Atlas of the Visible Human Male

National Library of Medicine

Atlas of the Visible Human Male

Reverse Engineering of the Human Body

VICTOR M. SPITZER, PH.D.
DAVID G. WHITLOCK, M.D., PH.D.
University of Colorado Health Sciences Center

JONES AND BARTLETT PUBLISHERS
Sudbury, Massachusetts
BOSTON LONDON SINGAPORE

Editorial, Sales, and Customer Service Offices

Jones and Bartlett Publishers, Inc.
40 Tall Pine Drive
Sudbury, MA 01776
508-443-5000
800-832-0034
info@jbpub.com
http://www.jbpub.com

Jones and Bartlett Publishers International
Barb House, Barb Mews
London W6 7PA
UK

Library of Congress Cataloging-in-Publication Data

Spitzer, Victor M.
Atlas of the visible human male : reverse engineering of the human body / Victor M. Spitzer , David G. Whitlock.
p. cm.
At head of title: National Library of Medicine.
Includes index.
ISBN 0-7637-0273-0 (hardcover : alk. paper). — ISBN 0-7637-0347-8 (pbk. : alk. paper)
1. Human anatomy—Atlases. 2. Tomography—Atlases. I. Whitlock, David G., 1924- . II. National Library of Medicine (U.S.) III. Title.
[DNLM: 1. Visible Human Project (National Library of Medicine (U.S.)) 2. Anatomy, Cross-Sectional—atlases. 3. Image Processing, Computer-Assisted. QS 17 S761a 1997]
QM25.S74 1997
611'.0022'2—dc21
DNLM/DLC
for Library of Congress 97-16572
CIP

Production Editor: Sue Michener
Editorial Production Service: Tower Graphics
Text Design, Cover Design, and Production: Marshall Henrichs
Text Printing: World Color
Cover Printing: Henry N. Sawyer Company
Binding: World Color (paperback), Acme Bookbinding (hardcover)

Printed in the United States of America
01 00 99 98 97 10 9 8 7 6 5 4 3 2 1

*Our foremost debt and highest acknowledgment are
to all people who have given their last perfect gift to society—their bodies.
Without these great gifts medical and health-care professionals cannot obtain
the rich understanding of the human body that underpins
all areas of medical knowledge.*

✦

To these generous people we dedicate this work.

✦

*To my father, one of these generous people,
I dedicate my small part in realizing his large dreams.*

VMS

Contents

ACKNOWLEDGMENTS viii

INTRODUCTION The Visible Human Project xi

PART I ORIGINAL IMAGES — Transverse 1

Head, 1001-1235; Neck 1160-1293; Upper Extremities, 1243-1908;
Thorax, 1293-1593; Abdomen, 1461-1812; Pelvis, 1812-2010;
Lower Extremities, 1861-1878

PART II RECONSTRUCTED IMAGES — Coronal 313

PART III RECONSTRUCTED IMAGES — Sagittal 409

PART IV Image Gallery 489

INDEX 509

ACKNOWLEDGMENTS

We are grateful to our University of Colorado department chairmen, Karl Pfenninger, MD, of Cellular and Structural Biology, and Michael Manco-Johnson, MD, of Radiology, for their collective vision in the late 1980s, which led to the teaming of the authors. They recognized that a superlative knowledge of anatomy was a crucial basis for medicine and that radiology had techniques and technology that could enormously benefit the computerization of anatomical knowledge. Their vision has been realized in the creation of the National Library of Medicine's The Visible Human Project™.

We are also deeply indebted to the National Library of Medicine (NLM) for providing the leadership and funds for the development of the Visible Human male database. Especially the Director, Donald Lindberg, MD, for his support and encouragement, and the NLM Boards of Directors, who approved this project and continue to support its development and use.

Michael Ackerman, the NLM project officer, deserves special commendation for his dedication to the project from its inception. We especially appreciate his efforts in facilitating special approvals for procedures not covered by government regulations, such as the expedient transport of dead bodies, and his insistence that review committees understand the pictures included in our proposals. We also gratefully acknowledge the contract work of Sharon Cummings and take solace in her concern that "her Library" was going to start cataloging corpses in the basement.

The acquisition, management, and preparation of the cadavers were also critical to the success of this project and for this, we offer our thanks to the State Anatomical Boards (SAB) of Colorado (Director, Michael Carry, PhD), Texas (Director, Andrew Payer, PhD), and Maryland (Director, Ronald S. Wade) for their cooperation and interest in the project. A special word of appreciation is also extended to Ed Molock and Mark Blatchley of the Colorado SAB for their round-the-clock efforts each time a cadaver was acquired. Also to Allen Tyler of the Texas SAB, thank you for spending many hours with one of the authors (VMS) waiting for the arrival of cadavers, during the embalming process, and even on a Lear Jet flight from Texas to Denver with one of the cadavers. Alan's companionship and experience in this uncommon situation was comforting at all times.

Radiological images, which were the basis for selection of the cadavers and later part of the Visible Human dataset, were acquired with expediency and expertise. The "emergency" procedures used to obtain radiographs, MR, and CT images, all within 24 to 36 hours and all with state-of-the-art equipment, required expert planning, cooperation, and hard work. We are grateful to Ann Scherzinger, PhD, for design and coordination of all the radiological imaging. Charles Ahrens, RT, provided radiographic imaging expertise and procedure development. His efforts and cooperation are especially noted. David Rubinstein, MD, put in many long evenings and early mornings to help with positioning of the cadavers and the acquisition and interpretation of the images. Paul Russ, MD, provided interpretations of the body images.

The long, night imaging hours, between regular, busy clinical days, for both the MR and CT imaging will always be remembered and appreciated. Cathy Gustafson, RT, Donna Callan, RT, and Deborah Singel, RT, put in many long nights on the MR scanner. Teresa Bonner, RT, supervised the CT imaging for all the cadavers and Thomas Hogan, RT, imaged the Visible Human Male. We are grateful to all these imagers for their last-minute response and dedication to the goals of the project.

Martha Pelster and Tim Butzer directed the support team and cut most of the 7,000 slices of the male and female specimens themselves. This team not only devised and tested freezing chambers and techniques and numerous cadaver handling systems, but also demonstrated the tenacity to perform the same function day after day. They were assisted by Helen Pelster, James Heath, Brian McNevin, Kate Finger, and Gregory Spitzer.

Helen Pelster, assisted by Gregory Spitzer, directed the image management, archiving, registration, and cataloging of the image database. Photography was overseen by James Heath. Control mechanisms for both the cryomacrotome and the cameras were designed and built by Charles Rush, MS, in collaboration with Karl Reinig, PhD. Numerous designs and redesigns of the equipment and protocol were successful because of their willingness, interest, and involvement along with support staff Bruce Burkhart and Dick Kennedy.

We also gratefully acknowledge Deborah Lehman, John Blackwell, and Renee Torgler for their work in the early portions of the contract and acknowledge that Renee was certainly willing, and in fact did, give the shirt off her back for the glory and goals of this work.

Of course, our gratitude for the participants in the Visible Human Male Project is quite different from our appreciation for the heroic efforts in bringing this book to fruition. Although numerous publishers showed little interest, Clayton Jones recognized the value of and believed in multimedia enough to not ignore the old and trusted media, the printed page, even for this entirely digital subject. It was his snap decision to commit (in less than one hour from presentation of our concept) to publish this book that has spurred our dedication to this effort. We also gratefully acknowledge Niel Patterson who shared confidence in our work and introduced us to Clayton. The constant facilitation and occasional badgering by our editors, Dave Phanco and Paul Lembo, could not have been done with more professionalism. The production staff of Jones and Bartlett, and the very special design influence of Marshall Henrichs, has been superb after they got over the normally encountered hurdles of what it was that these two crazy people in Colorado actually did.

For the detailed layout of all 250 labeled plates we are grateful to Raechel O'Kelly, Paul Sommers, Amy Schilling, and once again to Helen Pelster and Greg Spitzer for image management and database development. The 3D rendered images in the "Image Gallery" are only possible because of the teamwork of Zhonjun Fu and a team of 25 students and staff who have classified each and every visible structure in the body over the past 18 months. A few examples of this work are shown in the 3D pictures in the last part of the book. For these ray-traced renderings, which reflect both the color and texture of the original voxel data, we are indebted to Karl Reinig and Helen Pelster.

The images of simulated procedures, in the last part, reflect the exciting new directions that these fully segmented and classified data are facilitating. These images reflect thousands of hours of teamwork and cooperation from members of the University of Colorado Center for Human Simulation, in our effort to more fully utilize the potential of this extraordinary national resource. Karl Reinig, PhD, has led this simulation effort with the dedicated support of Tom Mahalik, PhD, Travis Johnson, and again Helen Pelster.

Our participation in this book certainly depended on previous support that laid the foundation for developing and assembling the technology necessary for the successful completion of the project. That includes a pivotal grant from the Frost Foundation, publishing support from Mosby-Yearbook, and equipment support from Hewlett Packard, Intel Corporation, Intergraph Computer Systems, Digital Equipment Corporation, Exabyte, Kodak, and Apple computer. Funding for all classification and segmentation of the Visible Human Male, which facilitates the 3D rendered images in the last part of this book, was provided by Gold Standard Multimedia. Research Systems Inc. generously provided the image processing software (IDL), which allowed rapid prototyping and development of image processing and handling of the Visible Human dataset.

Second to last, and most important, we express our continual appreciation of our wives and families for their support and understanding during the long hours involved in the preparation of this work. We also appreciate their technical assistance and occasional dominance. Last of all, we acknowledge all the people who have significantly contributed to the Visible Human Male Project at the University of Colorado and to this book but for some strange reason have slipped our collective minds.

ANDREAE VESALII
BRVXELLENSIS, SCHOLAE
medicorum Patauinæ professoris, de
Humani corporis fabrica
Libri septem.
CVM CAESAREAE
Maiest. Galliarum Regis, ac Senatus Veneti gra-
tia & priuilegio, ut in diplomatis eorundem continetur.

The hand colored title page to the first edition of the "De Humani Corporis Fabrica," 1543

INTRODUCTION

The Visible Human Male

The Visible Human Male is the most complete computerized database of the human body ever assembled. This book is your tour guide—taking you through the 1,878, one-millimeter thick transverse slices of an adult human male as well as sagittal and coronal reconstructions and representative clinical scans and 3D reconstructions. Called "the greatest contribution to anatomy since Vesalius's 1543 publication of De Humani Corporis Fabrica," the first collection of faithfully rendered drawings of human anatomy, these images, constituting 15 gigabytes of information, are the seeds for a growing medical revolution. With this atlas those already familiar with anatomy, such as radiologists and other health-care professionals, can follow structures through different views of the same body, seeing how an organ or muscle changes in space. Medical students can supplement their knowledge of gross anatomy by studying the labeled images. For the lay audience it is a window into the unknown, a way to visualize the body's mysterious machinery, and see what the professionals see. This dataset should change the way doctors are trained and surgery is practiced; it may help cut the cost of health care and allow physicians and other professionals to perform procedures unthinkable before. The National Library of Medicine's *The Visible Human Project*™ is an anatomy lesson for the twenty-first century.

Cadavers have been dissected since the thirteenth century, but it was not until the Renaissance that anatomists stopped searching for the seat of emotion or temperament locked within our organs and began seeing the body as a miraculous machine whose inner workings should be understood and mastered by physicians to help heal the sick. In 1405 the University of Bologna allowed up to twenty students to observe the dissection of a male corpse; thirty were permitted at a female's dissection. The pool of bodies for dissection consisted mainly of executed criminals who sometimes numbered only three or four a year, so simply put, there were never enough cadavers to go around. In the sixteenth century demand became so great that amphitheaters were built to accommodate the large audiences. Physicians, prominent members of the community, intellectuals, and the inevitable idle and rich snobs, in addition to medical students, gathered for the spectacle, which would last three to four days—a time frame determined by the lack of refrigeration. Later, audiences were made up of lay people as well as professionals, making it difficult to keep a sense of decorum, especially when people gathered in costume and sometimes became so rowdy that they would grab specimens right off the dissecting table.

Since then, anatomy has been studied in exquisite detail. Bodies have been dissected, cross-sectioned, photographed, and described ad infinitum. Yet those who have first-hand knowledge of the inside of a human body are still a highly select group. Most medical students only get to explore one cadaver, usually shared with at least three classmates. With *The Visible*

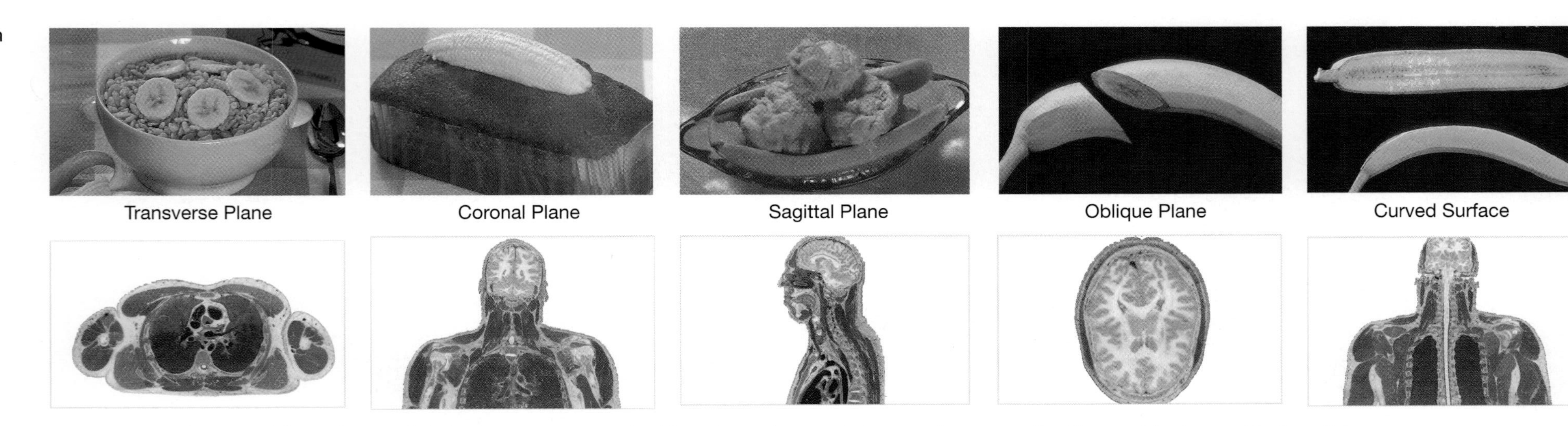
Transverse Plane
Coronal Plane
Sagittal Plane
Oblique Plane
Curved Surface

These views and their symmetry properties (though maybe not their names) are very familiar outside the world of medicine.

Human Project we expect to change that, enhancing the experience of cadaver dissection in medical education and providing virtual cadavers to anyone interested in the human body. The images in this book are today's version of a dissection in a public amphitheater, but ours is a global amphitheater able to serve an unlimited audience. The pieces we present can be reassembled in sections, just the brain, for example, or as a complete unit. It is what we like to call reverse engineering. Our man in the machine allows interested citizens of any age or profession to travel through and explore a real human body in its entirety on their own computer screens.

The Visible Human Project was conceived by the National Library of Medicine (NLM) in 1988 after bringing together representatives of eight medical centers working in the area of 3D anatomical visualization. Collectively these scientists decided that the NLM, until then a text-dominated repository of medical information, should create an image database of human anatomy. The idea, the dream really, was to obtain a uniform set of photographs and scans from one optimum specimen. Over one hundred medical schools vied for the job, but a proposal submitted by our school, the University of Colorado School of Medicine, won the contract which ultimately totaled $1.4 million for work on both the man and the woman. We inaugurated the project on September 1, 1991. The success of the Colorado proposal, it was later revealed, was not based on the superior writing skills of the authors, but rather on twenty previously captured pictures of a human abdomen that were submitted with the proposal. The pictures, each 7 megabytes or 3.5 million words, must have been worth a few million words.

To be of maximum clinical and educational value, it was decided that our virtual couple needed to be documented in several different formats commonly used by radiologists and other physicians. We decided on six: traditional X-rays and CT (computerized tomography) scans to optimally visualize bone, MRI (magnetic resonance imaging) for soft tissue, and three types of color photographs for definitive resolution. These would make up the multispectral database of images that would be captured by the most clinically useful imaging techniques currently available and one technique—photography—that would provide the ultimate standard of comparison.

Our first and most formidable task was to locate a specimen that could be taken as "normal" or representative of a large population. To achieve this goal, the NLM decided we should obtain a sample set of three cadavers and let a blue ribbon panel of experts select the "winner" based on radiological images of their remains. The success of this search for a healthy-appearing cadaver depended on resources (numbers of cadavers per year) beyond that of the State Anatomical Board of Colorado and we were very fortunate to be joined by the State Anatomical Board of Texas, in our original proposal, and later also by Maryland. State Anatomical Boards are organizations through which citizens can donate their bodies for medical research and training.

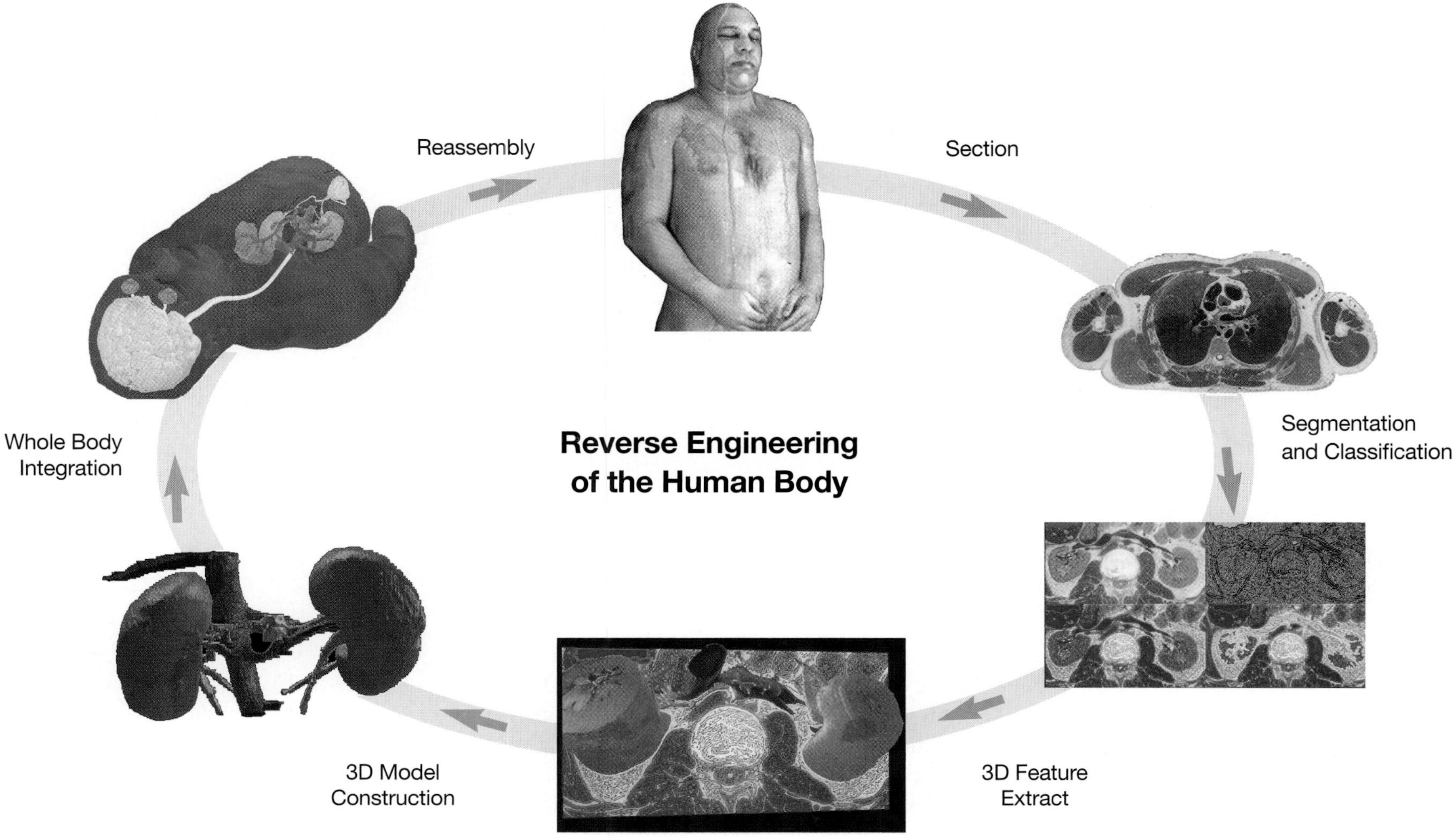

Reverse Engineering of the Human Body

Reassembly
Section
Segmentation and Classification
3D Feature Extract
3D Model Construction
Whole Body Integration

What we were searching for was someone 21- to 60-years of age who died without traumatic injuries or invasive or infectious disease. We got lucky. Some inmates on death row in Texas had decided to donate their bodies to science. They were young, relatively healthy men whose organs, tainted by lethal injection, were rendered unsuitable for transplant. Through screening of cadavers such as these as well as individuals obtained from other donations, our panel selected the body of a recently executed 38-year-old male. It was not lost on us that victims of execution, a population that taught anatomist's centuries ago, would be teaching us once again, perhaps in some way repaying society for their crimes.

Aside from some minor imperfections—no appendix, Number 14 tooth, and left testicle—this specimen fit all our criteria, which were that the body be: less than 6 feet tall, less than 22-inches wide, less than 14-inches from front to back, and normal height for weight. What accounted for these "ideal" human dimensions? Most were based on the limitations of equipment we were using either to image the body or slice it. Initially,

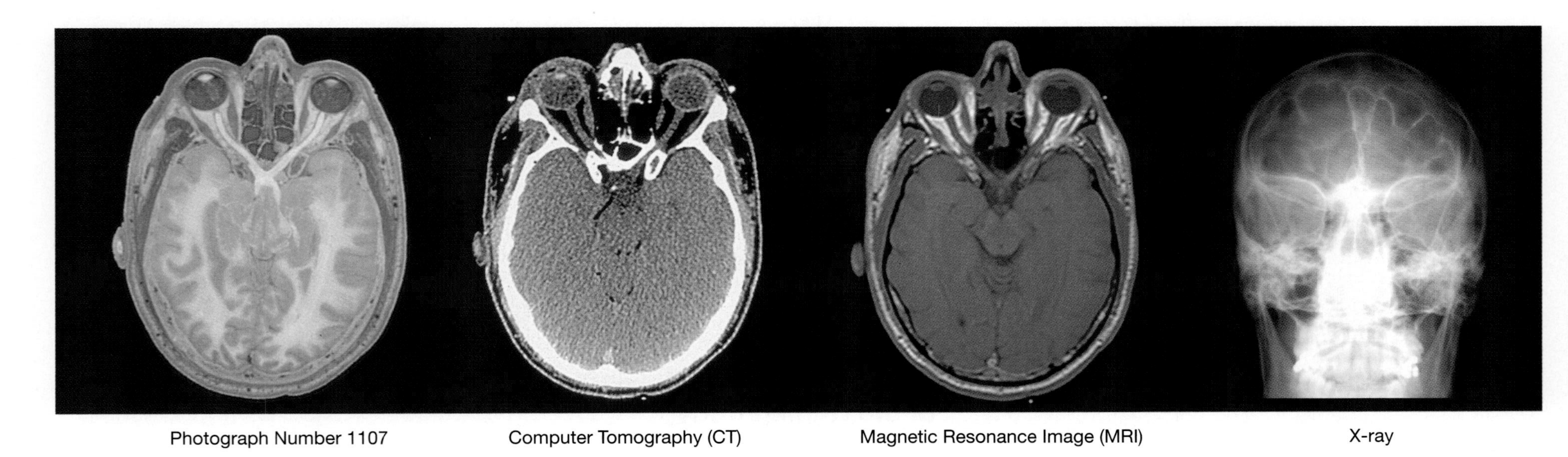

Photograph Number 1107 | Computer Tomography (CT) | Magnetic Resonance Image (MRI) | X-ray

The right-left symmetry is indicated by the bilateral structures in the head. The tilt of the head (proximity of the chin to the chest) is indicated by the length of the optic nerve in this single plane.

the ideal position was to be achieved by suspending the cadaver vertically while freezing it to preserve a "life-like" human form in the standing position. This required constructing a special freezing chamber for which the most practical height was no more than six feet. The process worked well for positioning internal viscera, because the body was suspended by the skull, underarms, and pelvis. But there were undesirable distortions in the areas of the cheeks, arms, and lower abdomen. After three or four attempts, this method was abandoned.

Other dimension limits had to do with the size of the cryomacrotome and other tools we used to slice the body. We were breaking new ground. According to published literature, what we did was unfeasible. Fortunately, no one told us that, so we experimented, sometimes erring, but improving our techniques every step of the way. Ultimately we opted to obtain the CT scans of the cadaver in the supine position and to hold the cadaver in that exact same position throughout the physical sectioning process. We did this by positioning the cadaver in a plywood box, adapted to mount on the CT gantry, and then filling all extraneous space in that confining box with a foaming agent called Alpha Cradle™ that hardens in 20 minutes. This is a product used in radiation therapy to construct a custom-fit cradle for patients to lie on during successive treatments. The foaming agent is not much different from foaming agents used for filling space in electronics packaging, with the exception that it generates so little heat that it does not burn the patient. A stable, low temperature was also an important issue for the cadaver because we were trying to preserve the tissue.

The standard technique for positioning a patient in the CT scanner includes a vertical positioning of the median sagittal plane which is obtained by positioning the ears in the same horizontal plane. This positioning is generally done on the CT scanner but for our cadaver the positioning was done in the laboratory. The success of the "positioning" can be judged by the symmetry of the body, in particular the head in the slices taken transverse to the long axis of the body. A particularly good example of this positioning is demonstrated by the symmetry of the optic nerves in Slice Number 1107. The length and symmetry of the optic nerves in the same slice also demonstrates a desirable tilt of the head from front to back.

Five of the 1871 transverse CT cross-sections

Proton density weighted
T1 weighted
T2 weighted
Five different depths

Magnetic resonance images (MRI) were obtained in transverse planes of the head and in coronal planes of the body.

When the NLM contract for the Visible Human was awarded, MR images of the body were not routine clinical practice, so the original contract called for MR images only of the head. By the time the Visible Human Male was acquired, body imaging was of clinical importance and we added it to our protocol. The initial contract also called for injection of the major vessels of the body with a colored latex (red in the arteries, blue in the veins) and iodine solution that could be seen in the MRI, CT scans, and physically cut sections. We injected five cadavers, and although this gave good results in the CT images, the solution leaked from the vessels in some areas of the body in three of the five cadavers. There is a long history of preparations using these kinds on injections (including latex, microfil, India ink, barium, etc.)—but they are rarely used on whole cadavers. Because it was unreasonable to assume that we would be able to prepare a leak-free specimen, vessel injection was eliminated from the protocol. Since MRI provided good delineation of the large vessels, these images were substituted for marking them by injection. The major drawback for MR images is simply that the body was not in the same position as used for CT scanning.

A $5,000 "backsaw" was commissioned from a local designer to partition the intact cadaver. We used it to section the cadaver into four blocks each less than 20 inches in height, the maximum height our cryomacrotome could handle. Concerned with future integration of data, we wanted our cuts to be in the exact same plane as the CT slices taken earlier and to make the saw kerf (the width of the saw cut) as thin as possible. The missing images in the Dataset represent anatomy destroyed by the saw. We will avoid this compromise in the future by expansion of the cryomacrotome to permit cutting the entire cadaver intact. The saw was manually operated with a single straight blade at 30,000 lb. of tension facilitating as straight a cut as possible. The major difficulty we had with the saw was when the blade encountered tissues of different resistance, in particular bone/soft tissue interfaces. We noted the problem and tried to overcome it by freezing the

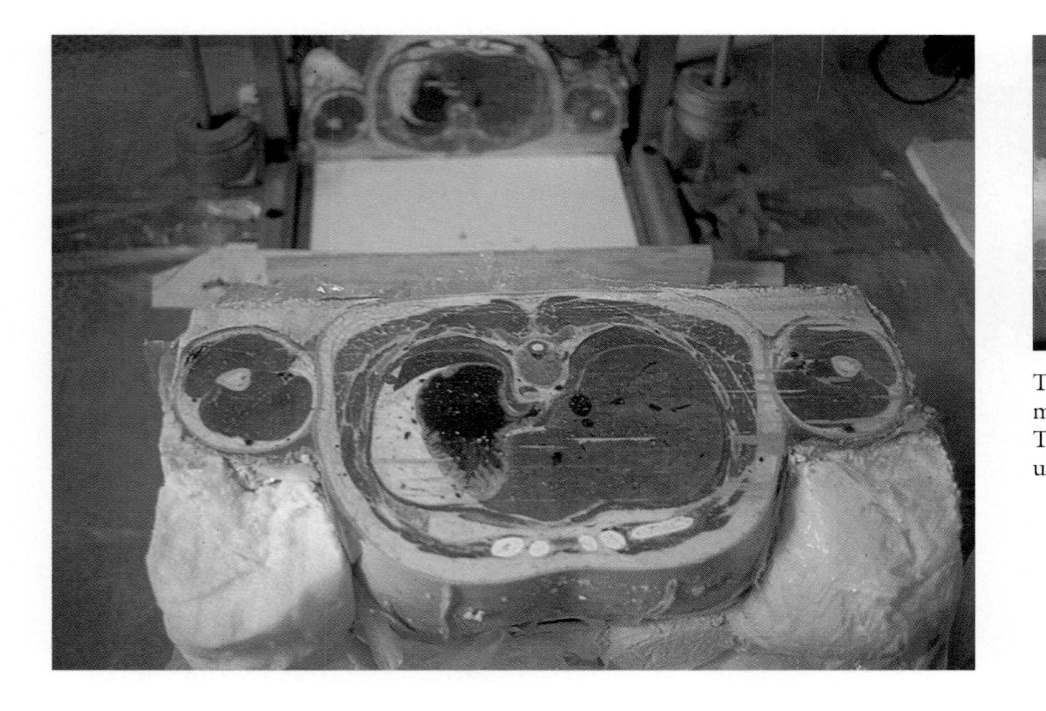

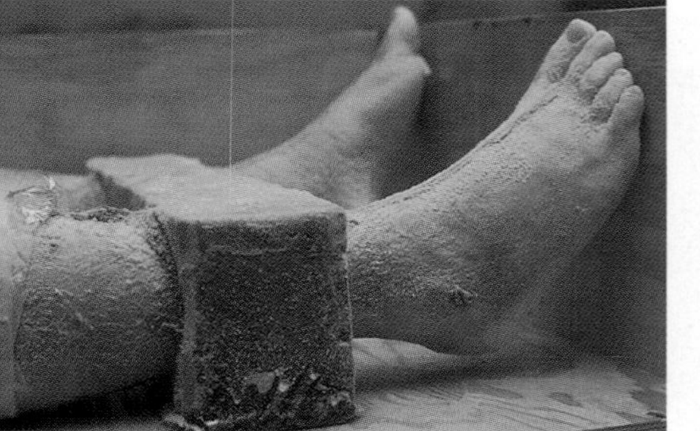

The legs of the cadaver were glued together to maintain their position during the cutting process. The "glue" was the same gelatin mixture later used to form the block to be sectioned.

cadaver to a lower temperature (−70° C) to equalize the tissue characteristics. Backsaw cuts made after packing the body in dry ice for two days prior to sectioning were most successful. We also carefully positioned the cut so that it went directly through the hard bone of the spine and not between vertebrae, which often deflected the blade and caused irregularities in the cut. The most difficult section to cut was that through the diaphragm and arms. What do you do with the arms once they are cut off? Cutting them separately and repositioning them at the correct locations in the computer database would have been extremely difficult, so we "glued" them to the abdomen and pelvis using colored gelatin, straight from our cafeteria, prior to sectioning the cadaver. To keep the legs in their relative positions we also used a "bridge" of gelatin between them before sectioning.

Six months after receiving the body, and after sectioning it into the blocks described above, we began imaging the 1-mm thick slices. These images were obtained by photographing the top surface of the block after removal of each successive millimeter by the cryomacrotome. No actual slices were retained, however, since we destroyed each documented slice by milling it down to get to the one underneath. A key element during the process was to keep all factors affecting the photography of the newly exposed surface constant, especially the lighting and specimen temperature. Uniformity of the lighting field was achieved with four polarized strobe lights. We optimized the consistency of the color of the top surface of the specimen by continuous cutting. We had two teams of two people each, cutting in shifts for as many as 12 hours a day. In a lab with no air conditioning during a hot Colorado summer, however, some cutting sessions were necessarily restricted to less than eight hours. Despite our progress, we were still somewhat limited by refrigeration, one of the same constraints encountered by fifteenth-century anatomists. Each slice, including cutting, preparation, photography, refreezing, and repositioning took anywhere from four to ten minutes.

The photography included image capture with three cameras, one digital and two conventional film, one 35-mm and one 70-mm. Refreezing the specimen was a "hand" operation. It entailed setting a tray of dry ice on the block surface for 20 seconds. The variable in the process was the surface preparation time. This included removing any debris from the surface, checking for tissue that was not cleanly cut (e.g., some tendons and fascia), spraying the surface with ethanol, applying a black mask over the bright white dry ice packing the block, attaching a Kodak photographic color strip (for

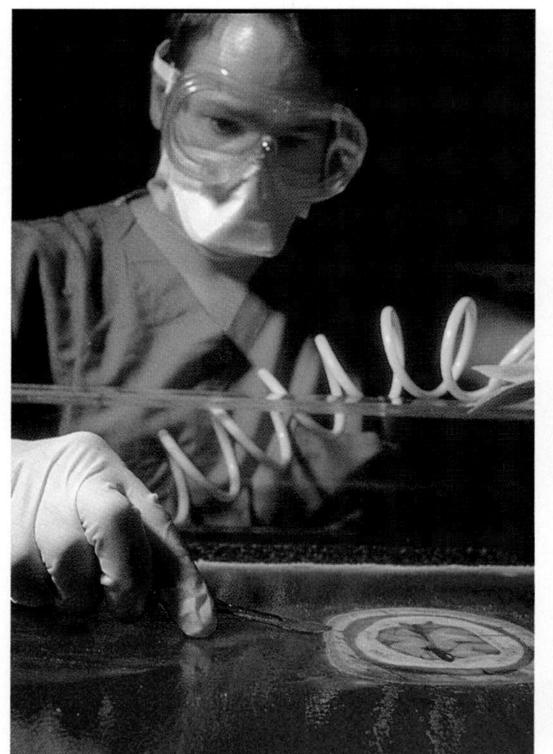

Surface preparation after removing each 1-mm slice includes removing all debris and checking for any "uncut" connective tissue.

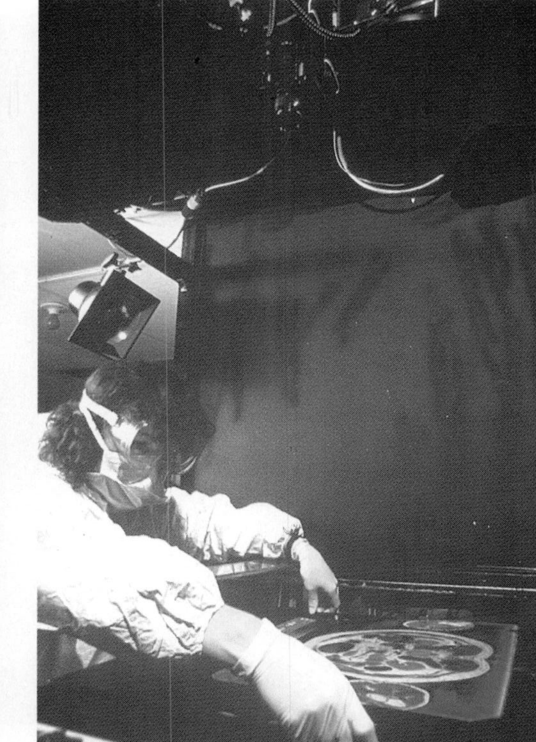

The black mask is placed over the freshly cut and prepared surface for photography. The three cameras used for image capture are at the top of the picture.

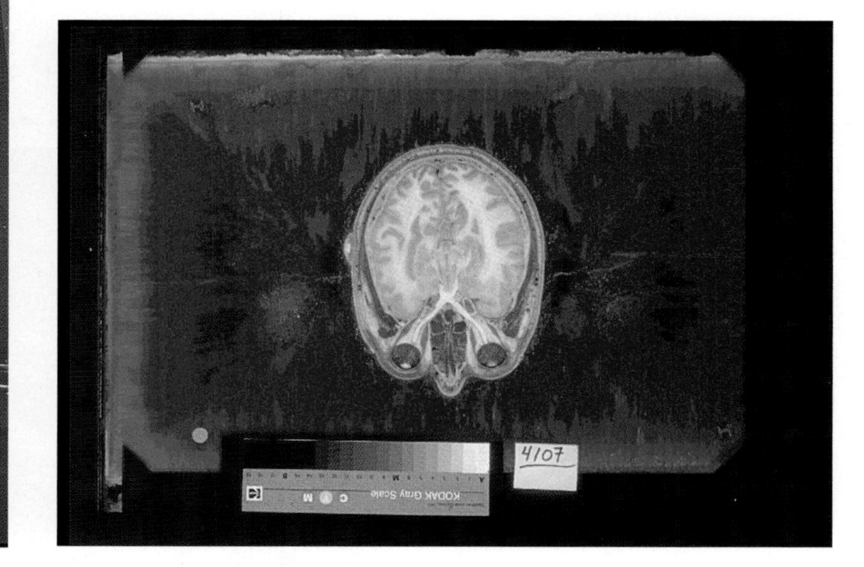
4107
KODAK Gray Scale

The image of the block surface as the camera viewed it. The color normalization strip is included at the bottom of each picture. The slice identifier indicates this was the 107th-mm from the top of the fourth block that was cut.

image color normalization), and assigning an internal slice number (written on a yellow Post-it Note™) indicating the block and the millimeter that each photograph represented (i.e., Slice 3105 is the 105th-mm cut in the third block). There were few typical days. Most components broke down at least once. A typical log entry:

> The first day is pretty good—It kinda looked like we would never get finished. First slice—power cord on 70-mm camera disconnected—took picture manually. Third slice—image capture computer stuck at "blue" image—reboot. Fifth slice—film counter on 70-mm camera malfunctions—fix programmed for the day. Sixth slice—computer stuck again— reboot. Slice 31—computer crashed again. Then things went well to slice 97.

The entire process took nearly nine months—four months of cutting time and the remainder for specimen preparation of each block.

The top and bottom few slices of each block were usually partial slices because the backsaw surface was not a perfect plane or the specimen section was not perfectly parallel to the cutting plane of the cryomacrotome. These partial slices and the 1.3-mm gap attributed to the saw kerf have been eliminated from the final, photographic image database. These missing slices account for the gaps at the diaphragm, upper thigh and just below the knee.

Our virtual male made his debut on November 28, 1994, at a meeting of the Radiological Society of North America in Chicago. Aware that the press might seize the opportunity to focus on the data's past—as housing for a convicted killer—and not its future—a revolution in the study of anatomy and surgical practice, we rehearsed our announcement carefully. Ironically the next day's headlines read, "Cannibal Dahmer Killed in Jail." By comparison, our news was tame. We stopped worrying.

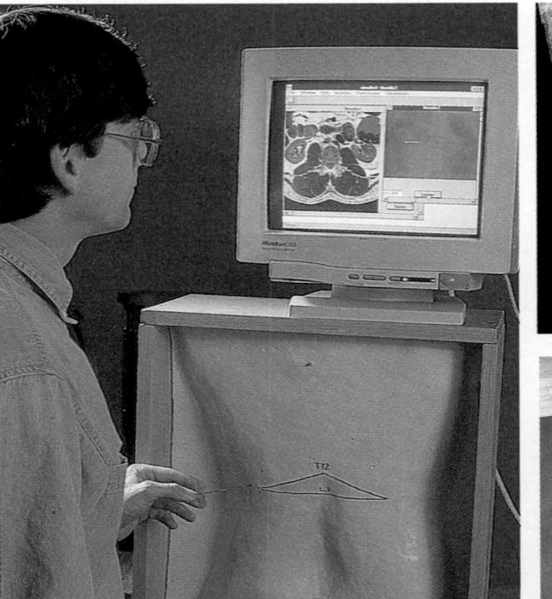

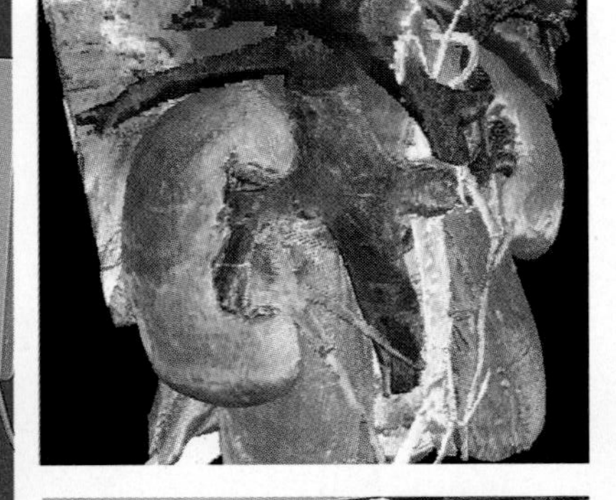

A celiac plexus nerve block is practiced with the plastic back, built from the data of the visible human (left). The student feels each tissue the needle encounters (upper right), through forces controlled by the Phantom, a haptic feedback interface from Sensable Technologies (lower right), and confirms the proper placement of the needle with a simulated radiograph showing its current position, as shown on the left side of monitor (upper left).

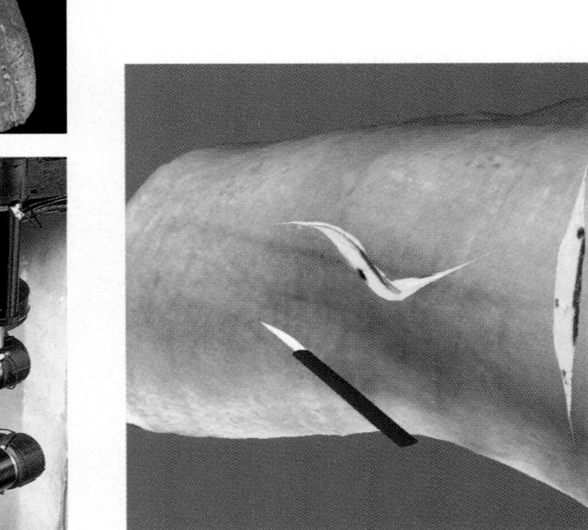

The student that just cut through the thigh of the Visible Human felt the cut as it was made. The cut reveals tissue defined by and taken from the Visible Human database.

With the release, over the Internet, of the *The Visible Human Project* by the National Library of Medicine, the images are now in the hands of medical researchers and technology designers all over the world. This dataset will facilitate the planning and rehearsal of medical interventional procedures prior to carrying them out in "real" operating or procedure rooms. Just as the aircraft industry provides simulators for its pilots, human body simulations are being constructed for training health-care personnel, including physicians, dentists, nurses, technicians, and EMTs. Col. Richard Satava, MD has predicted the impact of this new industry as "The use of the Visible Human database will fundamentally change the practice of medicine."

At the University of Colorado Center for Human Simulation, anesthesiologists use a haptic feedback device coupled to the classified data of the Visible Human abdomen to practice a celiac plexus block— an injection into a nerve center in the middle of the abdomen. With the haptic feedback device providing the interface to the computer, the student can "feel" the entire procedure. Before this simulator based on the Visible Human data was available, the procedure was performed by the student for the first time on a real patient. With the same haptic feedback device, this time coupled to the anatomy of the eye, we have developed a simulator for ophthalmologists to practice radial keratotomy; as they touch or cut the cornea, they feel and see the deformation they are causing. By coupling the same device to anatomical images of a tooth, dentists can experience, first-hand, the feeling and appearance of a tooth with carries and calculus. We can also mimic different stages in the progression of the disease. Simulators allow unlimited practice and rehearsal. These programs have value for the public sector as well, including patient education, general interest, and entertainment. Progress in this area is updated on our web site http://www.uchsc.edu/sm/chs.

We have embarked on other uses of this database in order to empower ourselves to more rapidly simulate the "new models" of humans, be the differences ever so slight, that keep coming off the assembly line. Our ultimate goal is to reverse engineer a single, or small number of models, of the human body in each fundamentally different category. We can then develop variations of these standard models to more closely match an individual of interest. This may mean a specific patient for the doctor, a small population of virtual heads for the helmet or headgear designer, an endless army of Size 6 bodies for the garment designer, or the ultimate stuntman for the movie producer. Our first model, presented here, is a middle-aged, Caucasian male. The second model, already sliced and imaged, is a middle-aged Caucasian female. We hope to expand on these categories with a variety of ages, shapes, and races. It is not unreasonable to extend this reverse engineering process to different species.

Available from the Internet (www.nlm.nih.gov/research/visible/visible_human.html) free of charge by agreeing to credit the NLM and send them a copy of any product produced, *The Visible Human Project* offers the medical and lay communities unlimited and previously unimaginable possibilities. We now have a renewable cadaver, a standardized patient, and a basis for digital populations of the future. Not only can we dissect it, we can put it back together again and start all over. Currently more than 700 licenses to use the dataset have been granted in more than 26 countries and applications are proliferating. The real future of medical education is not in the Visible Human Dataset itself but rather in the manipulation, distortion, and modification of the data to produce whole populations of virtual humans of every age, race, and pathology. In the past, we have provided visions of our theories and concepts through medical illustrations, photographs, and models, but each of these was a single representation in time. The Visible Human computerized database of the body, defined in three- or four-dimensional space, permits the viewing of our ideas in time—present, past, and future.

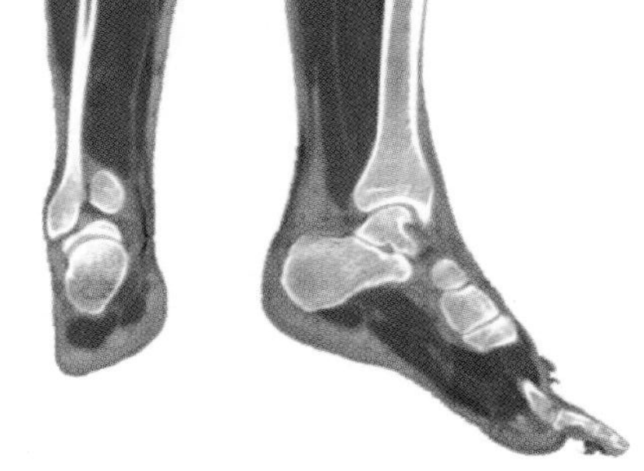

Transverse CT and corresponding anatomy planes through the ankle.

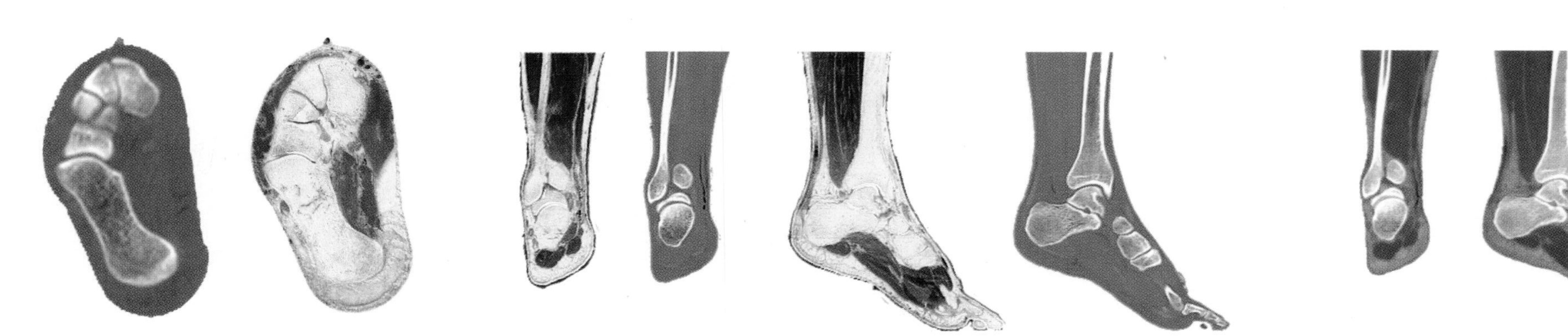

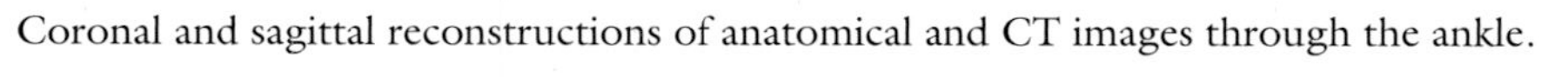

Coronal and sagittal reconstructions of anatomical and CT images through the ankle.

Overlay of the CT on the anatomical images showing alignment of the two modalities.

PART I

ORIGINAL IMAGES

Transverse

The transverse images are the original, digital pictures of the National Library of Medicine's The Visible Human Dataset™. Every picture (every millimeter) is included, from the head through the feet. These transverse planes are particularly useful when comparing bilateral anatomical structures. All images are presented in the standard radiological format, as though the viewer was standing at the foot of the bed and looking toward the patient's head.

The maximum resolution of images in the transverse plane is determined by the camera and field of view. All images in this part were acquired with a LEAF CCD camera. The pixel size is 0.32 millimeters square and the anatomy is contained in a matrix of 1760 x 1024 pixels. This pixel size and density is of similar resolution to the printing process used for the twelve-image arrays on the right-hand pages of this book.

The single labeled transverse image, on each left-hand page, was chosen from one of the twelve images, on the facing right-hand page. No attempt was made to keep labels in the same location from slice to slice. Bilateral structures are labeled on both sides when space permits. Labeling of the images generally follows naming conventions in *Gray's Anatomy,* 30th American edition (C. D. Clemente, editor, Lea & Febiger, 1985). Tissue classes are included when appropriate and abbreviations are used only when necessary. The word "bone" is used only with the carpal, tarsal, and skull bones. The word muscle is attached to all labeled muscles. The abbreviations used include: M., muscle; N., nerve; Lig., ligament; Trans., transverse; Post., posterior; and Ant., anterior. The twelve-image array permits anatomical structures to be followed easily, millimeter by millimeter, vertically through the body.

Every acquired transverse slice is included in this image collection. The only missing images are the three saw kerfs. An image relating to each saw kerf has been inserted in the image collection at these locations. The magnification used for the labeled images changes multiple times to accomidate the labels and to maximize the image size. The twelve images on the right-hand pages, change scale only once—at the base of the neck. Another image modification, in the labeled diagrams, occurs in the lower extremities where space has been removed between the legs and feet to bring them closer together. The original leg spacing is maintained for the twelve-image array pages.

Transverse
a_vm1012

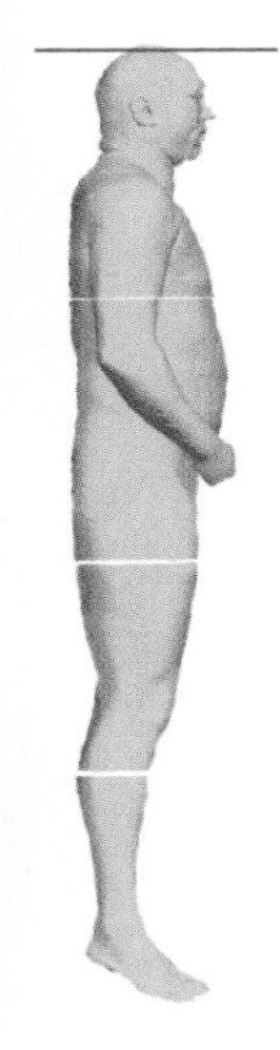

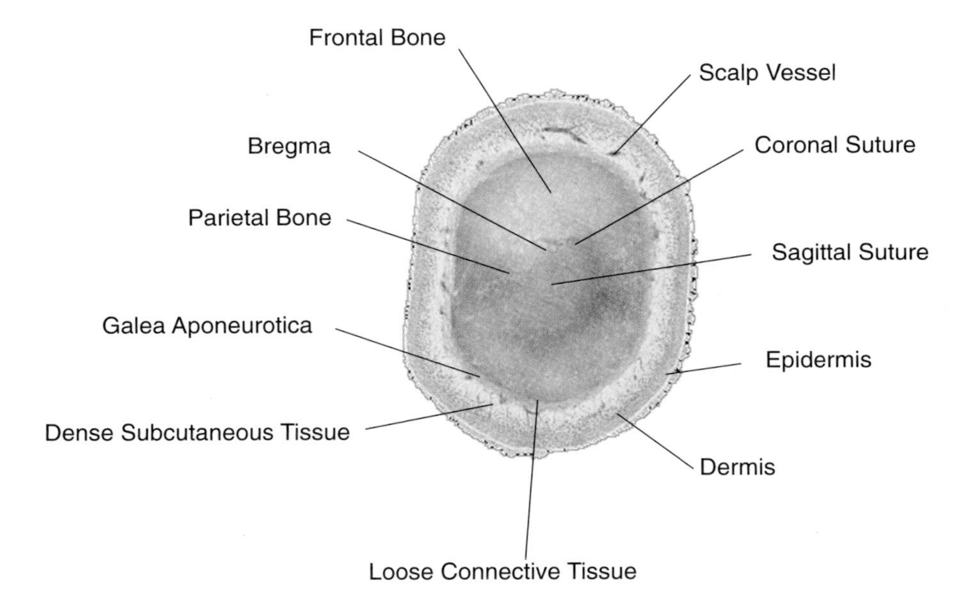
anterior
Frontal Bone
Scalp Vessel
Bregma
Coronal Suture
Parietal Bone
Sagittal Suture
Galea Aponeurotica
Epidermis
Dense Subcutaneous Tissue
Dermis
Loose Connective Tissue
right
left
posterior

a_vm1001	a_vm1002	a_vm1003
a_vm1004	a_vm1005	a_vm1006
a_vm1007	a_vm1008	a_vm1009
a_vm1010	a_vm1011	**a_vm1012**

Transverse
a_vm1018

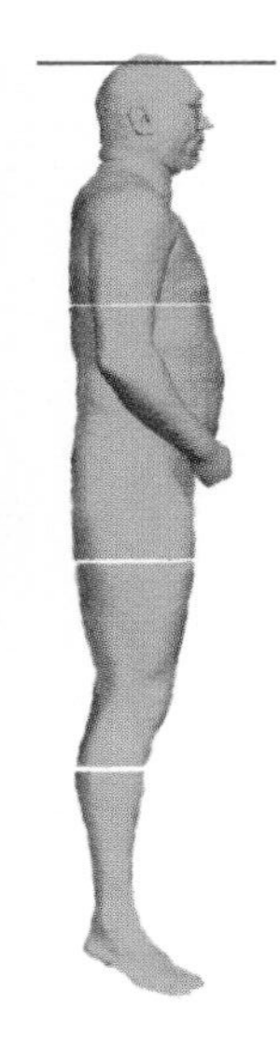

anterior

Bregma
Frontal Bone
Coronal Suture
Scalp Vessel
Diploe
Coronal Suture
Parietal Bone
Parietal Bone
Diploic Vessel
Sagittal Suture
Galea Aponeurotica
Outer Table
Dense Subcutaneous Tissue
Epidermis
Loose Connective Tissue
Dermis

right

left

posterior

a_vm1013

a_vm1014

a_vm1015

a_vm1016

a_vm1017

a_vm1018

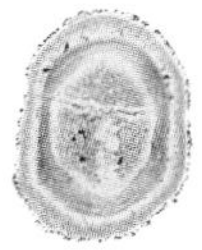

a_vm1019

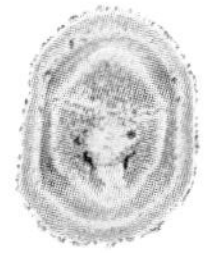

a_vm1020

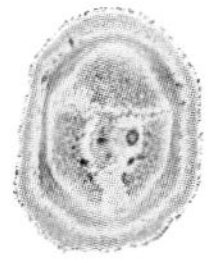

a_vm1021

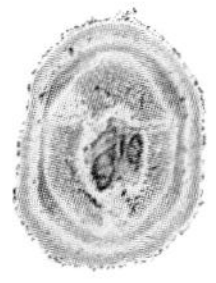

a_vm1022

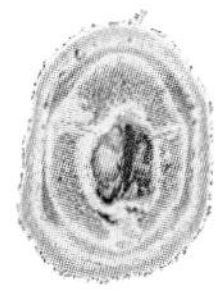

a_vm1023

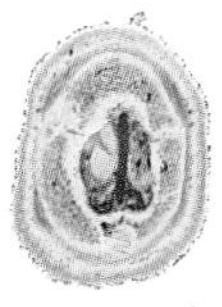

a_vm1024

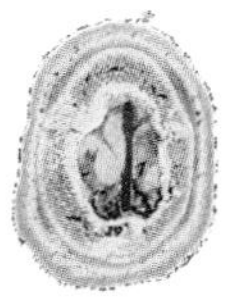

Transverse
a_vm1028

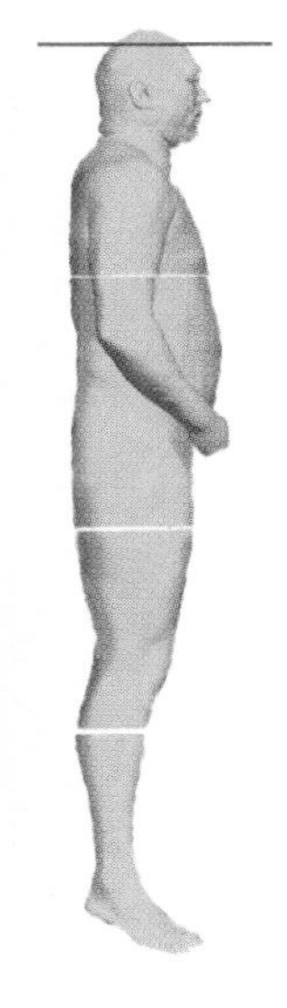

anterior

Lacrimal Lake

Supraorbital Vessel

Falx Cerebri

Subarachnoid Space

Coronal Suture

Superior Frontal Gyrus

Longitudinal Cerebral Fissure

Superior Frontal Sulcus

White Matter

Diploic Vessel

Parietal Bone

Gray Matter

Precentral Sulcus

Precentral Gyrus

Epidermis

Superior Sagittal Sinus

Dermis

Superior Sagittal Sinus

Frontal Bone

Loose Connective Tissue

Frontal Lobe

Coronal Suture

Parietal Bone

Dura Mater

Inner Table

Outer Table

Diploe

Galea Aponeurotica

Dense Connective Tissue

Sagittal Suture

right

left

posterior

a_vm1025

a_vm1026

a_vm1027

a_vm1028

a_vm1029

a_vm1030

a_vm1031

a_vm1032

a_vm1033

a_vm1034

a_vm1035

a_vm1036

Transverse
a_vm1048

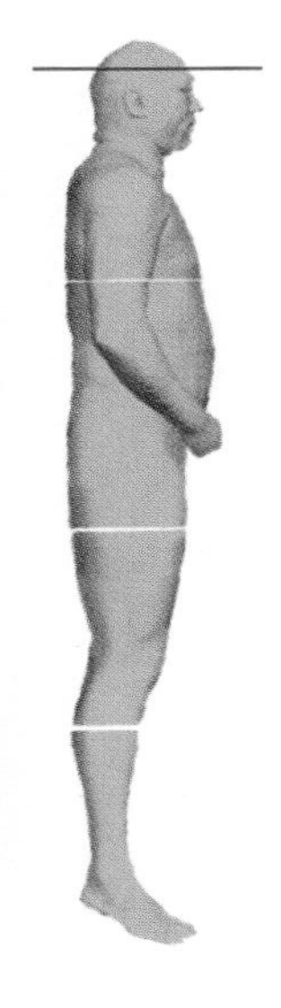

anterior
Frontal Sinus
Galea Aponeurotica
Superior Frontal Sulcus
Occipitofrontalis Muscle, Frontal Belly
Coronal Suture
Middle Frontal Gyrus
Diploic Vessel
Gray Matter
Temporalis Muscle
White Matter
Precentral Sulcus
Precentral Gyrus
Central Sulcus
Parietal Bone
Paracentral Lobule
Superior Cerebral Vessel
Outer Table
Diploe
Inner Table
Dura Mater
Precuneus
Superior Sagittal Sinus
Frontal Bone
Frontal Sinus
Frontal Pole
Falx Cerebri
Coronal Suture
Superior Frontal Gyrus
Superior Frontal Sulcus
Longitudinal Cerebral Fissure
Frontal Lobe
Lateral Fissure
Precentral Gyrus
Central Sulcus
Postcentral Gyrus
Postcentral Sulcus
Parietal Bone
Parietal Lobe
Superior Sagittal Sinus
Sagittal Suture
right
left
posterior

a_vm1037

a_vm1038

a_vm1039

a_vm1040

a_vm1041

a_vm1042

a_vm1043

a_vm1044

a_vm1045

a_vm1046

a_vm1047

a_vm1048

Transverse
a_vm1060

anterior

Supraorbital Vessel
Occipitofrontalis Muscle, Frontal Belly
Frontal Sinus
Frontal Bone
Superior Frontal Gyrus
Superior Frontal Sulcus
Skin
Middle Frontal Gyrus
Lateral Fissure
Temporalis Muscle
Cingulate Gyrus
Inner Table
Outer Table
Diploe
Paracentral Lobule
Diploic Vessel
Gray Matter
Skin
Parietal Bone
Superficial Cerebral Vessel
Interparietal Sulcus
Precuneus
Sagittal Suture
Superior Sagittal Sinus
Galea Aponeurotica
Frontal Sinus
Frontal Bone
Longitudinal Cerebral Fissure
Coronal Suture
Temporalis Muscle
Frontal Lobe
Lateral Fissure
White Matter
Temporal Bone
Gray Matter
Precentral Gyrus
Central Sulcus
Superficial Temporal Vessel
Falx Cerebri
Parietal Bone
Parietal Lobe
Parietooccipital Sulcus
Occipital Lobe
Occipital Vessel
Superior Sagittal Sinus
Dense Subcutaneous Tissue

right

left

posterior

a_vm1049

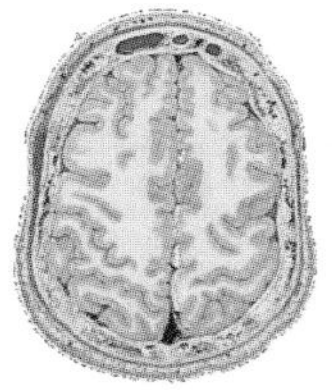

a_vm1050

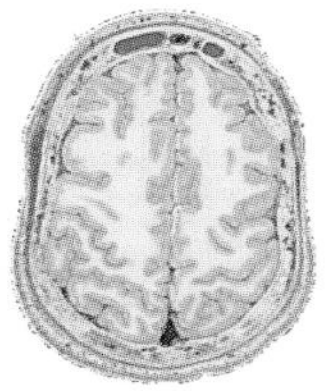

a_vm1051

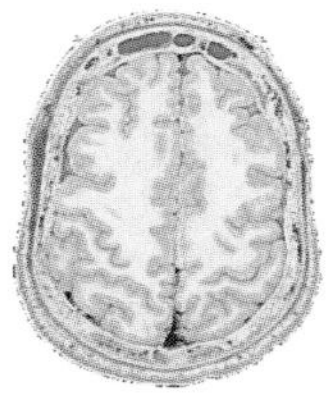

a_vm1052

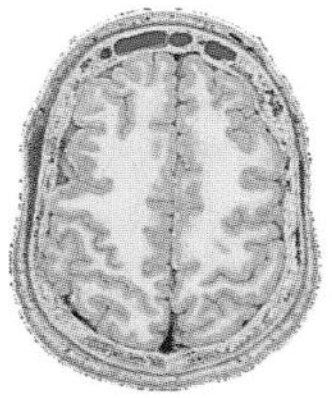

a_vm1053

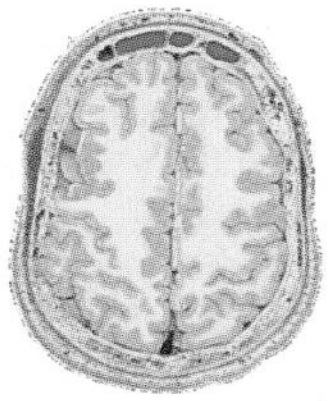

a_vm1054

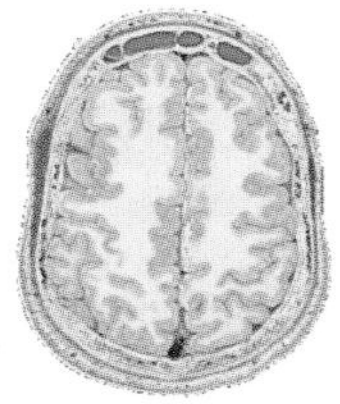

a_vm1055

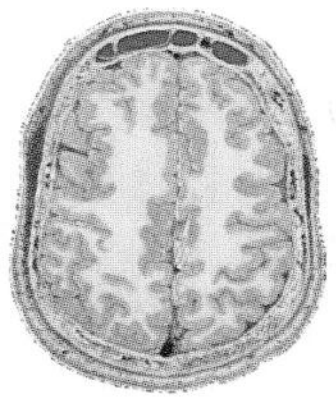

a_vm1056

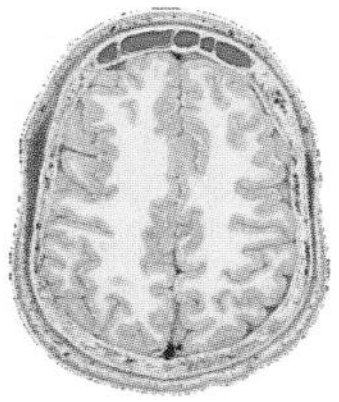

a_vm1057

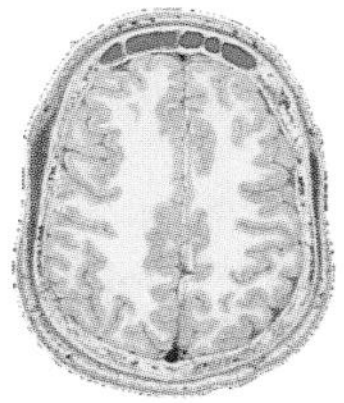

a_vm1058

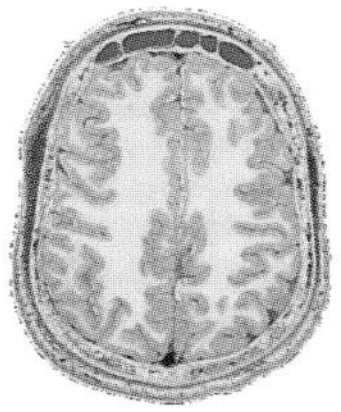

a_vm1059

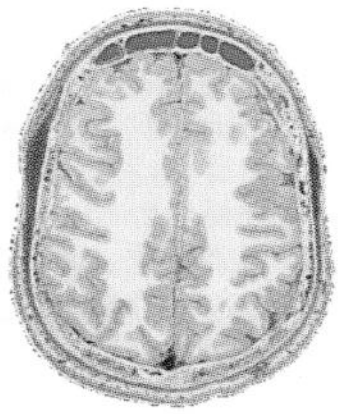

a_vm1060

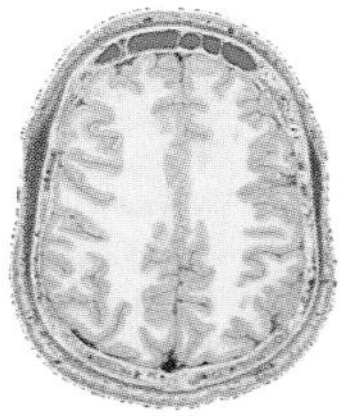

Transverse
a_vm1072

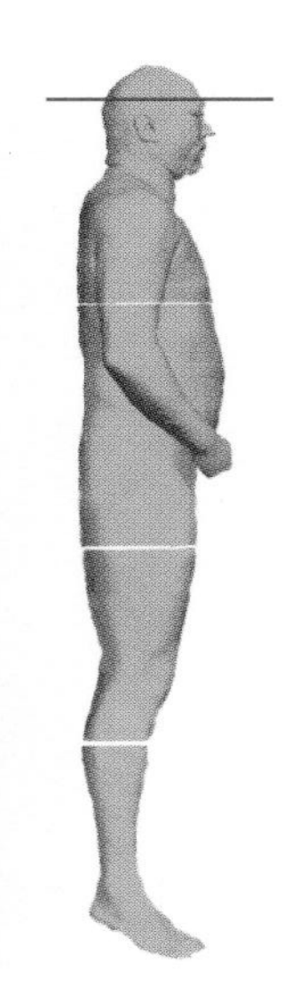

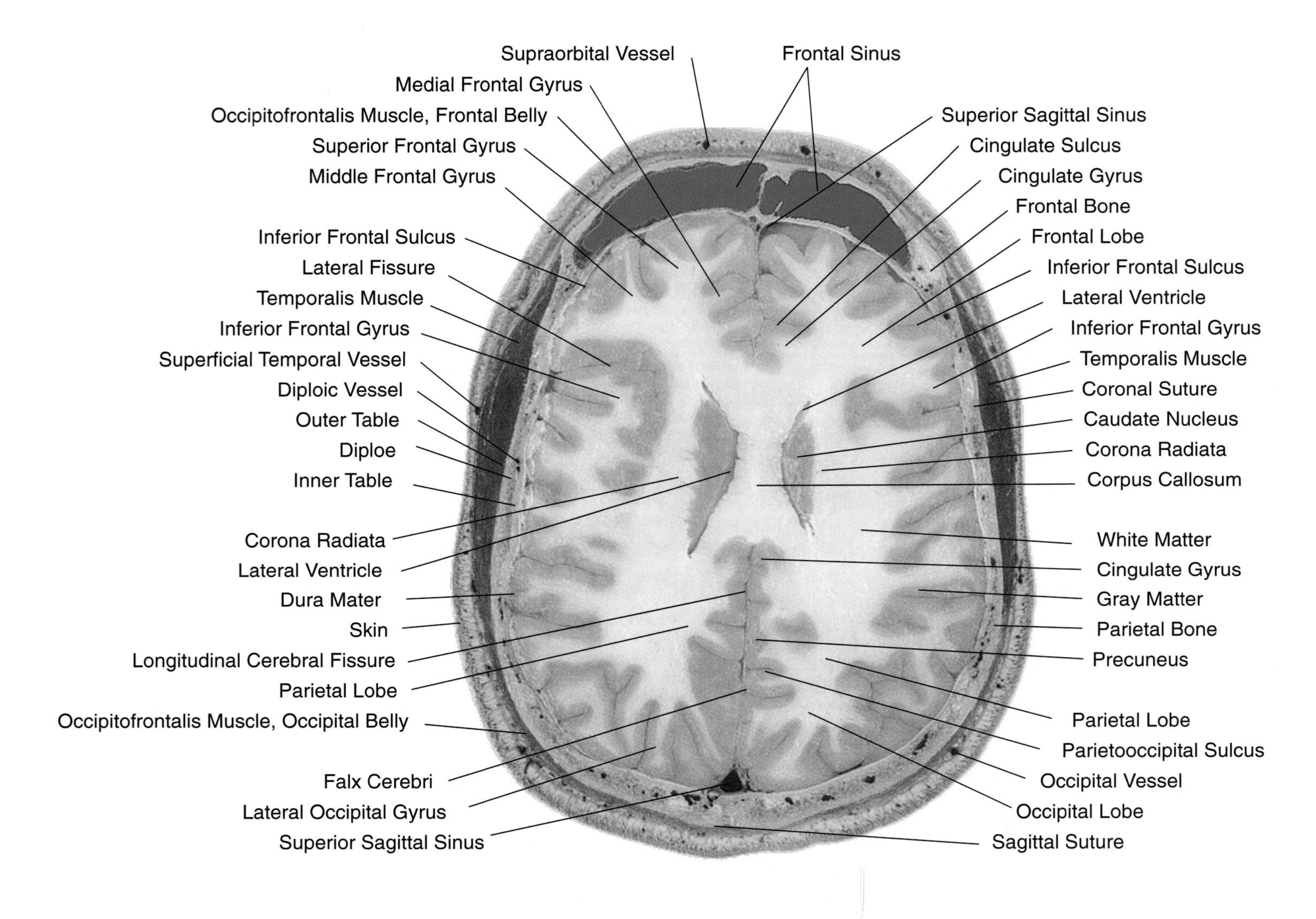
anterior
Supraorbital Vessel
Frontal Sinus
Medial Frontal Gyrus
Occipitofrontalis Muscle, Frontal Belly
Superior Frontal Gyrus
Middle Frontal Gyrus
Inferior Frontal Sulcus
Lateral Fissure
Temporalis Muscle
Inferior Frontal Gyrus
Superficial Temporal Vessel
Diploic Vessel
Outer Table
Diploe
Inner Table
Corona Radiata
Lateral Ventricle
Dura Mater
Skin
Longitudinal Cerebral Fissure
Parietal Lobe
Occipitofrontalis Muscle, Occipital Belly
Falx Cerebri
Lateral Occipital Gyrus
Superior Sagittal Sinus
Superior Sagittal Sinus
Cingulate Sulcus
Cingulate Gyrus
Frontal Bone
Frontal Lobe
Inferior Frontal Sulcus
Lateral Ventricle
Inferior Frontal Gyrus
Temporalis Muscle
Coronal Suture
Caudate Nucleus
Corona Radiata
Corpus Callosum
White Matter
Cingulate Gyrus
Gray Matter
Parietal Bone
Precuneus
Parietal Lobe
Parietooccipital Sulcus
Occipital Vessel
Occipital Lobe
Sagittal Suture
right
left
posterior

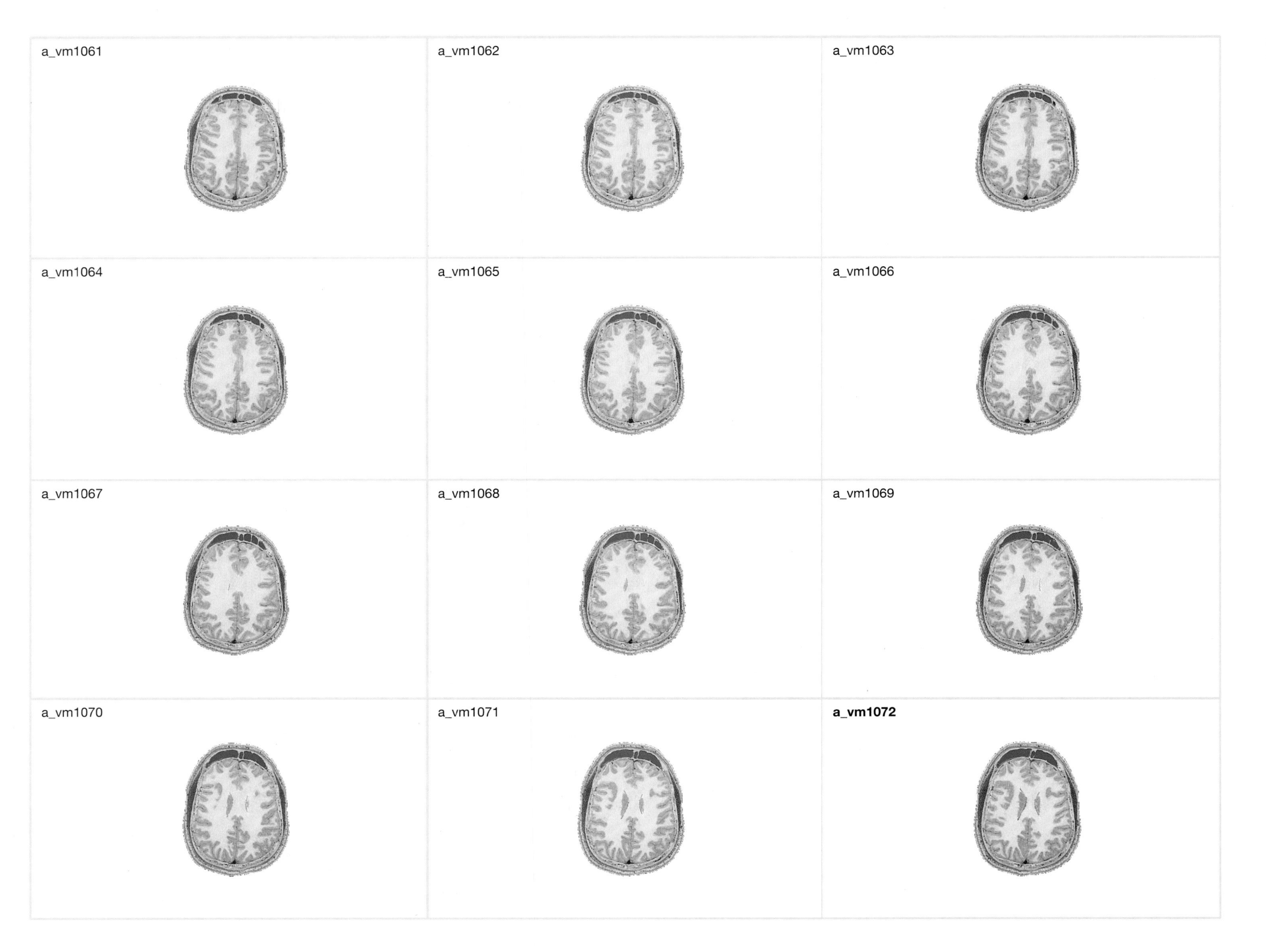
a_vm1061
a_vm1062
a_vm1063
a_vm1064
a_vm1065
a_vm1066
a_vm1067
a_vm1068
a_vm1069
a_vm1070
a_vm1071
a_vm1072

Transverse
a_vm1082

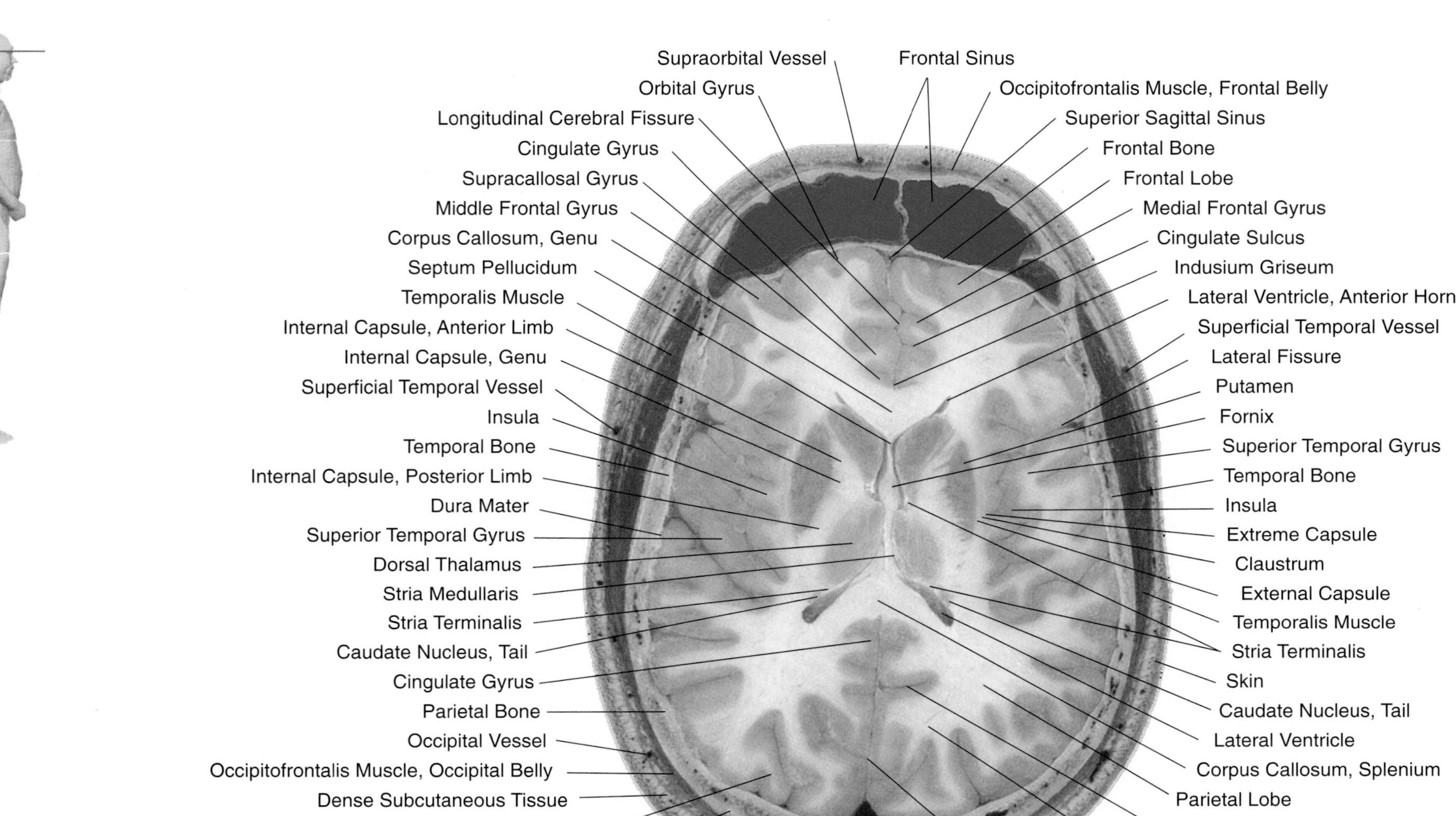

anterior
Supraorbital Vessel
Orbital Gyrus
Longitudinal Cerebral Fissure
Cingulate Gyrus
Supracallosal Gyrus
Middle Frontal Gyrus
Corpus Callosum, Genu
Septum Pellucidum
Temporalis Muscle
Internal Capsule, Anterior Limb
Internal Capsule, Genu
Superficial Temporal Vessel
Insula
Temporal Bone
Internal Capsule, Posterior Limb
Dura Mater
Superior Temporal Gyrus
Dorsal Thalamus
Stria Medullaris
Stria Terminalis
Caudate Nucleus, Tail
Cingulate Gyrus
Parietal Bone
Occipital Vessel
Occipitofrontalis Muscle, Occipital Belly
Dense Subcutaneous Tissue
Lateral Occipital Gyrus
Outer Table
Diploe
Inner Table
Lambdoidal Suture
Superior Sagittal Sinus
Frontal Sinus
Occipitofrontalis Muscle, Frontal Belly
Superior Sagittal Sinus
Frontal Bone
Frontal Lobe
Medial Frontal Gyrus
Cingulate Sulcus
Indusium Griseum
Lateral Ventricle, Anterior Horn
Superficial Temporal Vessel
Lateral Fissure
Putamen
Fornix
Superior Temporal Gyrus
Temporal Bone
Insula
Extreme Capsule
Claustrum
External Capsule
Temporalis Muscle
Stria Terminalis
Skin
Caudate Nucleus, Tail
Lateral Ventricle
Corpus Callosum, Splenium
Parietal Lobe
Parietooccipital Sulcus
Occipital Lobe
Occipital Vessel
Falx Cerebri
Occipital Bone
right
left
posterior

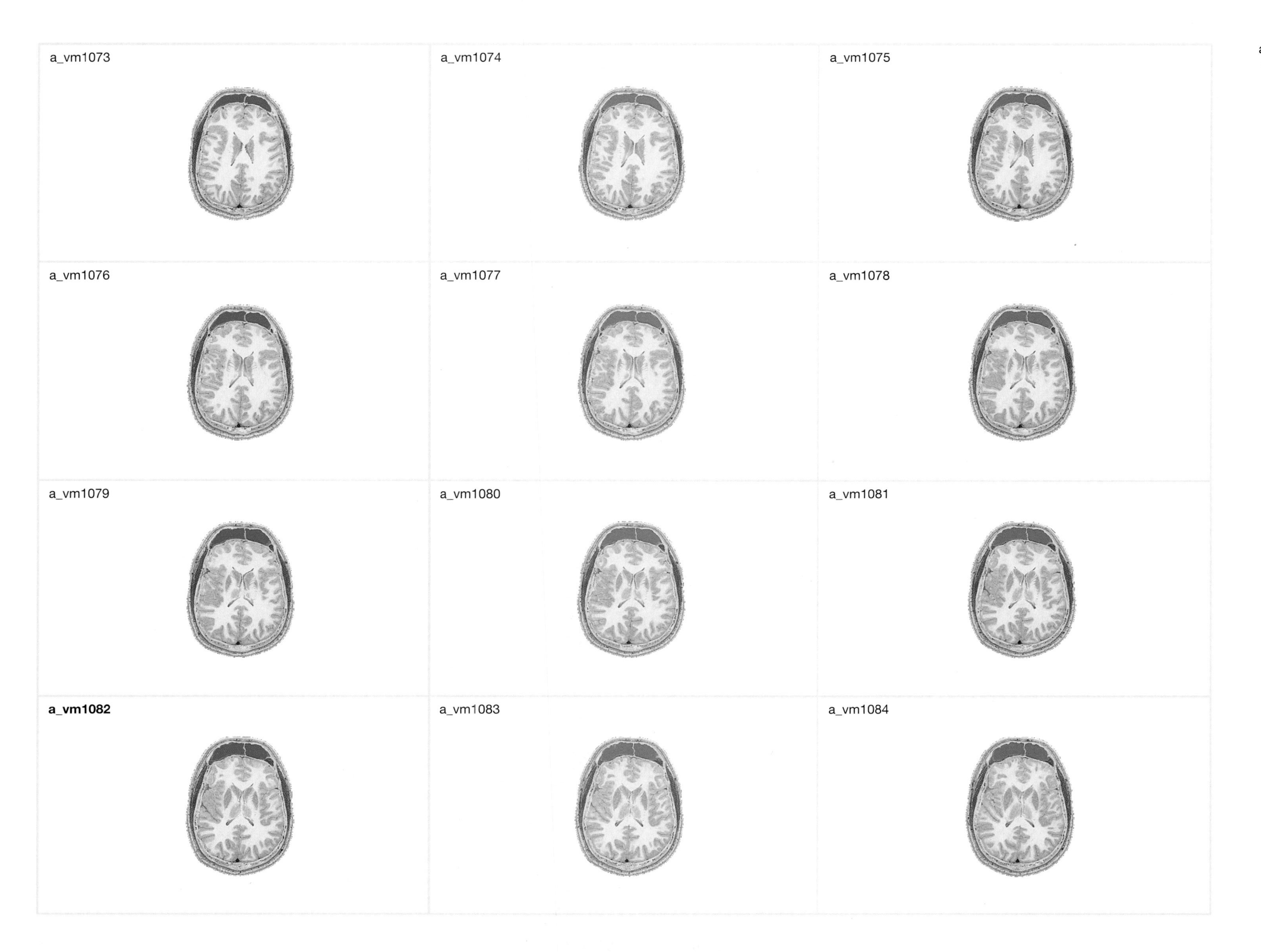
a_vm1073
a_vm1074
a_vm1075
a_vm1076
a_vm1077
a_vm1078
a_vm1079
a_vm1080
a_vm1081
a_vm1082
a_vm1083
a_vm1084

Transverse
a_vm1095

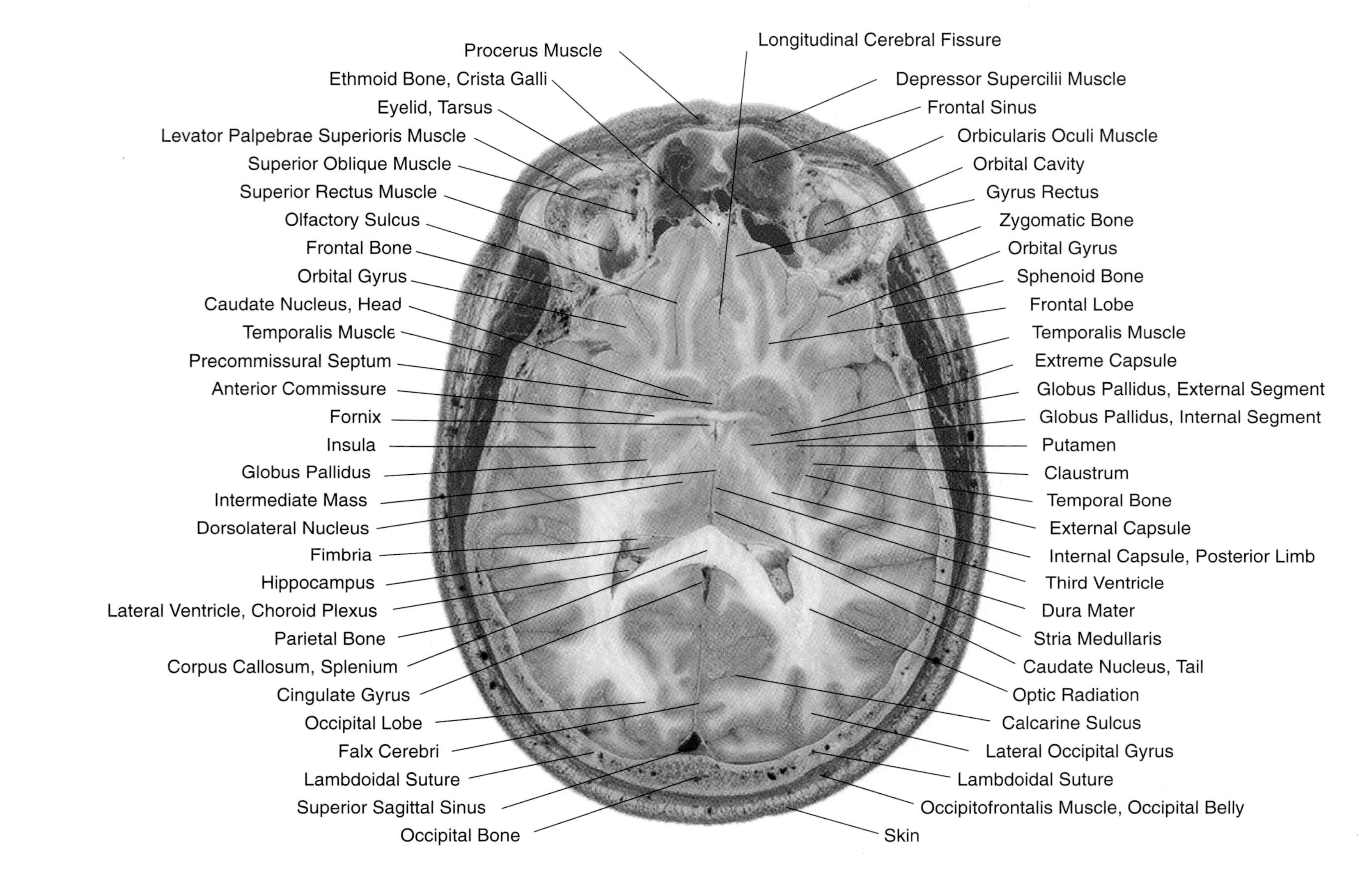

anterior
Procerus Muscle
Ethmoid Bone, Crista Galli
Eyelid, Tarsus
Levator Palpebrae Superioris Muscle
Superior Oblique Muscle
Superior Rectus Muscle
Olfactory Sulcus
Frontal Bone
Orbital Gyrus
Caudate Nucleus, Head
Temporalis Muscle
Precommissural Septum
Anterior Commissure
Fornix
Insula
Globus Pallidus
Intermediate Mass
Dorsolateral Nucleus
Fimbria
Hippocampus
Lateral Ventricle, Choroid Plexus
Parietal Bone
Corpus Callosum, Splenium
Cingulate Gyrus
Occipital Lobe
Falx Cerebri
Lambdoidal Suture
Superior Sagittal Sinus
Occipital Bone
Longitudinal Cerebral Fissure
Depressor Supercilii Muscle
Frontal Sinus
Orbicularis Oculi Muscle
Orbital Cavity
Gyrus Rectus
Zygomatic Bone
Orbital Gyrus
Sphenoid Bone
Frontal Lobe
Temporalis Muscle
Extreme Capsule
Globus Pallidus, External Segment
Globus Pallidus, Internal Segment
Putamen
Claustrum
Temporal Bone
External Capsule
Internal Capsule, Posterior Limb
Third Ventricle
Dura Mater
Stria Medullaris
Caudate Nucleus, Tail
Optic Radiation
Calcarine Sulcus
Lateral Occipital Gyrus
Lambdoidal Suture
Occipitofrontalis Muscle, Occipital Belly
Skin
right
left
posterior

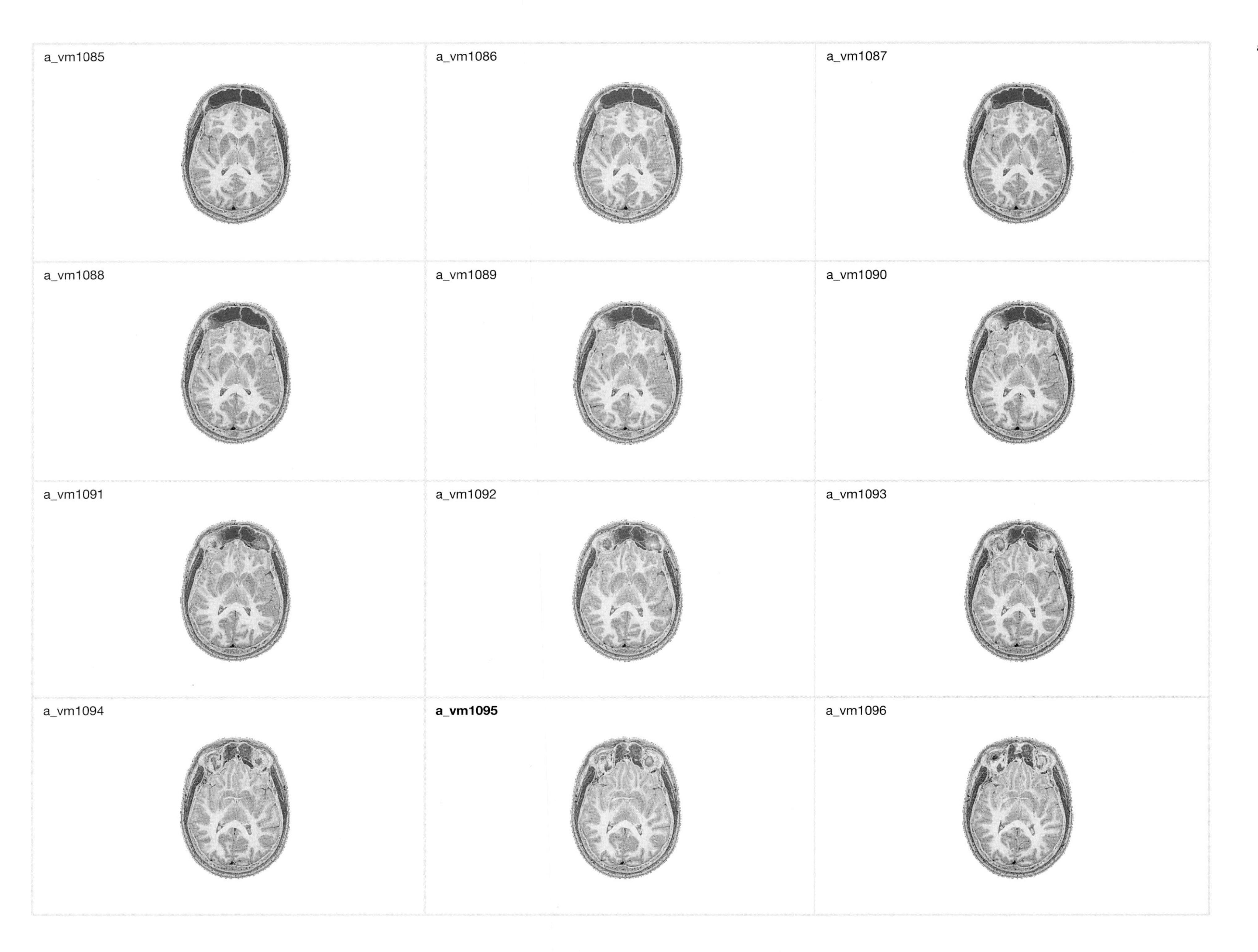
a_vm1085
a_vm1086
a_vm1087
a_vm1088
a_vm1089
a_vm1090
a_vm1091
a_vm1092
a_vm1093
a_vm1094
a_vm1095
a_vm1096

Transverse
a_vm1107

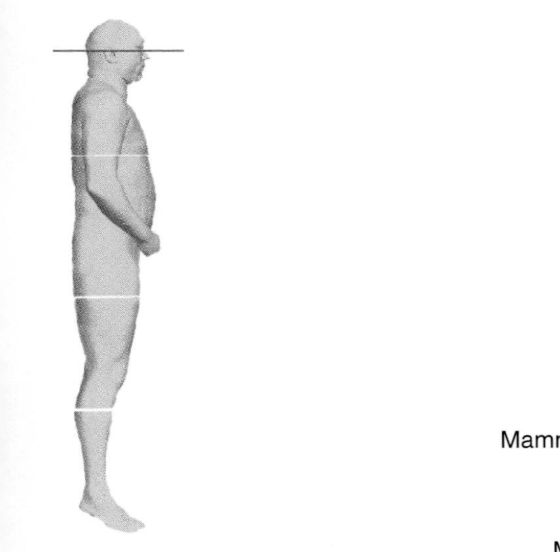

anterior
Nasal Bone
Sphenoid Sinus
Orbicularis Oculi Muscle
Tuber Cinereum
Optic Nerve
Hypothalamus
Zygomatic Bone
Sphenoid Bone
Rhinal Sulcus
Mammillothalamic Fasciculus
Temporalis Muscle
Substantia Nigra
Middle Temporal Gyrus
Habenulopeduncular Tract
Parietotemporopontine Fibers
Reticular Formation
Inferior Temporal Sulcus
Inferior Temporal Gyrus
Auricular Cartilage
Spinal Lemniscus
Superior Colliculus, Brachium
Superior Colliculus, Commissure
Collateral Sulcus
Tectum
Optic Radiation
Parietal Bone
Tentorium
Cerebellar Cortex
Lambdoidal Suture
Occipitofrontalis Muscle, Occipital Belly
Calcarine Sulcus
Occipital Bone
Lacrimal Bone
Maxillary Bone
Nasal Cavity
Ethmoid Bone
Middle Ethmoid Sinus
Medial Rectus Muscle
Lateral Rectus Muscle
Optic Chiasma
Superior Temporal Gyrus
Optic Tract
Uncus
Amygdala
Cerebral Cortex
Fornix
Third Ventricle
Cerebral Peduncle
Lateral Ventricle, Inferior Horn
Fimbria
Tapetum
Temporal Lobe
Hippocampus
Lateral Ventricle, Choroid Plexus
Medial Lemniscus
Hippocampal Sulcus
Parahippocampal Gyrus
Optic Radiation
Calcarine Sulcus
Lingual Gyrus
Lateral Occipital Gyrus
Occipital Lobe
Longitudinal Cerebral Fissure
Superior Sagittal Sinus
right
left
posterior

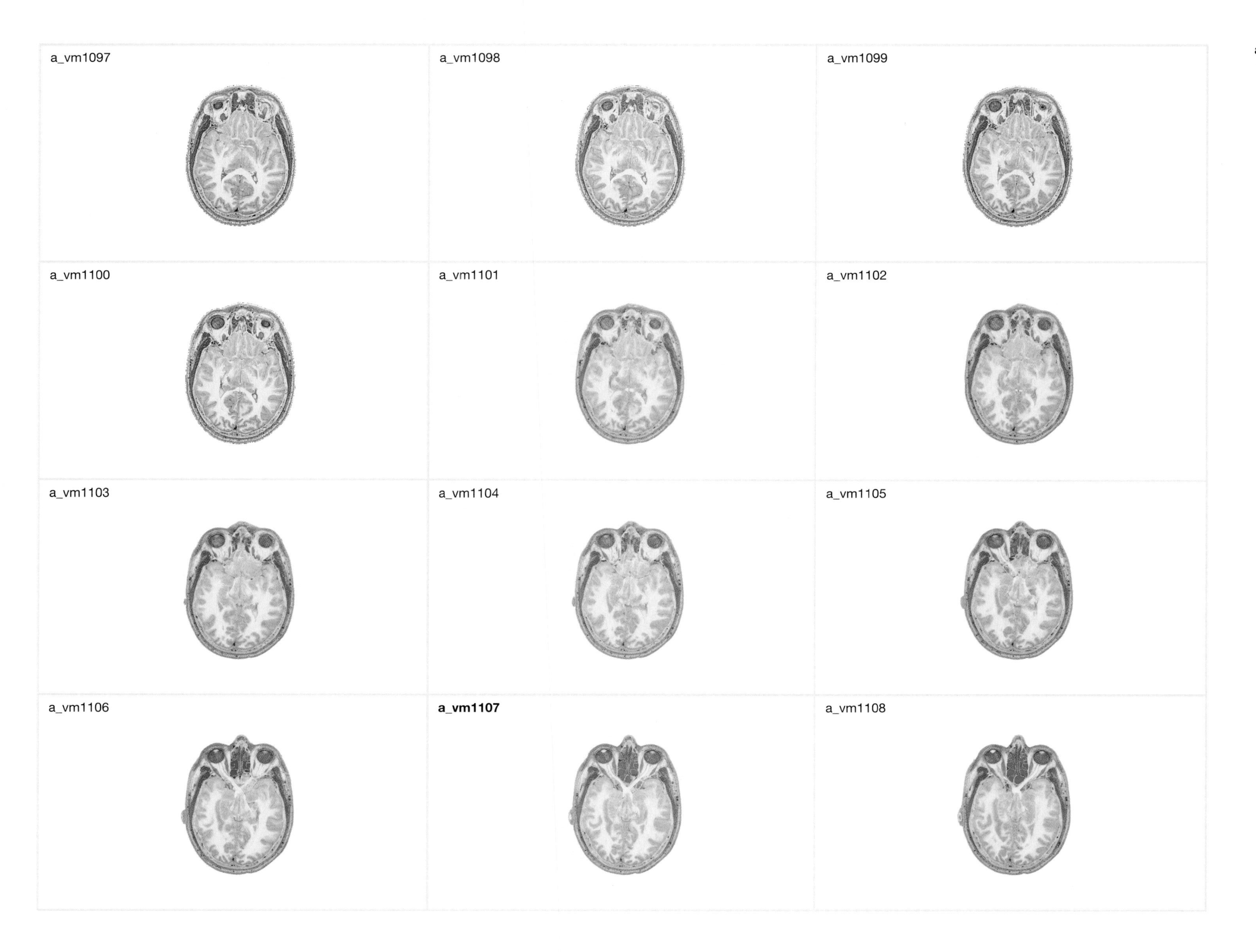
a_vm1097
a_vm1098
a_vm1099
a_vm1100
a_vm1101
a_vm1102
a_vm1103
a_vm1104
a_vm1105
a_vm1106
a_vm1107
a_vm1108

Transverse
a_vm1112

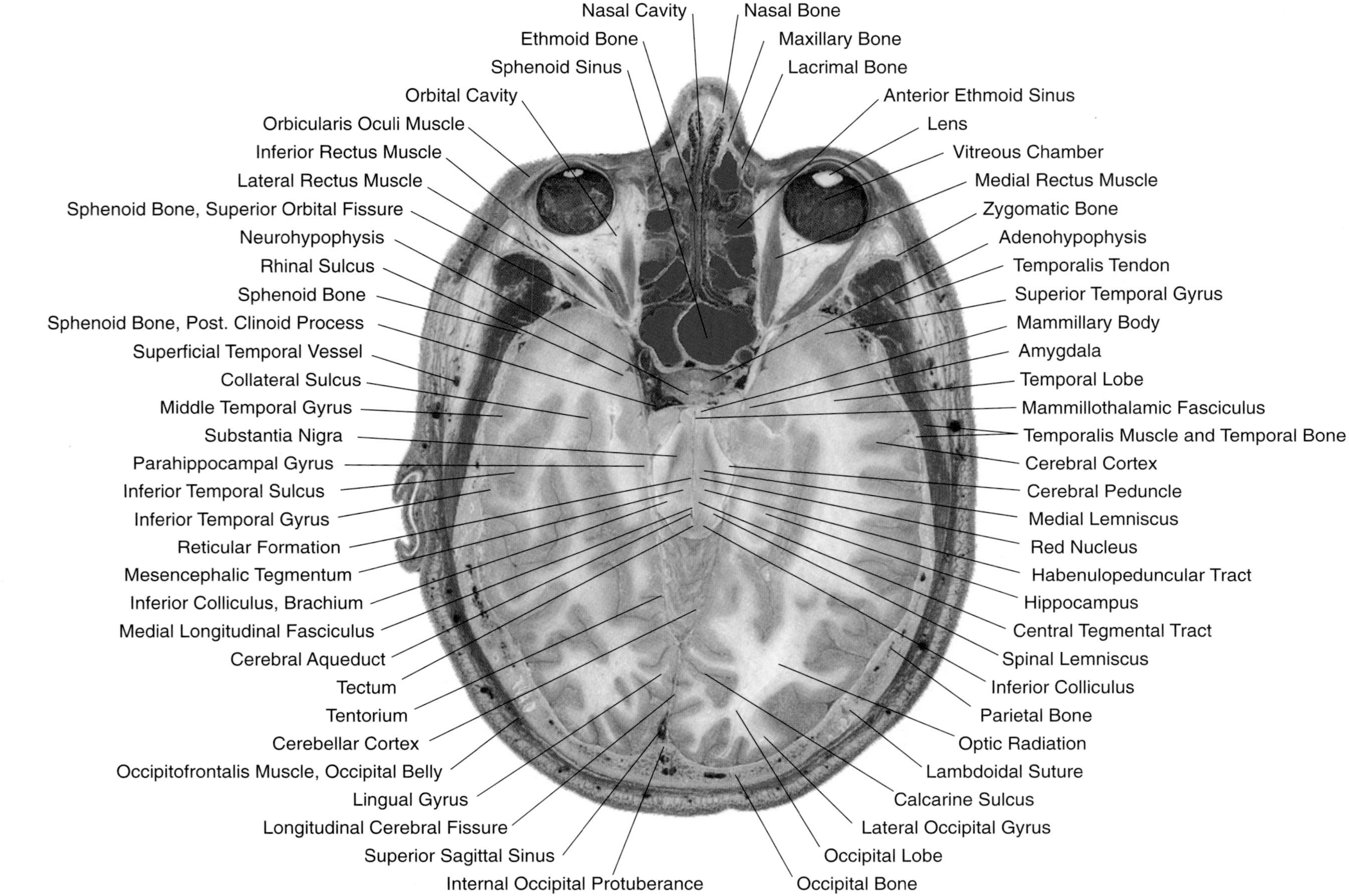
anterior
Nasal Cavity
Nasal Bone
Ethmoid Bone
Maxillary Bone
Sphenoid Sinus
Lacrimal Bone
Orbital Cavity
Anterior Ethmoid Sinus
Orbicularis Oculi Muscle
Lens
Inferior Rectus Muscle
Vitreous Chamber
Lateral Rectus Muscle
Medial Rectus Muscle
Sphenoid Bone, Superior Orbital Fissure
Zygomatic Bone
Neurohypophysis
Adenohypophysis
Rhinal Sulcus
Temporalis Tendon
Sphenoid Bone
Superior Temporal Gyrus
Sphenoid Bone, Post. Clinoid Process
Mammillary Body
Superficial Temporal Vessel
Amygdala
Collateral Sulcus
Temporal Lobe
Middle Temporal Gyrus
Mammillothalamic Fasciculus
Substantia Nigra
Temporalis Muscle and Temporal Bone
Parahippocampal Gyrus
Cerebral Cortex
Inferior Temporal Sulcus
Cerebral Peduncle
Inferior Temporal Gyrus
Medial Lemniscus
Reticular Formation
Red Nucleus
Mesencephalic Tegmentum
Habenulopeduncular Tract
Inferior Colliculus, Brachium
Hippocampus
Medial Longitudinal Fasciculus
Central Tegmental Tract
Cerebral Aqueduct
Spinal Lemniscus
Tectum
Inferior Colliculus
Tentorium
Parietal Bone
Cerebellar Cortex
Optic Radiation
Occipitofrontalis Muscle, Occipital Belly
Lambdoidal Suture
Lingual Gyrus
Calcarine Sulcus
Longitudinal Cerebral Fissure
Lateral Occipital Gyrus
right
left
Superior Sagittal Sinus
Occipital Lobe
Internal Occipital Protuberance
Occipital Bone
posterior

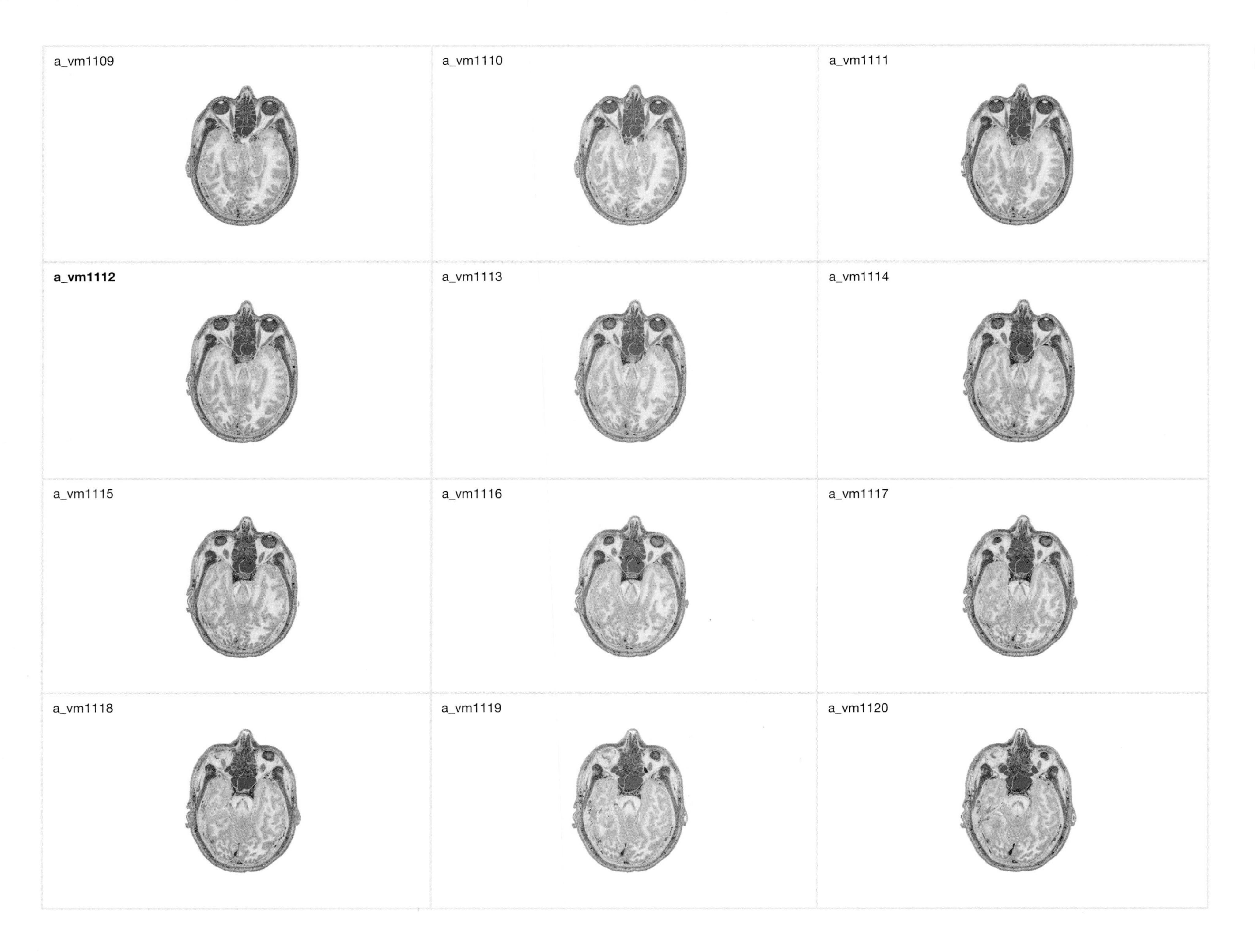
a_vm1109
a_vm1110
a_vm1111
a_vm1112
a_vm1113
a_vm1114
a_vm1115
a_vm1116
a_vm1117
a_vm1118
a_vm1119
a_vm1120

Transverse
a_vm1122

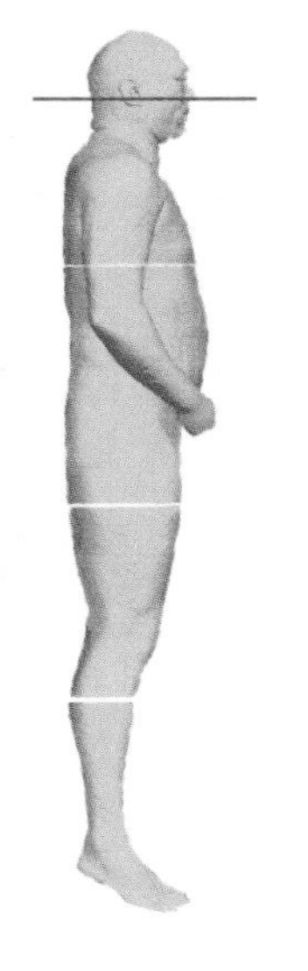

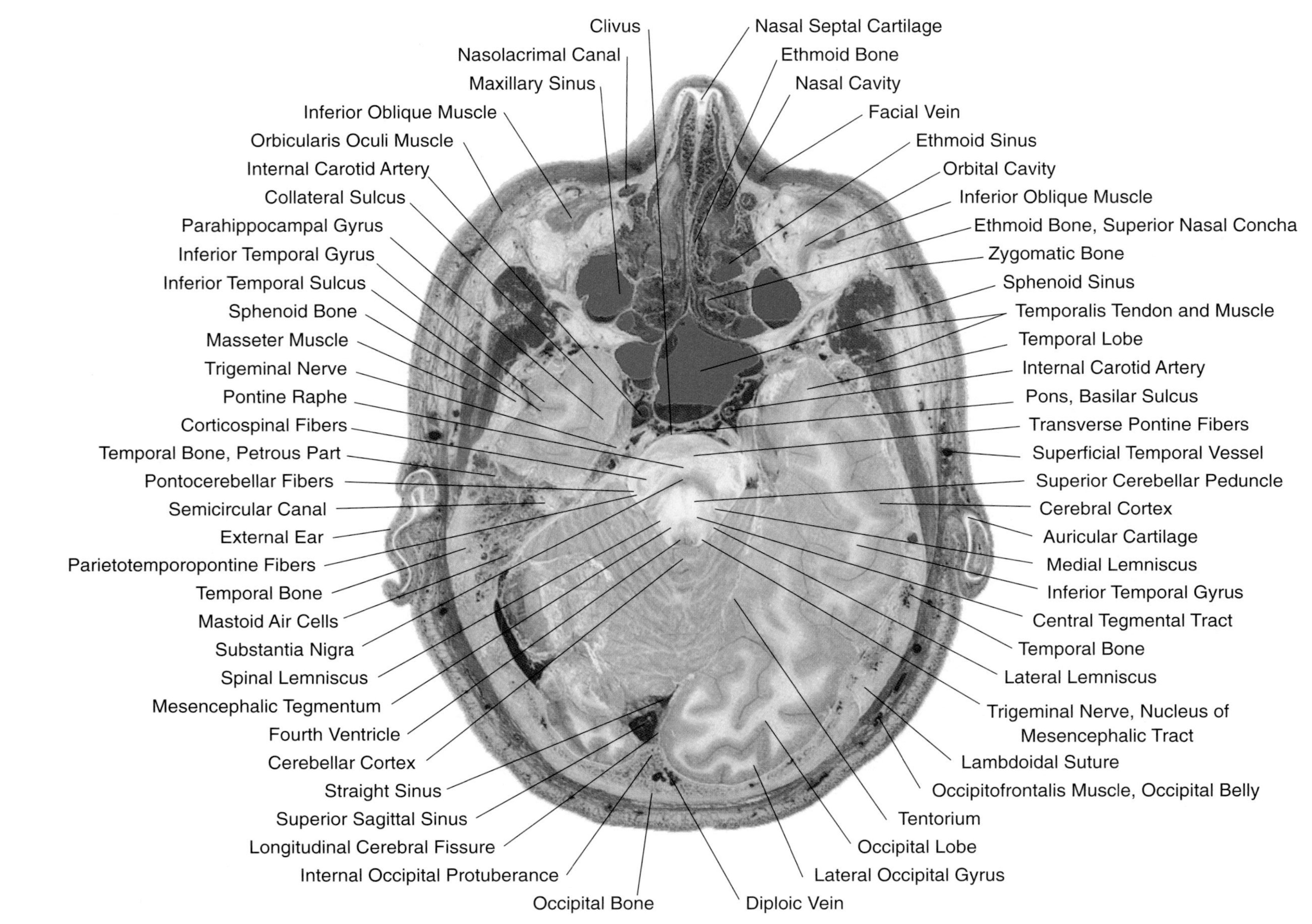
anterior
Clivus
Nasolacrimal Canal
Maxillary Sinus
Inferior Oblique Muscle
Orbicularis Oculi Muscle
Internal Carotid Artery
Collateral Sulcus
Parahippocampal Gyrus
Inferior Temporal Gyrus
Inferior Temporal Sulcus
Sphenoid Bone
Masseter Muscle
Trigeminal Nerve
Pontine Raphe
Corticospinal Fibers
Temporal Bone, Petrous Part
Pontocerebellar Fibers
Semicircular Canal
External Ear
Parietotemporopontine Fibers
Temporal Bone
Mastoid Air Cells
Substantia Nigra
Spinal Lemniscus
Mesencephalic Tegmentum
Fourth Ventricle
Cerebellar Cortex
Straight Sinus
Superior Sagittal Sinus
Longitudinal Cerebral Fissure
Internal Occipital Protuberance
Occipital Bone
Nasal Septal Cartilage
Ethmoid Bone
Nasal Cavity
Facial Vein
Ethmoid Sinus
Orbital Cavity
Inferior Oblique Muscle
Ethmoid Bone, Superior Nasal Concha
Zygomatic Bone
Sphenoid Sinus
Temporalis Tendon and Muscle
Temporal Lobe
Internal Carotid Artery
Pons, Basilar Sulcus
Transverse Pontine Fibers
Superficial Temporal Vessel
Superior Cerebellar Peduncle
Cerebral Cortex
Auricular Cartilage
Medial Lemniscus
Inferior Temporal Gyrus
Central Tegmental Tract
Temporal Bone
Lateral Lemniscus
Trigeminal Nerve, Nucleus of Mesencephalic Tract
Lambdoidal Suture
Occipitofrontalis Muscle, Occipital Belly
Tentorium
Occipital Lobe
Lateral Occipital Gyrus
Diploic Vein
right
left
posterior

a_vm1121

a_vm1122

a_vm1123

a_vm1124

a_vm1125

a_vm1126

a_vm1127

a_vm1128

a_vm1129

a_vm1130

a_vm1131

a_vm1132

Transverse
a_vm1134

anterior
Nasal Septal Cartilage
Nasal Cavity
Nasolacrimal Canal
Pontine Nuclei
Sphenoid Bone
Corticospinal Fibers
Temporalis Muscle
Zygomaticus Major Muscle
Masseter Muscle
Pontocerebellar Fibers
Temporal Bone, Auditory Tube
Temporal Bone, Tubercle
Temporomandibular Joint
Mandible, Condyle
Sphenoid Bone, Foramen Spinosum
Superficial Temporal Vessel
Carotid Canal
External Acoustic Meatus
Middle Cerebellar Peduncle
Temporal Bone
Locus Ceruleus
External Ear
Mastoid Air Cells
Transverse Sinus
Superior Cerebellar Peduncle
Cerebellum, Fastigial Nucleus
Cerebellum, Emboliform Nucleus
Cerebellar Cortex
Lateral Cerebellar Hemisphere
Diploic Vessel
Semispinalis Capitis Muscle
Loose Connective Tissue
Occipital Bone, Outer Table
Cerebellum, Vermis
Ethmoid Bone, Middle Nasal Concha
Maxillary Bone
Levator Labii Superioris Muscle
Occipital Bone, Basal Portion
Maxillary Sinus
Orbicularis Oculi Muscle
Pons, Basilar Sulcus
Lateral Pterygoid Muscle
Mandible, Coronoid Process
Internal Carotid Artery
Pontine Raphe
Transverse Pontine Fibers
Masseter Muscle
Retromandibular Vein
Medial Lemniscus
Superficial Temporal Vein
Tympanic Cavity
Cochlea
Auricular Cartilage
Temporal Bone
Lateral Lemniscus
Pons, Tegmentum
Central Tegmental Tract
Mastoid Air Cells
Spinal Lemniscus
Medial Longitudinal Fasciculus
Lambdoidal Suture
Fourth Ventricle
Cerebellum, Globose Nucleus
Artifact
Internal Occipital Protuberance
Dense Connective Tissue
Occipital Bone
External Occipital Protuberance
right
left
posterior

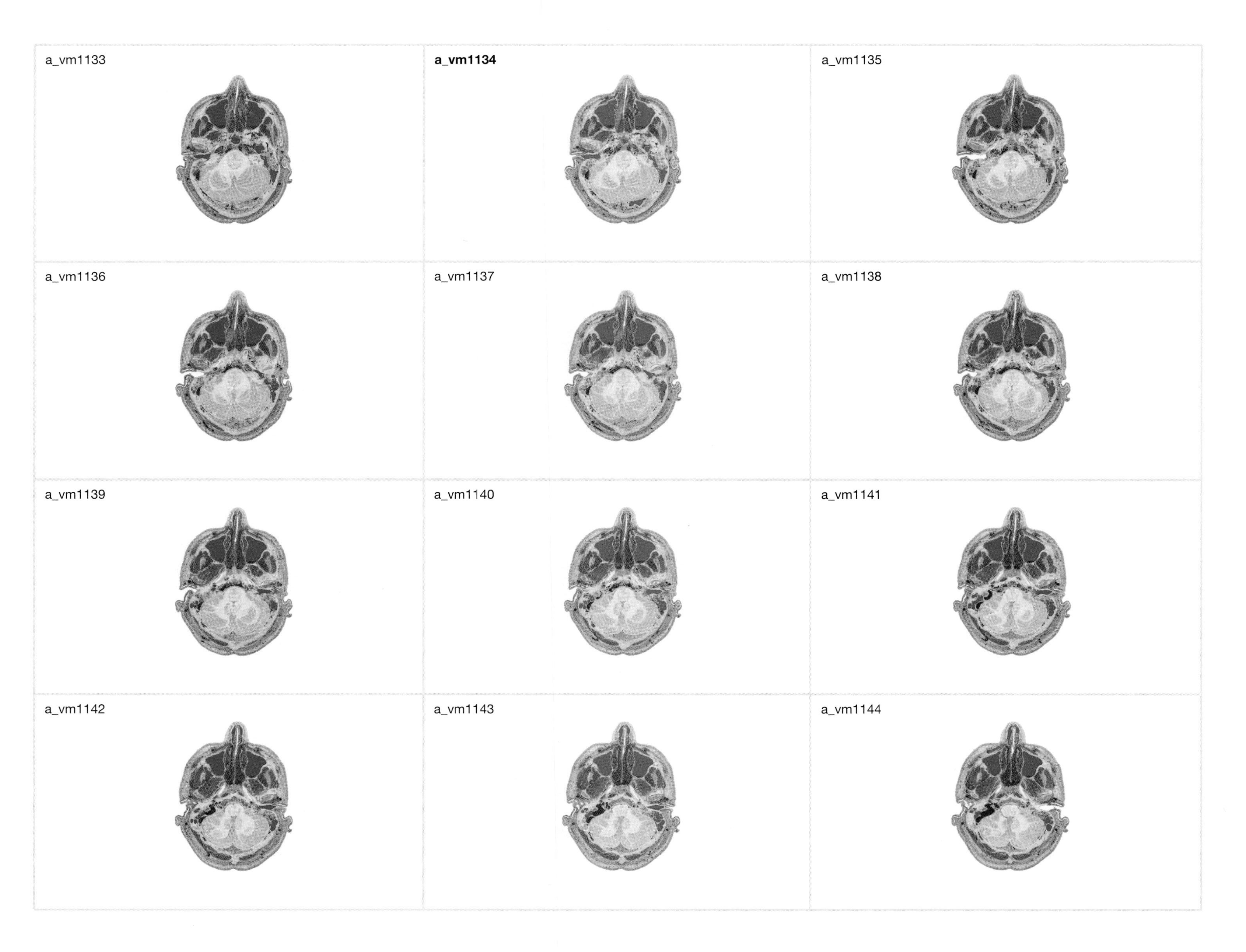
a_vm1133
a_vm1134
a_vm1135
a_vm1136
a_vm1137
a_vm1138
a_vm1139
a_vm1140
a_vm1141
a_vm1142
a_vm1143
a_vm1144

Transverse
a_vm1150

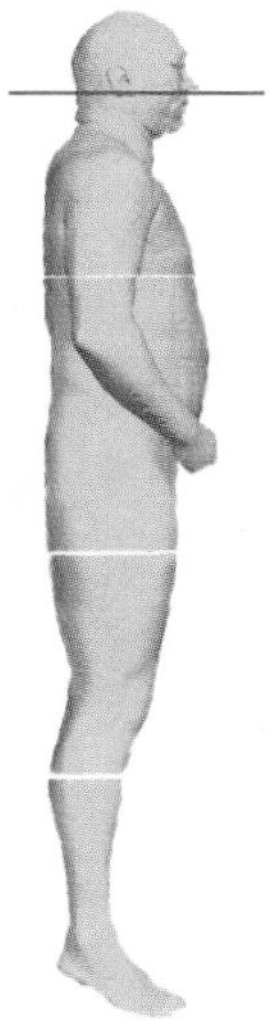

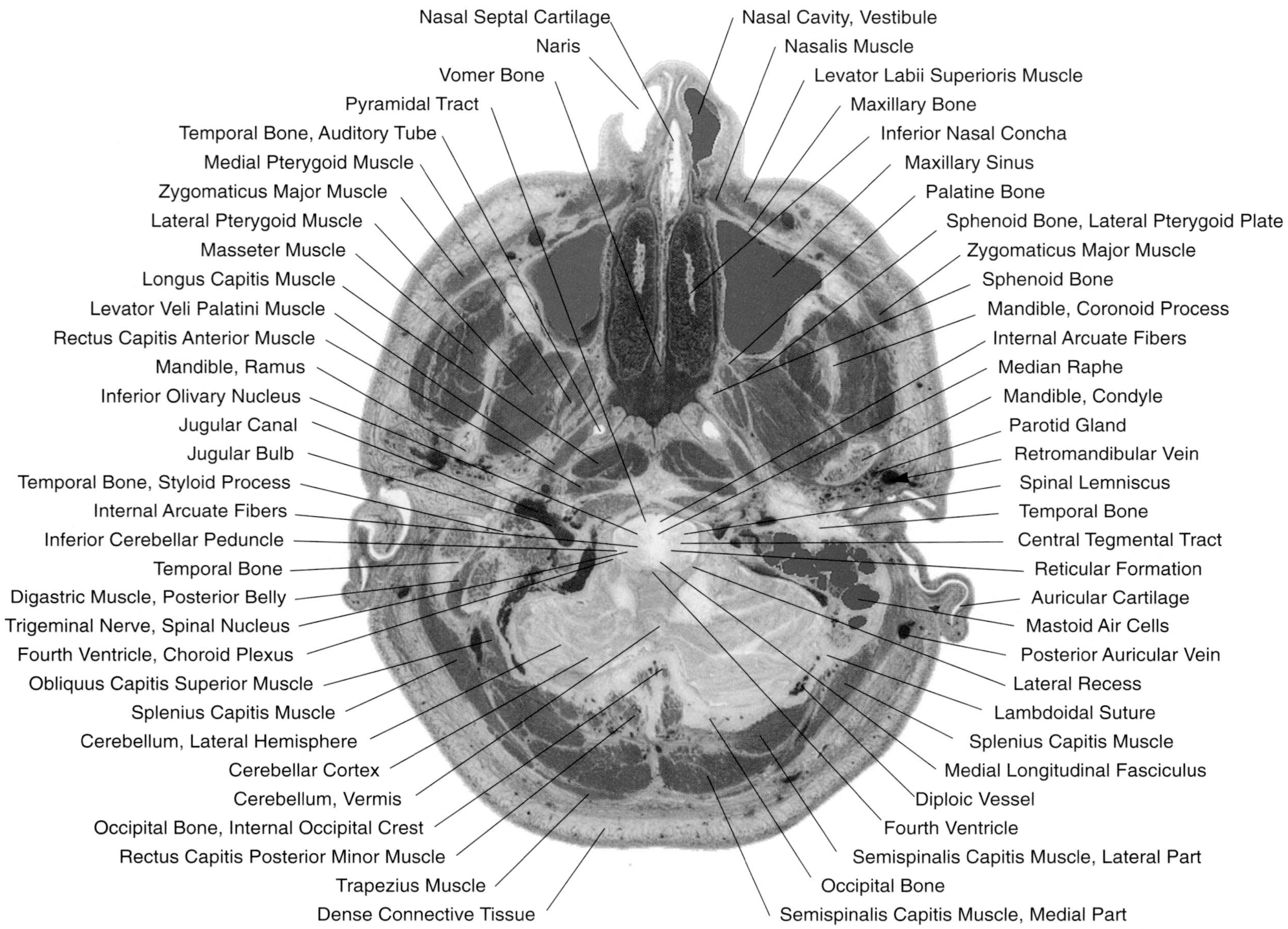

anterior

Nasal Septal Cartilage
Naris
Vomer Bone
Pyramidal Tract
Temporal Bone, Auditory Tube
Medial Pterygoid Muscle
Zygomaticus Major Muscle
Lateral Pterygoid Muscle
Masseter Muscle
Longus Capitis Muscle
Levator Veli Palatini Muscle
Rectus Capitis Anterior Muscle
Mandible, Ramus
Inferior Olivary Nucleus
Jugular Canal
Jugular Bulb
Temporal Bone, Styloid Process
Internal Arcuate Fibers
Inferior Cerebellar Peduncle
Temporal Bone
Digastric Muscle, Posterior Belly
Trigeminal Nerve, Spinal Nucleus
Fourth Ventricle, Choroid Plexus
Obliquus Capitis Superior Muscle
Splenius Capitis Muscle
Cerebellum, Lateral Hemisphere
Cerebellar Cortex
Cerebellum, Vermis
Occipital Bone, Internal Occipital Crest
Rectus Capitis Posterior Minor Muscle
Trapezius Muscle
Dense Connective Tissue

Nasal Cavity, Vestibule
Nasalis Muscle
Levator Labii Superioris Muscle
Maxillary Bone
Inferior Nasal Concha
Maxillary Sinus
Palatine Bone
Sphenoid Bone, Lateral Pterygoid Plate
Zygomaticus Major Muscle
Sphenoid Bone
Mandible, Coronoid Process
Internal Arcuate Fibers
Median Raphe
Mandible, Condyle
Parotid Gland
Retromandibular Vein
Spinal Lemniscus
Temporal Bone
Central Tegmental Tract
Reticular Formation
Auricular Cartilage
Mastoid Air Cells
Posterior Auricular Vein
Lateral Recess
Lambdoidal Suture
Splenius Capitis Muscle
Medial Longitudinal Fasciculus
Diploic Vessel
Fourth Ventricle
Semispinalis Capitis Muscle, Lateral Part
Occipital Bone
Semispinalis Capitis Muscle, Medial Part

right

left

posterior

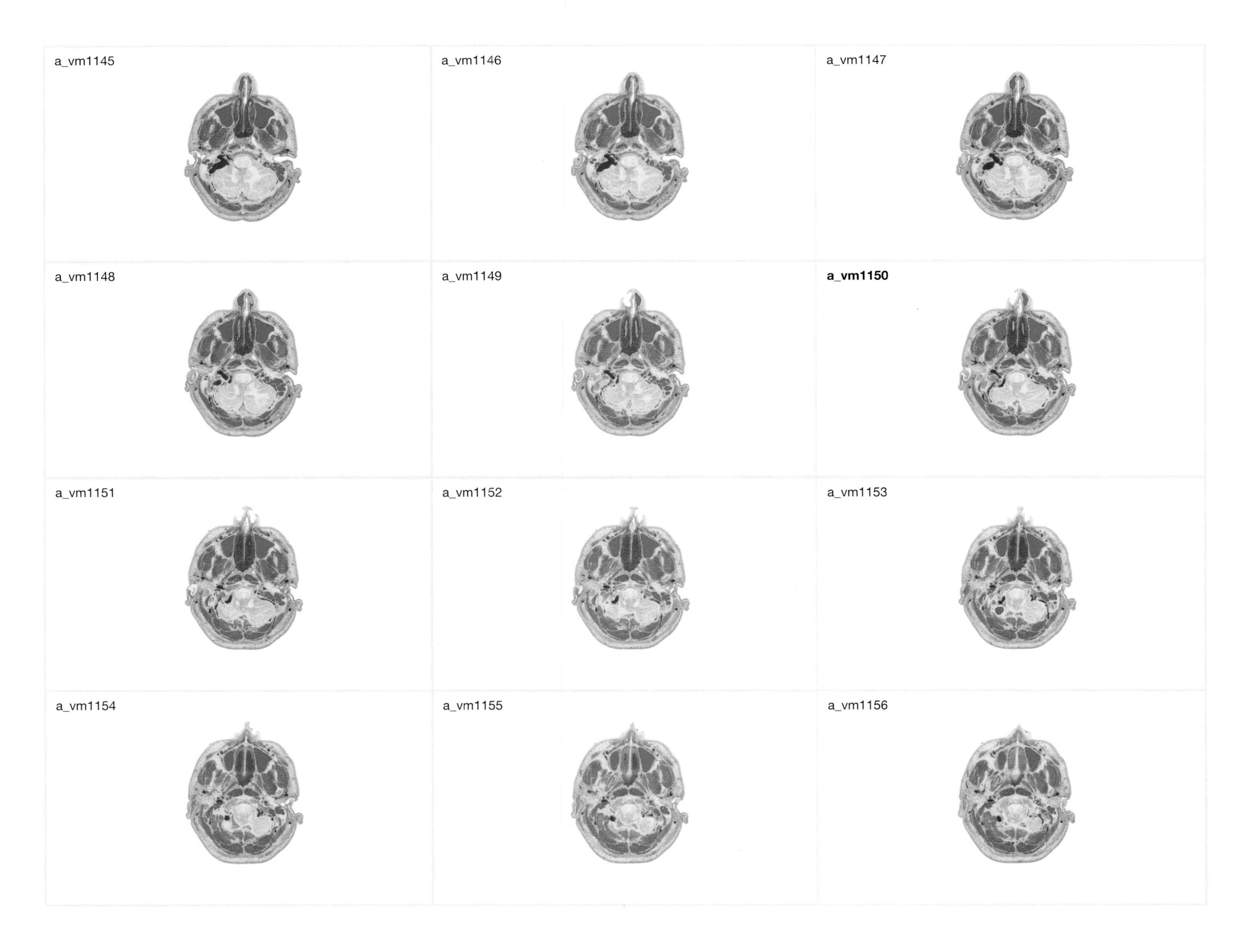
a_vm1145
a_vm1146
a_vm1147
a_vm1148
a_vm1149
a_vm1150
a_vm1151
a_vm1152
a_vm1153
a_vm1154
a_vm1155
a_vm1156

Transverse
a_vm1160

anterior
Hard Palate
Levator Labii Superioris Muscle
Levator Anguli Oris Muscle
Facial Vein
Zygomaticus Major Muscle
Palatine Bone
Uvula Muscle
Levator Veli Palatini Muscle
Longus Capitis Muscle
C1, Atlas, Anterior Arch
Mandible, Ramus
Stylohyoid Muscle
Retromandibular Vein
Parotid Gland
C1, Atlas
Internal Jugular Vein
C1, Atlas, Superior Articular Facet
Digastric Muscle, Posterior Belly
C2, Axis, Dens
Pyramidal Tract
Sternocleidomastoid Muscle
Longissimus Capitis Muscle
Obliquus Capitis Superior Muscle
Deep Cervical Veins
Trigeminal Nerve, Spinal Nucleus
Reticular Formation
Gracilis Nucleus
Posterior Atlantooccipital Membrane
Cisterna Magna
Depressor Septi Muscle
Anterior Nasal Spine
Maxillary Bone
Facial Vein
Maxillary Sinus
Sphenoid Bone, Lateral Pterygoid Plate
Temporalis Muscle
Lateral Pterygoid Muscle
Masseter Muscle
Medial Pterygoid Muscle
Mandible, Ramus
Occipital Condyle
Internal Carotid Artery
Retromandibular Vein
Temporal Bone, Styloid Process
Rectus Capitis Lateralis Muscle
Medulla Oblongata, Raphe
Auricular Cartilage
Mastoid Air Cells
Pyramids, Decussation
Posterior Auricular Vein
Obliquus Capitis Superior Muscle
Cuneate Fasciculus
Gracilis Fasciculus
Splenius Capitis Muscle
Rectus Capitis Posterior Major Muscle
Rectus Capitis Posterior Minor Muscle
Semispinalis Capitis Muscle
Trapezius Muscle
right
left
posterior

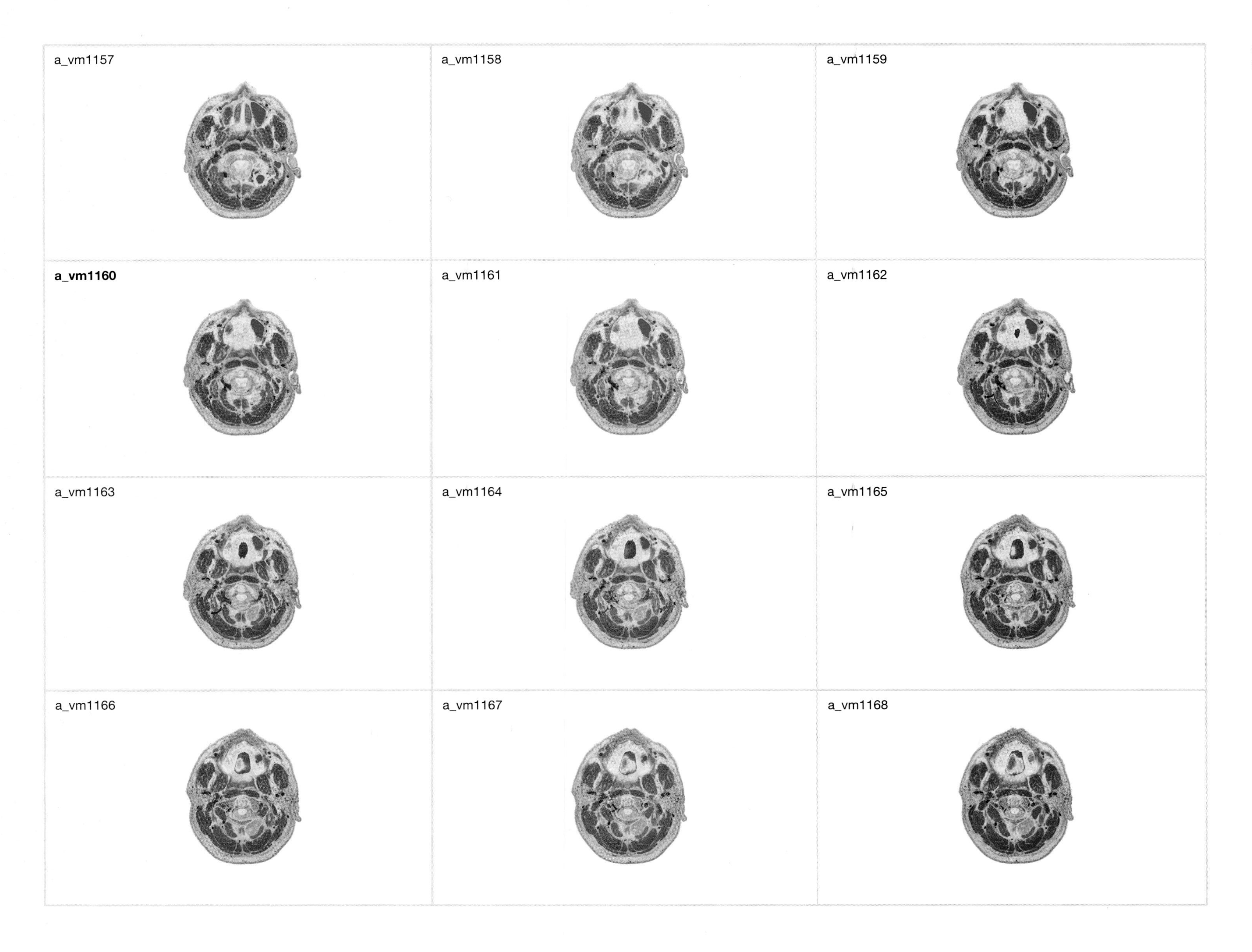
a_vm1157
a_vm1158
a_vm1159
a_vm1160
a_vm1161
a_vm1162
a_vm1163
a_vm1164
a_vm1165
a_vm1166
a_vm1167
a_vm1168

Transverse
a_vm1178

anterior

Upper Incisor Tooth
Upper Lip
Upper Canine Tooth
Tongue
Upper Premolar Teeth
Vestibule
Medial Lingual Raphe
Oral Cavity
Uvula
Nasopharynx
Palatine Tonsil
Stylopharyngeus Muscle
Styloglossus Muscle
Retromandibular Vein
Parotid Gland
Stylohyoid Muscle
Internal Carotid Artery
Digastric Muscle, Posterior Belly
Internal Jugular Vein
Pharyngeal Constrictor Muscle
C2, Axis, Body
Sternocleidomastoid Muscle
Longissimus Capitis Muscle
Obliquus Capitis Inferior Muscle
Splenius Capitis Muscle
Cisterna Magna
Deep Cervical Vein
Semispinalis Capitis Muscle, Lateral Part
Rectus Capitis Posterior Major Muscle
Semispinalis Capitis Muscle, Medial Part

Orbicularis Oris Muscle
Maxillary Bone
Vestibule
Depressor Anguli Oris Muscle
Zygomaticus Major Muscle
Dental Restoration
Facial Vein
Buccinator Muscle
Buccal Fat Pad
Masseter Muscle
Vallate Papilla
Palatoglossus Muscle
Medial Pterygoid Muscle
Mandible, Ramus
Palatopharyngeus Muscle
Retromandibular Vein
Temporal Bone, Styloid Process
Parotid Gland
Internal Carotid Artery
Posterior Auricular Vein
Levator Scapulae Muscle
Longus Capitis Muscle
Vertebral Artery
C2, Axis, Superior Articular Facet
C2, Dorsal Root Ganglion
Spinal Cord
Deep Cervical Fascia
Occipital Vessel
Trapezius Muscle
Artifact

right

left

posterior

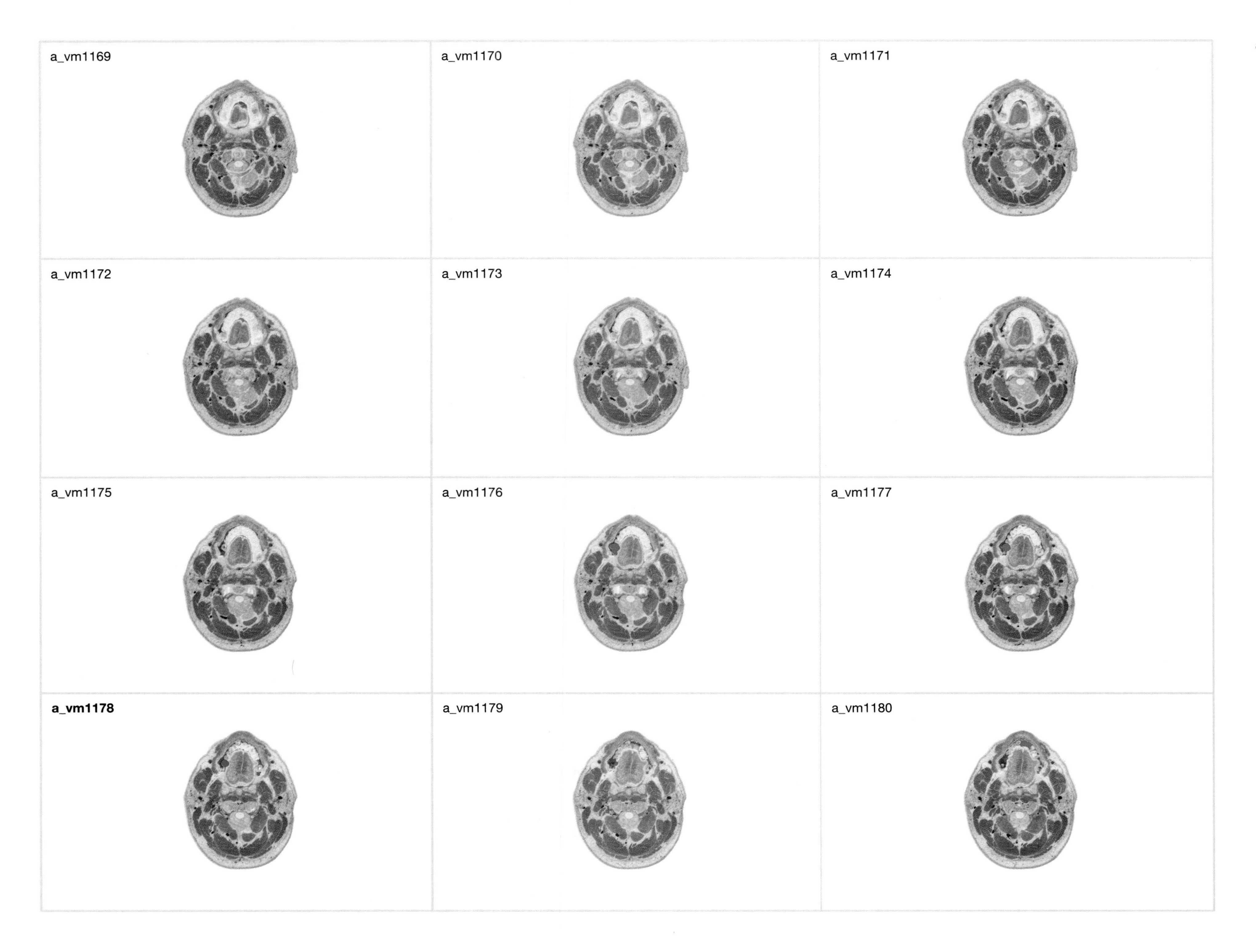
a_vm1169
a_vm1170
a_vm1171
a_vm1172
a_vm1173
a_vm1174
a_vm1175
a_vm1176
a_vm1177
a_vm1178
a_vm1179
a_vm1180

Transverse
a_vm1188

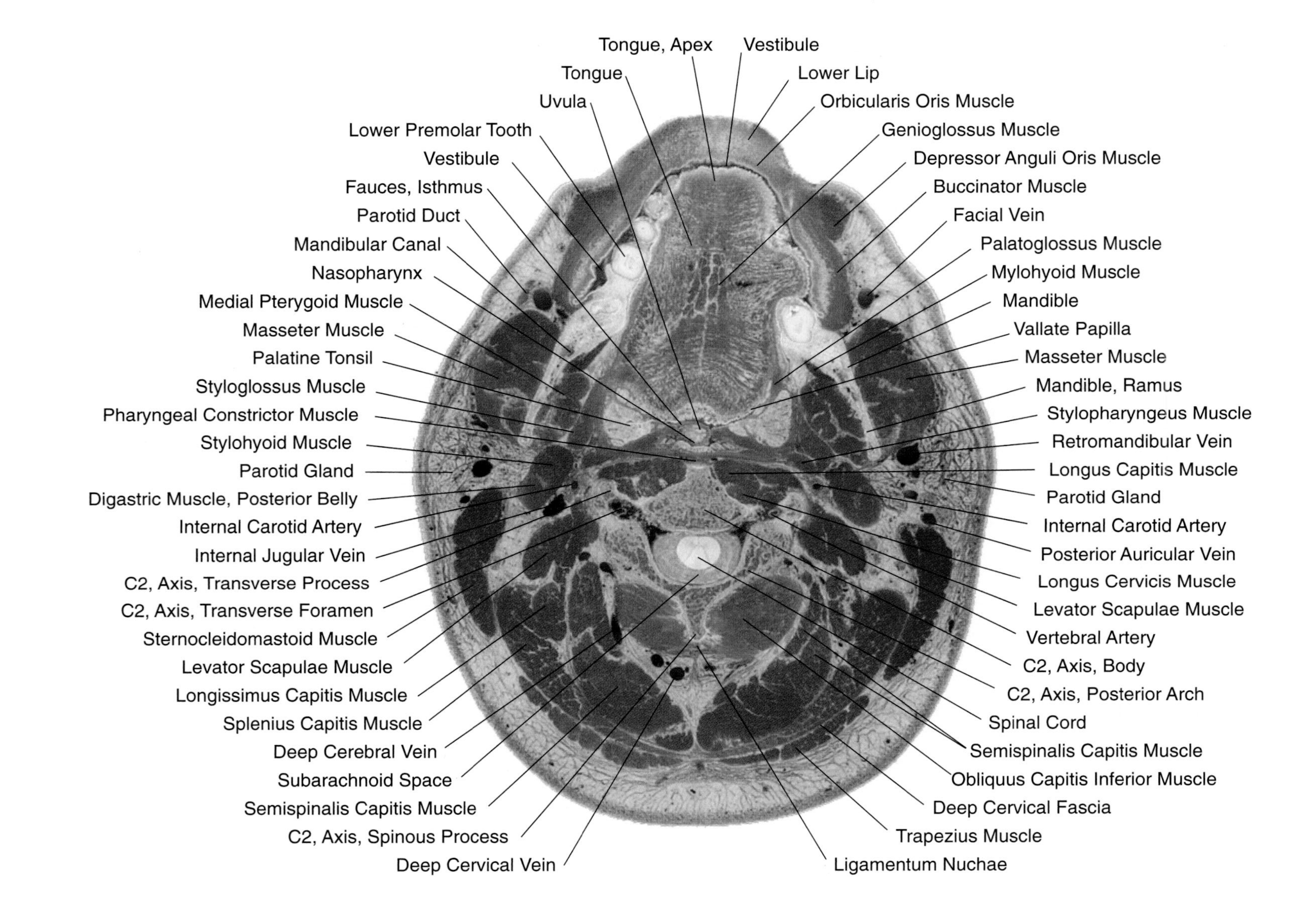
anterior
Tongue, Apex
Vestibule
Tongue
Lower Lip
Uvula
Orbicularis Oris Muscle
Lower Premolar Tooth
Genioglossus Muscle
Vestibule
Depressor Anguli Oris Muscle
Fauces, Isthmus
Buccinator Muscle
Parotid Duct
Facial Vein
Mandibular Canal
Palatoglossus Muscle
Nasopharynx
Mylohyoid Muscle
Medial Pterygoid Muscle
Mandible
Masseter Muscle
Vallate Papilla
Palatine Tonsil
Masseter Muscle
Styloglossus Muscle
Mandible, Ramus
Pharyngeal Constrictor Muscle
Stylopharyngeus Muscle
Stylohyoid Muscle
Retromandibular Vein
Parotid Gland
Longus Capitis Muscle
Digastric Muscle, Posterior Belly
Parotid Gland
Internal Carotid Artery
Internal Carotid Artery
Internal Jugular Vein
Posterior Auricular Vein
C2, Axis, Transverse Process
Longus Cervicis Muscle
C2, Axis, Transverse Foramen
Levator Scapulae Muscle
Sternocleidomastoid Muscle
Vertebral Artery
Levator Scapulae Muscle
C2, Axis, Body
Longissimus Capitis Muscle
C2, Axis, Posterior Arch
Splenius Capitis Muscle
Spinal Cord
Deep Cerebral Vein
Semispinalis Capitis Muscle
Subarachnoid Space
Obliquus Capitis Inferior Muscle
Semispinalis Capitis Muscle
Deep Cervical Fascia
C2, Axis, Spinous Process
Trapezius Muscle
Deep Cervical Vein
Ligamentum Nuchae
right
left
posterior

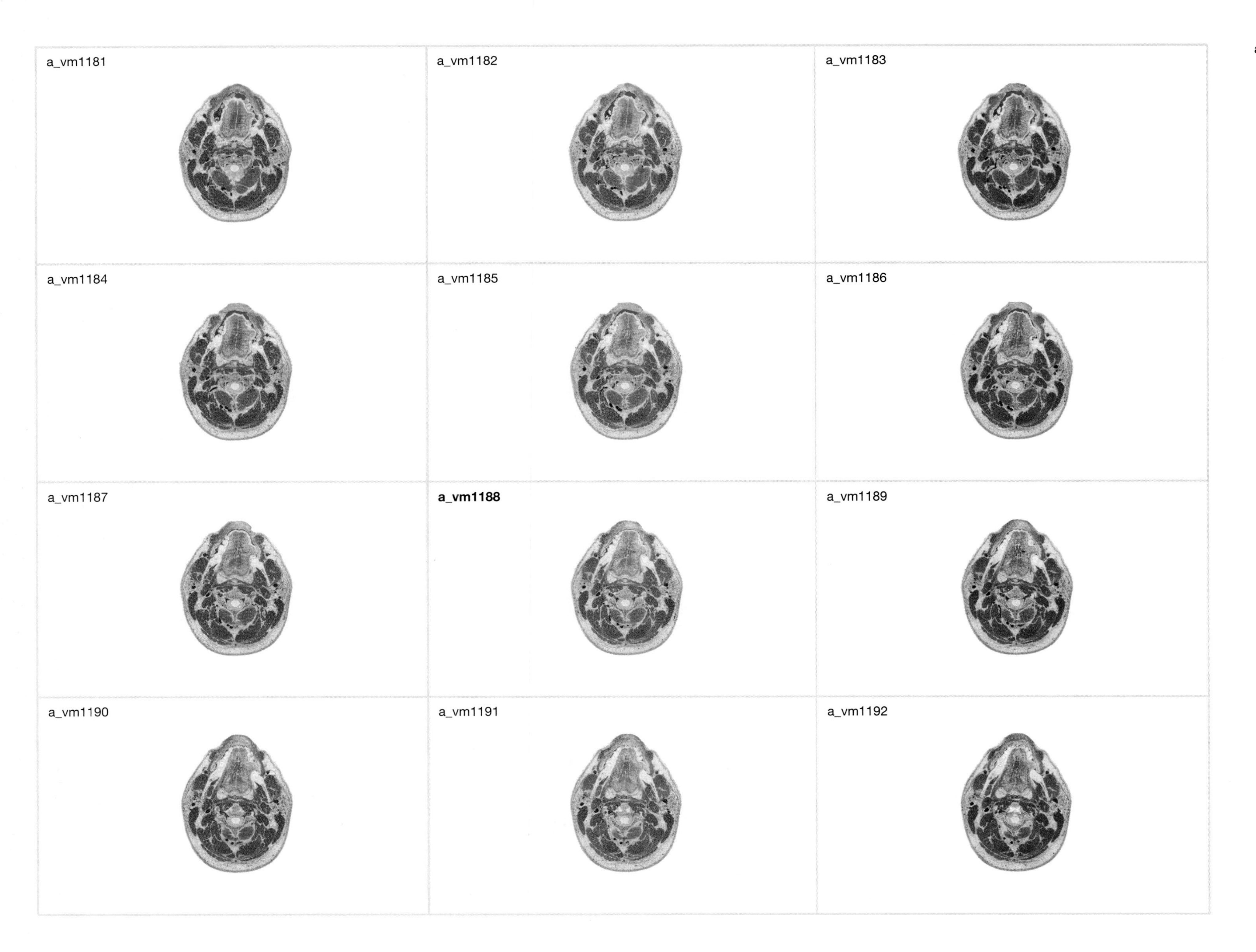
a_vm1181
a_vm1182
a_vm1183
a_vm1184
a_vm1185
a_vm1186
a_vm1187
a_vm1188
a_vm1189
a_vm1190
a_vm1191
a_vm1192

Transverse
a_vm1197

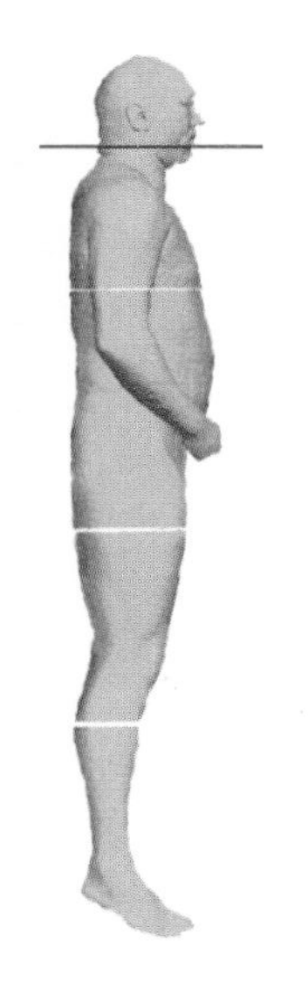

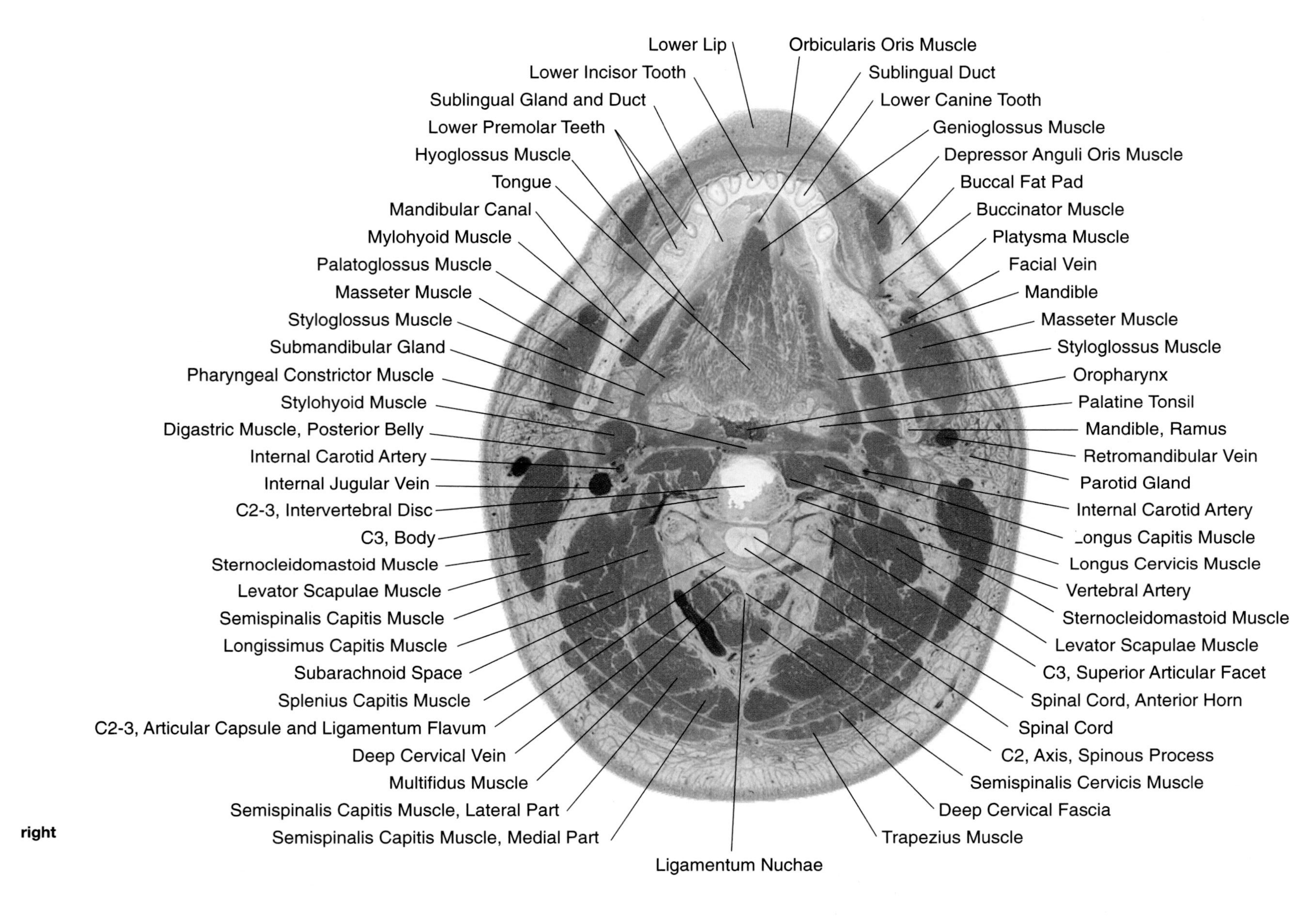
anterior
Lower Lip
Lower Incisor Tooth
Sublingual Gland and Duct
Lower Premolar Teeth
Hyoglossus Muscle
Tongue
Mandibular Canal
Mylohyoid Muscle
Palatoglossus Muscle
Masseter Muscle
Styloglossus Muscle
Submandibular Gland
Pharyngeal Constrictor Muscle
Stylohyoid Muscle
Digastric Muscle, Posterior Belly
Internal Carotid Artery
Internal Jugular Vein
C2-3, Intervertebral Disc
C3, Body
Sternocleidomastoid Muscle
Levator Scapulae Muscle
Semispinalis Capitis Muscle
Longissimus Capitis Muscle
Subarachnoid Space
Splenius Capitis Muscle
C2-3, Articular Capsule and Ligamentum Flavum
Deep Cervical Vein
Multifidus Muscle
Semispinalis Capitis Muscle, Lateral Part
Semispinalis Capitis Muscle, Medial Part
Ligamentum Nuchae
Orbicularis Oris Muscle
Sublingual Duct
Lower Canine Tooth
Genioglossus Muscle
Depressor Anguli Oris Muscle
Buccal Fat Pad
Buccinator Muscle
Platysma Muscle
Facial Vein
Mandible
Masseter Muscle
Styloglossus Muscle
Oropharynx
Palatine Tonsil
Mandible, Ramus
Retromandibular Vein
Parotid Gland
Internal Carotid Artery
Longus Capitis Muscle
Longus Cervicis Muscle
Vertebral Artery
Sternocleidomastoid Muscle
Levator Scapulae Muscle
C3, Superior Articular Facet
Spinal Cord, Anterior Horn
Spinal Cord
C2, Axis, Spinous Process
Semispinalis Cervicis Muscle
Deep Cervical Fascia
Trapezius Muscle
right
left
posterior

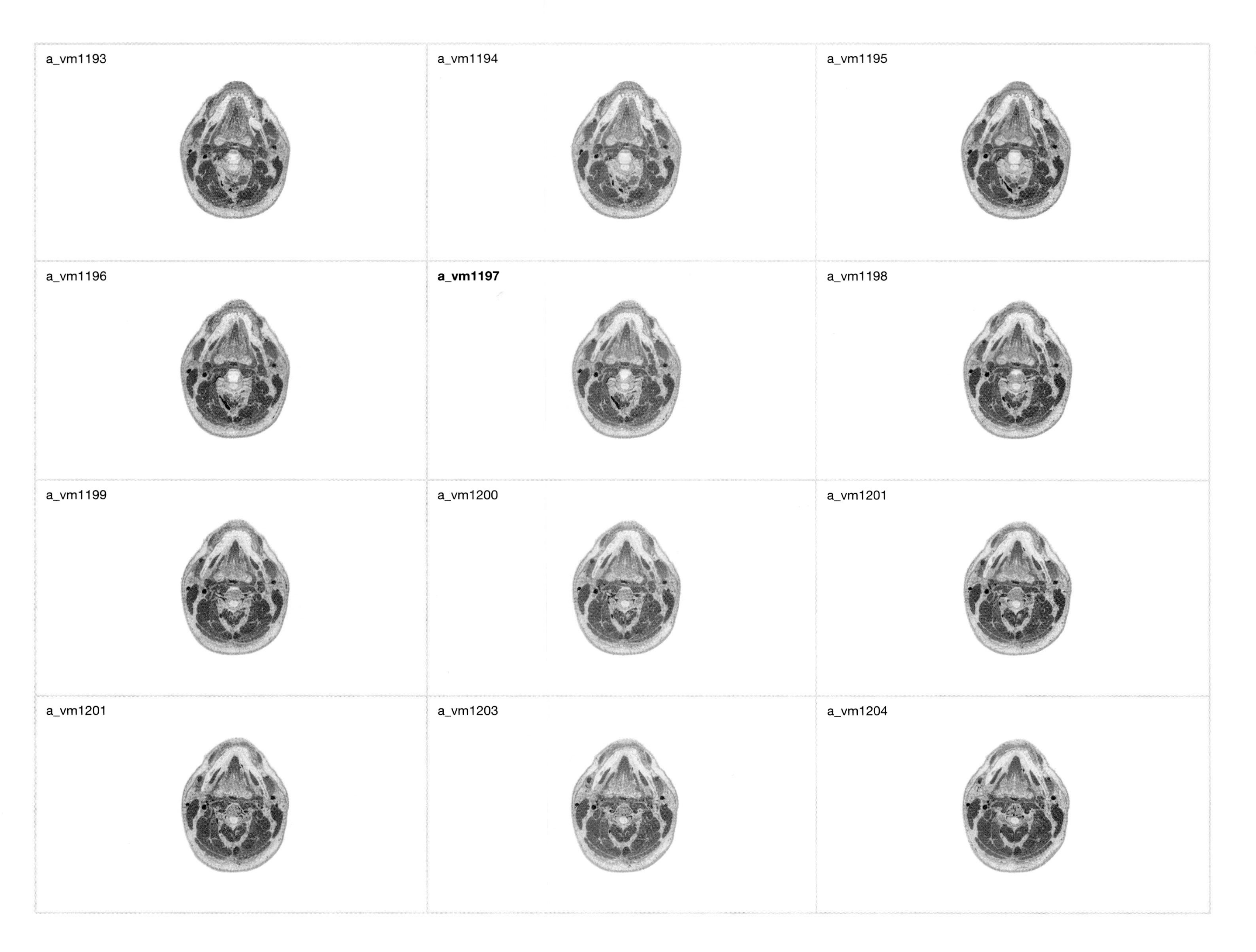
a_vm1193
a_vm1194
a_vm1195
a_vm1196
a_vm1197
a_vm1198
a_vm1199
a_vm1200
a_vm1201
a_vm1201
a_vm1203
a_vm1204

Transverse
a_vm1212

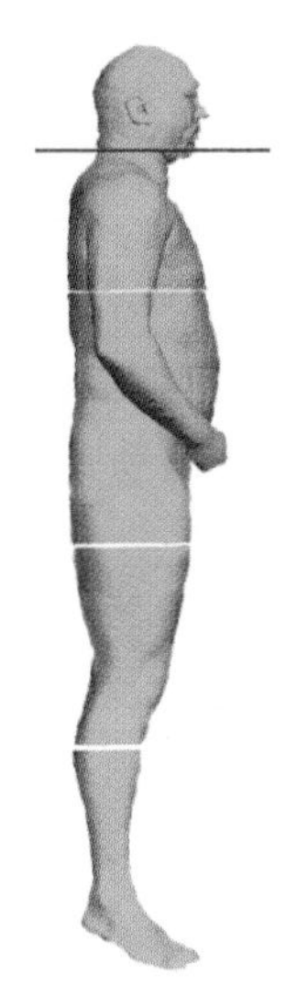

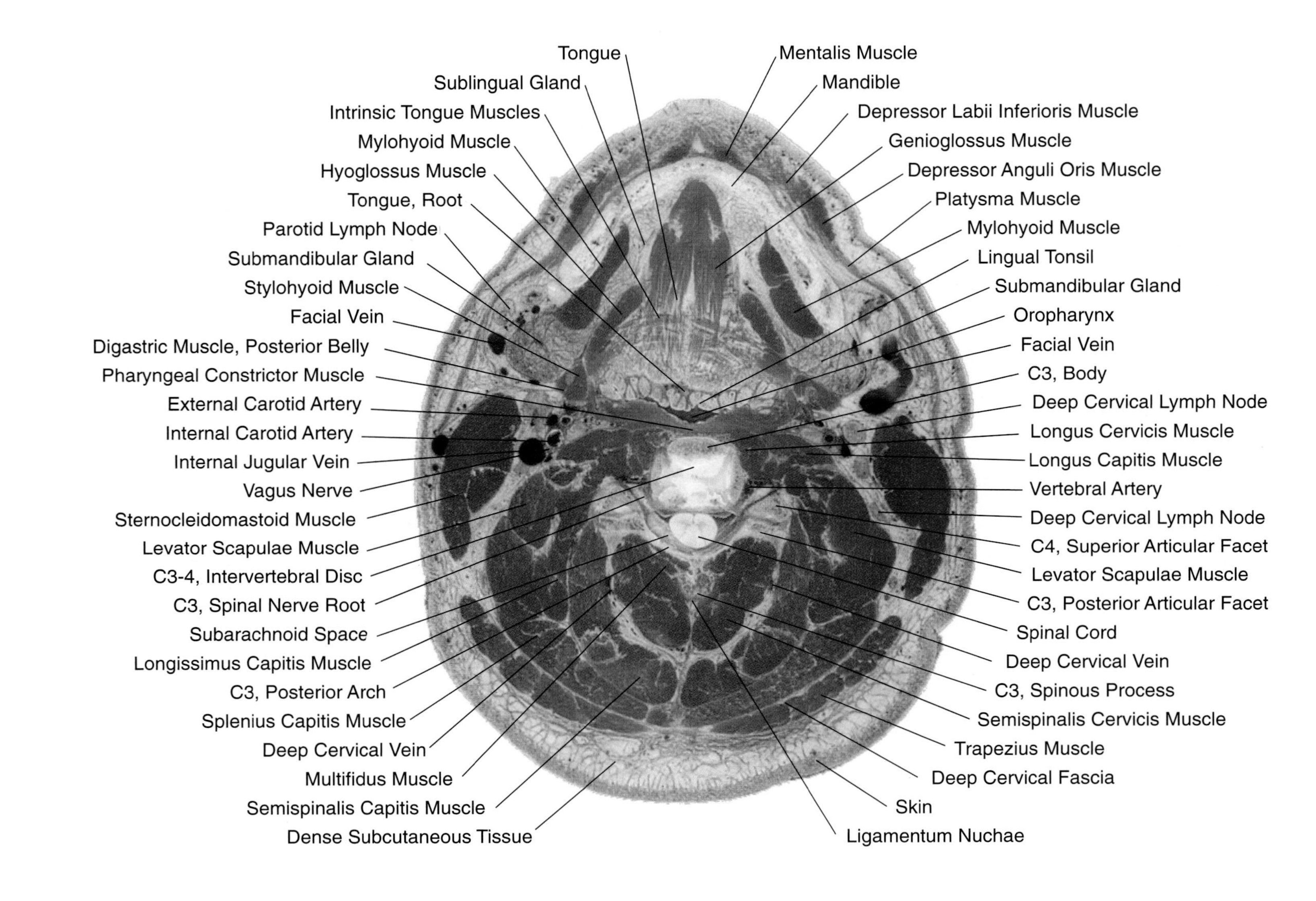
anterior
Tongue
Sublingual Gland
Intrinsic Tongue Muscles
Mylohyoid Muscle
Hyoglossus Muscle
Tongue, Root
Parotid Lymph Node
Submandibular Gland
Stylohyoid Muscle
Facial Vein
Digastric Muscle, Posterior Belly
Pharyngeal Constrictor Muscle
External Carotid Artery
Internal Carotid Artery
Internal Jugular Vein
Vagus Nerve
Sternocleidomastoid Muscle
Levator Scapulae Muscle
C3-4, Intervertebral Disc
C3, Spinal Nerve Root
Subarachnoid Space
Longissimus Capitis Muscle
C3, Posterior Arch
Splenius Capitis Muscle
Deep Cervical Vein
Multifidus Muscle
Semispinalis Capitis Muscle
Dense Subcutaneous Tissue
Mentalis Muscle
Mandible
Depressor Labii Inferioris Muscle
Genioglossus Muscle
Depressor Anguli Oris Muscle
Platysma Muscle
Mylohyoid Muscle
Lingual Tonsil
Submandibular Gland
Oropharynx
Facial Vein
C3, Body
Deep Cervical Lymph Node
Longus Cervicis Muscle
Longus Capitis Muscle
Vertebral Artery
Deep Cervical Lymph Node
C4, Superior Articular Facet
Levator Scapulae Muscle
C3, Posterior Articular Facet
Spinal Cord
Deep Cervical Vein
C3, Spinous Process
Semispinalis Cervicis Muscle
Trapezius Muscle
Deep Cervical Fascia
Skin
Ligamentum Nuchae
right
left
posterior

a_vm1205

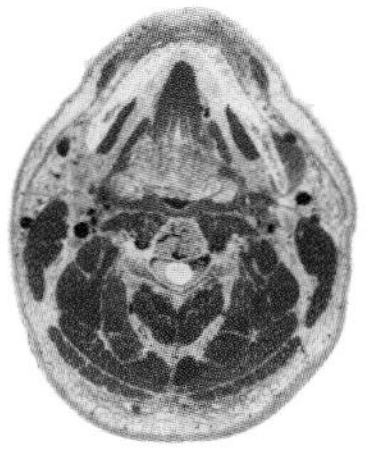

a_vm1206

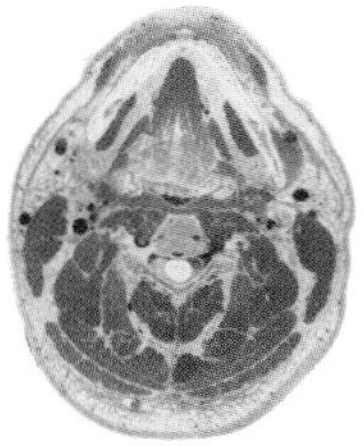

a_vm1207

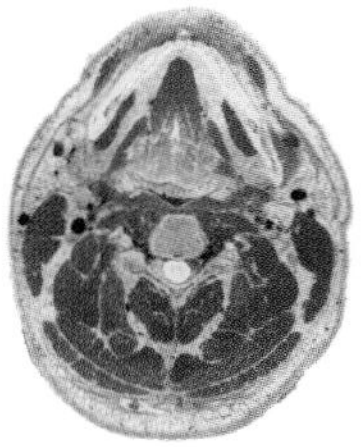

a_vm1208

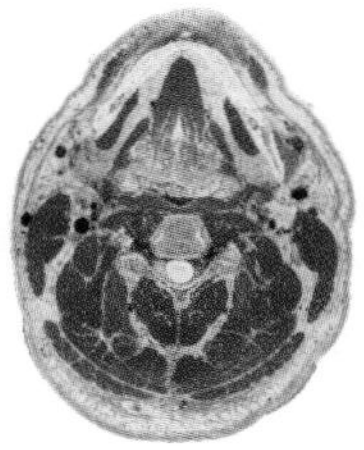

a_vm1209

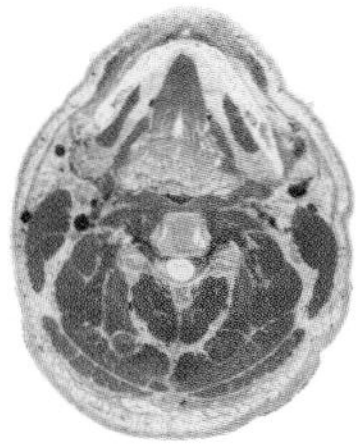

a_vm1210

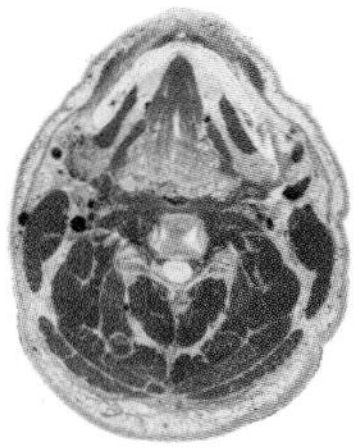

a_vm1211

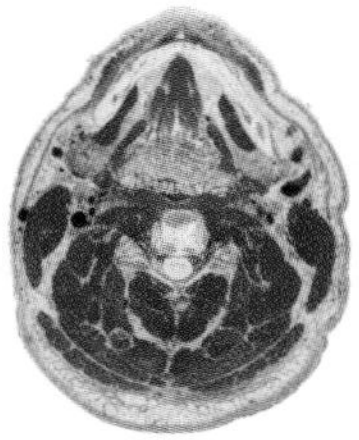

a_vm1212

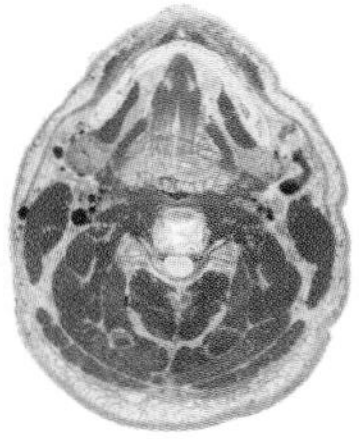

a_vm1213

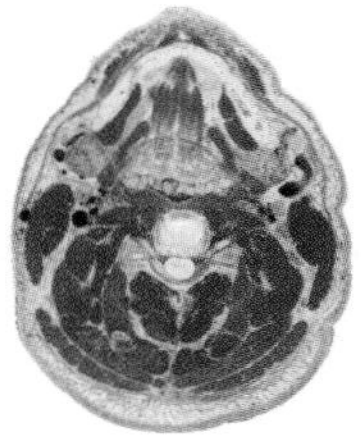

a_vm1214

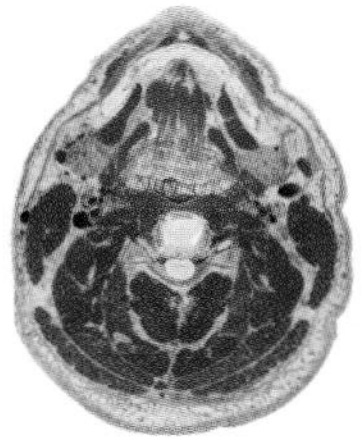

a_vm1215

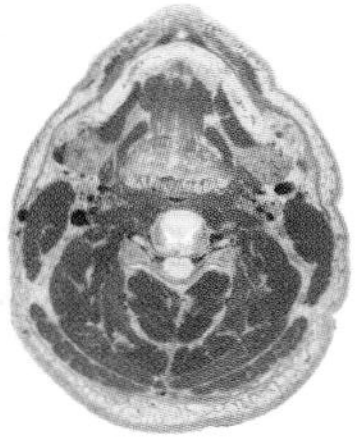

a_vm1216

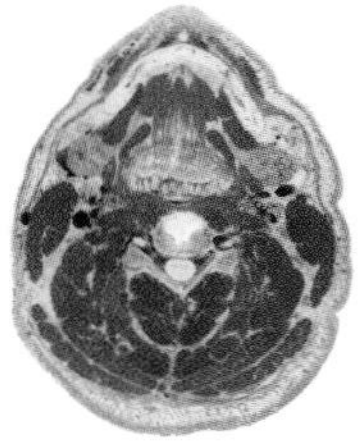

Transverse
a_vm1218

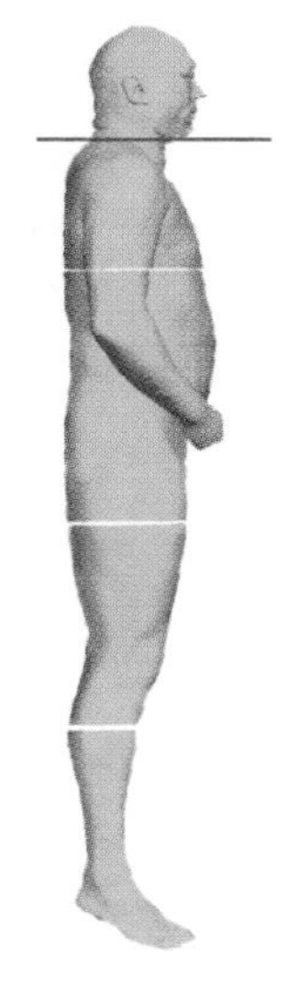

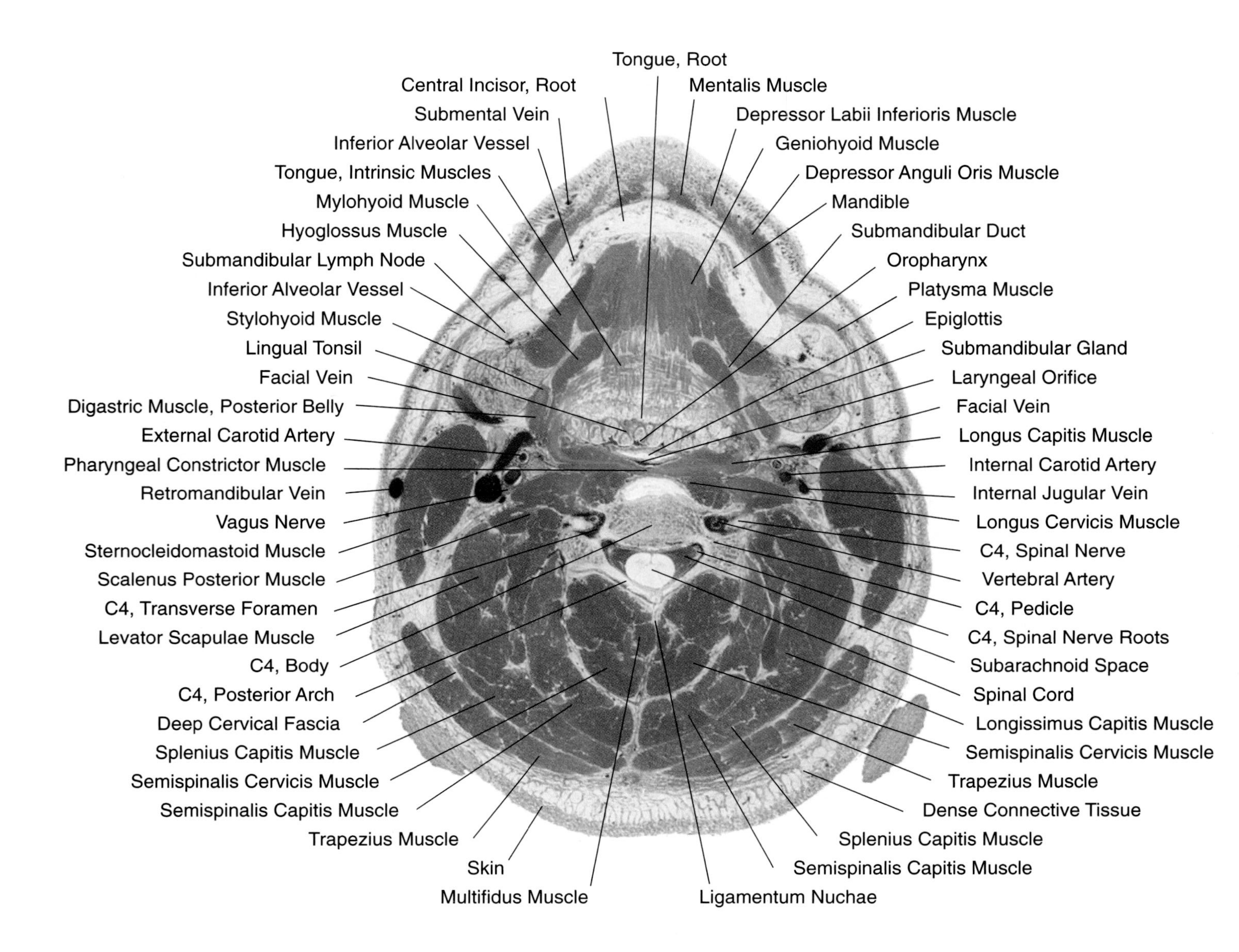
anterior
Tongue, Root
Central Incisor, Root
Mentalis Muscle
Submental Vein
Depressor Labii Inferioris Muscle
Inferior Alveolar Vessel
Geniohyoid Muscle
Tongue, Intrinsic Muscles
Depressor Anguli Oris Muscle
Mylohyoid Muscle
Mandible
Hyoglossus Muscle
Submandibular Duct
Submandibular Lymph Node
Oropharynx
Inferior Alveolar Vessel
Platysma Muscle
Stylohyoid Muscle
Epiglottis
Lingual Tonsil
Submandibular Gland
Facial Vein
Laryngeal Orifice
Digastric Muscle, Posterior Belly
Facial Vein
External Carotid Artery
Longus Capitis Muscle
Pharyngeal Constrictor Muscle
Internal Carotid Artery
Retromandibular Vein
Internal Jugular Vein
Vagus Nerve
Longus Cervicis Muscle
Sternocleidomastoid Muscle
C4, Spinal Nerve
Scalenus Posterior Muscle
Vertebral Artery
C4, Transverse Foramen
C4, Pedicle
Levator Scapulae Muscle
C4, Spinal Nerve Roots
C4, Body
Subarachnoid Space
C4, Posterior Arch
Spinal Cord
Deep Cervical Fascia
Longissimus Capitis Muscle
Splenius Capitis Muscle
Semispinalis Cervicis Muscle
Semispinalis Cervicis Muscle
Trapezius Muscle
Semispinalis Capitis Muscle
Dense Connective Tissue
Trapezius Muscle
Splenius Capitis Muscle
Skin
Semispinalis Capitis Muscle
Multifidus Muscle
Ligamentum Nuchae
right
left
posterior

a_vm1217

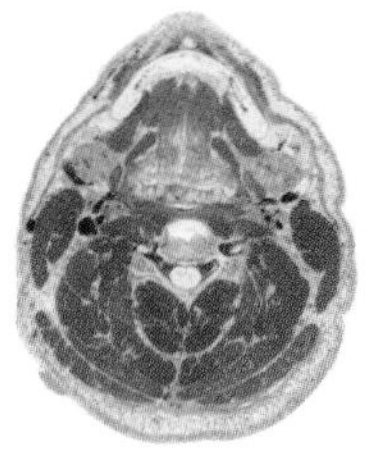

a_vm1218

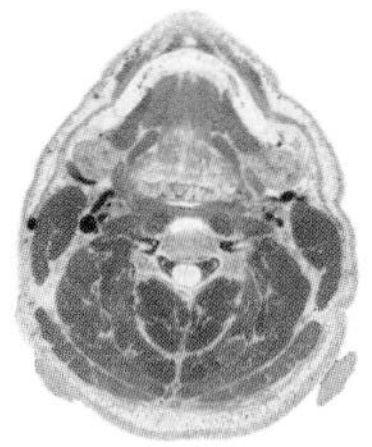

a_vm1219

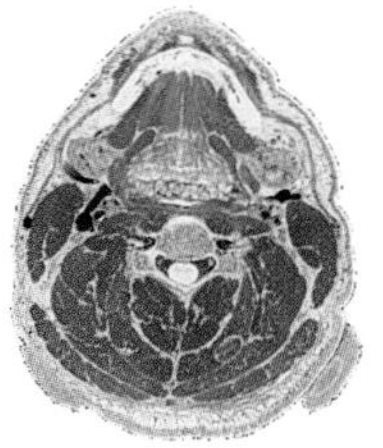

a_vm1220

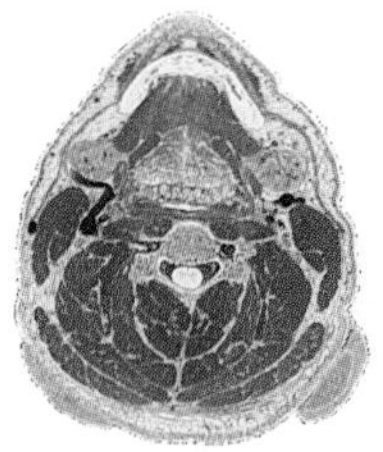

a_vm1221

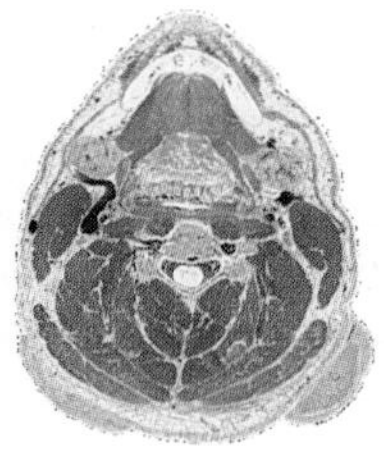

a_vm1222

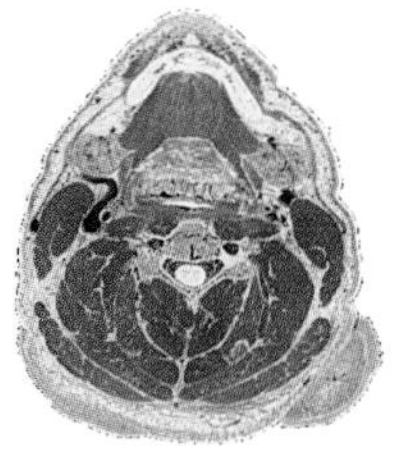

a_vm1223

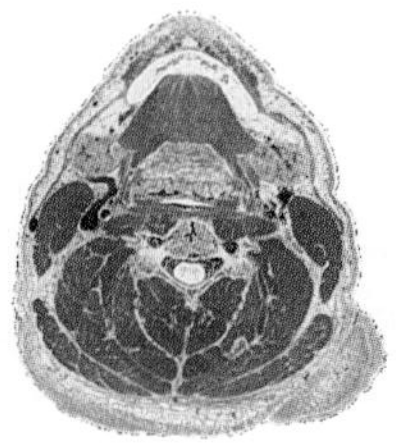

a_vm1224

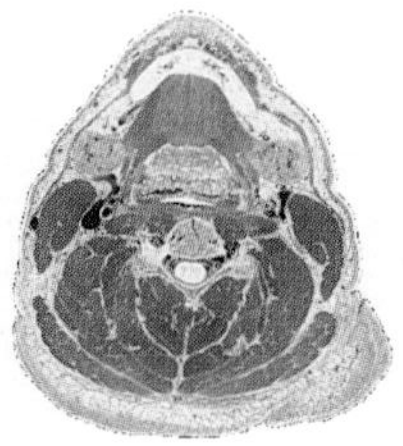

a_vm1225

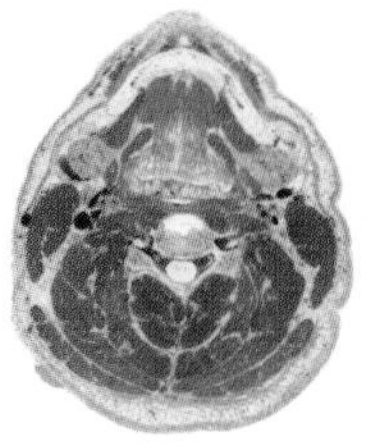

a_vm1226

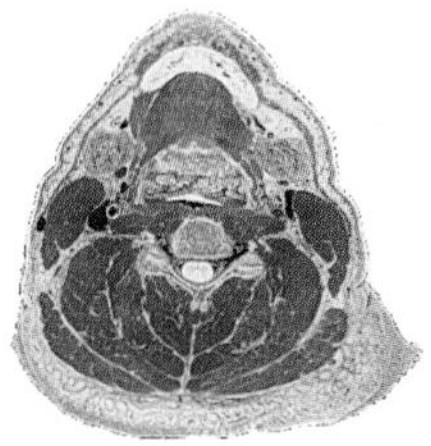

a_vm1227

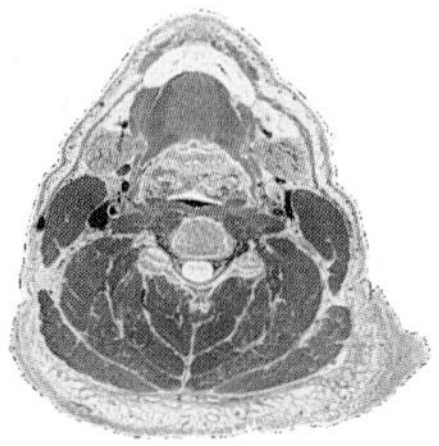

a_vm1228

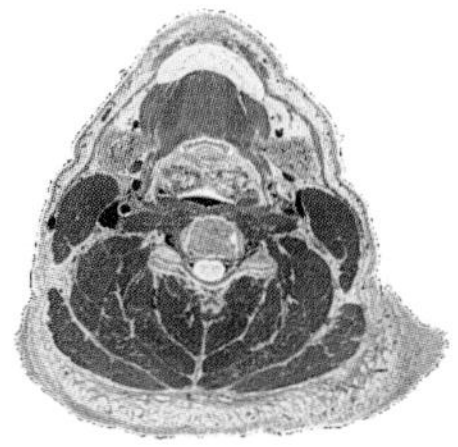

Transverse
a_vm1235

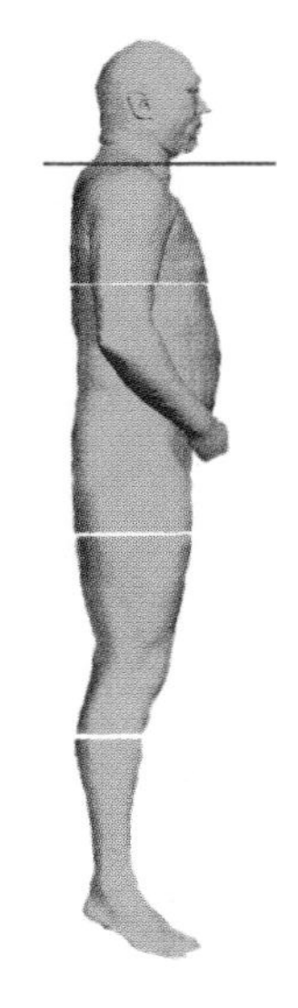

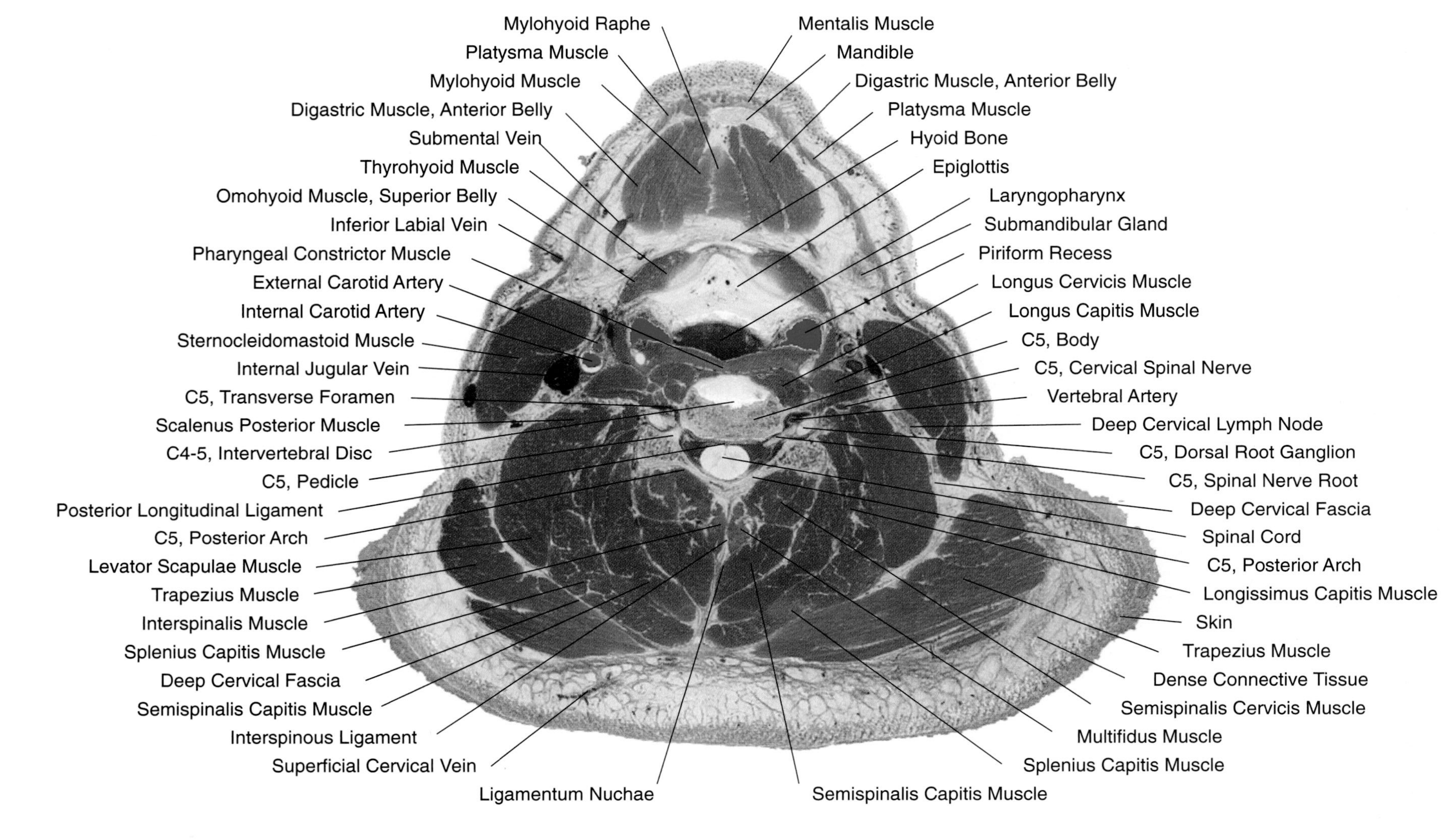
anterior
Mylohyoid Raphe
Platysma Muscle
Mylohyoid Muscle
Digastric Muscle, Anterior Belly
Submental Vein
Thyrohyoid Muscle
Omohyoid Muscle, Superior Belly
Inferior Labial Vein
Pharyngeal Constrictor Muscle
External Carotid Artery
Internal Carotid Artery
Sternocleidomastoid Muscle
Internal Jugular Vein
C5, Transverse Foramen
Scalenus Posterior Muscle
C4-5, Intervertebral Disc
C5, Pedicle
Posterior Longitudinal Ligament
C5, Posterior Arch
Levator Scapulae Muscle
Trapezius Muscle
Interspinalis Muscle
Splenius Capitis Muscle
Deep Cervical Fascia
Semispinalis Capitis Muscle
Interspinous Ligament
Superficial Cervical Vein
Ligamentum Nuchae
Mentalis Muscle
Mandible
Digastric Muscle, Anterior Belly
Platysma Muscle
Hyoid Bone
Epiglottis
Laryngopharynx
Submandibular Gland
Piriform Recess
Longus Cervicis Muscle
Longus Capitis Muscle
C5, Body
C5, Cervical Spinal Nerve
Vertebral Artery
Deep Cervical Lymph Node
C5, Dorsal Root Ganglion
C5, Spinal Nerve Root
Deep Cervical Fascia
Spinal Cord
C5, Posterior Arch
Longissimus Capitis Muscle
Skin
Trapezius Muscle
Dense Connective Tissue
Semispinalis Cervicis Muscle
Multifidus Muscle
Splenius Capitis Muscle
Semispinalis Capitis Muscle
right
left
posterior

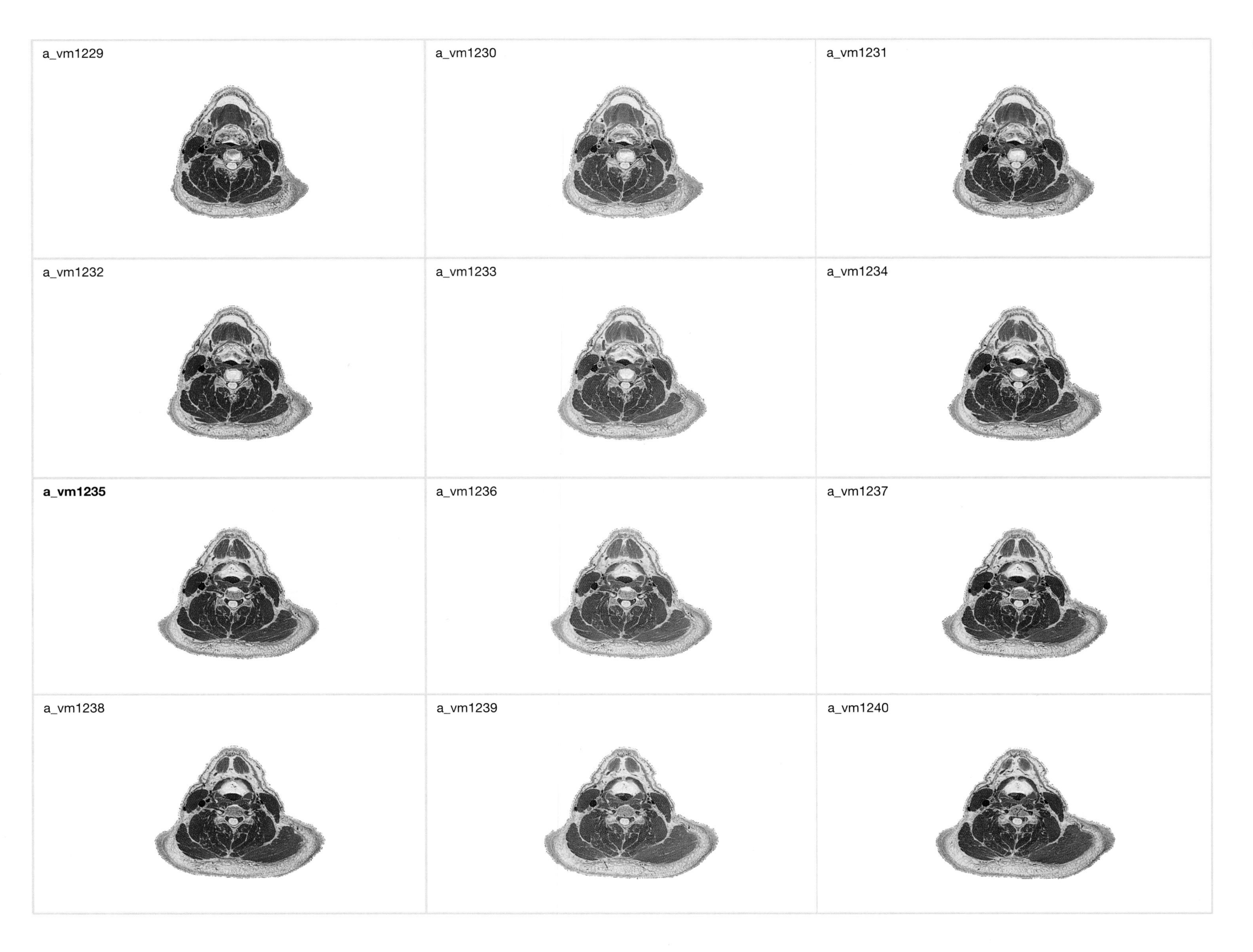
a_vm1229
a_vm1230
a_vm1231
a_vm1232
a_vm1233
a_vm1234
a_vm1235
a_vm1236
a_vm1237
a_vm1238
a_vm1239
a_vm1240

Transverse
a_vm1243

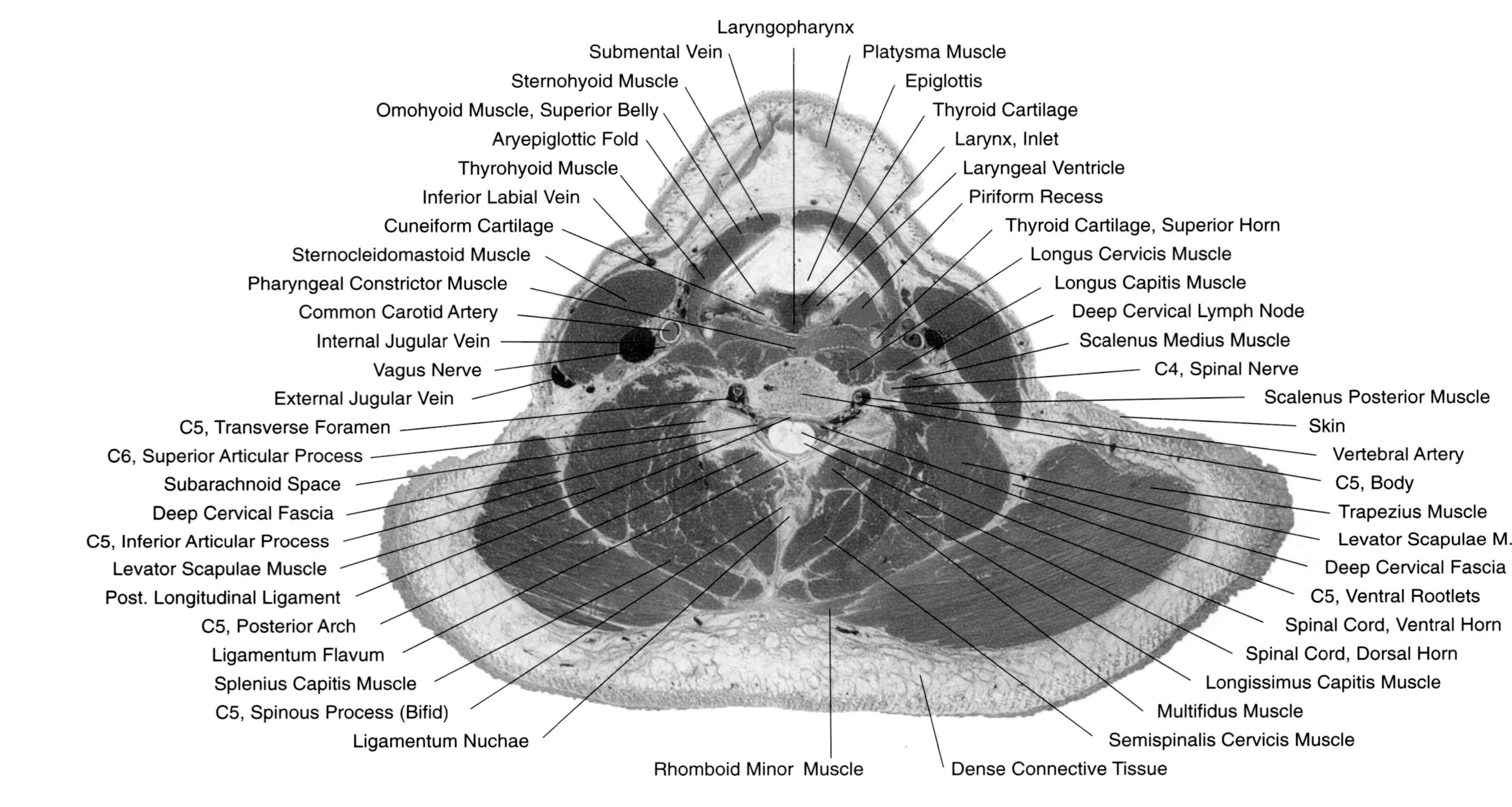
anterior
Laryngopharynx
Submental Vein
Sternohyoid Muscle
Omohyoid Muscle, Superior Belly
Aryepiglottic Fold
Thyrohyoid Muscle
Inferior Labial Vein
Cuneiform Cartilage
Sternocleidomastoid Muscle
Pharyngeal Constrictor Muscle
Common Carotid Artery
Internal Jugular Vein
Vagus Nerve
External Jugular Vein
C5, Transverse Foramen
C6, Superior Articular Process
Subarachnoid Space
Deep Cervical Fascia
C5, Inferior Articular Process
Levator Scapulae Muscle
Post. Longitudinal Ligament
C5, Posterior Arch
Ligamentum Flavum
Splenius Capitis Muscle
C5, Spinous Process (Bifid)
Ligamentum Nuchae
Rhomboid Minor Muscle
Platysma Muscle
Epiglottis
Thyroid Cartilage
Larynx, Inlet
Laryngeal Ventricle
Piriform Recess
Thyroid Cartilage, Superior Horn
Longus Cervicis Muscle
Longus Capitis Muscle
Deep Cervical Lymph Node
Scalenus Medius Muscle
C4, Spinal Nerve
Scalenus Posterior Muscle
Skin
Vertebral Artery
C5, Body
Trapezius Muscle
Levator Scapulae M.
Deep Cervical Fascia
C5, Ventral Rootlets
Spinal Cord, Ventral Horn
Spinal Cord, Dorsal Horn
Longissimus Capitis Muscle
Multifidus Muscle
Semispinalis Cervicis Muscle
Dense Connective Tissue
right
left
posterior

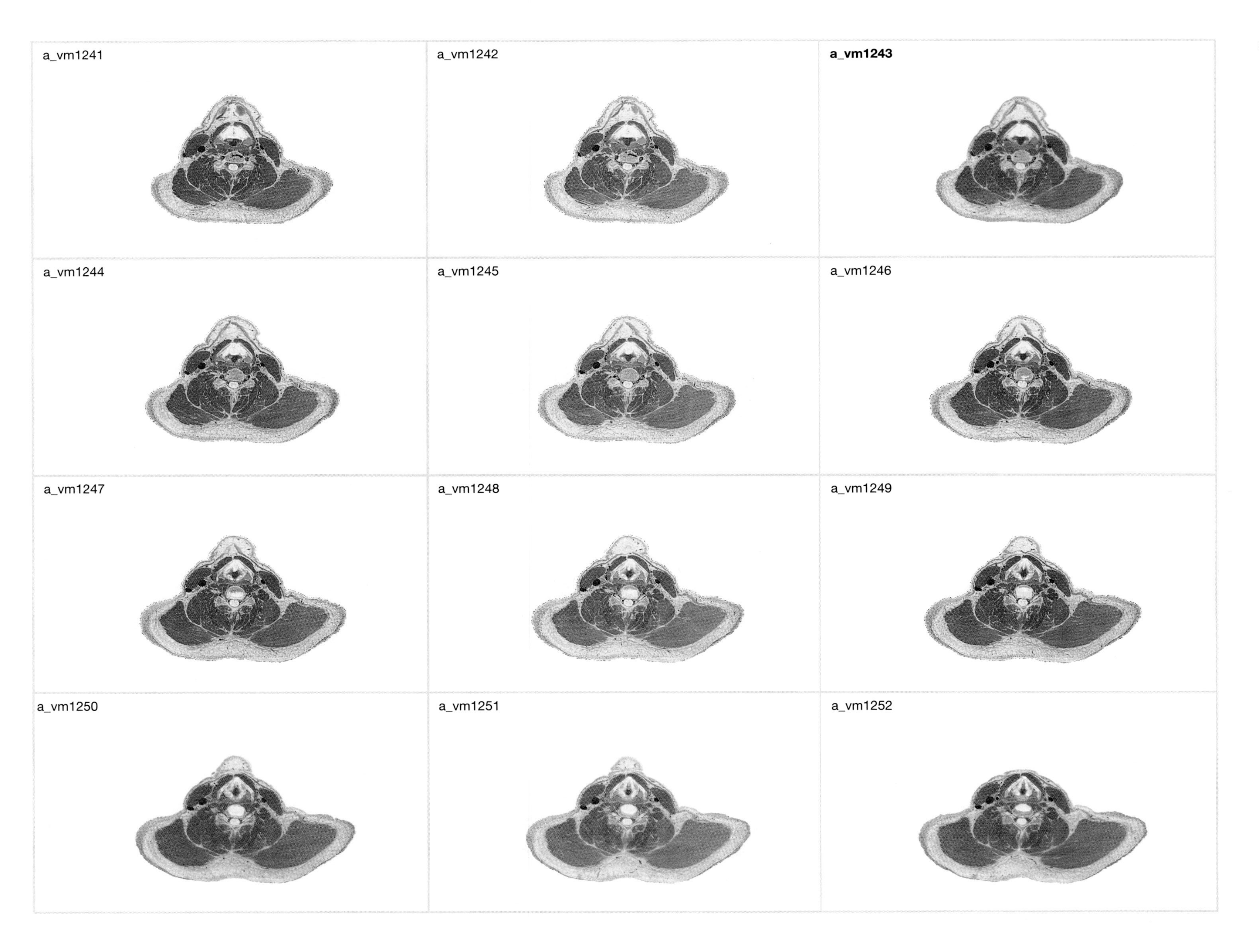
a_vm1241
a_vm1242
a_vm1243
a_vm1244
a_vm1245
a_vm1246
a_vm1247
a_vm1248
a_vm1249
a_vm1250
a_vm1251
a_vm1252

Transverse
a_vm1260

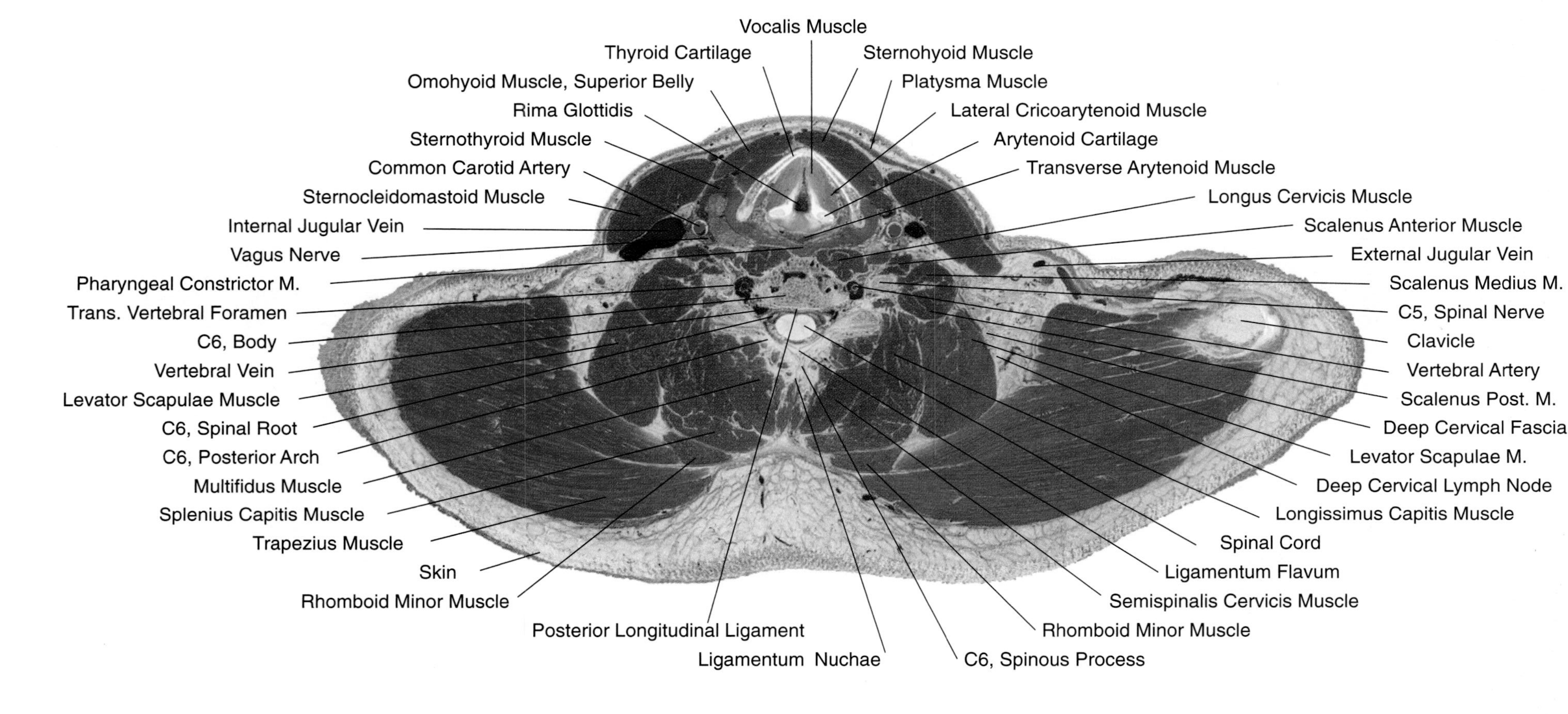
anterior
Vocalis Muscle
Thyroid Cartilage
Sternohyoid Muscle
Omohyoid Muscle, Superior Belly
Platysma Muscle
Rima Glottidis
Lateral Cricoarytenoid Muscle
Sternothyroid Muscle
Arytenoid Cartilage
Common Carotid Artery
Transverse Arytenoid Muscle
Sternocleidomastoid Muscle
Longus Cervicis Muscle
Internal Jugular Vein
Scalenus Anterior Muscle
Vagus Nerve
External Jugular Vein
Pharyngeal Constrictor M.
Scalenus Medius M.
Trans. Vertebral Foramen
C5, Spinal Nerve
C6, Body
Clavicle
Vertebral Vein
Vertebral Artery
Levator Scapulae Muscle
Scalenus Post. M.
C6, Spinal Root
Deep Cervical Fascia
C6, Posterior Arch
Levator Scapulae M.
Multifidus Muscle
Deep Cervical Lymph Node
Splenius Capitis Muscle
Longissimus Capitis Muscle
Trapezius Muscle
Spinal Cord
Skin
Ligamentum Flavum
Rhomboid Minor Muscle
Semispinalis Cervicis Muscle
Posterior Longitudinal Ligament
Rhomboid Minor Muscle
Ligamentum Nuchae
C6, Spinous Process
right
left
posterior

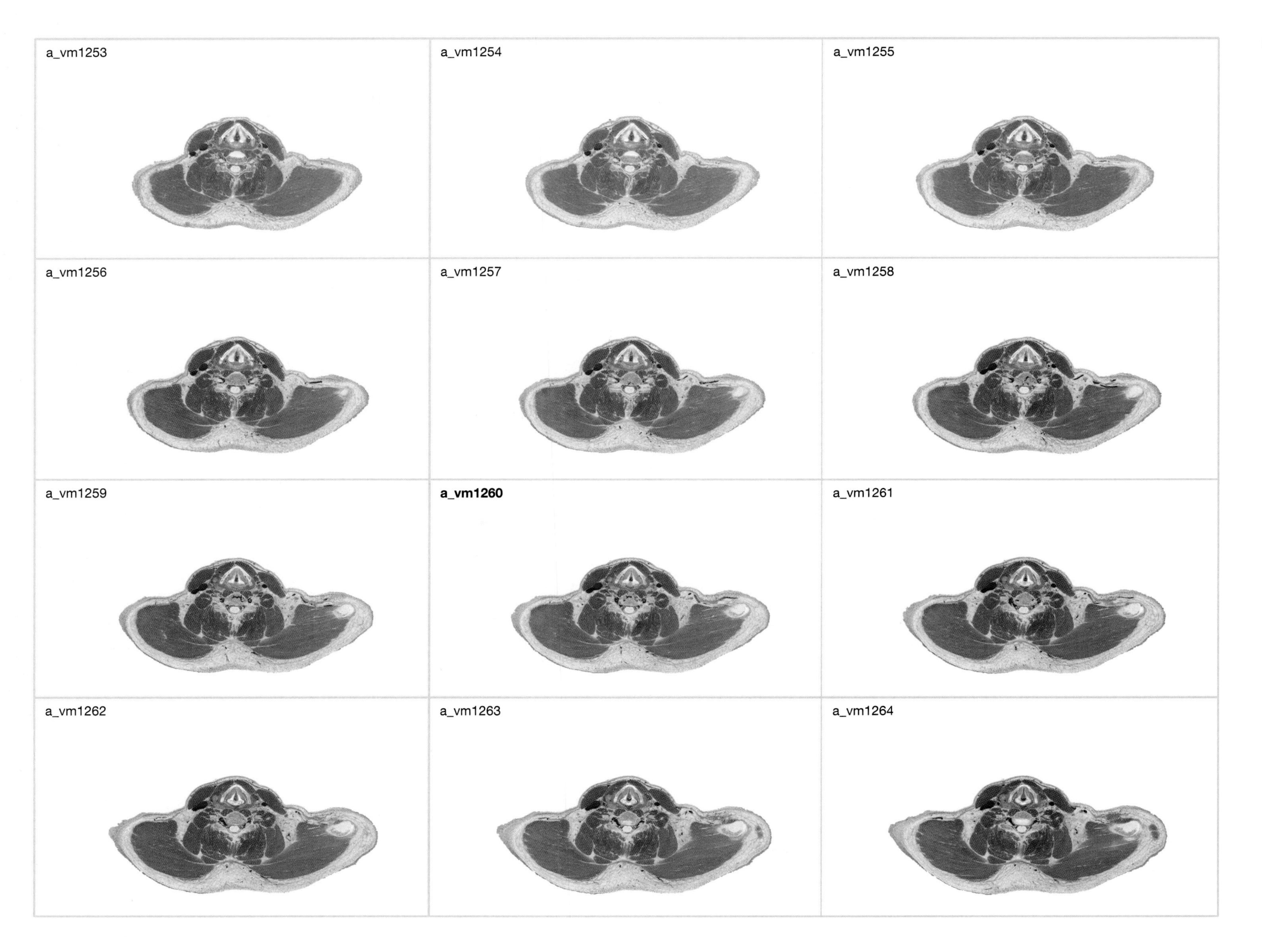
a_vm1253
a_vm1254
a_vm1255
a_vm1256
a_vm1257
a_vm1258
a_vm1259
a_vm1260
a_vm1261
a_vm1262
a_vm1263
a_vm1264

Transverse
a_vm1272

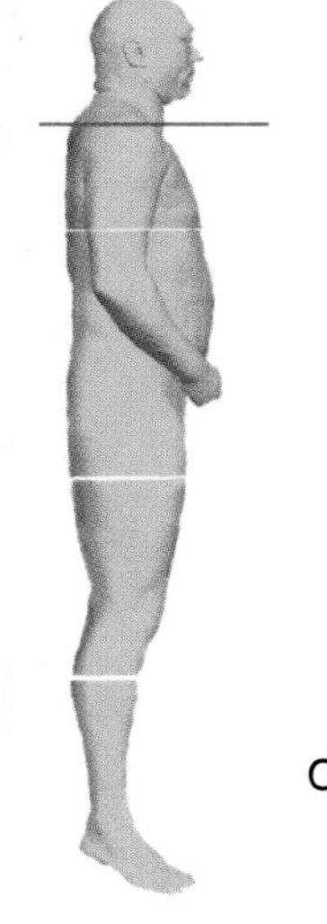

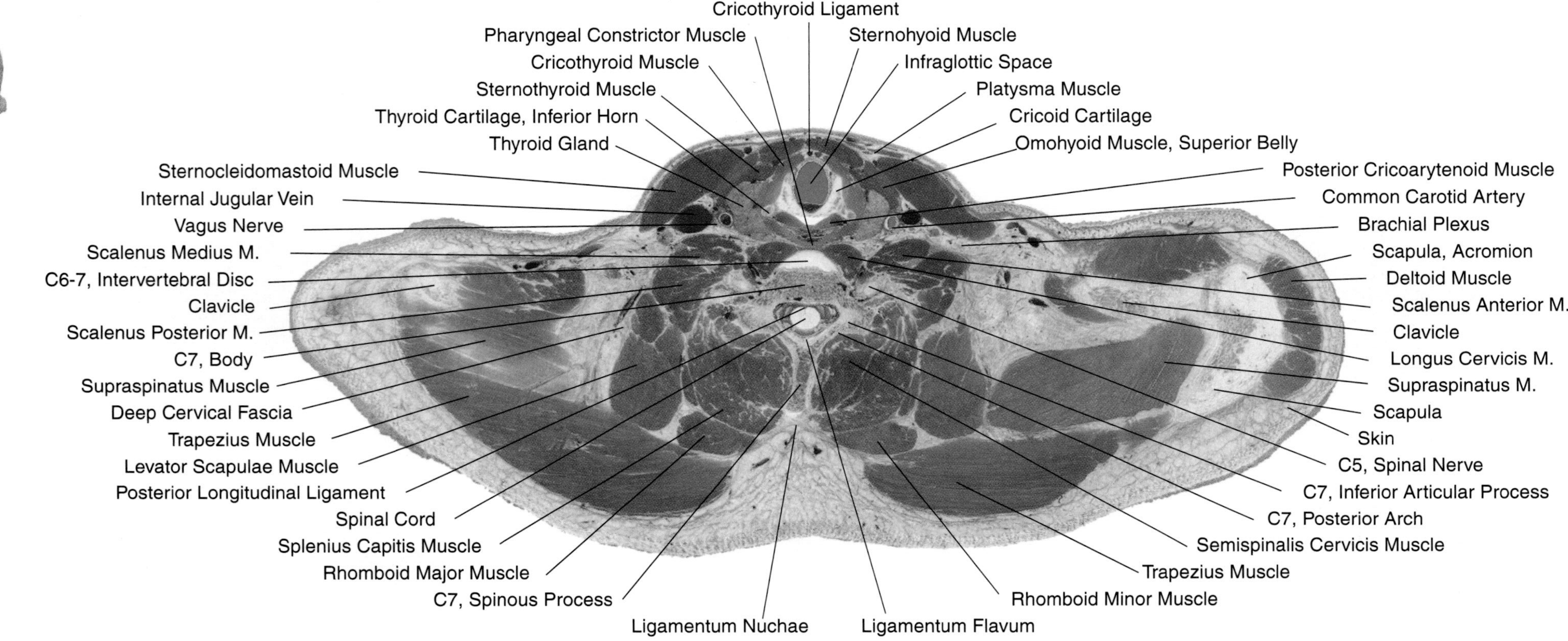
anterior
Cricothyroid Ligament
Pharyngeal Constrictor Muscle
Sternohyoid Muscle
Cricothyroid Muscle
Infraglottic Space
Sternothyroid Muscle
Platysma Muscle
Thyroid Cartilage, Inferior Horn
Cricoid Cartilage
Thyroid Gland
Omohyoid Muscle, Superior Belly
Sternocleidomastoid Muscle
Posterior Cricoarytenoid Muscle
Internal Jugular Vein
Common Carotid Artery
Vagus Nerve
Brachial Plexus
Scalenus Medius M.
Scapula, Acromion
C6-7, Intervertebral Disc
Deltoid Muscle
Clavicle
Scalenus Anterior M.
Scalenus Posterior M.
Clavicle
C7, Body
Longus Cervicis M.
Supraspinatus Muscle
Supraspinatus M.
Deep Cervical Fascia
Scapula
Trapezius Muscle
Skin
Levator Scapulae Muscle
C5, Spinal Nerve
Posterior Longitudinal Ligament
C7, Inferior Articular Process
Spinal Cord
C7, Posterior Arch
Splenius Capitis Muscle
Semispinalis Cervicis Muscle
Rhomboid Major Muscle
Trapezius Muscle
C7, Spinous Process
Rhomboid Minor Muscle
Ligamentum Nuchae
Ligamentum Flavum
right
left
posterior

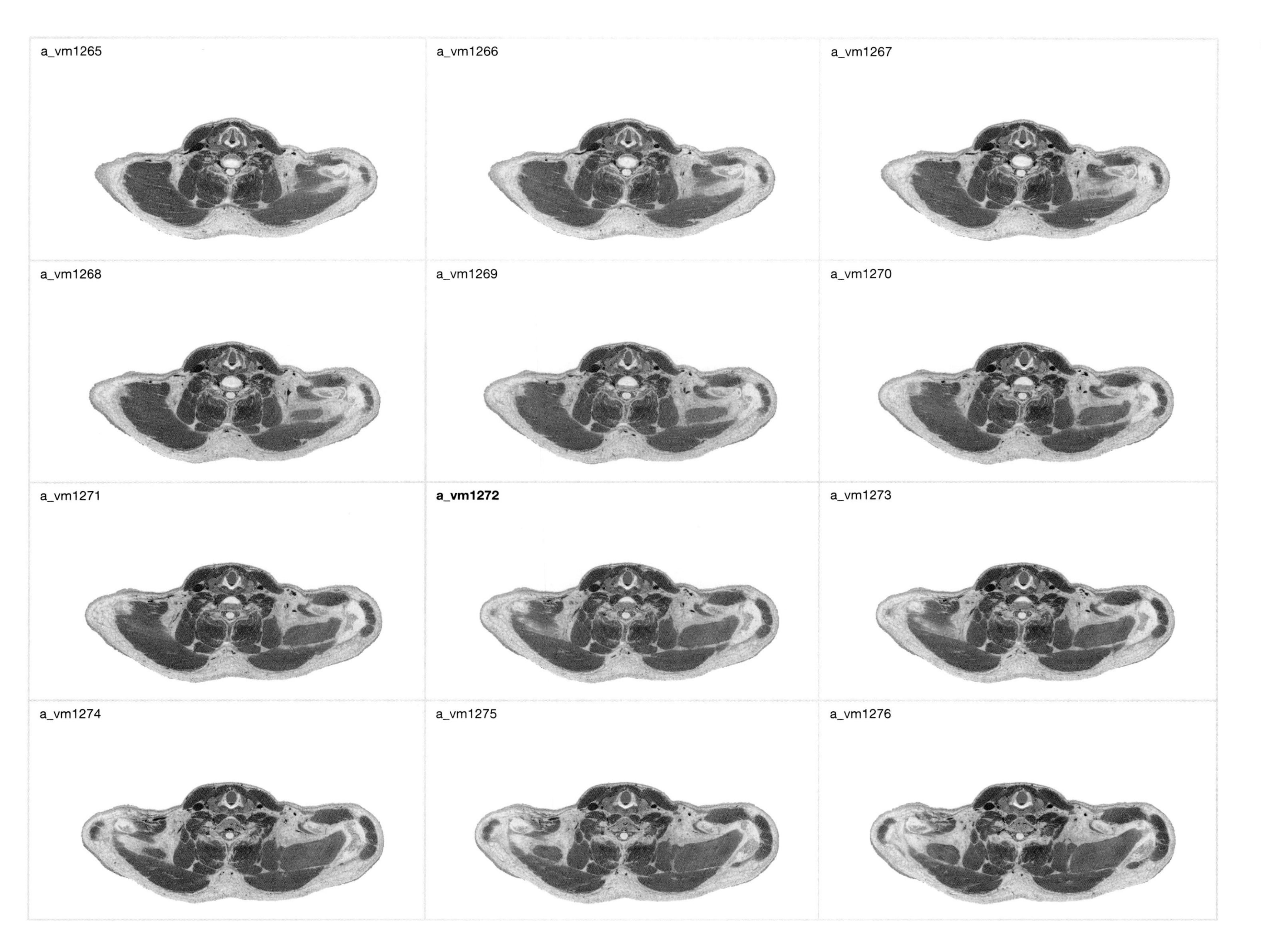
a_vm1265
a_vm1266
a_vm1267
a_vm1268
a_vm1269
a_vm1270
a_vm1271
a_vm1272
a_vm1273
a_vm1274
a_vm1275
a_vm1276

Transverse
a_vm1317

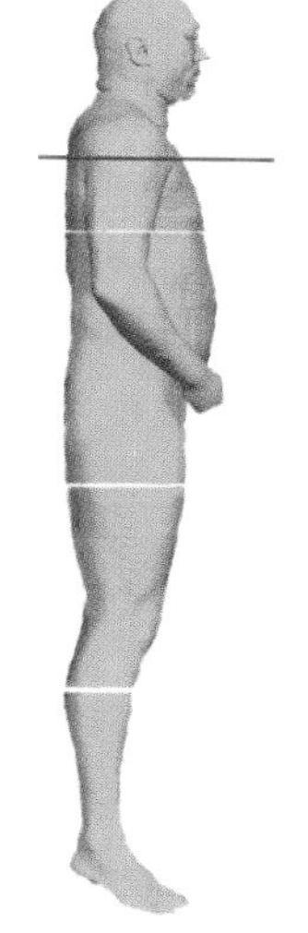

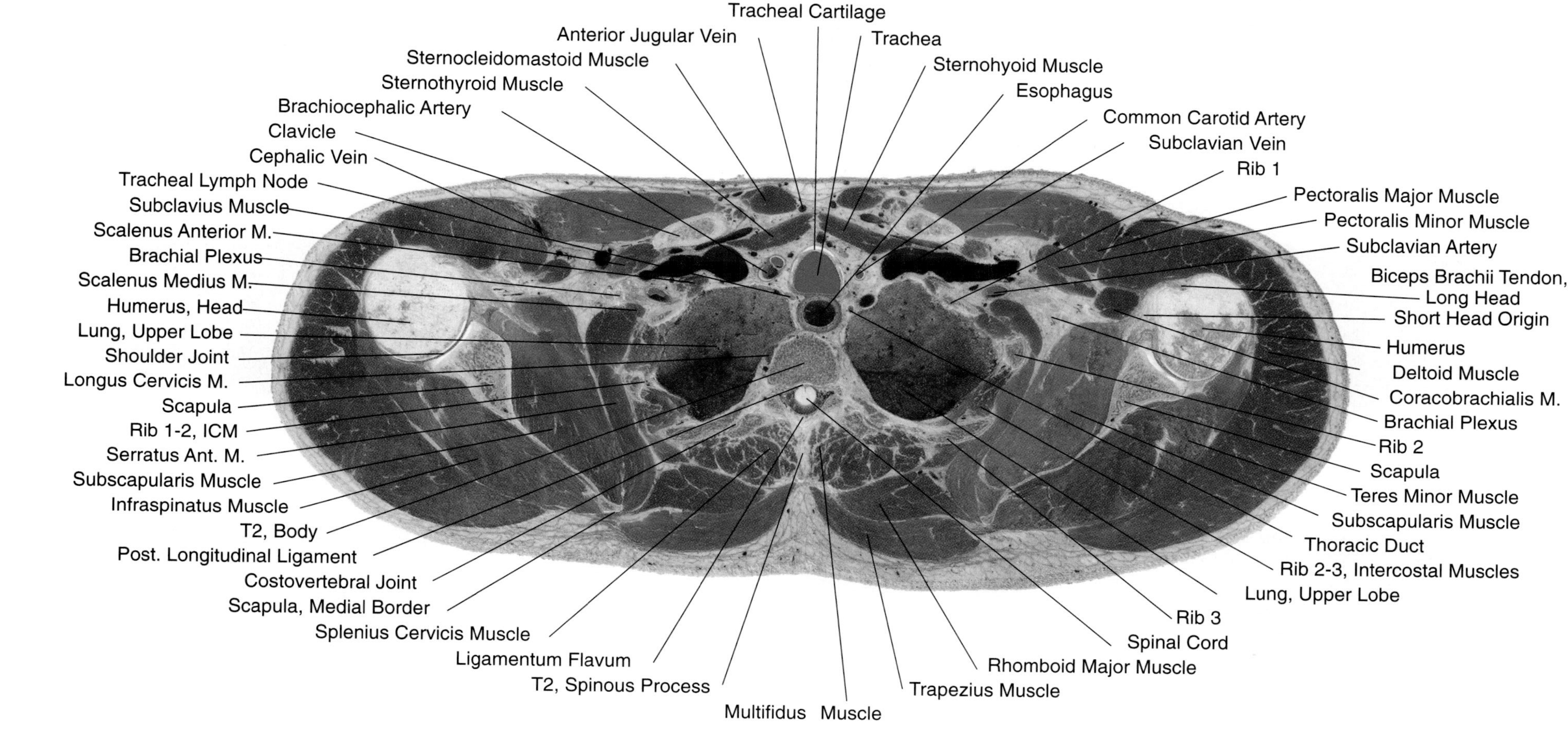
anterior
Tracheal Cartilage
Anterior Jugular Vein
Trachea
Sternocleidomastoid Muscle
Sternohyoid Muscle
Sternothyroid Muscle
Esophagus
Brachiocephalic Artery
Common Carotid Artery
Clavicle
Subclavian Vein
Cephalic Vein
Rib 1
Tracheal Lymph Node
Pectoralis Major Muscle
Subclavius Muscle
Pectoralis Minor Muscle
Scalenus Anterior M.
Subclavian Artery
Brachial Plexus
Biceps Brachii Tendon, Long Head
Scalenus Medius M.
Short Head Origin
Humerus, Head
Humerus
Lung, Upper Lobe
Deltoid Muscle
Shoulder Joint
Coracobrachialis M.
Longus Cervicis M.
Brachial Plexus
Scapula
Rib 2
Rib 1-2, ICM
Scapula
Serratus Ant. M.
Teres Minor Muscle
Subscapularis Muscle
Subscapularis Muscle
Infraspinatus Muscle
Thoracic Duct
T2, Body
Rib 2-3, Intercostal Muscles
Post. Longitudinal Ligament
Lung, Upper Lobe
Costovertebral Joint
Rib 3
Scapula, Medial Border
Spinal Cord
Splenius Cervicis Muscle
Rhomboid Major Muscle
Ligamentum Flavum
Trapezius Muscle
T2, Spinous Process
Multifidus Muscle
right
left
posterior

ICM=Intercostal Muscles

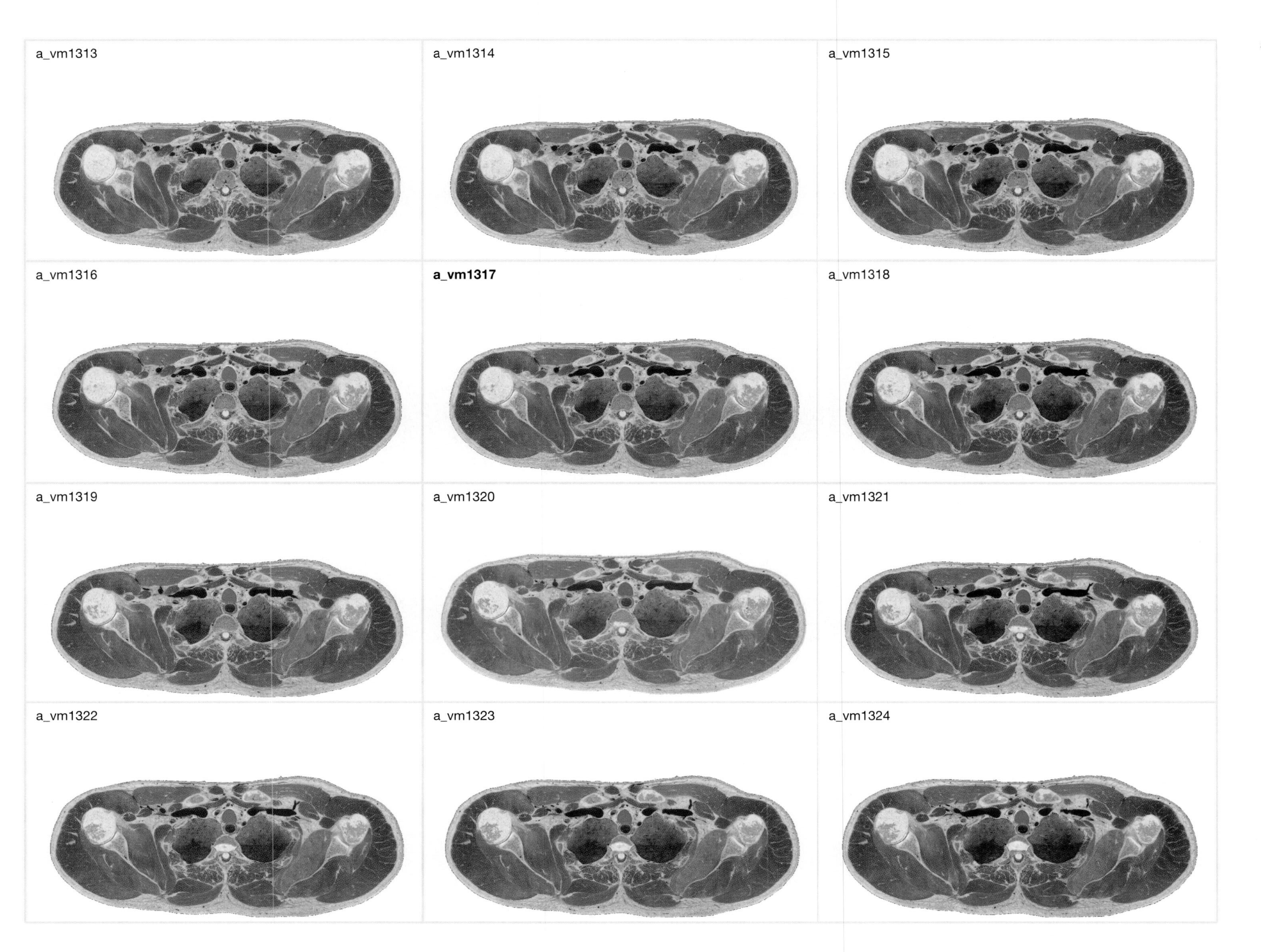
a_vm1313
a_vm1314
a_vm1315
a_vm1316
a_vm1317
a_vm1318
a_vm1319
a_vm1320
a_vm1321
a_vm1322
a_vm1323
a_vm1324

Transverse
a_vm1326

anterior

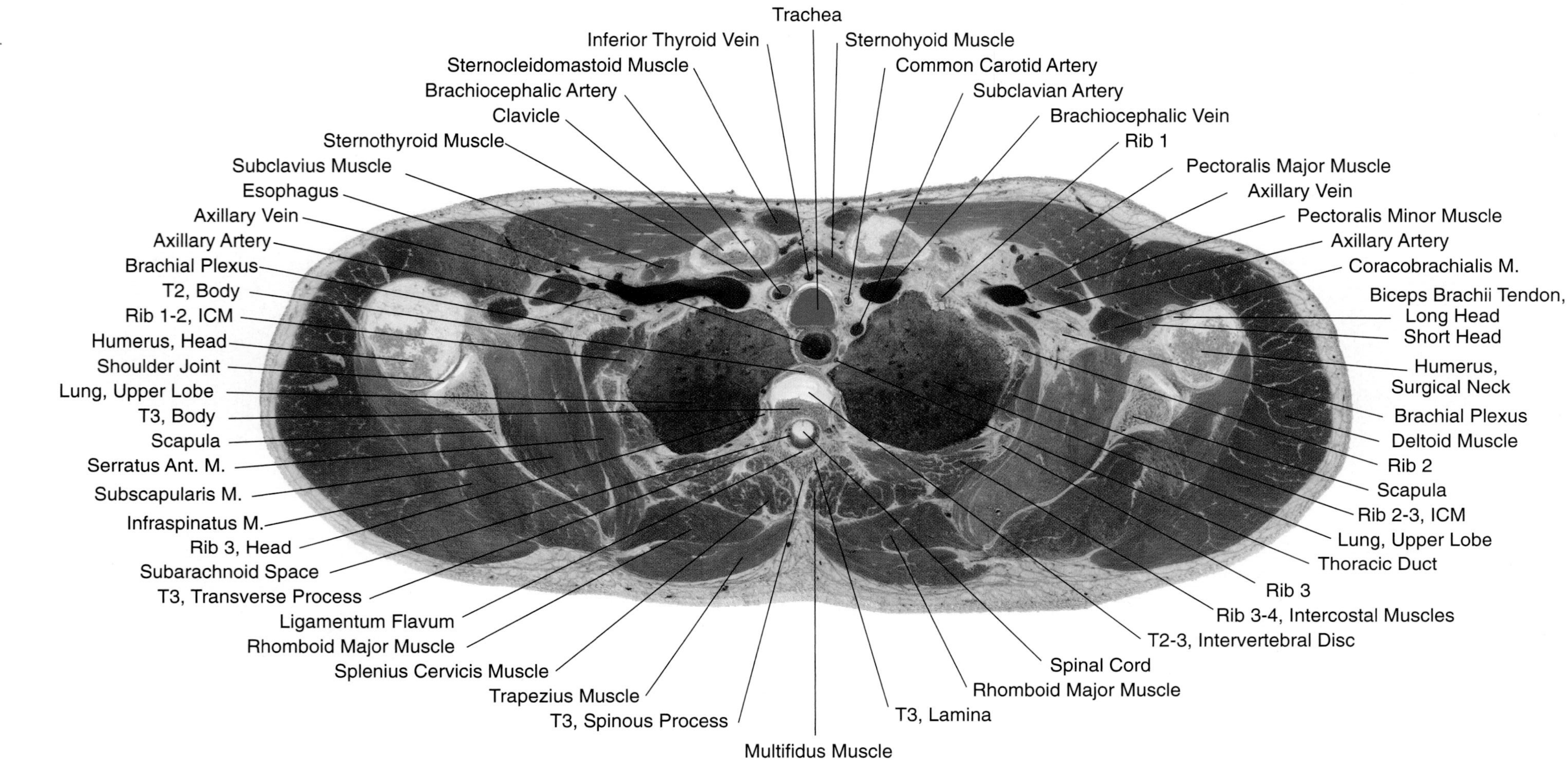
Trachea
Inferior Thyroid Vein
Sternohyoid Muscle
Sternocleidomastoid Muscle
Common Carotid Artery
Brachiocephalic Artery
Subclavian Artery
Clavicle
Brachiocephalic Vein
Sternothyroid Muscle
Rib 1
Subclavius Muscle
Pectoralis Major Muscle
Esophagus
Axillary Vein
Axillary Vein
Pectoralis Minor Muscle
Axillary Artery
Axillary Artery
Brachial Plexus
Coracobrachialis M.
T2, Body
Biceps Brachii Tendon,
Long Head
Short Head
Rib 1-2, ICM
Humerus, Head
Shoulder Joint
Humerus,
Surgical Neck
Lung, Upper Lobe
T3, Body
Brachial Plexus
Scapula
Deltoid Muscle
Serratus Ant. M.
Rib 2
Subscapularis M.
Scapula
Infraspinatus M.
Rib 2-3, ICM
Rib 3, Head
Lung, Upper Lobe
Subarachnoid Space
Thoracic Duct
T3, Transverse Process
Rib 3
Ligamentum Flavum
Rib 3-4, Intercostal Muscles
Rhomboid Major Muscle
T2-3, Intervertebral Disc
Splenius Cervicis Muscle
Spinal Cord
Trapezius Muscle
Rhomboid Major Muscle
T3, Spinous Process
T3, Lamina
Multifidus Muscle

right

left

ICM=Intercostal Muscles

posterior

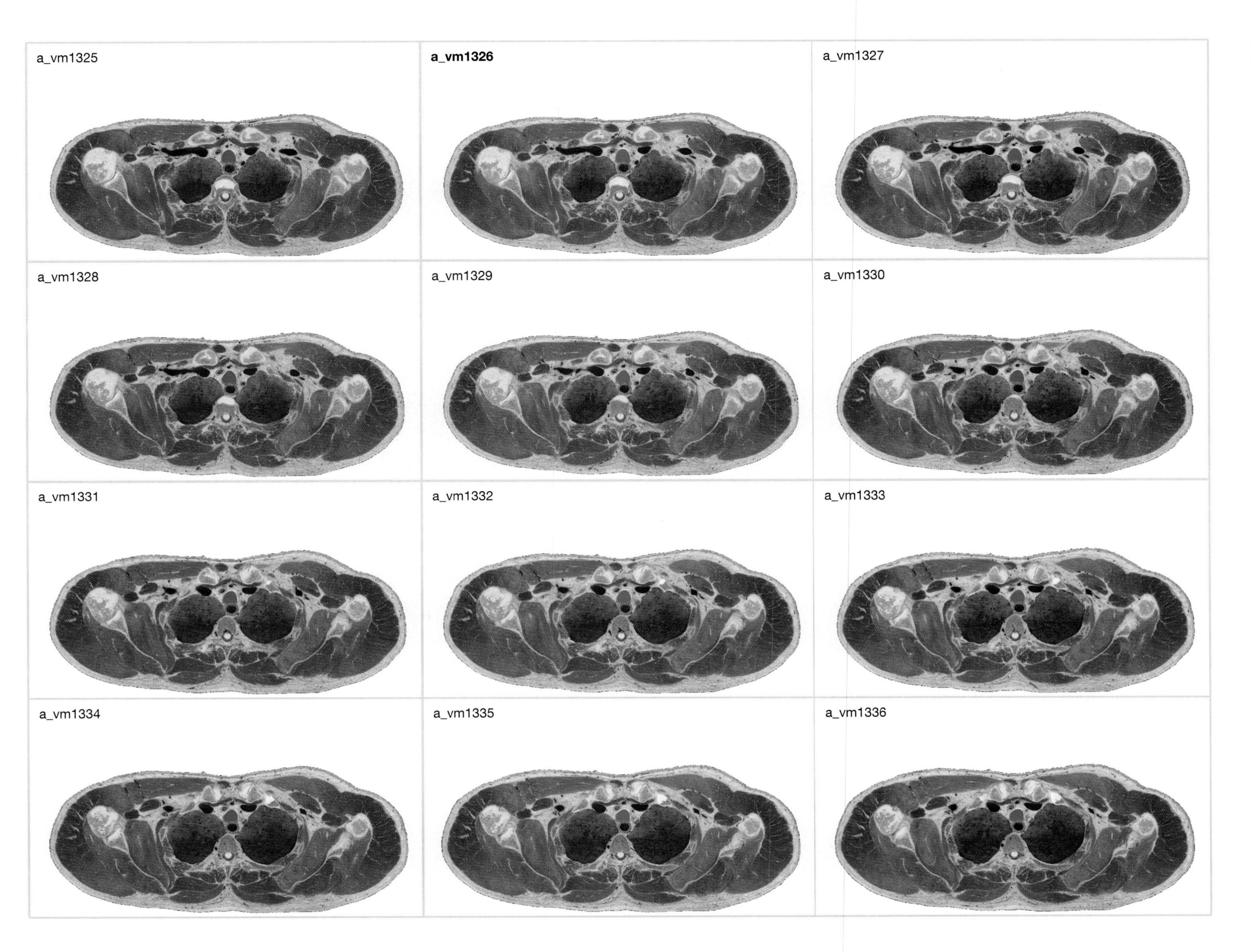
a_vm1325
a_vm1326
a_vm1327
a_vm1328
a_vm1329
a_vm1330
a_vm1331
a_vm1332
a_vm1333
a_vm1334
a_vm1335
a_vm1336

Transverse
a_vm1340

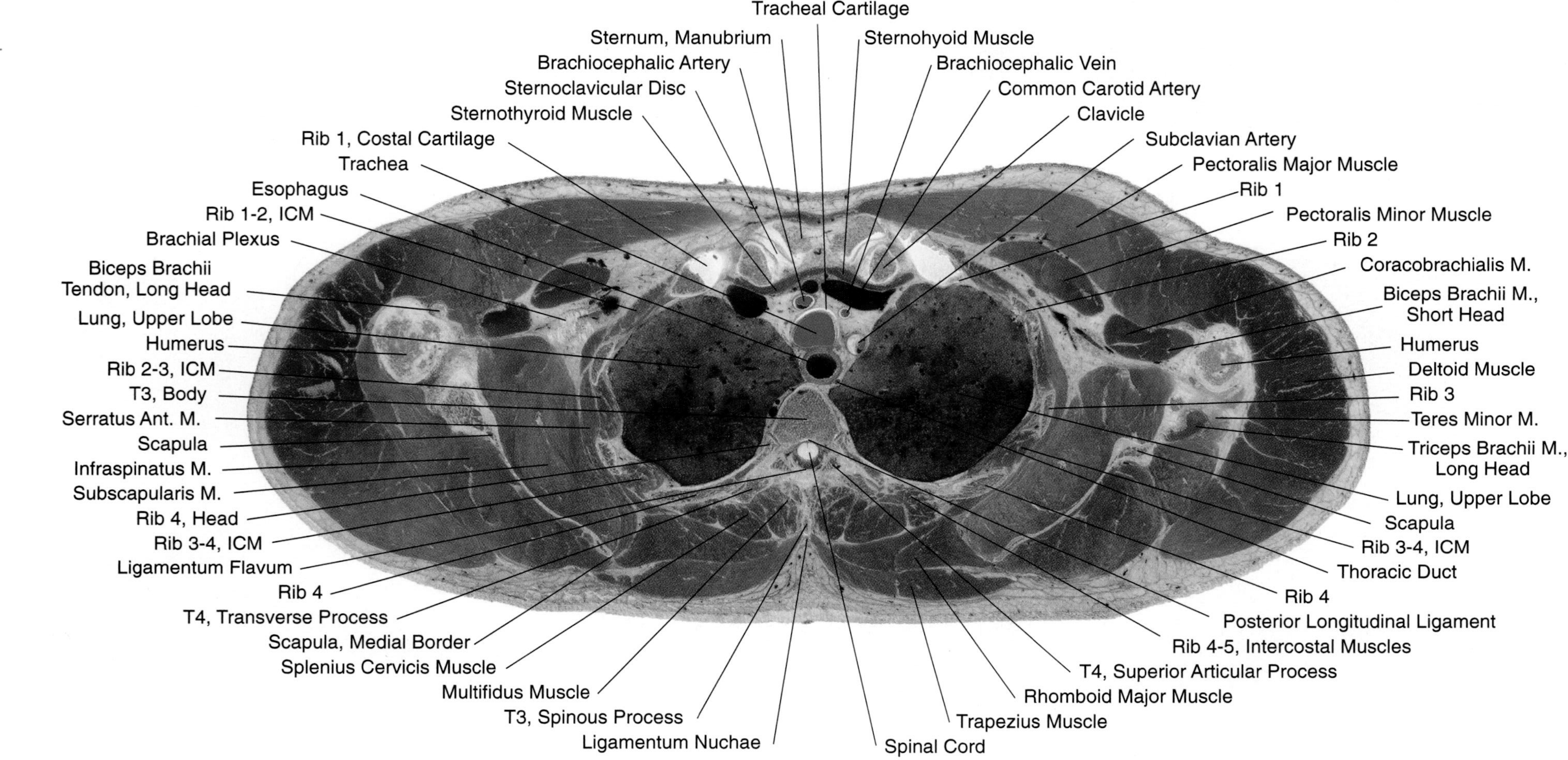

anterior
Tracheal Cartilage
Sternum, Manubrium
Sternohyoid Muscle
Brachiocephalic Artery
Brachiocephalic Vein
Sternoclavicular Disc
Common Carotid Artery
Sternothyroid Muscle
Clavicle
Rib 1, Costal Cartilage
Subclavian Artery
Trachea
Pectoralis Major Muscle
Esophagus
Rib 1
Rib 1-2, ICM
Pectoralis Minor Muscle
Brachial Plexus
Rib 2
Coracobrachialis M.
Biceps Brachii
Tendon, Long Head
Biceps Brachii M.,
Short Head
Lung, Upper Lobe
Humerus
Humerus
Deltoid Muscle
Rib 2-3, ICM
T3, Body
Rib 3
Serratus Ant. M.
Teres Minor M.
Scapula
Triceps Brachii M.,
Long Head
Infraspinatus M.
Subscapularis M.
Lung, Upper Lobe
Rib 4, Head
Scapula
Rib 3-4, ICM
Rib 3-4, ICM
Ligamentum Flavum
Thoracic Duct
Rib 4
Rib 4
T4, Transverse Process
Posterior Longitudinal Ligament
Scapula, Medial Border
Rib 4-5, Intercostal Muscles
Splenius Cervicis Muscle
T4, Superior Articular Process
Multifidus Muscle
Rhomboid Major Muscle
T3, Spinous Process
Trapezius Muscle
Ligamentum Nuchae
Spinal Cord
right
left
posterior

ICM=Intercostal Muscles

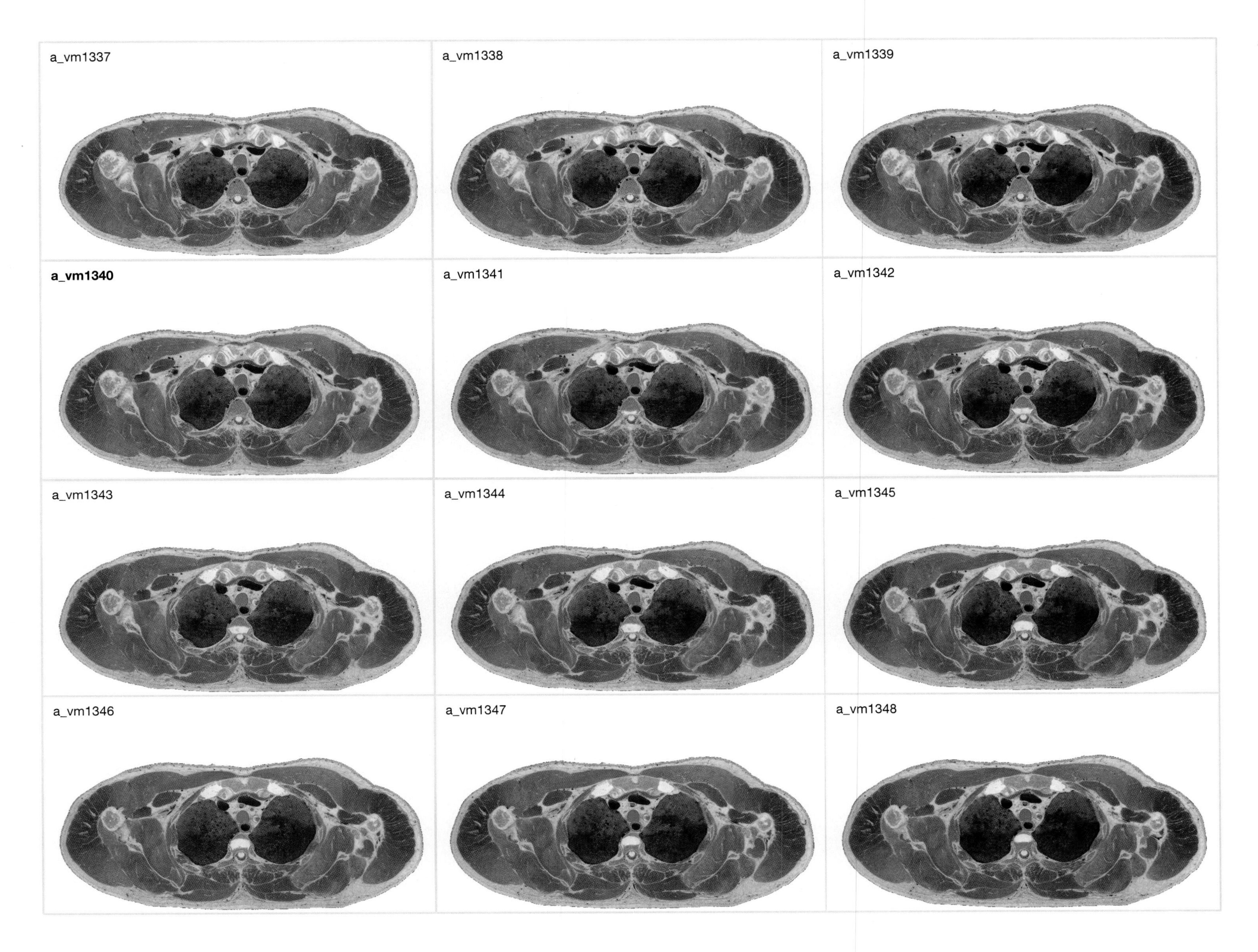
a_vm1337
a_vm1338
a_vm1339
a_vm1340
a_vm1341
a_vm1342
a_vm1343
a_vm1344
a_vm1345
a_vm1346
a_vm1347
a_vm1348

Transverse
a_vm1352

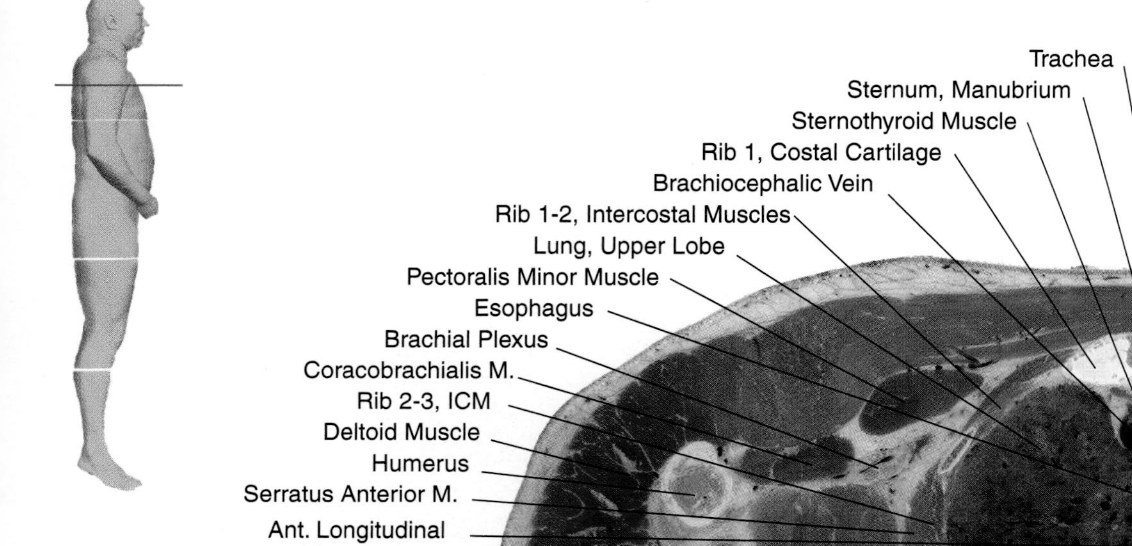

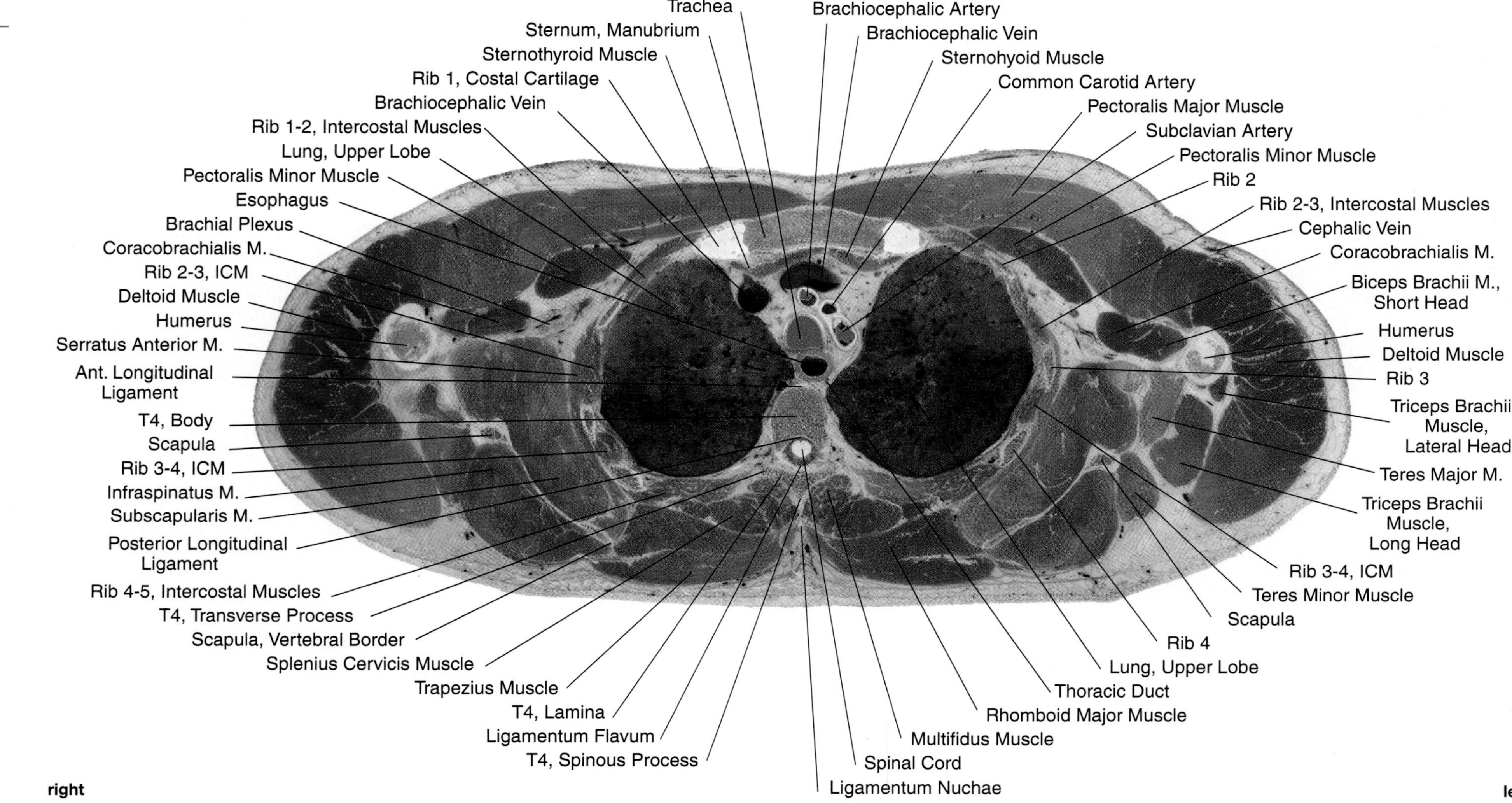

anterior
Trachea
Sternum, Manubrium
Sternothyroid Muscle
Rib 1, Costal Cartilage
Brachiocephalic Vein
Rib 1-2, Intercostal Muscles
Lung, Upper Lobe
Pectoralis Minor Muscle
Esophagus
Brachial Plexus
Coracobrachialis M.
Rib 2-3, ICM
Deltoid Muscle
Humerus
Serratus Anterior M.
Ant. Longitudinal Ligament
T4, Body
Scapula
Rib 3-4, ICM
Infraspinatus M.
Subscapularis M.
Posterior Longitudinal Ligament
Rib 4-5, Intercostal Muscles
T4, Transverse Process
Scapula, Vertebral Border
Splenius Cervicis Muscle
Trapezius Muscle
T4, Lamina
Ligamentum Flavum
T4, Spinous Process
Brachiocephalic Artery
Brachiocephalic Vein
Sternohyoid Muscle
Common Carotid Artery
Pectoralis Major Muscle
Subclavian Artery
Pectoralis Minor Muscle
Rib 2
Rib 2-3, Intercostal Muscles
Cephalic Vein
Coracobrachialis M.
Biceps Brachii M., Short Head
Humerus
Deltoid Muscle
Rib 3
Triceps Brachii Muscle, Lateral Head
Teres Major M.
Triceps Brachii Muscle, Long Head
Rib 3-4, ICM
Teres Minor Muscle
Scapula
Rib 4
Lung, Upper Lobe
Thoracic Duct
Rhomboid Major Muscle
Multifidus Muscle
Spinal Cord
Ligamentum Nuchae
right
left
posterior

ICM=Intercostal Muscles

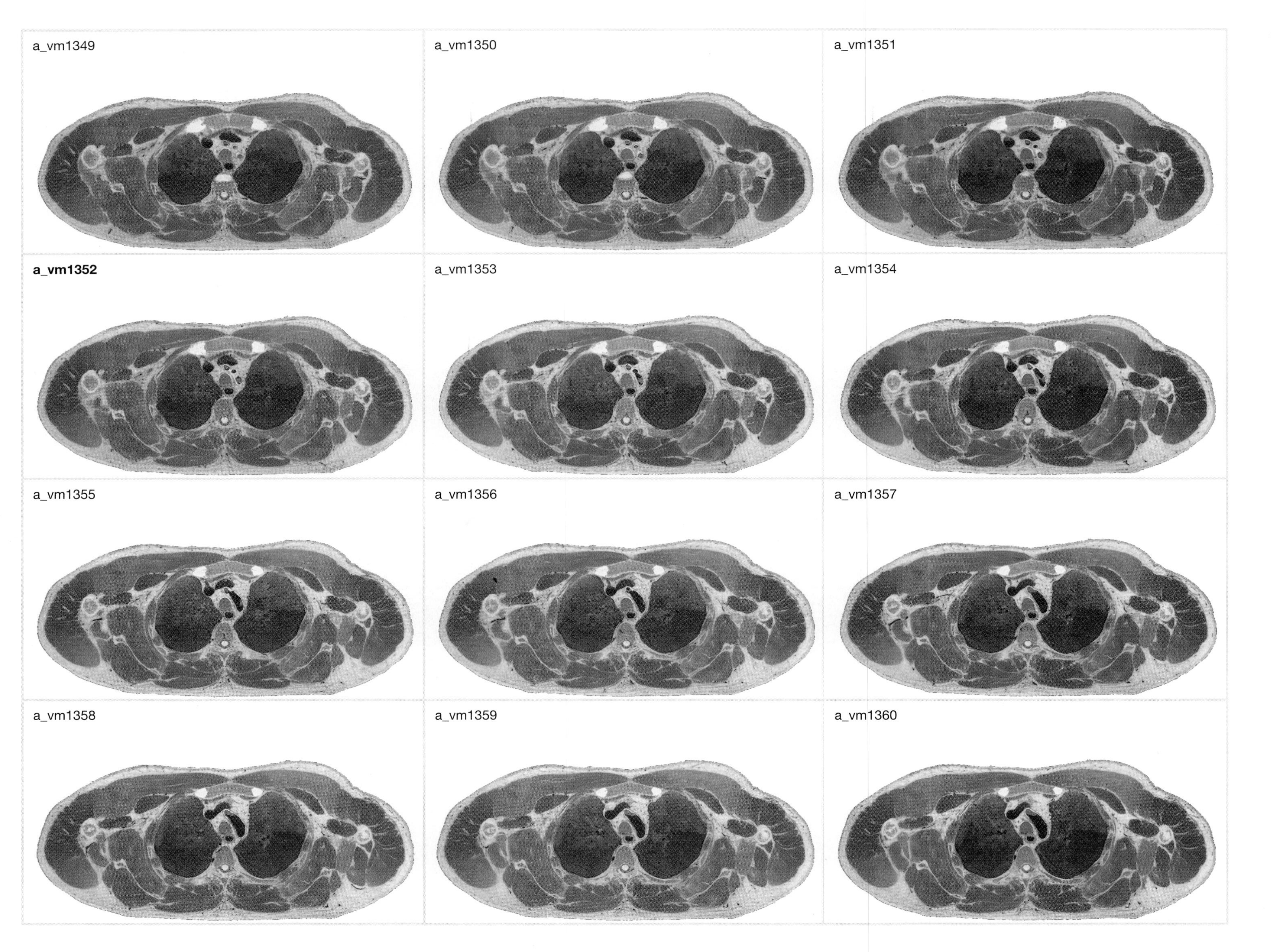
a_vm1349
a_vm1350
a_vm1351
a_vm1352
a_vm1353
a_vm1354
a_vm1355
a_vm1356
a_vm1357
a_vm1358
a_vm1359
a_vm1360

Transverse
a_vm1362

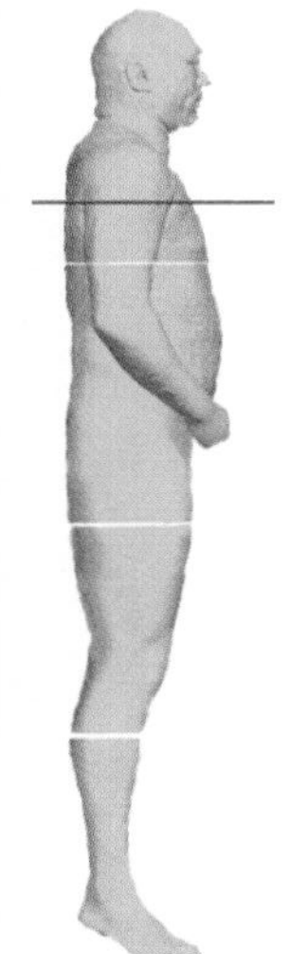

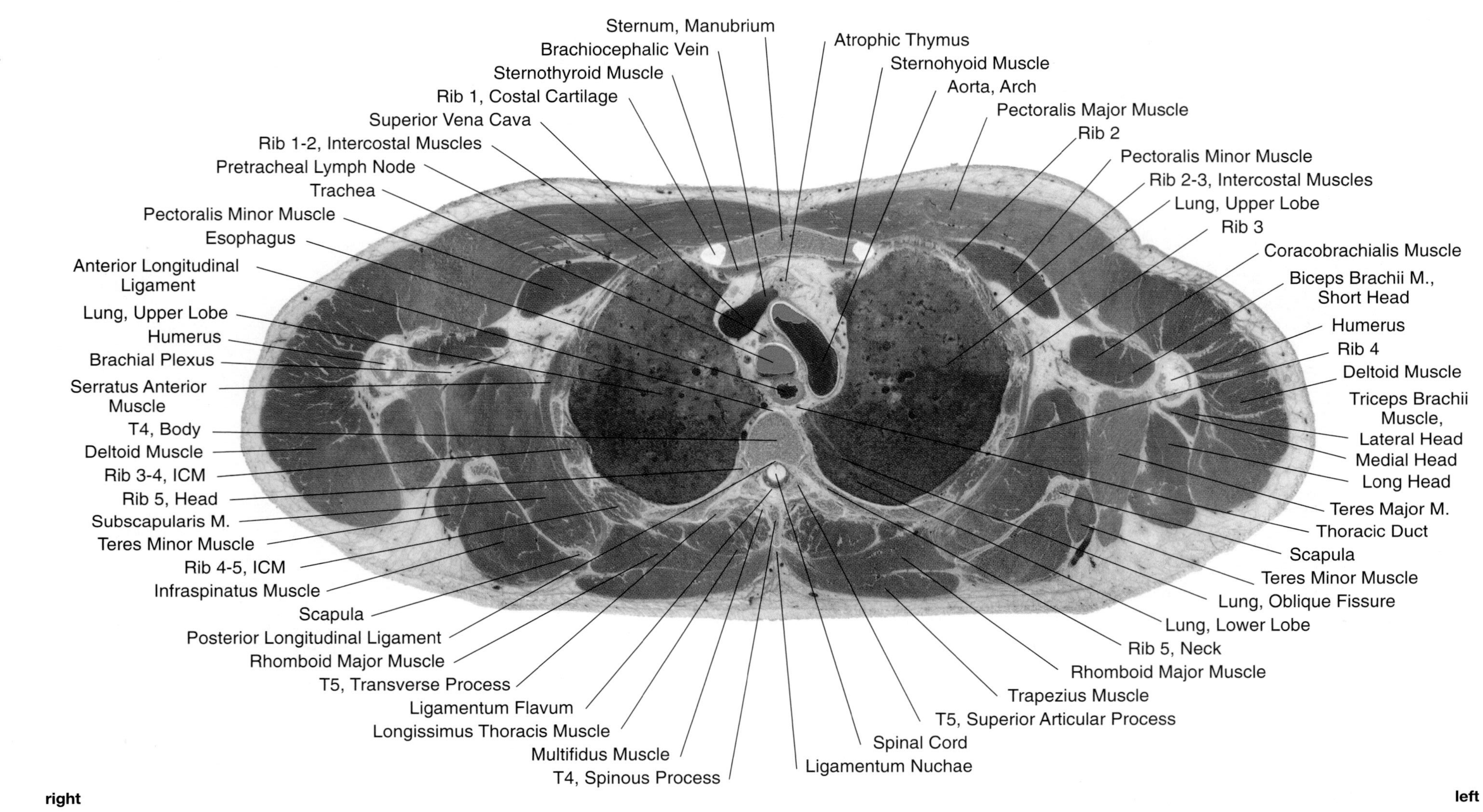
anterior
Sternum, Manubrium
Brachiocephalic Vein
Sternothyroid Muscle
Rib 1, Costal Cartilage
Superior Vena Cava
Rib 1-2, Intercostal Muscles
Pretracheal Lymph Node
Trachea
Pectoralis Minor Muscle
Esophagus
Anterior Longitudinal Ligament
Lung, Upper Lobe
Humerus
Brachial Plexus
Serratus Anterior Muscle
T4, Body
Deltoid Muscle
Rib 3-4, ICM
Rib 5, Head
Subscapularis M.
Teres Minor Muscle
Rib 4-5, ICM
Infraspinatus Muscle
Scapula
Posterior Longitudinal Ligament
Rhomboid Major Muscle
T5, Transverse Process
Ligamentum Flavum
Longissimus Thoracis Muscle
Multifidus Muscle
T4, Spinous Process
Atrophic Thymus
Sternohyoid Muscle
Aorta, Arch
Pectoralis Major Muscle
Rib 2
Pectoralis Minor Muscle
Rib 2-3, Intercostal Muscles
Lung, Upper Lobe
Rib 3
Coracobrachialis Muscle
Biceps Brachii M., Short Head
Humerus
Rib 4
Deltoid Muscle
Triceps Brachii Muscle,
Lateral Head
Medial Head
Long Head
Teres Major M.
Thoracic Duct
Scapula
Teres Minor Muscle
Lung, Oblique Fissure
Lung, Lower Lobe
Rib 5, Neck
Rhomboid Major Muscle
Trapezius Muscle
T5, Superior Articular Process
Spinal Cord
Ligamentum Nuchae
right
left
posterior

ICM=Intercostal Muscles

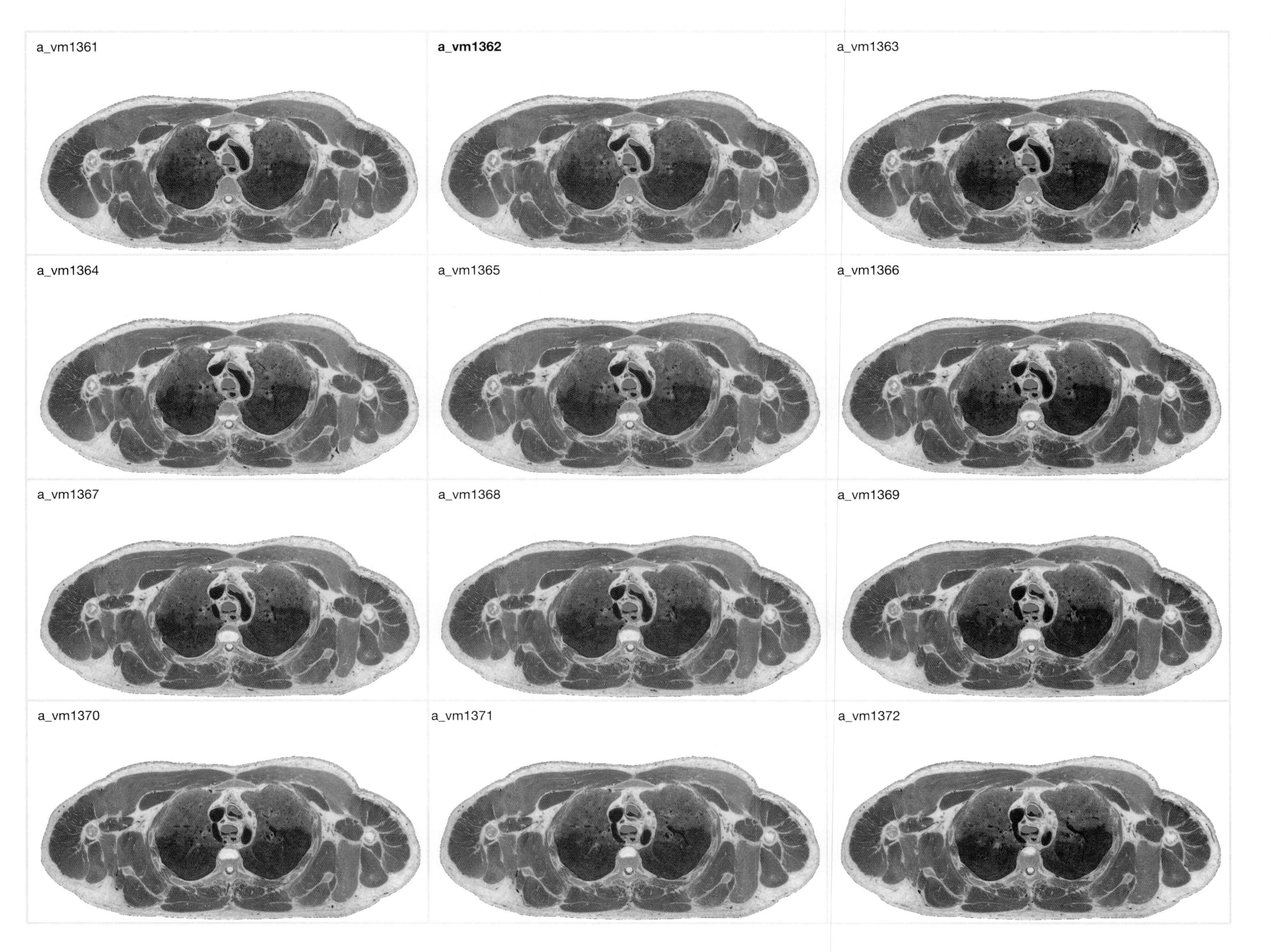
a_vm1361
a_vm1362
a_vm1363
a_vm1364
a_vm1365
a_vm1366
a_vm1367
a_vm1368
a_vm1369
a_vm1370
a_vm1371
a_vm1372

Transverse
a_vm1381

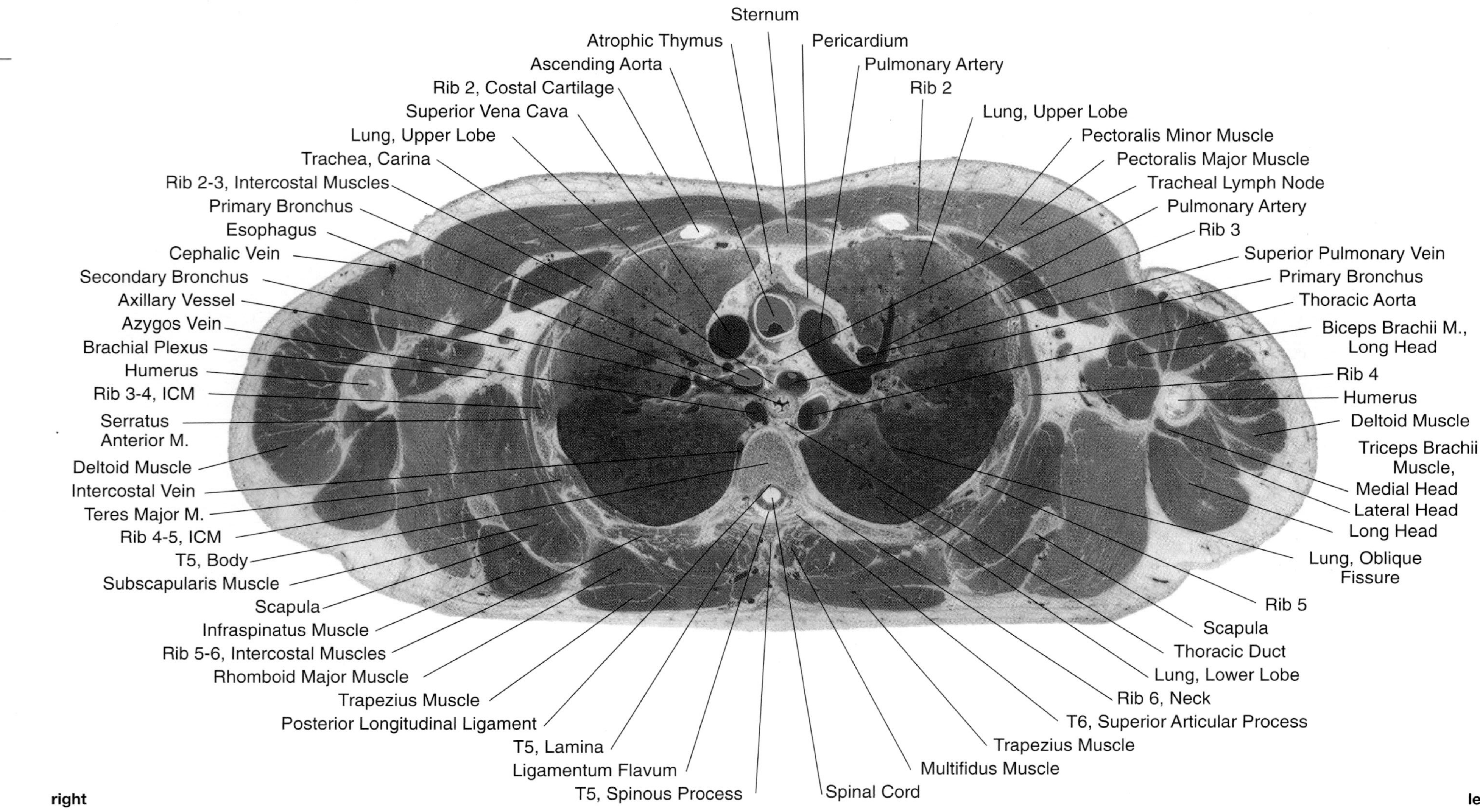

anterior
Sternum
Atrophic Thymus
Pericardium
Ascending Aorta
Pulmonary Artery
Rib 2, Costal Cartilage
Rib 2
Superior Vena Cava
Lung, Upper Lobe
Lung, Upper Lobe
Pectoralis Minor Muscle
Trachea, Carina
Pectoralis Major Muscle
Rib 2-3, Intercostal Muscles
Tracheal Lymph Node
Primary Bronchus
Pulmonary Artery
Esophagus
Rib 3
Cephalic Vein
Superior Pulmonary Vein
Secondary Bronchus
Primary Bronchus
Axillary Vessel
Thoracic Aorta
Azygos Vein
Biceps Brachii M., Long Head
Brachial Plexus
Humerus
Rib 4
Rib 3-4, ICM
Humerus
Serratus Anterior M.
Deltoid Muscle
Deltoid Muscle
Triceps Brachii Muscle,
Intercostal Vein
Medial Head
Teres Major M.
Lateral Head
Rib 4-5, ICM
Long Head
T5, Body
Lung, Oblique Fissure
Subscapularis Muscle
Scapula
Rib 5
Infraspinatus Muscle
Scapula
Rib 5-6, Intercostal Muscles
Thoracic Duct
Rhomboid Major Muscle
Lung, Lower Lobe
Trapezius Muscle
Rib 6, Neck
Posterior Longitudinal Ligament
T6, Superior Articular Process
T5, Lamina
Trapezius Muscle
Ligamentum Flavum
Multifidus Muscle
T5, Spinous Process
Spinal Cord
right
left
posterior

ICM=Intercostal Muscles

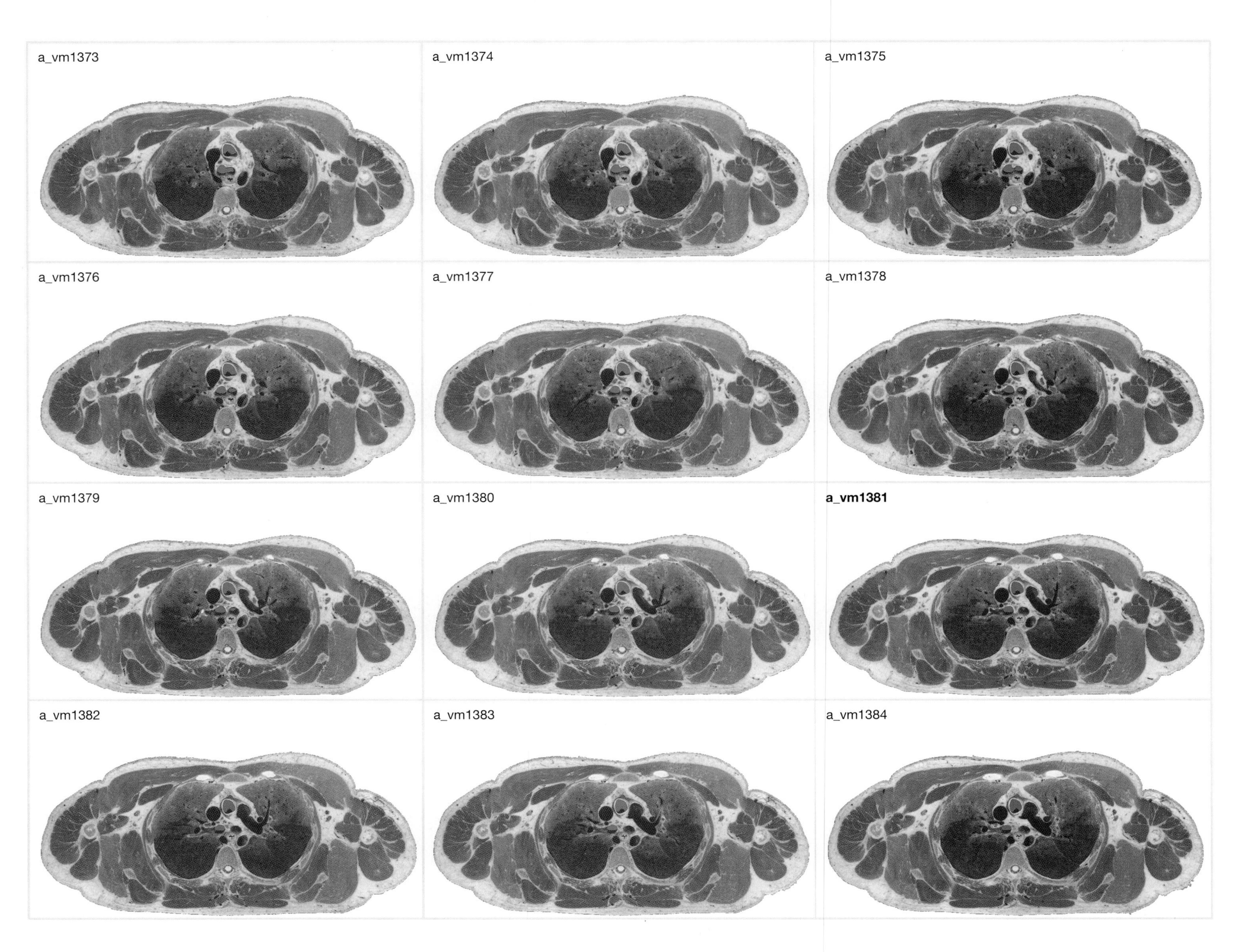
a_vm1373
a_vm1374
a_vm1375
a_vm1376
a_vm1377
a_vm1378
a_vm1379
a_vm1380
a_vm1381
a_vm1382
a_vm1383
a_vm1384

Transverse
a_vm1388

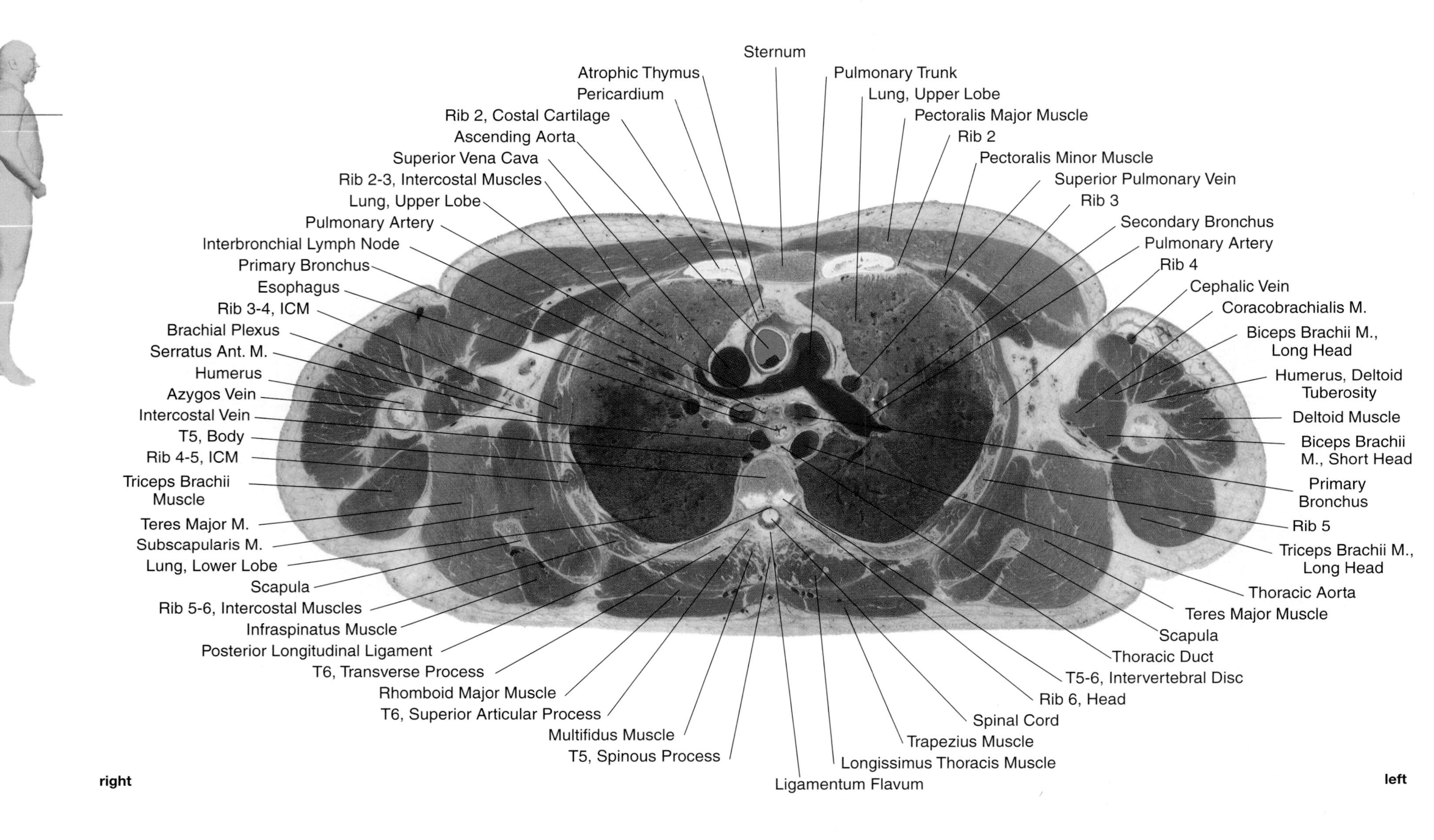
anterior
Sternum
Atrophic Thymus
Pericardium
Rib 2, Costal Cartilage
Ascending Aorta
Superior Vena Cava
Rib 2-3, Intercostal Muscles
Lung, Upper Lobe
Pulmonary Artery
Interbronchial Lymph Node
Primary Bronchus
Esophagus
Rib 3-4, ICM
Brachial Plexus
Serratus Ant. M.
Humerus
Azygos Vein
Intercostal Vein
T5, Body
Rib 4-5, ICM
Triceps Brachii Muscle
Teres Major M.
Subscapularis M.
Lung, Lower Lobe
Scapula
Rib 5-6, Intercostal Muscles
Infraspinatus Muscle
Posterior Longitudinal Ligament
T6, Transverse Process
Rhomboid Major Muscle
T6, Superior Articular Process
Multifidus Muscle
T5, Spinous Process
Ligamentum Flavum
Longissimus Thoracis Muscle
Trapezius Muscle
Spinal Cord
Rib 6, Head
T5-6, Intervertebral Disc
Thoracic Duct
Scapula
Teres Major Muscle
Thoracic Aorta
Triceps Brachii M., Long Head
Rib 5
Primary Bronchus
Biceps Brachii M., Short Head
Deltoid Muscle
Humerus, Deltoid Tuberosity
Biceps Brachii M., Long Head
Coracobrachialis M.
Cephalic Vein
Rib 4
Pulmonary Artery
Secondary Bronchus
Rib 3
Superior Pulmonary Vein
Pectoralis Minor Muscle
Rib 2
Pectoralis Major Muscle
Lung, Upper Lobe
Pulmonary Trunk
right
left
posterior

ICM=Intercostal Muscles

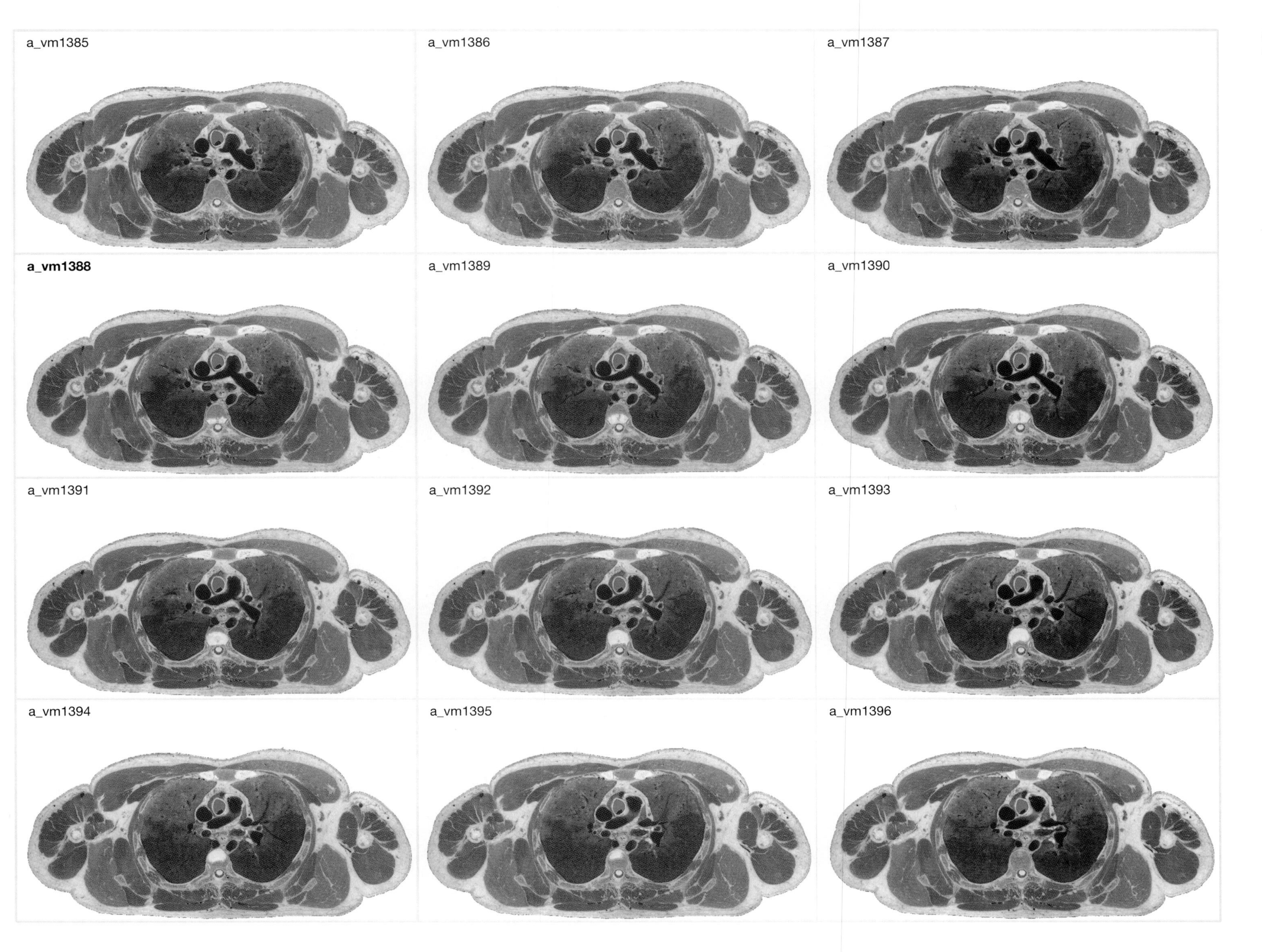
a_vm1385
a_vm1386
a_vm1387
a_vm1388
a_vm1389
a_vm1390
a_vm1391
a_vm1392
a_vm1393
a_vm1394
a_vm1395
a_vm1396

Transverse
a_vm1405

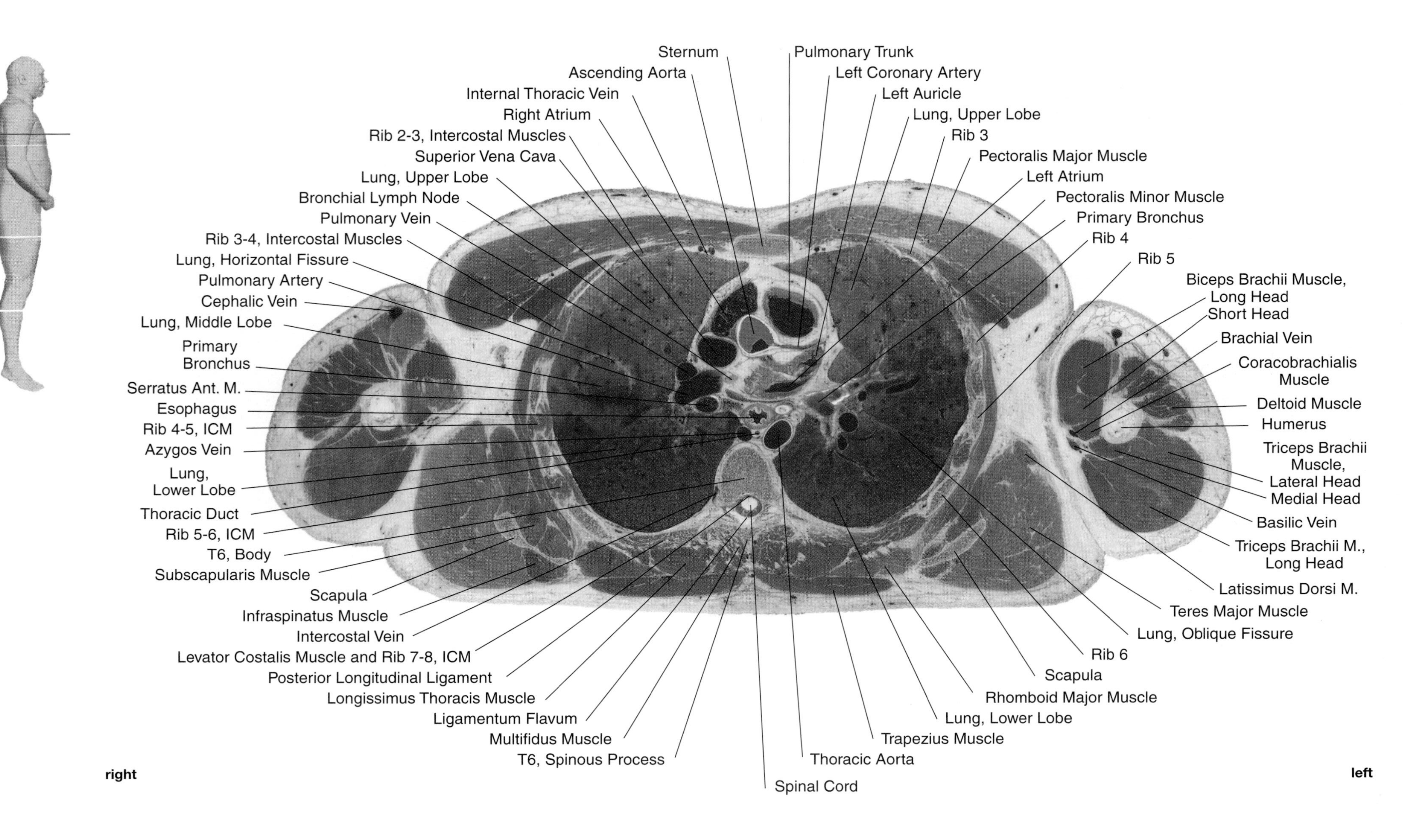

anterior
Sternum
Ascending Aorta
Internal Thoracic Vein
Right Atrium
Rib 2-3, Intercostal Muscles
Superior Vena Cava
Lung, Upper Lobe
Bronchial Lymph Node
Pulmonary Vein
Rib 3-4, Intercostal Muscles
Lung, Horizontal Fissure
Pulmonary Artery
Cephalic Vein
Lung, Middle Lobe
Primary Bronchus
Serratus Ant. M.
Esophagus
Rib 4-5, ICM
Azygos Vein
Lung, Lower Lobe
Thoracic Duct
Rib 5-6, ICM
T6, Body
Subscapularis Muscle
Scapula
Infraspinatus Muscle
Intercostal Vein
Levator Costalis Muscle and Rib 7-8, ICM
Posterior Longitudinal Ligament
Longissimus Thoracis Muscle
Ligamentum Flavum
Multifidus Muscle
T6, Spinous Process
Spinal Cord
Thoracic Aorta
Trapezius Muscle
Lung, Lower Lobe
Rhomboid Major Muscle
Scapula
Rib 6
Lung, Oblique Fissure
Teres Major Muscle
Latissimus Dorsi M.
Triceps Brachii M., Long Head
Basilic Vein
Medial Head
Triceps Brachii Muscle, Lateral Head
Humerus
Deltoid Muscle
Coracobrachialis Muscle
Brachial Vein
Short Head
Biceps Brachii Muscle, Long Head
Rib 5
Rib 4
Primary Bronchus
Pectoralis Minor Muscle
Left Atrium
Pectoralis Major Muscle
Rib 3
Lung, Upper Lobe
Left Auricle
Left Coronary Artery
Pulmonary Trunk
right
left
posterior

ICM=Intercostal Muscles

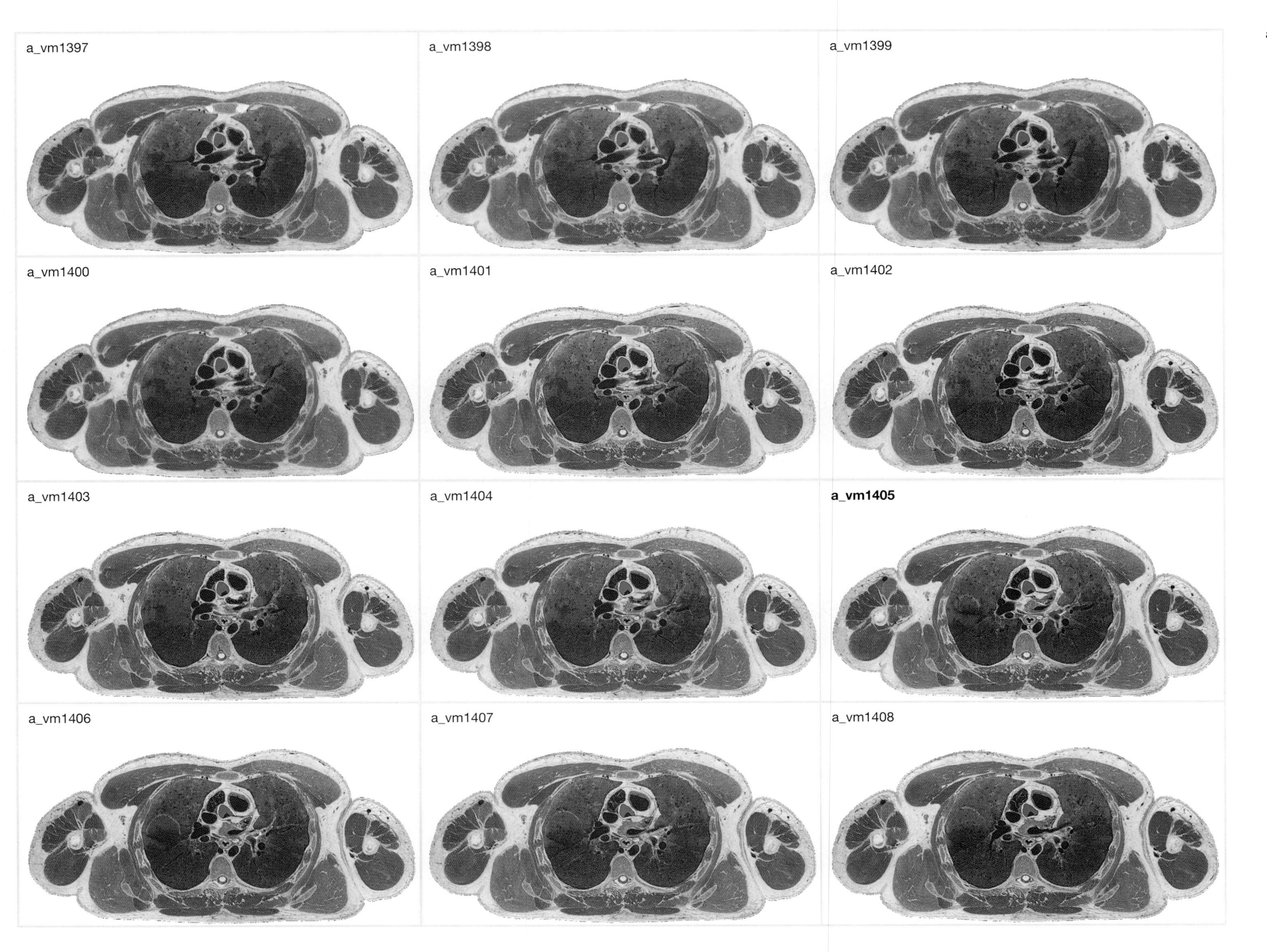
a_vm1397
a_vm1398
a_vm1399
a_vm1400
a_vm1401
a_vm1402
a_vm1403
a_vm1404
a_vm1405
a_vm1406
a_vm1407
a_vm1408

Transverse
a_vm1420

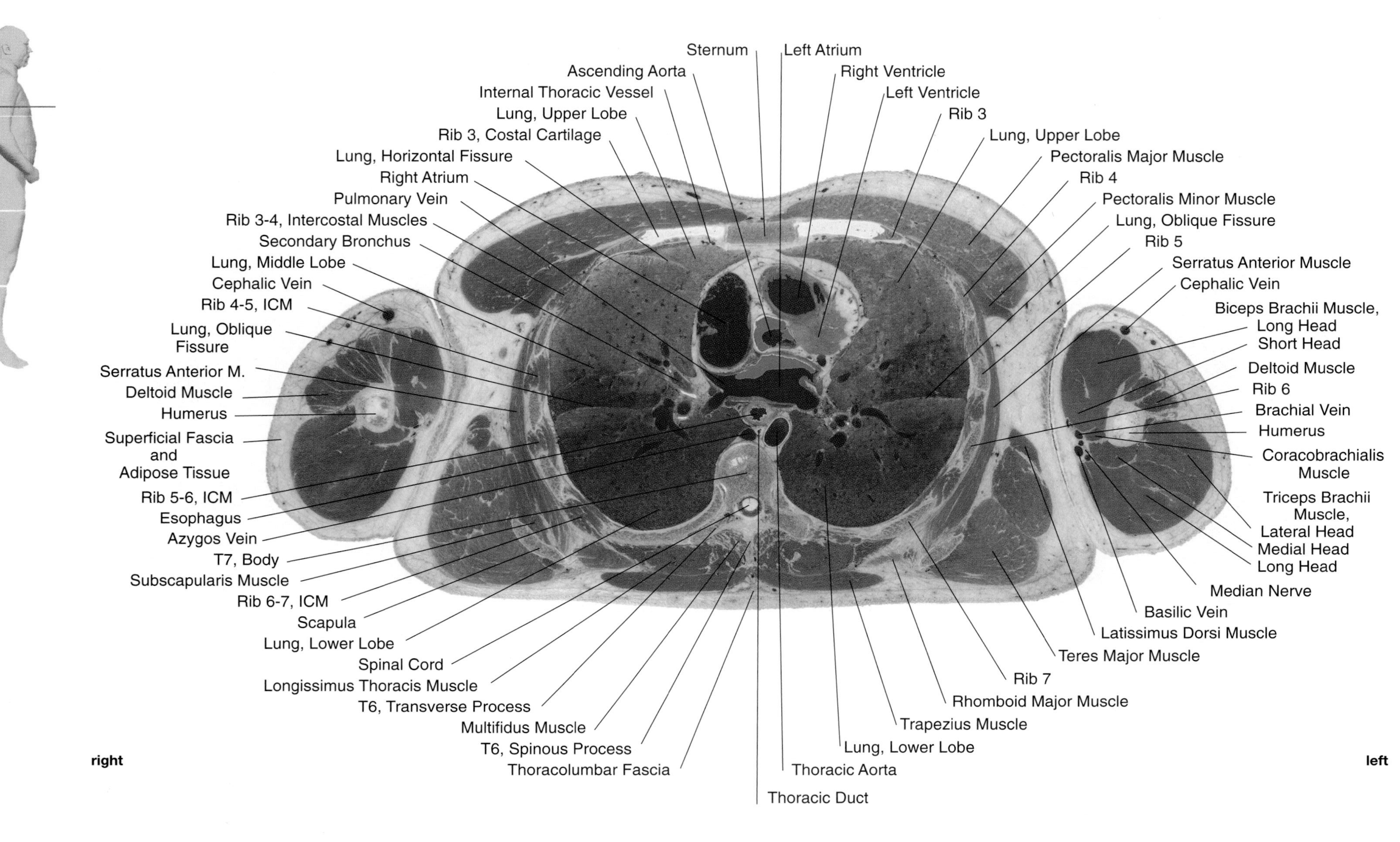

anterior
Sternum
Left Atrium
Ascending Aorta
Right Ventricle
Internal Thoracic Vessel
Left Ventricle
Lung, Upper Lobe
Rib 3
Rib 3, Costal Cartilage
Lung, Upper Lobe
Lung, Horizontal Fissure
Pectoralis Major Muscle
Right Atrium
Rib 4
Pulmonary Vein
Pectoralis Minor Muscle
Rib 3-4, Intercostal Muscles
Lung, Oblique Fissure
Secondary Bronchus
Rib 5
Lung, Middle Lobe
Serratus Anterior Muscle
Cephalic Vein
Cephalic Vein
Rib 4-5, ICM
Biceps Brachii Muscle,
Long Head
Short Head
Lung, Oblique
Fissure
Serratus Anterior M.
Deltoid Muscle
Deltoid Muscle
Rib 6
Humerus
Brachial Vein
Humerus
Superficial Fascia
and
Adipose Tissue
Coracobrachialis
Muscle
Rib 5-6, ICM
Triceps Brachii
Muscle,
Lateral Head
Medial Head
Long Head
Esophagus
Azygos Vein
T7, Body
Subscapularis Muscle
Median Nerve
Rib 6-7, ICM
Basilic Vein
Scapula
Latissimus Dorsi Muscle
Lung, Lower Lobe
Teres Major Muscle
Spinal Cord
Rib 7
Longissimus Thoracis Muscle
Rhomboid Major Muscle
T6, Transverse Process
Trapezius Muscle
Multifidus Muscle
Lung, Lower Lobe
T6, Spinous Process
Thoracolumbar Fascia
Thoracic Aorta
Thoracic Duct
right
left
posterior

ICM=Intercostal Muscles

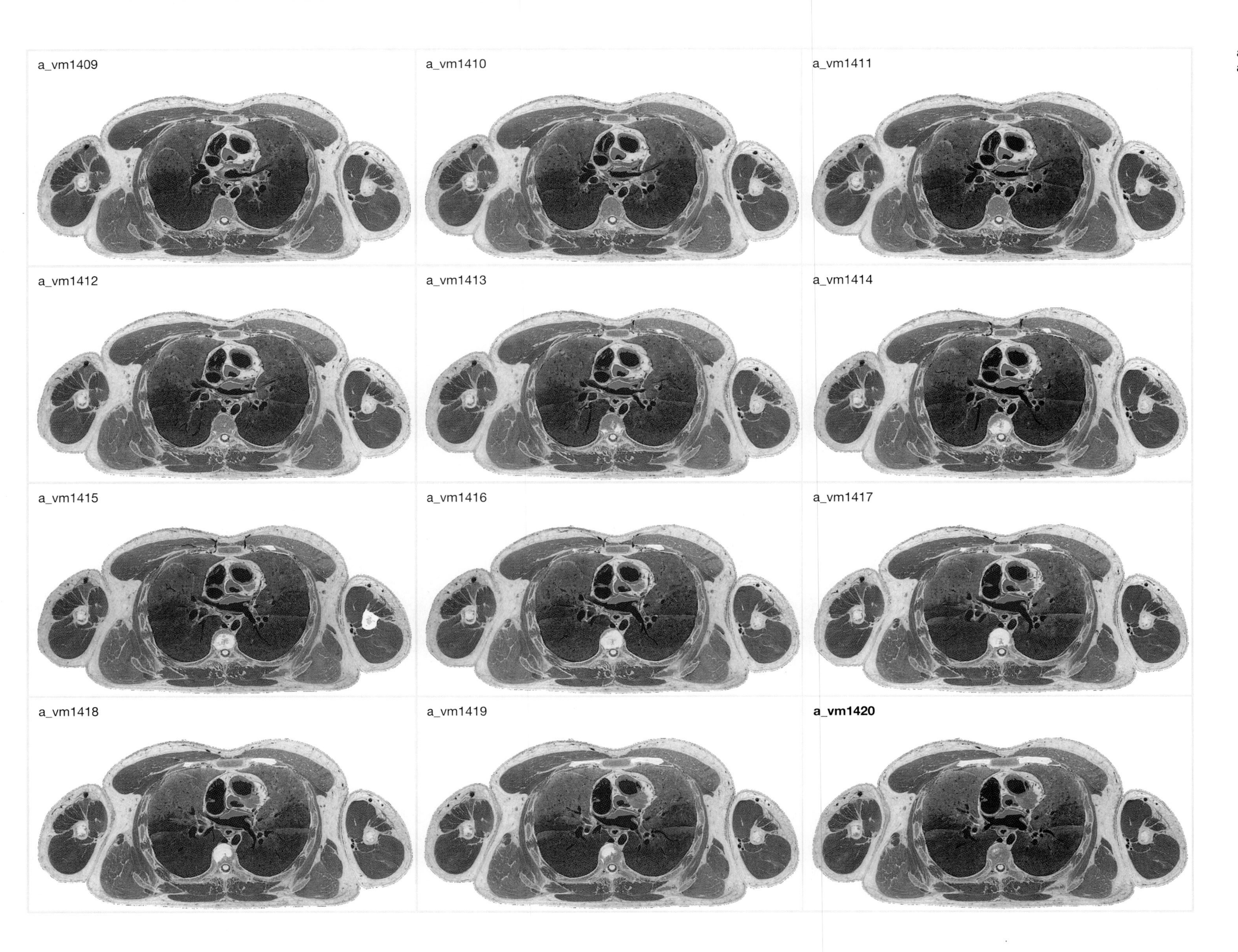
a_vm1409
a_vm1410
a_vm1411
a_vm1412
a_vm1413
a_vm1414
a_vm1415
a_vm1416
a_vm1417
a_vm1418
a_vm1419
a_vm1420

Transverse
a_vm1428

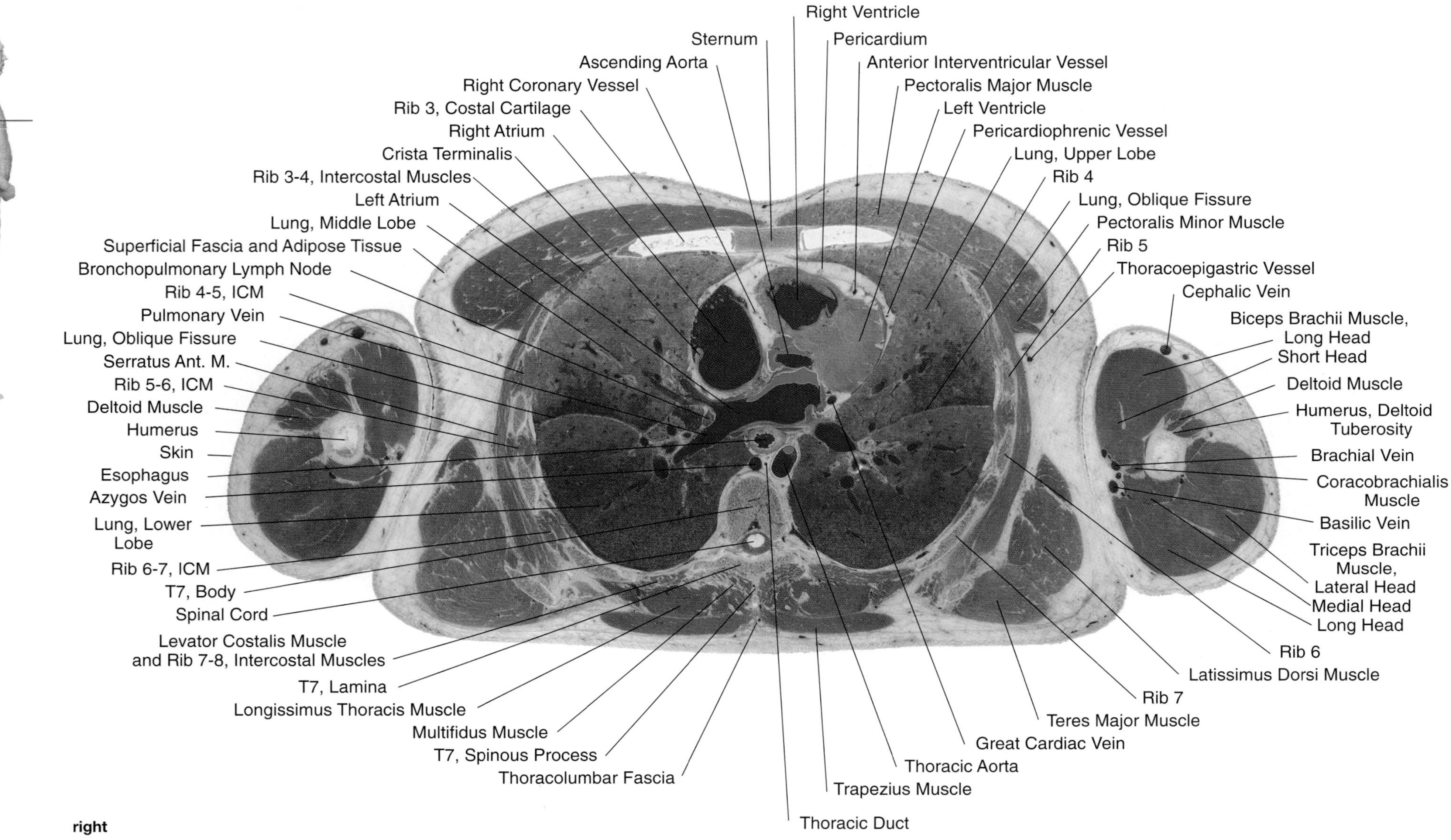
anterior
Right Ventricle
Sternum
Pericardium
Ascending Aorta
Anterior Interventricular Vessel
Right Coronary Vessel
Pectoralis Major Muscle
Rib 3, Costal Cartilage
Left Ventricle
Right Atrium
Pericardiophrenic Vessel
Crista Terminalis
Lung, Upper Lobe
Rib 3-4, Intercostal Muscles
Rib 4
Left Atrium
Lung, Oblique Fissure
Lung, Middle Lobe
Pectoralis Minor Muscle
Superficial Fascia and Adipose Tissue
Rib 5
Bronchopulmonary Lymph Node
Thoracoepigastric Vessel
Rib 4-5, ICM
Cephalic Vein
Pulmonary Vein
Biceps Brachii Muscle, Long Head
Lung, Oblique Fissure
Short Head
Serratus Ant. M.
Rib 5-6, ICM
Deltoid Muscle
Deltoid Muscle
Humerus, Deltoid Tuberosity
Humerus
Skin
Brachial Vein
Esophagus
Coracobrachialis Muscle
Azygos Vein
Basilic Vein
Lung, Lower Lobe
Triceps Brachii Muscle, Lateral Head
Rib 6-7, ICM
Medial Head
T7, Body
Long Head
Spinal Cord
Rib 6
Levator Costalis Muscle and Rib 7-8, Intercostal Muscles
Latissimus Dorsi Muscle
T7, Lamina
Rib 7
Longissimus Thoracis Muscle
Teres Major Muscle
Multifidus Muscle
Great Cardiac Vein
T7, Spinous Process
Thoracic Aorta
Thoracolumbar Fascia
Trapezius Muscle
Thoracic Duct
right
left
posterior

ICM=Intercostal Muscles

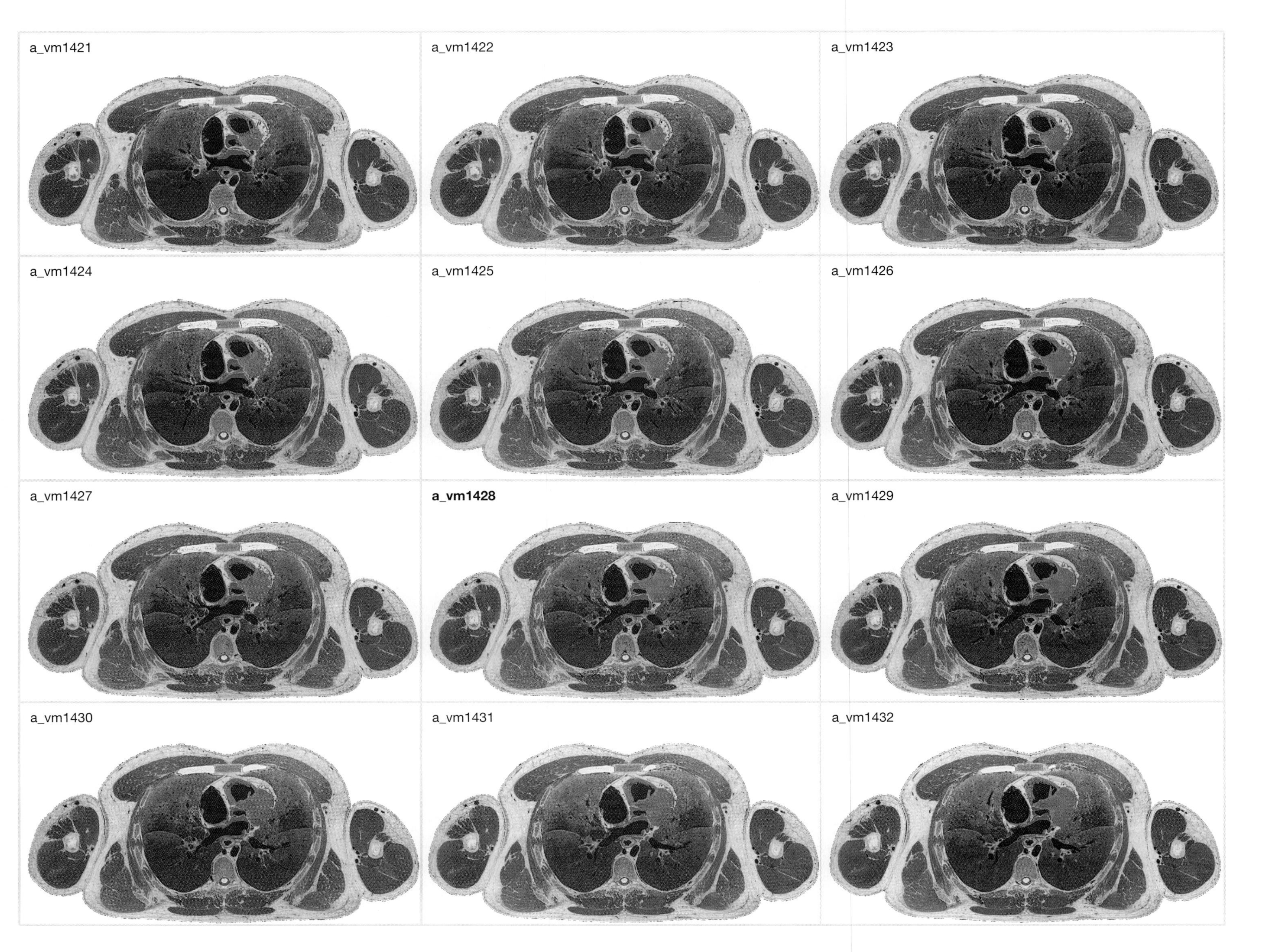
a_vm1421
a_vm1422
a_vm1423
a_vm1424
a_vm1425
a_vm1426
a_vm1427
a_vm1428
a_vm1429
a_vm1430
a_vm1431
a_vm1432

Transverse
a_vm1436

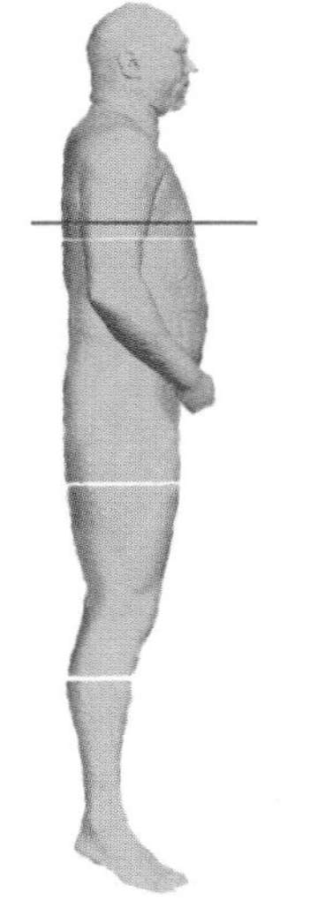

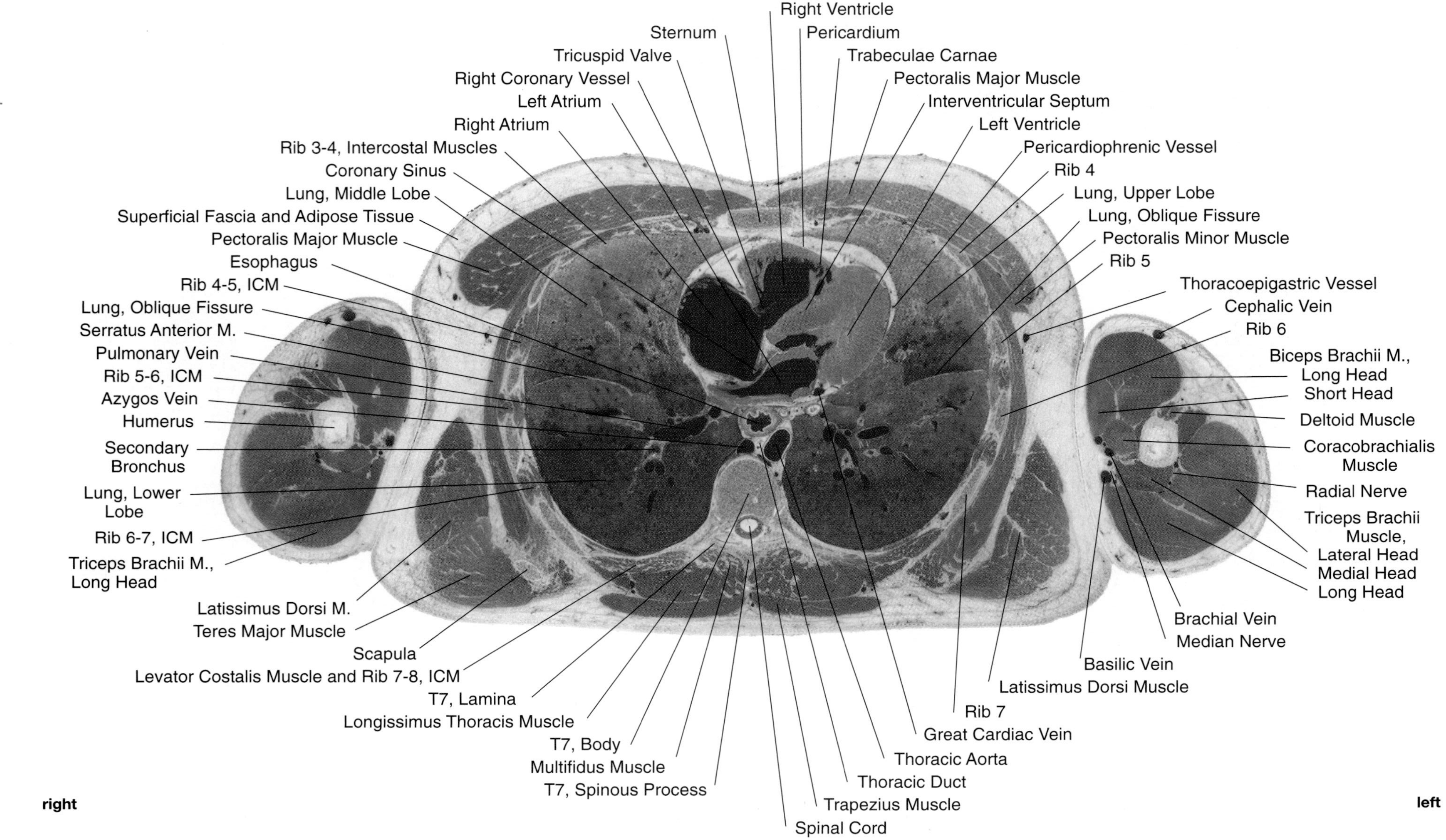
anterior
Right Ventricle
Sternum
Pericardium
Tricuspid Valve
Trabeculae Carnae
Right Coronary Vessel
Pectoralis Major Muscle
Left Atrium
Interventricular Septum
Right Atrium
Left Ventricle
Rib 3-4, Intercostal Muscles
Pericardiophrenic Vessel
Coronary Sinus
Rib 4
Lung, Middle Lobe
Lung, Upper Lobe
Superficial Fascia and Adipose Tissue
Lung, Oblique Fissure
Pectoralis Major Muscle
Pectoralis Minor Muscle
Esophagus
Rib 5
Rib 4-5, ICM
Thoracoepigastric Vessel
Lung, Oblique Fissure
Cephalic Vein
Serratus Anterior M.
Rib 6
Pulmonary Vein
Biceps Brachii M.,
Long Head
Rib 5-6, ICM
Short Head
Azygos Vein
Deltoid Muscle
Humerus
Secondary
Bronchus
Coracobrachialis
Muscle
Lung, Lower
Lobe
Radial Nerve
Rib 6-7, ICM
Triceps Brachii
Muscle,
Lateral Head
Medial Head
Long Head
Triceps Brachii M.,
Long Head
Latissimus Dorsi M.
Brachial Vein
Teres Major Muscle
Median Nerve
Scapula
Basilic Vein
Levator Costalis Muscle and Rib 7-8, ICM
Latissimus Dorsi Muscle
T7, Lamina
Rib 7
Longissimus Thoracis Muscle
Great Cardiac Vein
T7, Body
Thoracic Aorta
Multifidus Muscle
Thoracic Duct
T7, Spinous Process
Trapezius Muscle
Spinal Cord
right
left
posterior

ICM=Intercostal Muscles

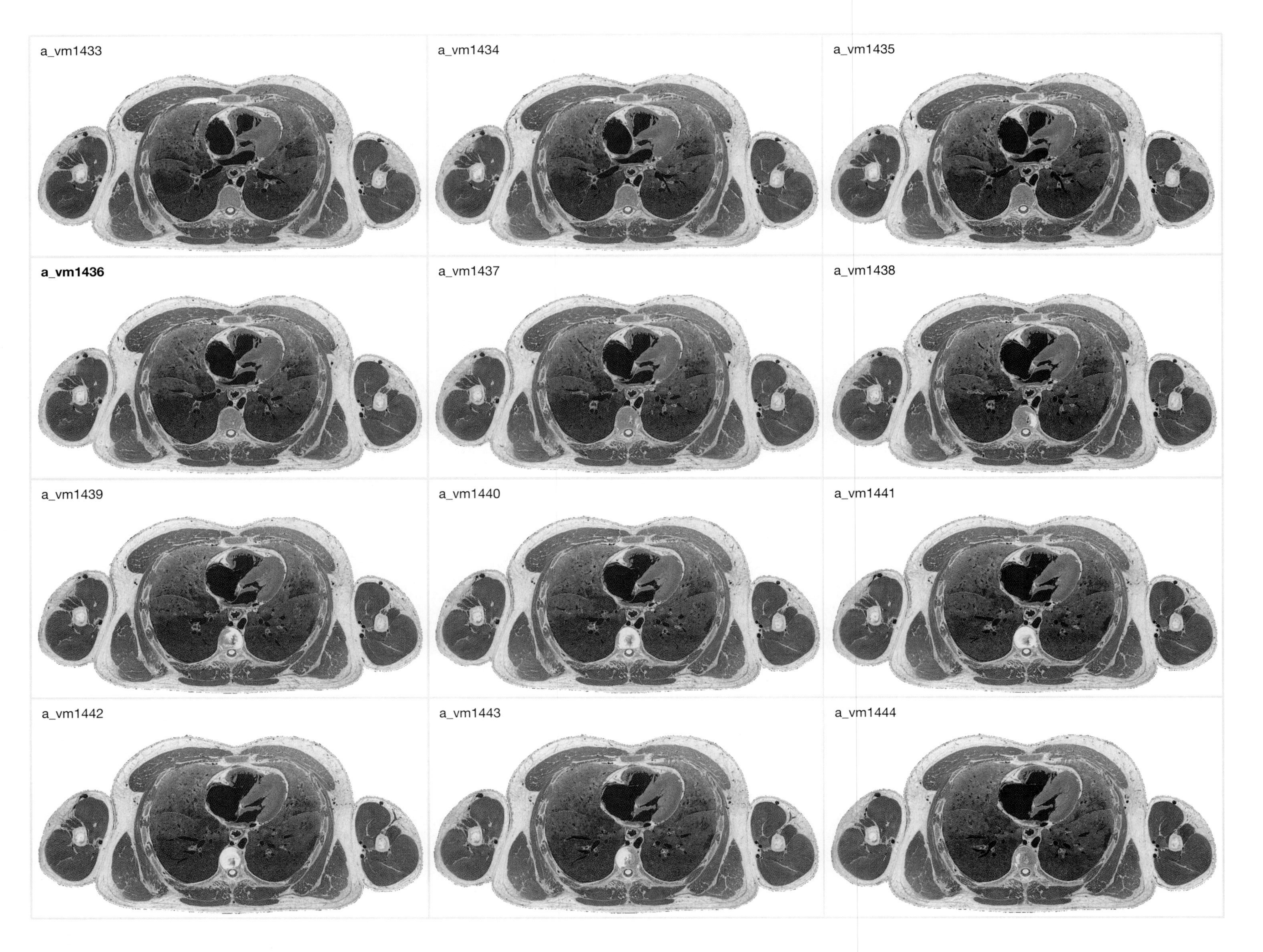
a_vm1433
a_vm1434
a_vm1435
a_vm1436
a_vm1437
a_vm1438
a_vm1439
a_vm1440
a_vm1441
a_vm1442
a_vm1443
a_vm1444

Transverse
a_vm1452

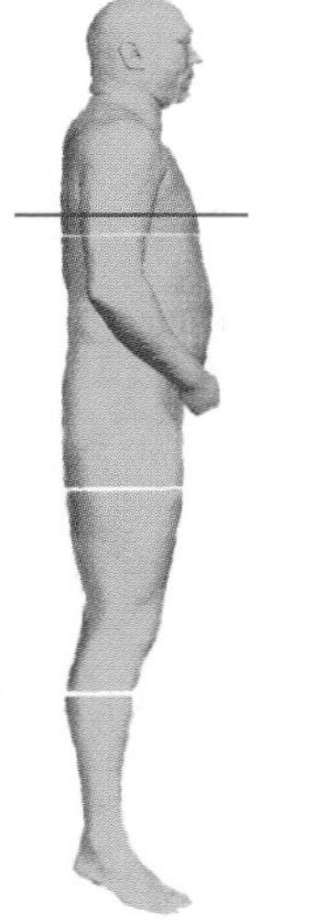

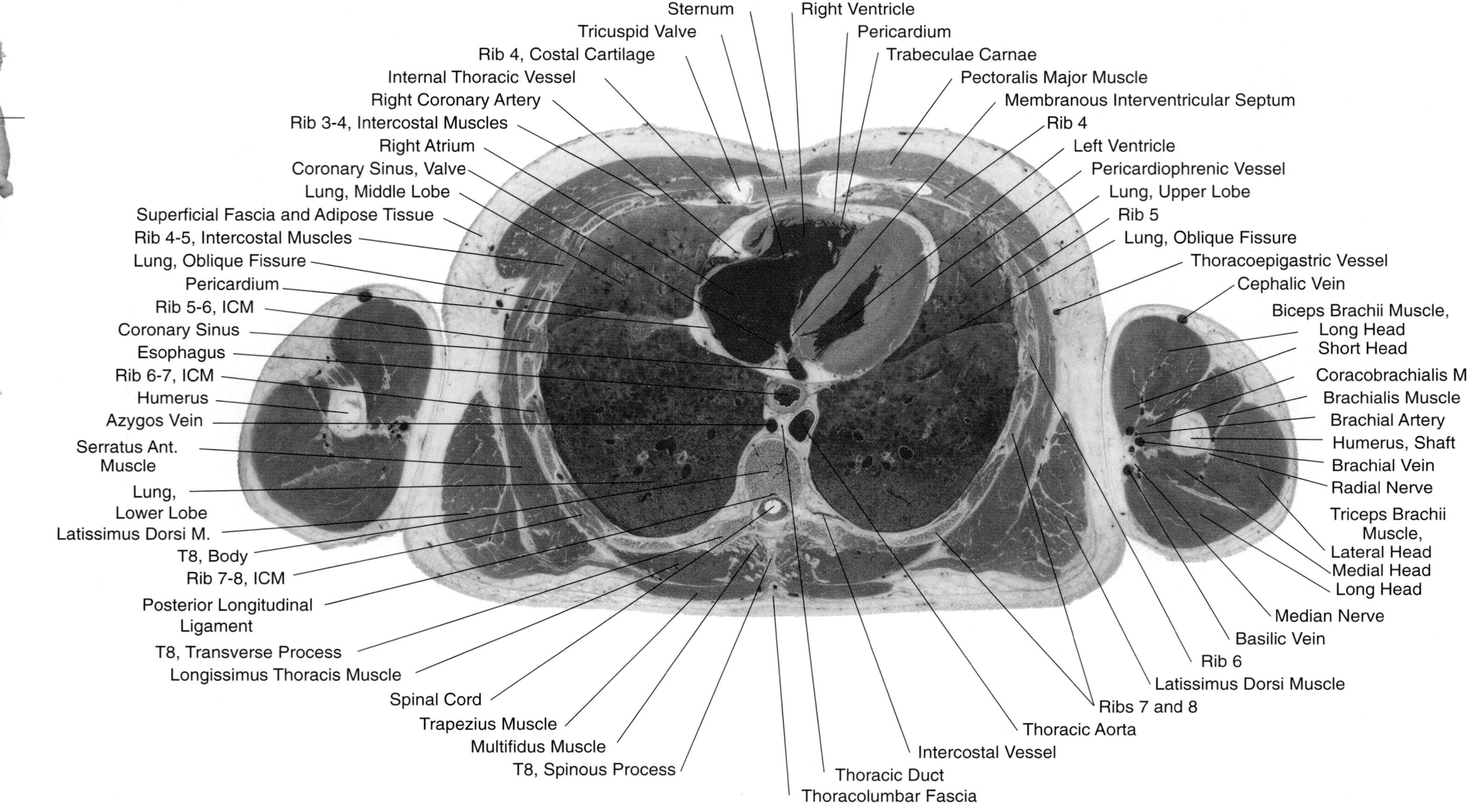
anterior
Sternum
Tricuspid Valve
Rib 4, Costal Cartilage
Internal Thoracic Vessel
Right Coronary Artery
Rib 3-4, Intercostal Muscles
Right Atrium
Coronary Sinus, Valve
Lung, Middle Lobe
Superficial Fascia and Adipose Tissue
Rib 4-5, Intercostal Muscles
Lung, Oblique Fissure
Pericardium
Rib 5-6, ICM
Coronary Sinus
Esophagus
Rib 6-7, ICM
Humerus
Azygos Vein
Serratus Ant. Muscle
Lung, Lower Lobe
Latissimus Dorsi M.
T8, Body
Rib 7-8, ICM
Posterior Longitudinal Ligament
T8, Transverse Process
Longissimus Thoracis Muscle
Spinal Cord
Trapezius Muscle
Multifidus Muscle
T8, Spinous Process
Right Ventricle
Pericardium
Trabeculae Carnae
Pectoralis Major Muscle
Membranous Interventricular Septum
Rib 4
Left Ventricle
Pericardiophrenic Vessel
Lung, Upper Lobe
Rib 5
Lung, Oblique Fissure
Thoracoepigastric Vessel
Cephalic Vein
Biceps Brachii Muscle, Long Head
Short Head
Coracobrachialis M.
Brachialis Muscle
Brachial Artery
Humerus, Shaft
Brachial Vein
Radial Nerve
Triceps Brachii Muscle, Lateral Head
Medial Head
Long Head
Median Nerve
Basilic Vein
Rib 6
Latissimus Dorsi Muscle
Ribs 7 and 8
Thoracic Aorta
Intercostal Vessel
Thoracic Duct
Thoracolumbar Fascia
right
left
posterior

ICM=Intercostal Muscles

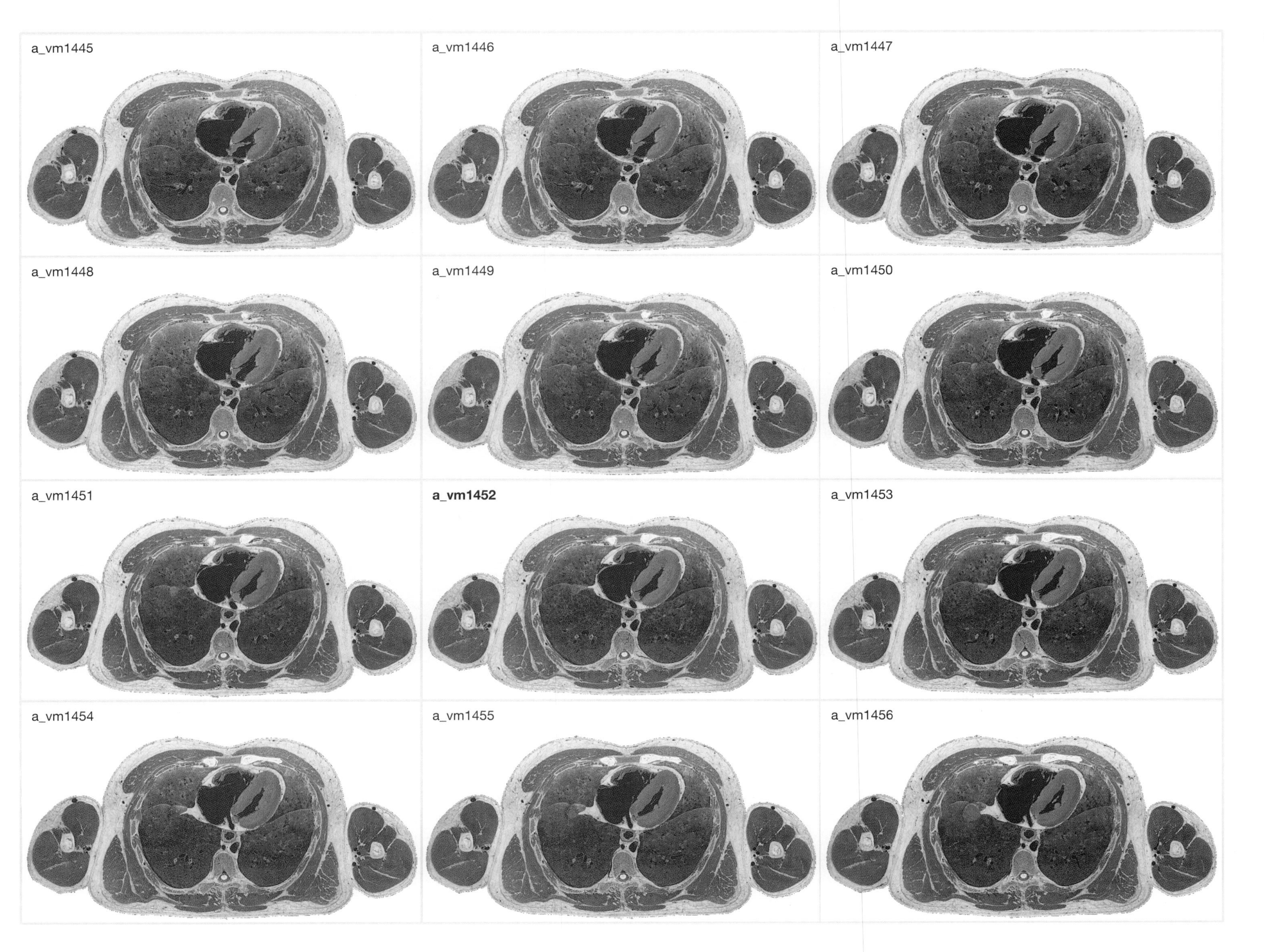
a_vm1445
a_vm1446
a_vm1447
a_vm1448
a_vm1449
a_vm1450
a_vm1451
a_vm1452
a_vm1453
a_vm1454
a_vm1455
a_vm1456

Transverse
a_vm1461

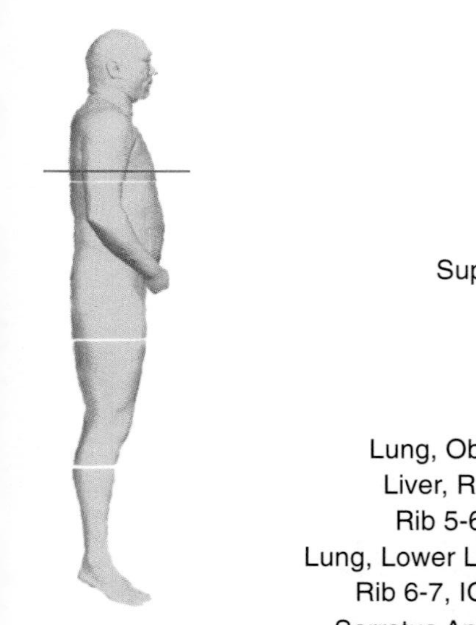

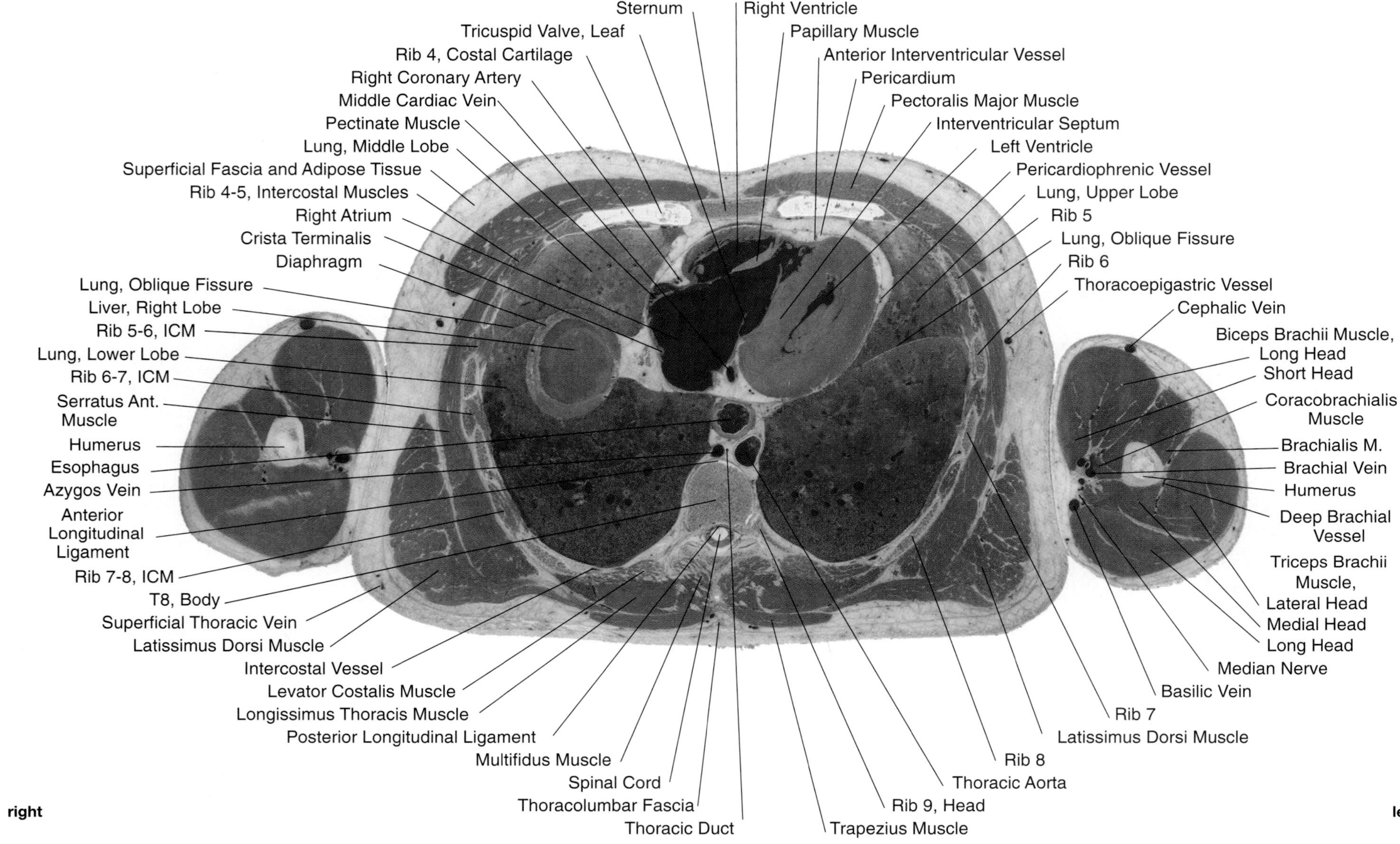
anterior
Sternum
Tricuspid Valve, Leaf
Rib 4, Costal Cartilage
Right Coronary Artery
Middle Cardiac Vein
Pectinate Muscle
Lung, Middle Lobe
Superficial Fascia and Adipose Tissue
Rib 4-5, Intercostal Muscles
Right Atrium
Crista Terminalis
Diaphragm
Lung, Oblique Fissure
Liver, Right Lobe
Rib 5-6, ICM
Lung, Lower Lobe
Rib 6-7, ICM
Serratus Ant. Muscle
Humerus
Esophagus
Azygos Vein
Anterior Longitudinal Ligament
Rib 7-8, ICM
T8, Body
Superficial Thoracic Vein
Latissimus Dorsi Muscle
Intercostal Vessel
Levator Costalis Muscle
Longissimus Thoracis Muscle
Posterior Longitudinal Ligament
Multifidus Muscle
Spinal Cord
Thoracolumbar Fascia
Thoracic Duct
Right Ventricle
Papillary Muscle
Anterior Interventricular Vessel
Pericardium
Pectoralis Major Muscle
Interventricular Septum
Left Ventricle
Pericardiophrenic Vessel
Lung, Upper Lobe
Rib 5
Lung, Oblique Fissure
Rib 6
Thoracoepigastric Vessel
Cephalic Vein
Biceps Brachii Muscle, Long Head
Short Head
Coracobrachialis Muscle
Brachialis M.
Brachial Vein
Humerus
Deep Brachial Vessel
Triceps Brachii Muscle,
Lateral Head
Medial Head
Long Head
Median Nerve
Basilic Vein
Rib 7
Latissimus Dorsi Muscle
Rib 8
Thoracic Aorta
Rib 9, Head
Trapezius Muscle
right
left
posterior

ICM=Intercostal Muscles

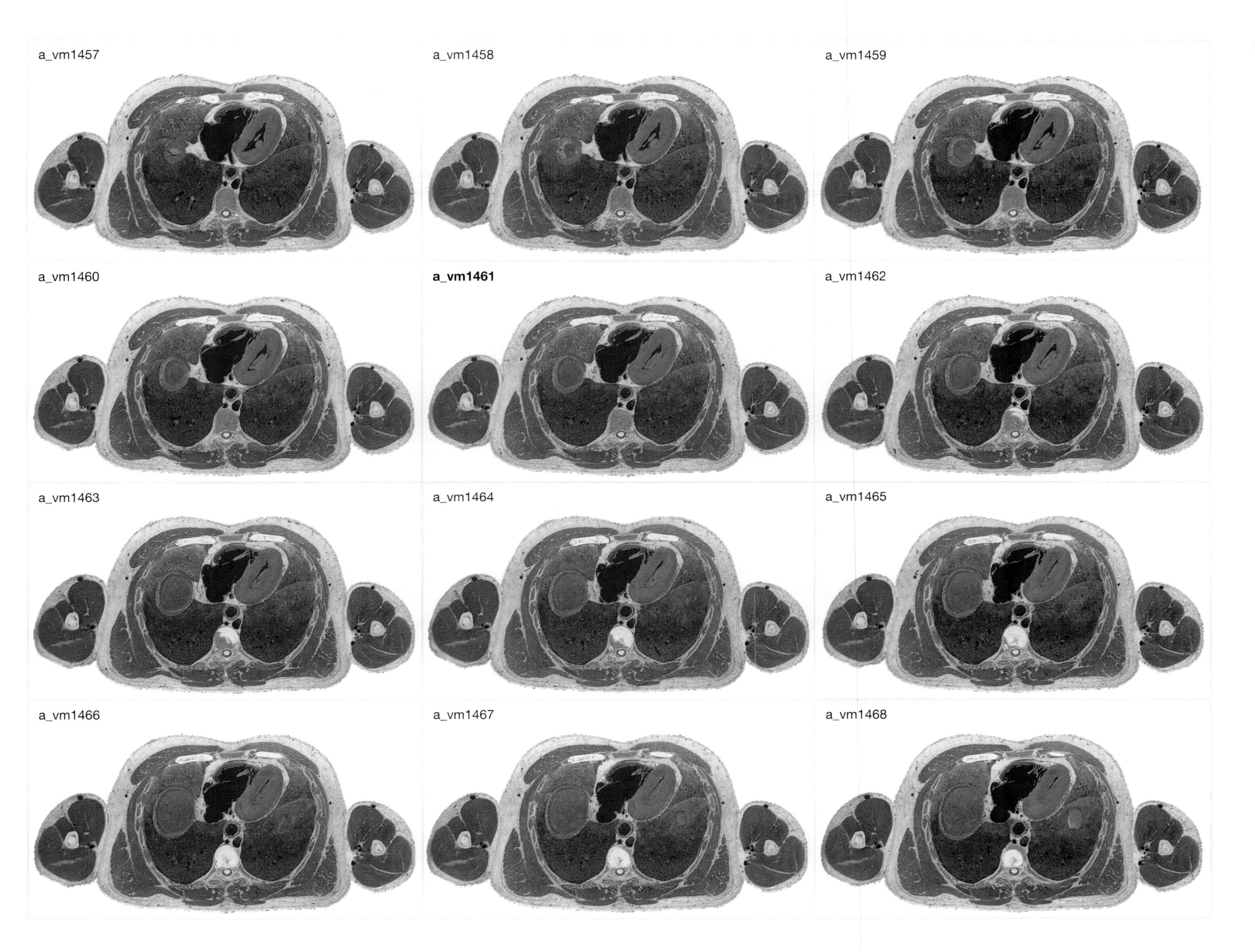
a_vm1457
a_vm1458
a_vm1459
a_vm1460
a_vm1461
a_vm1462
a_vm1463
a_vm1464
a_vm1465
a_vm1466
a_vm1467
a_vm1468

Transverse
a_vm1510

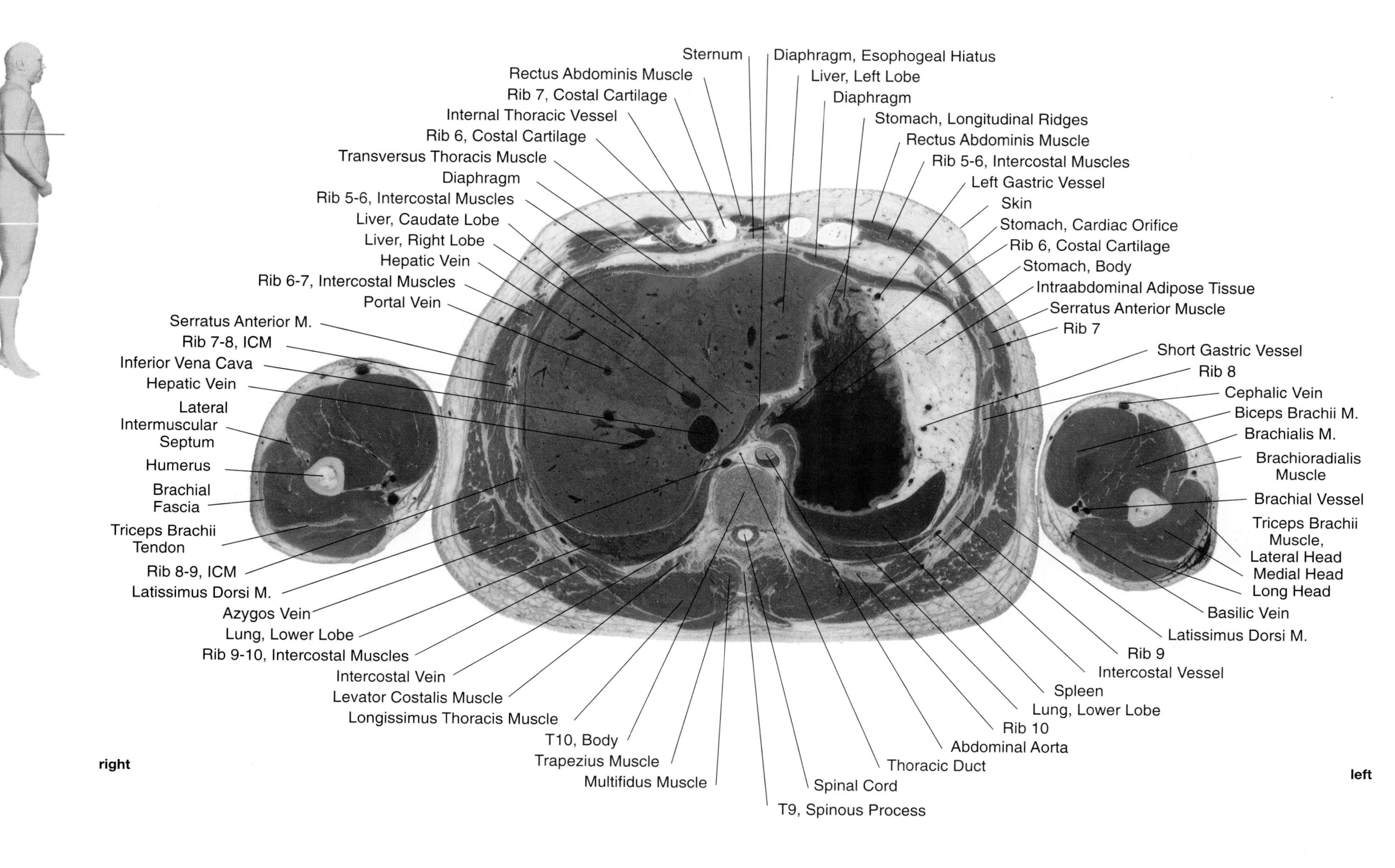
anterior
Sternum
Rectus Abdominis Muscle
Rib 7, Costal Cartilage
Internal Thoracic Vessel
Rib 6, Costal Cartilage
Transversus Thoracis Muscle
Diaphragm
Rib 5-6, Intercostal Muscles
Liver, Caudate Lobe
Liver, Right Lobe
Hepatic Vein
Rib 6-7, Intercostal Muscles
Portal Vein
Serratus Anterior M.
Rib 7-8, ICM
Inferior Vena Cava
Hepatic Vein
Lateral Intermuscular Septum
Humerus
Brachial Fascia
Triceps Brachii Tendon
Rib 8-9, ICM
Latissimus Dorsi M.
Azygos Vein
Lung, Lower Lobe
Rib 9-10, Intercostal Muscles
Intercostal Vein
Levator Costalis Muscle
Longissimus Thoracis Muscle
T10, Body
Trapezius Muscle
Multifidus Muscle
Diaphragm, Esophogeal Hiatus
Liver, Left Lobe
Diaphragm
Stomach, Longitudinal Ridges
Rectus Abdominis Muscle
Rib 5-6, Intercostal Muscles
Left Gastric Vessel
Skin
Stomach, Cardiac Orifice
Rib 6, Costal Cartilage
Stomach, Body
Intraabdominal Adipose Tissue
Serratus Anterior Muscle
Rib 7
Short Gastric Vessel
Rib 8
Cephalic Vein
Biceps Brachii M.
Brachialis M.
Brachioradialis Muscle
Brachial Vessel
Triceps Brachii Muscle,
Lateral Head
Medial Head
Long Head
Basilic Vein
Latissimus Dorsi M.
Rib 9
Intercostal Vessel
Spleen
Lung, Lower Lobe
Rib 10
Abdominal Aorta
Thoracic Duct
Spinal Cord
T9, Spinous Process
right
left
posterior

ICM=Intercostal Muscles

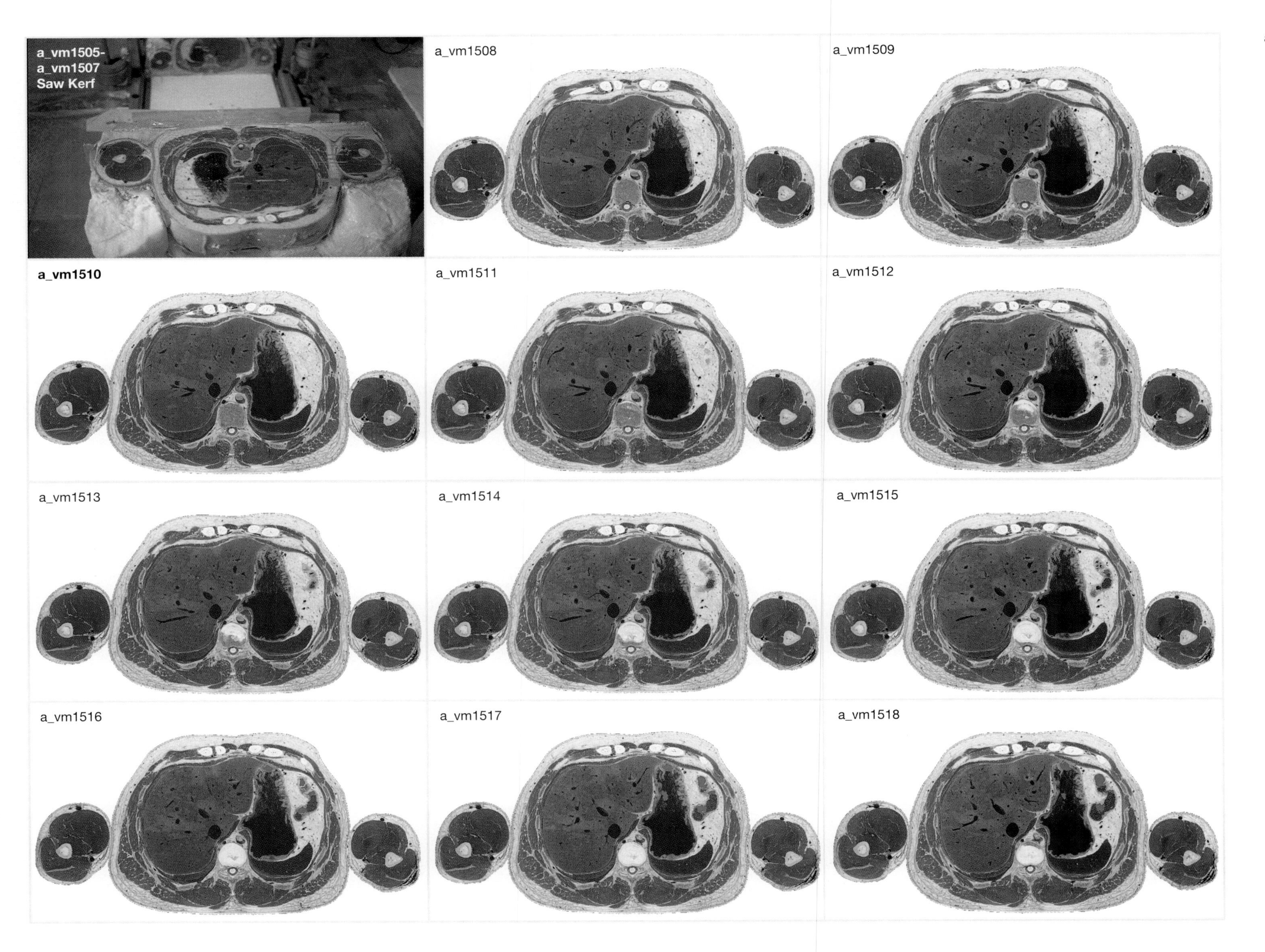
a_vm1505-
a_vm1507
Saw Kerf
a_vm1508
a_vm1509
a_vm1510
a_vm1511
a_vm1512
a_vm1513
a_vm1514
a_vm1515
a_vm1516
a_vm1517
a_vm1518

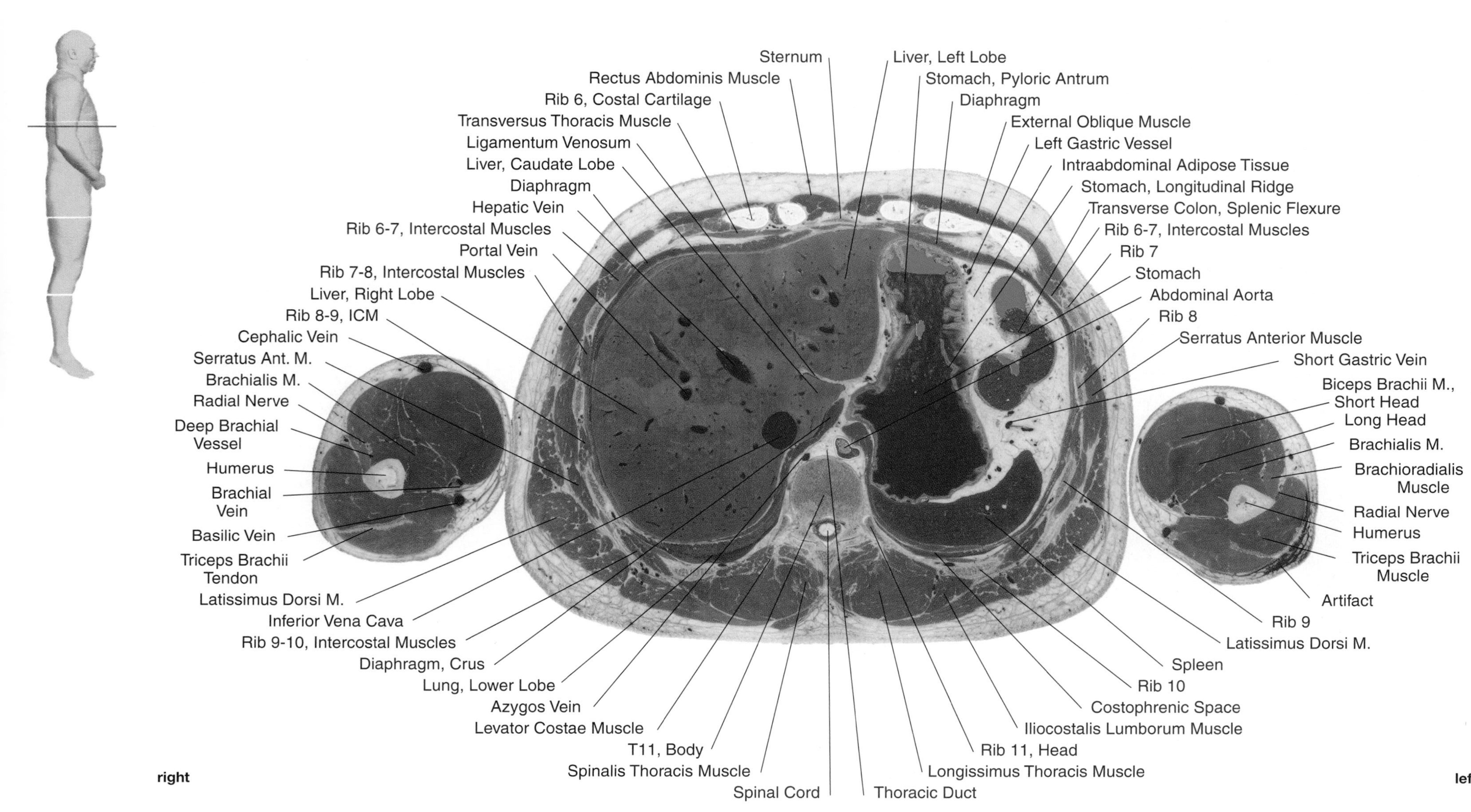
anterior
Sternum
Liver, Left Lobe
Rectus Abdominis Muscle
Stomach, Pyloric Antrum
Rib 6, Costal Cartilage
Diaphragm
Transversus Thoracis Muscle
External Oblique Muscle
Ligamentum Venosum
Left Gastric Vessel
Liver, Caudate Lobe
Intraabdominal Adipose Tissue
Diaphragm
Stomach, Longitudinal Ridge
Hepatic Vein
Transverse Colon, Splenic Flexure
Rib 6-7, Intercostal Muscles
Rib 6-7, Intercostal Muscles
Portal Vein
Rib 7
Rib 7-8, Intercostal Muscles
Stomach
Liver, Right Lobe
Abdominal Aorta
Rib 8-9, ICM
Rib 8
Cephalic Vein
Serratus Anterior Muscle
Serratus Ant. M.
Short Gastric Vein
Brachialis M.
Biceps Brachii M.,
Short Head
Radial Nerve
Long Head
Deep Brachial
Vessel
Brachialis M.
Humerus
Brachioradialis
Muscle
Brachial
Vein
Radial Nerve
Basilic Vein
Humerus
Triceps Brachii
Tendon
Triceps Brachii
Muscle
Latissimus Dorsi M.
Artifact
Inferior Vena Cava
Rib 9
Rib 9-10, Intercostal Muscles
Latissimus Dorsi M.
Diaphragm, Crus
Spleen
Lung, Lower Lobe
Rib 10
Azygos Vein
Costophrenic Space
Levator Costae Muscle
Iliocostalis Lumborum Muscle
T11, Body
Rib 11, Head
right
Spinalis Thoracis Muscle
Longissimus Thoracis Muscle
left
Spinal Cord
Thoracic Duct
posterior

ICM=Intercostal Muscles

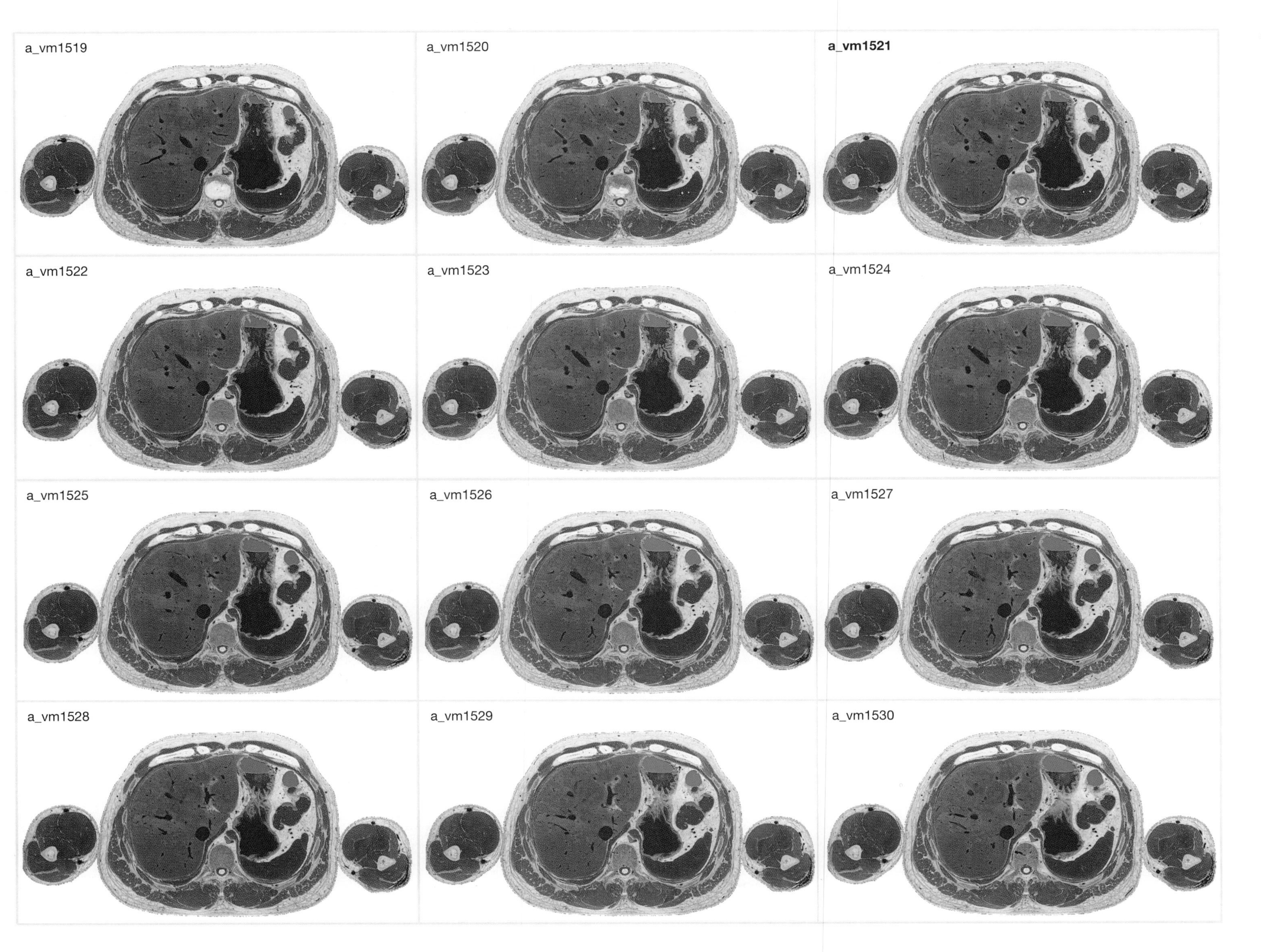
a_vm1519
a_vm1520
a_vm1521
a_vm1522
a_vm1523
a_vm1524
a_vm1525
a_vm1526
a_vm1527
a_vm1528
a_vm1529
a_vm1530

Transverse
a_vm1539

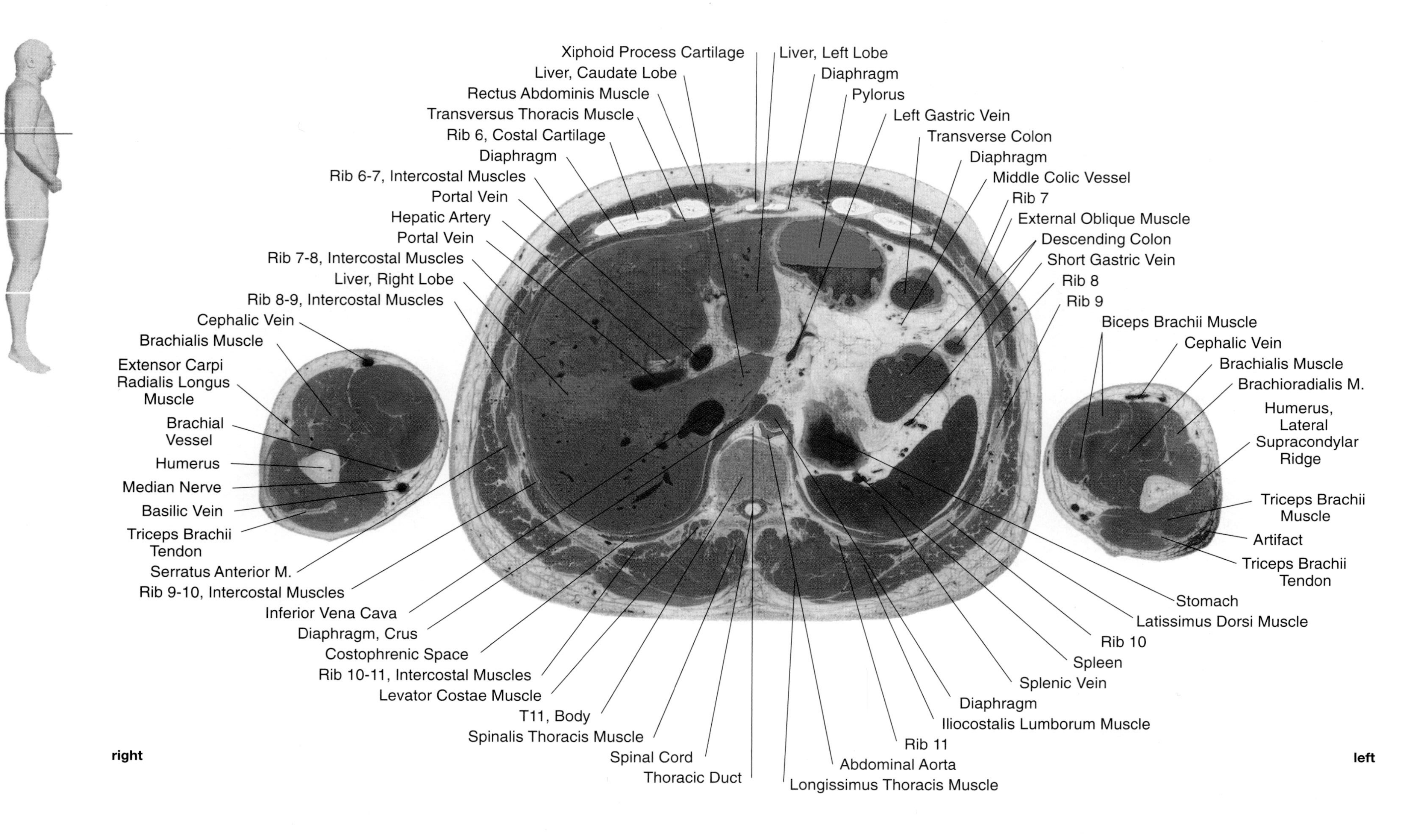
anterior
Xiphoid Process Cartilage
Liver, Caudate Lobe
Rectus Abdominis Muscle
Transversus Thoracis Muscle
Rib 6, Costal Cartilage
Diaphragm
Rib 6-7, Intercostal Muscles
Portal Vein
Hepatic Artery
Portal Vein
Rib 7-8, Intercostal Muscles
Liver, Right Lobe
Rib 8-9, Intercostal Muscles
Cephalic Vein
Brachialis Muscle
Extensor Carpi Radialis Longus Muscle
Brachial Vessel
Humerus
Median Nerve
Basilic Vein
Triceps Brachii Tendon
Serratus Anterior M.
Rib 9-10, Intercostal Muscles
Inferior Vena Cava
Diaphragm, Crus
Costophrenic Space
Rib 10-11, Intercostal Muscles
Levator Costae Muscle
T11, Body
Spinalis Thoracis Muscle
Spinal Cord
Thoracic Duct
Liver, Left Lobe
Diaphragm
Pylorus
Left Gastric Vein
Transverse Colon
Diaphragm
Middle Colic Vessel
Rib 7
External Oblique Muscle
Descending Colon
Short Gastric Vein
Rib 8
Rib 9
Biceps Brachii Muscle
Cephalic Vein
Brachialis Muscle
Brachioradialis M.
Humerus, Lateral Supracondylar Ridge
Triceps Brachii Muscle
Artifact
Triceps Brachii Tendon
Stomach
Latissimus Dorsi Muscle
Rib 10
Spleen
Splenic Vein
Diaphragm
Iliocostalis Lumborum Muscle
Rib 11
Abdominal Aorta
Longissimus Thoracis Muscle
right
left
posterior

ICM=Intercostal Muscles

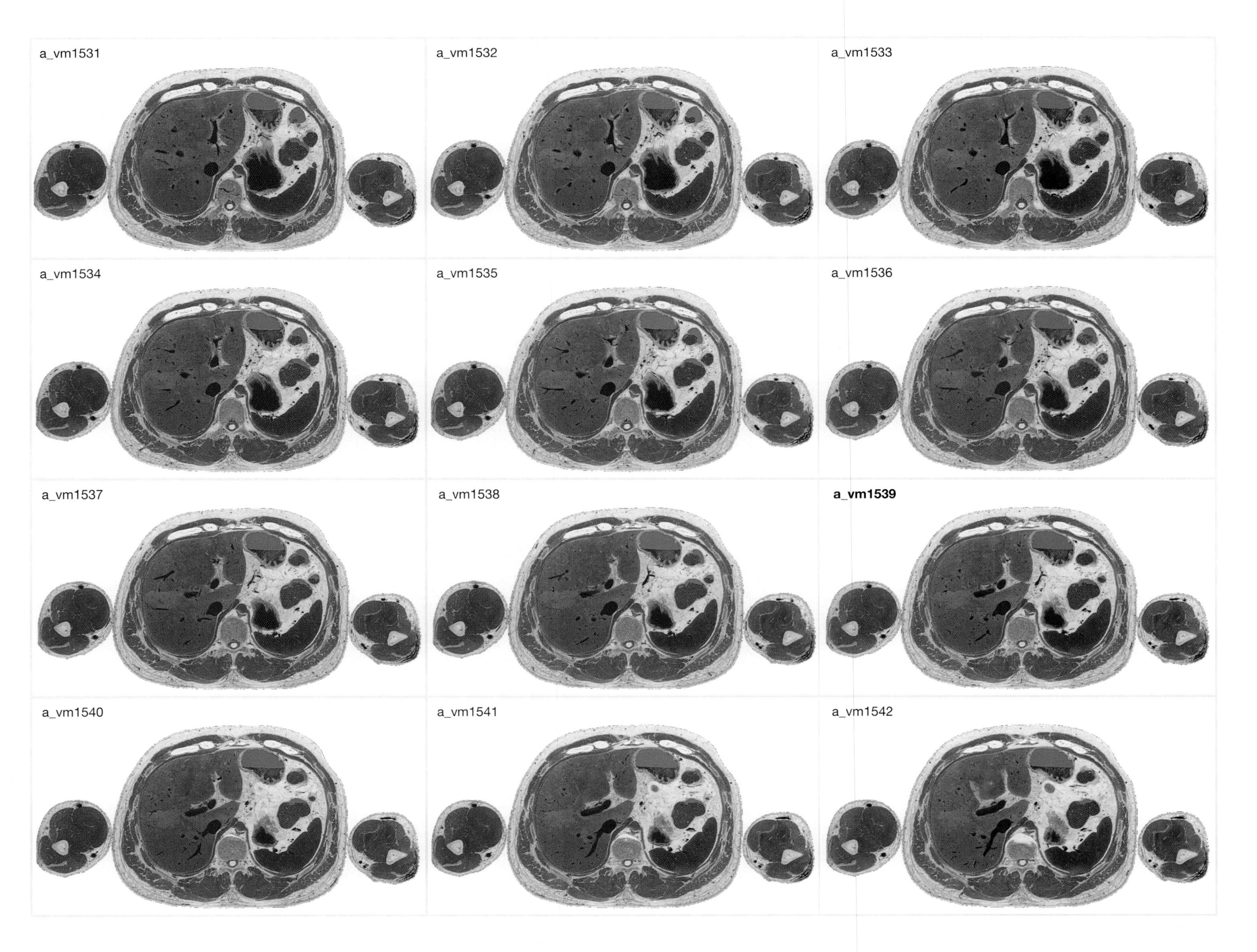
a_vm1531
a_vm1532
a_vm1533
a_vm1534
a_vm1535
a_vm1536
a_vm1537
a_vm1538
a_vm1539
a_vm1540
a_vm1541
a_vm1542

Transverse
a_vm1549

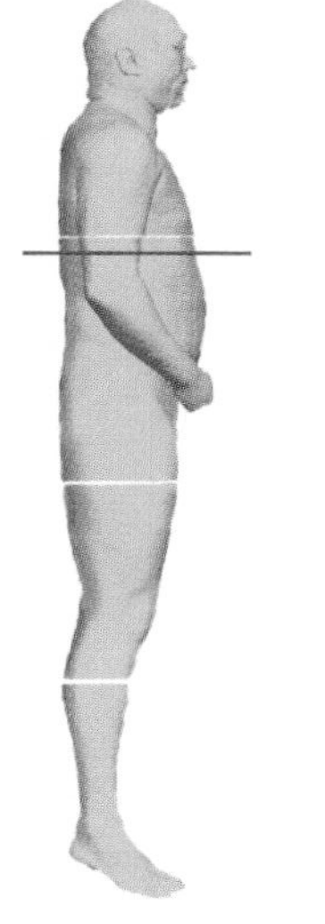

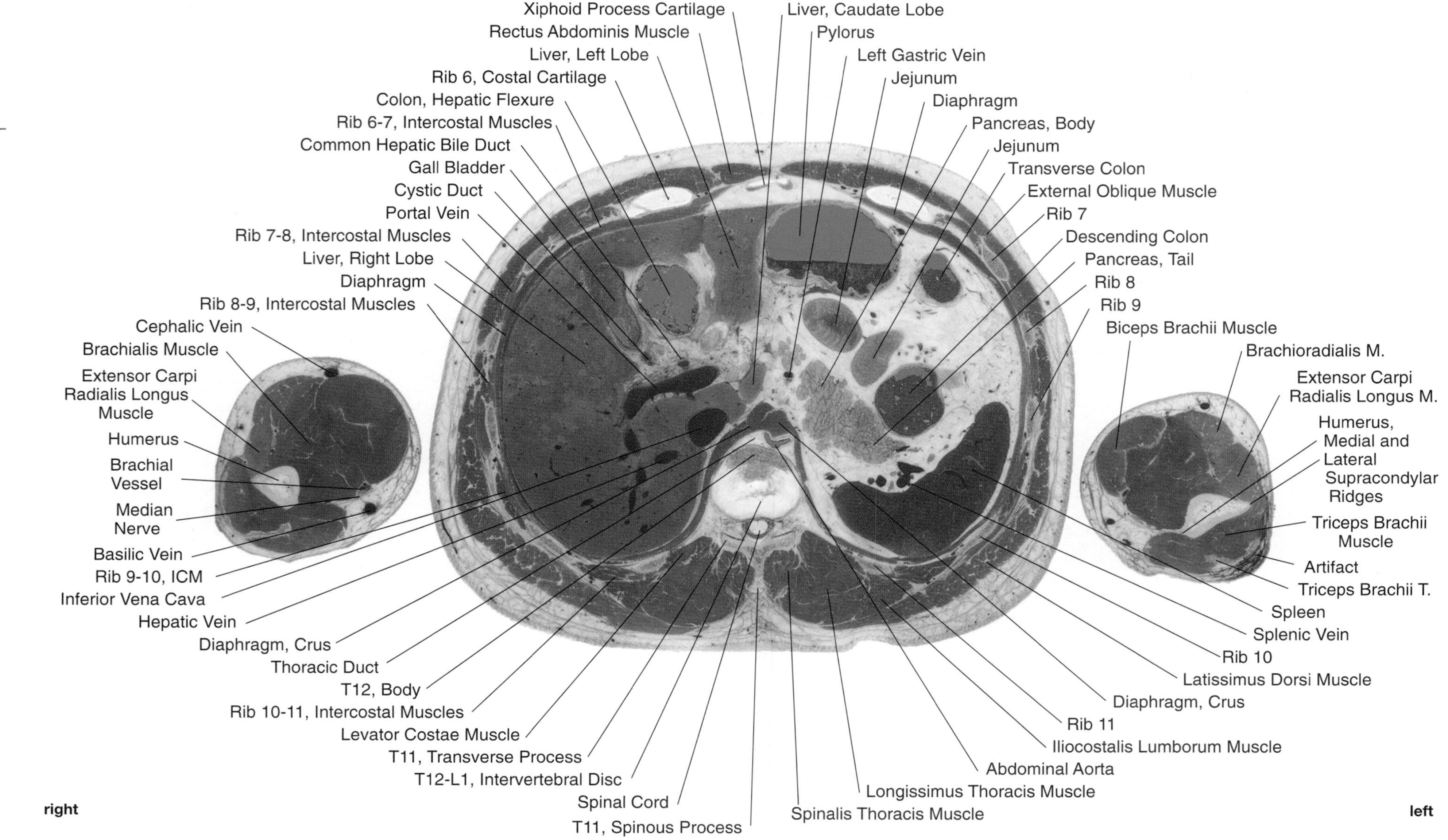
anterior
Xiphoid Process Cartilage
Rectus Abdominis Muscle
Liver, Left Lobe
Rib 6, Costal Cartilage
Colon, Hepatic Flexure
Rib 6-7, Intercostal Muscles
Common Hepatic Bile Duct
Gall Bladder
Cystic Duct
Portal Vein
Rib 7-8, Intercostal Muscles
Liver, Right Lobe
Diaphragm
Rib 8-9, Intercostal Muscles
Cephalic Vein
Brachialis Muscle
Extensor Carpi Radialis Longus Muscle
Humerus
Brachial Vessel
Median Nerve
Basilic Vein
Rib 9-10, ICM
Inferior Vena Cava
Hepatic Vein
Diaphragm, Crus
Thoracic Duct
T12, Body
Rib 10-11, Intercostal Muscles
Levator Costae Muscle
T11, Transverse Process
T12-L1, Intervertebral Disc
Spinal Cord
T11, Spinous Process
Liver, Caudate Lobe
Pylorus
Left Gastric Vein
Jejunum
Diaphragm
Pancreas, Body
Jejunum
Transverse Colon
External Oblique Muscle
Rib 7
Descending Colon
Pancreas, Tail
Rib 8
Rib 9
Biceps Brachii Muscle
Brachioradialis M.
Extensor Carpi Radialis Longus M.
Humerus, Medial and Lateral Supracondylar Ridges
Triceps Brachii Muscle
Artifact
Triceps Brachii T.
Spleen
Splenic Vein
Rib 10
Latissimus Dorsi Muscle
Diaphragm, Crus
Rib 11
Iliocostalis Lumborum Muscle
Abdominal Aorta
Longissimus Thoracis Muscle
Spinalis Thoracis Muscle
right
left
posterior

ICM=Intercostal Muscles

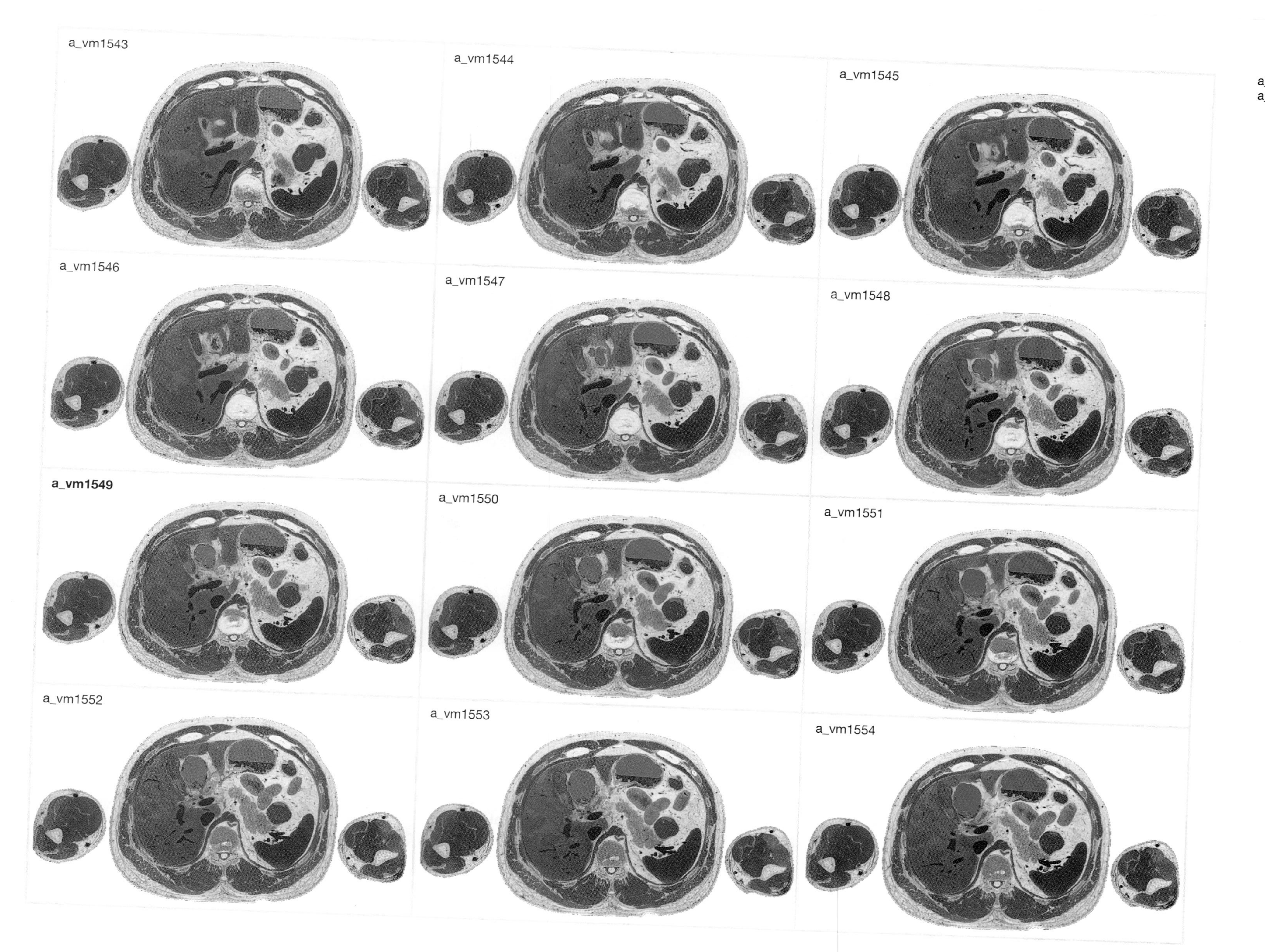
a_vm1543
a_vm1544
a_vm1545
a_vm1546
a_vm1547
a_vm1548
a_vm1549
a_vm1550
a_vm1551
a_vm1552
a_vm1553
a_vm1554

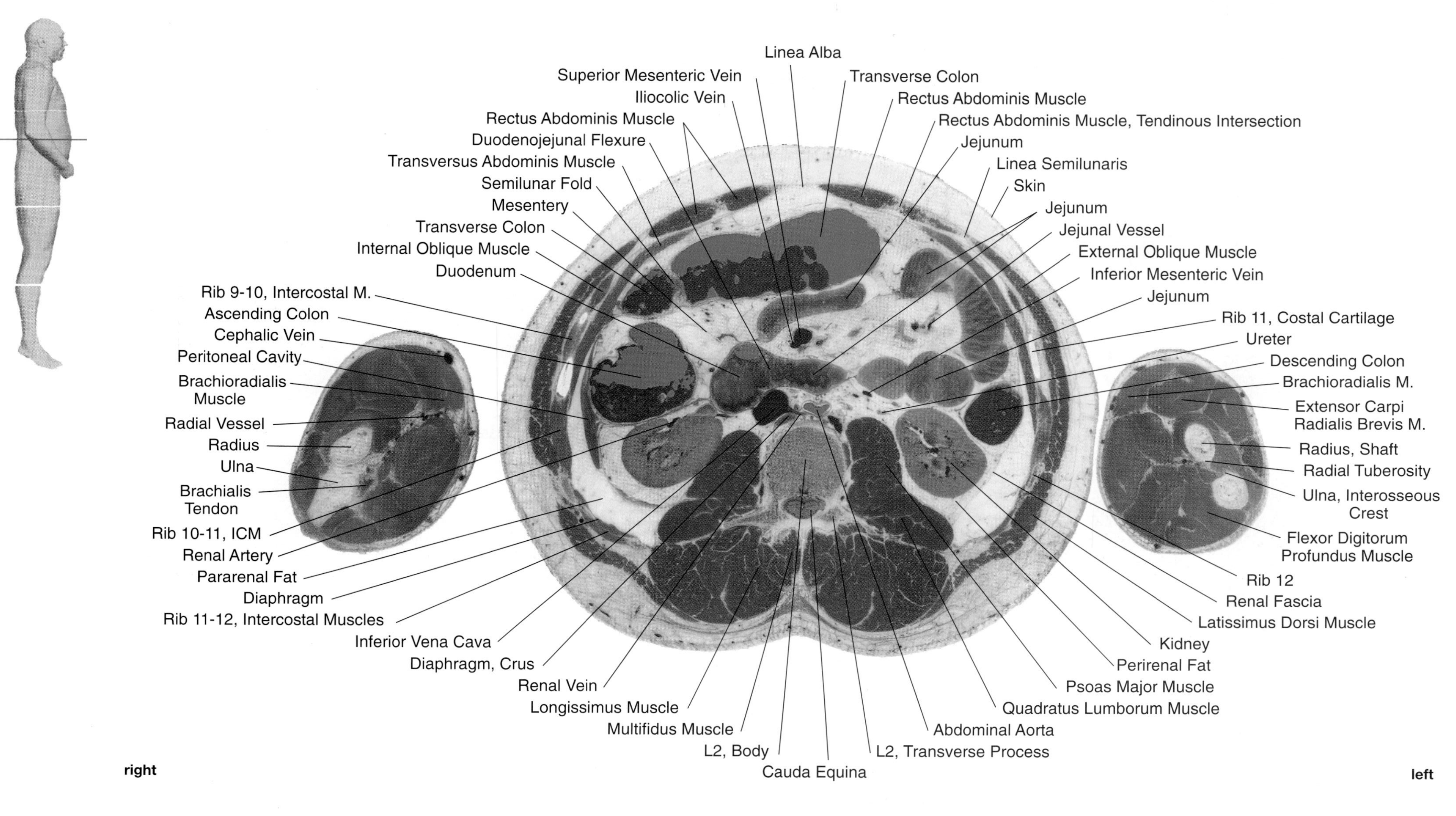
anterior
Linea Alba
Superior Mesenteric Vein
Iliocolic Vein
Rectus Abdominis Muscle
Duodenojejunal Flexure
Transversus Abdominis Muscle
Semilunar Fold
Mesentery
Transverse Colon
Internal Oblique Muscle
Duodenum
Rib 9-10, Intercostal M.
Ascending Colon
Cephalic Vein
Peritoneal Cavity
Brachioradialis Muscle
Radial Vessel
Radius
Ulna
Brachialis Tendon
Rib 10-11, ICM
Renal Artery
Pararenal Fat
Diaphragm
Rib 11-12, Intercostal Muscles
Inferior Vena Cava
Diaphragm, Crus
Renal Vein
Longissimus Muscle
Multifidus Muscle
L2, Body
Cauda Equina
L2, Transverse Process
Abdominal Aorta
Quadratus Lumborum Muscle
Psoas Major Muscle
Perirenal Fat
Kidney
Latissimus Dorsi Muscle
Renal Fascia
Rib 12
Flexor Digitorum Profundus Muscle
Ulna, Interosseous Crest
Radial Tuberosity
Radius, Shaft
Extensor Carpi Radialis Brevis M.
Brachioradialis M.
Descending Colon
Ureter
Rib 11, Costal Cartilage
Jejunum
Inferior Mesenteric Vein
External Oblique Muscle
Jejunal Vessel
Jejunum
Skin
Linea Semilunaris
Jejunum
Rectus Abdominis Muscle, Tendinous Intersection
Rectus Abdominis Muscle
Transverse Colon
right
left
posterior

ICM=Intercostal Muscles

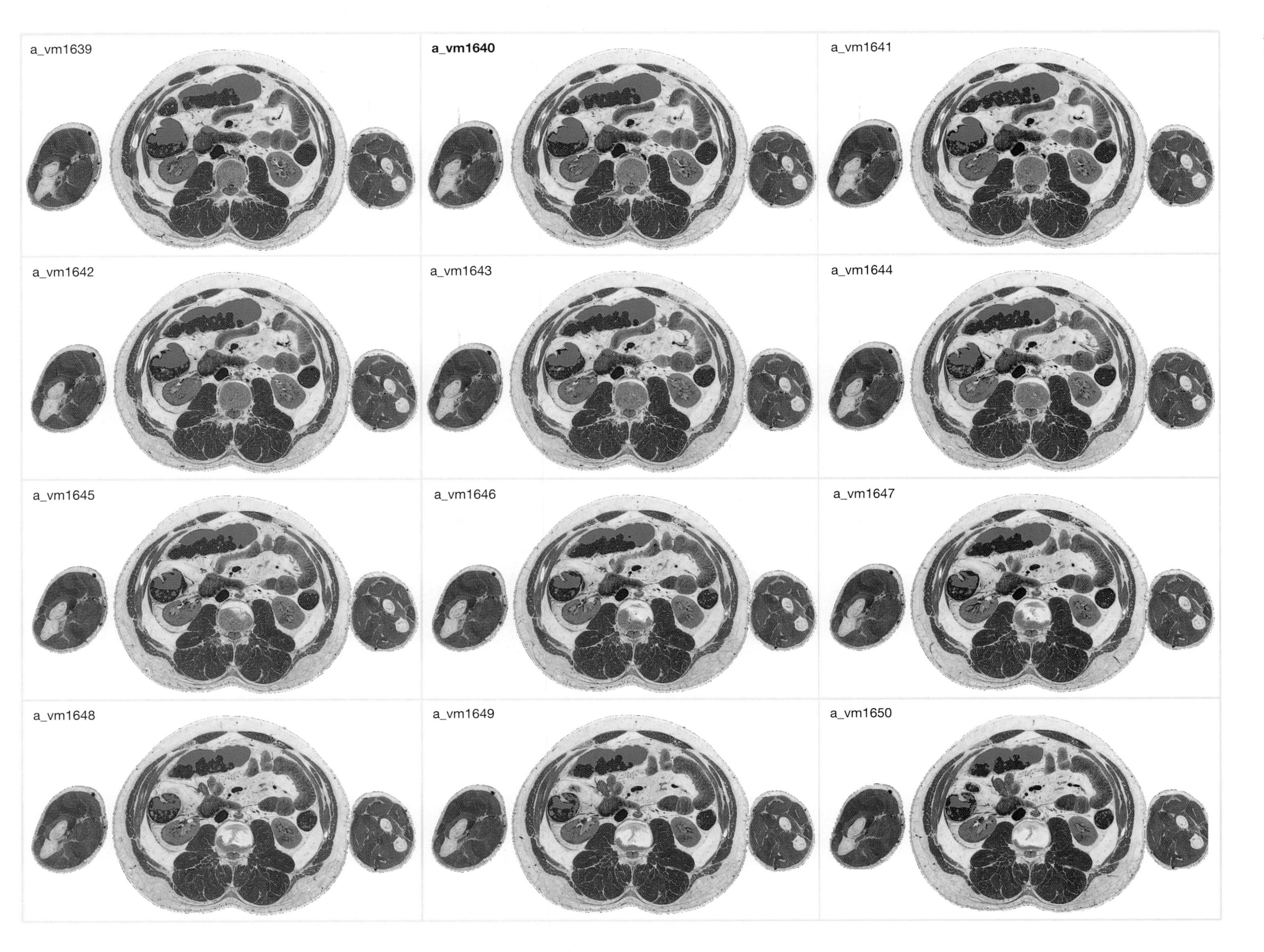
a_vm1639
a_vm1640
a_vm1641
a_vm1642
a_vm1643
a_vm1644
a_vm1645
a_vm1646
a_vm1647
a_vm1648
a_vm1649
a_vm1650

Transverse
a_vm1657

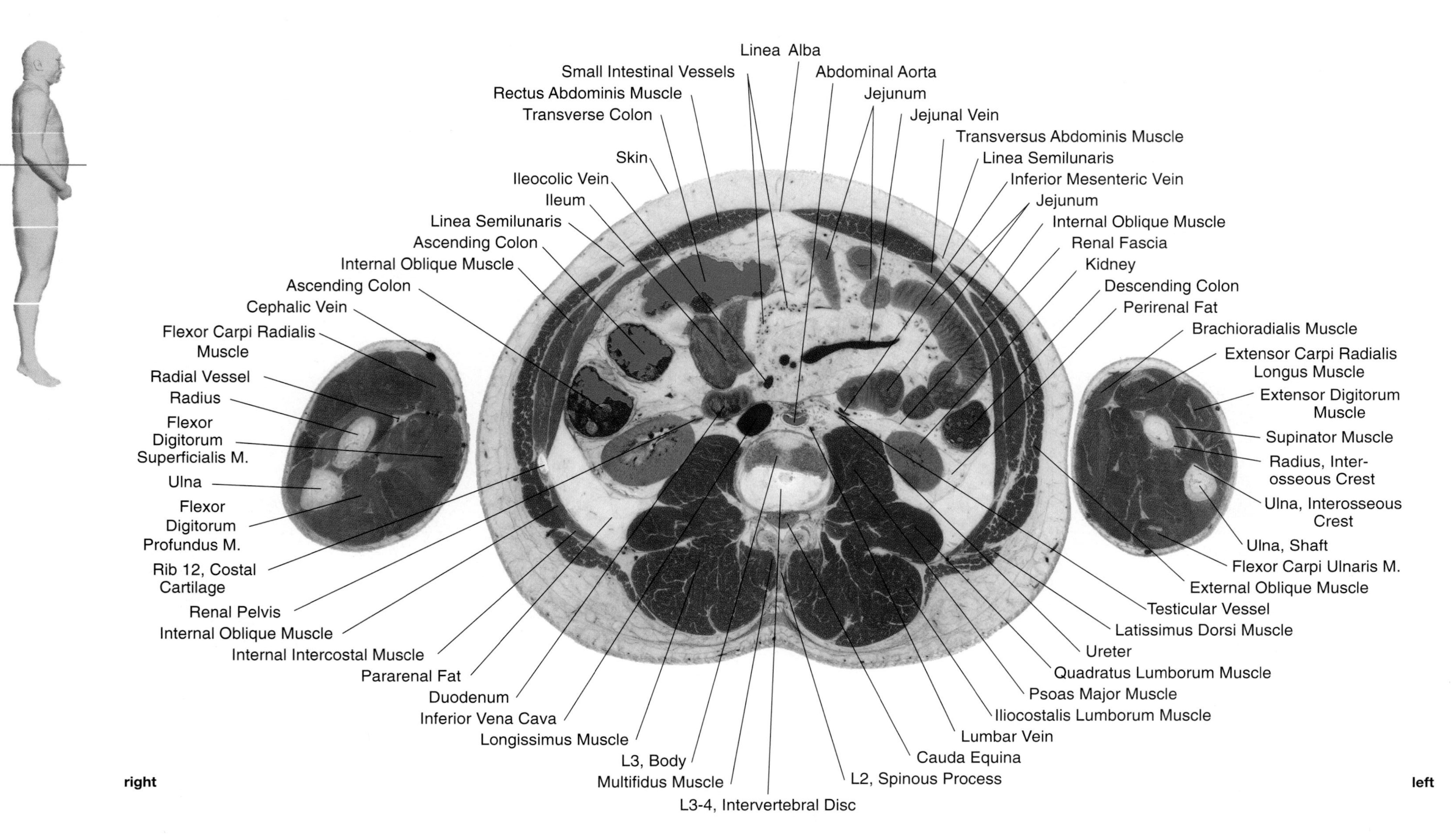
anterior
Linea Alba
Small Intestinal Vessels
Abdominal Aorta
Rectus Abdominis Muscle
Jejunum
Transverse Colon
Jejunal Vein
Transversus Abdominis Muscle
Skin
Linea Semilunaris
Ileocolic Vein
Inferior Mesenteric Vein
Ileum
Jejunum
Linea Semilunaris
Internal Oblique Muscle
Ascending Colon
Renal Fascia
Internal Oblique Muscle
Kidney
Ascending Colon
Descending Colon
Cephalic Vein
Perirenal Fat
Flexor Carpi Radialis Muscle
Brachioradialis Muscle
Extensor Carpi Radialis Longus Muscle
Radial Vessel
Radius
Extensor Digitorum Muscle
Flexor Digitorum Superficialis M.
Supinator Muscle
Radius, Inter-osseous Crest
Ulna
Ulna, Interosseous Crest
Flexor Digitorum Profundus M.
Ulna, Shaft
Flexor Carpi Ulnaris M.
Rib 12, Costal Cartilage
External Oblique Muscle
Testicular Vessel
Renal Pelvis
Latissimus Dorsi Muscle
Internal Oblique Muscle
Ureter
Internal Intercostal Muscle
Quadratus Lumborum Muscle
Pararenal Fat
Psoas Major Muscle
Duodenum
Iliocostalis Lumborum Muscle
Inferior Vena Cava
Lumbar Vein
Longissimus Muscle
Cauda Equina
L3, Body
L2, Spinous Process
Multifidus Muscle
L3-4, Intervertebral Disc
right
left
posterior

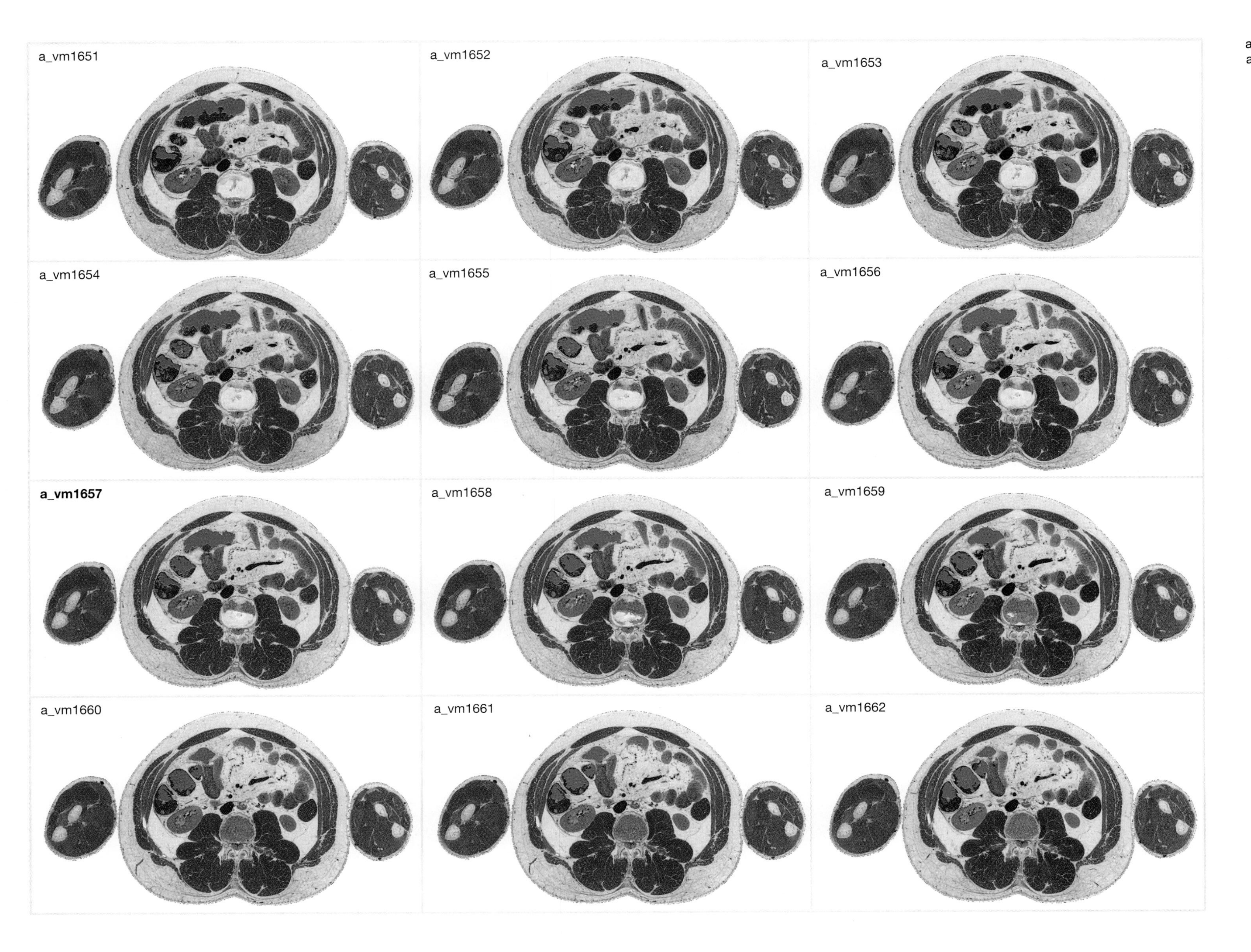
a_vm1651
a_vm1652
a_vm1653
a_vm1654
a_vm1655
a_vm1656
a_vm1657
a_vm1658
a_vm1659
a_vm1660
a_vm1661
a_vm1662

Transverse
a_vm1671

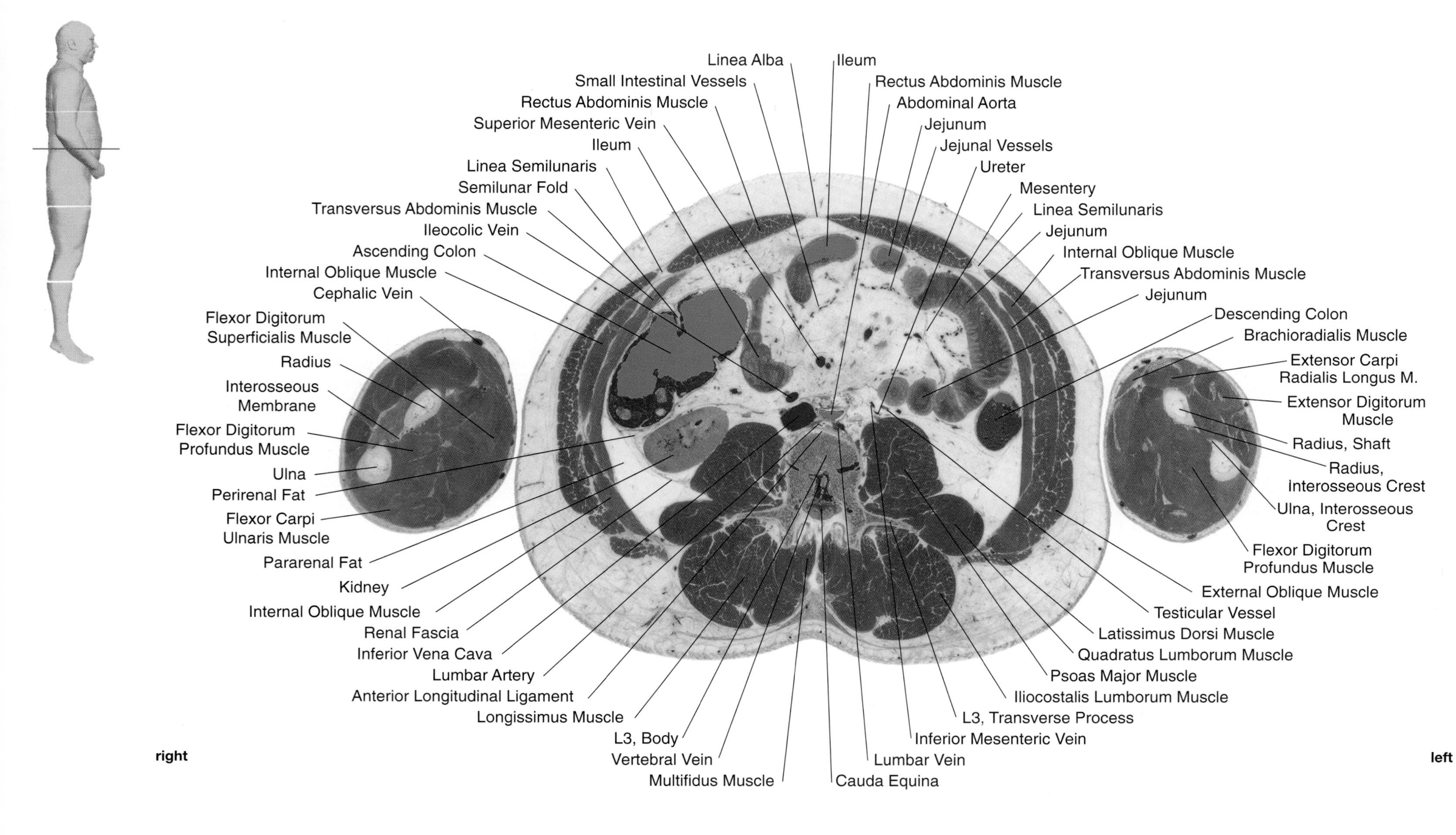
anterior
Linea Alba
Small Intestinal Vessels
Rectus Abdominis Muscle
Superior Mesenteric Vein
Ileum
Linea Semilunaris
Semilunar Fold
Transversus Abdominis Muscle
Ileocolic Vein
Ascending Colon
Internal Oblique Muscle
Cephalic Vein
Flexor Digitorum Superficialis Muscle
Radius
Interosseous Membrane
Flexor Digitorum Profundus Muscle
Ulna
Perirenal Fat
Flexor Carpi Ulnaris Muscle
Pararenal Fat
Kidney
Internal Oblique Muscle
Renal Fascia
Inferior Vena Cava
Lumbar Artery
Anterior Longitudinal Ligament
Longissimus Muscle
L3, Body
Vertebral Vein
Multifidus Muscle
Ileum
Rectus Abdominis Muscle
Abdominal Aorta
Jejunum
Jejunal Vessels
Ureter
Mesentery
Linea Semilunaris
Jejunum
Internal Oblique Muscle
Transversus Abdominis Muscle
Jejunum
Descending Colon
Brachioradialis Muscle
Extensor Carpi Radialis Longus M.
Extensor Digitorum Muscle
Radius, Shaft
Radius, Interosseous Crest
Ulna, Interosseous Crest
Flexor Digitorum Profundus Muscle
External Oblique Muscle
Testicular Vessel
Latissimus Dorsi Muscle
Quadratus Lumborum Muscle
Psoas Major Muscle
Iliocostalis Lumborum Muscle
L3, Transverse Process
Inferior Mesenteric Vein
Lumbar Vein
Cauda Equina
right
left
posterior

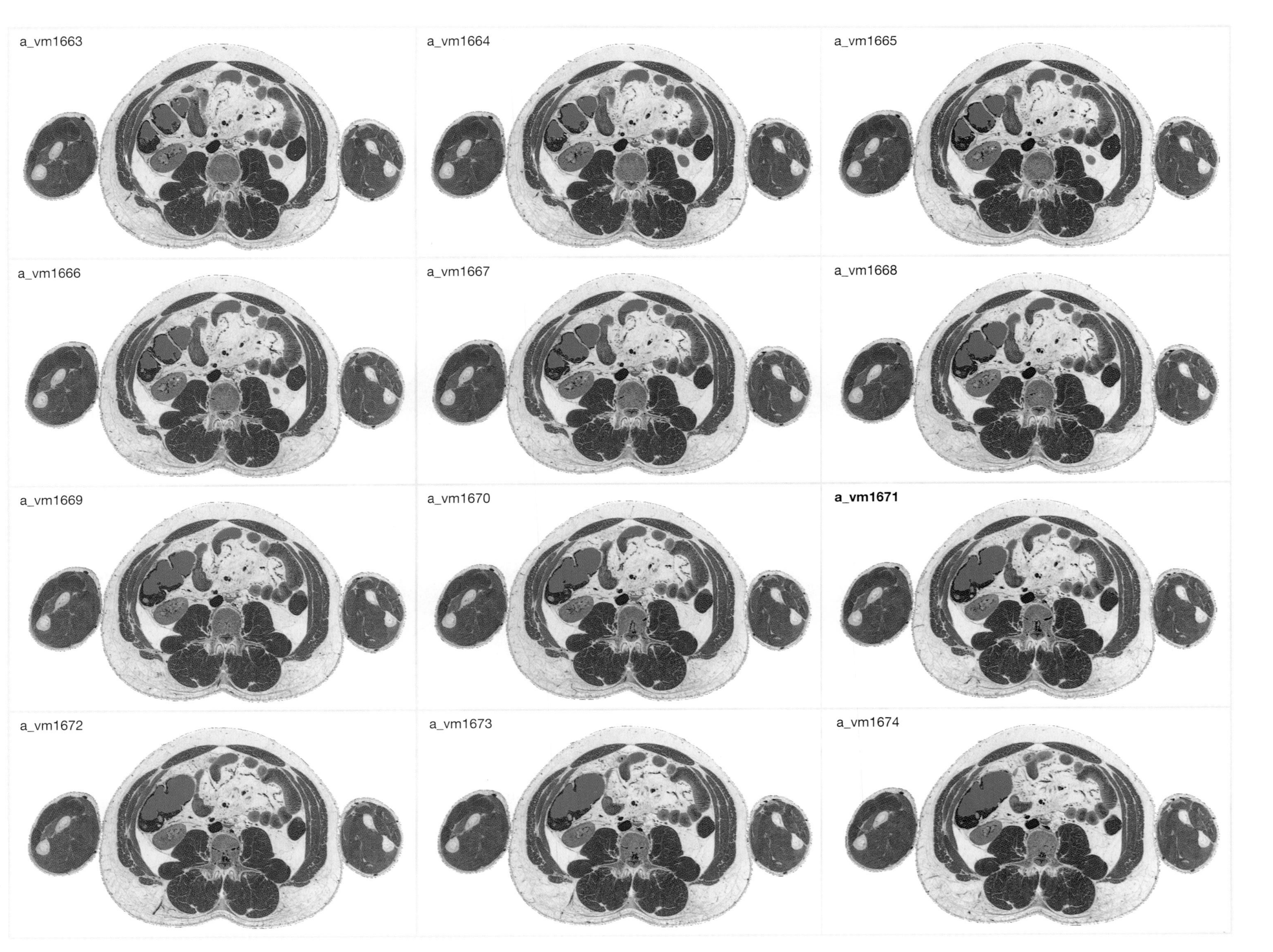
a_vm1663
a_vm1664
a_vm1665
a_vm1666
a_vm1667
a_vm1668
a_vm1669
a_vm1670
a_vm1671
a_vm1672
a_vm1673
a_vm1674

Transverse
a_vm1740

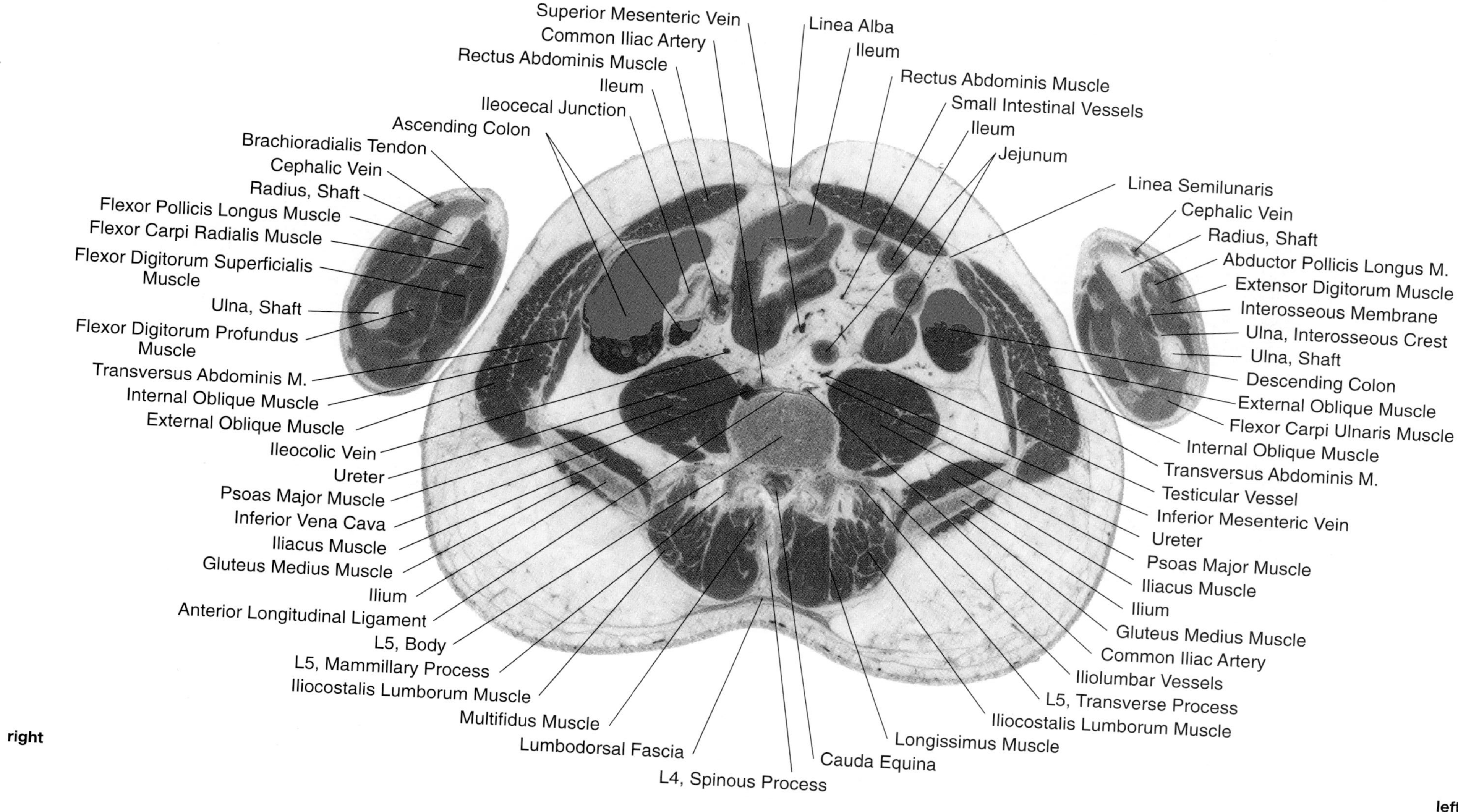
anterior
posterior
right
left
Superior Mesenteric Vein
Common Iliac Artery
Rectus Abdominis Muscle
Ileum
Ileocecal Junction
Ascending Colon
Brachioradialis Tendon
Cephalic Vein
Radius, Shaft
Flexor Pollicis Longus Muscle
Flexor Carpi Radialis Muscle
Flexor Digitorum Superficialis Muscle
Ulna, Shaft
Flexor Digitorum Profundus Muscle
Transversus Abdominis M.
Internal Oblique Muscle
External Oblique Muscle
Ileocolic Vein
Ureter
Psoas Major Muscle
Inferior Vena Cava
Iliacus Muscle
Gluteus Medius Muscle
Ilium
Anterior Longitudinal Ligament
L5, Body
L5, Mammillary Process
Iliocostalis Lumborum Muscle
Multifidus Muscle
Lumbodorsal Fascia
L4, Spinous Process
Cauda Equina
Longissimus Muscle
Iliocostalis Lumborum Muscle
L5, Transverse Process
Iliolumbar Vessels
Common Iliac Artery
Gluteus Medius Muscle
Ilium
Iliacus Muscle
Psoas Major Muscle
Ureter
Inferior Mesenteric Vein
Testicular Vessel
Transversus Abdominis M.
Internal Oblique Muscle
Flexor Carpi Ulnaris Muscle
External Oblique Muscle
Descending Colon
Ulna, Shaft
Ulna, Interosseous Crest
Interosseous Membrane
Extensor Digitorum Muscle
Abductor Pollicis Longus M.
Radius, Shaft
Cephalic Vein
Linea Semilunaris
Jejunum
Ileum
Small Intestinal Vessels
Rectus Abdominis Muscle
Ileum
Linea Alba

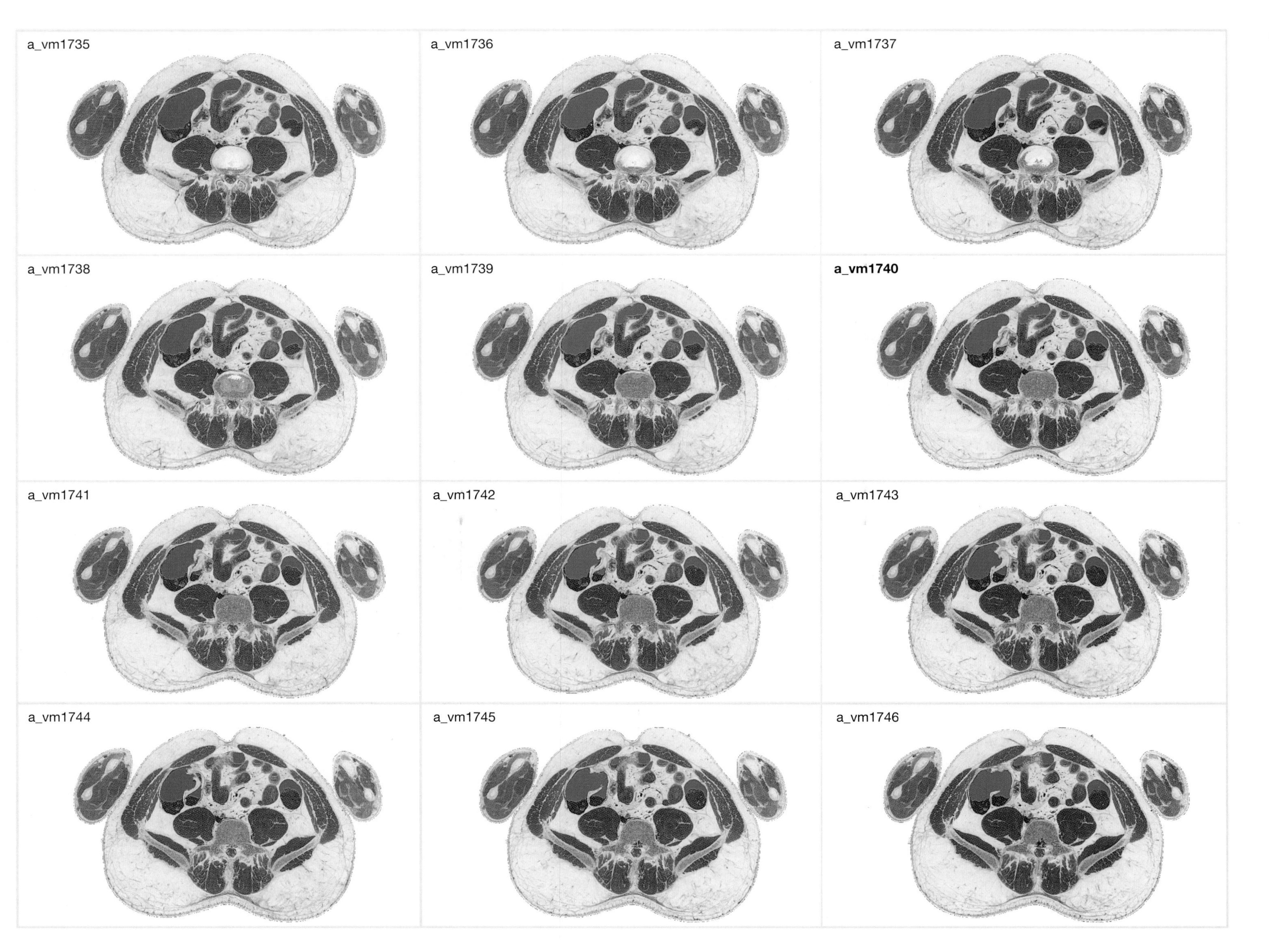
a_vm1735
a_vm1736
a_vm1737
a_vm1738
a_vm1739
a_vm1740
a_vm1741
a_vm1742
a_vm1743
a_vm1744
a_vm1745
a_vm1746

Transverse
a_vm1750

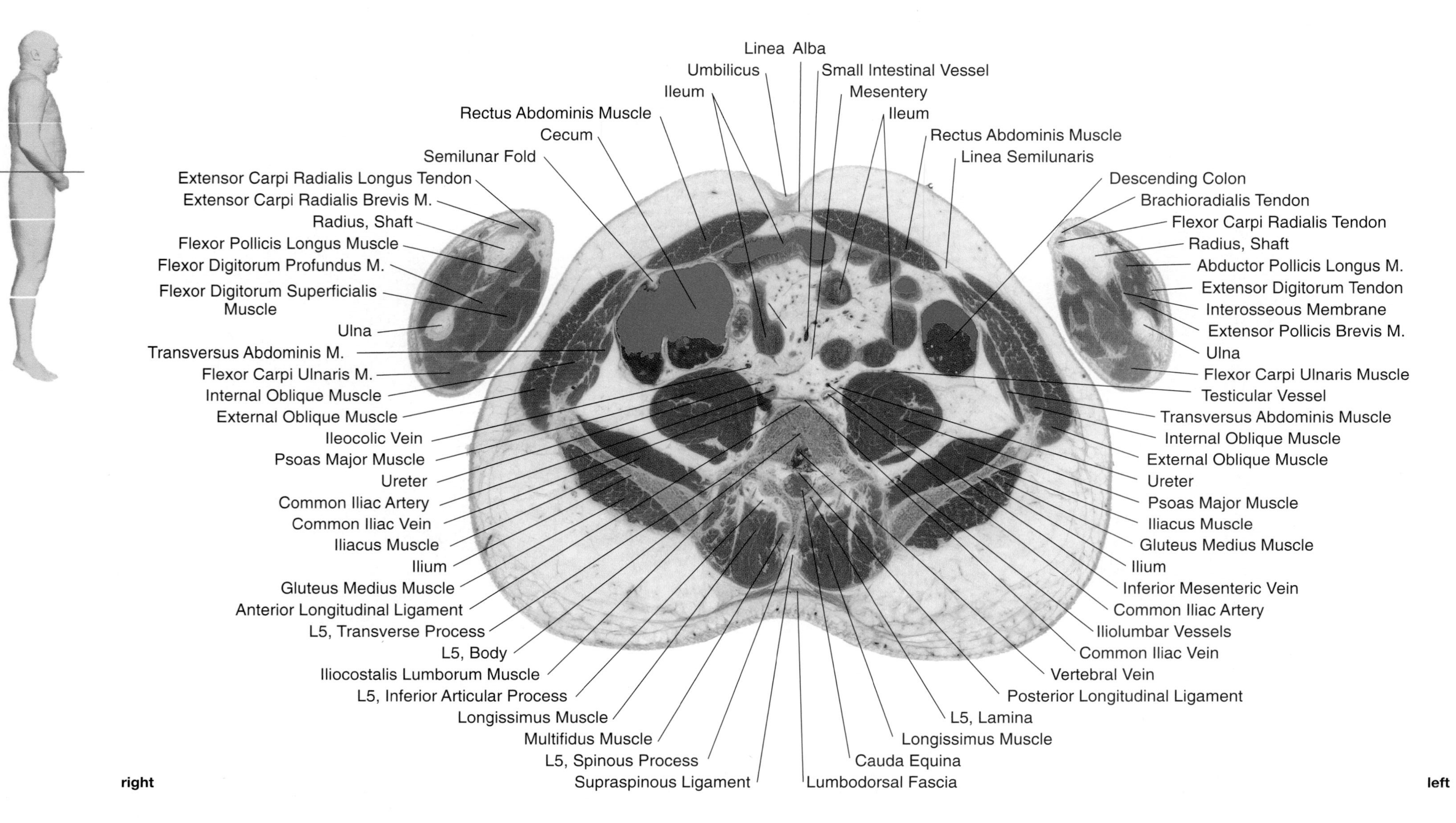
anterior
Linea Alba
Umbilicus
Small Intestinal Vessel
Ileum
Mesentery
Ileum
Rectus Abdominis Muscle
Cecum
Rectus Abdominis Muscle
Semilunar Fold
Linea Semilunaris
Extensor Carpi Radialis Longus Tendon
Descending Colon
Extensor Carpi Radialis Brevis M.
Brachioradialis Tendon
Radius, Shaft
Flexor Carpi Radialis Tendon
Flexor Pollicis Longus Muscle
Radius, Shaft
Flexor Digitorum Profundus M.
Abductor Pollicis Longus M.
Flexor Digitorum Superficialis Muscle
Extensor Digitorum Tendon
Interosseous Membrane
Extensor Pollicis Brevis M.
Ulna
Ulna
Transversus Abdominis M.
Flexor Carpi Ulnaris M.
Flexor Carpi Ulnaris Muscle
Internal Oblique Muscle
Testicular Vessel
External Oblique Muscle
Transversus Abdominis Muscle
Ileocolic Vein
Internal Oblique Muscle
Psoas Major Muscle
External Oblique Muscle
Ureter
Ureter
Common Iliac Artery
Psoas Major Muscle
Common Iliac Vein
Iliacus Muscle
Iliacus Muscle
Gluteus Medius Muscle
Ilium
Ilium
Gluteus Medius Muscle
Inferior Mesenteric Vein
Anterior Longitudinal Ligament
Common Iliac Artery
L5, Transverse Process
Iliolumbar Vessels
L5, Body
Common Iliac Vein
Iliocostalis Lumborum Muscle
Vertebral Vein
L5, Inferior Articular Process
Posterior Longitudinal Ligament
Longissimus Muscle
L5, Lamina
Multifidus Muscle
Longissimus Muscle
L5, Spinous Process
Cauda Equina
Supraspinous Ligament
Lumbodorsal Fascia
right
left
posterior

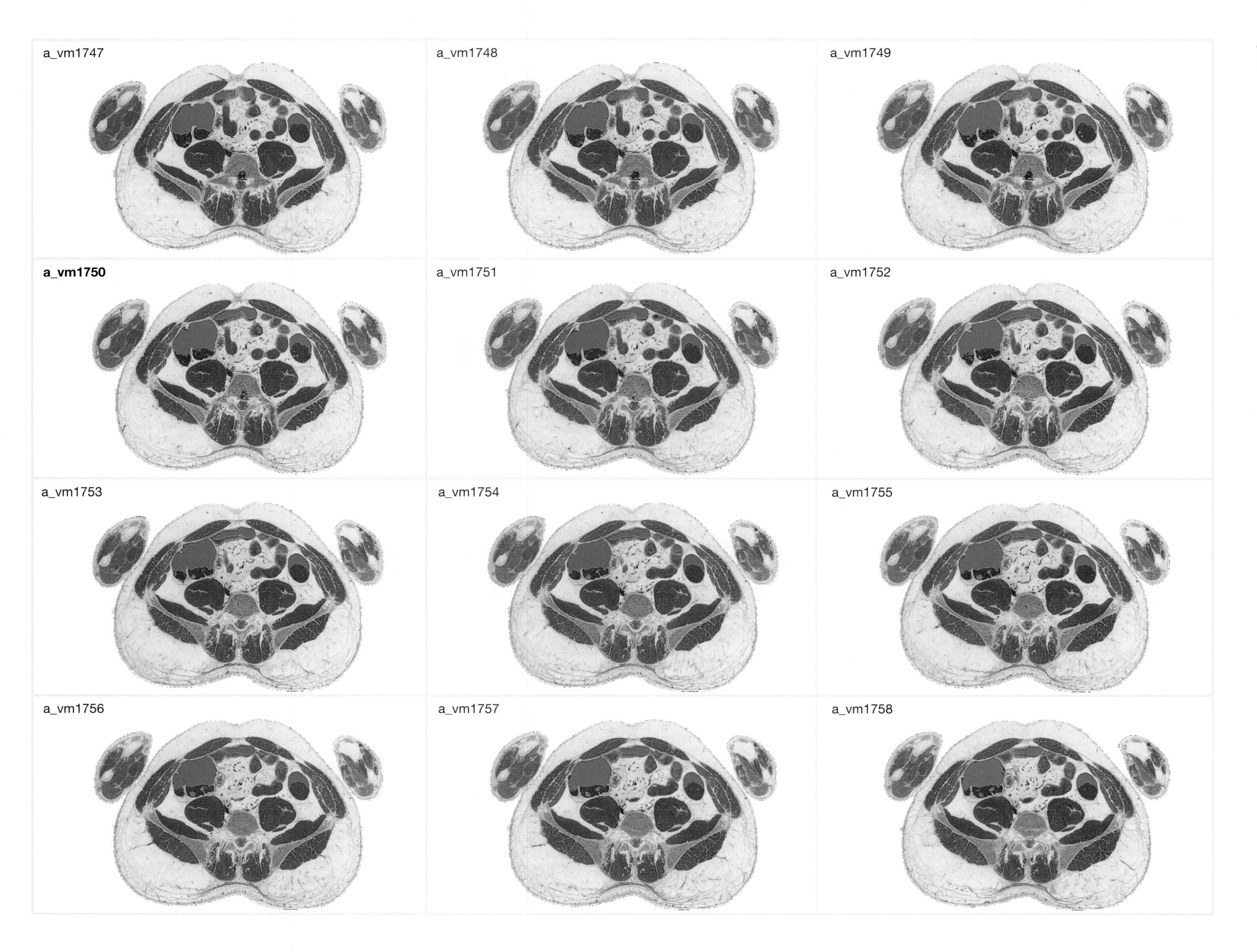
a_vm1747
a_vm1748
a_vm1749
a_vm1750
a_vm1751
a_vm1752
a_vm1753
a_vm1754
a_vm1755
a_vm1756
a_vm1757
a_vm1758

Transverse
a_vm1762

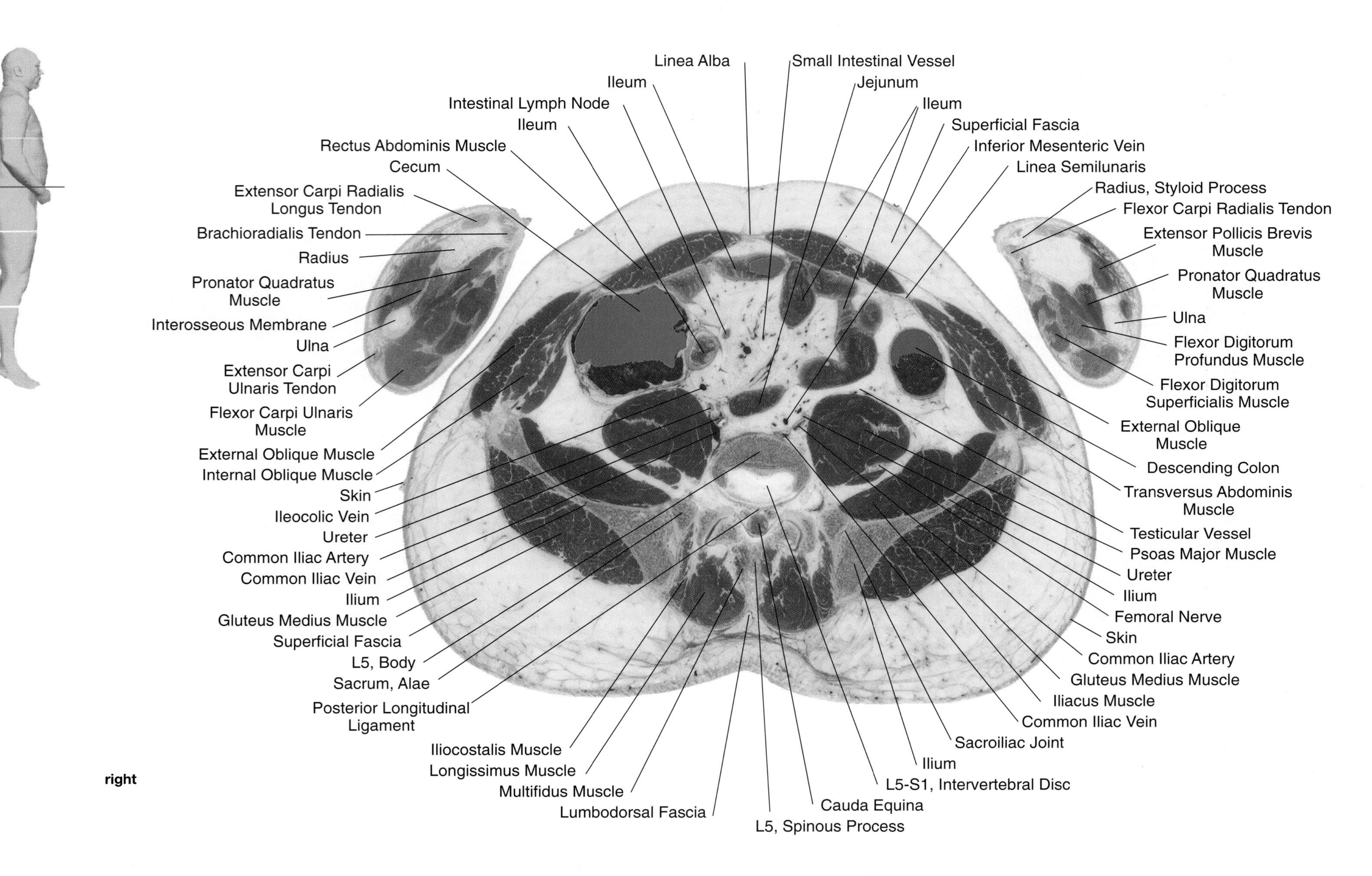

anterior
posterior
right
left
Linea Alba
Small Intestinal Vessel
Ileum
Jejunum
Intestinal Lymph Node
Ileum
Ileum
Superficial Fascia
Rectus Abdominis Muscle
Inferior Mesenteric Vein
Cecum
Linea Semilunaris
Extensor Carpi Radialis Longus Tendon
Radius, Styloid Process
Flexor Carpi Radialis Tendon
Brachioradialis Tendon
Extensor Pollicis Brevis Muscle
Radius
Pronator Quadratus Muscle
Pronator Quadratus Muscle
Ulna
Interosseous Membrane
Flexor Digitorum Profundus Muscle
Ulna
Extensor Carpi Ulnaris Tendon
Flexor Digitorum Superficialis Muscle
Flexor Carpi Ulnaris Muscle
External Oblique Muscle
External Oblique Muscle
Descending Colon
Internal Oblique Muscle
Transversus Abdominis Muscle
Skin
Ileocolic Vein
Testicular Vessel
Ureter
Psoas Major Muscle
Common Iliac Artery
Ureter
Common Iliac Vein
Ilium
Ilium
Femoral Nerve
Gluteus Medius Muscle
Skin
Superficial Fascia
Common Iliac Artery
L5, Body
Gluteus Medius Muscle
Sacrum, Alae
Iliacus Muscle
Posterior Longitudinal Ligament
Common Iliac Vein
Sacroiliac Joint
Iliocostalis Muscle
Ilium
Longissimus Muscle
L5-S1, Intervertebral Disc
Multifidus Muscle
Cauda Equina
Lumbodorsal Fascia
L5, Spinous Process

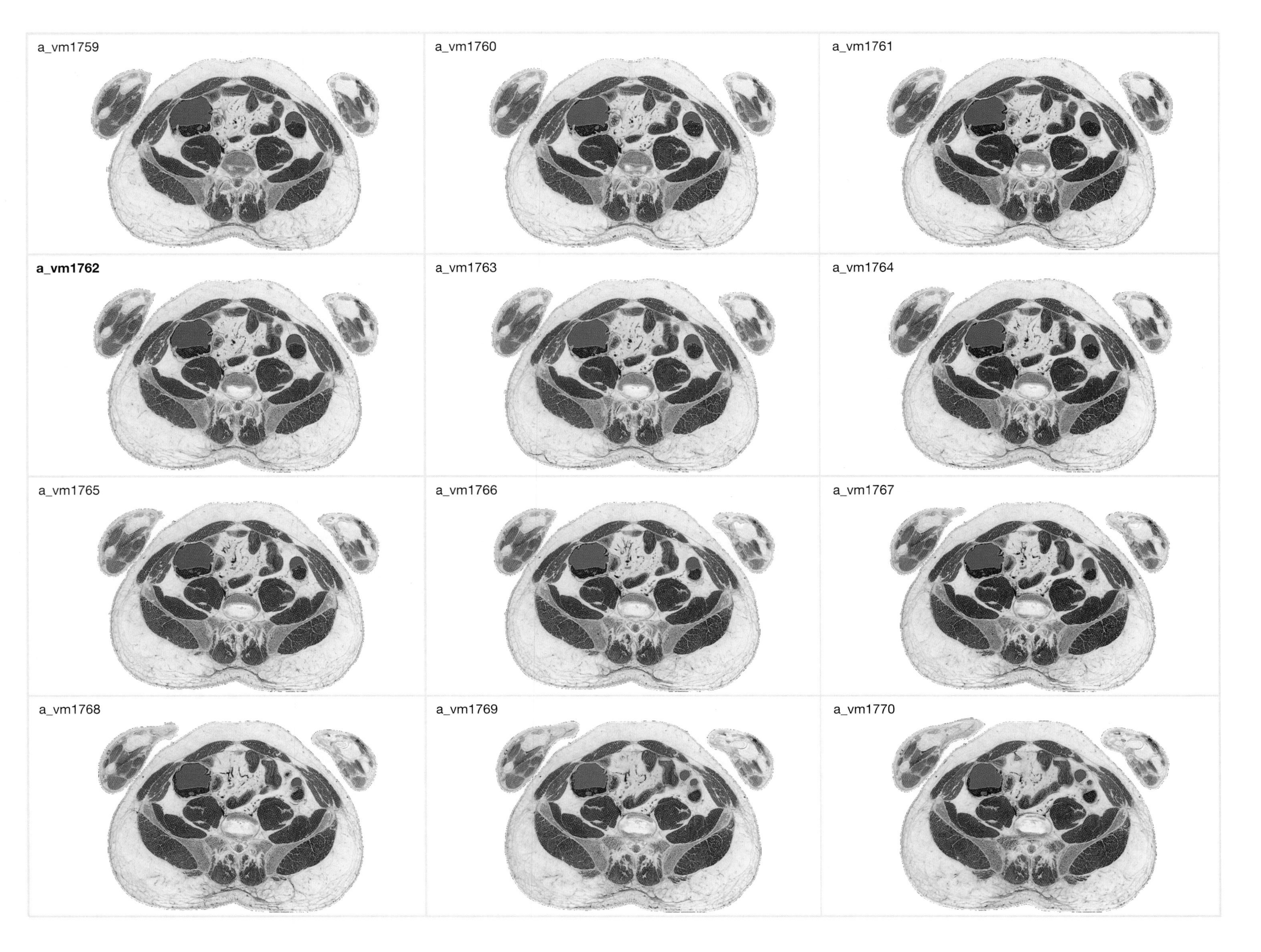
a_vm1759
a_vm1760
a_vm1761
a_vm1762
a_vm1763
a_vm1764
a_vm1765
a_vm1766
a_vm1767
a_vm1768
a_vm1769
a_vm1770

Transverse
a_vm1778

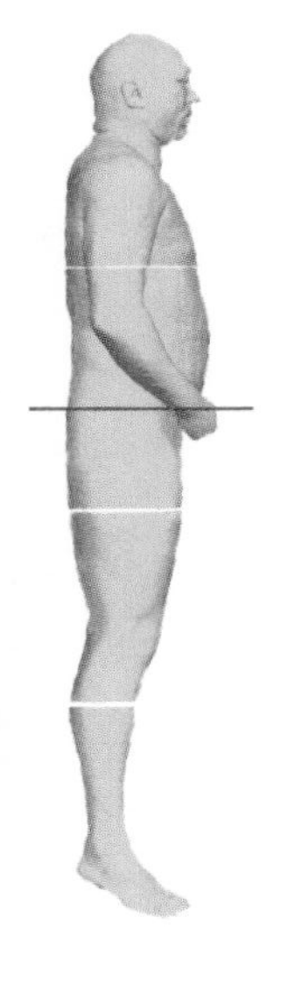

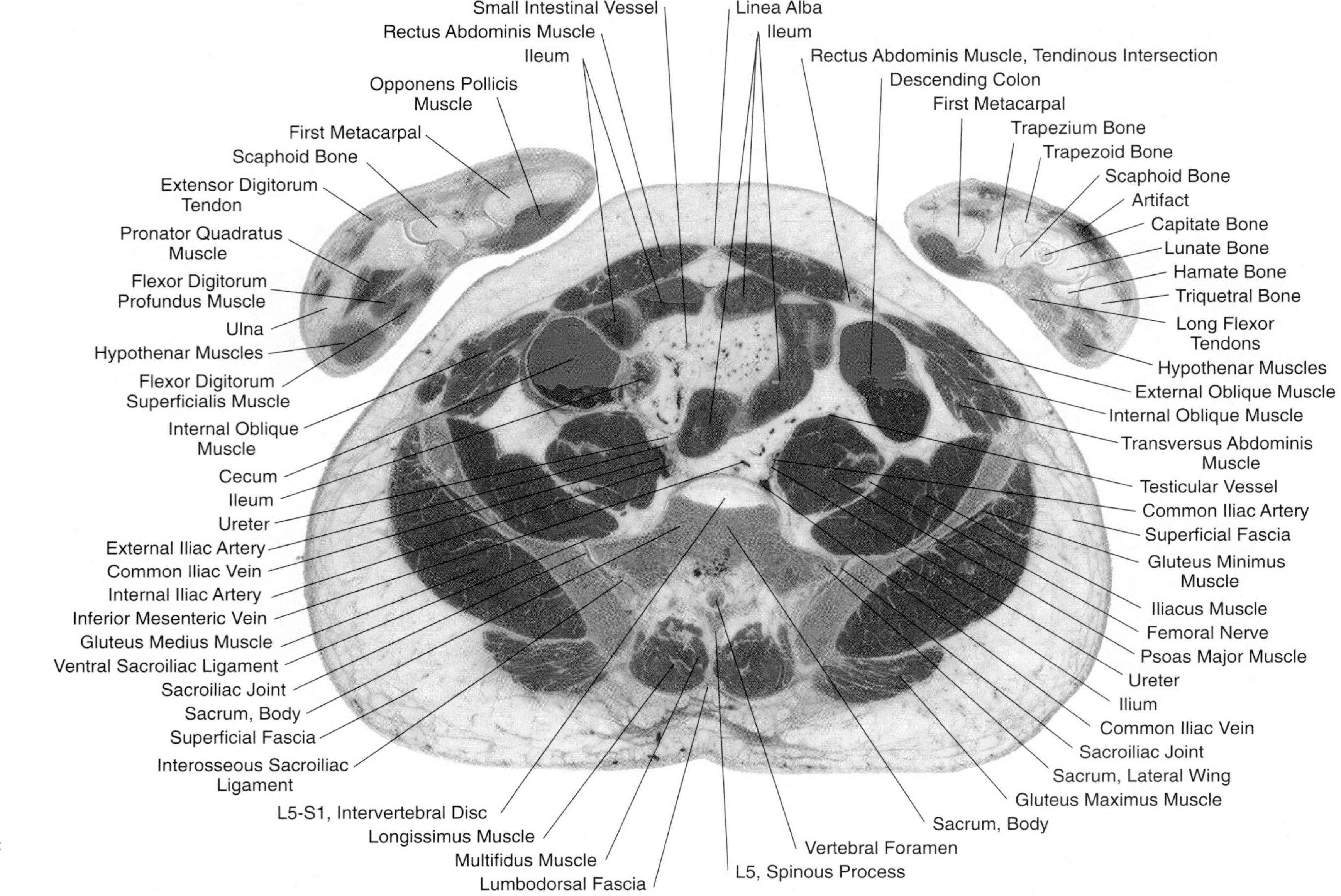
anterior
posterior
right
left
Small Intestinal Vessel
Rectus Abdominis Muscle
Ileum
Opponens Pollicis Muscle
First Metacarpal
Scaphoid Bone
Extensor Digitorum Tendon
Pronator Quadratus Muscle
Flexor Digitorum Profundus Muscle
Ulna
Hypothenar Muscles
Flexor Digitorum Superficialis Muscle
Internal Oblique Muscle
Cecum
Ileum
Ureter
External Iliac Artery
Common Iliac Vein
Internal Iliac Artery
Inferior Mesenteric Vein
Gluteus Medius Muscle
Ventral Sacroiliac Ligament
Sacroiliac Joint
Sacrum, Body
Superficial Fascia
Interosseous Sacroiliac Ligament
L5-S1, Intervertebral Disc
Longissimus Muscle
Multifidus Muscle
Lumbodorsal Fascia
Linea Alba
Ileum
Rectus Abdominis Muscle, Tendinous Intersection
Descending Colon
First Metacarpal
Trapezium Bone
Trapezoid Bone
Scaphoid Bone
Artifact
Capitate Bone
Lunate Bone
Hamate Bone
Triquetral Bone
Long Flexor Tendons
Hypothenar Muscles
External Oblique Muscle
Internal Oblique Muscle
Transversus Abdominis Muscle
Testicular Vessel
Common Iliac Artery
Superficial Fascia
Gluteus Minimus Muscle
Iliacus Muscle
Femoral Nerve
Psoas Major Muscle
Ureter
Ilium
Common Iliac Vein
Sacroiliac Joint
Sacrum, Lateral Wing
Gluteus Maximus Muscle
Sacrum, Body
Vertebral Foramen
L5, Spinous Process

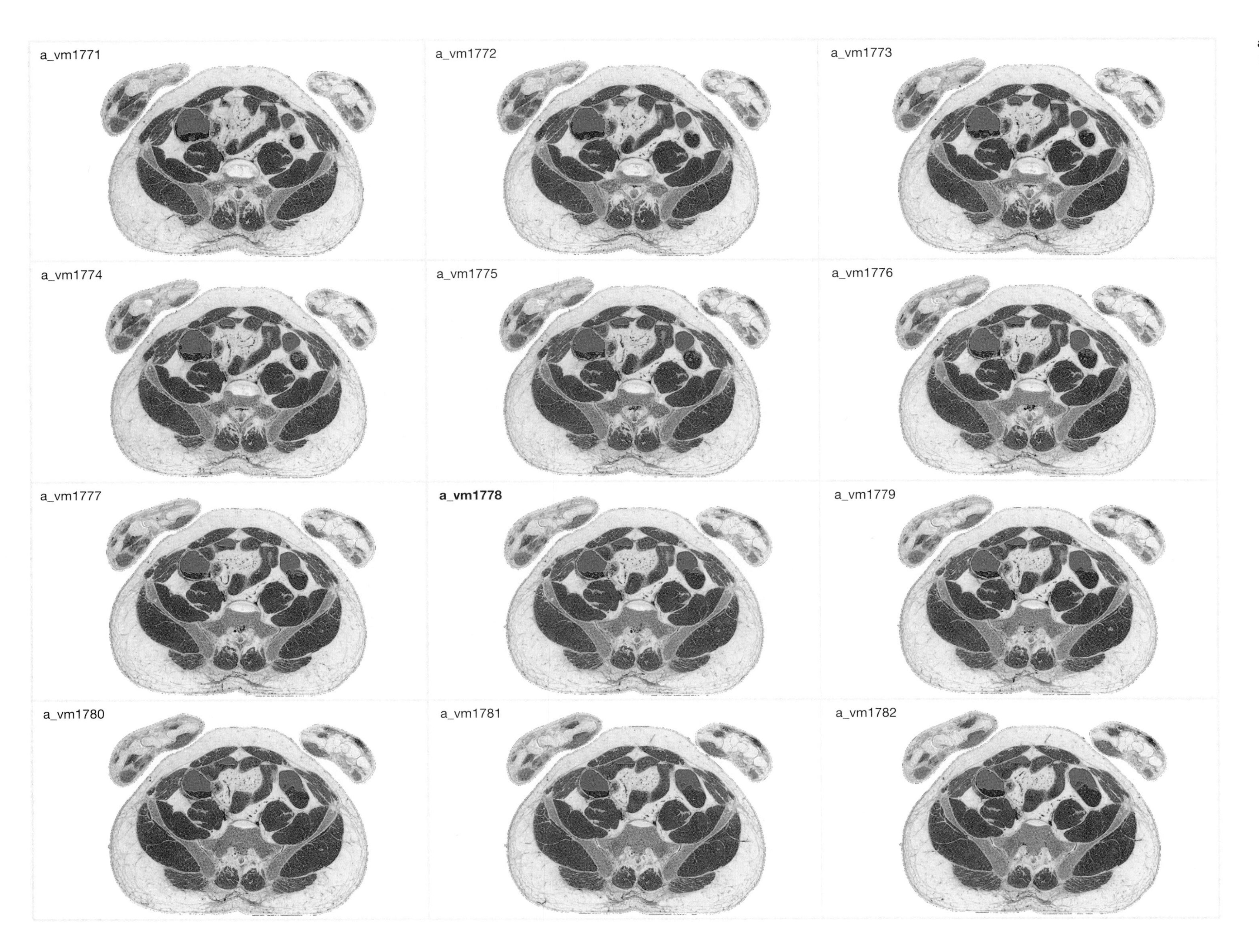
a_vm1771
a_vm1772
a_vm1773
a_vm1774
a_vm1775
a_vm1776
a_vm1777
a_vm1778
a_vm1779
a_vm1780
a_vm1781
a_vm1782

Transverse
a_vm1789

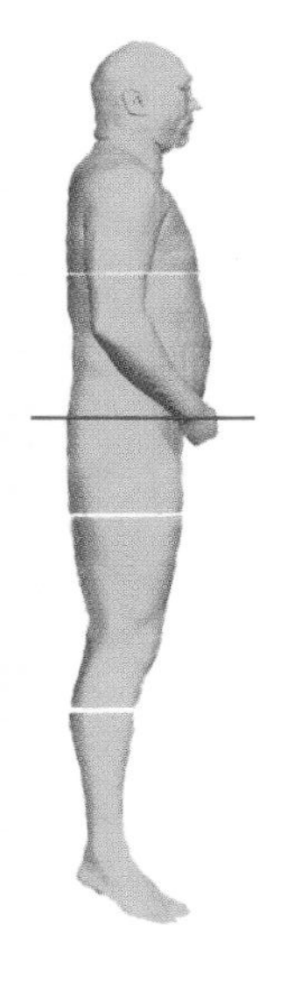

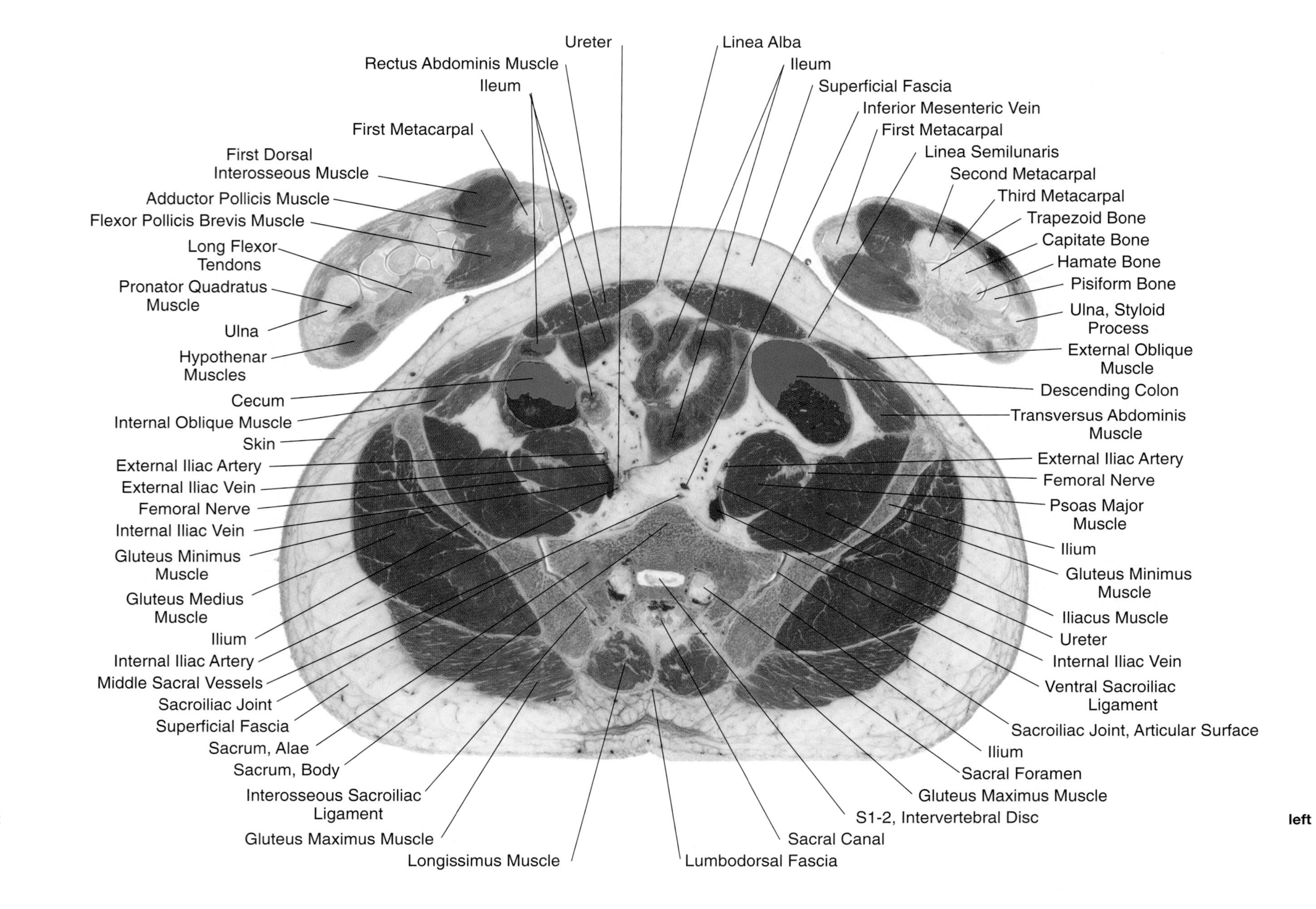

anterior
Ureter
Rectus Abdominis Muscle
Ileum
First Metacarpal
First Dorsal Interosseous Muscle
Adductor Pollicis Muscle
Flexor Pollicis Brevis Muscle
Long Flexor Tendons
Pronator Quadratus Muscle
Ulna
Hypothenar Muscles
Cecum
Internal Oblique Muscle
Skin
External Iliac Artery
External Iliac Vein
Femoral Nerve
Internal Iliac Vein
Gluteus Minimus Muscle
Gluteus Medius Muscle
Ilium
Internal Iliac Artery
Middle Sacral Vessels
Sacroiliac Joint
Superficial Fascia
Sacrum, Alae
Sacrum, Body
Interosseous Sacroiliac Ligament
Gluteus Maximus Muscle
Longissimus Muscle
Linea Alba
Ileum
Superficial Fascia
Inferior Mesenteric Vein
First Metacarpal
Linea Semilunaris
Second Metacarpal
Third Metacarpal
Trapezoid Bone
Capitate Bone
Hamate Bone
Pisiform Bone
Ulna, Styloid Process
External Oblique Muscle
Descending Colon
Transversus Abdominis Muscle
External Iliac Artery
Femoral Nerve
Psoas Major Muscle
Ilium
Gluteus Minimus Muscle
Iliacus Muscle
Ureter
Internal Iliac Vein
Ventral Sacroiliac Ligament
Sacroiliac Joint, Articular Surface
Ilium
Sacral Foramen
Gluteus Maximus Muscle
S1-2, Intervertebral Disc
Sacral Canal
Lumbodorsal Fascia
right
left
posterior

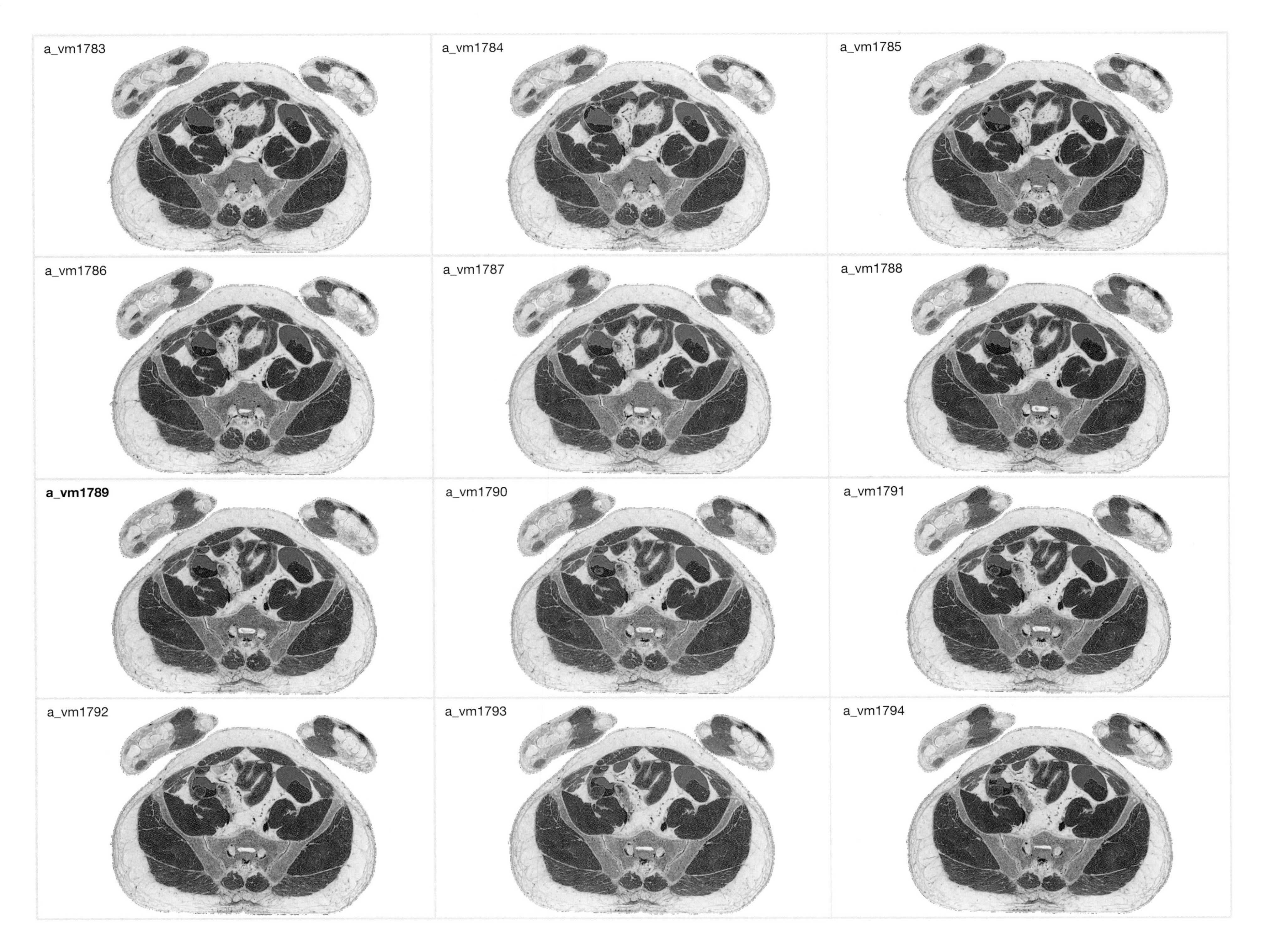
a_vm1783
a_vm1784
a_vm1785
a_vm1786
a_vm1787
a_vm1788
a_vm1789
a_vm1790
a_vm1791
a_vm1792
a_vm1793
a_vm1794

Transverse
a_vm1800

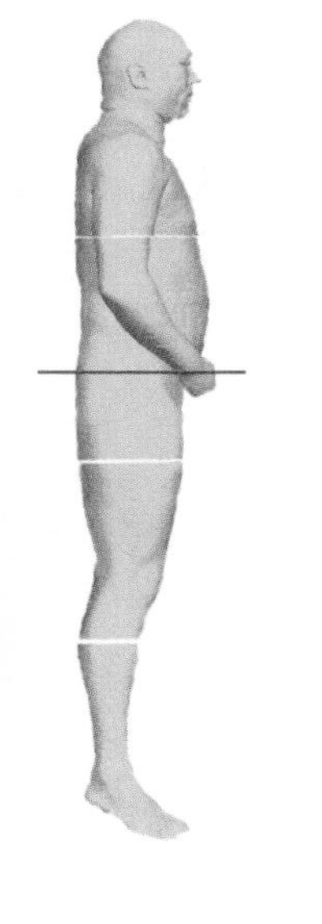

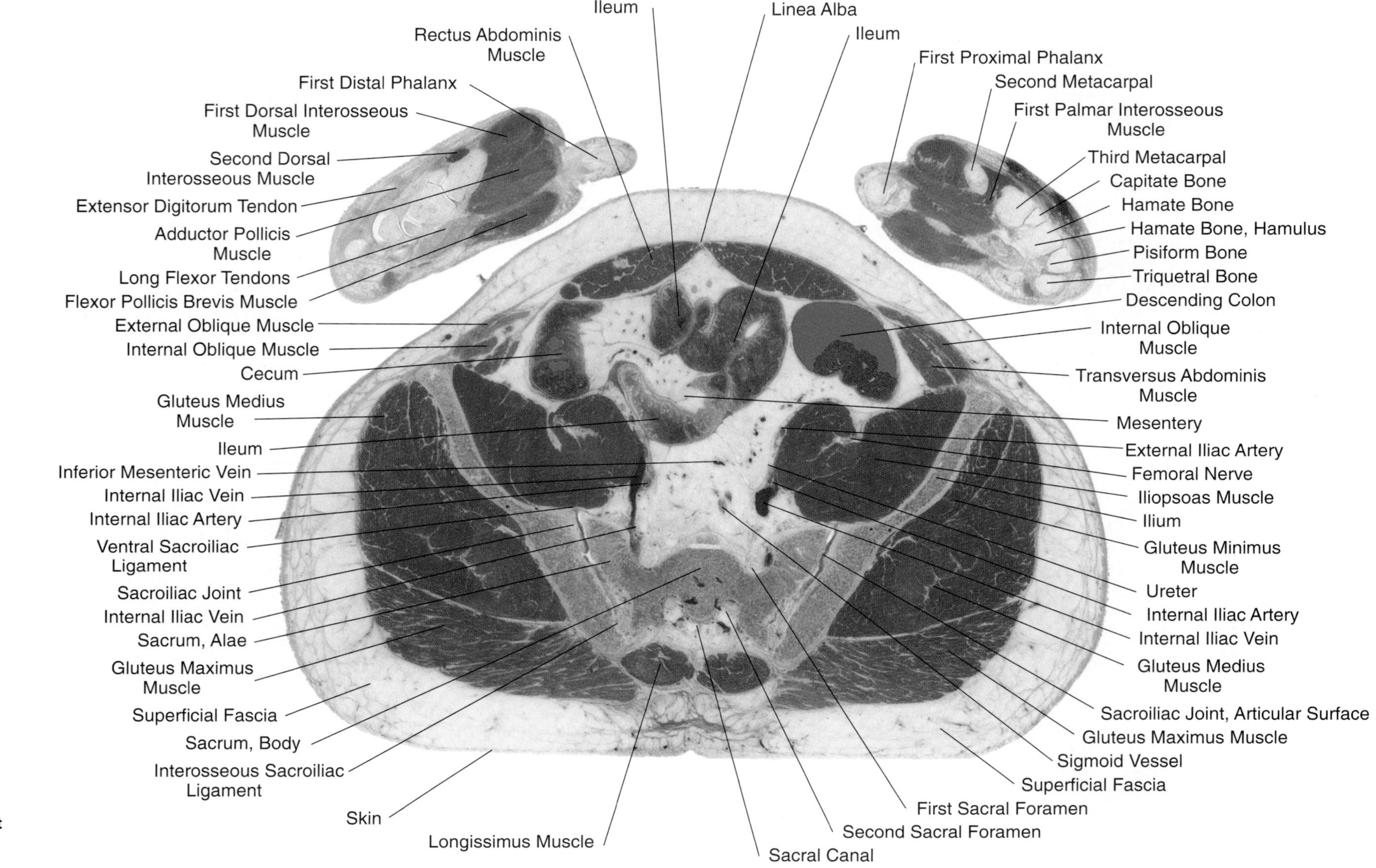
anterior
Ileum
Linea Alba
Ileum
Rectus Abdominis Muscle
First Distal Phalanx
First Dorsal Interosseous Muscle
Second Dorsal Interosseous Muscle
Extensor Digitorum Tendon
Adductor Pollicis Muscle
Long Flexor Tendons
Flexor Pollicis Brevis Muscle
External Oblique Muscle
Internal Oblique Muscle
Cecum
Gluteus Medius Muscle
Ileum
Inferior Mesenteric Vein
Internal Iliac Vein
Internal Iliac Artery
Ventral Sacroiliac Ligament
Sacroiliac Joint
Internal Iliac Vein
Sacrum, Alae
Gluteus Maximus Muscle
Superficial Fascia
Sacrum, Body
Interosseous Sacroiliac Ligament
Skin
Longissimus Muscle
Sacral Canal
Second Sacral Foramen
First Sacral Foramen
Superficial Fascia
Sigmoid Vessel
Gluteus Maximus Muscle
Sacroiliac Joint, Articular Surface
Gluteus Medius Muscle
Internal Iliac Vein
Internal Iliac Artery
Ureter
Gluteus Minimus Muscle
Ilium
Iliopsoas Muscle
Femoral Nerve
External Iliac Artery
Mesentery
Transversus Abdominis Muscle
Internal Oblique Muscle
Descending Colon
Triquetral Bone
Pisiform Bone
Hamate Bone, Hamulus
Hamate Bone
Capitate Bone
Third Metacarpal
First Palmar Interosseous Muscle
Second Metacarpal
First Proximal Phalanx
right
left
posterior

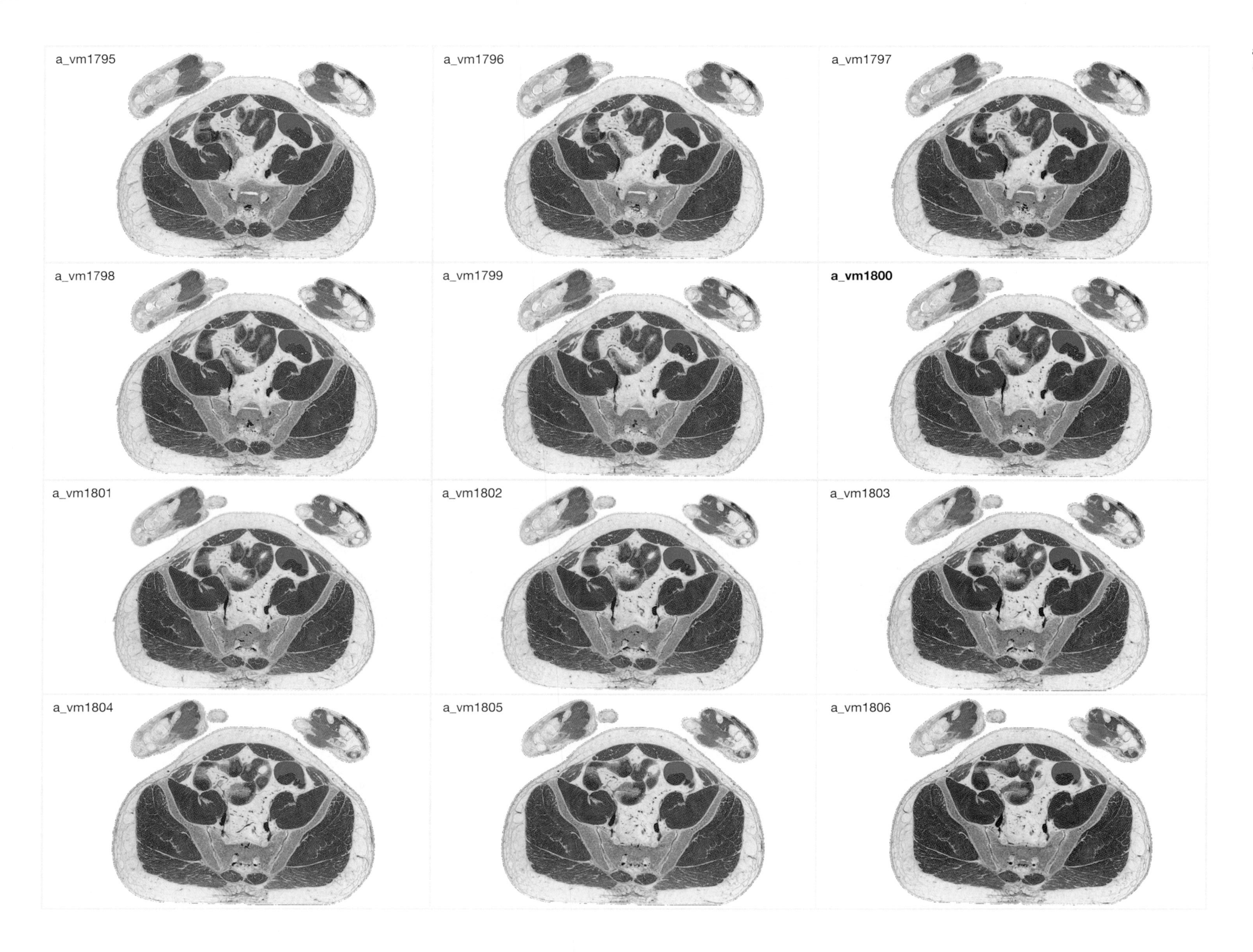
a_vm1795
a_vm1796
a_vm1797
a_vm1798
a_vm1799
a_vm1800
a_vm1801
a_vm1802
a_vm1803
a_vm1804
a_vm1805
a_vm1806

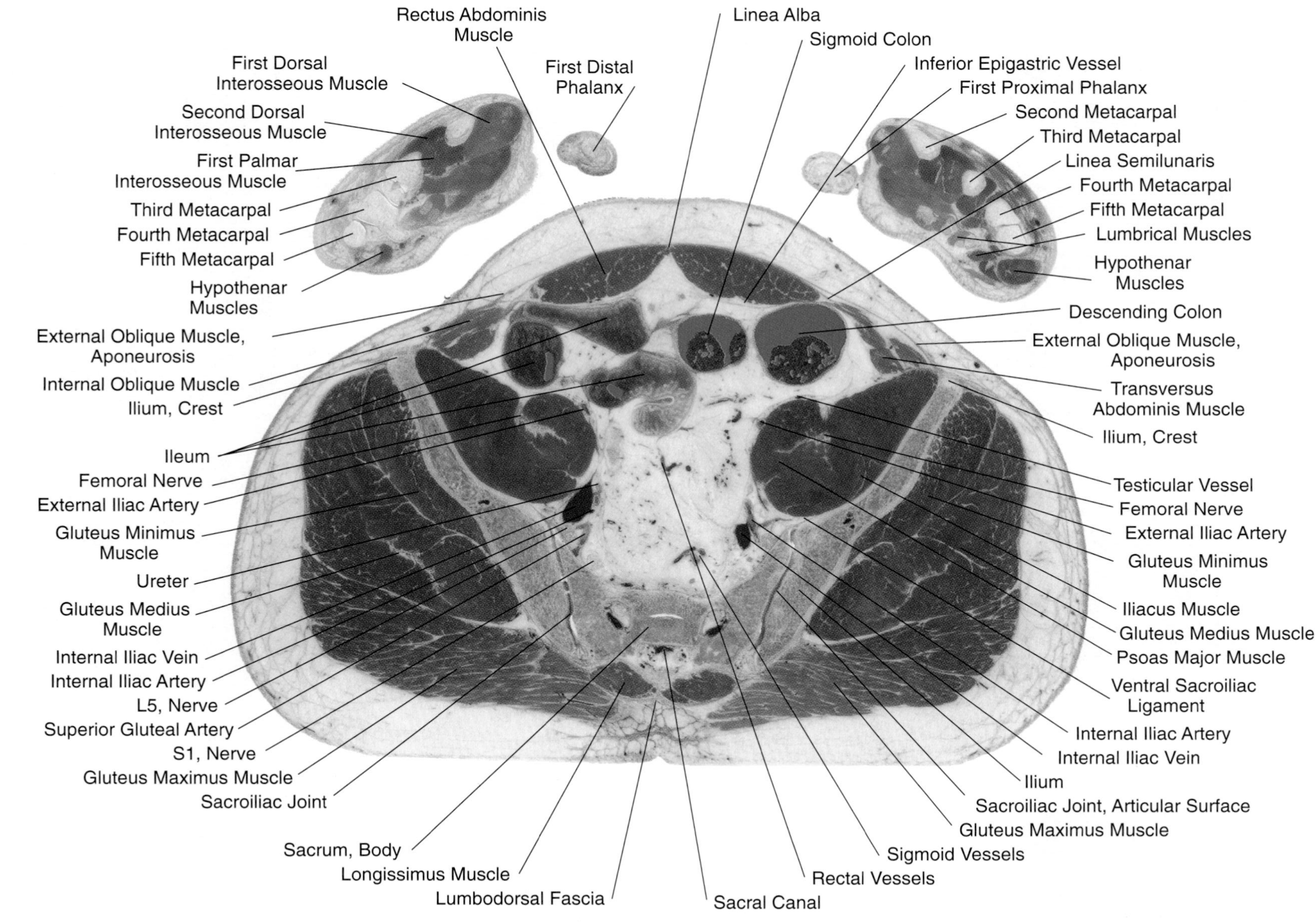
anterior
Rectus Abdominis Muscle
First Distal Phalanx
Linea Alba
Sigmoid Colon
Inferior Epigastric Vessel
First Proximal Phalanx
Second Metacarpal
Third Metacarpal
Linea Semilunaris
Fourth Metacarpal
Fifth Metacarpal
Lumbrical Muscles
Hypothenar Muscles
Descending Colon
External Oblique Muscle, Aponeurosis
Transversus Abdominis Muscle
Ilium, Crest
Testicular Vessel
Femoral Nerve
External Iliac Artery
Gluteus Minimus Muscle
Iliacus Muscle
Gluteus Medius Muscle
Psoas Major Muscle
Ventral Sacroiliac Ligament
Internal Iliac Artery
Internal Iliac Vein
Ilium
Sacroiliac Joint, Articular Surface
Gluteus Maximus Muscle
Sigmoid Vessels
Rectal Vessels
Sacral Canal
First Dorsal Interosseous Muscle
Second Dorsal Interosseous Muscle
First Palmar Interosseous Muscle
Third Metacarpal
Fourth Metacarpal
Fifth Metacarpal
Hypothenar Muscles
External Oblique Muscle, Aponeurosis
Internal Oblique Muscle
Ilium, Crest
Ileum
Femoral Nerve
External Iliac Artery
Gluteus Minimus Muscle
Ureter
Gluteus Medius Muscle
Internal Iliac Vein
Internal Iliac Artery
L5, Nerve
Superior Gluteal Artery
S1, Nerve
Gluteus Maximus Muscle
Sacroiliac Joint
Sacrum, Body
Longissimus Muscle
Lumbodorsal Fascia
right
left
posterior

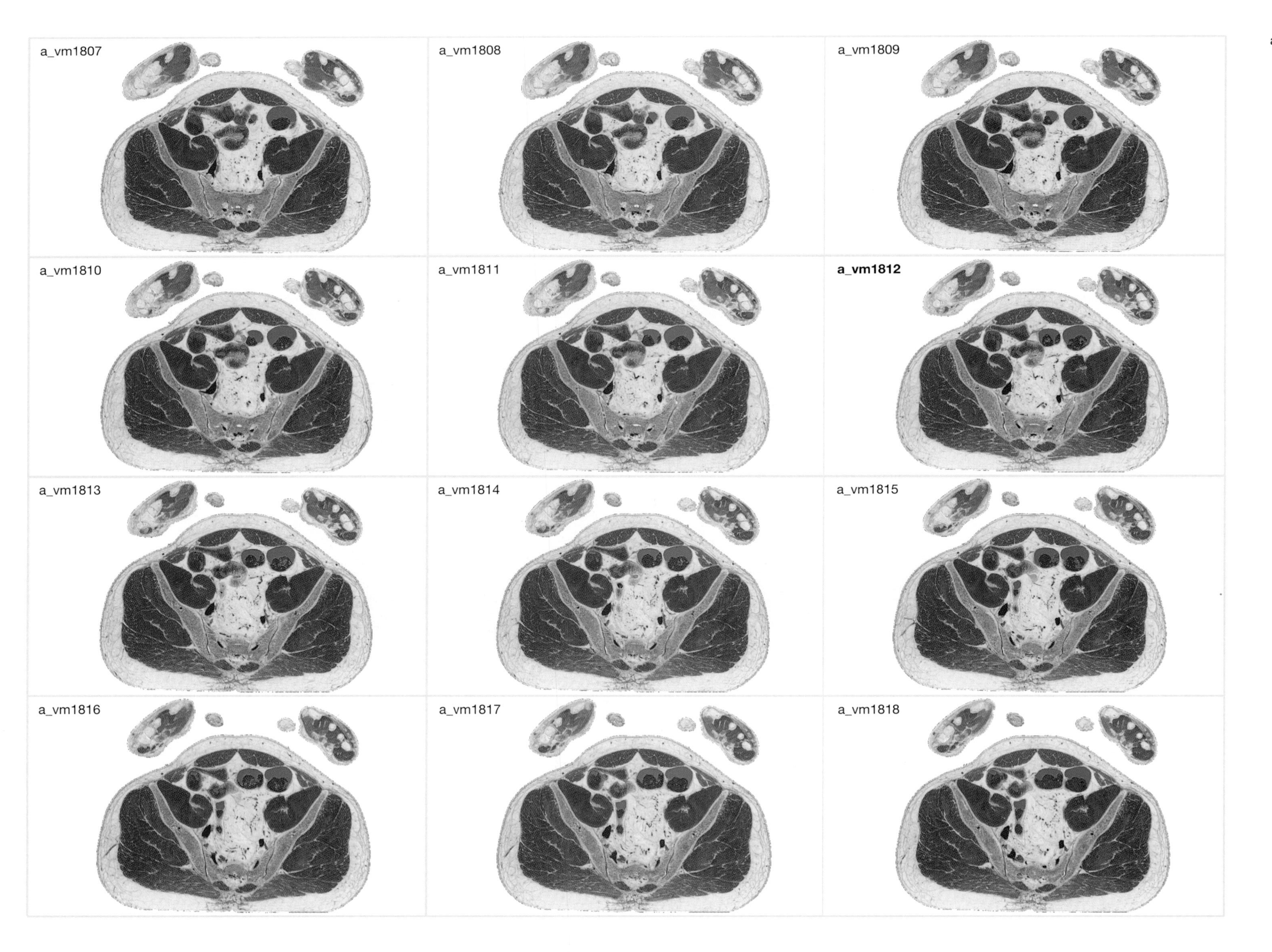
a_vm1807
a_vm1808
a_vm1809
a_vm1810
a_vm1811
a_vm1812
a_vm1813
a_vm1814
a_vm1815
a_vm1816
a_vm1817
a_vm1818

Transverse
a_vm1829

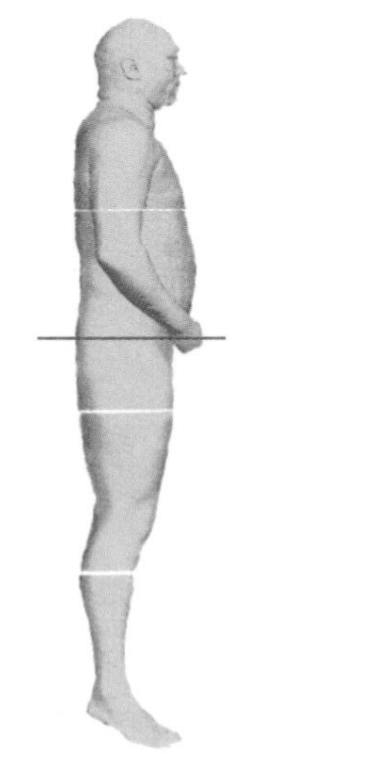

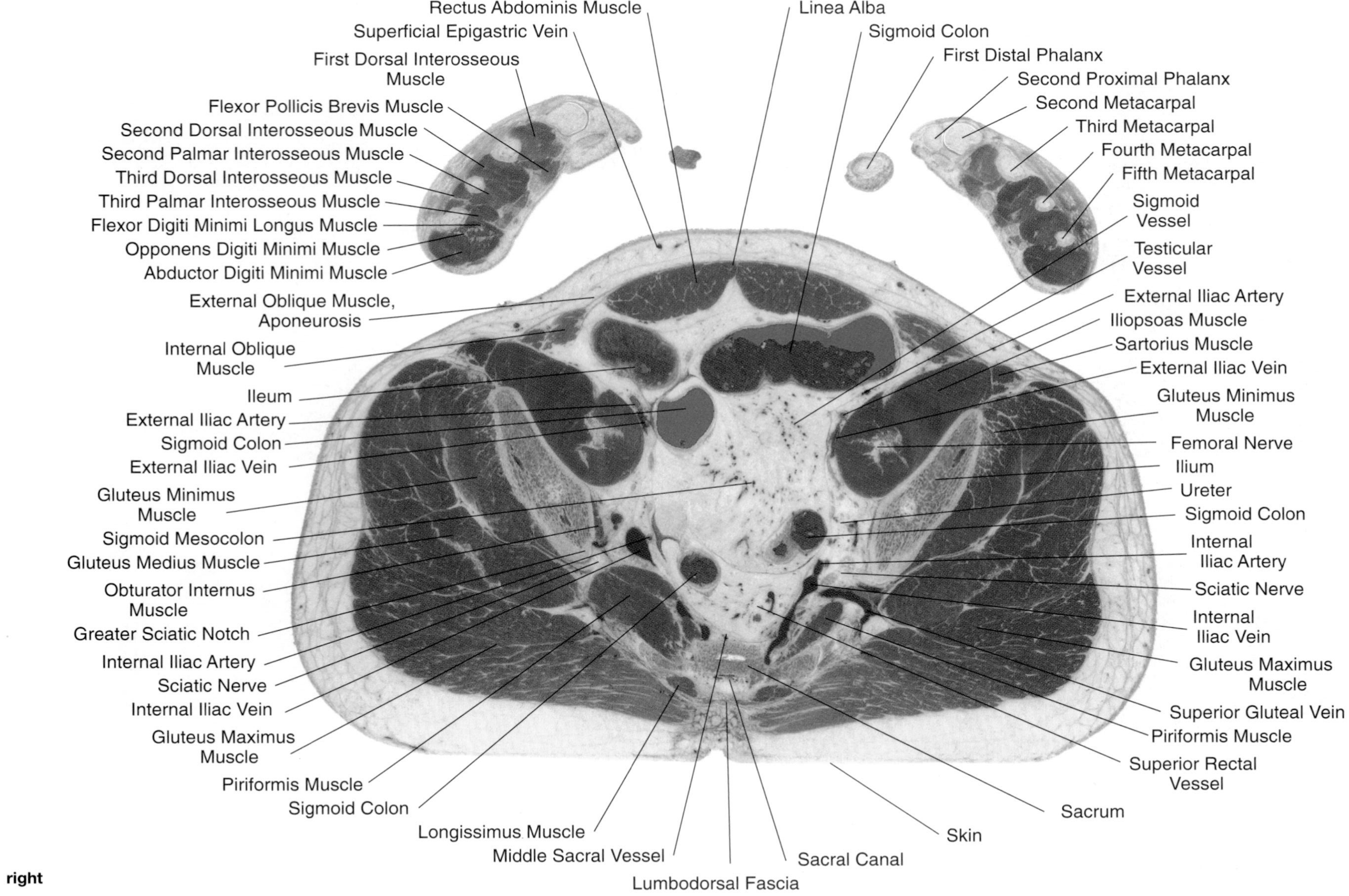
anterior
Rectus Abdominis Muscle
Superficial Epigastric Vein
First Dorsal Interosseous Muscle
Flexor Pollicis Brevis Muscle
Second Dorsal Interosseous Muscle
Second Palmar Interosseous Muscle
Third Dorsal Interosseous Muscle
Third Palmar Interosseous Muscle
Flexor Digiti Minimi Longus Muscle
Opponens Digiti Minimi Muscle
Abductor Digiti Minimi Muscle
External Oblique Muscle, Aponeurosis
Internal Oblique Muscle
Ileum
External Iliac Artery
Sigmoid Colon
External Iliac Vein
Gluteus Minimus Muscle
Sigmoid Mesocolon
Gluteus Medius Muscle
Obturator Internus Muscle
Greater Sciatic Notch
Internal Iliac Artery
Sciatic Nerve
Internal Iliac Vein
Gluteus Maximus Muscle
Piriformis Muscle
Sigmoid Colon
Longissimus Muscle
Middle Sacral Vessel
Lumbodorsal Fascia
Sacral Canal
Skin
Sacrum
Linea Alba
Sigmoid Colon
First Distal Phalanx
Second Proximal Phalanx
Second Metacarpal
Third Metacarpal
Fourth Metacarpal
Fifth Metacarpal
Sigmoid Vessel
Testicular Vessel
External Iliac Artery
Iliopsoas Muscle
Sartorius Muscle
External Iliac Vein
Gluteus Minimus Muscle
Femoral Nerve
Ilium
Ureter
Sigmoid Colon
Internal Iliac Artery
Sciatic Nerve
Internal Iliac Vein
Gluteus Maximus Muscle
Superior Gluteal Vein
Piriformis Muscle
Superior Rectal Vessel
right
left
posterior

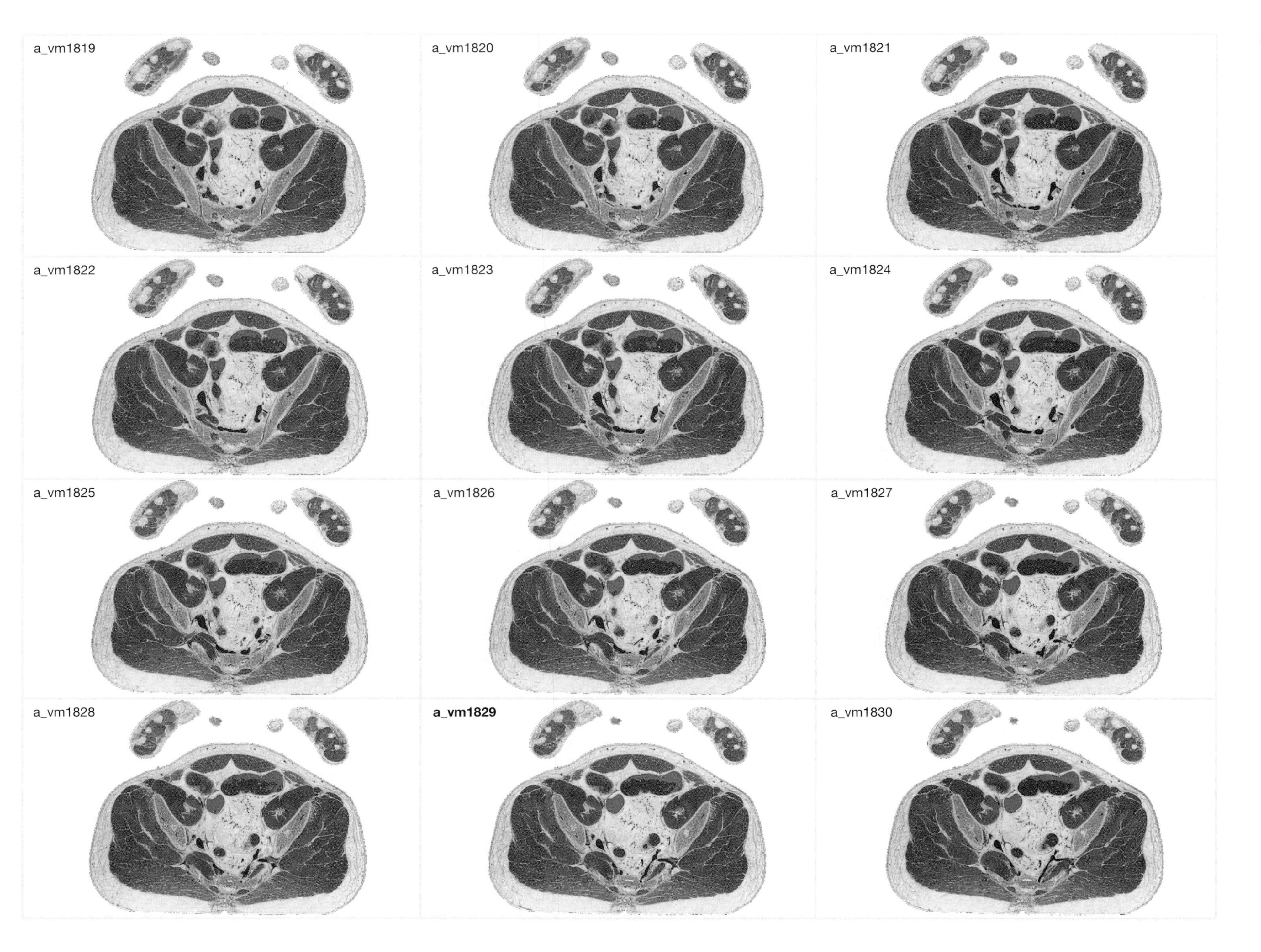
a_vm1819
a_vm1820
a_vm1821
a_vm1822
a_vm1823
a_vm1824
a_vm1825
a_vm1826
a_vm1827
a_vm1828
a_vm1829
a_vm1830

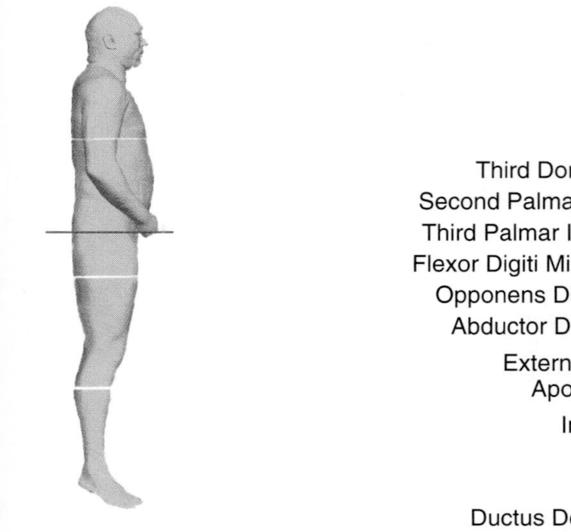

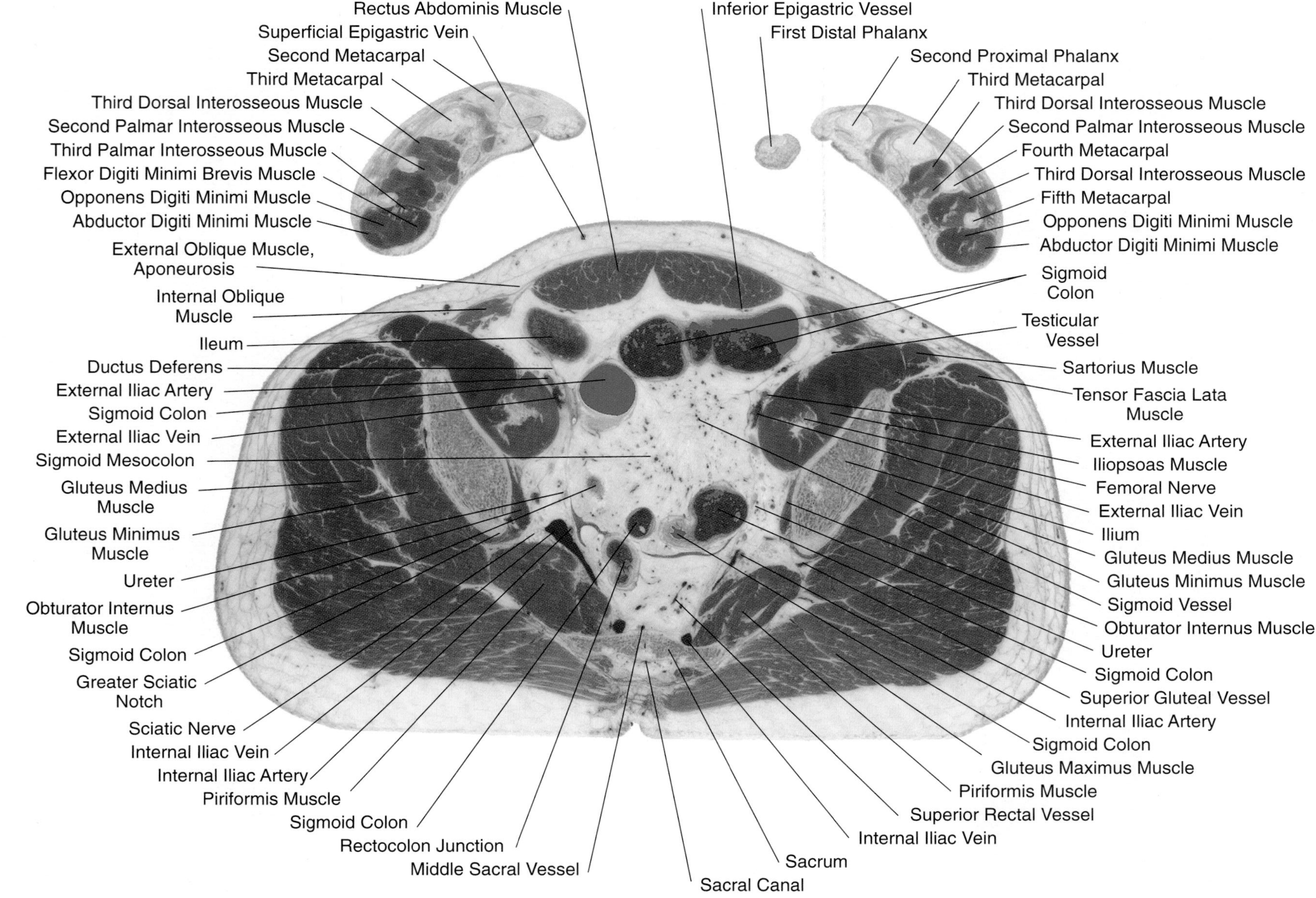
anterior
Rectus Abdominis Muscle
Superficial Epigastric Vein
Second Metacarpal
Third Metacarpal
Third Dorsal Interosseous Muscle
Second Palmar Interosseous Muscle
Third Palmar Interosseous Muscle
Flexor Digiti Minimi Brevis Muscle
Opponens Digiti Minimi Muscle
Abductor Digiti Minimi Muscle
External Oblique Muscle, Aponeurosis
Internal Oblique Muscle
Ileum
Ductus Deferens
External Iliac Artery
Sigmoid Colon
External Iliac Vein
Sigmoid Mesocolon
Gluteus Medius Muscle
Gluteus Minimus Muscle
Ureter
Obturator Internus Muscle
Sigmoid Colon
Greater Sciatic Notch
Sciatic Nerve
Internal Iliac Vein
Internal Iliac Artery
Piriformis Muscle
Sigmoid Colon
Rectocolon Junction
Middle Sacral Vessel
Inferior Epigastric Vessel
First Distal Phalanx
Second Proximal Phalanx
Third Metacarpal
Third Dorsal Interosseous Muscle
Second Palmar Interosseous Muscle
Fourth Metacarpal
Third Dorsal Interosseous Muscle
Fifth Metacarpal
Opponens Digiti Minimi Muscle
Abductor Digiti Minimi Muscle
Sigmoid Colon
Testicular Vessel
Sartorius Muscle
Tensor Fascia Lata Muscle
External Iliac Artery
Iliopsoas Muscle
Femoral Nerve
External Iliac Vein
Ilium
Gluteus Medius Muscle
Gluteus Minimus Muscle
Sigmoid Vessel
Obturator Internus Muscle
Ureter
Sigmoid Colon
Superior Gluteal Vessel
Internal Iliac Artery
Sigmoid Colon
Gluteus Maximus Muscle
Piriformis Muscle
Superior Rectal Vessel
Internal Iliac Vein
Sacrum
Sacral Canal
right
left
posterior

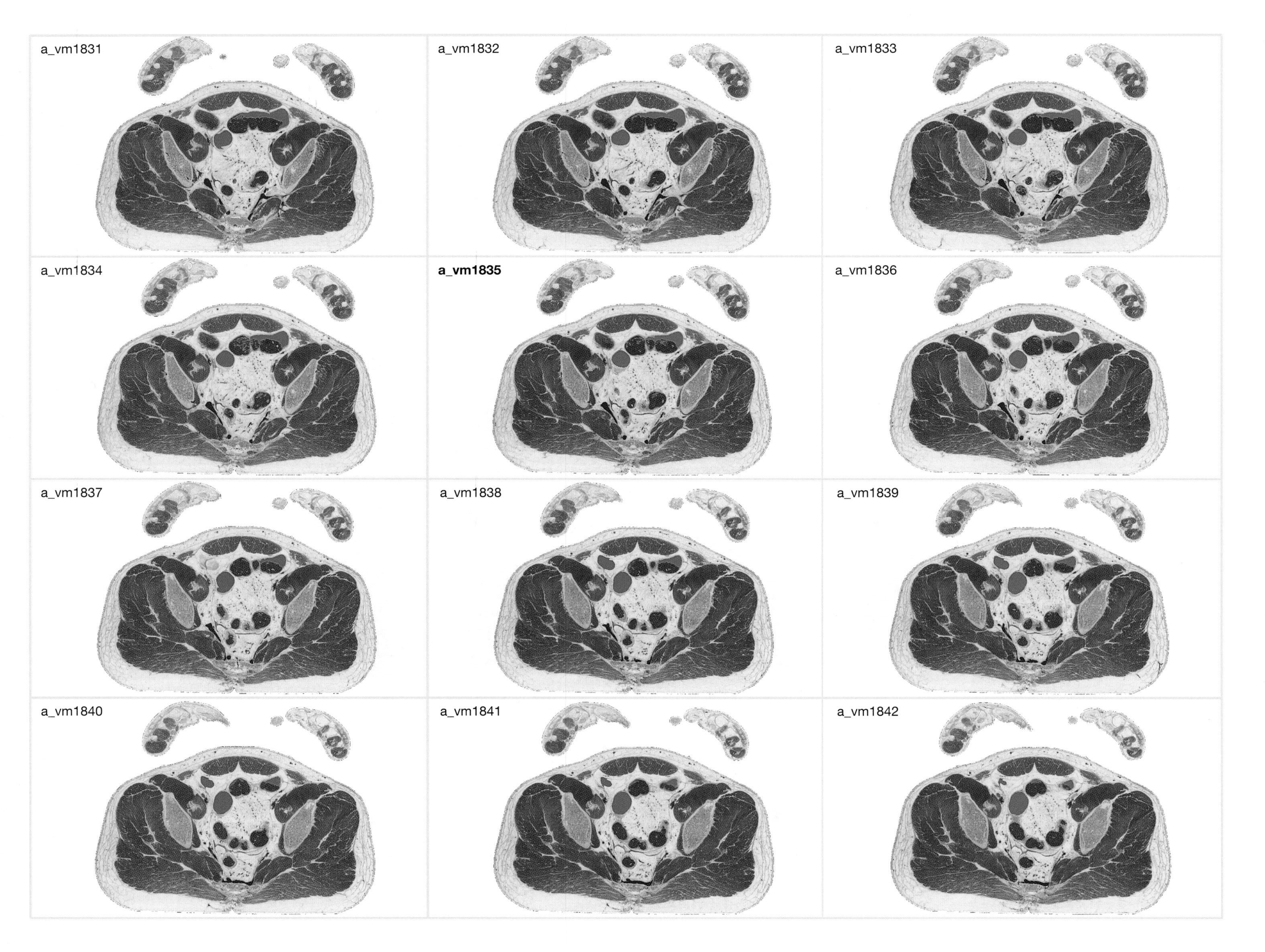
a_vm1831
a_vm1832
a_vm1833
a_vm1834
a_vm1835
a_vm1836
a_vm1837
a_vm1838
a_vm1839
a_vm1840
a_vm1841
a_vm1842

Transverse
a_vm1850

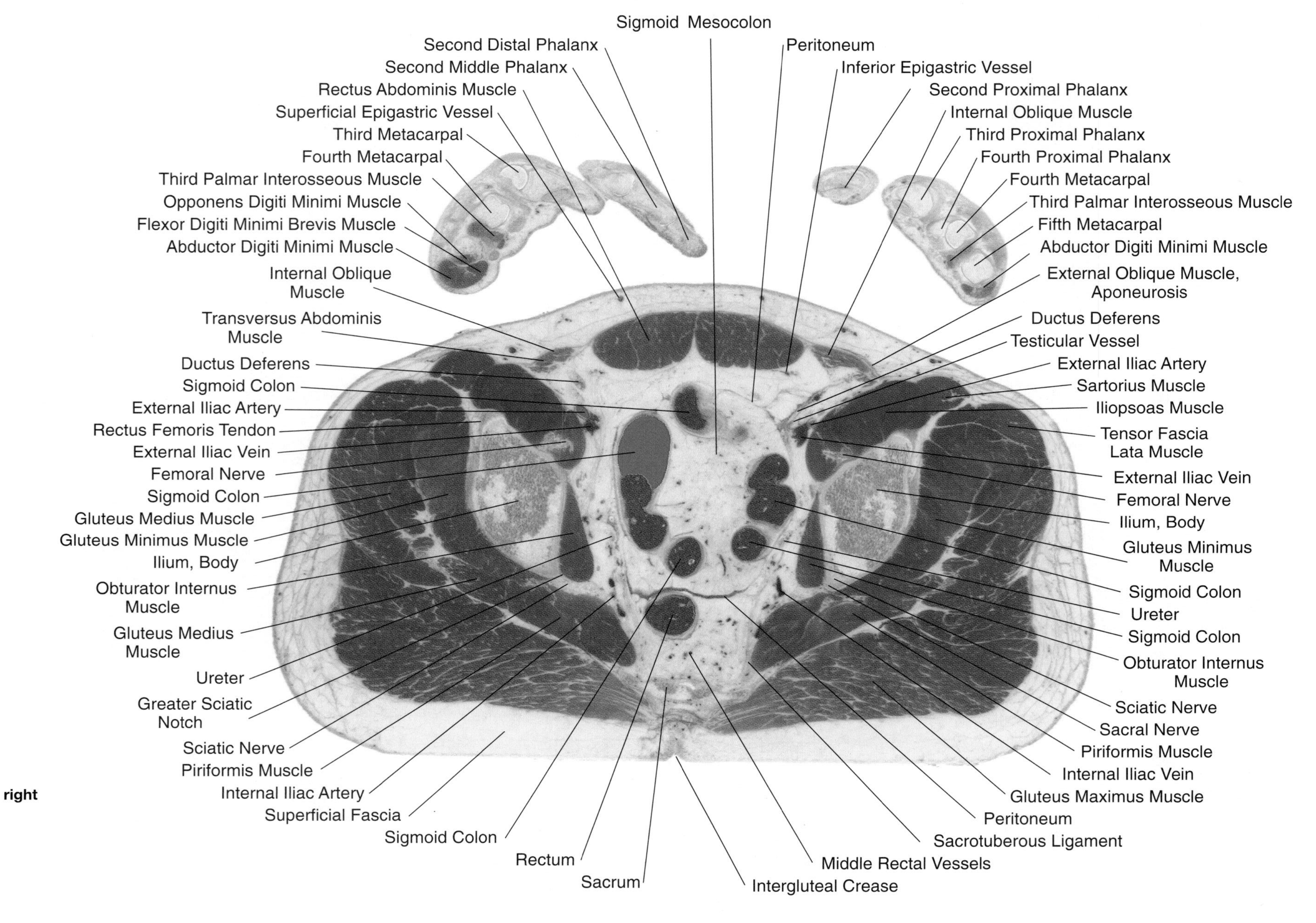

anterior
Sigmoid Mesocolon
Second Distal Phalanx
Second Middle Phalanx
Rectus Abdominis Muscle
Superficial Epigastric Vessel
Third Metacarpal
Fourth Metacarpal
Third Palmar Interosseous Muscle
Opponens Digiti Minimi Muscle
Flexor Digiti Minimi Brevis Muscle
Abductor Digiti Minimi Muscle
Internal Oblique Muscle
Transversus Abdominis Muscle
Ductus Deferens
Sigmoid Colon
External Iliac Artery
Rectus Femoris Tendon
External Iliac Vein
Femoral Nerve
Sigmoid Colon
Gluteus Medius Muscle
Gluteus Minimus Muscle
Ilium, Body
Obturator Internus Muscle
Gluteus Medius Muscle
Ureter
Greater Sciatic Notch
Sciatic Nerve
Piriformis Muscle
Internal Iliac Artery
Superficial Fascia
Sigmoid Colon
Rectum
Sacrum
Peritoneum
Inferior Epigastric Vessel
Second Proximal Phalanx
Internal Oblique Muscle
Third Proximal Phalanx
Fourth Proximal Phalanx
Fourth Metacarpal
Third Palmar Interosseous Muscle
Fifth Metacarpal
Abductor Digiti Minimi Muscle
External Oblique Muscle, Aponeurosis
Ductus Deferens
Testicular Vessel
External Iliac Artery
Sartorius Muscle
Iliopsoas Muscle
Tensor Fascia Lata Muscle
External Iliac Vein
Femoral Nerve
Ilium, Body
Gluteus Minimus Muscle
Sigmoid Colon
Ureter
Sigmoid Colon
Obturator Internus Muscle
Sciatic Nerve
Sacral Nerve
Piriformis Muscle
Internal Iliac Vein
Gluteus Maximus Muscle
Peritoneum
Sacrotuberous Ligament
Middle Rectal Vessels
Intergluteal Crease
right
left
posterior

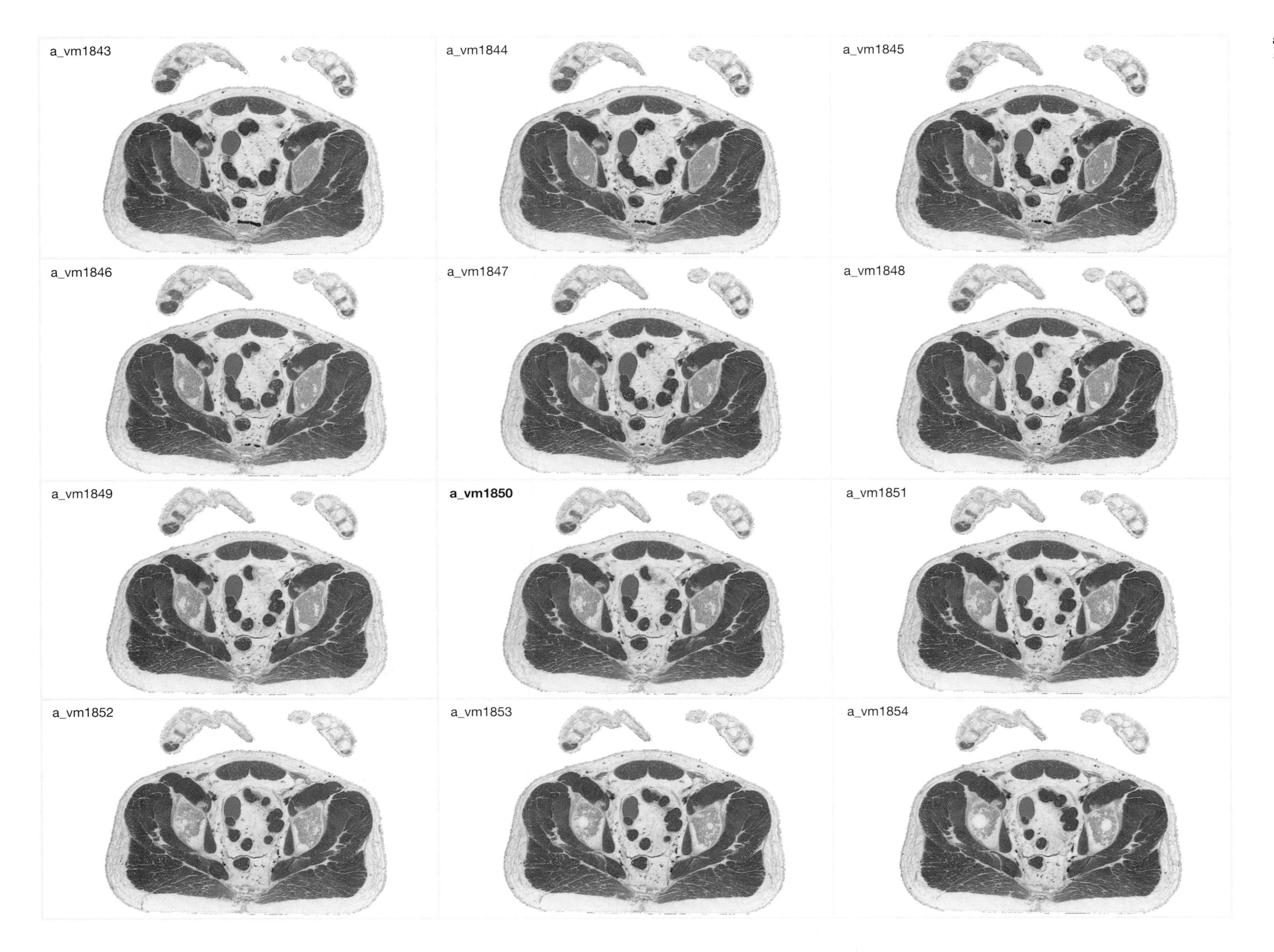
a_vm1843
a_vm1844
a_vm1845
a_vm1846
a_vm1847
a_vm1848
a_vm1849
a_vm1850
a_vm1851
a_vm1852
a_vm1853
a_vm1854

Transverse
a_vm1861

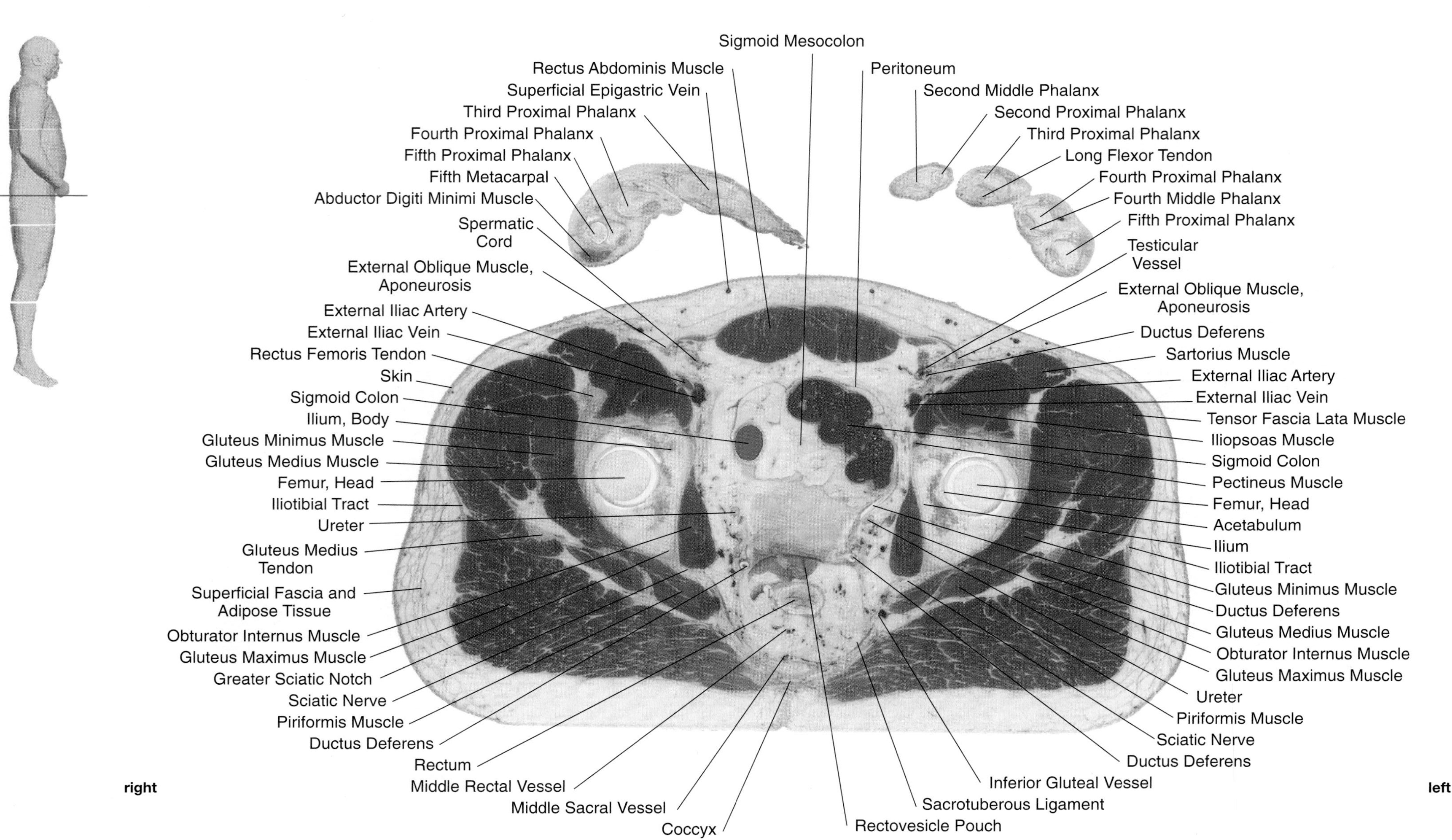
anterior
Sigmoid Mesocolon
Rectus Abdominis Muscle
Superficial Epigastric Vein
Third Proximal Phalanx
Fourth Proximal Phalanx
Fifth Proximal Phalanx
Fifth Metacarpal
Abductor Digiti Minimi Muscle
Spermatic Cord
External Oblique Muscle, Aponeurosis
External Iliac Artery
External Iliac Vein
Rectus Femoris Tendon
Skin
Sigmoid Colon
Ilium, Body
Gluteus Minimus Muscle
Gluteus Medius Muscle
Femur, Head
Iliotibial Tract
Ureter
Gluteus Medius Tendon
Superficial Fascia and Adipose Tissue
Obturator Internus Muscle
Gluteus Maximus Muscle
Greater Sciatic Notch
Sciatic Nerve
Piriformis Muscle
Ductus Deferens
Rectum
Middle Rectal Vessel
Middle Sacral Vessel
Coccyx
Peritoneum
Second Middle Phalanx
Second Proximal Phalanx
Third Proximal Phalanx
Long Flexor Tendon
Fourth Proximal Phalanx
Fourth Middle Phalanx
Fifth Proximal Phalanx
Testicular Vessel
External Oblique Muscle, Aponeurosis
Ductus Deferens
Sartorius Muscle
External Iliac Artery
External Iliac Vein
Tensor Fascia Lata Muscle
Iliopsoas Muscle
Sigmoid Colon
Pectineus Muscle
Femur, Head
Acetabulum
Ilium
Iliotibial Tract
Gluteus Minimus Muscle
Ductus Deferens
Gluteus Medius Muscle
Obturator Internus Muscle
Gluteus Maximus Muscle
Ureter
Piriformis Muscle
Sciatic Nerve
Ductus Deferens
Inferior Gluteal Vessel
Sacrotuberous Ligament
Rectovesicle Pouch
right
left
posterior

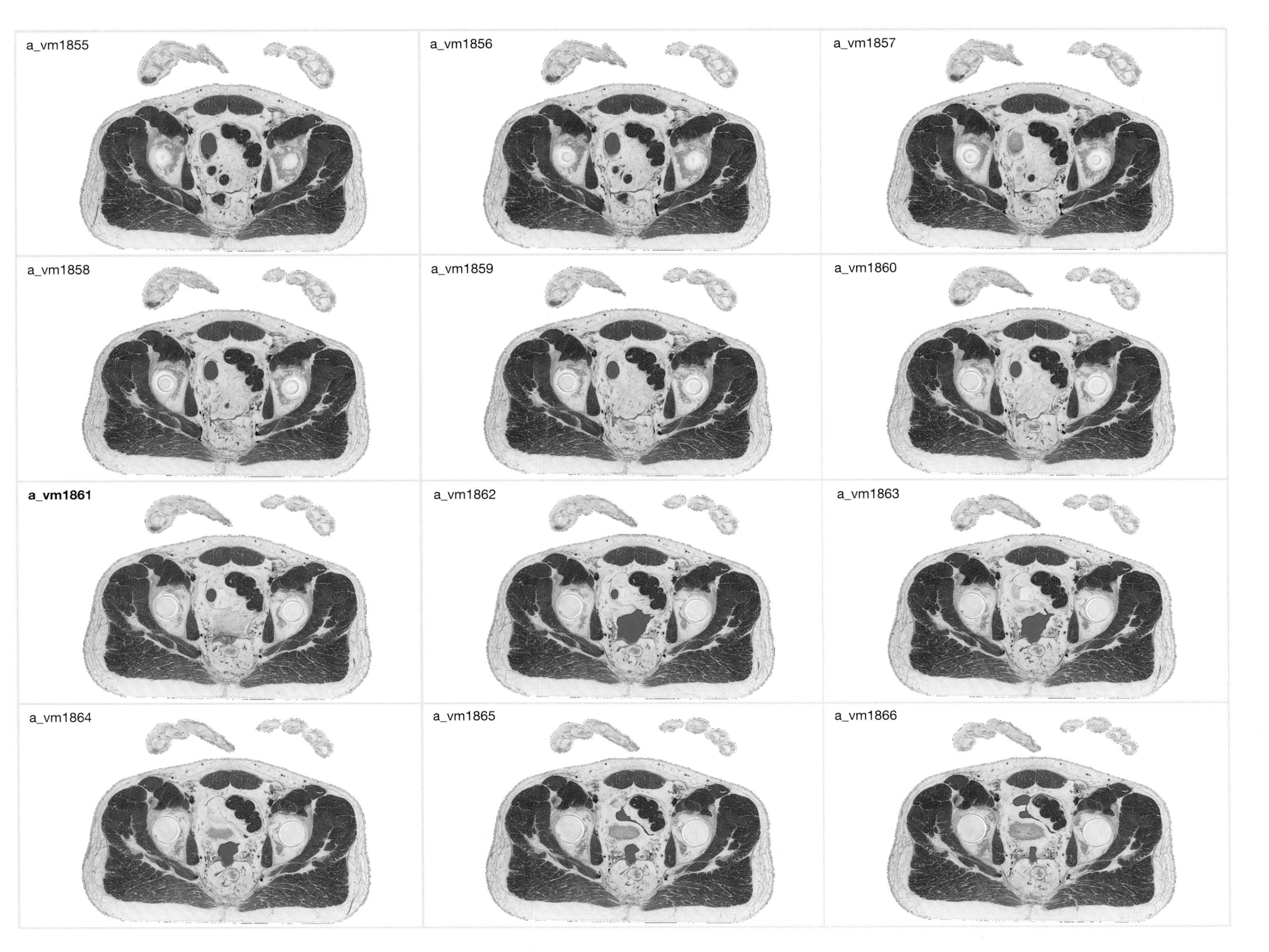
a_vm1855
a_vm1856
a_vm1857
a_vm1858
a_vm1859
a_vm1860
a_vm1861
a_vm1862
a_vm1863
a_vm1864
a_vm1865
a_vm1866

Transverse
a_vm1873

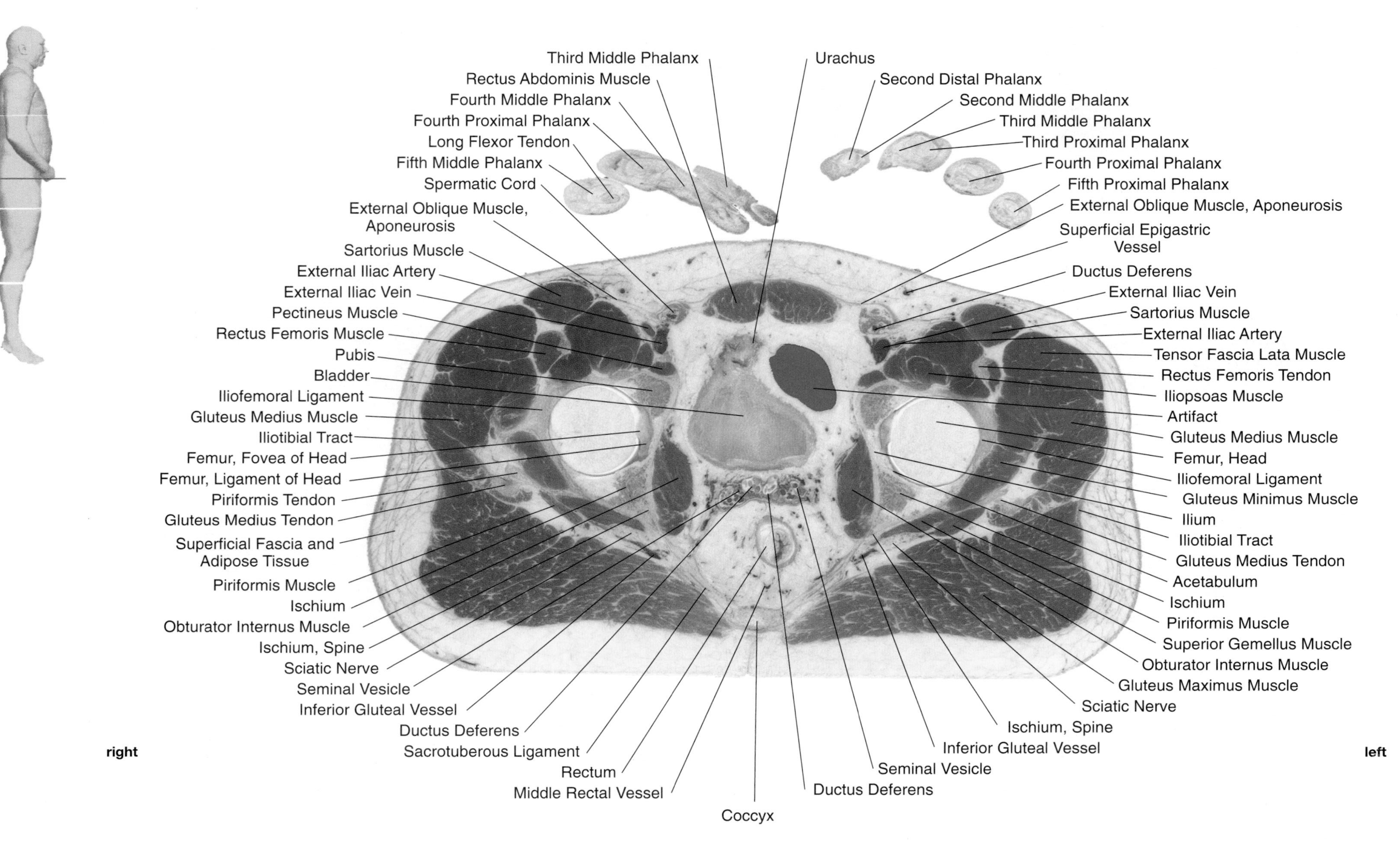
anterior
Third Middle Phalanx
Rectus Abdominis Muscle
Fourth Middle Phalanx
Fourth Proximal Phalanx
Long Flexor Tendon
Fifth Middle Phalanx
Spermatic Cord
External Oblique Muscle, Aponeurosis
Sartorius Muscle
External Iliac Artery
External Iliac Vein
Pectineus Muscle
Rectus Femoris Muscle
Pubis
Bladder
Iliofemoral Ligament
Gluteus Medius Muscle
Iliotibial Tract
Femur, Fovea of Head
Femur, Ligament of Head
Piriformis Tendon
Gluteus Medius Tendon
Superficial Fascia and Adipose Tissue
Piriformis Muscle
Ischium
Obturator Internus Muscle
Ischium, Spine
Sciatic Nerve
Seminal Vesicle
Inferior Gluteal Vessel
Ductus Deferens
Sacrotuberous Ligament
Rectum
Middle Rectal Vessel
Coccyx
Urachus
Second Distal Phalanx
Second Middle Phalanx
Third Middle Phalanx
Third Proximal Phalanx
Fourth Proximal Phalanx
Fifth Proximal Phalanx
External Oblique Muscle, Aponeurosis
Superficial Epigastric Vessel
Ductus Deferens
External Iliac Vein
Sartorius Muscle
External Iliac Artery
Tensor Fascia Lata Muscle
Rectus Femoris Tendon
Iliopsoas Muscle
Artifact
Gluteus Medius Muscle
Femur, Head
Iliofemoral Ligament
Gluteus Minimus Muscle
Ilium
Iliotibial Tract
Gluteus Medius Tendon
Acetabulum
Ischium
Piriformis Muscle
Superior Gemellus Muscle
Obturator Internus Muscle
Gluteus Maximus Muscle
Sciatic Nerve
Ischium, Spine
Inferior Gluteal Vessel
Seminal Vesicle
Ductus Deferens
right
left
posterior

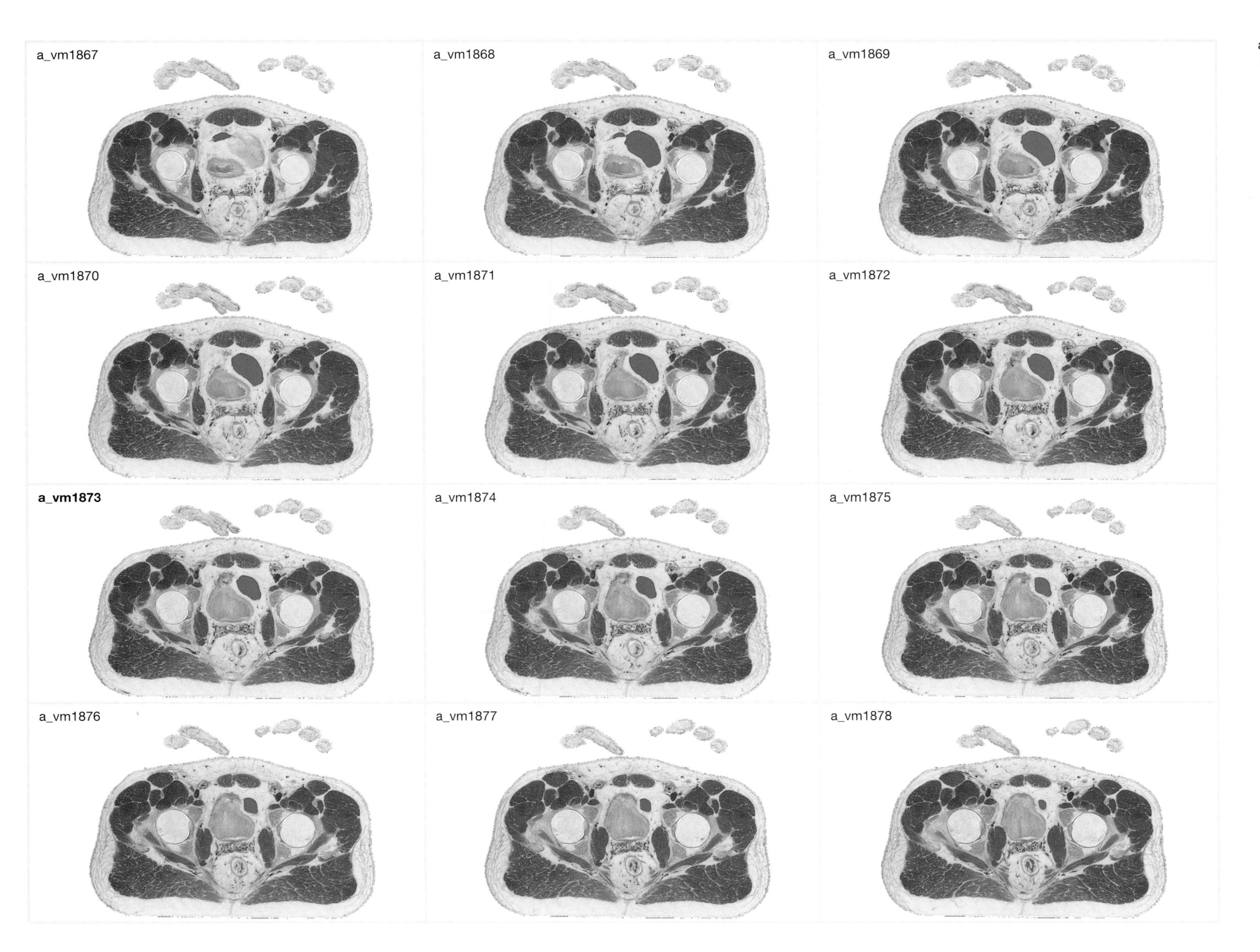
a_vm1867
a_vm1868
a_vm1869
a_vm1870
a_vm1871
a_vm1872
a_vm1873
a_vm1874
a_vm1875
a_vm1876
a_vm1877
a_vm1878

Transverse
a_vm1886

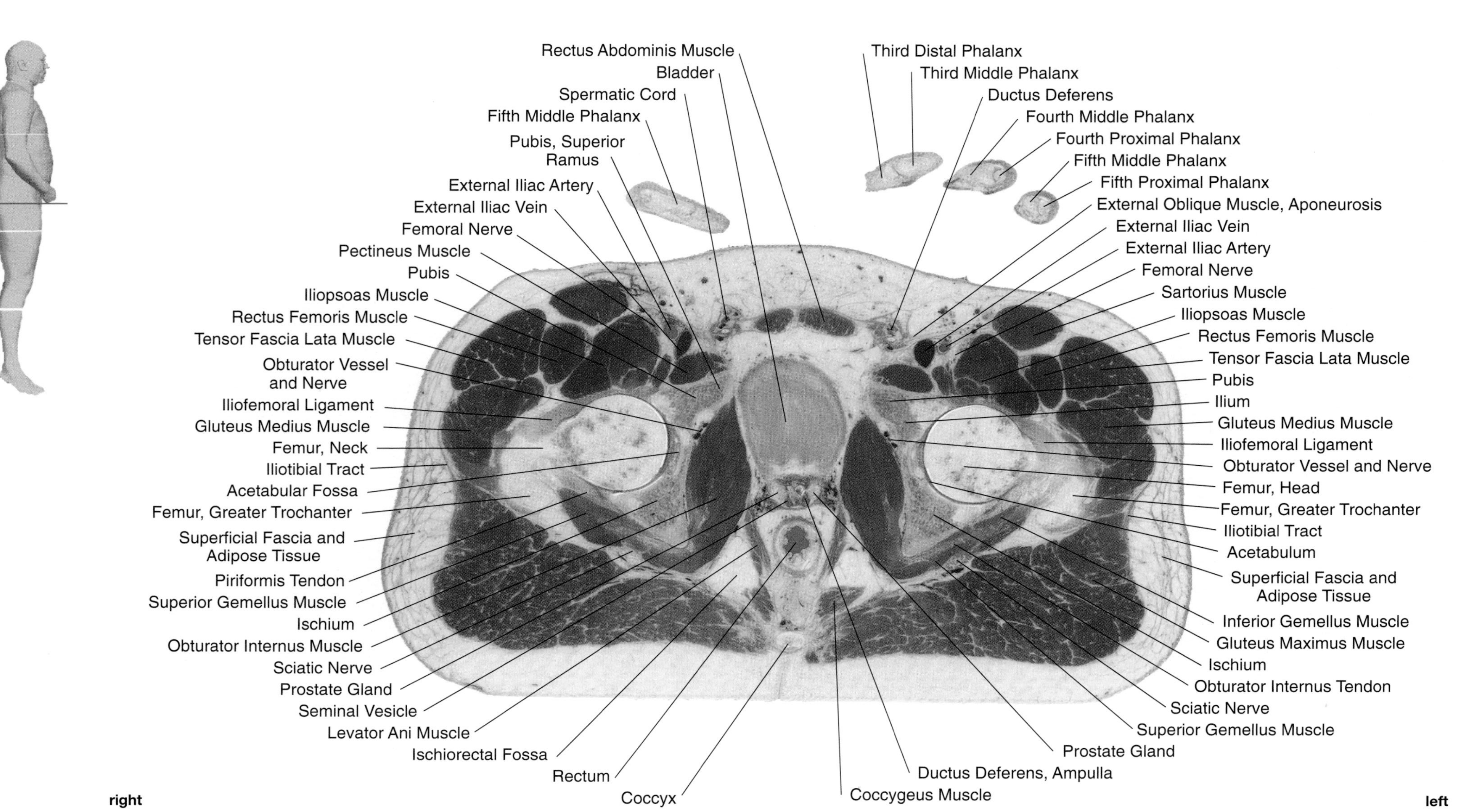

anterior
Rectus Abdominis Muscle
Bladder
Spermatic Cord
Fifth Middle Phalanx
Pubis, Superior Ramus
External Iliac Artery
External Iliac Vein
Femoral Nerve
Pectineus Muscle
Pubis
Iliopsoas Muscle
Rectus Femoris Muscle
Tensor Fascia Lata Muscle
Obturator Vessel and Nerve
Iliofemoral Ligament
Gluteus Medius Muscle
Femur, Neck
Iliotibial Tract
Acetabular Fossa
Femur, Greater Trochanter
Superficial Fascia and Adipose Tissue
Piriformis Tendon
Superior Gemellus Muscle
Ischium
Obturator Internus Muscle
Sciatic Nerve
Prostate Gland
Seminal Vesicle
Levator Ani Muscle
Ischiorectal Fossa
Rectum
Coccyx
Third Distal Phalanx
Third Middle Phalanx
Ductus Deferens
Fourth Middle Phalanx
Fourth Proximal Phalanx
Fifth Middle Phalanx
Fifth Proximal Phalanx
External Oblique Muscle, Aponeurosis
External Iliac Vein
External Iliac Artery
Femoral Nerve
Sartorius Muscle
Iliopsoas Muscle
Rectus Femoris Muscle
Tensor Fascia Lata Muscle
Pubis
Ilium
Gluteus Medius Muscle
Iliofemoral Ligament
Obturator Vessel and Nerve
Femur, Head
Femur, Greater Trochanter
Iliotibial Tract
Acetabulum
Superficial Fascia and Adipose Tissue
Inferior Gemellus Muscle
Gluteus Maximus Muscle
Ischium
Obturator Internus Tendon
Sciatic Nerve
Superior Gemellus Muscle
Prostate Gland
Ductus Deferens, Ampulla
Coccygeus Muscle
right
left
posterior

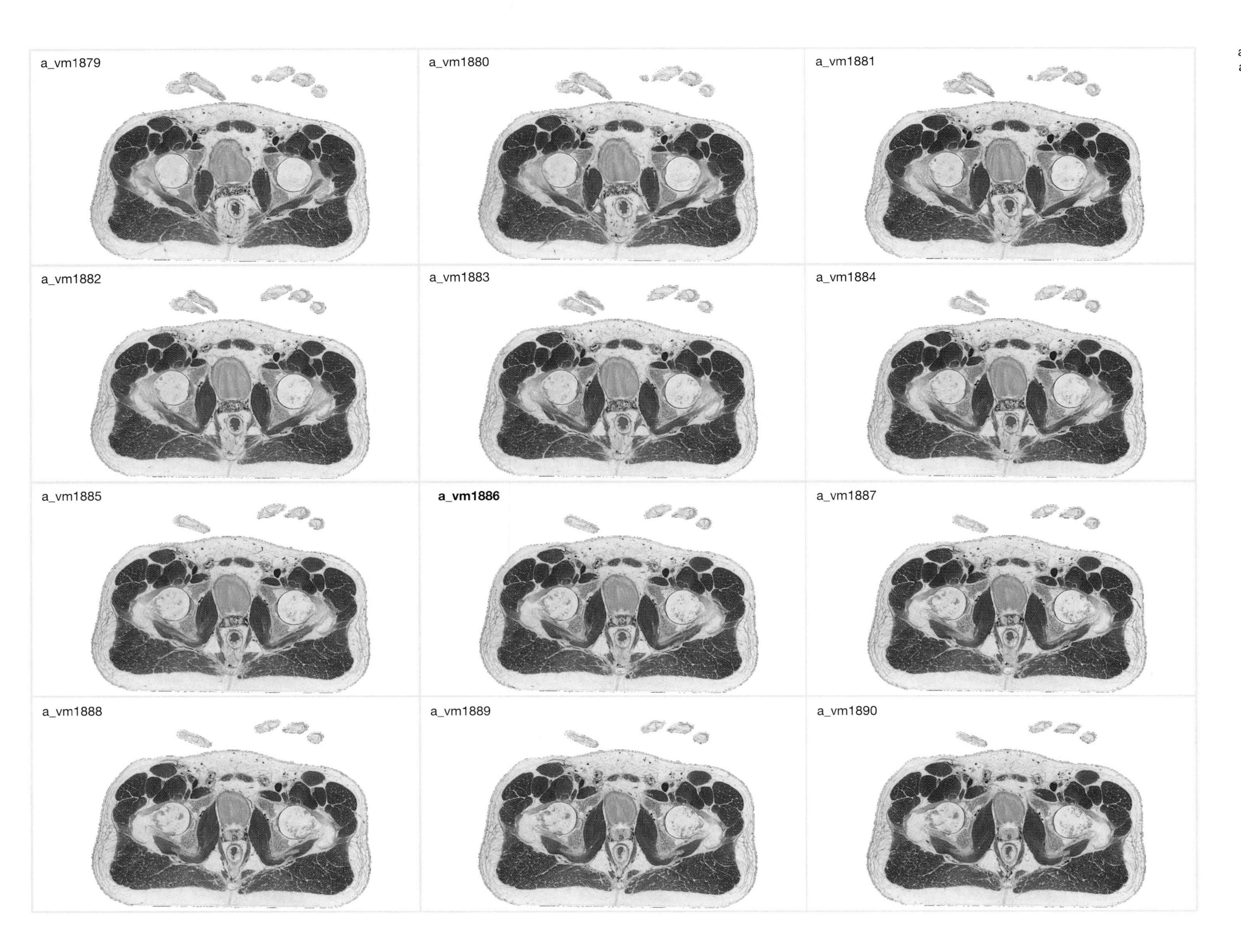
a_vm1879
a_vm1880
a_vm1881
a_vm1882
a_vm1883
a_vm1884
a_vm1885
a_vm1886
a_vm1887
a_vm1888
a_vm1889
a_vm1890

Transverse
a_vm1897

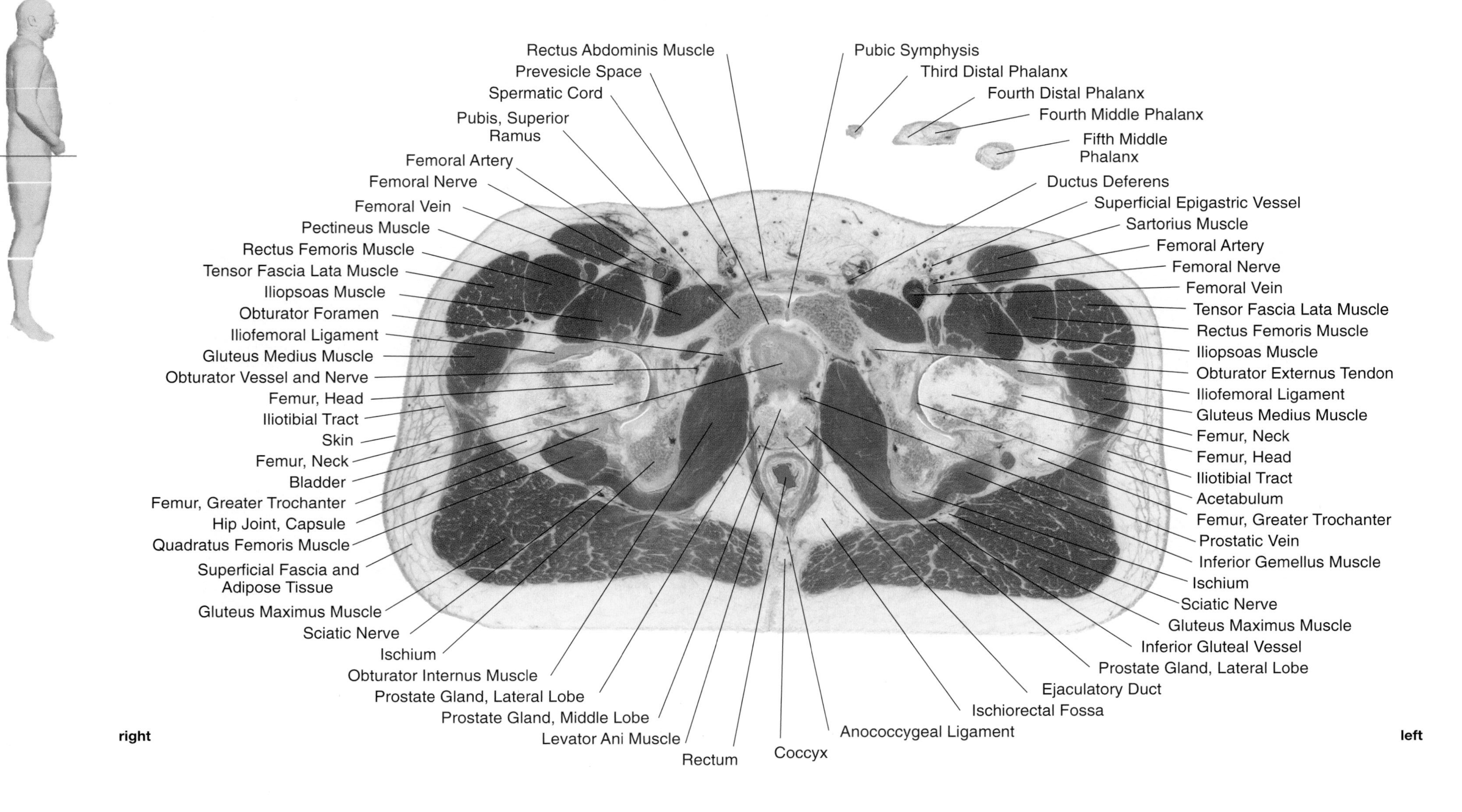
anterior
Rectus Abdominis Muscle
Prevesicle Space
Spermatic Cord
Pubis, Superior Ramus
Femoral Artery
Femoral Nerve
Femoral Vein
Pectineus Muscle
Rectus Femoris Muscle
Tensor Fascia Lata Muscle
Iliopsoas Muscle
Obturator Foramen
Iliofemoral Ligament
Gluteus Medius Muscle
Obturator Vessel and Nerve
Femur, Head
Iliotibial Tract
Skin
Femur, Neck
Bladder
Femur, Greater Trochanter
Hip Joint, Capsule
Quadratus Femoris Muscle
Superficial Fascia and Adipose Tissue
Gluteus Maximus Muscle
Sciatic Nerve
Ischium
Obturator Internus Muscle
Prostate Gland, Lateral Lobe
Prostate Gland, Middle Lobe
Levator Ani Muscle
Rectum
Coccyx
Pubic Symphysis
Third Distal Phalanx
Fourth Distal Phalanx
Fourth Middle Phalanx
Fifth Middle Phalanx
Ductus Deferens
Superficial Epigastric Vessel
Sartorius Muscle
Femoral Artery
Femoral Nerve
Femoral Vein
Tensor Fascia Lata Muscle
Rectus Femoris Muscle
Iliopsoas Muscle
Obturator Externus Tendon
Iliofemoral Ligament
Gluteus Medius Muscle
Femur, Neck
Femur, Head
Iliotibial Tract
Acetabulum
Femur, Greater Trochanter
Prostatic Vein
Inferior Gemellus Muscle
Ischium
Sciatic Nerve
Gluteus Maximus Muscle
Inferior Gluteal Vessel
Prostate Gland, Lateral Lobe
Ejaculatory Duct
Ischiorectal Fossa
Anococcygeal Ligament
right
left
posterior

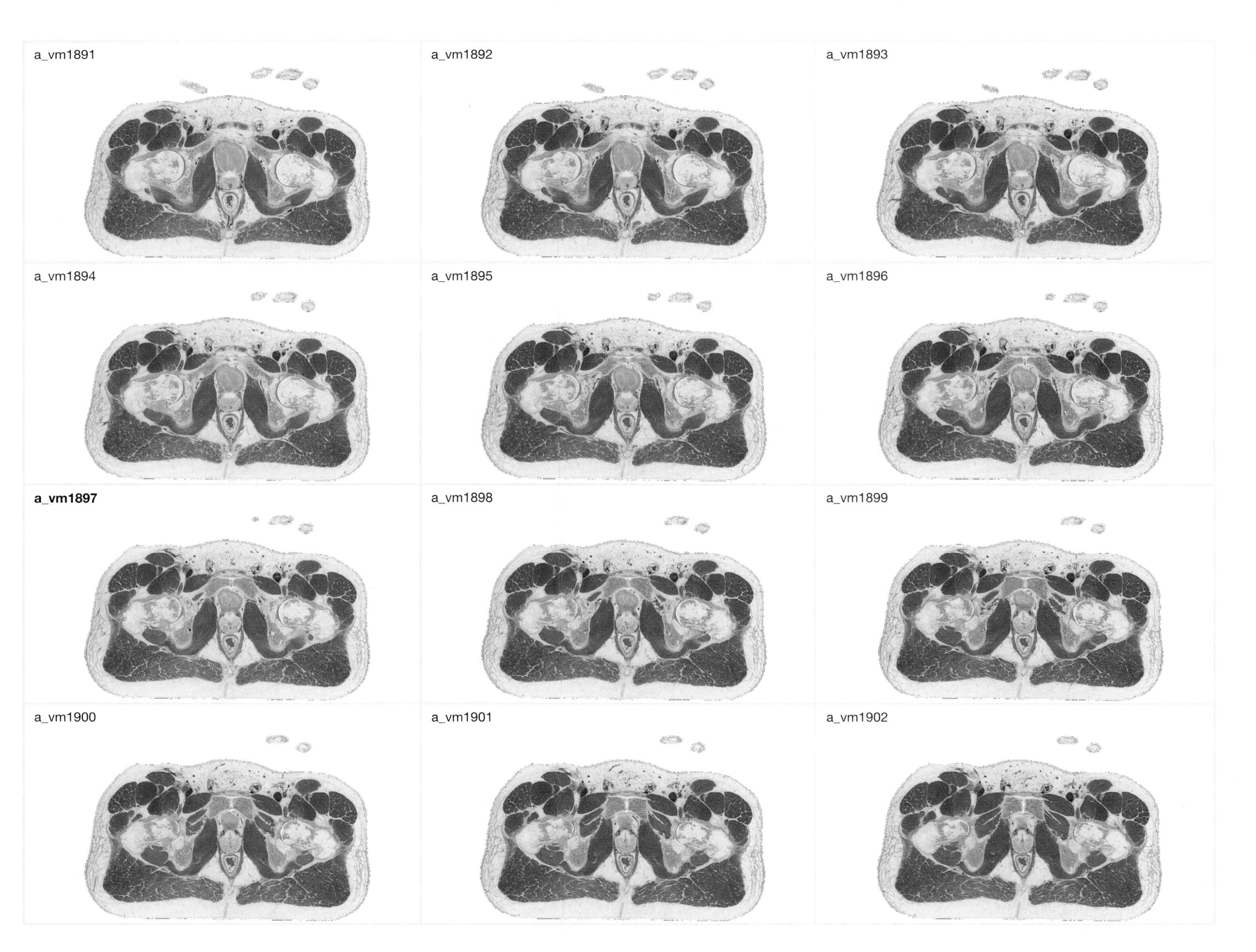
a_vm1891
a_vm1892
a_vm1893
a_vm1894
a_vm1895
a_vm1896
a_vm1897
a_vm1898
a_vm1899
a_vm1900
a_vm1901
a_vm1902

Transverse
a_vm1908

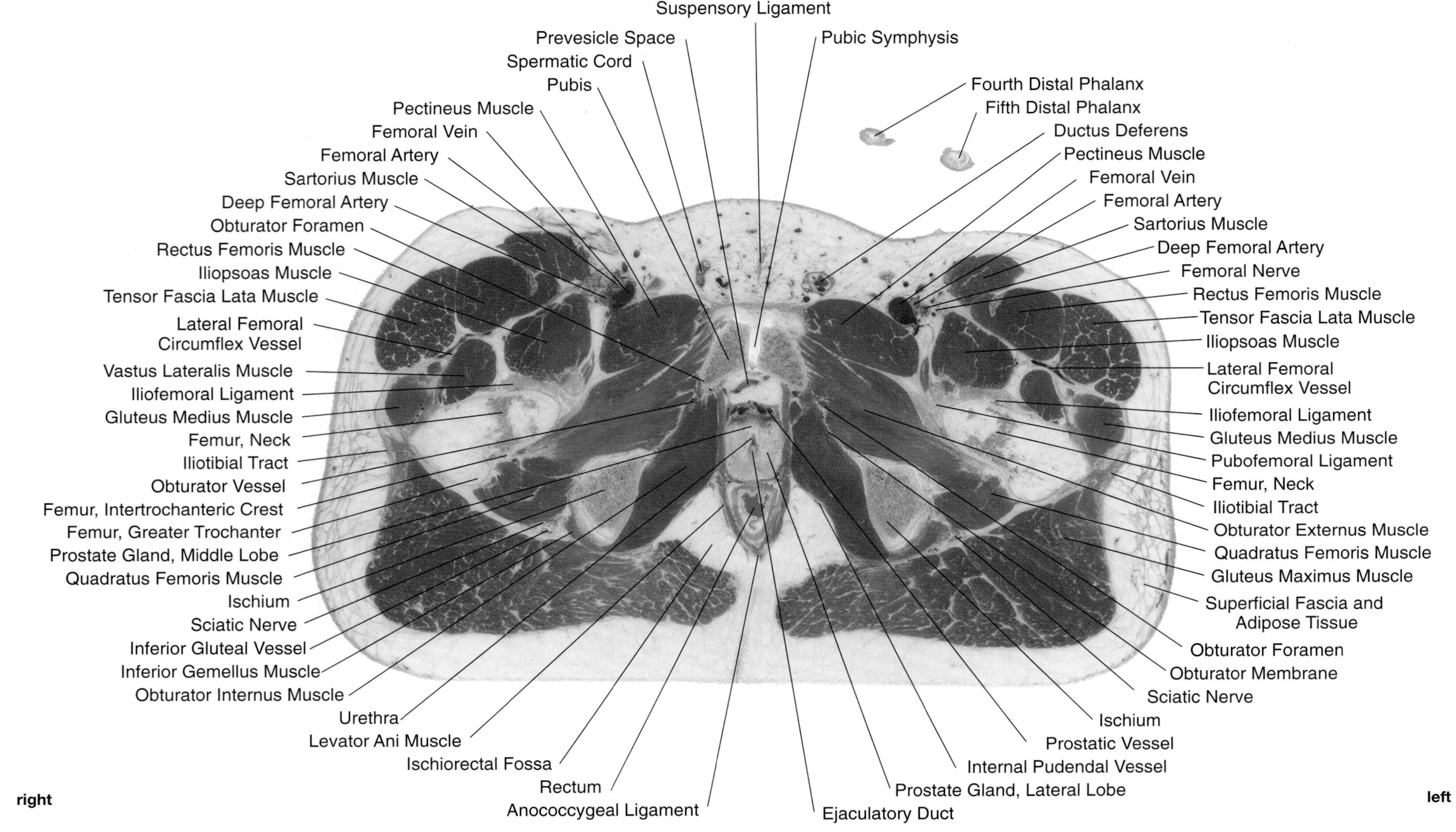

anterior
posterior
right
left
Suspensory Ligament
Prevesicle Space
Spermatic Cord
Pubis
Pubic Symphysis
Pectineus Muscle
Femoral Vein
Femoral Artery
Sartorius Muscle
Deep Femoral Artery
Obturator Foramen
Rectus Femoris Muscle
Iliopsoas Muscle
Tensor Fascia Lata Muscle
Lateral Femoral Circumflex Vessel
Vastus Lateralis Muscle
Iliofemoral Ligament
Gluteus Medius Muscle
Femur, Neck
Iliotibial Tract
Obturator Vessel
Femur, Intertrochanteric Crest
Femur, Greater Trochanter
Prostate Gland, Middle Lobe
Quadratus Femoris Muscle
Ischium
Sciatic Nerve
Inferior Gluteal Vessel
Inferior Gemellus Muscle
Obturator Internus Muscle
Urethra
Levator Ani Muscle
Ischiorectal Fossa
Rectum
Anococcygeal Ligament
Fourth Distal Phalanx
Fifth Distal Phalanx
Ductus Deferens
Pectineus Muscle
Femoral Vein
Femoral Artery
Sartorius Muscle
Deep Femoral Artery
Femoral Nerve
Rectus Femoris Muscle
Tensor Fascia Lata Muscle
Iliopsoas Muscle
Lateral Femoral Circumflex Vessel
Iliofemoral Ligament
Gluteus Medius Muscle
Pubofemoral Ligament
Femur, Neck
Iliotibial Tract
Obturator Externus Muscle
Quadratus Femoris Muscle
Gluteus Maximus Muscle
Superficial Fascia and Adipose Tissue
Obturator Foramen
Obturator Membrane
Sciatic Nerve
Ischium
Prostatic Vessel
Internal Pudendal Vessel
Prostate Gland, Lateral Lobe
Ejaculatory Duct

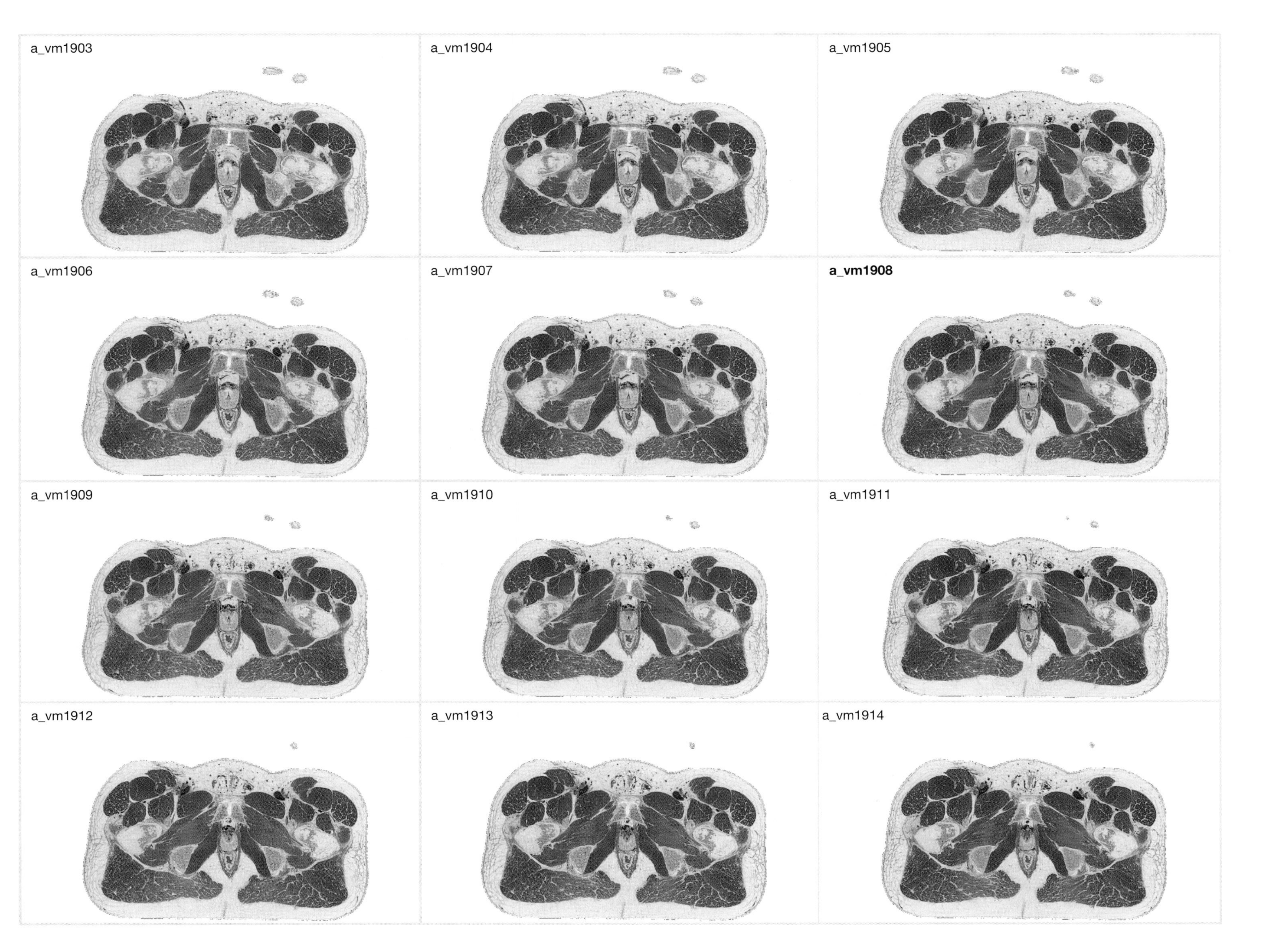
a_vm1903
a_vm1904
a_vm1905
a_vm1906
a_vm1907
a_vm1908
a_vm1909
a_vm1910
a_vm1911
a_vm1912
a_vm1913
a_vm1914

Transverse
a_vm1922

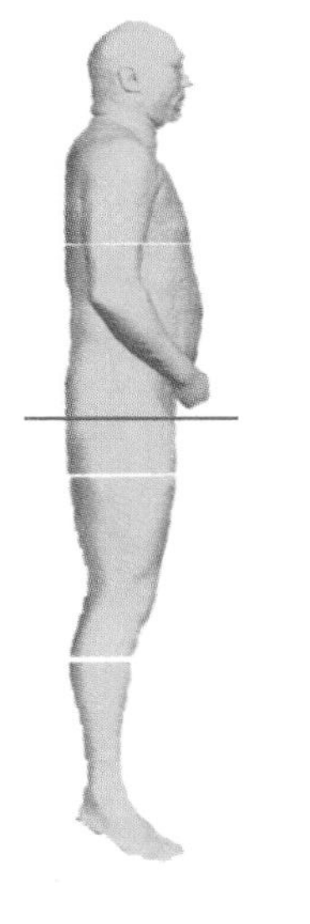

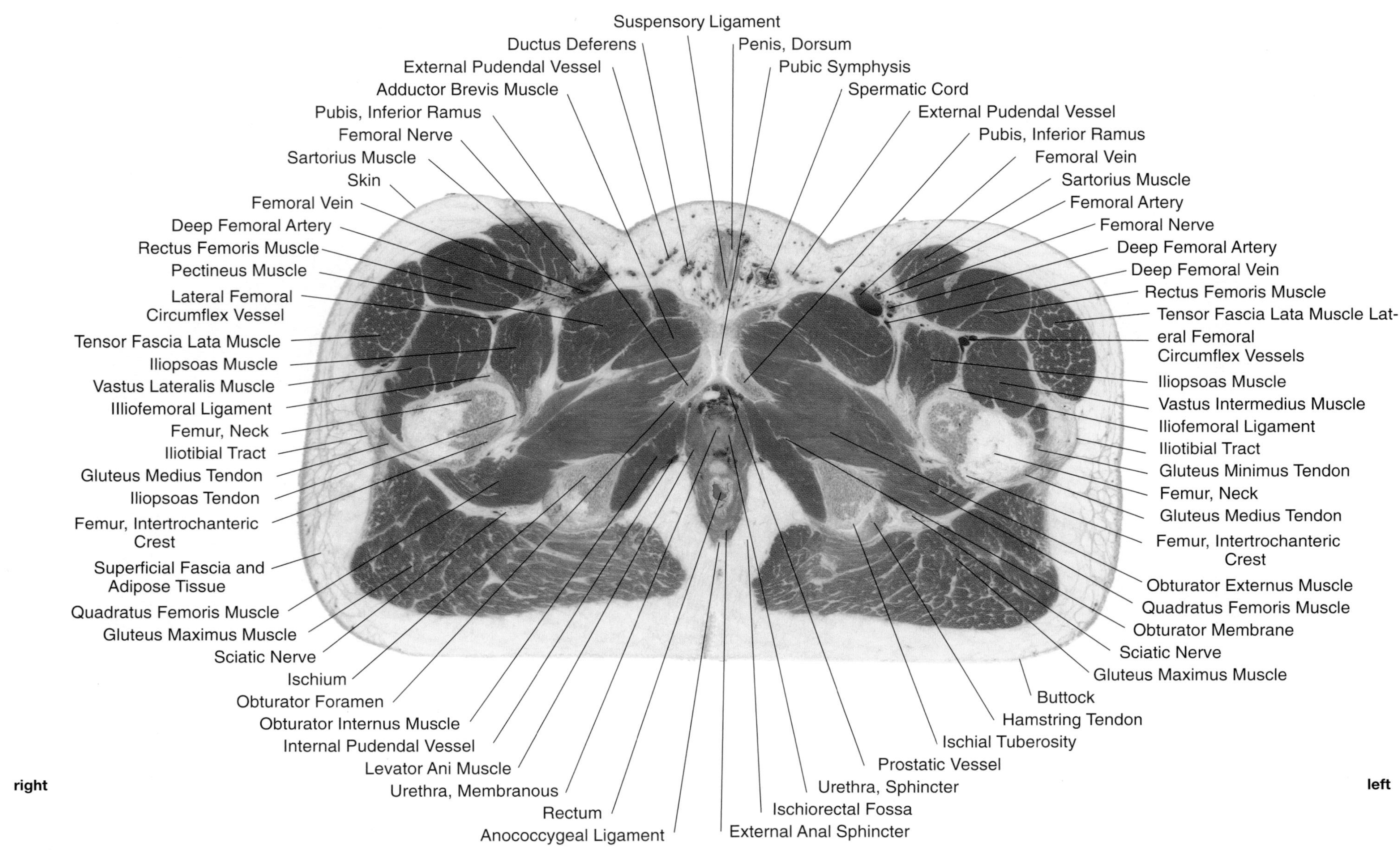

anterior
Suspensory Ligament
Ductus Deferens
Penis, Dorsum
External Pudendal Vessel
Pubic Symphysis
Adductor Brevis Muscle
Spermatic Cord
Pubis, Inferior Ramus
External Pudendal Vessel
Femoral Nerve
Pubis, Inferior Ramus
Sartorius Muscle
Femoral Vein
Skin
Sartorius Muscle
Femoral Vein
Femoral Artery
Deep Femoral Artery
Femoral Nerve
Rectus Femoris Muscle
Deep Femoral Artery
Pectineus Muscle
Deep Femoral Vein
Lateral Femoral Circumflex Vessel
Rectus Femoris Muscle
Tensor Fascia Lata Muscle
Tensor Fascia Lata Muscle Lat-eral Femoral Circumflex Vessels
Iliopsoas Muscle
Iliopsoas Muscle
Vastus Lateralis Muscle
Vastus Intermedius Muscle
Illiofemoral Ligament
Iliofemoral Ligament
Femur, Neck
Iliotibial Tract
Iliotibial Tract
Gluteus Minimus Tendon
Gluteus Medius Tendon
Femur, Neck
Iliopsoas Tendon
Gluteus Medius Tendon
Femur, Intertrochanteric Crest
Femur, Intertrochanteric Crest
Superficial Fascia and Adipose Tissue
Obturator Externus Muscle
Quadratus Femoris Muscle
Quadratus Femoris Muscle
Gluteus Maximus Muscle
Obturator Membrane
Sciatic Nerve
Sciatic Nerve
Ischium
Gluteus Maximus Muscle
Obturator Foramen
Buttock
Obturator Internus Muscle
Hamstring Tendon
Internal Pudendal Vessel
Ischial Tuberosity
Levator Ani Muscle
Prostatic Vessel
Urethra, Membranous
Urethra, Sphincter
Rectum
Ischiorectal Fossa
Anococcygeal Ligament
External Anal Sphincter
right
left
posterior

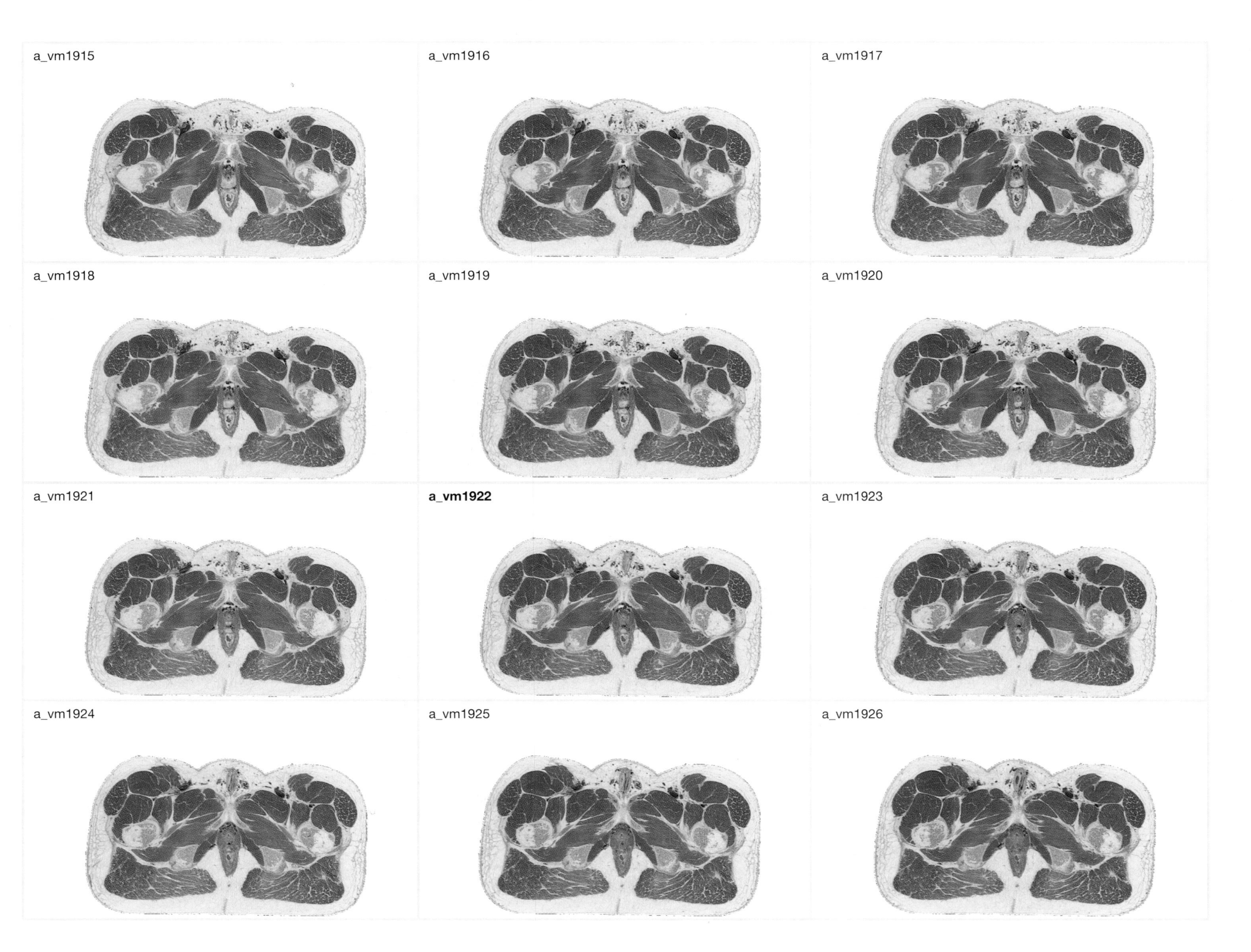
a_vm1915
a_vm1916
a_vm1917
a_vm1918
a_vm1919
a_vm1920
a_vm1921
a_vm1922
a_vm1923
a_vm1924
a_vm1925
a_vm1926

Transverse
a_vm1933

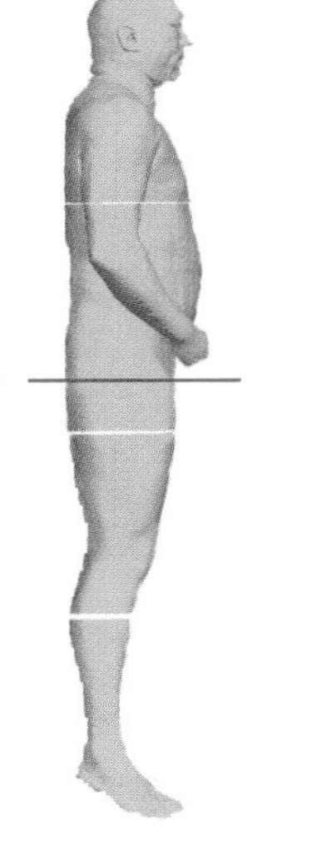

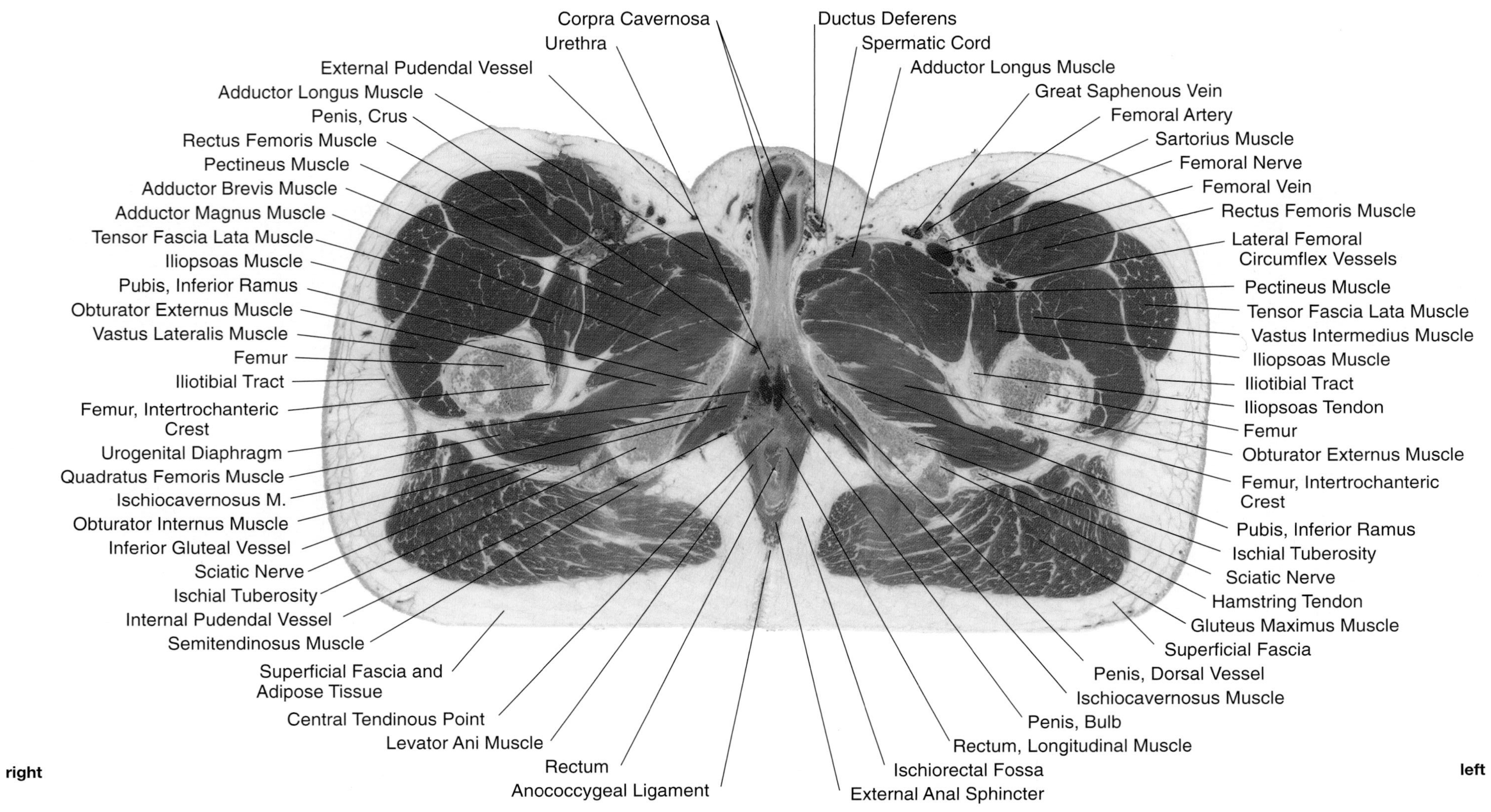

anterior
posterior
right
left
Corpra Cavernosa
Urethra
External Pudendal Vessel
Adductor Longus Muscle
Penis, Crus
Rectus Femoris Muscle
Pectineus Muscle
Adductor Brevis Muscle
Adductor Magnus Muscle
Tensor Fascia Lata Muscle
Iliopsoas Muscle
Pubis, Inferior Ramus
Obturator Externus Muscle
Vastus Lateralis Muscle
Femur
Iliotibial Tract
Femur, Intertrochanteric Crest
Urogenital Diaphragm
Quadratus Femoris Muscle
Ischiocavernosus M.
Obturator Internus Muscle
Inferior Gluteal Vessel
Sciatic Nerve
Ischial Tuberosity
Internal Pudendal Vessel
Semitendinosus Muscle
Superficial Fascia and Adipose Tissue
Central Tendinous Point
Levator Ani Muscle
Rectum
Anococcygeal Ligament
Ductus Deferens
Spermatic Cord
Adductor Longus Muscle
Great Saphenous Vein
Femoral Artery
Sartorius Muscle
Femoral Nerve
Femoral Vein
Rectus Femoris Muscle
Lateral Femoral Circumflex Vessels
Pectineus Muscle
Tensor Fascia Lata Muscle
Vastus Intermedius Muscle
Iliopsoas Muscle
Iliotibial Tract
Iliopsoas Tendon
Femur
Obturator Externus Muscle
Femur, Intertrochanteric Crest
Pubis, Inferior Ramus
Ischial Tuberosity
Sciatic Nerve
Hamstring Tendon
Gluteus Maximus Muscle
Superficial Fascia
Penis, Dorsal Vessel
Ischiocavernosus Muscle
Penis, Bulb
Rectum, Longitudinal Muscle
Ischiorectal Fossa
External Anal Sphincter

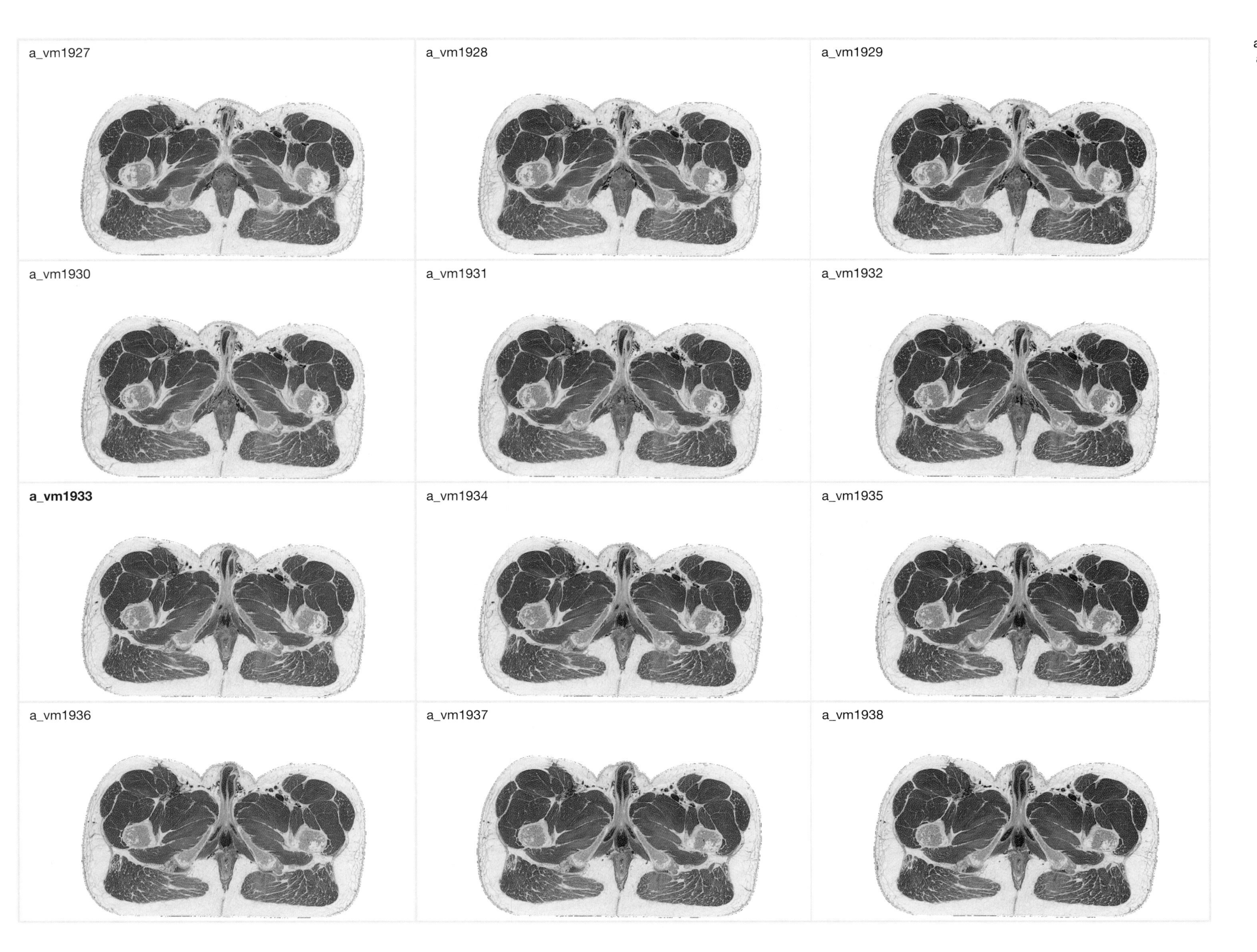
a_vm1927
a_vm1928
a_vm1929
a_vm1930
a_vm1931
a_vm1932
a_vm1933
a_vm1934
a_vm1935
a_vm1936
a_vm1937
a_vm1938

Transverse
a_vm1940

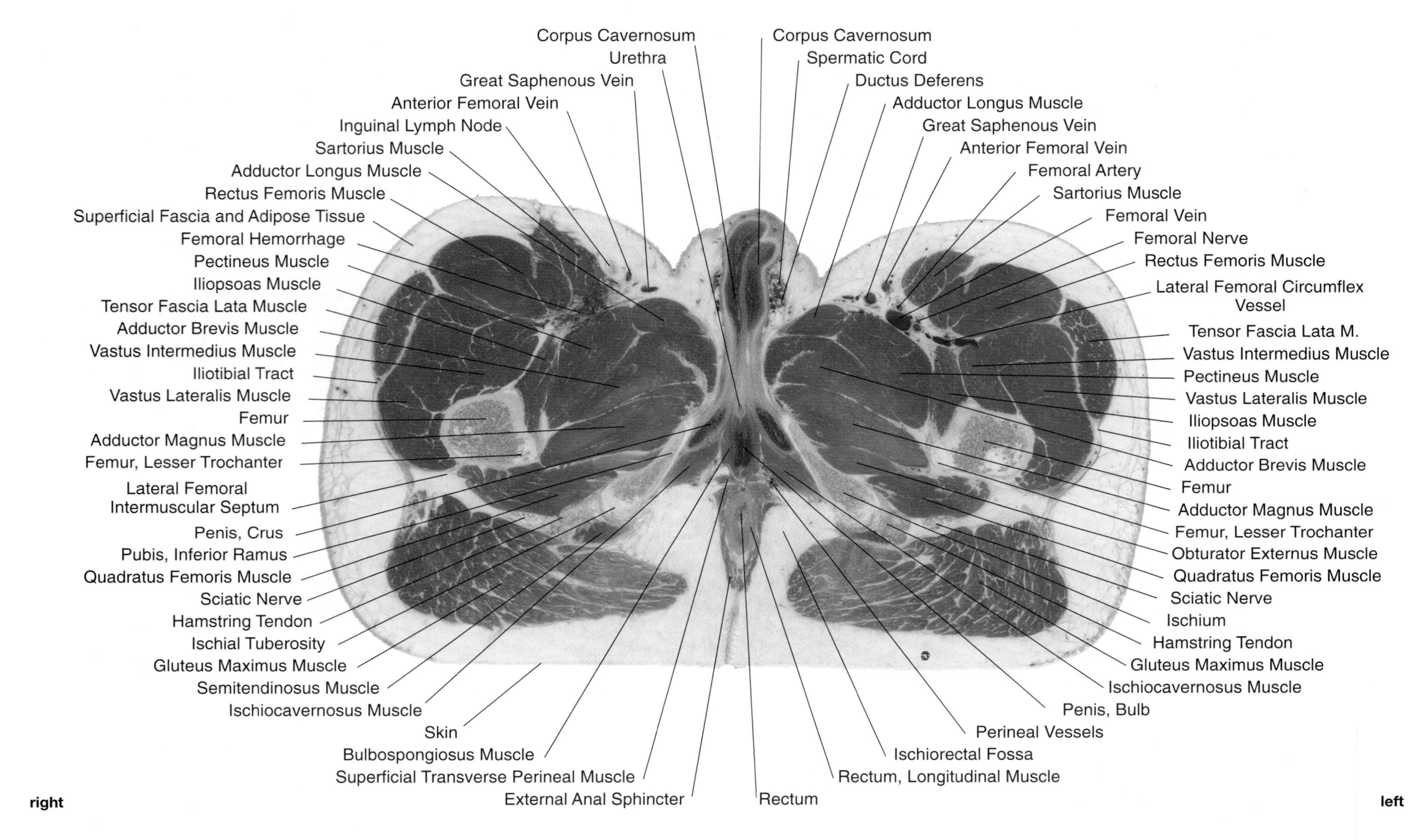

anterior
Corpus Cavernosum
Urethra
Great Saphenous Vein
Anterior Femoral Vein
Inguinal Lymph Node
Sartorius Muscle
Adductor Longus Muscle
Rectus Femoris Muscle
Superficial Fascia and Adipose Tissue
Femoral Hemorrhage
Pectineus Muscle
Iliopsoas Muscle
Tensor Fascia Lata Muscle
Adductor Brevis Muscle
Vastus Intermedius Muscle
Iliotibial Tract
Vastus Lateralis Muscle
Femur
Adductor Magnus Muscle
Femur, Lesser Trochanter
Lateral Femoral Intermuscular Septum
Penis, Crus
Pubis, Inferior Ramus
Quadratus Femoris Muscle
Sciatic Nerve
Hamstring Tendon
Ischial Tuberosity
Gluteus Maximus Muscle
Semitendinosus Muscle
Ischiocavernosus Muscle
Skin
Bulbospongiosus Muscle
Superficial Transverse Perineal Muscle
External Anal Sphincter
Corpus Cavernosum
Spermatic Cord
Ductus Deferens
Adductor Longus Muscle
Great Saphenous Vein
Anterior Femoral Vein
Femoral Artery
Sartorius Muscle
Femoral Vein
Femoral Nerve
Rectus Femoris Muscle
Lateral Femoral Circumflex Vessel
Tensor Fascia Lata M.
Vastus Intermedius Muscle
Pectineus Muscle
Vastus Lateralis Muscle
Iliopsoas Muscle
Iliotibial Tract
Adductor Brevis Muscle
Femur
Adductor Magnus Muscle
Femur, Lesser Trochanter
Obturator Externus Muscle
Quadratus Femoris Muscle
Sciatic Nerve
Ischium
Hamstring Tendon
Gluteus Maximus Muscle
Ischiocavernosus Muscle
Penis, Bulb
Perineal Vessels
Ischiorectal Fossa
Rectum, Longitudinal Muscle
Rectum
right
left
posterior

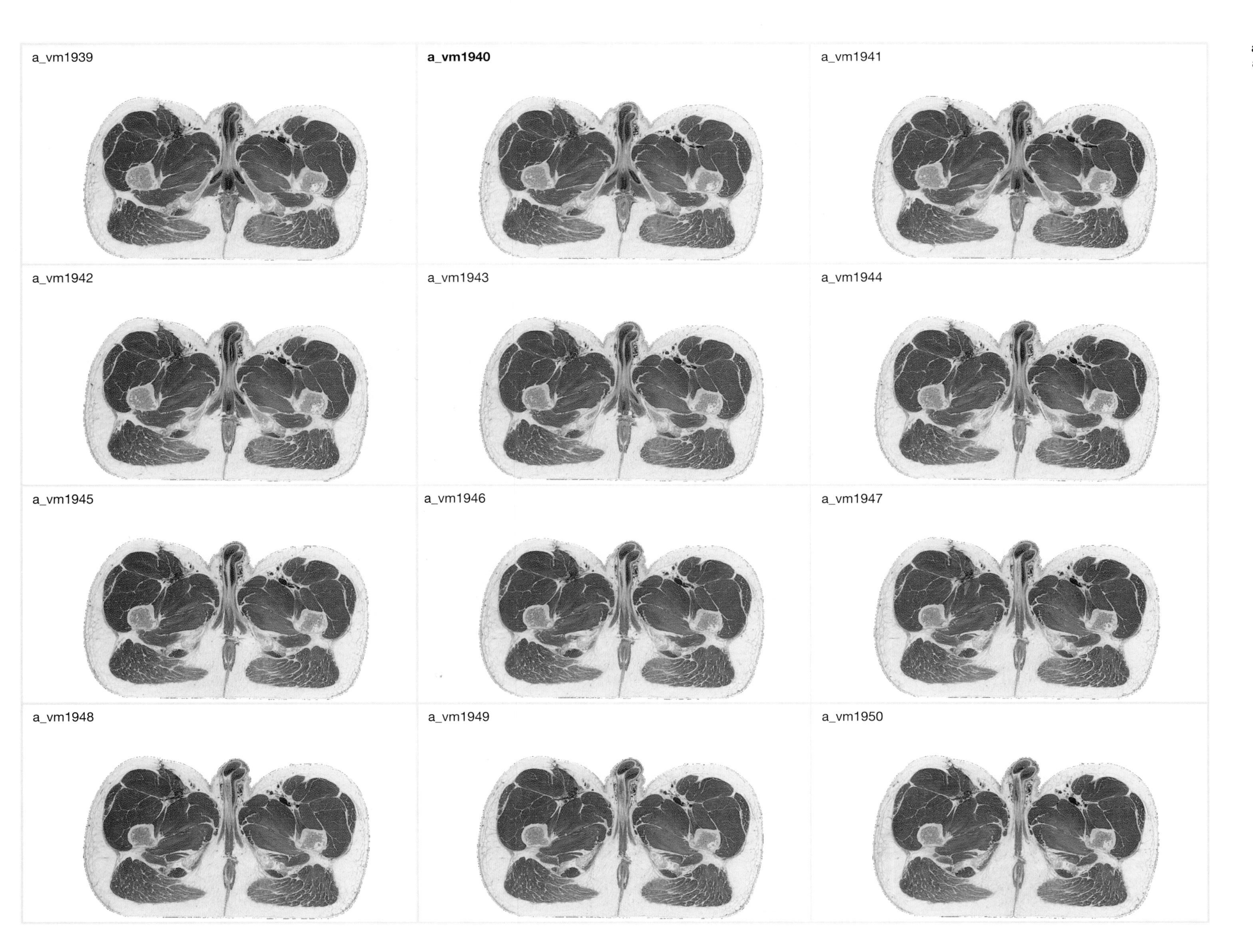
a_vm1939
a_vm1940
a_vm1941
a_vm1942
a_vm1943
a_vm1944
a_vm1945
a_vm1946
a_vm1947
a_vm1948
a_vm1949
a_vm1950

Transverse
a_vm1951

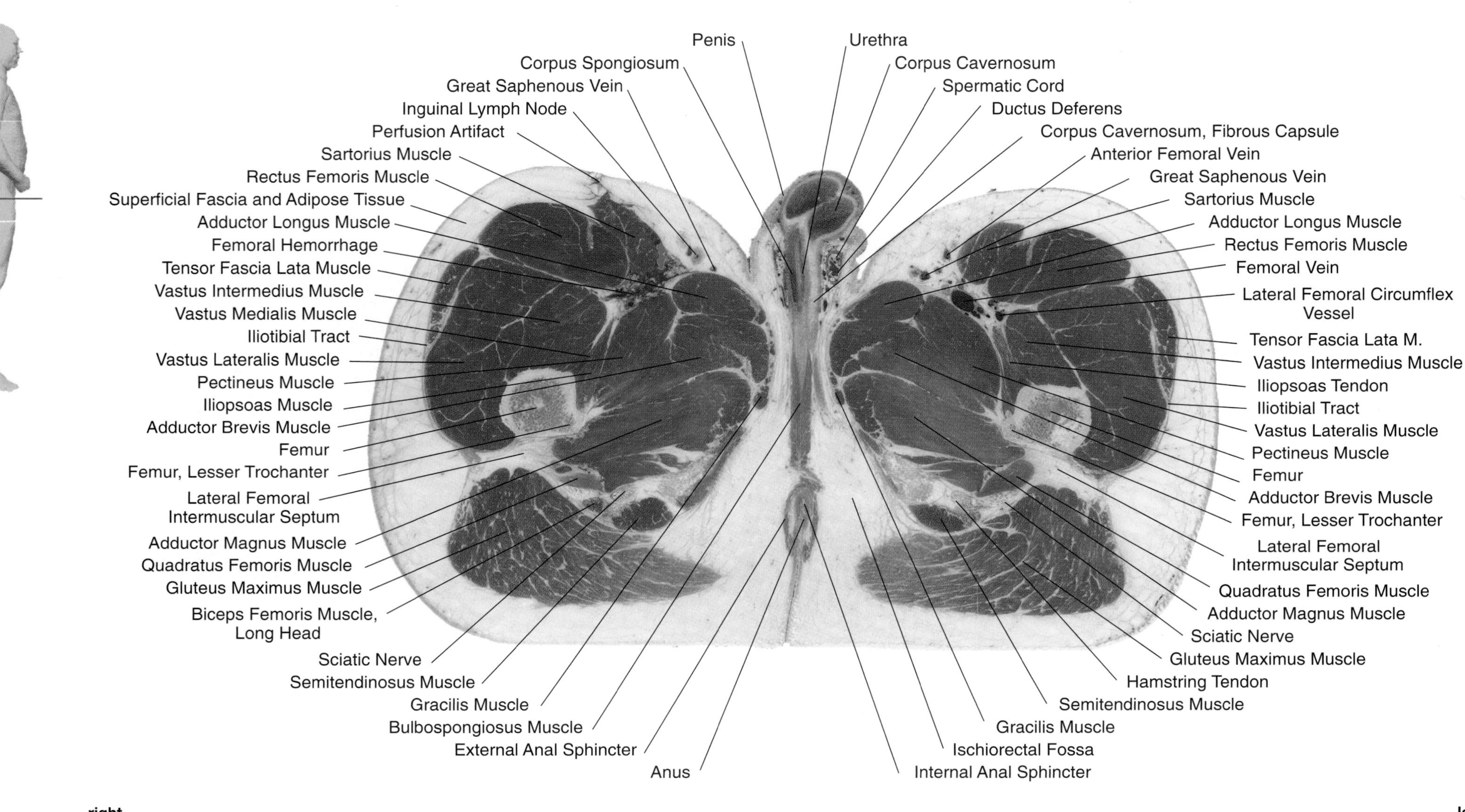
anterior
Penis
Corpus Spongiosum
Great Saphenous Vein
Inguinal Lymph Node
Perfusion Artifact
Sartorius Muscle
Rectus Femoris Muscle
Superficial Fascia and Adipose Tissue
Adductor Longus Muscle
Femoral Hemorrhage
Tensor Fascia Lata Muscle
Vastus Intermedius Muscle
Vastus Medialis Muscle
Iliotibial Tract
Vastus Lateralis Muscle
Pectineus Muscle
Iliopsoas Muscle
Adductor Brevis Muscle
Femur
Femur, Lesser Trochanter
Lateral Femoral Intermuscular Septum
Adductor Magnus Muscle
Quadratus Femoris Muscle
Gluteus Maximus Muscle
Biceps Femoris Muscle, Long Head
Sciatic Nerve
Semitendinosus Muscle
Gracilis Muscle
Bulbospongiosus Muscle
External Anal Sphincter
Anus
Urethra
Corpus Cavernosum
Spermatic Cord
Ductus Deferens
Corpus Cavernosum, Fibrous Capsule
Anterior Femoral Vein
Great Saphenous Vein
Sartorius Muscle
Adductor Longus Muscle
Rectus Femoris Muscle
Femoral Vein
Lateral Femoral Circumflex Vessel
Tensor Fascia Lata M.
Vastus Intermedius Muscle
Iliopsoas Tendon
Iliotibial Tract
Vastus Lateralis Muscle
Pectineus Muscle
Femur
Adductor Brevis Muscle
Femur, Lesser Trochanter
Lateral Femoral Intermuscular Septum
Quadratus Femoris Muscle
Adductor Magnus Muscle
Sciatic Nerve
Gluteus Maximus Muscle
Hamstring Tendon
Semitendinosus Muscle
Gracilis Muscle
Ischiorectal Fossa
Internal Anal Sphincter
right
left
posterior

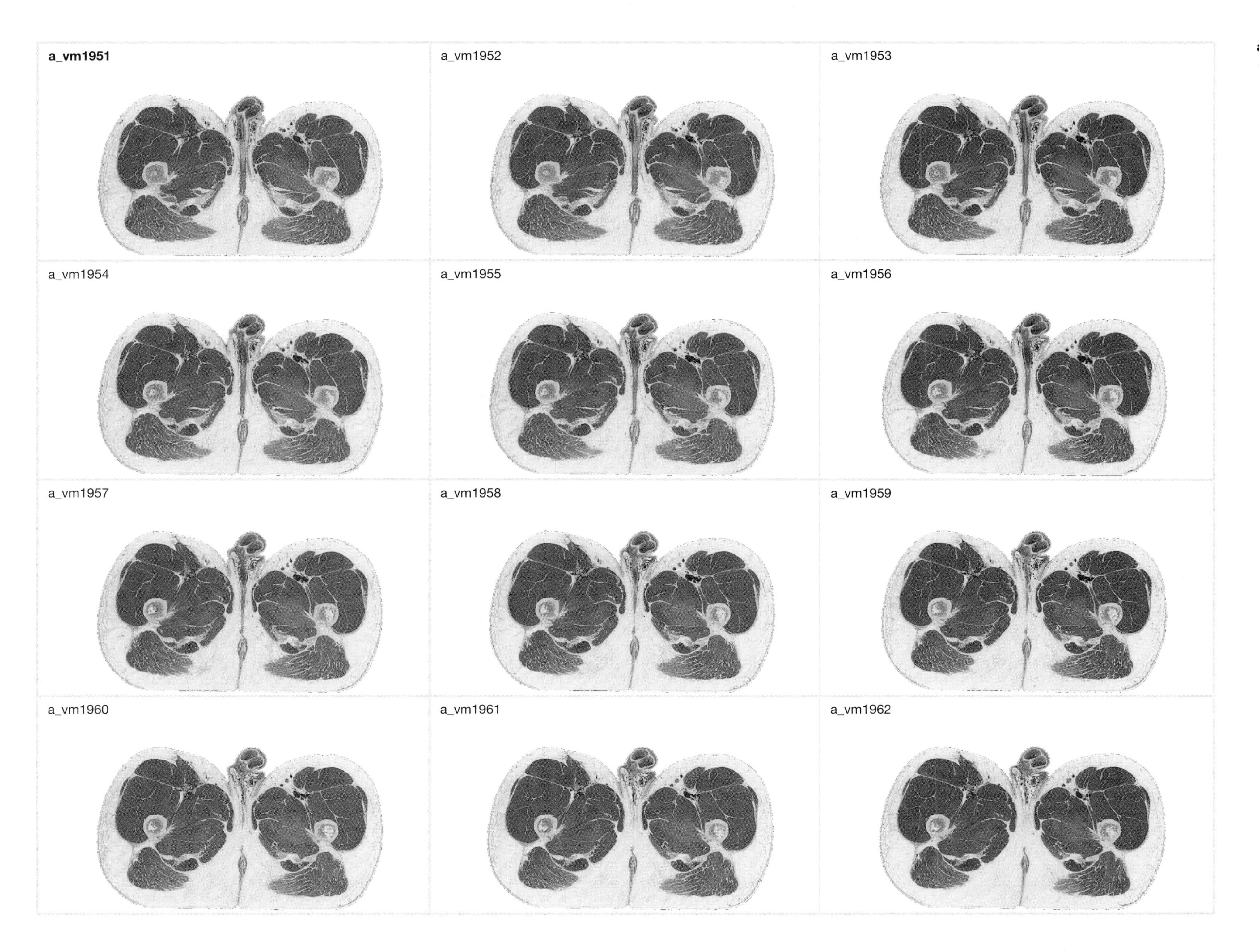
a_vm1951
a_vm1952
a_vm1953
a_vm1954
a_vm1955
a_vm1956
a_vm1957
a_vm1958
a_vm1959
a_vm1960
a_vm1961
a_vm1962

Transverse
a_vm1964

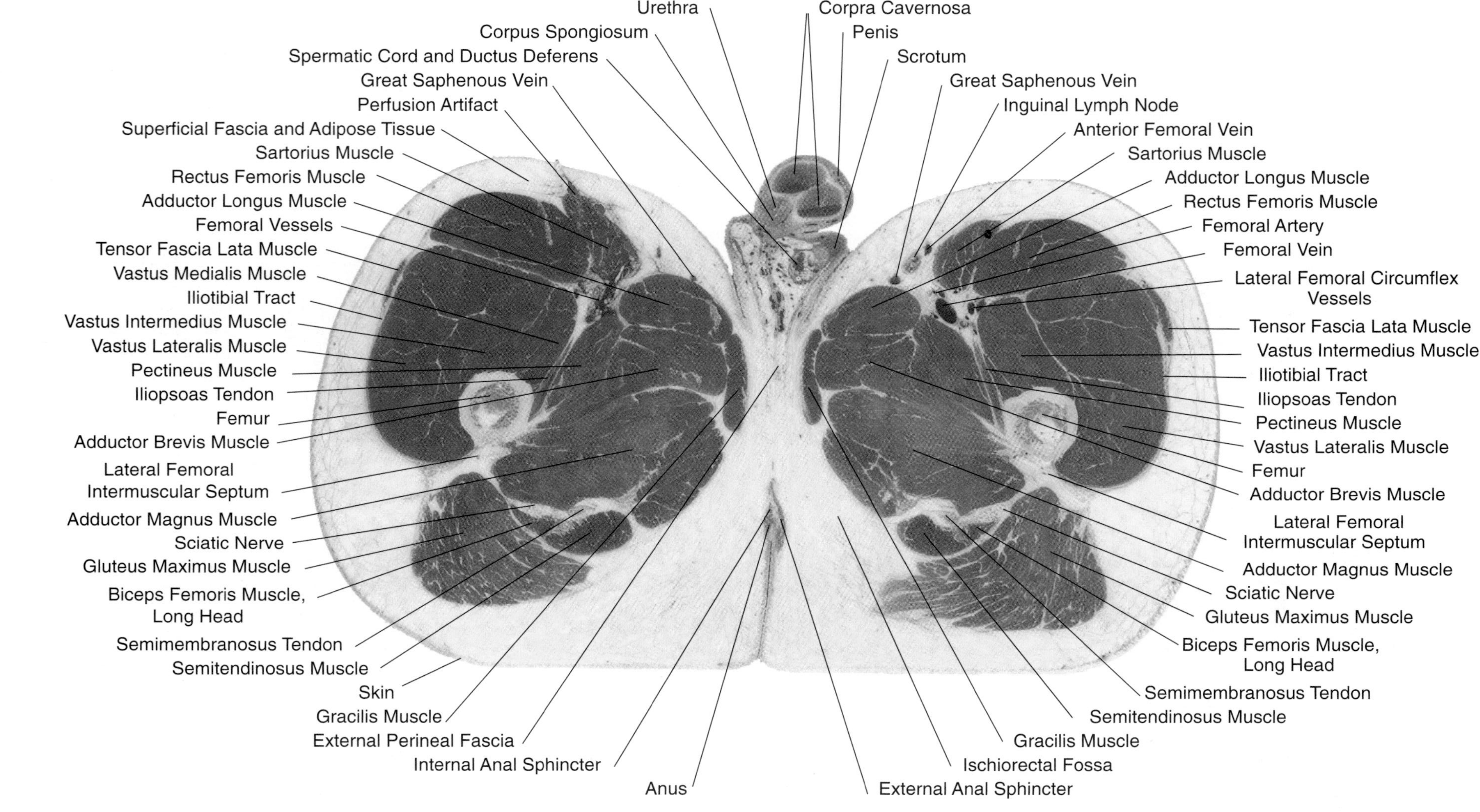
anterior
Urethra
Corpus Spongiosum
Spermatic Cord and Ductus Deferens
Great Saphenous Vein
Perfusion Artifact
Superficial Fascia and Adipose Tissue
Sartorius Muscle
Rectus Femoris Muscle
Adductor Longus Muscle
Femoral Vessels
Tensor Fascia Lata Muscle
Vastus Medialis Muscle
Iliotibial Tract
Vastus Intermedius Muscle
Vastus Lateralis Muscle
Pectineus Muscle
Iliopsoas Tendon
Femur
Adductor Brevis Muscle
Lateral Femoral Intermuscular Septum
Adductor Magnus Muscle
Sciatic Nerve
Gluteus Maximus Muscle
Biceps Femoris Muscle, Long Head
Semimembranosus Tendon
Semitendinosus Muscle
Skin
Gracilis Muscle
External Perineal Fascia
Internal Anal Sphincter
Anus
Corpra Cavernosa
Penis
Scrotum
Great Saphenous Vein
Inguinal Lymph Node
Anterior Femoral Vein
Sartorius Muscle
Adductor Longus Muscle
Rectus Femoris Muscle
Femoral Artery
Femoral Vein
Lateral Femoral Circumflex Vessels
Tensor Fascia Lata Muscle
Vastus Intermedius Muscle
Iliotibial Tract
Iliopsoas Tendon
Pectineus Muscle
Vastus Lateralis Muscle
Femur
Adductor Brevis Muscle
Lateral Femoral Intermuscular Septum
Adductor Magnus Muscle
Sciatic Nerve
Gluteus Maximus Muscle
Biceps Femoris Muscle, Long Head
Semimembranosus Tendon
Semitendinosus Muscle
Gracilis Muscle
Ischiorectal Fossa
External Anal Sphincter
right
left
posterior

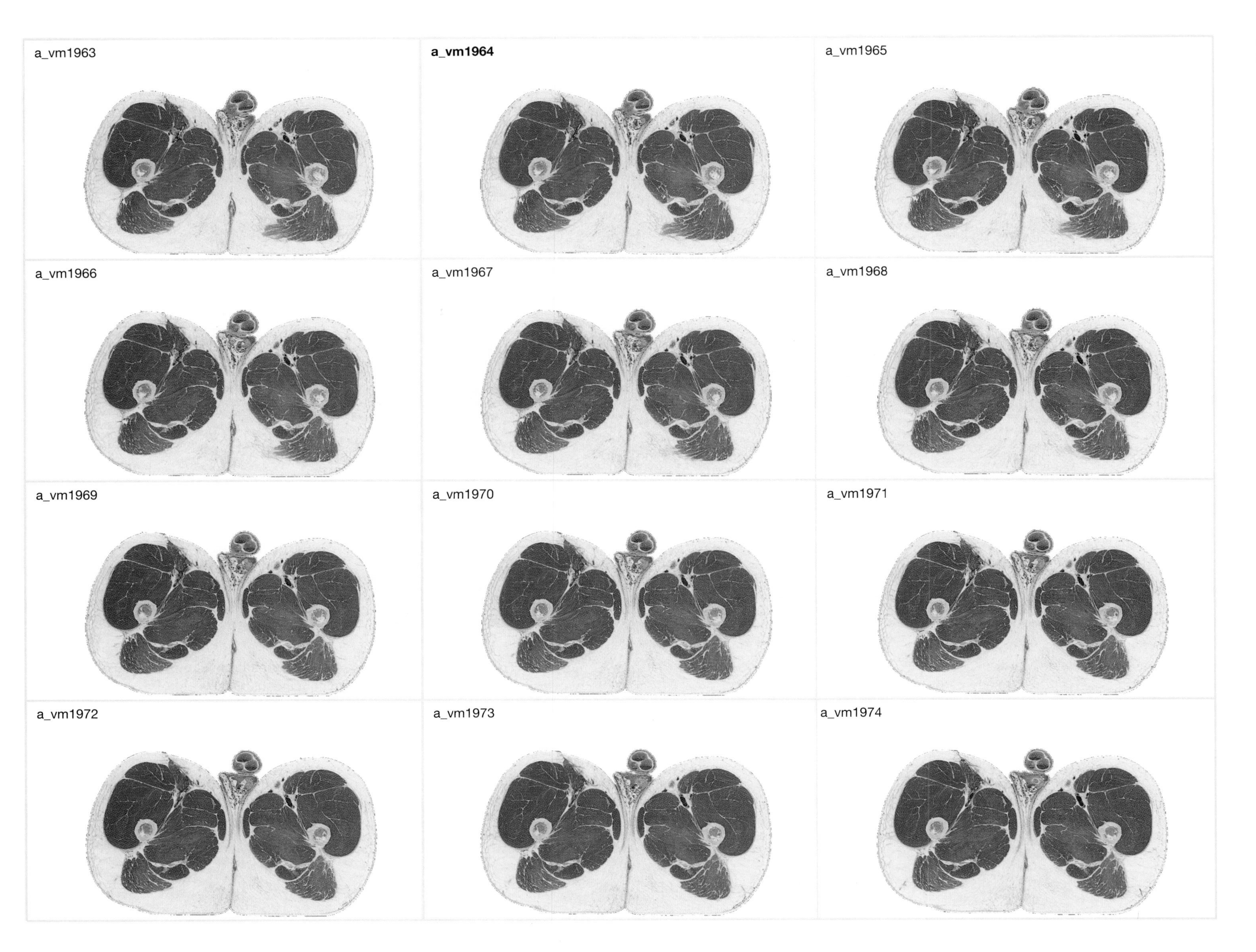
a_vm1963
a_vm1964
a_vm1965
a_vm1966
a_vm1967
a_vm1968
a_vm1969
a_vm1970
a_vm1971
a_vm1972
a_vm1973
a_vm1974

Transverse
a_vm1980

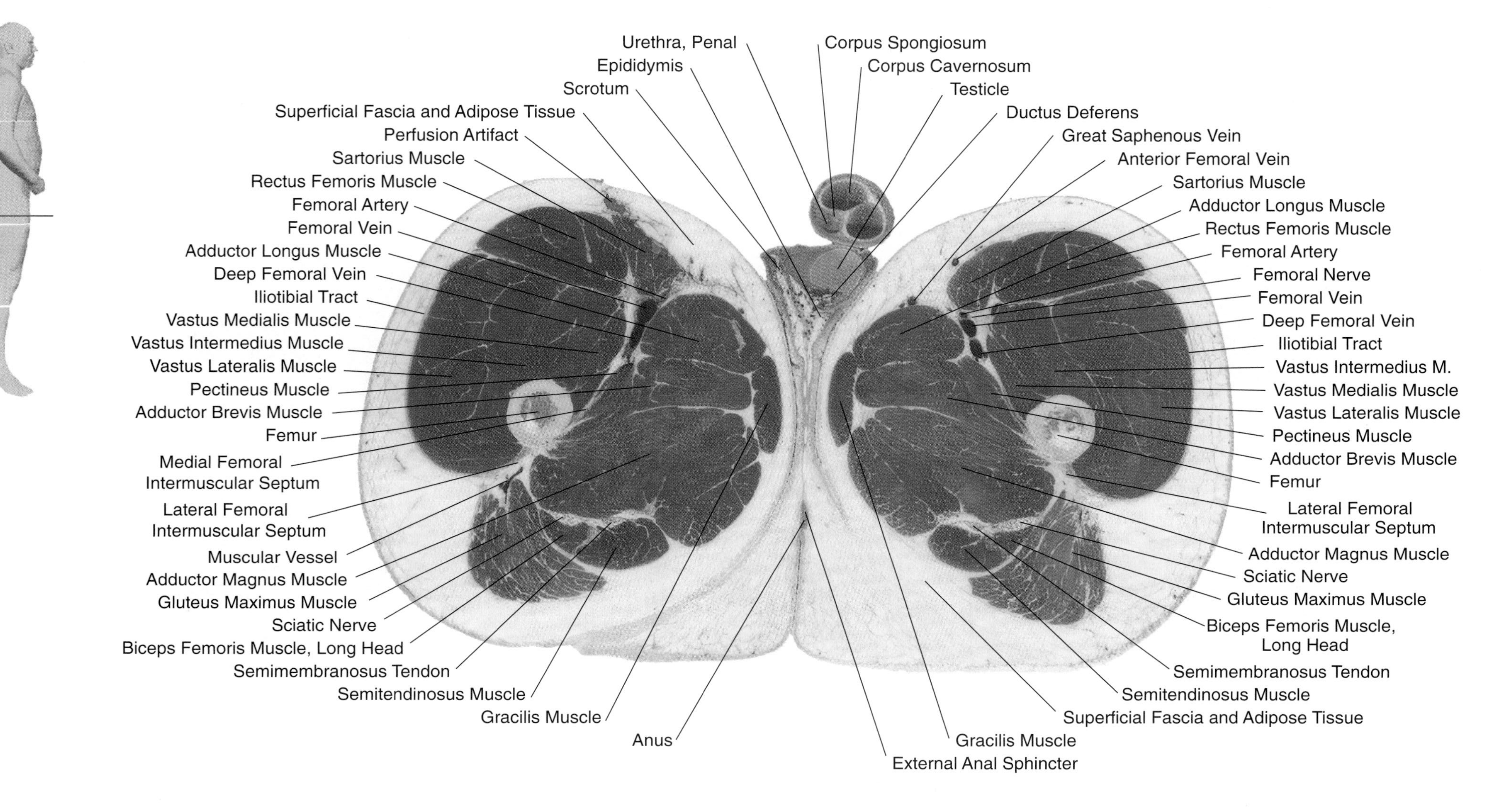

anterior
Urethra, Penal
Epididymis
Scrotum
Superficial Fascia and Adipose Tissue
Perfusion Artifact
Sartorius Muscle
Rectus Femoris Muscle
Femoral Artery
Femoral Vein
Adductor Longus Muscle
Deep Femoral Vein
Iliotibial Tract
Vastus Medialis Muscle
Vastus Intermedius Muscle
Vastus Lateralis Muscle
Pectineus Muscle
Adductor Brevis Muscle
Femur
Medial Femoral Intermuscular Septum
Lateral Femoral Intermuscular Septum
Muscular Vessel
Adductor Magnus Muscle
Gluteus Maximus Muscle
Sciatic Nerve
Biceps Femoris Muscle, Long Head
Semimembranosus Tendon
Semitendinosus Muscle
Gracilis Muscle
Anus
Corpus Spongiosum
Corpus Cavernosum
Testicle
Ductus Deferens
Great Saphenous Vein
Anterior Femoral Vein
Sartorius Muscle
Adductor Longus Muscle
Rectus Femoris Muscle
Femoral Artery
Femoral Nerve
Femoral Vein
Deep Femoral Vein
Iliotibial Tract
Vastus Intermedius M.
Vastus Medialis Muscle
Vastus Lateralis Muscle
Pectineus Muscle
Adductor Brevis Muscle
Femur
Lateral Femoral Intermuscular Septum
Adductor Magnus Muscle
Sciatic Nerve
Gluteus Maximus Muscle
Biceps Femoris Muscle, Long Head
Semimembranosus Tendon
Semitendinosus Muscle
Superficial Fascia and Adipose Tissue
Gracilis Muscle
External Anal Sphincter
right
left
posterior

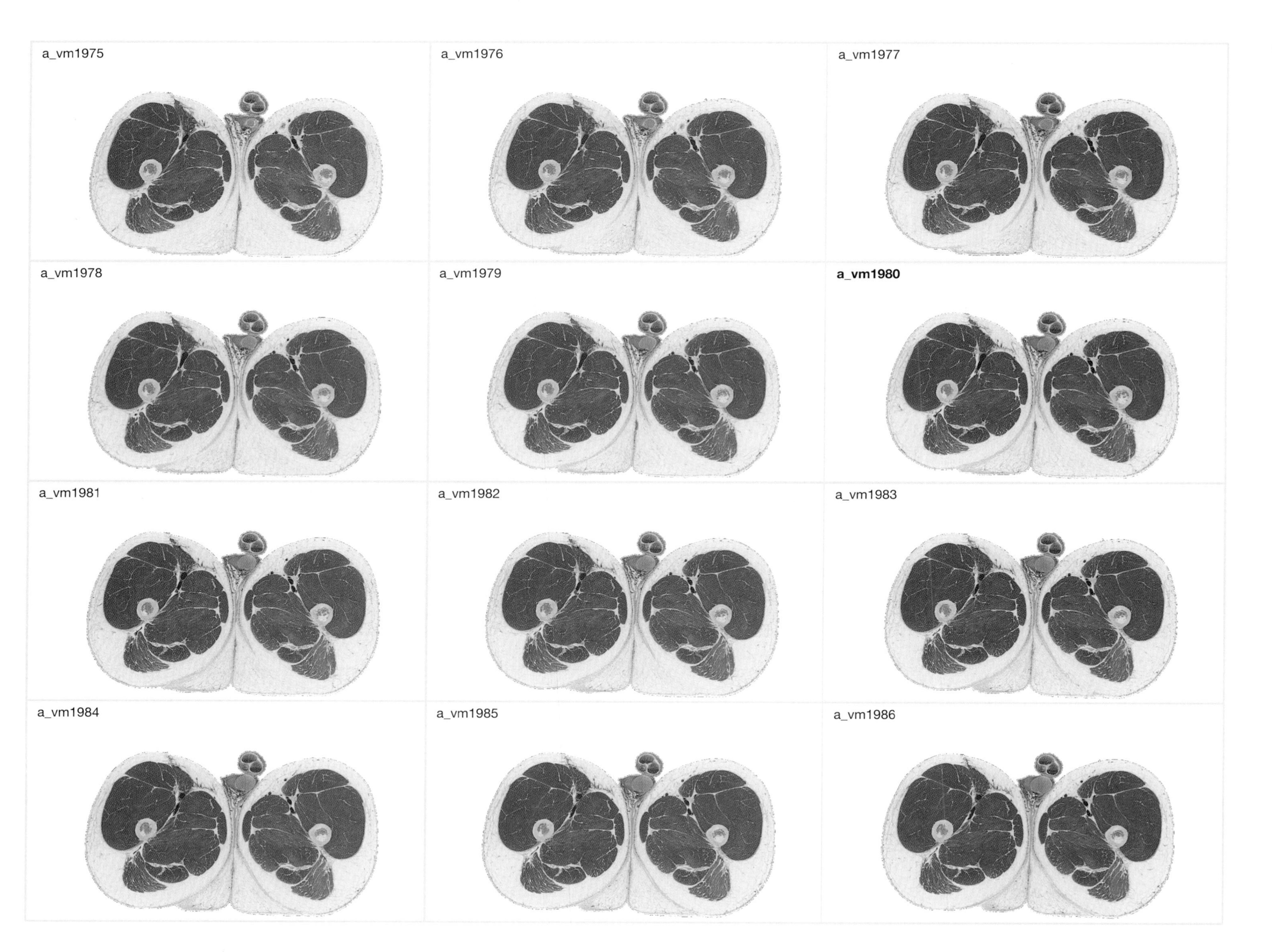
a_vm1975
a_vm1976
a_vm1977
a_vm1978
a_vm1979
a_vm1980
a_vm1981
a_vm1982
a_vm1983
a_vm1984
a_vm1985
a_vm1986

Transverse
a_vm1992

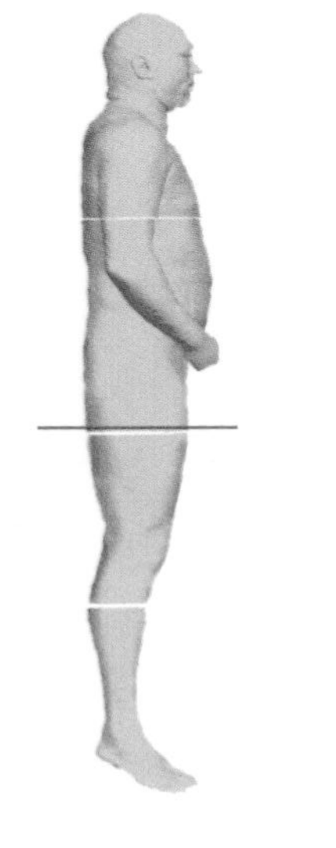

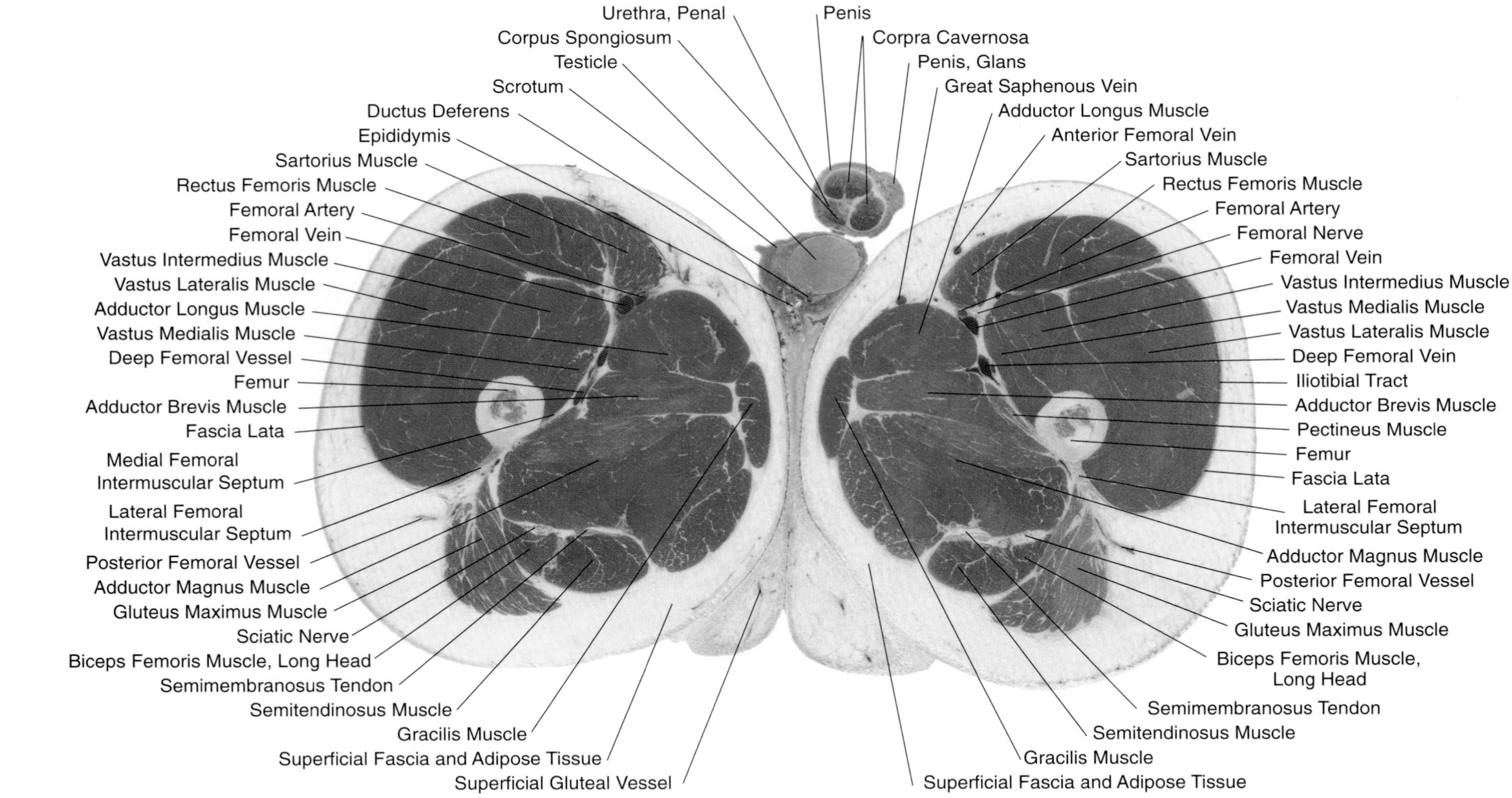
anterior
Urethra, Penal
Corpus Spongiosum
Testicle
Scrotum
Ductus Deferens
Epididymis
Sartorius Muscle
Rectus Femoris Muscle
Femoral Artery
Femoral Vein
Vastus Intermedius Muscle
Vastus Lateralis Muscle
Adductor Longus Muscle
Vastus Medialis Muscle
Deep Femoral Vessel
Femur
Adductor Brevis Muscle
Fascia Lata
Medial Femoral Intermuscular Septum
Lateral Femoral Intermuscular Septum
Posterior Femoral Vessel
Adductor Magnus Muscle
Gluteus Maximus Muscle
Sciatic Nerve
Biceps Femoris Muscle, Long Head
Semimembranosus Tendon
Semitendinosus Muscle
Gracilis Muscle
Superficial Fascia and Adipose Tissue
Superficial Gluteal Vessel
Penis
Corpra Cavernosa
Penis, Glans
Great Saphenous Vein
Adductor Longus Muscle
Anterior Femoral Vein
Sartorius Muscle
Rectus Femoris Muscle
Femoral Artery
Femoral Nerve
Femoral Vein
Vastus Intermedius Muscle
Vastus Medialis Muscle
Vastus Lateralis Muscle
Deep Femoral Vein
Iliotibial Tract
Adductor Brevis Muscle
Pectineus Muscle
Femur
Fascia Lata
Lateral Femoral Intermuscular Septum
Adductor Magnus Muscle
Posterior Femoral Vessel
Sciatic Nerve
Gluteus Maximus Muscle
Biceps Femoris Muscle, Long Head
Semimembranosus Tendon
Semitendinosus Muscle
Gracilis Muscle
Superficial Fascia and Adipose Tissue
right
left
posterior

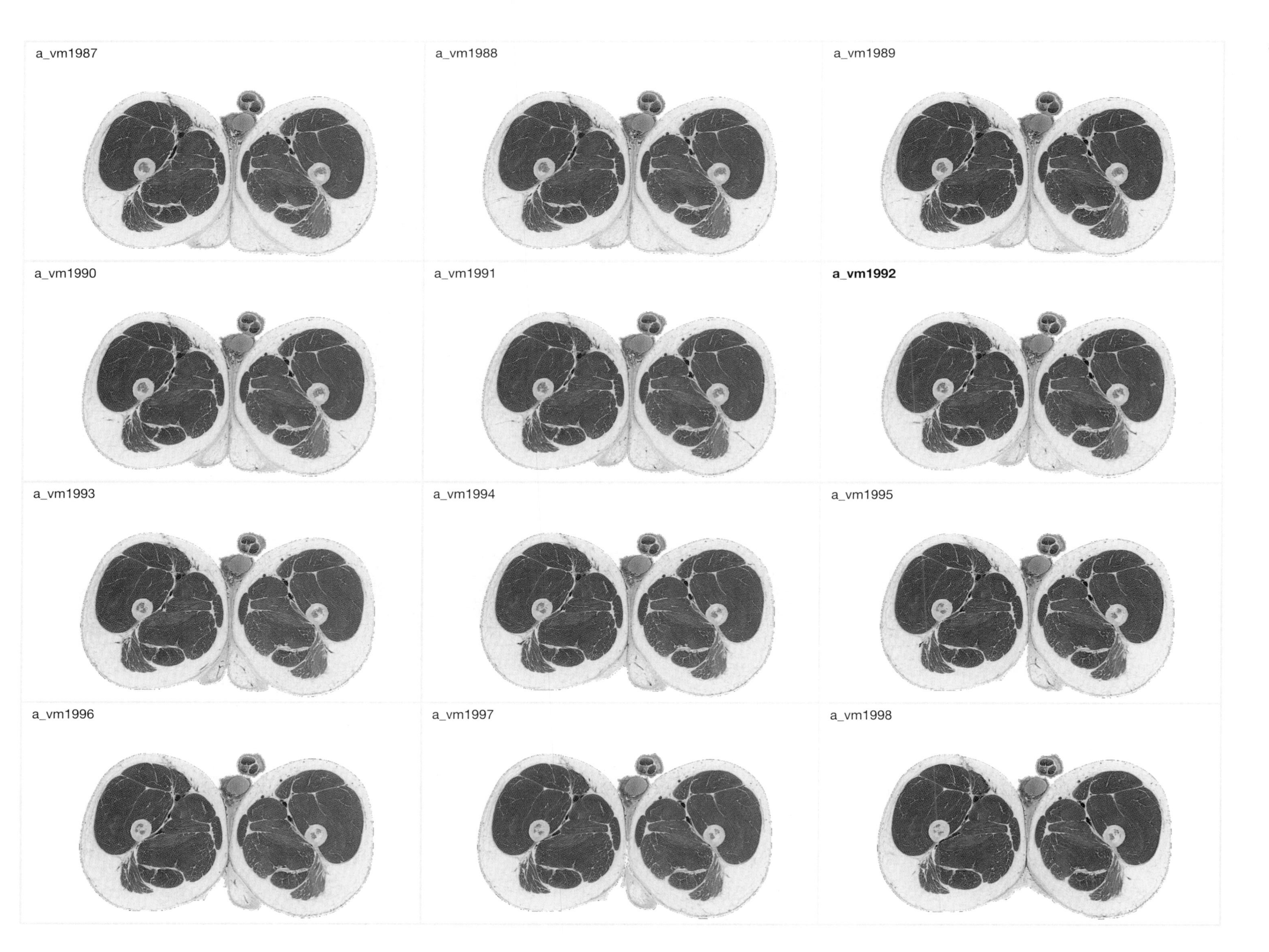
a_vm1987
a_vm1988
a_vm1989
a_vm1990
a_vm1991
a_vm1992
a_vm1993
a_vm1994
a_vm1995
a_vm1996
a_vm1997
a_vm1998

Transverse
a_vm2010

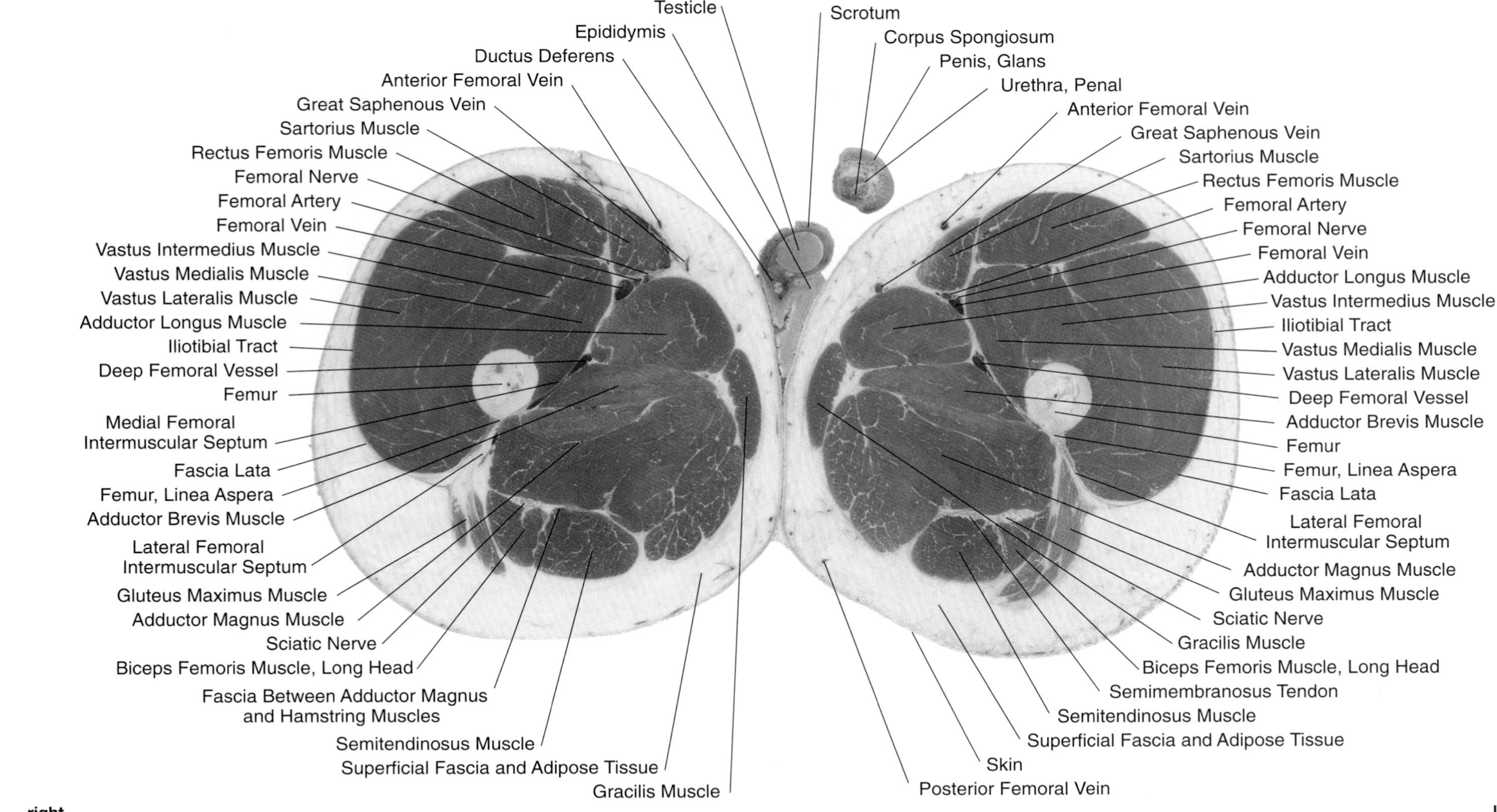
anterior
Testicle
Epididymis
Ductus Deferens
Anterior Femoral Vein
Great Saphenous Vein
Sartorius Muscle
Rectus Femoris Muscle
Femoral Nerve
Femoral Artery
Femoral Vein
Vastus Intermedius Muscle
Vastus Medialis Muscle
Vastus Lateralis Muscle
Adductor Longus Muscle
Iliotibial Tract
Deep Femoral Vessel
Femur
Medial Femoral Intermuscular Septum
Fascia Lata
Femur, Linea Aspera
Adductor Brevis Muscle
Lateral Femoral Intermuscular Septum
Gluteus Maximus Muscle
Adductor Magnus Muscle
Sciatic Nerve
Biceps Femoris Muscle, Long Head
Fascia Between Adductor Magnus and Hamstring Muscles
Semitendinosus Muscle
Superficial Fascia and Adipose Tissue
Gracilis Muscle
Scrotum
Corpus Spongiosum
Penis, Glans
Urethra, Penal
Anterior Femoral Vein
Great Saphenous Vein
Sartorius Muscle
Rectus Femoris Muscle
Femoral Artery
Femoral Nerve
Femoral Vein
Adductor Longus Muscle
Vastus Intermedius Muscle
Iliotibial Tract
Vastus Medialis Muscle
Vastus Lateralis Muscle
Deep Femoral Vessel
Adductor Brevis Muscle
Femur
Femur, Linea Aspera
Fascia Lata
Lateral Femoral Intermuscular Septum
Adductor Magnus Muscle
Gluteus Maximus Muscle
Sciatic Nerve
Gracilis Muscle
Biceps Femoris Muscle, Long Head
Semimembranosus Tendon
Semitendinosus Muscle
Superficial Fascia and Adipose Tissue
Skin
Posterior Femoral Vein
right
left
posterior

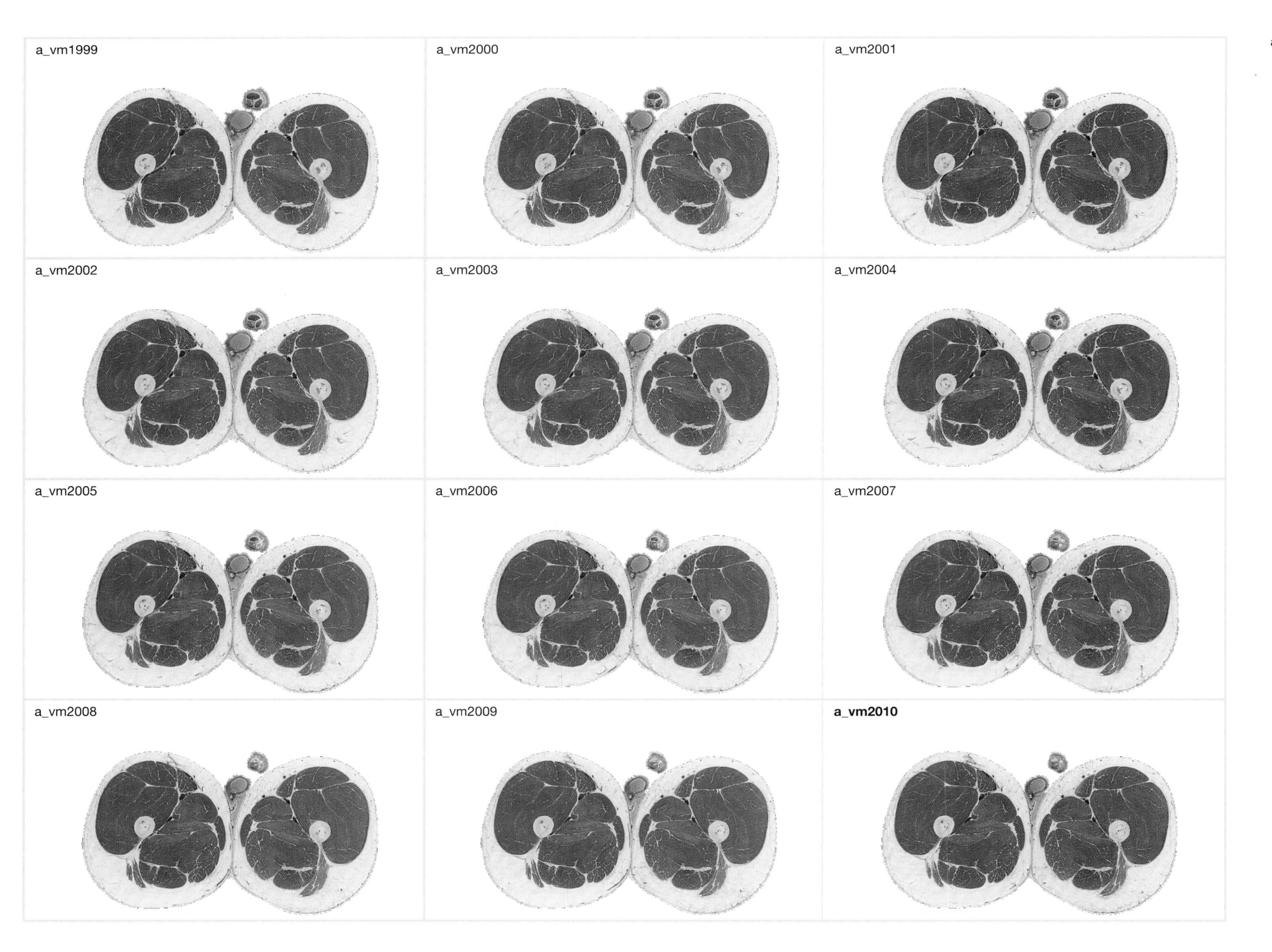
a_vm1999
a_vm2000
a_vm2001
a_vm2002
a_vm2003
a_vm2004
a_vm2005
a_vm2006
a_vm2007
a_vm2008
a_vm2009
a_vm2010

Transverse
a_vm2026

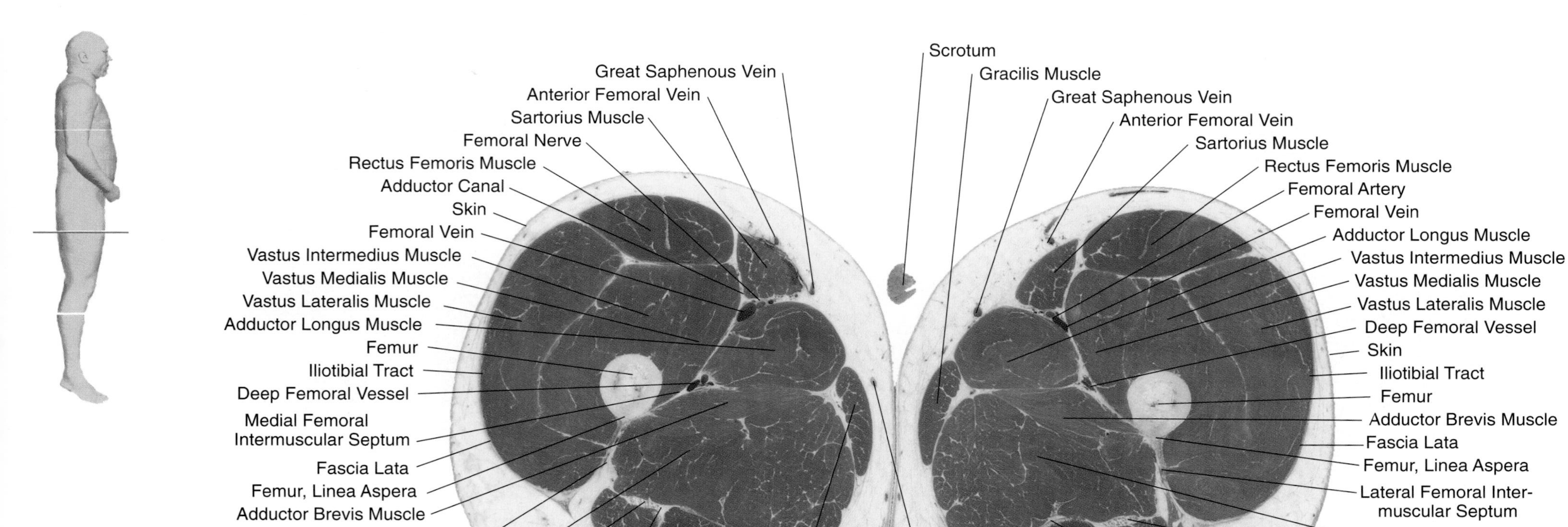

anterior
Great Saphenous Vein
Anterior Femoral Vein
Sartorius Muscle
Femoral Nerve
Rectus Femoris Muscle
Adductor Canal
Skin
Femoral Vein
Vastus Intermedius Muscle
Vastus Medialis Muscle
Vastus Lateralis Muscle
Adductor Longus Muscle
Femur
Iliotibial Tract
Deep Femoral Vessel
Medial Femoral
Intermuscular Septum
Fascia Lata
Femur, Linea Aspera
Adductor Brevis Muscle
Lateral Femoral
Intermuscular Septum
Adductor Magnus Muscle
Sciatic Nerve
Biceps Femoris Muscle, Long Head
Fascia Between Adductor Magnus
and Hamstring Muscles
Semitendinosus Muscle
Gracilis Muscle
Posterior Femoral Vein
Scrotum
Gracilis Muscle
Great Saphenous Vein
Anterior Femoral Vein
Sartorius Muscle
Rectus Femoris Muscle
Femoral Artery
Femoral Vein
Adductor Longus Muscle
Vastus Intermedius Muscle
Vastus Medialis Muscle
Vastus Lateralis Muscle
Deep Femoral Vessel
Skin
Iliotibial Tract
Femur
Adductor Brevis Muscle
Fascia Lata
Femur, Linea Aspera
Lateral Femoral Inter-
muscular Septum
Adductor Magnus Muscle
Sciatic Nerve
Gluteus Maximus Muscle
Biceps Femoris M., Long Head
Fascia Between Adductor Magnus
and Hamstring Muscles
Semitendinosus Muscle
Semimembranosus Muscle
Posterior Femoral Vein
Accessory Saphenous Vein
right
left
posterior

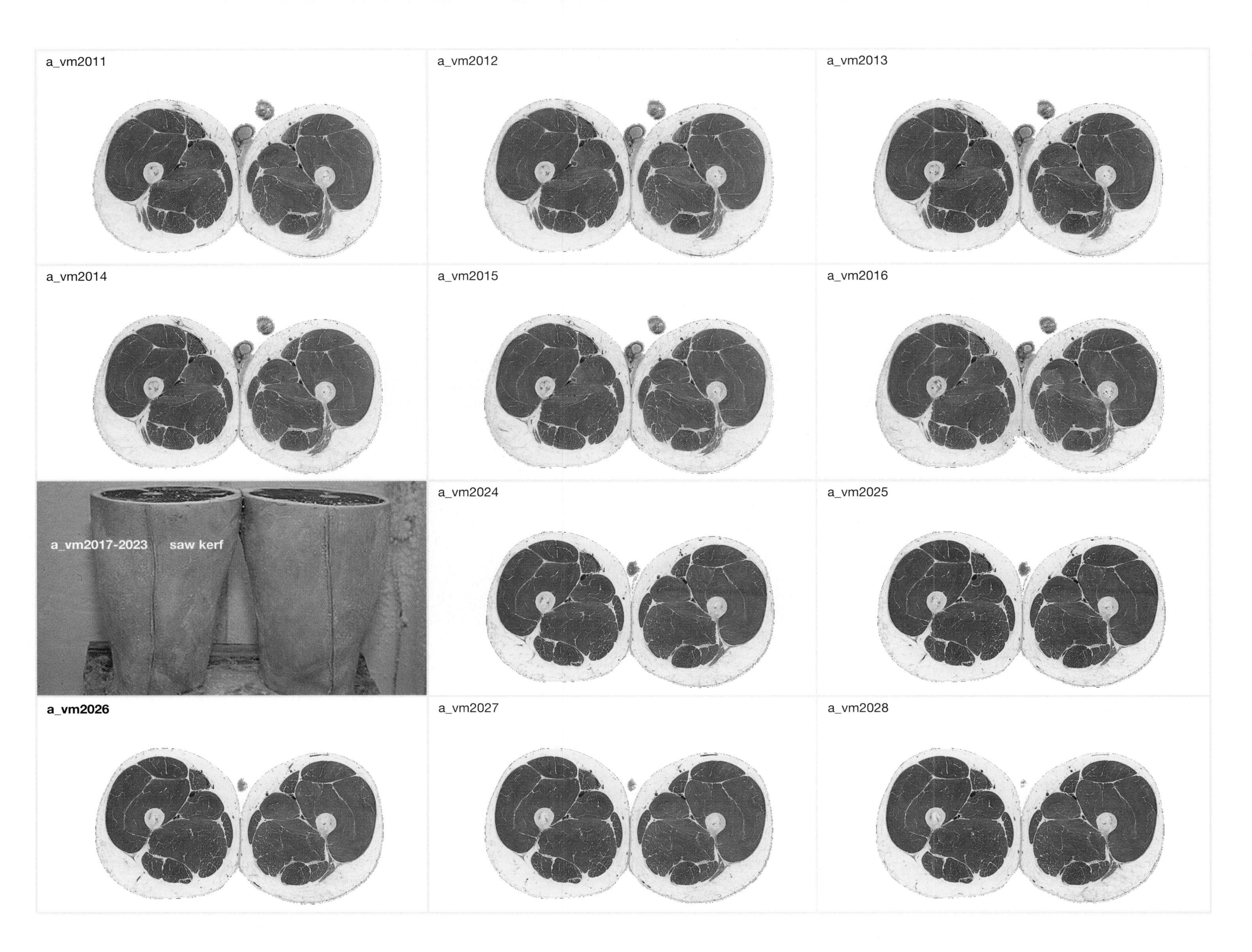
a_vm2011
a_vm2012
a_vm2013
a_vm2014
a_vm2015
a_vm2016
a_vm2017-2023
saw kerf
a_vm2024
a_vm2025
a_vm2026
a_vm2027
a_vm2028

Transverse
a_vm2036

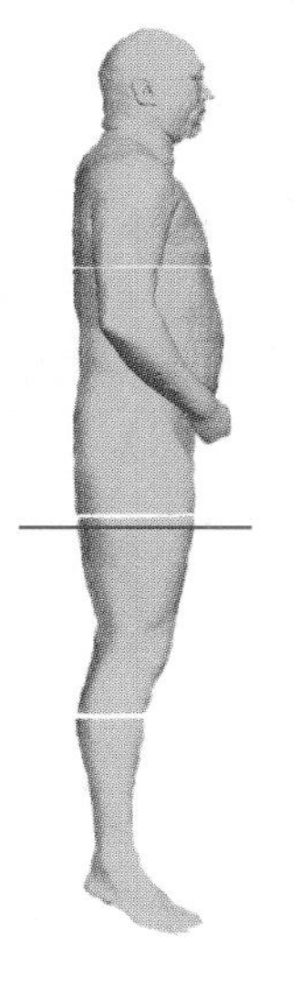

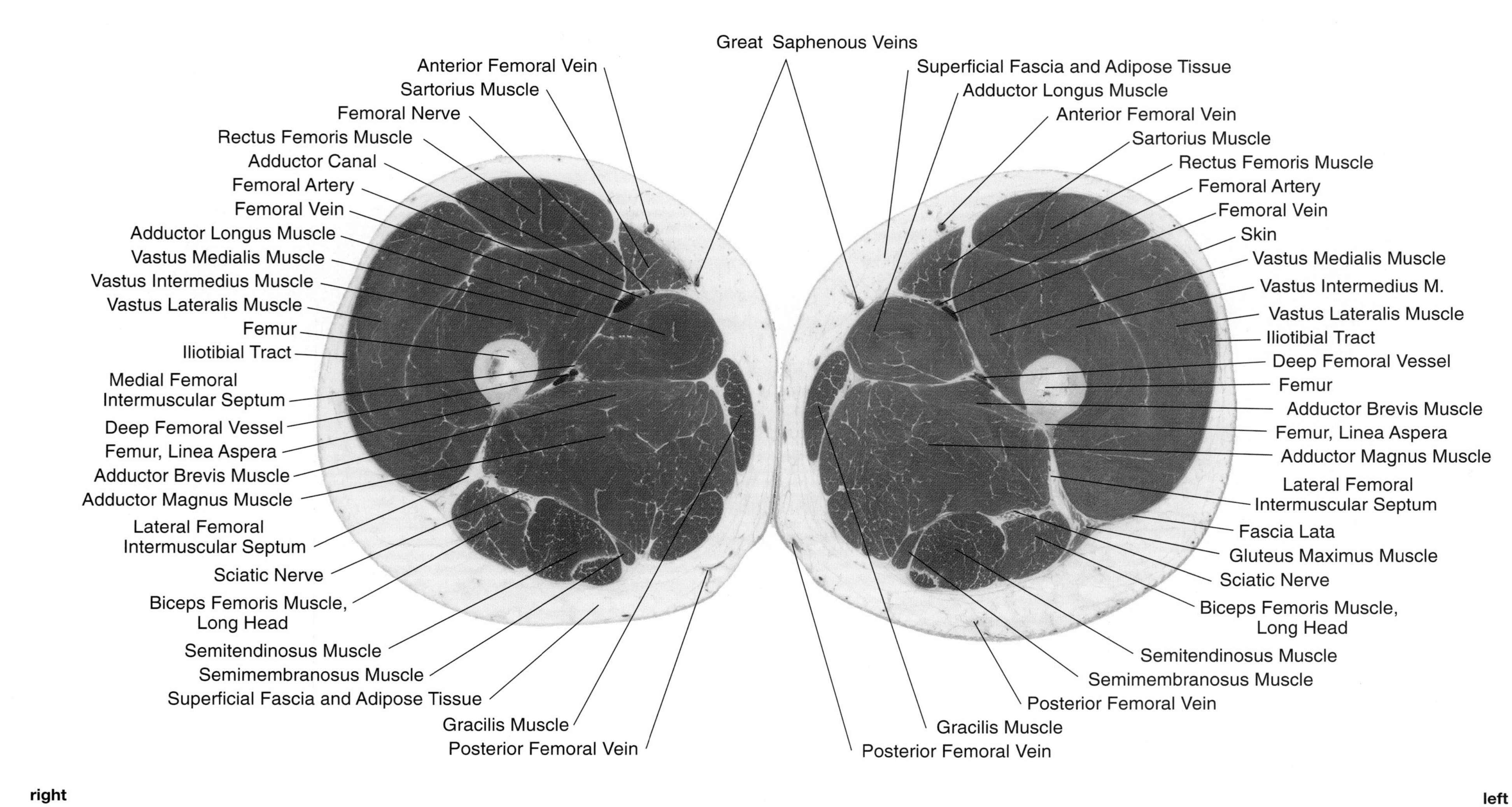
anterior
Great Saphenous Veins
Anterior Femoral Vein
Sartorius Muscle
Femoral Nerve
Rectus Femoris Muscle
Adductor Canal
Femoral Artery
Femoral Vein
Adductor Longus Muscle
Vastus Medialis Muscle
Vastus Intermedius Muscle
Vastus Lateralis Muscle
Femur
Iliotibial Tract
Medial Femoral Intermuscular Septum
Deep Femoral Vessel
Femur, Linea Aspera
Adductor Brevis Muscle
Adductor Magnus Muscle
Lateral Femoral Intermuscular Septum
Sciatic Nerve
Biceps Femoris Muscle, Long Head
Semitendinosus Muscle
Semimembranosus Muscle
Superficial Fascia and Adipose Tissue
Gracilis Muscle
Posterior Femoral Vein
Superficial Fascia and Adipose Tissue
Adductor Longus Muscle
Anterior Femoral Vein
Sartorius Muscle
Rectus Femoris Muscle
Femoral Artery
Femoral Vein
Skin
Vastus Medialis Muscle
Vastus Intermedius M.
Vastus Lateralis Muscle
Iliotibial Tract
Deep Femoral Vessel
Femur
Adductor Brevis Muscle
Femur, Linea Aspera
Adductor Magnus Muscle
Lateral Femoral Intermuscular Septum
Fascia Lata
Gluteus Maximus Muscle
Sciatic Nerve
Biceps Femoris Muscle, Long Head
Semitendinosus Muscle
Semimembranosus Muscle
Posterior Femoral Vein
Gracilis Muscle
Posterior Femoral Vein
right
left
posterior

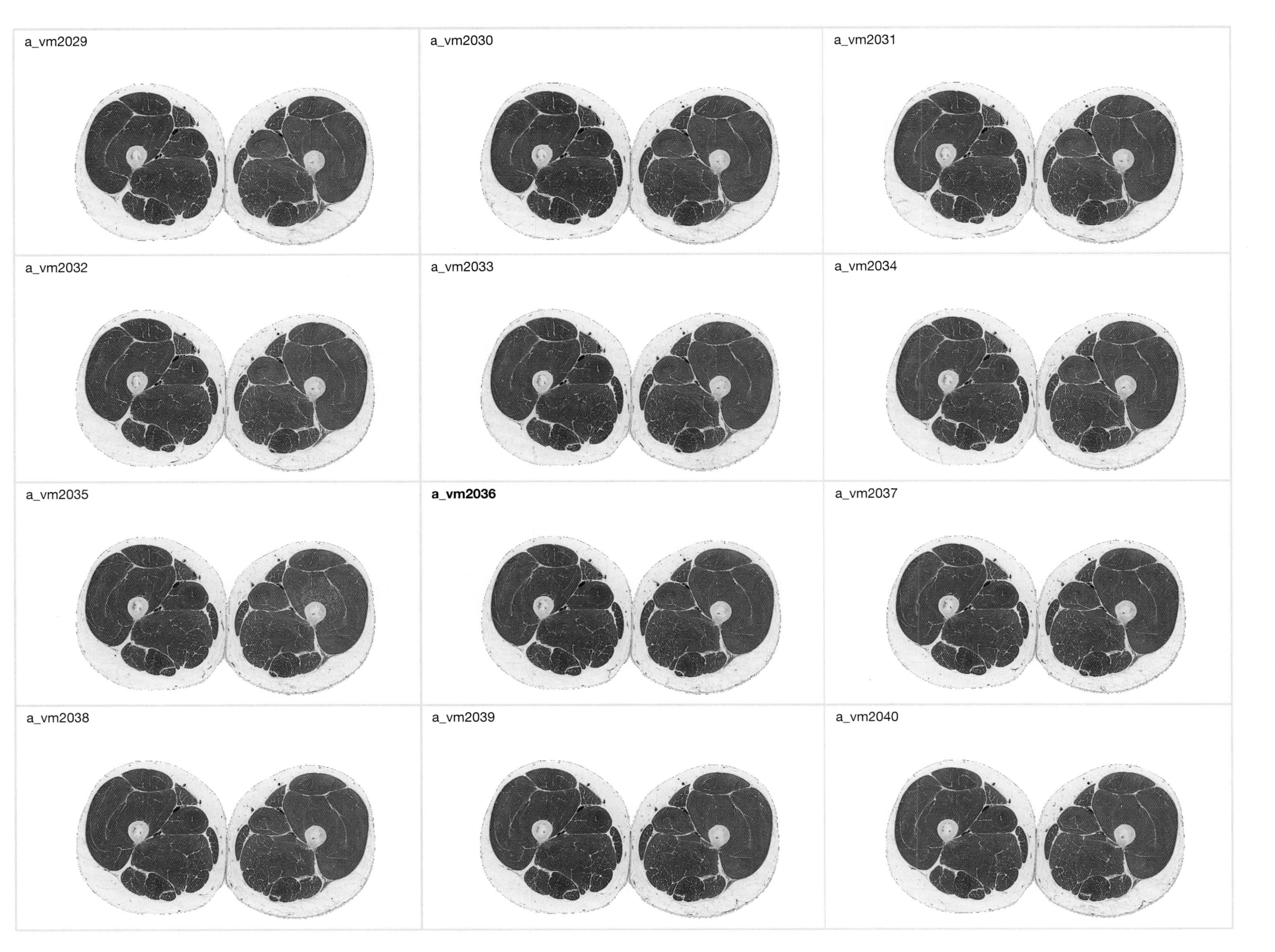
a_vm2029
a_vm2030
a_vm2031
a_vm2032
a_vm2033
a_vm2034
a_vm2035
a_vm2036
a_vm2037
a_vm2038
a_vm2039
a_vm2040

Transverse
a_cm2046

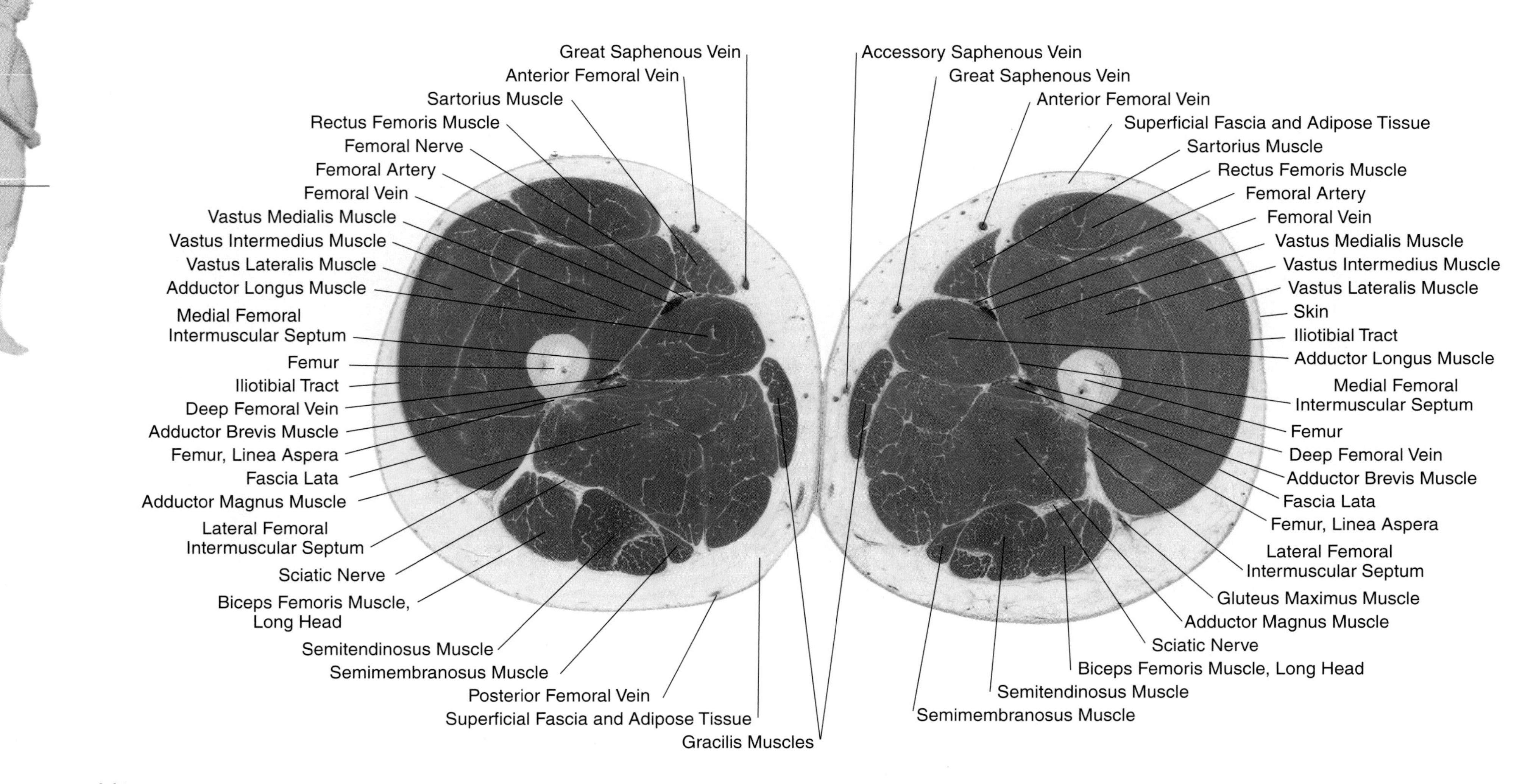
anterior
Great Saphenous Vein
Anterior Femoral Vein
Sartorius Muscle
Rectus Femoris Muscle
Femoral Nerve
Femoral Artery
Femoral Vein
Vastus Medialis Muscle
Vastus Intermedius Muscle
Vastus Lateralis Muscle
Adductor Longus Muscle
Medial Femoral Intermuscular Septum
Femur
Iliotibial Tract
Deep Femoral Vein
Adductor Brevis Muscle
Femur, Linea Aspera
Fascia Lata
Adductor Magnus Muscle
Lateral Femoral Intermuscular Septum
Sciatic Nerve
Biceps Femoris Muscle, Long Head
Semitendinosus Muscle
Semimembranosus Muscle
Posterior Femoral Vein
Superficial Fascia and Adipose Tissue
Gracilis Muscles
Accessory Saphenous Vein
Great Saphenous Vein
Anterior Femoral Vein
Superficial Fascia and Adipose Tissue
Sartorius Muscle
Rectus Femoris Muscle
Femoral Artery
Femoral Vein
Vastus Medialis Muscle
Vastus Intermedius Muscle
Vastus Lateralis Muscle
Skin
Iliotibial Tract
Adductor Longus Muscle
Medial Femoral Intermuscular Septum
Femur
Deep Femoral Vein
Adductor Brevis Muscle
Fascia Lata
Femur, Linea Aspera
Lateral Femoral Intermuscular Septum
Gluteus Maximus Muscle
Adductor Magnus Muscle
Sciatic Nerve
Biceps Femoris Muscle, Long Head
Semitendinosus Muscle
Semimembranosus Muscle
right
left
posterior

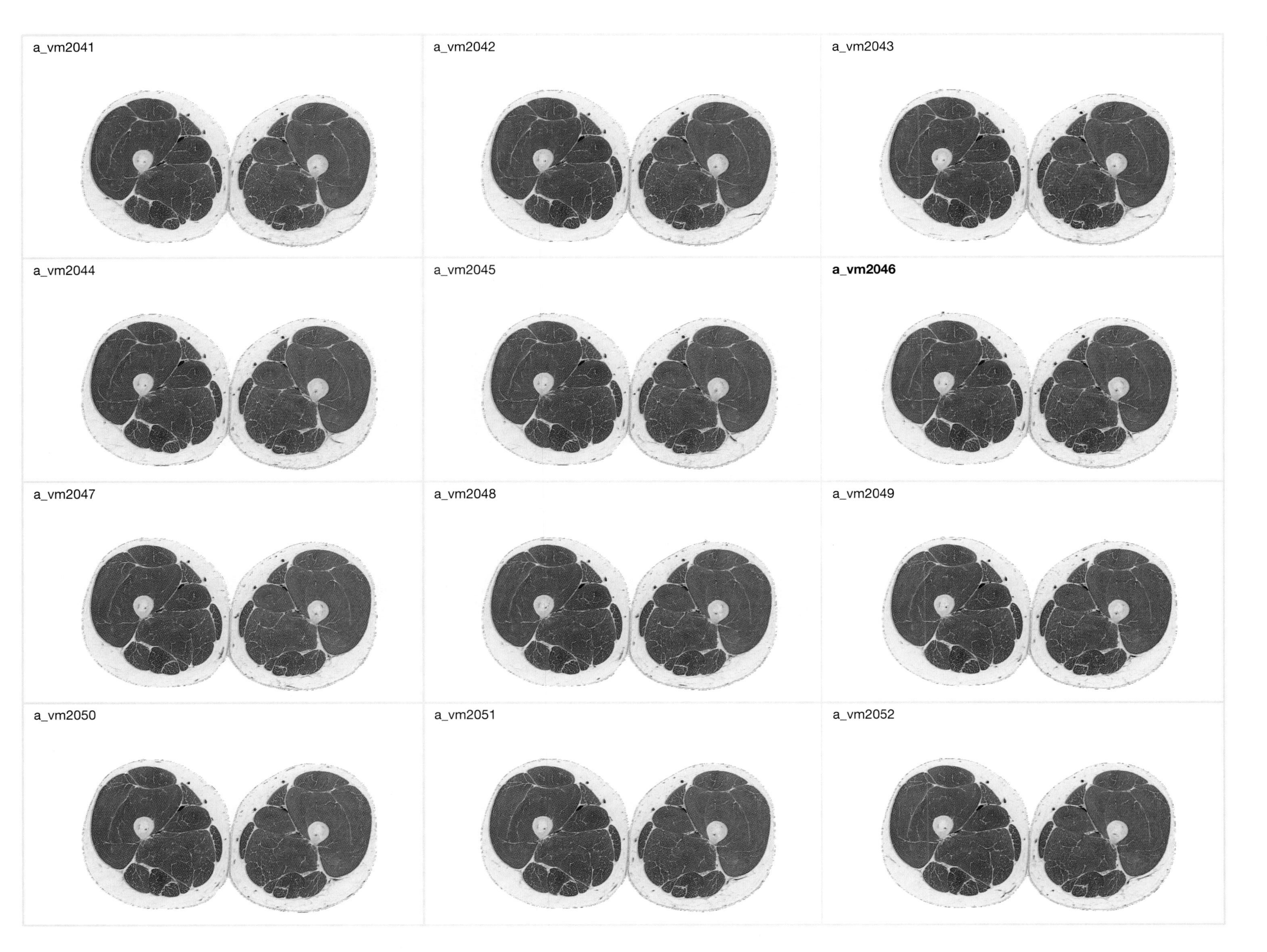
a_vm2041
a_vm2042
a_vm2043
a_vm2044
a_vm2045
a_vm2046
a_vm2047
a_vm2048
a_vm2049
a_vm2050
a_vm2051
a_vm2052

Transverse
a_vm2058

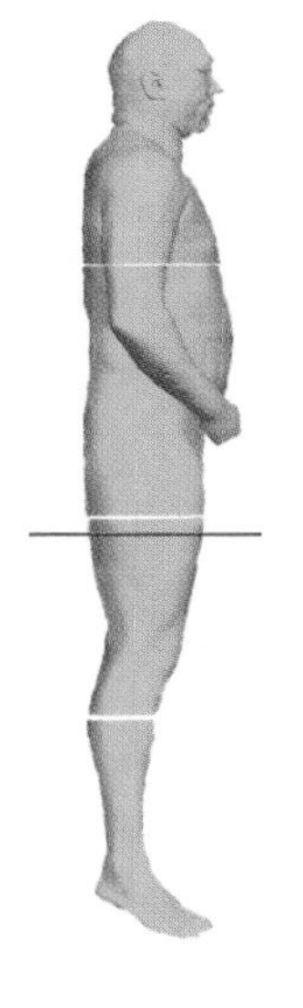

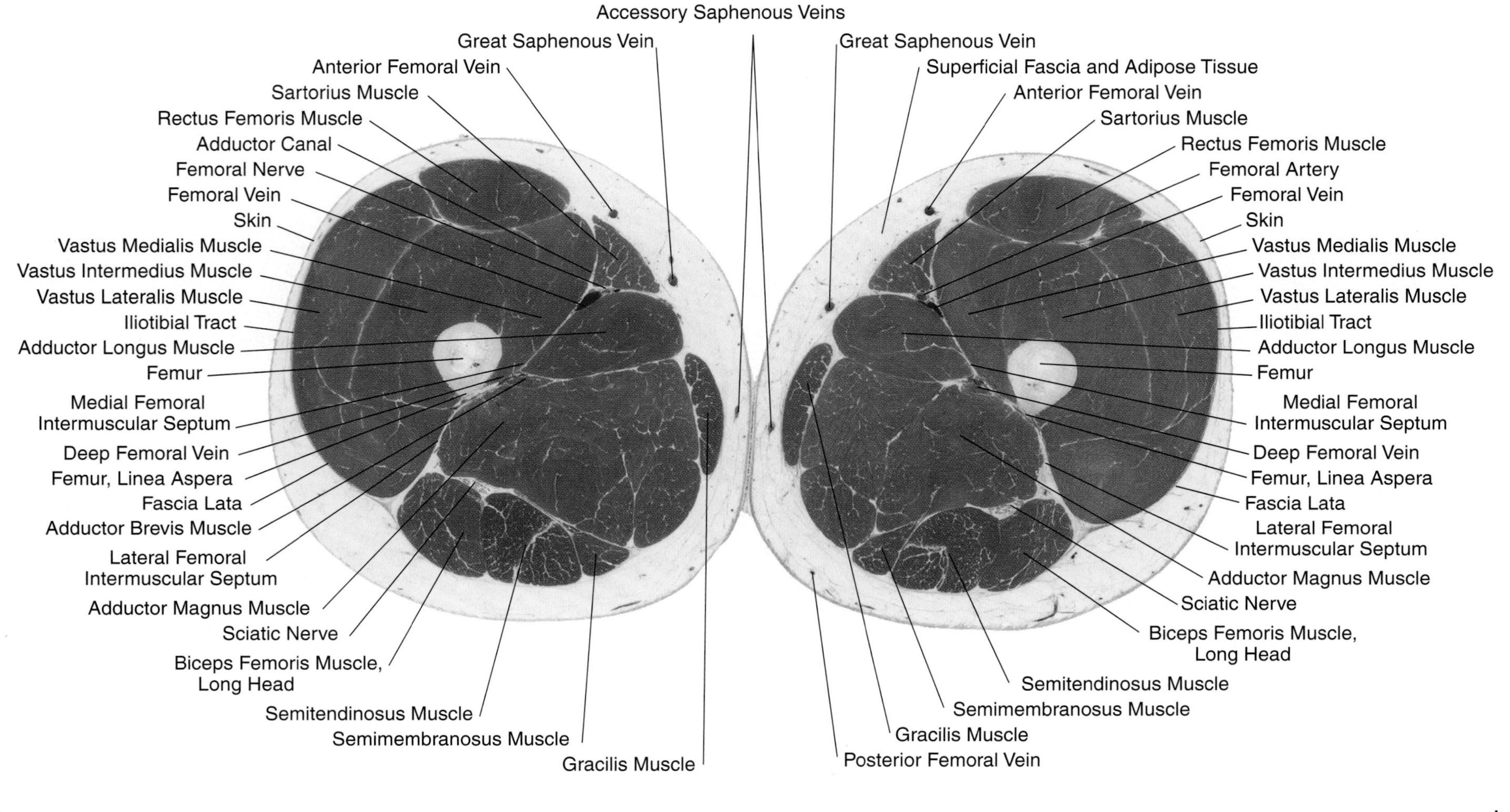
anterior
Accessory Saphenous Veins
Great Saphenous Vein
Anterior Femoral Vein
Sartorius Muscle
Rectus Femoris Muscle
Adductor Canal
Femoral Nerve
Femoral Vein
Skin
Vastus Medialis Muscle
Vastus Intermedius Muscle
Vastus Lateralis Muscle
Iliotibial Tract
Adductor Longus Muscle
Femur
Medial Femoral Intermuscular Septum
Deep Femoral Vein
Femur, Linea Aspera
Fascia Lata
Adductor Brevis Muscle
Lateral Femoral Intermuscular Septum
Adductor Magnus Muscle
Sciatic Nerve
Biceps Femoris Muscle, Long Head
Semitendinosus Muscle
Semimembranosus Muscle
Gracilis Muscle
Great Saphenous Vein
Superficial Fascia and Adipose Tissue
Anterior Femoral Vein
Sartorius Muscle
Rectus Femoris Muscle
Femoral Artery
Femoral Vein
Skin
Vastus Medialis Muscle
Vastus Intermedius Muscle
Vastus Lateralis Muscle
Iliotibial Tract
Adductor Longus Muscle
Femur
Medial Femoral Intermuscular Septum
Deep Femoral Vein
Femur, Linea Aspera
Fascia Lata
Lateral Femoral Intermuscular Septum
Adductor Magnus Muscle
Sciatic Nerve
Biceps Femoris Muscle, Long Head
Semitendinosus Muscle
Semimembranosus Muscle
Gracilis Muscle
Posterior Femoral Vein
right
left
posterior

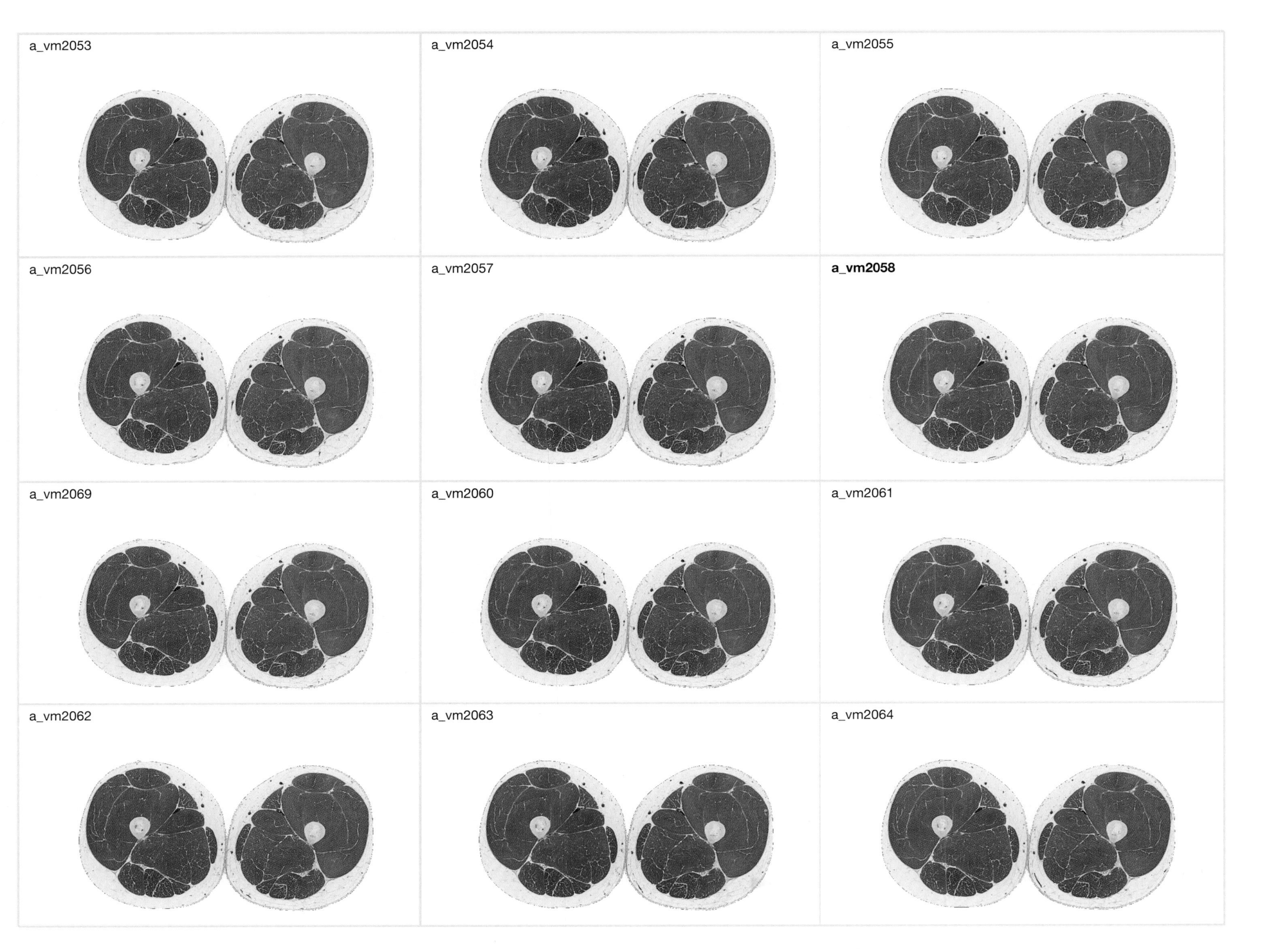
a_vm2053
a_vm2054
a_vm2055
a_vm2056
a_vm2057
a_vm2058
a_vm2069
a_vm2060
a_vm2061
a_vm2062
a_vm2063
a_vm2064

Transverse
a_vm2070

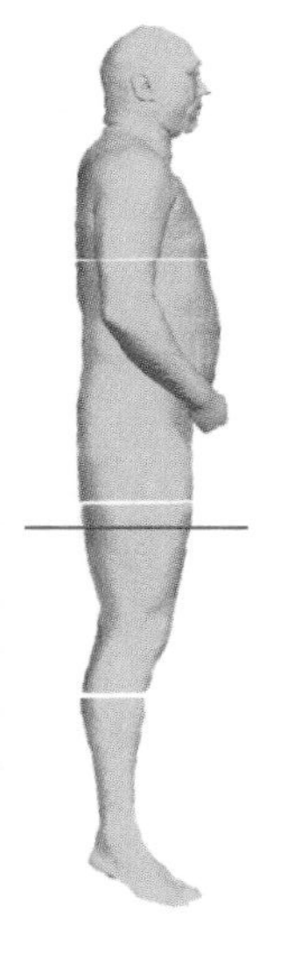

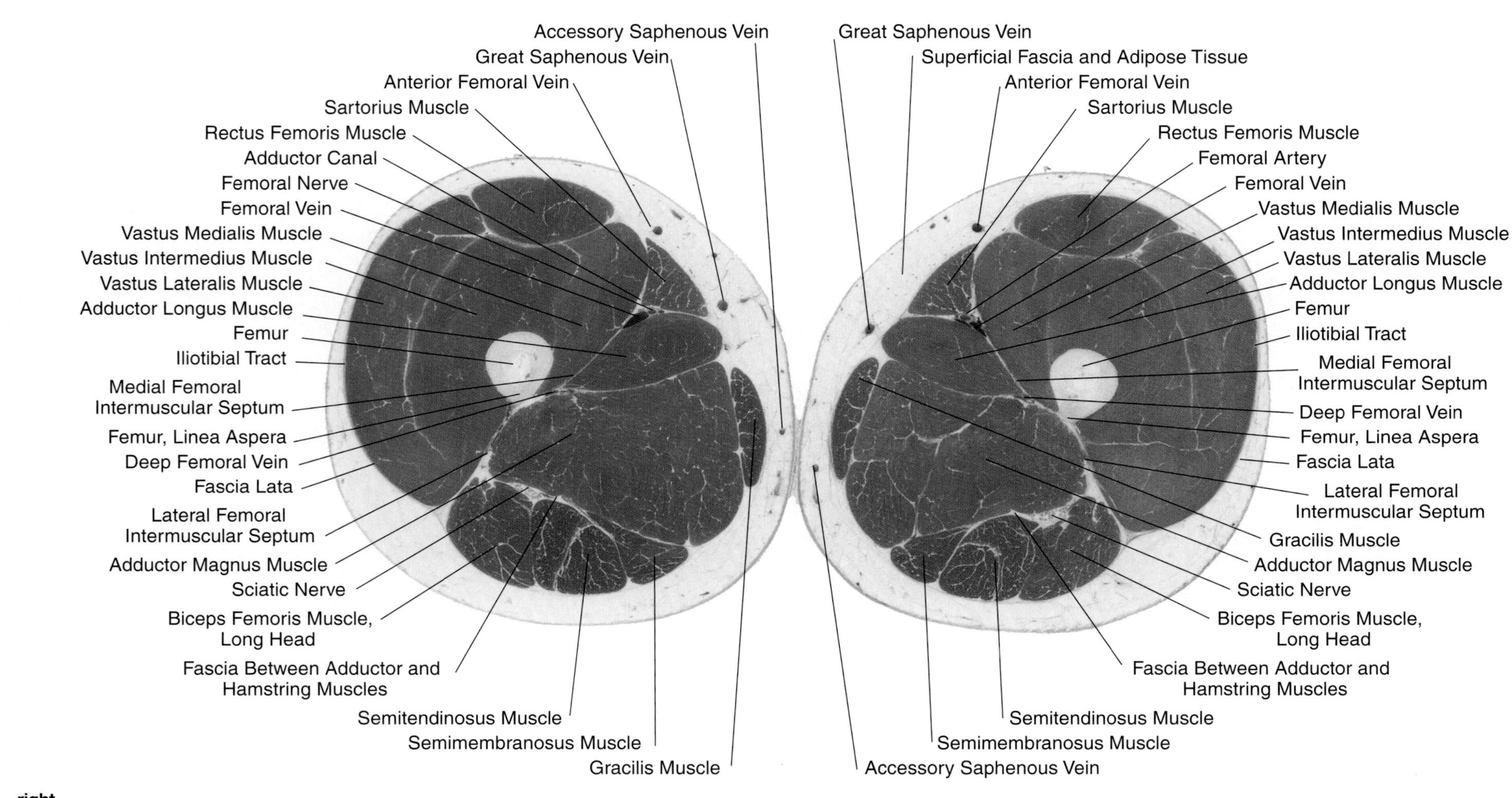
anterior
Accessory Saphenous Vein
Great Saphenous Vein
Anterior Femoral Vein
Sartorius Muscle
Rectus Femoris Muscle
Adductor Canal
Femoral Nerve
Femoral Vein
Vastus Medialis Muscle
Vastus Intermedius Muscle
Vastus Lateralis Muscle
Adductor Longus Muscle
Femur
Iliotibial Tract
Medial Femoral Intermuscular Septum
Femur, Linea Aspera
Deep Femoral Vein
Fascia Lata
Lateral Femoral Intermuscular Septum
Adductor Magnus Muscle
Sciatic Nerve
Biceps Femoris Muscle, Long Head
Fascia Between Adductor and Hamstring Muscles
Semitendinosus Muscle
Semimembranosus Muscle
Gracilis Muscle
Great Saphenous Vein
Superficial Fascia and Adipose Tissue
Anterior Femoral Vein
Sartorius Muscle
Rectus Femoris Muscle
Femoral Artery
Femoral Vein
Vastus Medialis Muscle
Vastus Intermedius Muscle
Vastus Lateralis Muscle
Adductor Longus Muscle
Femur
Iliotibial Tract
Medial Femoral Intermuscular Septum
Deep Femoral Vein
Femur, Linea Aspera
Fascia Lata
Lateral Femoral Intermuscular Septum
Gracilis Muscle
Adductor Magnus Muscle
Sciatic Nerve
Biceps Femoris Muscle, Long Head
Fascia Between Adductor and Hamstring Muscles
Semitendinosus Muscle
Semimembranosus Muscle
Accessory Saphenous Vein
right
left
posterior

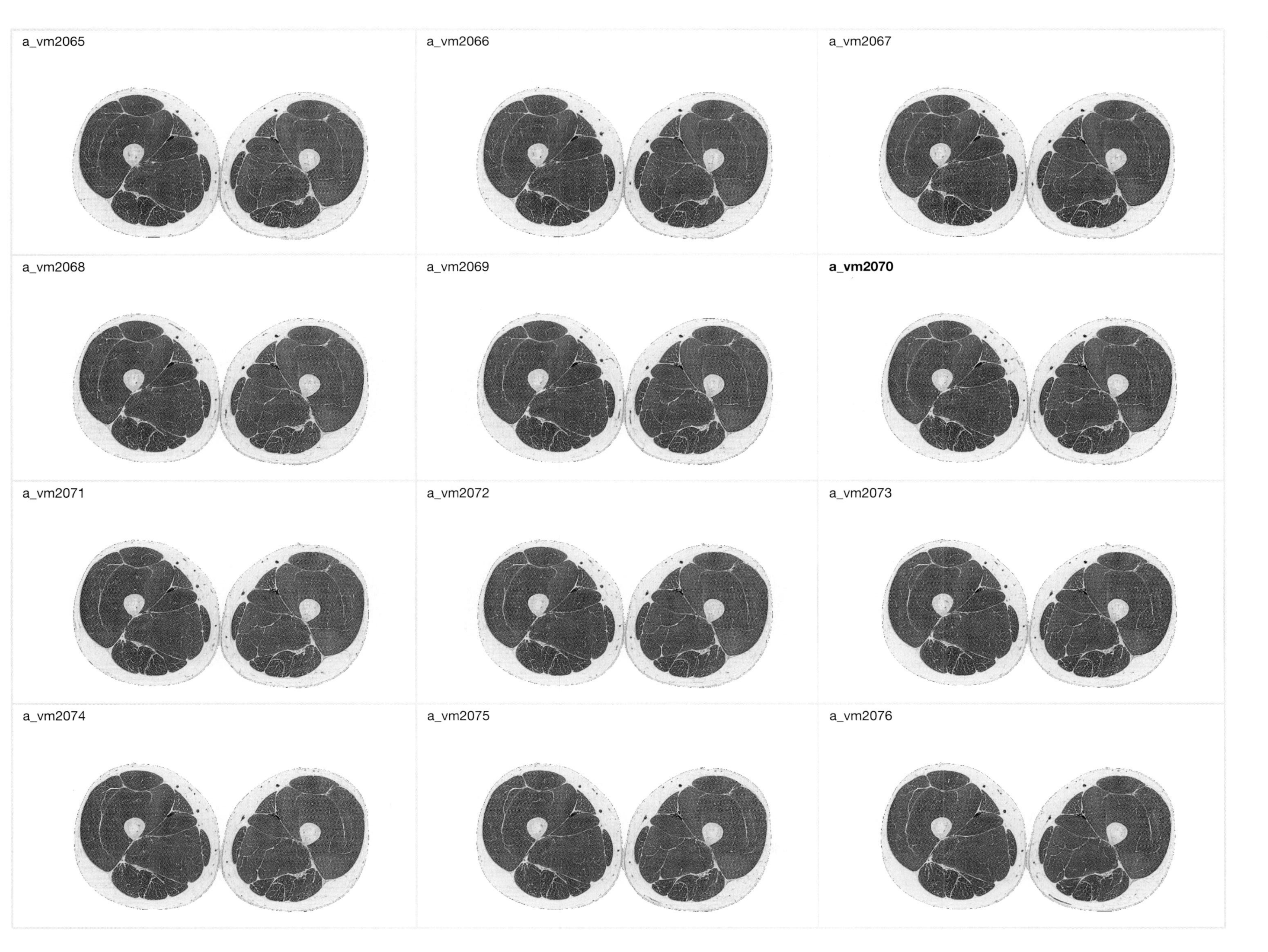
a_vm2065
a_vm2066
a_vm2067
a_vm2068
a_vm2069
a_vm2070
a_vm2071
a_vm2072
a_vm2073
a_vm2074
a_vm2075
a_vm2076

Transverse
a_vm2082

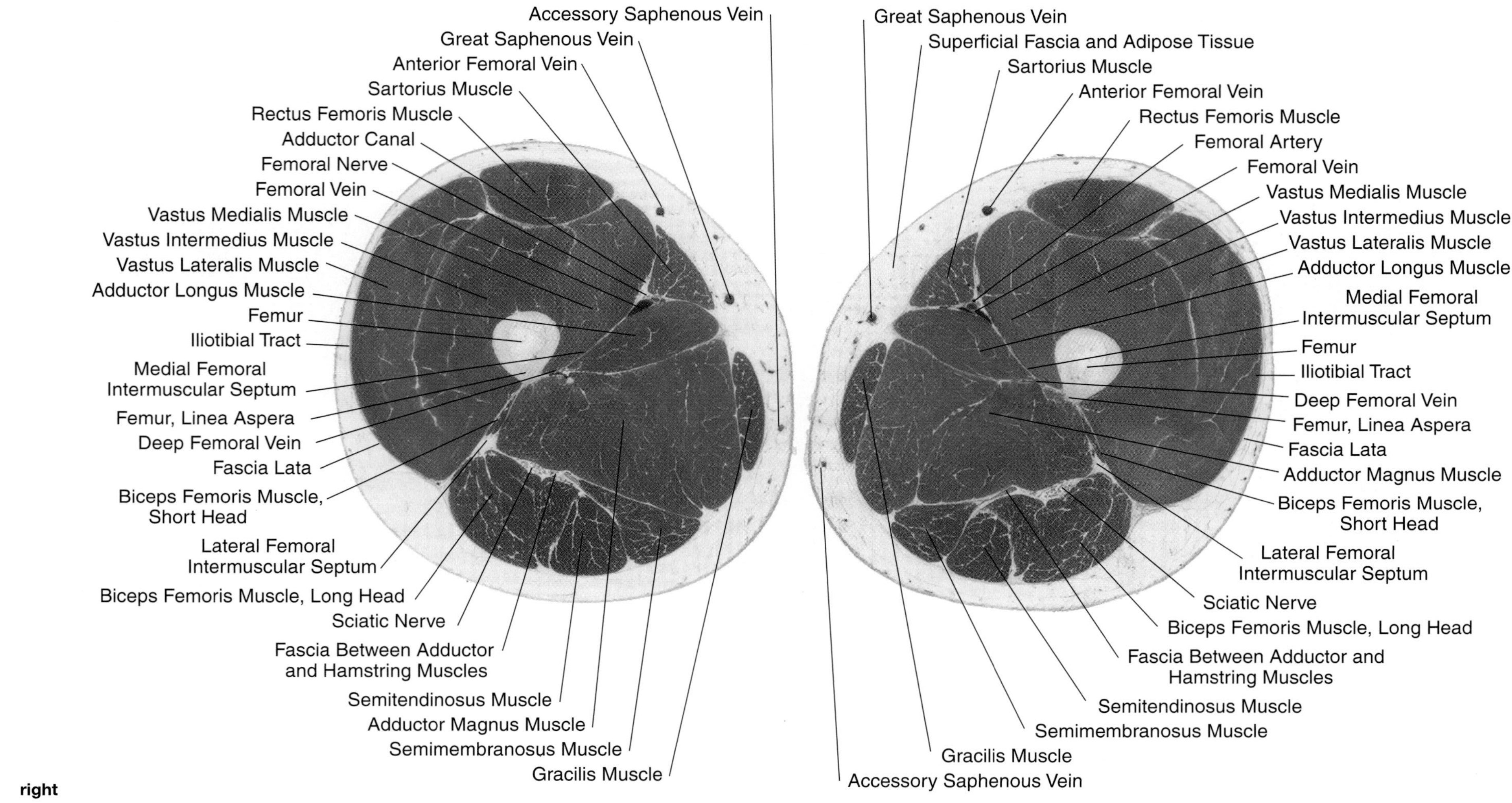
anterior
Accessory Saphenous Vein
Great Saphenous Vein
Anterior Femoral Vein
Sartorius Muscle
Rectus Femoris Muscle
Adductor Canal
Femoral Nerve
Femoral Vein
Vastus Medialis Muscle
Vastus Intermedius Muscle
Vastus Lateralis Muscle
Adductor Longus Muscle
Femur
Iliotibial Tract
Medial Femoral Intermuscular Septum
Femur, Linea Aspera
Deep Femoral Vein
Fascia Lata
Biceps Femoris Muscle, Short Head
Lateral Femoral Intermuscular Septum
Biceps Femoris Muscle, Long Head
Sciatic Nerve
Fascia Between Adductor and Hamstring Muscles
Semitendinosus Muscle
Adductor Magnus Muscle
Semimembranosus Muscle
Gracilis Muscle
Great Saphenous Vein
Superficial Fascia and Adipose Tissue
Sartorius Muscle
Anterior Femoral Vein
Rectus Femoris Muscle
Femoral Artery
Femoral Vein
Vastus Medialis Muscle
Vastus Intermedius Muscle
Vastus Lateralis Muscle
Adductor Longus Muscle
Medial Femoral Intermuscular Septum
Femur
Iliotibial Tract
Deep Femoral Vein
Femur, Linea Aspera
Fascia Lata
Adductor Magnus Muscle
Biceps Femoris Muscle, Short Head
Lateral Femoral Intermuscular Septum
Sciatic Nerve
Biceps Femoris Muscle, Long Head
Fascia Between Adductor and Hamstring Muscles
Semitendinosus Muscle
Semimembranosus Muscle
Gracilis Muscle
Accessory Saphenous Vein
right
left
posterior

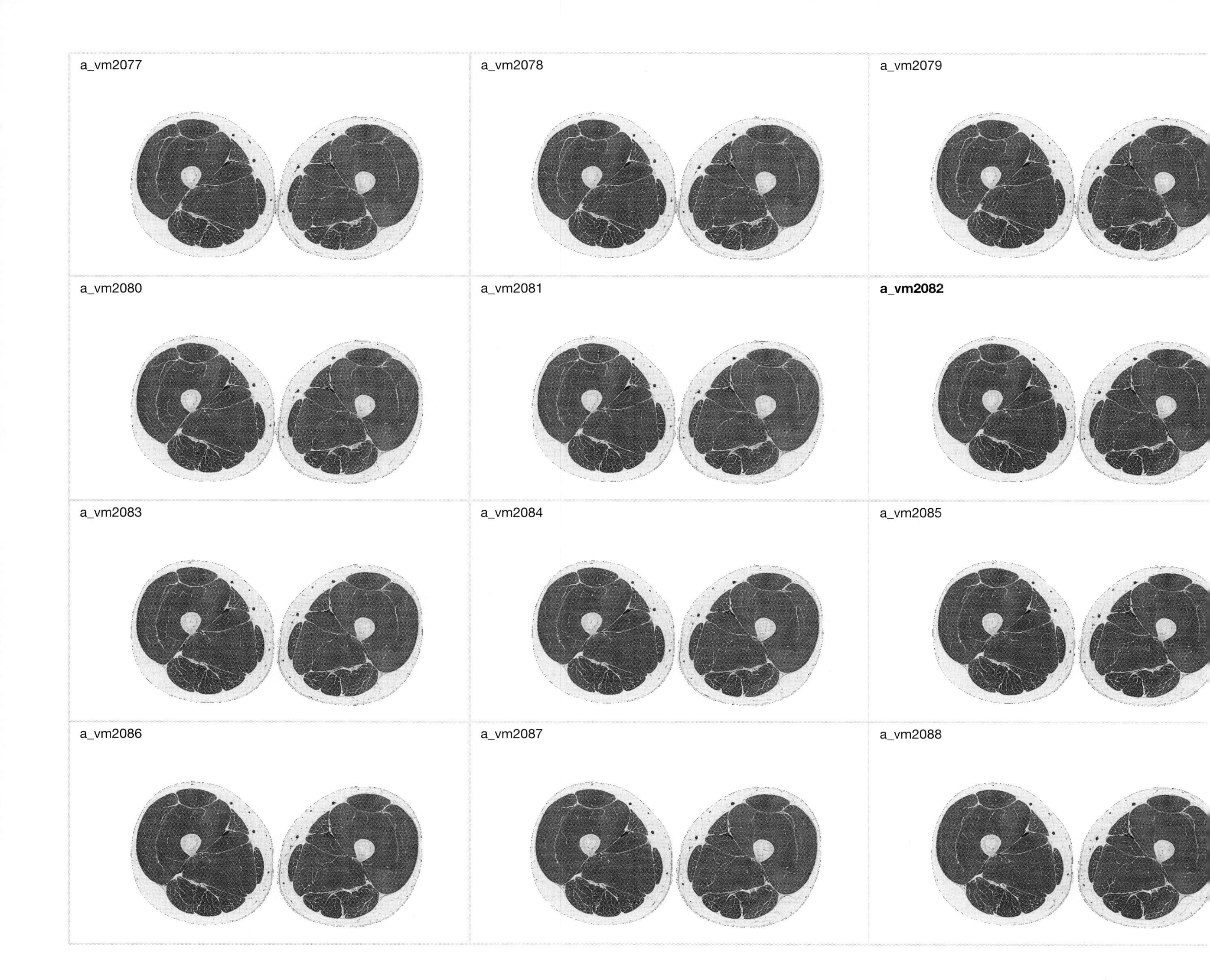
a_vm2077
a_vm2078
a_vm2079
a_vm2080
a_vm2081
a_vm2082
a_vm2083
a_vm2084
a_vm2085
a_vm2086
a_vm2087
a_vm2088

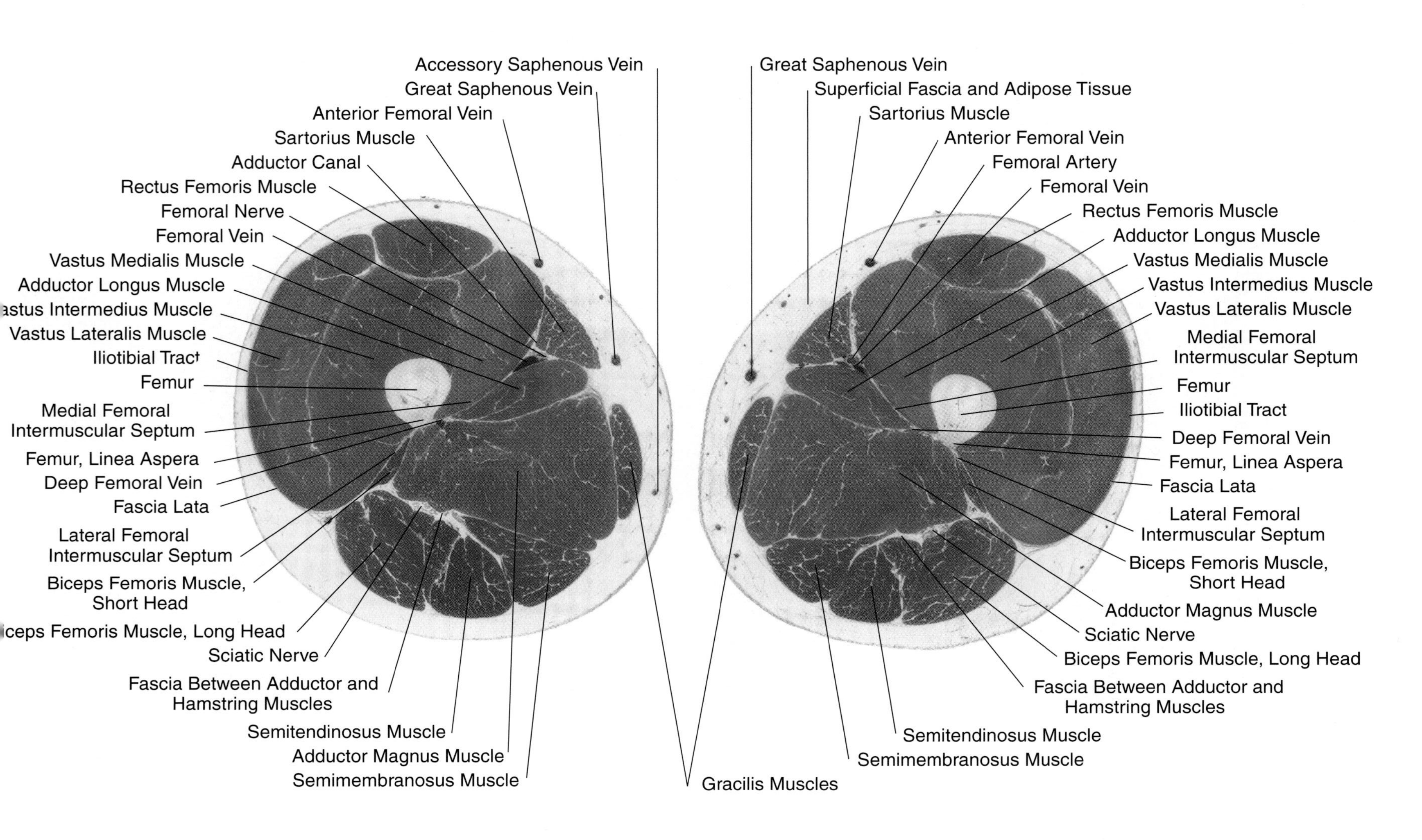
anterior
Accessory Saphenous Vein
Great Saphenous Vein
Anterior Femoral Vein
Sartorius Muscle
Adductor Canal
Rectus Femoris Muscle
Femoral Nerve
Femoral Vein
Vastus Medialis Muscle
Adductor Longus Muscle
astus Intermedius Muscle
Vastus Lateralis Muscle
Iliotibial Tract
Femur
Medial Femoral
Intermuscular Septum
Femur, Linea Aspera
Deep Femoral Vein
Fascia Lata
Lateral Femoral
Intermuscular Septum
Biceps Femoris Muscle,
Short Head
iceps Femoris Muscle, Long Head
Sciatic Nerve
Fascia Between Adductor and
Hamstring Muscles
Semitendinosus Muscle
Adductor Magnus Muscle
Semimembranosus Muscle
Gracilis Muscles
Great Saphenous Vein
Superficial Fascia and Adipose Tissue
Sartorius Muscle
Anterior Femoral Vein
Femoral Artery
Femoral Vein
Rectus Femoris Muscle
Adductor Longus Muscle
Vastus Medialis Muscle
Vastus Intermedius Muscle
Vastus Lateralis Muscle
Medial Femoral
Intermuscular Septum
Femur
Iliotibial Tract
Deep Femoral Vein
Femur, Linea Aspera
Fascia Lata
Lateral Femoral
Intermuscular Septum
Biceps Femoris Muscle,
Short Head
Adductor Magnus Muscle
Sciatic Nerve
Biceps Femoris Muscle, Long Head
Fascia Between Adductor and
Hamstring Muscles
Semitendinosus Muscle
Semimembranosus Muscle
left
posterior

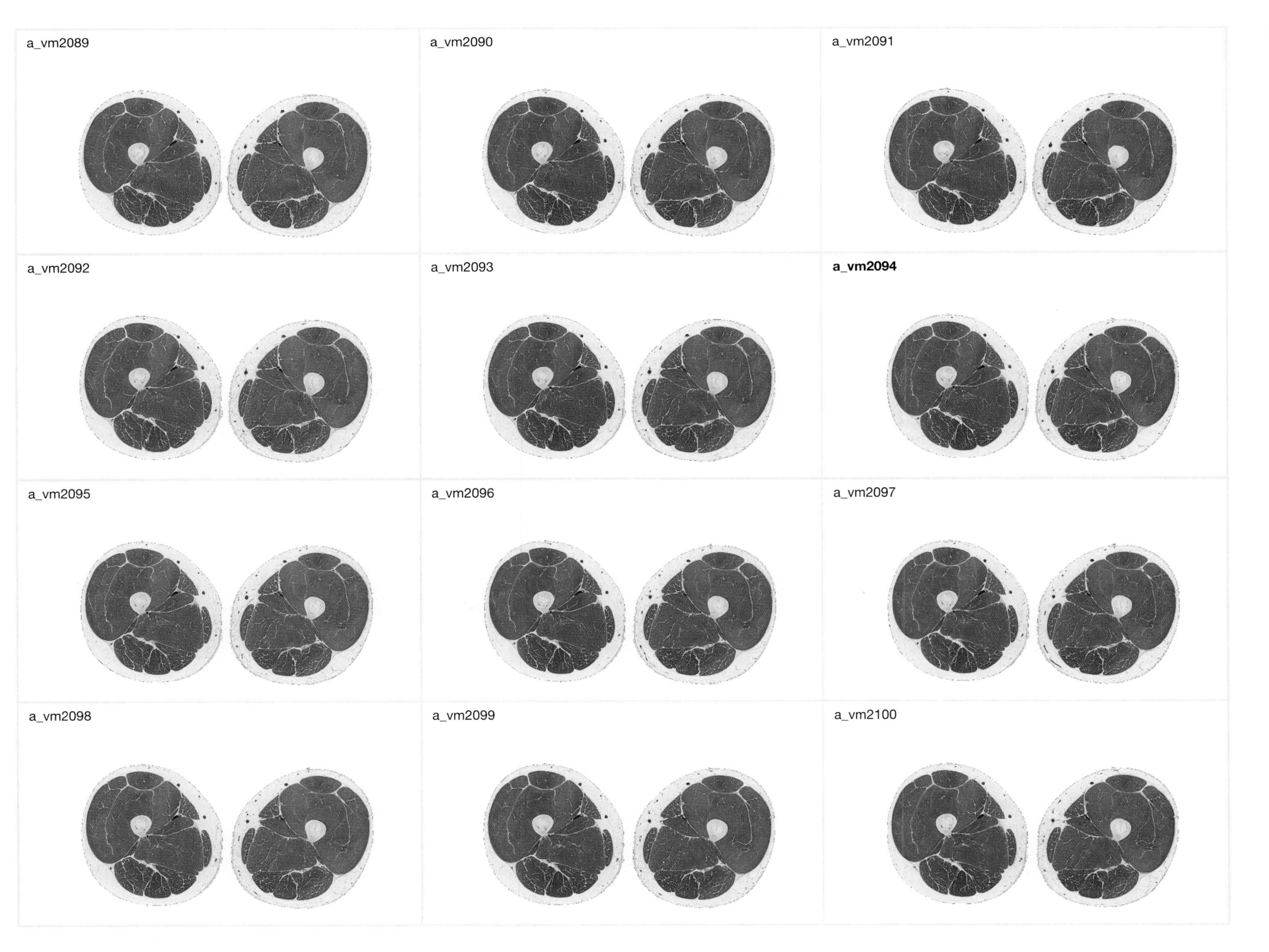
a_vm2089
a_vm2090
a_vm2091
a_vm2092
a_vm2093
a_vm2094
a_vm2095
a_vm2096
a_vm2097
a_vm2098
a_vm2099
a_vm2100

Transverse
a_vm2106

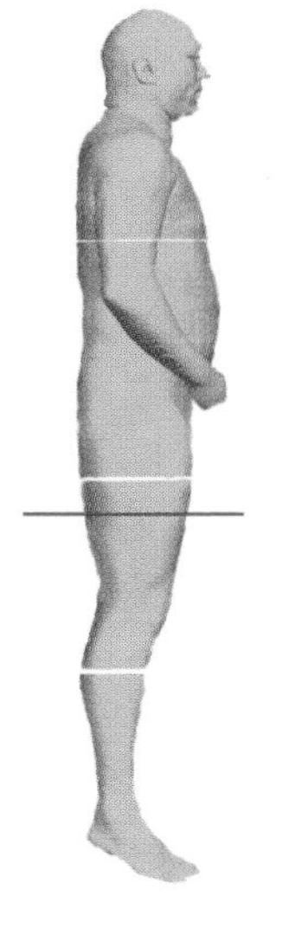

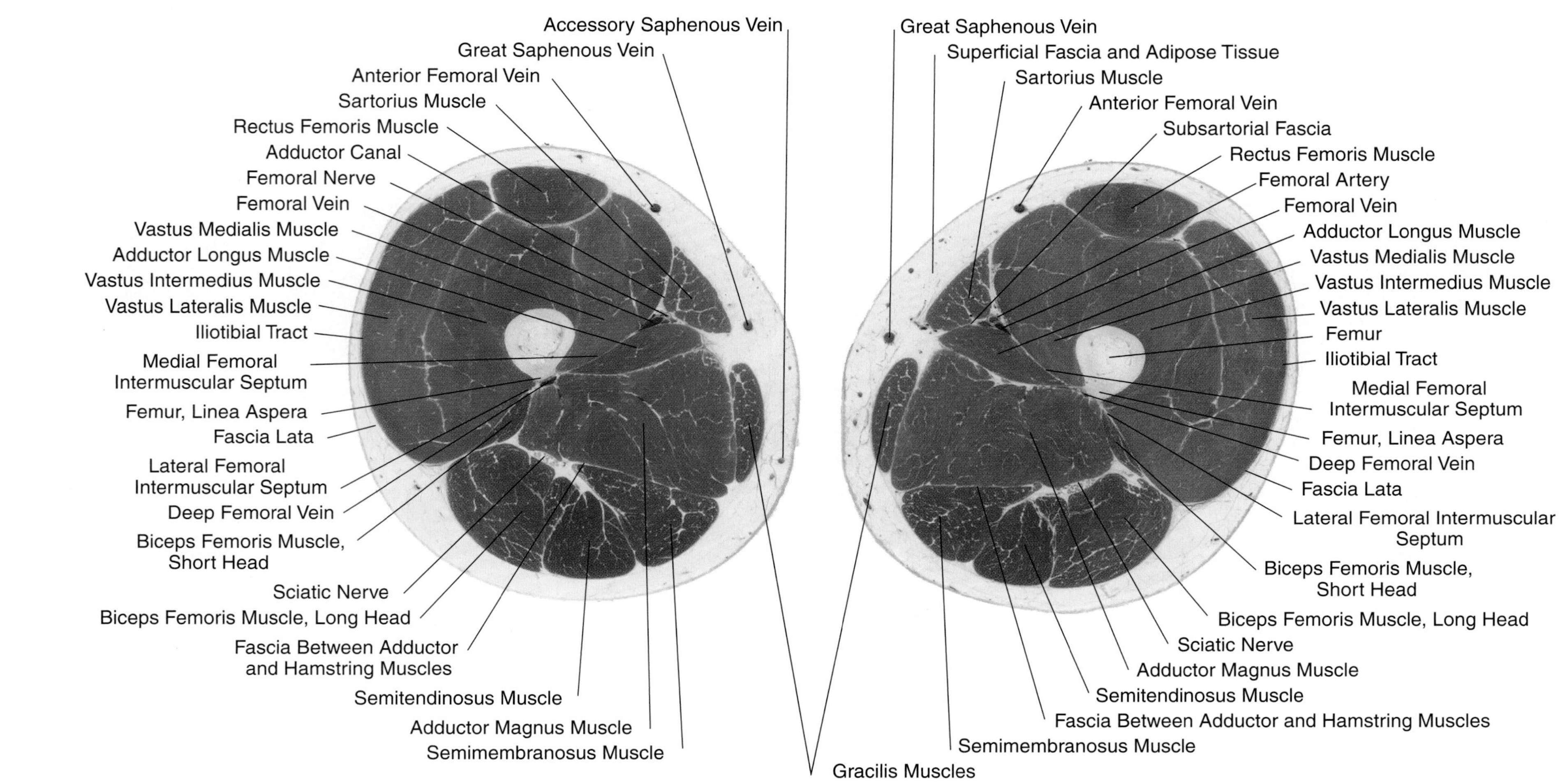
anterior
Accessory Saphenous Vein
Great Saphenous Vein
Anterior Femoral Vein
Sartorius Muscle
Rectus Femoris Muscle
Adductor Canal
Femoral Nerve
Femoral Vein
Vastus Medialis Muscle
Adductor Longus Muscle
Vastus Intermedius Muscle
Vastus Lateralis Muscle
Iliotibial Tract
Medial Femoral Intermuscular Septum
Femur, Linea Aspera
Fascia Lata
Lateral Femoral Intermuscular Septum
Deep Femoral Vein
Biceps Femoris Muscle, Short Head
Sciatic Nerve
Biceps Femoris Muscle, Long Head
Fascia Between Adductor and Hamstring Muscles
Semitendinosus Muscle
Adductor Magnus Muscle
Semimembranosus Muscle
Gracilis Muscles
Great Saphenous Vein
Superficial Fascia and Adipose Tissue
Sartorius Muscle
Anterior Femoral Vein
Subsartorial Fascia
Rectus Femoris Muscle
Femoral Artery
Femoral Vein
Adductor Longus Muscle
Vastus Medialis Muscle
Vastus Intermedius Muscle
Vastus Lateralis Muscle
Femur
Iliotibial Tract
Medial Femoral Intermuscular Septum
Femur, Linea Aspera
Deep Femoral Vein
Fascia Lata
Lateral Femoral Intermuscular Septum
Biceps Femoris Muscle, Short Head
Biceps Femoris Muscle, Long Head
Sciatic Nerve
Adductor Magnus Muscle
Semitendinosus Muscle
Fascia Between Adductor and Hamstring Muscles
Semimembranosus Muscle
right
left
posterior

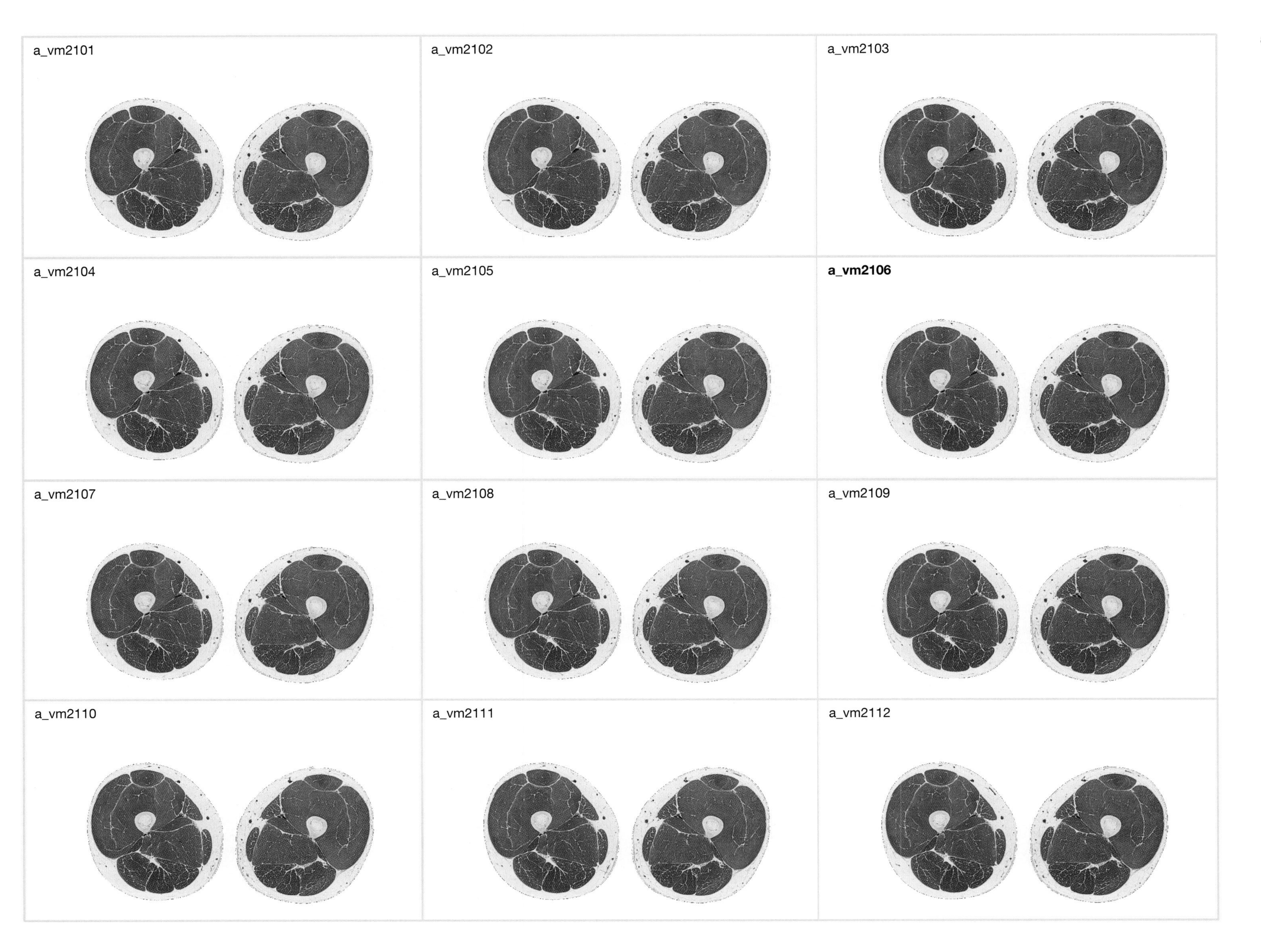
a_vm2101
a_vm2102
a_vm2103
a_vm2104
a_vm2105
a_vm2106
a_vm2107
a_vm2108
a_vm2109
a_vm2110
a_vm2111
a_vm2112

Transverse
a_vm2118

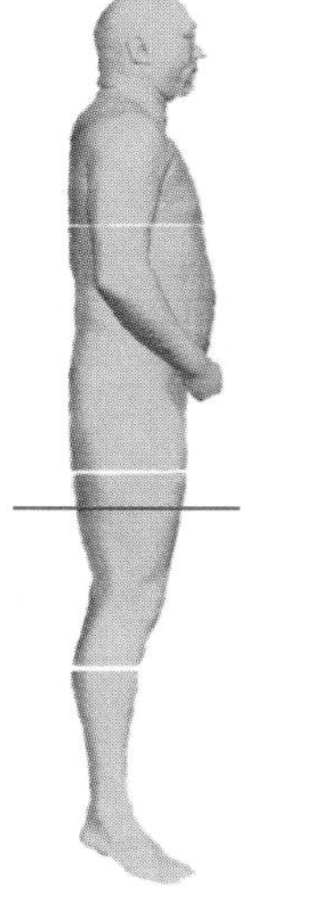

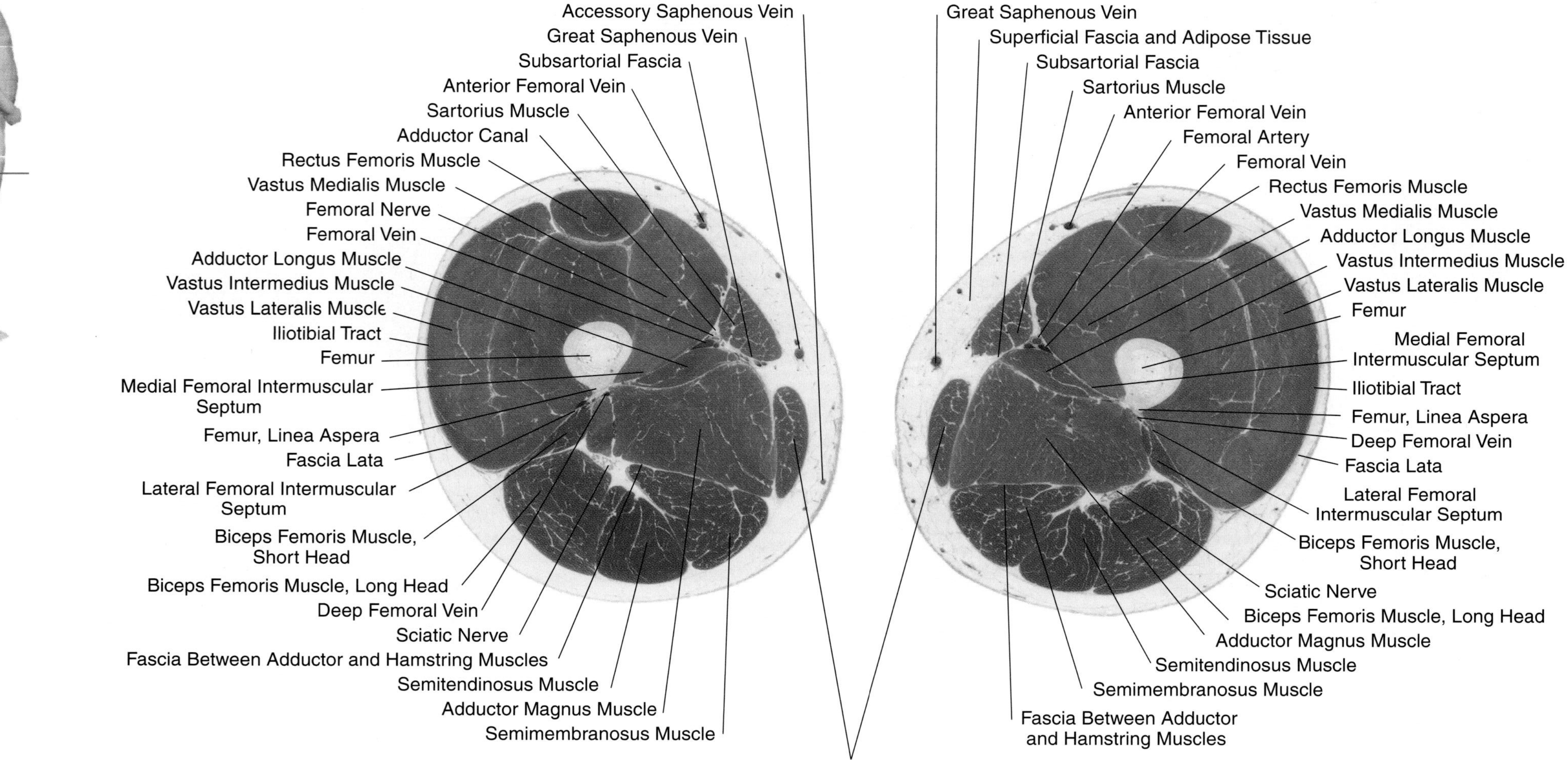
anterior
Accessory Saphenous Vein
Great Saphenous Vein
Subsartorial Fascia
Anterior Femoral Vein
Sartorius Muscle
Adductor Canal
Rectus Femoris Muscle
Vastus Medialis Muscle
Femoral Nerve
Femoral Vein
Adductor Longus Muscle
Vastus Intermedius Muscle
Vastus Lateralis Muscle
Iliotibial Tract
Femur
Medial Femoral Intermuscular Septum
Femur, Linea Aspera
Fascia Lata
Lateral Femoral Intermuscular Septum
Biceps Femoris Muscle, Short Head
Biceps Femoris Muscle, Long Head
Deep Femoral Vein
Sciatic Nerve
Fascia Between Adductor and Hamstring Muscles
Semitendinosus Muscle
Adductor Magnus Muscle
Semimembranosus Muscle
Gracilis Muscles
Great Saphenous Vein
Superficial Fascia and Adipose Tissue
Subsartorial Fascia
Sartorius Muscle
Anterior Femoral Vein
Femoral Artery
Femoral Vein
Rectus Femoris Muscle
Vastus Medialis Muscle
Adductor Longus Muscle
Vastus Intermedius Muscle
Vastus Lateralis Muscle
Femur
Medial Femoral Intermuscular Septum
Iliotibial Tract
Femur, Linea Aspera
Deep Femoral Vein
Fascia Lata
Lateral Femoral Intermuscular Septum
Biceps Femoris Muscle, Short Head
Sciatic Nerve
Biceps Femoris Muscle, Long Head
Adductor Magnus Muscle
Semitendinosus Muscle
Semimembranosus Muscle
Fascia Between Adductor and Hamstring Muscles
right
left
posterior

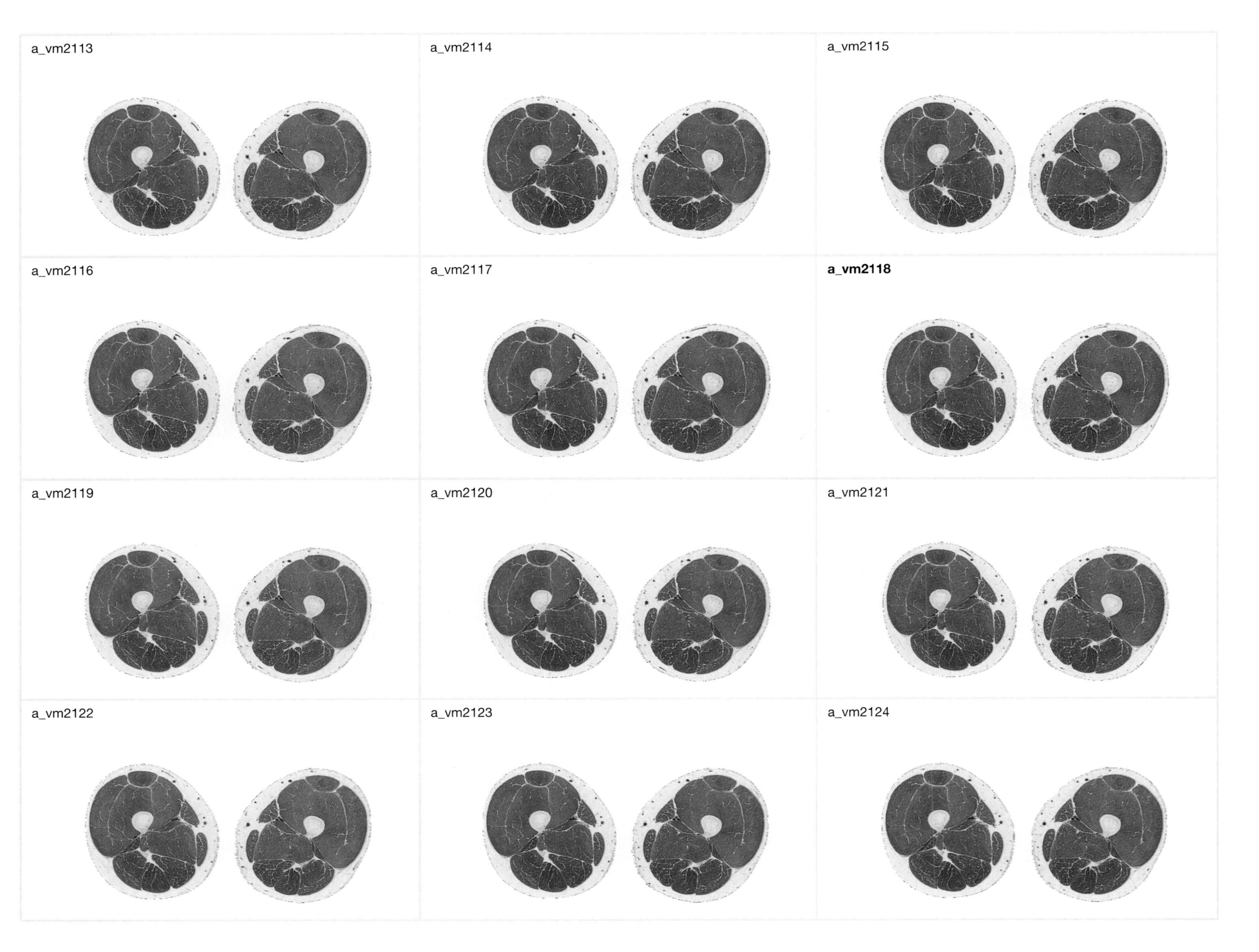
a_vm2113
a_vm2114
a_vm2115
a_vm2116
a_vm2117
a_vm2118
a_vm2119
a_vm2120
a_vm2121
a_vm2122
a_vm2123
a_vm2124

Transverse
a_vm2130

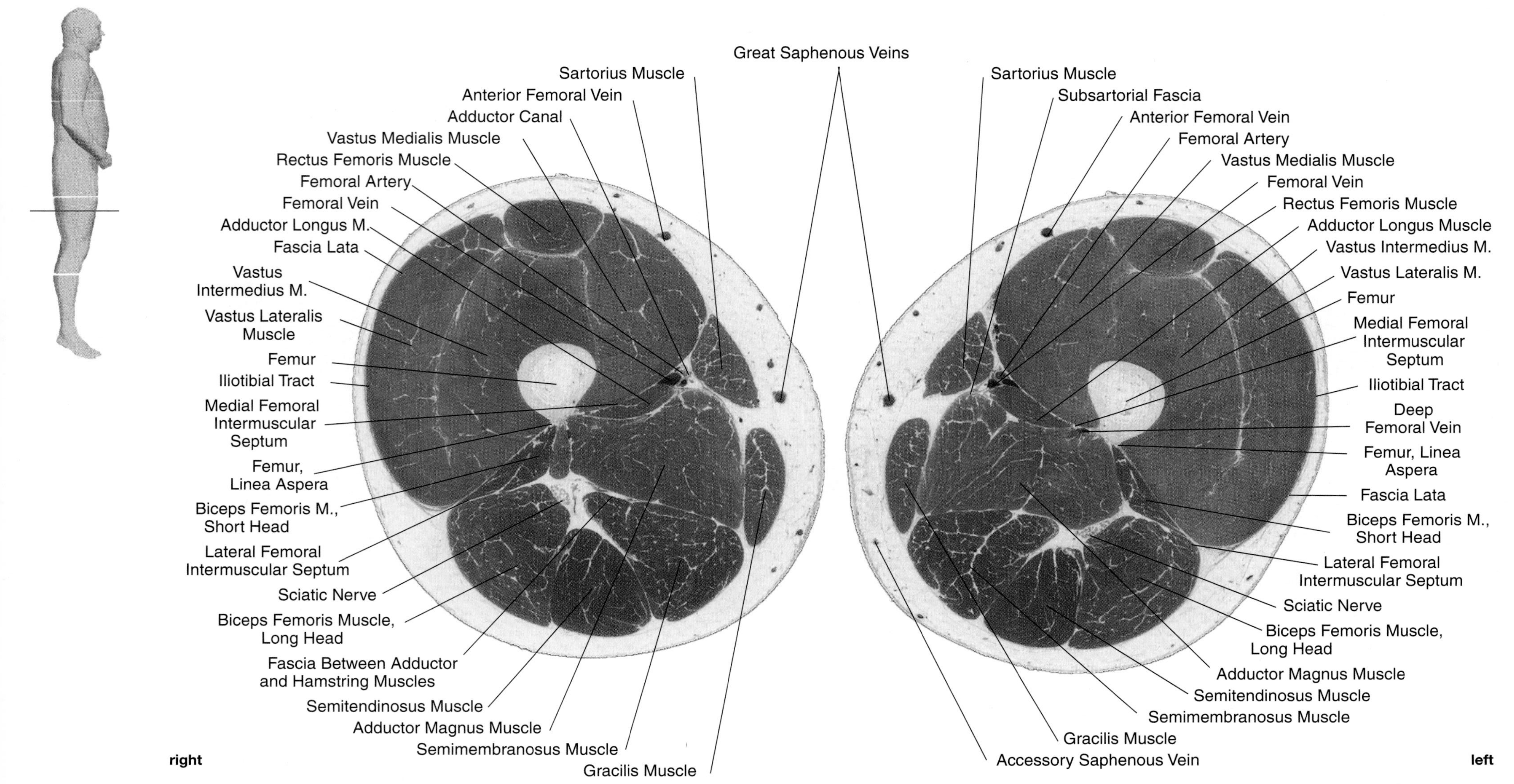

anterior
Great Saphenous Veins
Sartorius Muscle
Anterior Femoral Vein
Adductor Canal
Vastus Medialis Muscle
Rectus Femoris Muscle
Femoral Artery
Femoral Vein
Adductor Longus M.
Fascia Lata
Vastus Intermedius M.
Vastus Lateralis Muscle
Femur
Iliotibial Tract
Medial Femoral Intermuscular Septum
Femur, Linea Aspera
Biceps Femoris M., Short Head
Lateral Femoral Intermuscular Septum
Sciatic Nerve
Biceps Femoris Muscle, Long Head
Fascia Between Adductor and Hamstring Muscles
Semitendinosus Muscle
Adductor Magnus Muscle
Semimembranosus Muscle
Gracilis Muscle
right
Sartorius Muscle
Subsartorial Fascia
Anterior Femoral Vein
Femoral Artery
Vastus Medialis Muscle
Femoral Vein
Rectus Femoris Muscle
Adductor Longus Muscle
Vastus Intermedius M.
Vastus Lateralis M.
Femur
Medial Femoral Intermuscular Septum
Iliotibial Tract
Deep Femoral Vein
Femur, Linea Aspera
Fascia Lata
Biceps Femoris M., Short Head
Lateral Femoral Intermuscular Septum
Sciatic Nerve
Biceps Femoris Muscle, Long Head
Adductor Magnus Muscle
Semitendinosus Muscle
Semimembranosus Muscle
Gracilis Muscle
Accessory Saphenous Vein
left
posterior

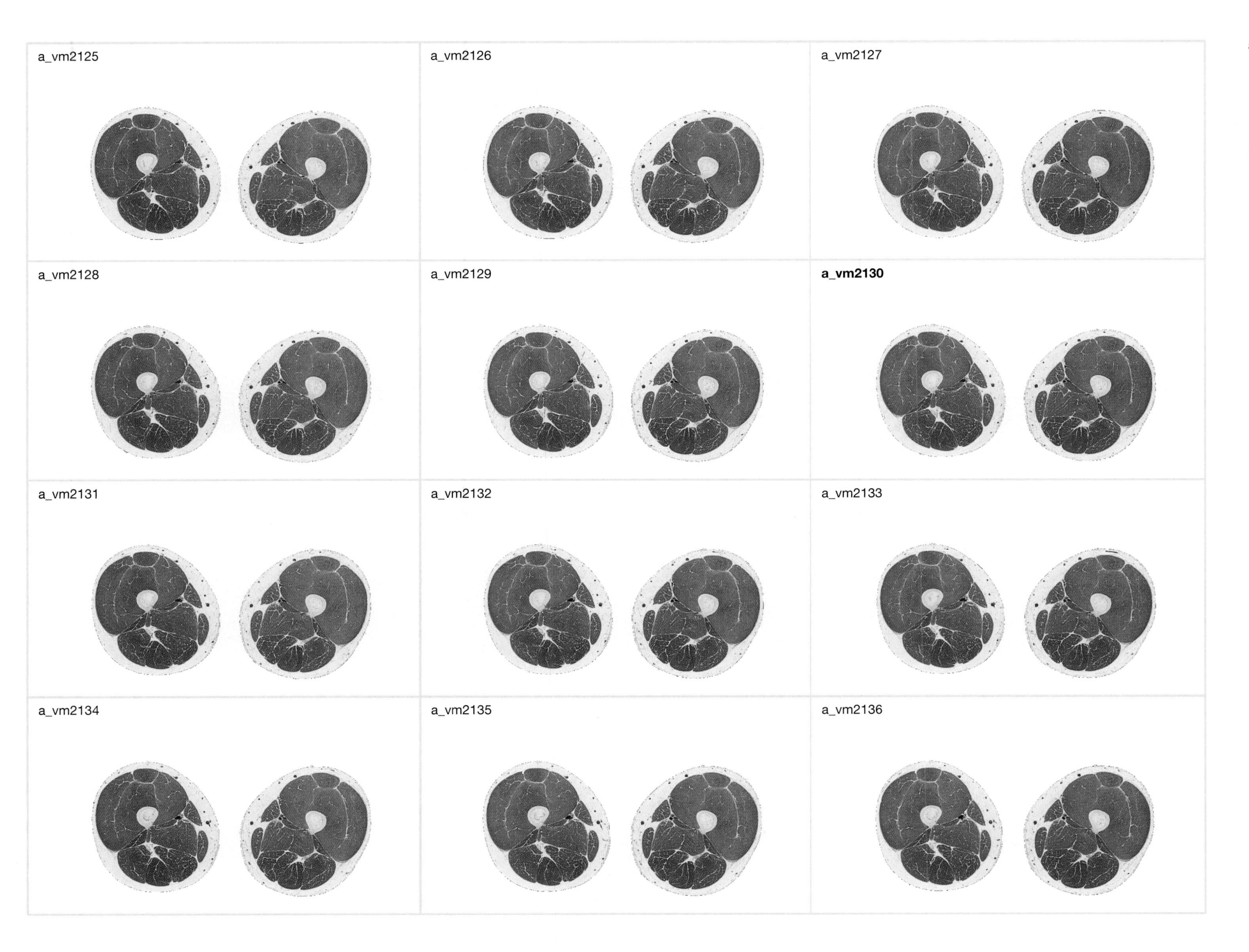
a_vm2125
a_vm2126
a_vm2127
a_vm2128
a_vm2129
a_vm2130
a_vm2131
a_vm2132
a_vm2133
a_vm2134
a_vm2135
a_vm2136

Transverse
a_vm2142

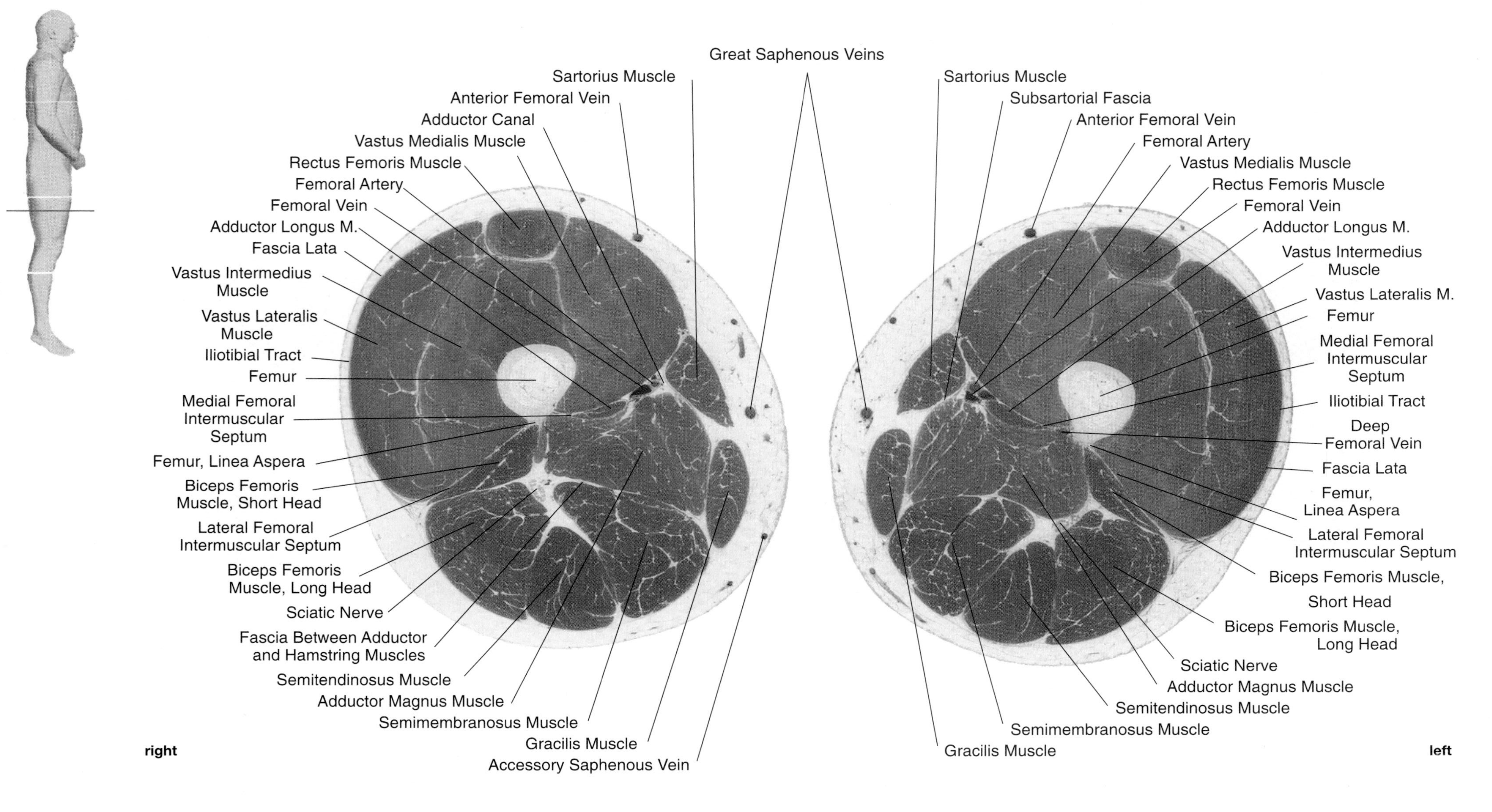

anterior
Great Saphenous Veins
Sartorius Muscle
Anterior Femoral Vein
Adductor Canal
Vastus Medialis Muscle
Rectus Femoris Muscle
Femoral Artery
Femoral Vein
Adductor Longus M.
Fascia Lata
Vastus Intermedius Muscle
Vastus Lateralis Muscle
Iliotibial Tract
Femur
Medial Femoral Intermuscular Septum
Femur, Linea Aspera
Biceps Femoris Muscle, Short Head
Lateral Femoral Intermuscular Septum
Biceps Femoris Muscle, Long Head
Sciatic Nerve
Fascia Between Adductor and Hamstring Muscles
Semitendinosus Muscle
Adductor Magnus Muscle
Semimembranosus Muscle
Gracilis Muscle
Accessory Saphenous Vein
right
Sartorius Muscle
Subsartorial Fascia
Anterior Femoral Vein
Femoral Artery
Vastus Medialis Muscle
Rectus Femoris Muscle
Femoral Vein
Adductor Longus M.
Vastus Intermedius Muscle
Vastus Lateralis M.
Femur
Medial Femoral Intermuscular Septum
Iliotibial Tract
Deep Femoral Vein
Fascia Lata
Femur, Linea Aspera
Lateral Femoral Intermuscular Septum
Biceps Femoris Muscle, Short Head
Biceps Femoris Muscle, Long Head
Sciatic Nerve
Adductor Magnus Muscle
Semitendinosus Muscle
Semimembranosus Muscle
Gracilis Muscle
left
posterior

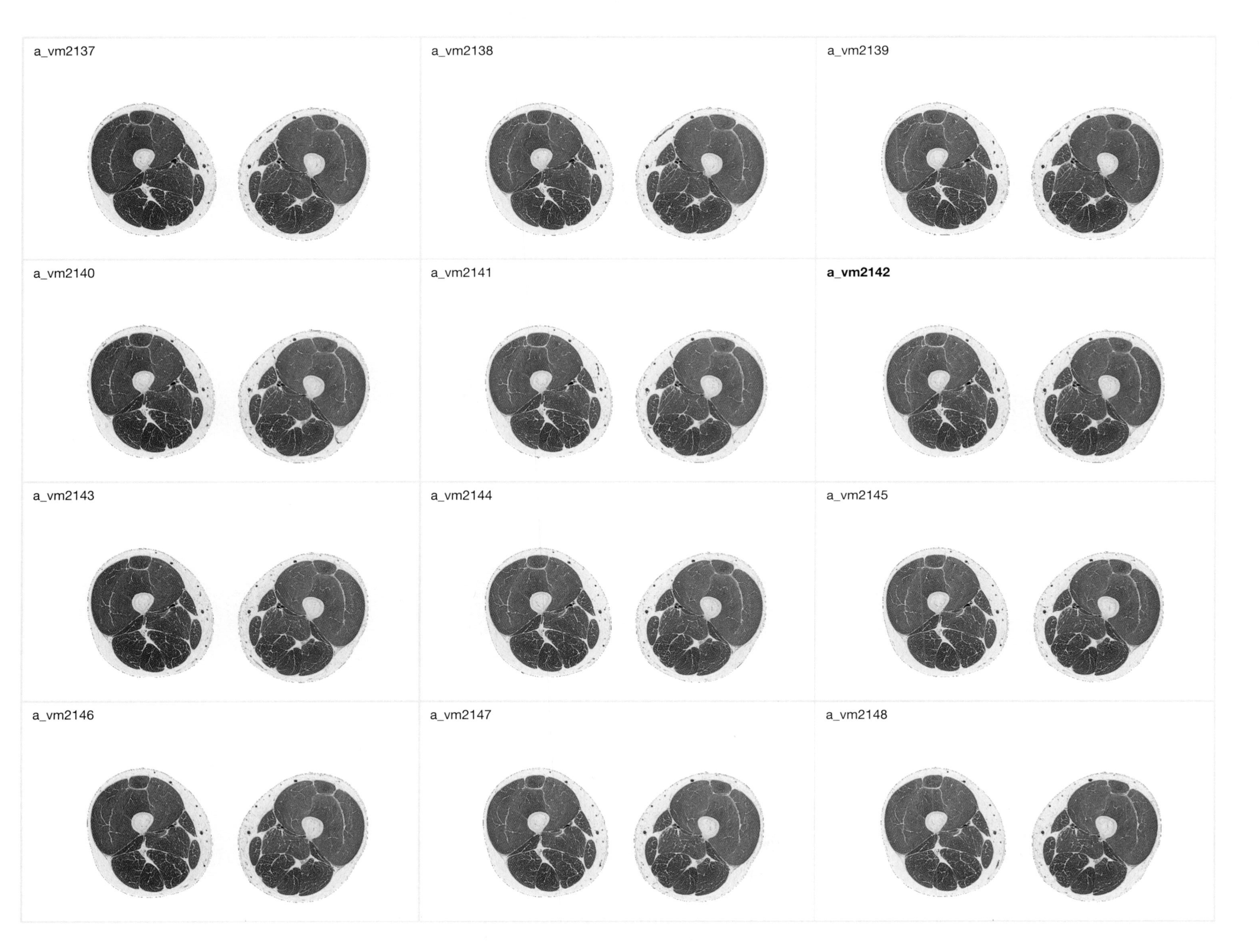
a_vm2137
a_vm2138
a_vm2139
a_vm2140
a_vm2141
a_vm2142
a_vm2143
a_vm2144
a_vm2145
a_vm2146
a_vm2147
a_vm2148

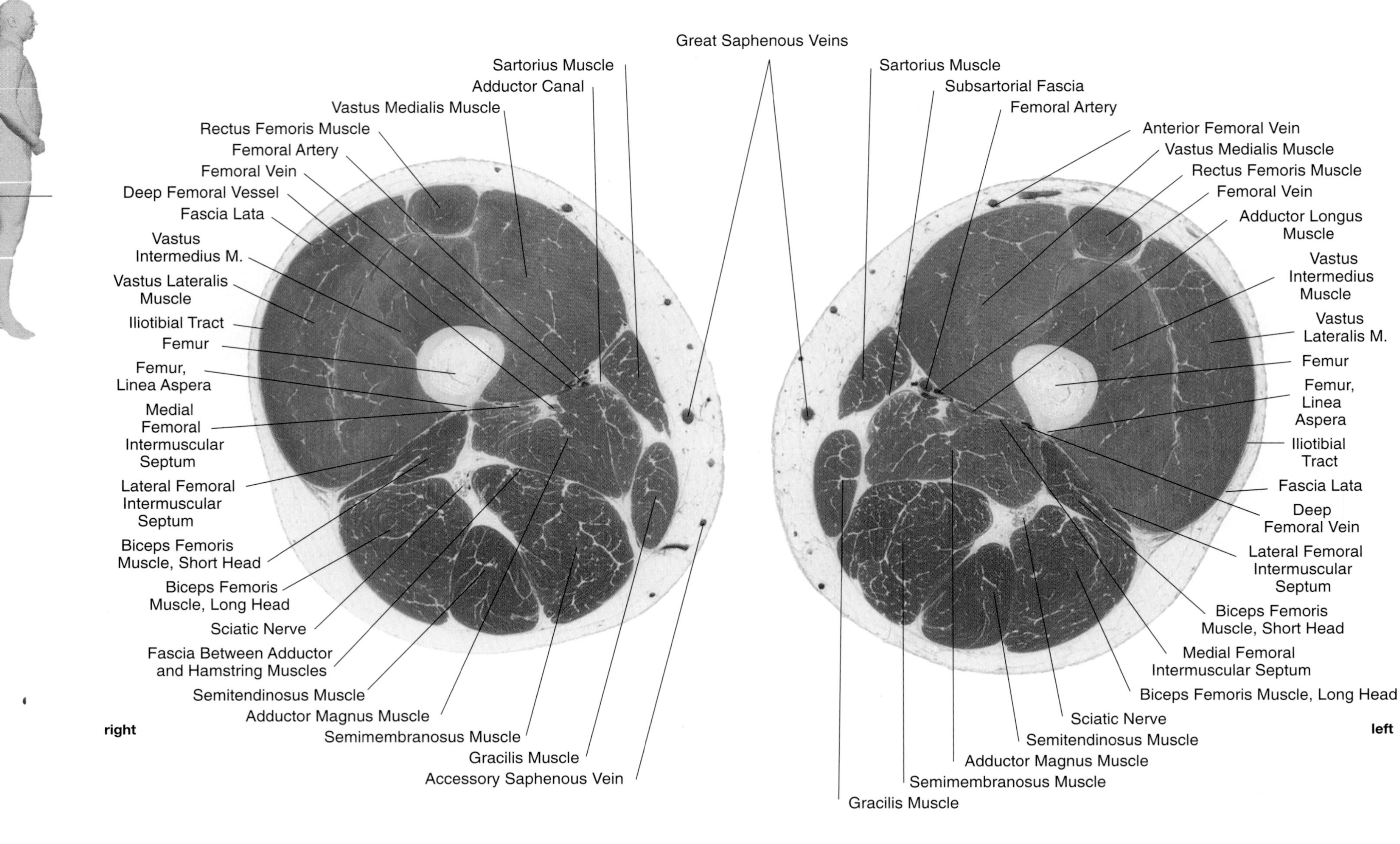
Transverse
a_vm2154
anterior
Great Saphenous Veins
Sartorius Muscle
Adductor Canal
Vastus Medialis Muscle
Rectus Femoris Muscle
Femoral Artery
Femoral Vein
Deep Femoral Vessel
Fascia Lata
Vastus Intermedius M.
Vastus Lateralis Muscle
Iliotibial Tract
Femur
Femur, Linea Aspera
Medial Femoral Intermuscular Septum
Lateral Femoral Intermuscular Septum
Biceps Femoris Muscle, Short Head
Biceps Femoris Muscle, Long Head
Sciatic Nerve
Fascia Between Adductor and Hamstring Muscles
Semitendinosus Muscle
Adductor Magnus Muscle
Semimembranosus Muscle
Gracilis Muscle
Accessory Saphenous Vein
right
Sartorius Muscle
Subsartorial Fascia
Femoral Artery
Anterior Femoral Vein
Vastus Medialis Muscle
Rectus Femoris Muscle
Femoral Vein
Adductor Longus Muscle
Vastus Intermedius Muscle
Vastus Lateralis M.
Femur
Femur, Linea Aspera
Iliotibial Tract
Fascia Lata
Deep Femoral Vein
Lateral Femoral Intermuscular Septum
Biceps Femoris Muscle, Short Head
Medial Femoral Intermuscular Septum
Biceps Femoris Muscle, Long Head
Sciatic Nerve
Semitendinosus Muscle
Adductor Magnus Muscle
Semimembranosus Muscle
Gracilis Muscle
left
posterior

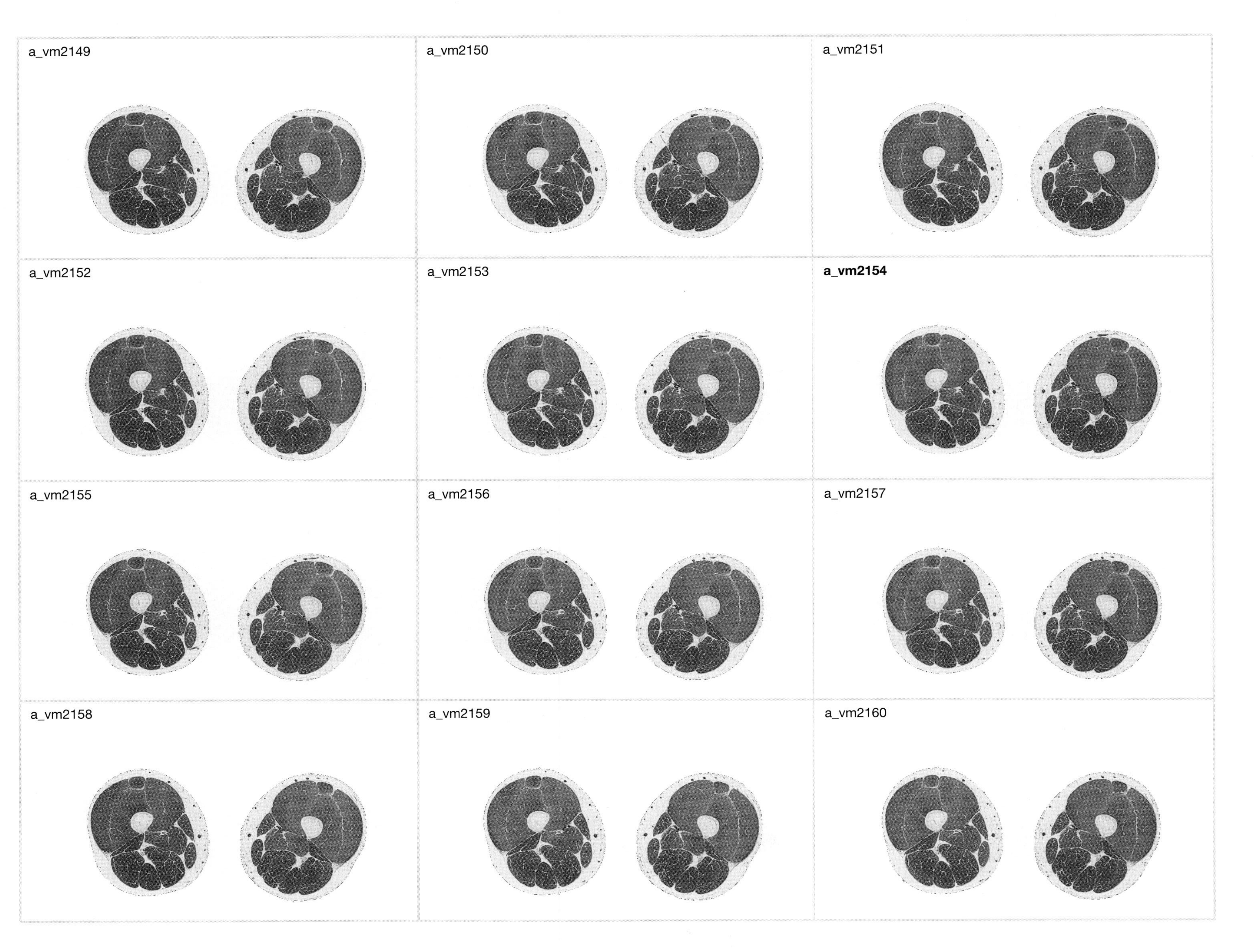
a_vm2149
a_vm2150
a_vm2151
a_vm2152
a_vm2153
a_vm2154
a_vm2155
a_vm2156
a_vm2157
a_vm2158
a_vm2159
a_vm2160

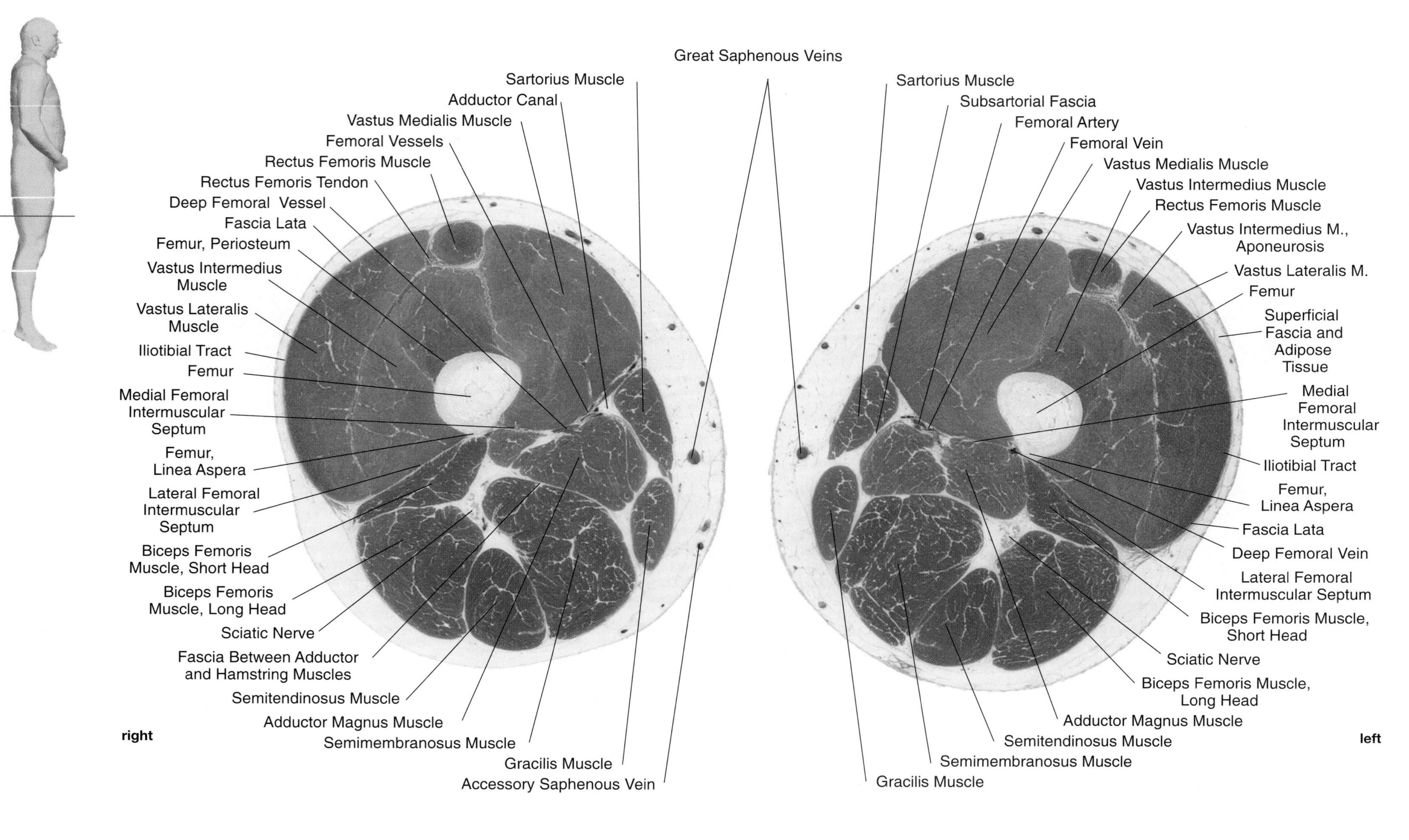
anterior
Great Saphenous Veins
Sartorius Muscle
Adductor Canal
Vastus Medialis Muscle
Femoral Vessels
Rectus Femoris Muscle
Rectus Femoris Tendon
Deep Femoral Vessel
Fascia Lata
Femur, Periosteum
Vastus Intermedius Muscle
Vastus Lateralis Muscle
Iliotibial Tract
Femur
Medial Femoral Intermuscular Septum
Femur, Linea Aspera
Lateral Femoral Intermuscular Septum
Biceps Femoris Muscle, Short Head
Biceps Femoris Muscle, Long Head
Sciatic Nerve
Fascia Between Adductor and Hamstring Muscles
Semitendinosus Muscle
Adductor Magnus Muscle
Semimembranosus Muscle
Gracilis Muscle
Accessory Saphenous Vein
right
Sartorius Muscle
Subsartorial Fascia
Femoral Artery
Femoral Vein
Vastus Medialis Muscle
Vastus Intermedius Muscle
Rectus Femoris Muscle
Vastus Intermedius M., Aponeurosis
Vastus Lateralis M.
Femur
Superficial Fascia and Adipose Tissue
Medial Femoral Intermuscular Septum
Iliotibial Tract
Femur, Linea Aspera
Fascia Lata
Deep Femoral Vein
Lateral Femoral Intermuscular Septum
Biceps Femoris Muscle, Short Head
Sciatic Nerve
Biceps Femoris Muscle, Long Head
Adductor Magnus Muscle
Semitendinosus Muscle
Semimembranosus Muscle
Gracilis Muscle
left
posterior

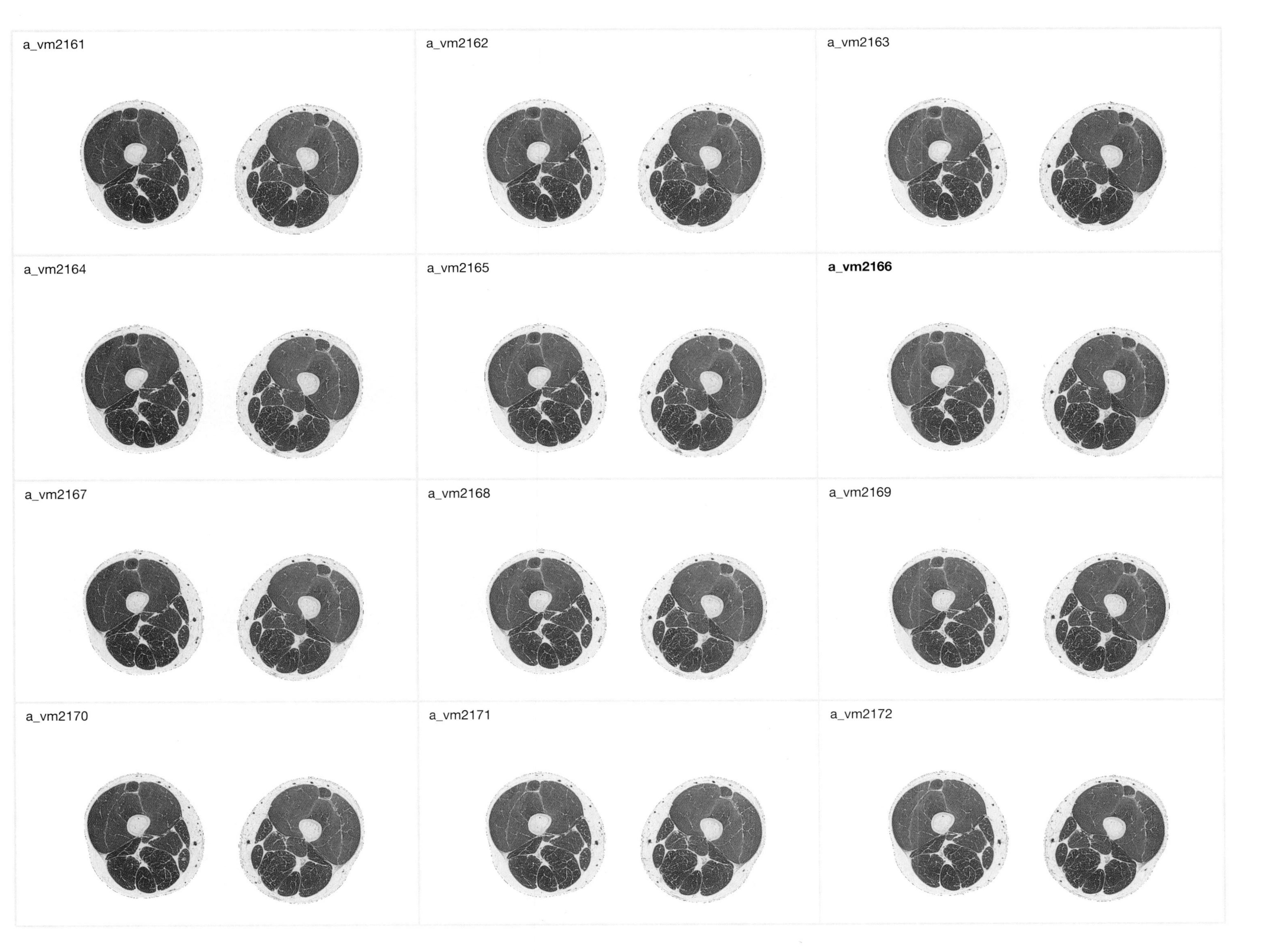
a_vm2161
a_vm2162
a_vm2163
a_vm2164
a_vm2165
a_vm2166
a_vm2167
a_vm2168
a_vm2169
a_vm2170
a_vm2171
a_vm2172

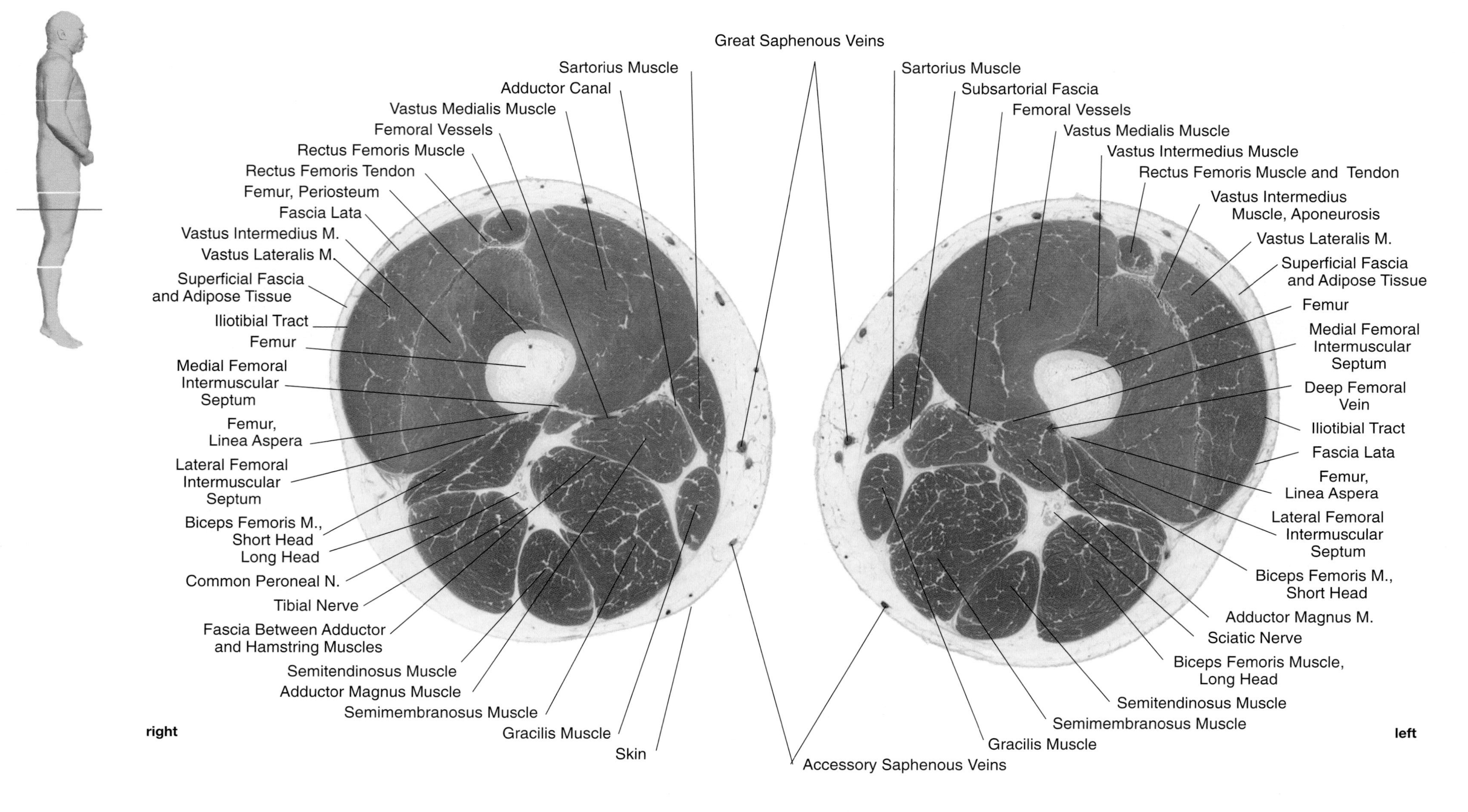
Transverse
a_vm2178
anterior
Great Saphenous Veins
Sartorius Muscle
Adductor Canal
Vastus Medialis Muscle
Femoral Vessels
Rectus Femoris Muscle
Rectus Femoris Tendon
Femur, Periosteum
Fascia Lata
Vastus Intermedius M.
Vastus Lateralis M.
Superficial Fascia and Adipose Tissue
Iliotibial Tract
Femur
Medial Femoral Intermuscular Septum
Femur, Linea Aspera
Lateral Femoral Intermuscular Septum
Biceps Femoris M., Short Head
Long Head
Common Peroneal N.
Tibial Nerve
Fascia Between Adductor and Hamstring Muscles
Semitendinosus Muscle
Adductor Magnus Muscle
Semimembranosus Muscle
Gracilis Muscle
Skin
right
Sartorius Muscle
Subsartorial Fascia
Femoral Vessels
Vastus Medialis Muscle
Vastus Intermedius Muscle
Rectus Femoris Muscle and Tendon
Vastus Intermedius Muscle, Aponeurosis
Vastus Lateralis M.
Superficial Fascia and Adipose Tissue
Femur
Medial Femoral Intermuscular Septum
Deep Femoral Vein
Iliotibial Tract
Fascia Lata
Femur, Linea Aspera
Lateral Femoral Intermuscular Septum
Biceps Femoris M., Short Head
Adductor Magnus M.
Sciatic Nerve
Biceps Femoris Muscle, Long Head
Semitendinosus Muscle
Semimembranosus Muscle
Gracilis Muscle
Accessory Saphenous Veins
left
posterior

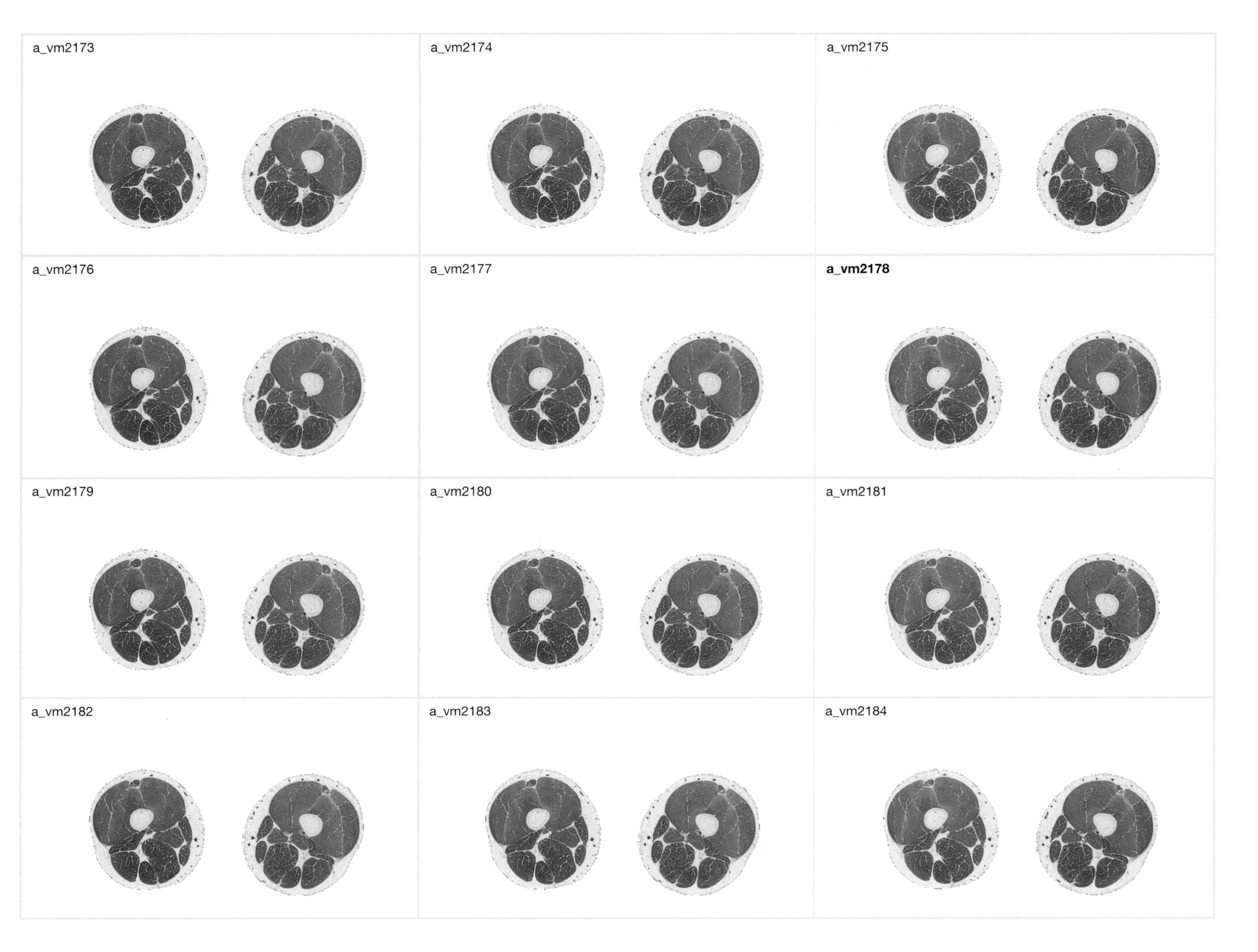
a_vm2173
a_vm2174
a_vm2175
a_vm2176
a_vm2177
a_vm2178
a_vm2179
a_vm2180
a_vm2181
a_vm2182
a_vm2183
a_vm2184

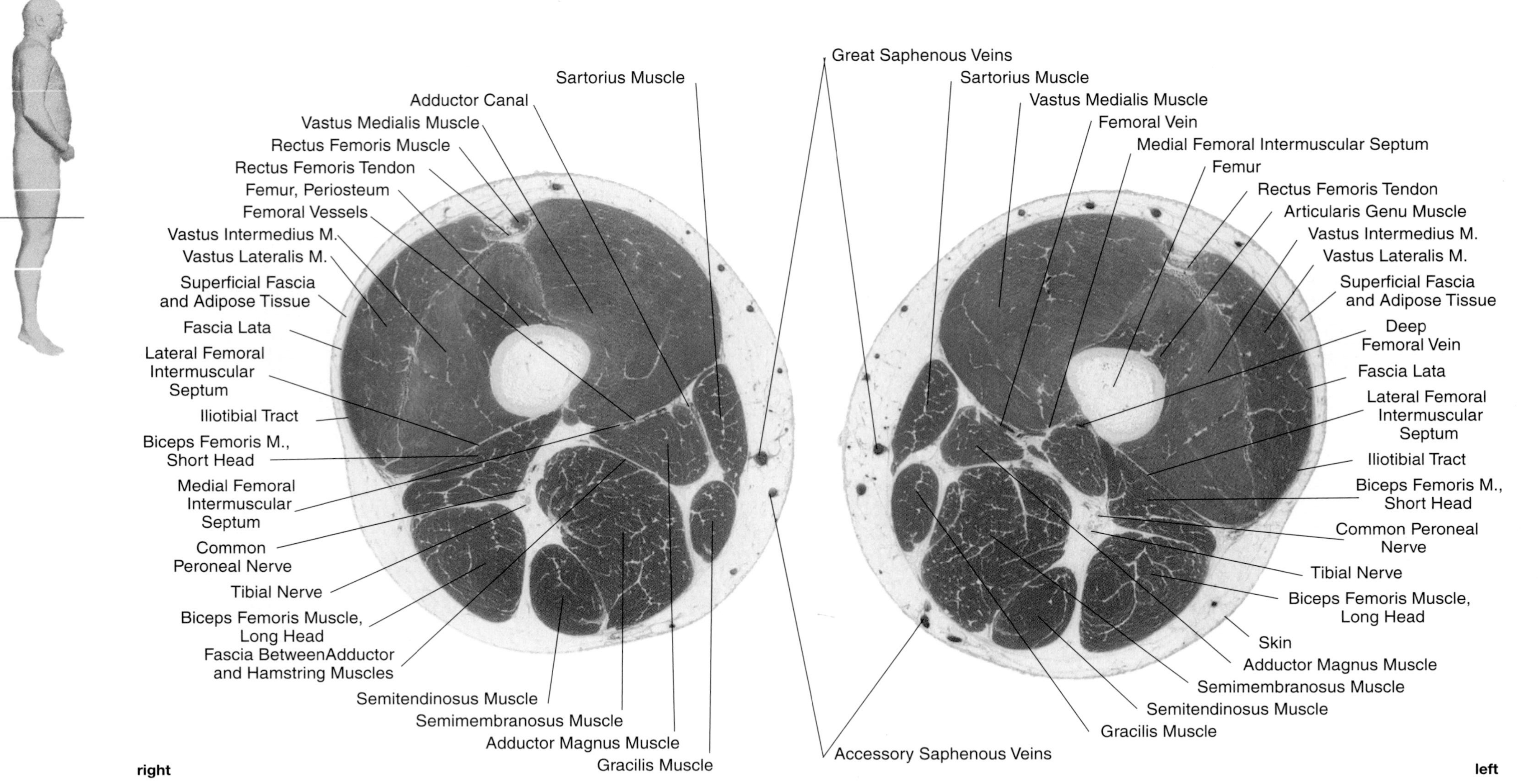
anterior
Sartorius Muscle
Adductor Canal
Vastus Medialis Muscle
Rectus Femoris Muscle
Rectus Femoris Tendon
Femur, Periosteum
Femoral Vessels
Vastus Intermedius M.
Vastus Lateralis M.
Superficial Fascia
and Adipose Tissue
Fascia Lata
Lateral Femoral
Intermuscular
Septum
Iliotibial Tract
Biceps Femoris M.,
Short Head
Medial Femoral
Intermuscular
Septum
Common
Peroneal Nerve
Tibial Nerve
Biceps Femoris Muscle,
Long Head
Fascia BetweenAdductor
and Hamstring Muscles
Semitendinosus Muscle
Semimembranosus Muscle
Adductor Magnus Muscle
Gracilis Muscle
right
Great Saphenous Veins
Sartorius Muscle
Vastus Medialis Muscle
Femoral Vein
Medial Femoral Intermuscular Septum
Femur
Rectus Femoris Tendon
Articularis Genu Muscle
Vastus Intermedius M.
Vastus Lateralis M.
Superficial Fascia
and Adipose Tissue
Deep
Femoral Vein
Fascia Lata
Lateral Femoral
Intermuscular
Septum
Iliotibial Tract
Biceps Femoris M.,
Short Head
Common Peroneal
Nerve
Tibial Nerve
Biceps Femoris Muscle,
Long Head
Skin
Adductor Magnus Muscle
Semimembranosus Muscle
Semitendinosus Muscle
Gracilis Muscle
Accessory Saphenous Veins
left
posterior

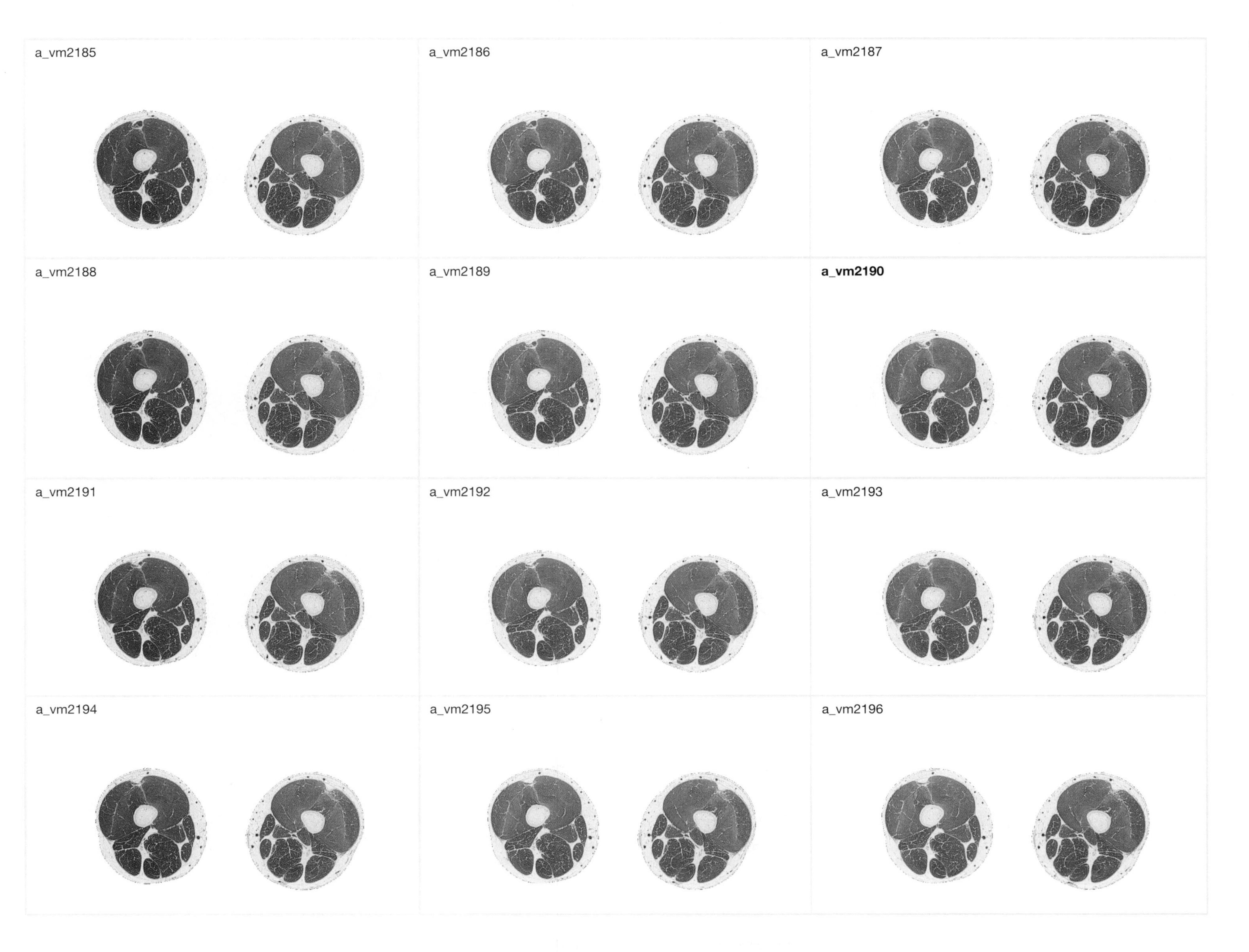
a_vm2185
a_vm2186
a_vm2187
a_vm2188
a_vm2189
a_vm2190
a_vm2191
a_vm2192
a_vm2193
a_vm2194
a_vm2195
a_vm2196

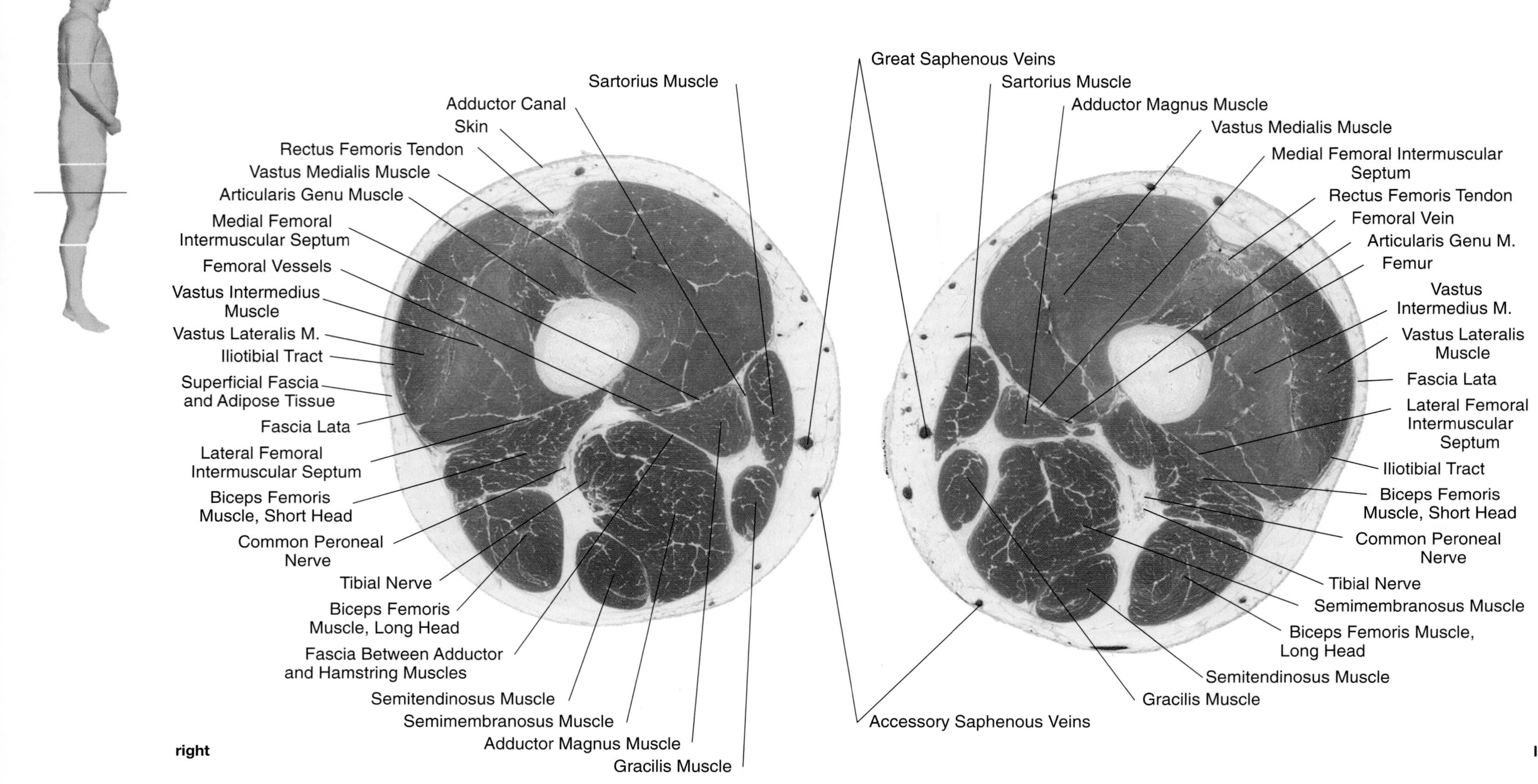
Transverse
a_vm2202
anterior
Great Saphenous Veins
Sartorius Muscle
Adductor Canal
Skin
Rectus Femoris Tendon
Vastus Medialis Muscle
Articularis Genu Muscle
Medial Femoral Intermuscular Septum
Femoral Vessels
Vastus Intermedius Muscle
Vastus Lateralis M.
Iliotibial Tract
Superficial Fascia and Adipose Tissue
Fascia Lata
Lateral Femoral Intermuscular Septum
Biceps Femoris Muscle, Short Head
Common Peroneal Nerve
Tibial Nerve
Biceps Femoris Muscle, Long Head
Fascia Between Adductor and Hamstring Muscles
Semitendinosus Muscle
Semimembranosus Muscle
Adductor Magnus Muscle
Gracilis Muscle
right
Sartorius Muscle
Adductor Magnus Muscle
Vastus Medialis Muscle
Medial Femoral Intermuscular Septum
Rectus Femoris Tendon
Femoral Vein
Articularis Genu M.
Femur
Vastus Intermedius M.
Vastus Lateralis Muscle
Fascia Lata
Lateral Femoral Intermuscular Septum
Iliotibial Tract
Biceps Femoris Muscle, Short Head
Common Peroneal Nerve
Tibial Nerve
Semimembranosus Muscle
Biceps Femoris Muscle, Long Head
Semitendinosus Muscle
Gracilis Muscle
Accessory Saphenous Veins
left
posterior

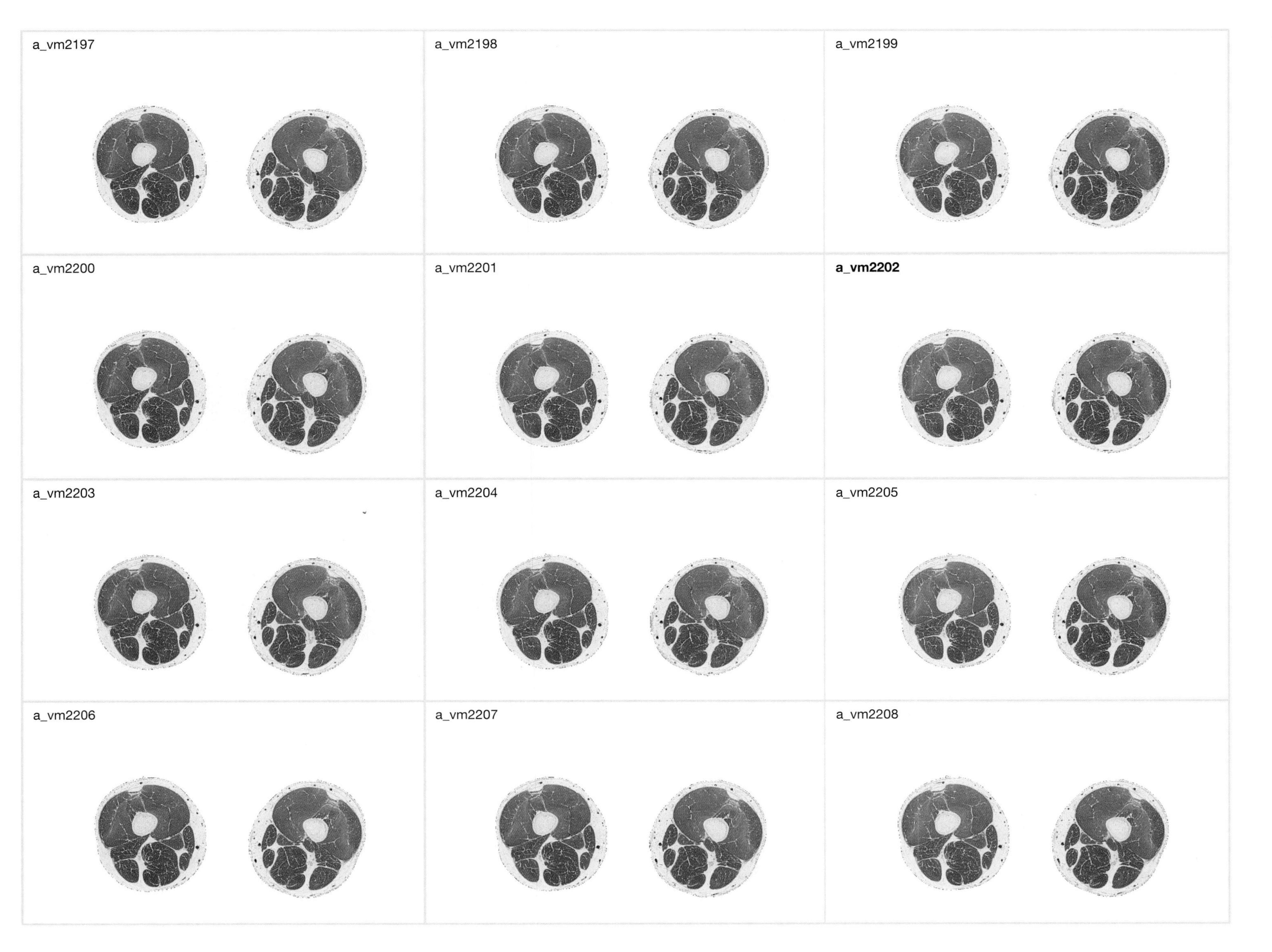
a_vm2197
a_vm2198
a_vm2199
a_vm2200
a_vm2201
a_vm2202
a_vm2203
a_vm2204
a_vm2205
a_vm2206
a_vm2207
a_vm2208

Transverse
a_vm2214

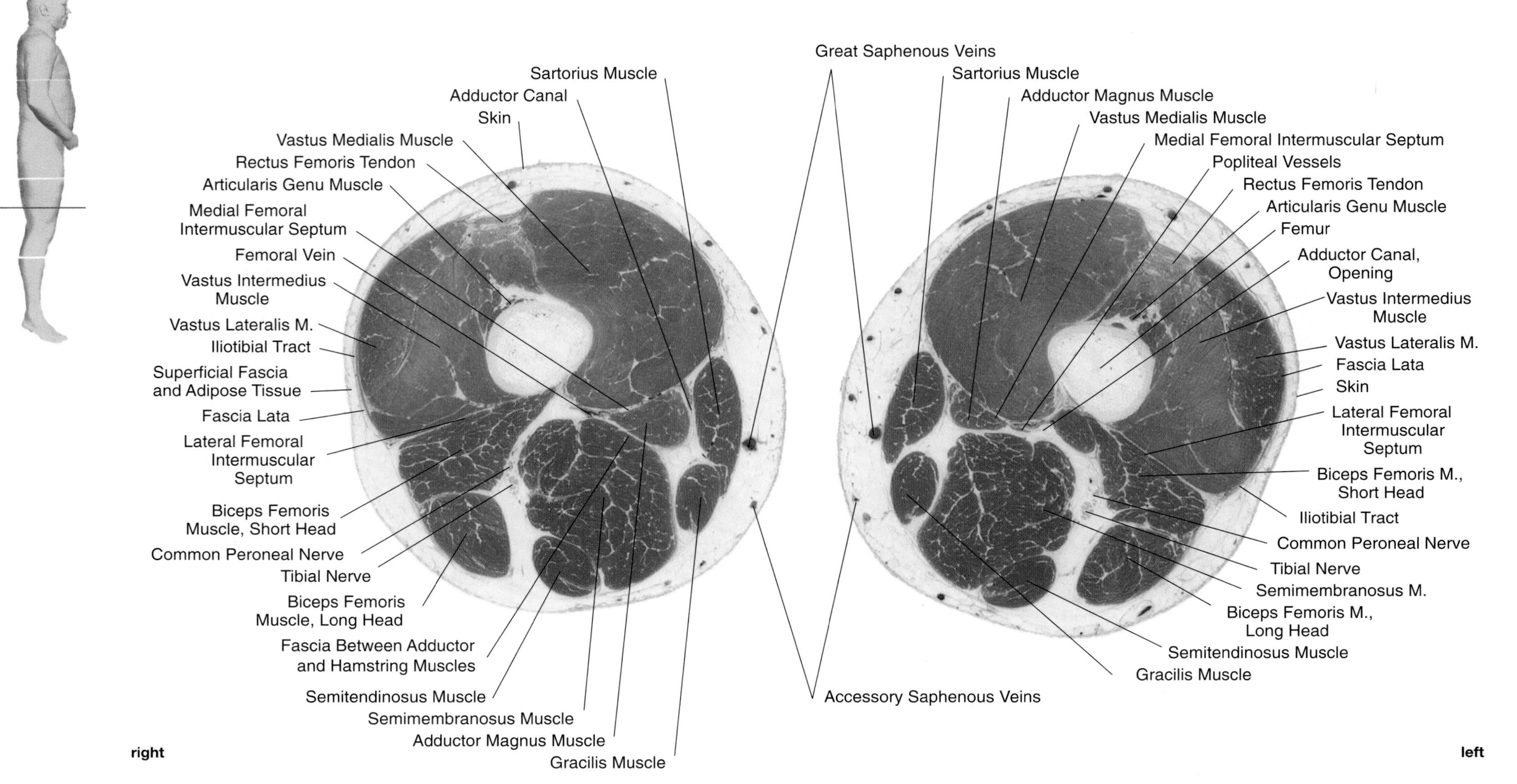

anterior
Sartorius Muscle
Adductor Canal
Skin
Vastus Medialis Muscle
Rectus Femoris Tendon
Articularis Genu Muscle
Medial Femoral Intermuscular Septum
Femoral Vein
Vastus Intermedius Muscle
Vastus Lateralis M.
Iliotibial Tract
Superficial Fascia and Adipose Tissue
Fascia Lata
Lateral Femoral Intermuscular Septum
Biceps Femoris Muscle, Short Head
Common Peroneal Nerve
Tibial Nerve
Biceps Femoris Muscle, Long Head
Fascia Between Adductor and Hamstring Muscles
Semitendinosus Muscle
Semimembranosus Muscle
Adductor Magnus Muscle
Gracilis Muscle
Great Saphenous Veins
Accessory Saphenous Veins
Sartorius Muscle
Adductor Magnus Muscle
Vastus Medialis Muscle
Medial Femoral Intermuscular Septum
Popliteal Vessels
Rectus Femoris Tendon
Articularis Genu Muscle
Femur
Adductor Canal, Opening
Vastus Intermedius Muscle
Vastus Lateralis M.
Fascia Lata
Skin
Lateral Femoral Intermuscular Septum
Biceps Femoris M., Short Head
Iliotibial Tract
Common Peroneal Nerve
Tibial Nerve
Semimembranosus M.
Biceps Femoris M., Long Head
Semitendinosus Muscle
Gracilis Muscle
right
left
posterior

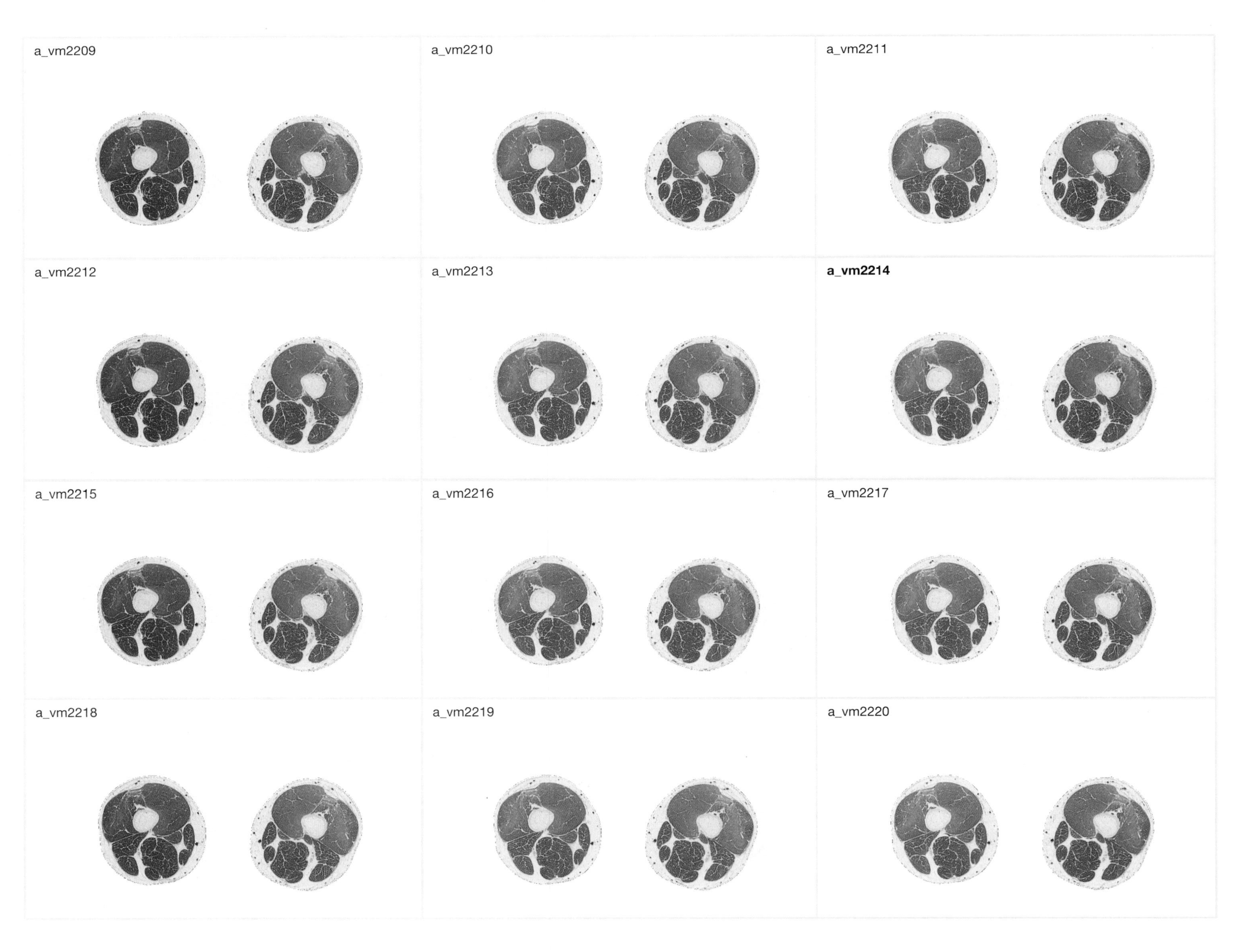
a_vm2209
a_vm2210
a_vm2211
a_vm2212
a_vm2213
a_vm2214
a_vm2215
a_vm2216
a_vm2217
a_vm2218
a_vm2219
a_vm2220

Transverse
a_vm2226

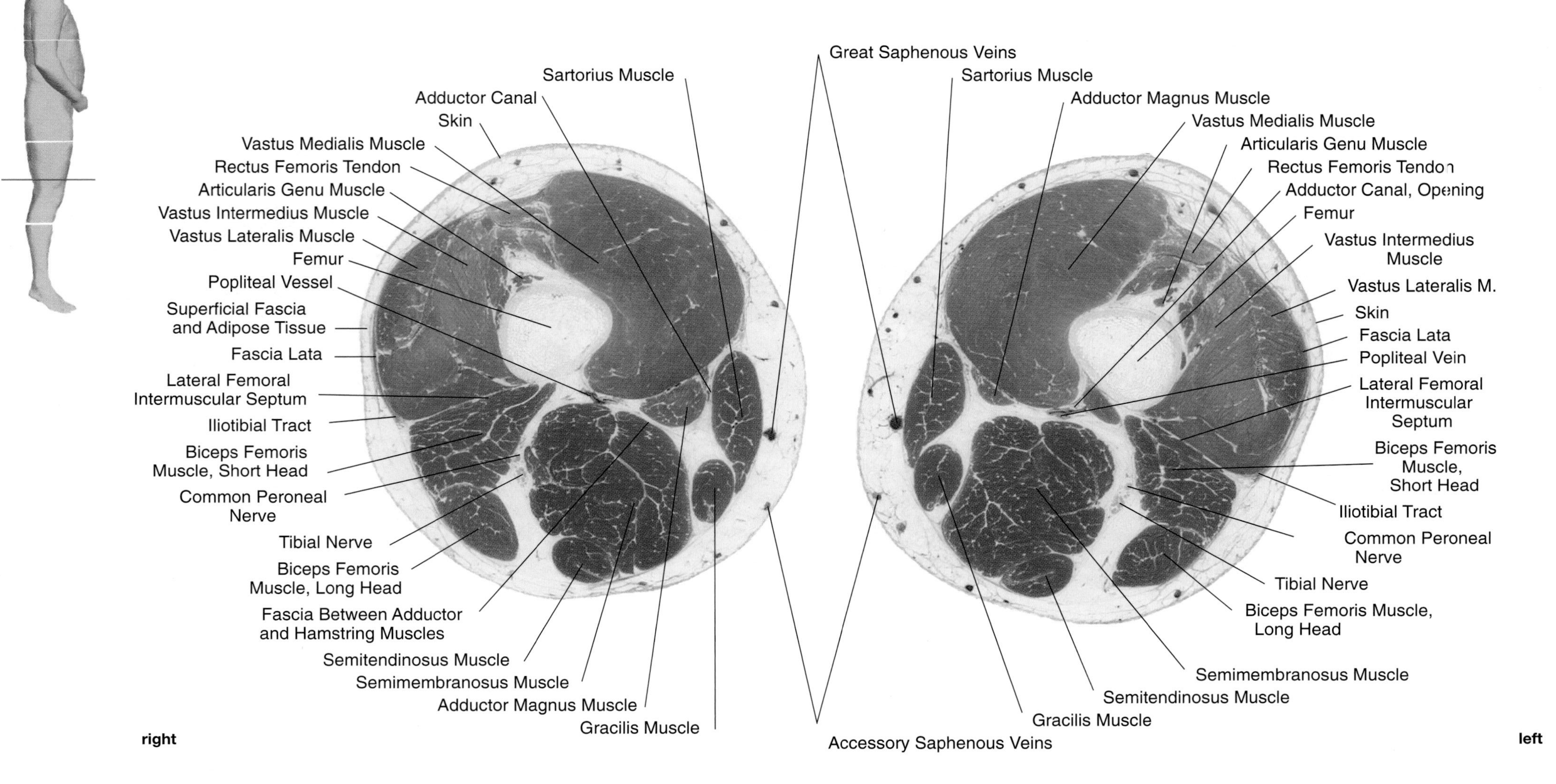

anterior
Sartorius Muscle
Adductor Canal
Skin
Vastus Medialis Muscle
Rectus Femoris Tendon
Articularis Genu Muscle
Vastus Intermedius Muscle
Vastus Lateralis Muscle
Femur
Popliteal Vessel
Superficial Fascia and Adipose Tissue
Fascia Lata
Lateral Femoral Intermuscular Septum
Iliotibial Tract
Biceps Femoris Muscle, Short Head
Common Peroneal Nerve
Tibial Nerve
Biceps Femoris Muscle, Long Head
Fascia Between Adductor and Hamstring Muscles
Semitendinosus Muscle
Semimembranosus Muscle
Adductor Magnus Muscle
Gracilis Muscle
right
Great Saphenous Veins
Accessory Saphenous Veins
Sartorius Muscle
Adductor Magnus Muscle
Vastus Medialis Muscle
Articularis Genu Muscle
Rectus Femoris Tendon
Adductor Canal, Opening
Femur
Vastus Intermedius Muscle
Vastus Lateralis M.
Skin
Fascia Lata
Popliteal Vein
Lateral Femoral Intermuscular Septum
Biceps Femoris Muscle, Short Head
Iliotibial Tract
Common Peroneal Nerve
Tibial Nerve
Biceps Femoris Muscle, Long Head
Semimembranosus Muscle
Semitendinosus Muscle
Gracilis Muscle
left
posterior

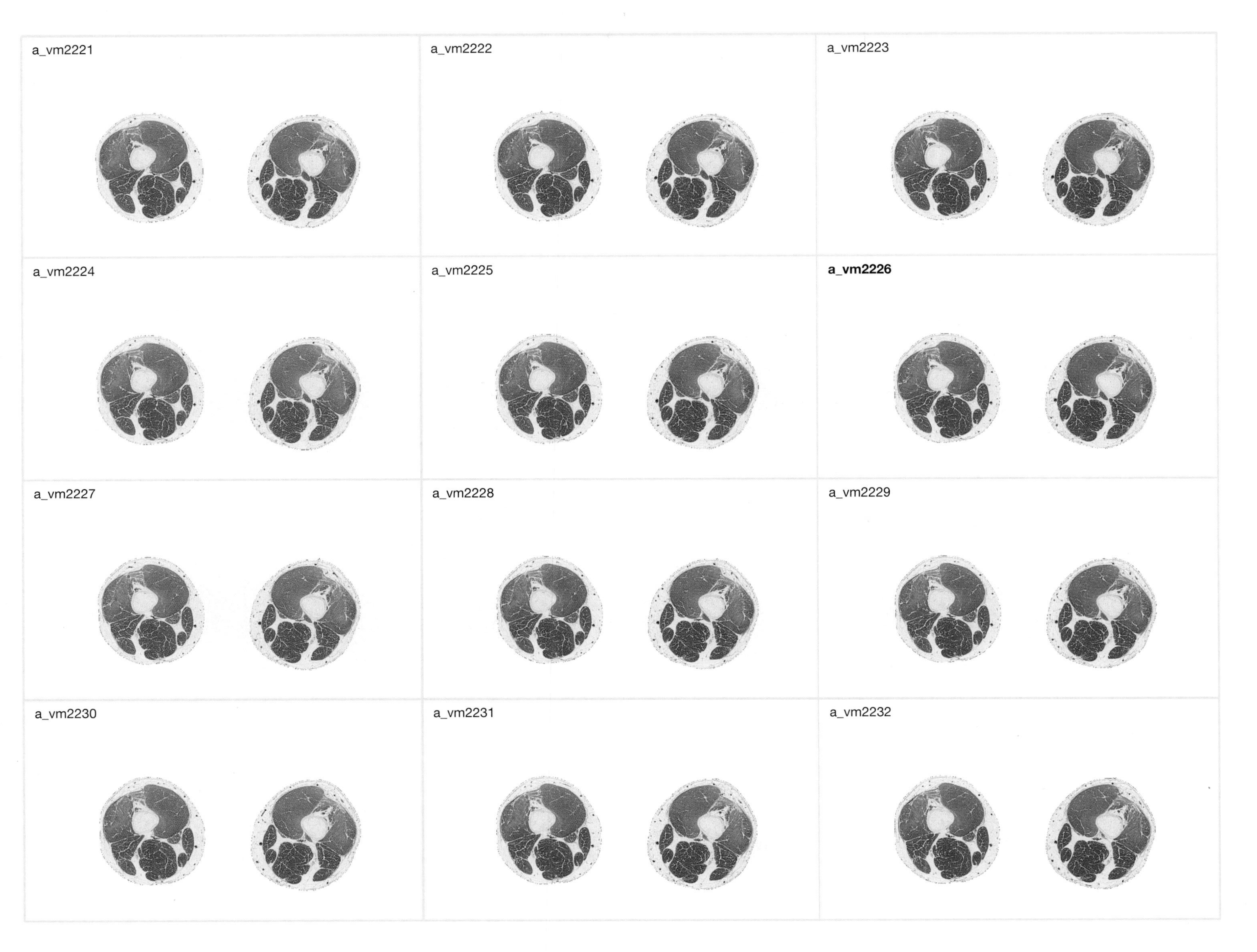
a_vm2221
a_vm2222
a_vm2223
a_vm2224
a_vm2225
a_vm2226
a_vm2227
a_vm2228
a_vm2229
a_vm2230
a_vm2231
a_vm2232

Transverse
a_vm2238

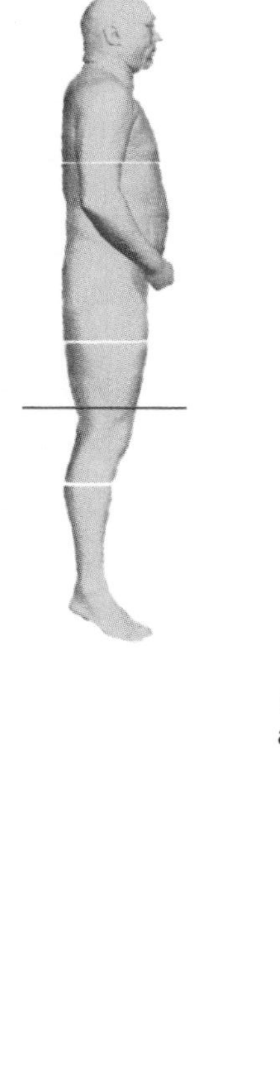

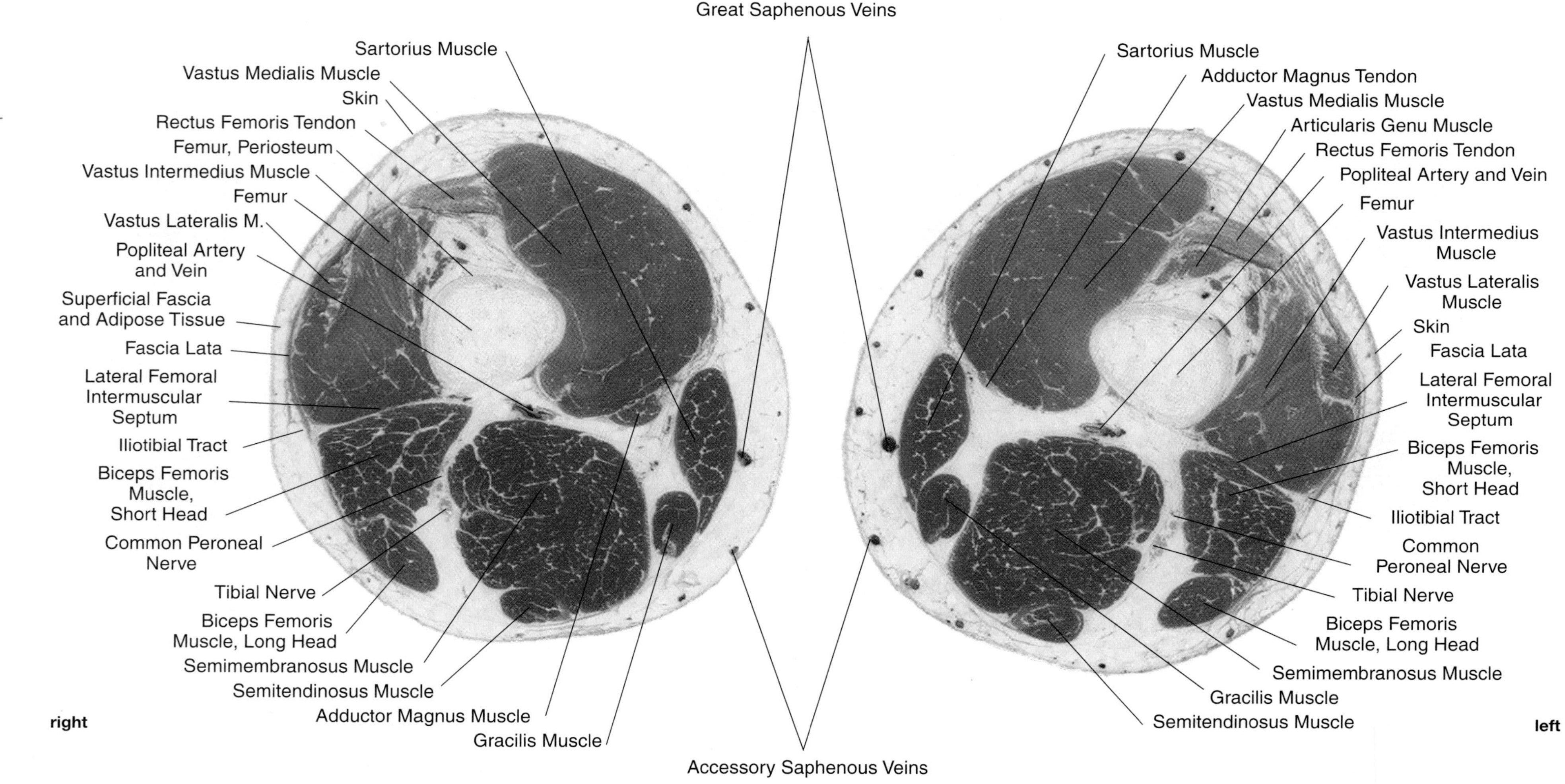
anterior
Great Saphenous Veins
Sartorius Muscle
Vastus Medialis Muscle
Skin
Rectus Femoris Tendon
Femur, Periosteum
Vastus Intermedius Muscle
Femur
Vastus Lateralis M.
Popliteal Artery and Vein
Superficial Fascia and Adipose Tissue
Fascia Lata
Lateral Femoral Intermuscular Septum
Iliotibial Tract
Biceps Femoris Muscle, Short Head
Common Peroneal Nerve
Tibial Nerve
Biceps Femoris Muscle, Long Head
Semimembranosus Muscle
Semitendinosus Muscle
Adductor Magnus Muscle
Gracilis Muscle
right
Sartorius Muscle
Adductor Magnus Tendon
Vastus Medialis Muscle
Articularis Genu Muscle
Rectus Femoris Tendon
Popliteal Artery and Vein
Femur
Vastus Intermedius Muscle
Vastus Lateralis Muscle
Skin
Fascia Lata
Lateral Femoral Intermuscular Septum
Biceps Femoris Muscle, Short Head
Iliotibial Tract
Common Peroneal Nerve
Tibial Nerve
Biceps Femoris Muscle, Long Head
Semimembranosus Muscle
Gracilis Muscle
Semitendinosus Muscle
left
Accessory Saphenous Veins
posterior

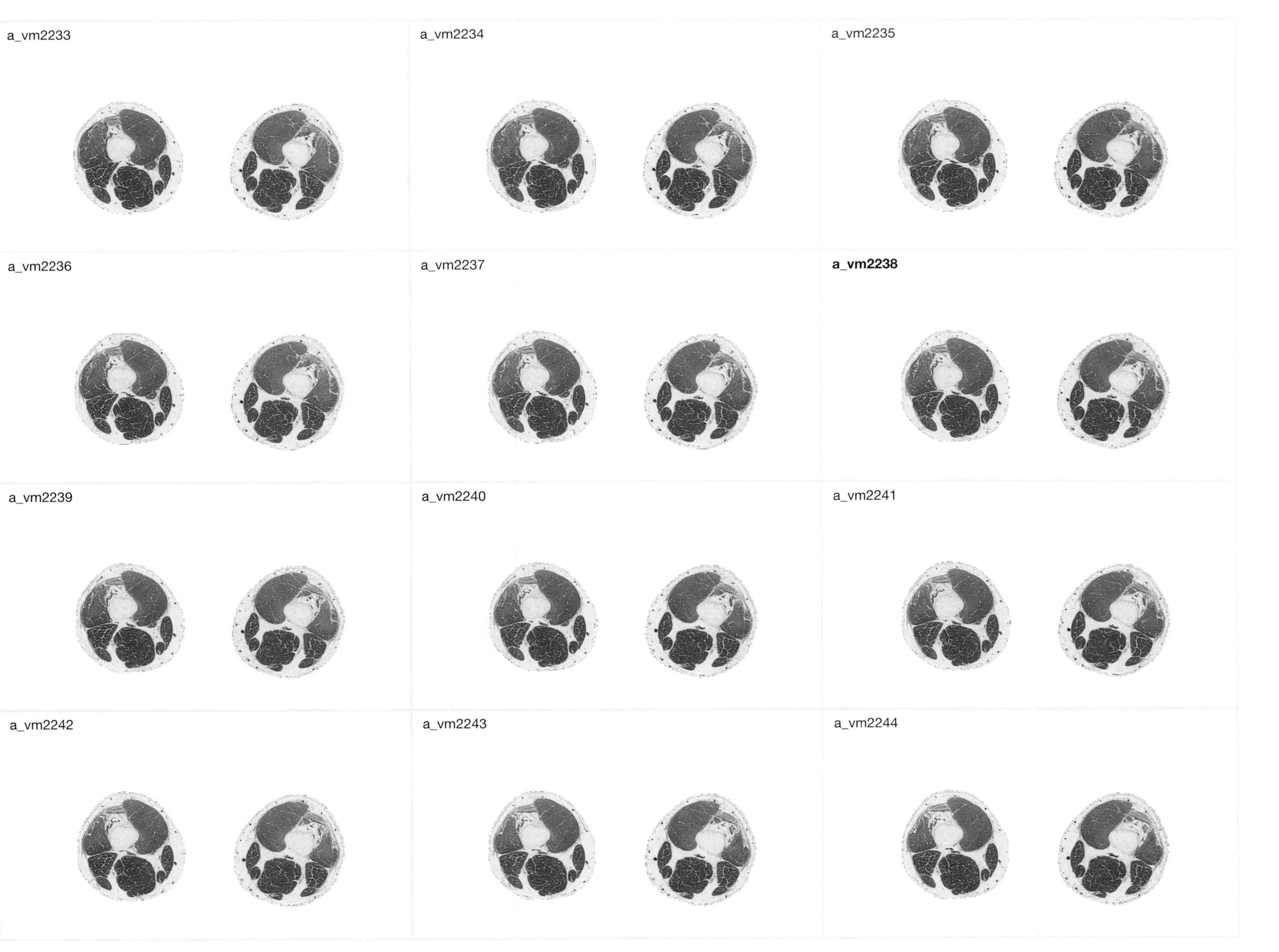
a_vm2233
a_vm2234
a_vm2235
a_vm2236
a_vm2237
a_vm2238
a_vm2239
a_vm2240
a_vm2241
a_vm2242
a_vm2243
a_vm2244

Transverse
a_vm2250

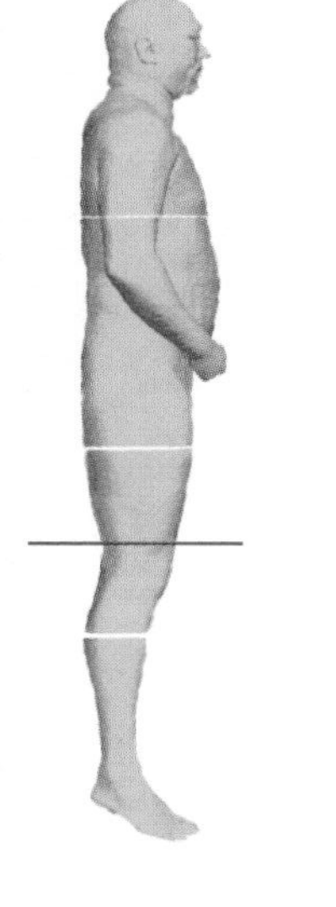

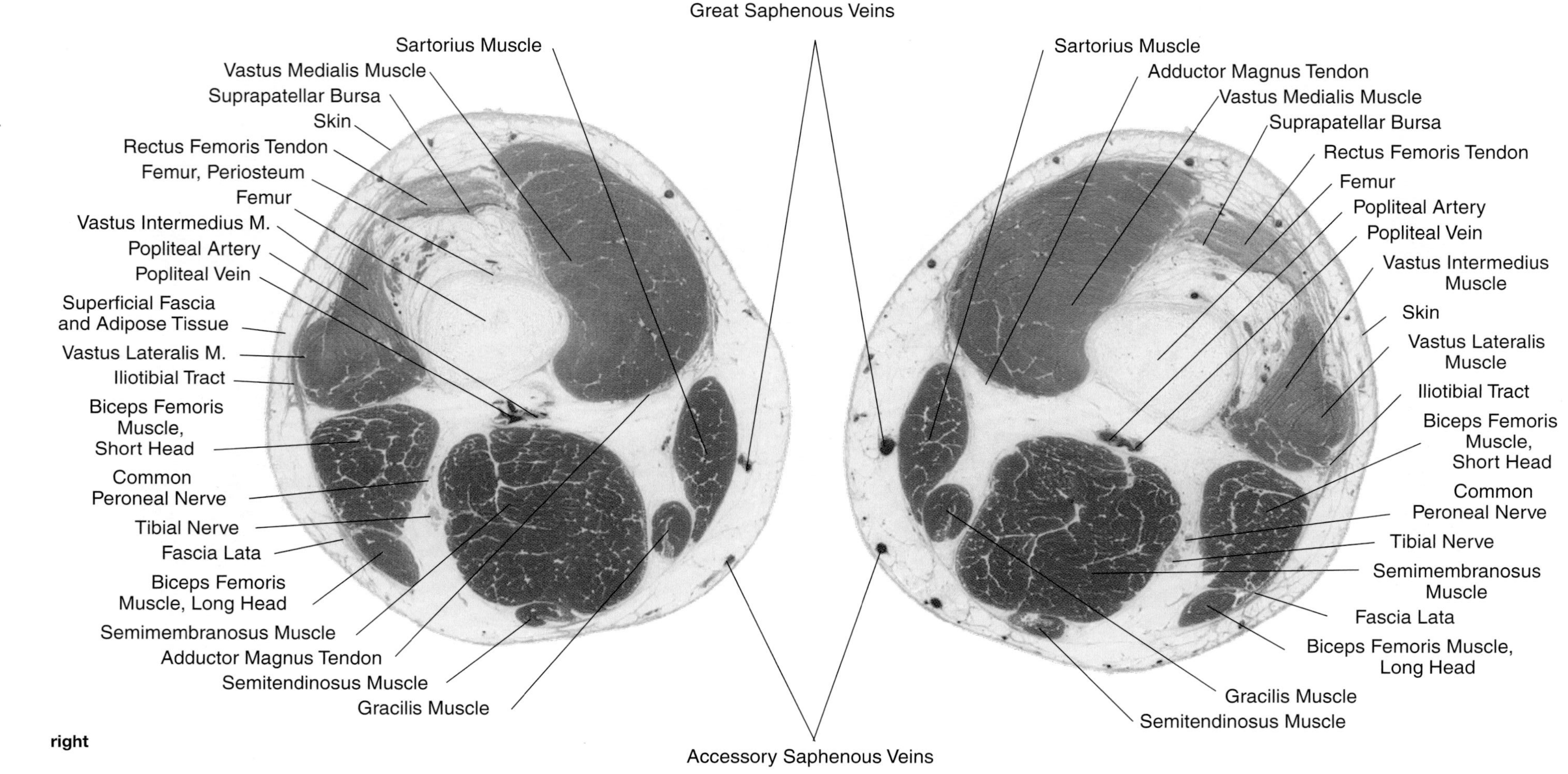
anterior
Great Saphenous Veins
Sartorius Muscle
Vastus Medialis Muscle
Suprapatellar Bursa
Skin
Rectus Femoris Tendon
Femur, Periosteum
Femur
Vastus Intermedius M.
Popliteal Artery
Popliteal Vein
Superficial Fascia and Adipose Tissue
Vastus Lateralis M.
Iliotibial Tract
Biceps Femoris Muscle, Short Head
Common Peroneal Nerve
Tibial Nerve
Fascia Lata
Biceps Femoris Muscle, Long Head
Semimembranosus Muscle
Adductor Magnus Tendon
Semitendinosus Muscle
Gracilis Muscle
Sartorius Muscle
Adductor Magnus Tendon
Vastus Medialis Muscle
Suprapatellar Bursa
Rectus Femoris Tendon
Femur
Popliteal Artery
Popliteal Vein
Vastus Intermedius Muscle
Skin
Vastus Lateralis Muscle
Iliotibial Tract
Biceps Femoris Muscle, Short Head
Common Peroneal Nerve
Tibial Nerve
Semimembranosus Muscle
Fascia Lata
Biceps Femoris Muscle, Long Head
Gracilis Muscle
Semitendinosus Muscle
right
Accessory Saphenous Veins
left
posterior

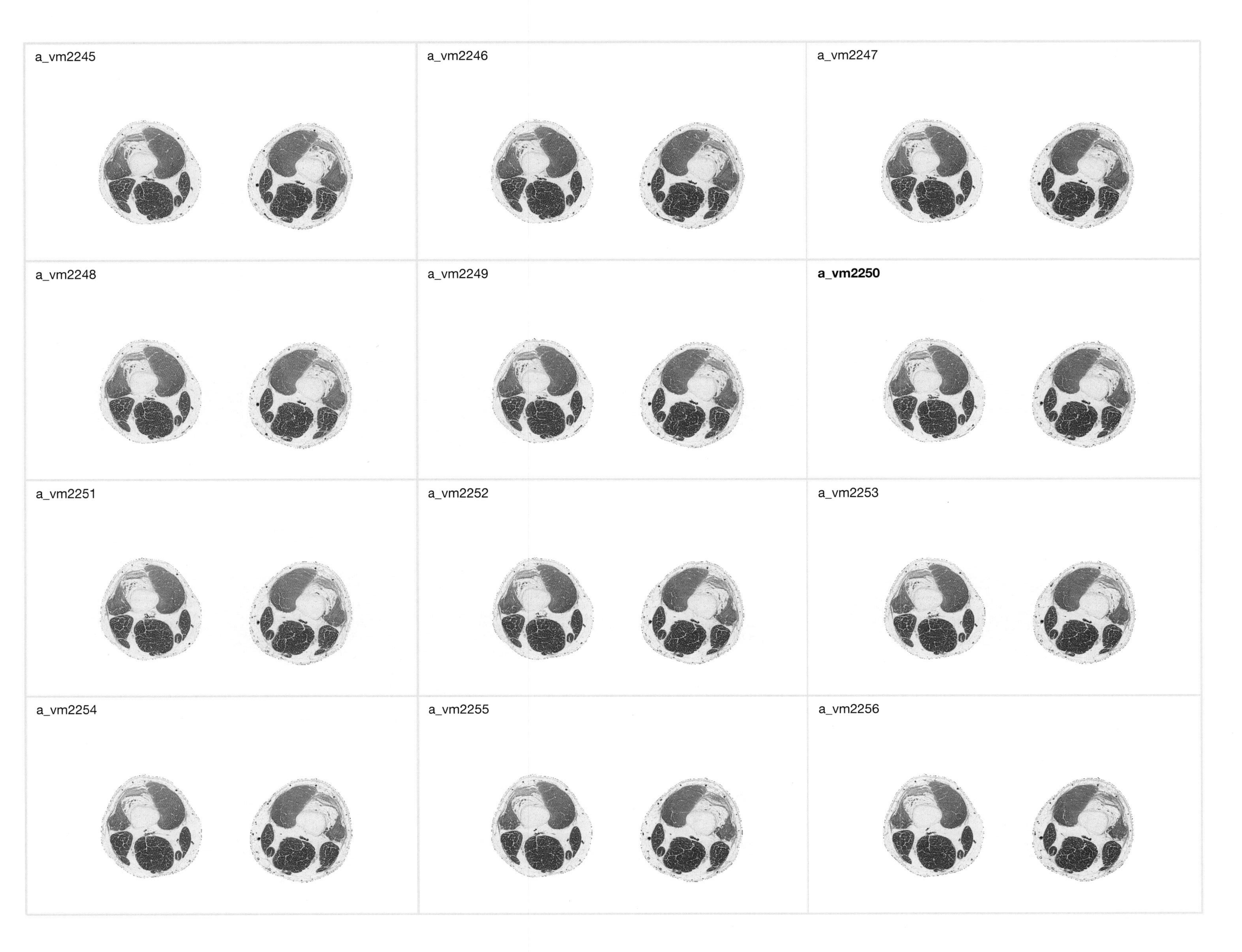
a_vm2245
a_vm2246
a_vm2247
a_vm2248
a_vm2249
a_vm2250
a_vm2251
a_vm2252
a_vm2253
a_vm2254
a_vm2255
a_vm2256

Transverse
a_vm2262

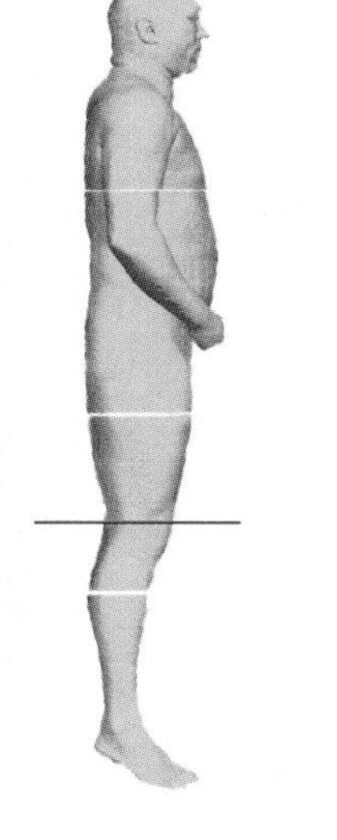

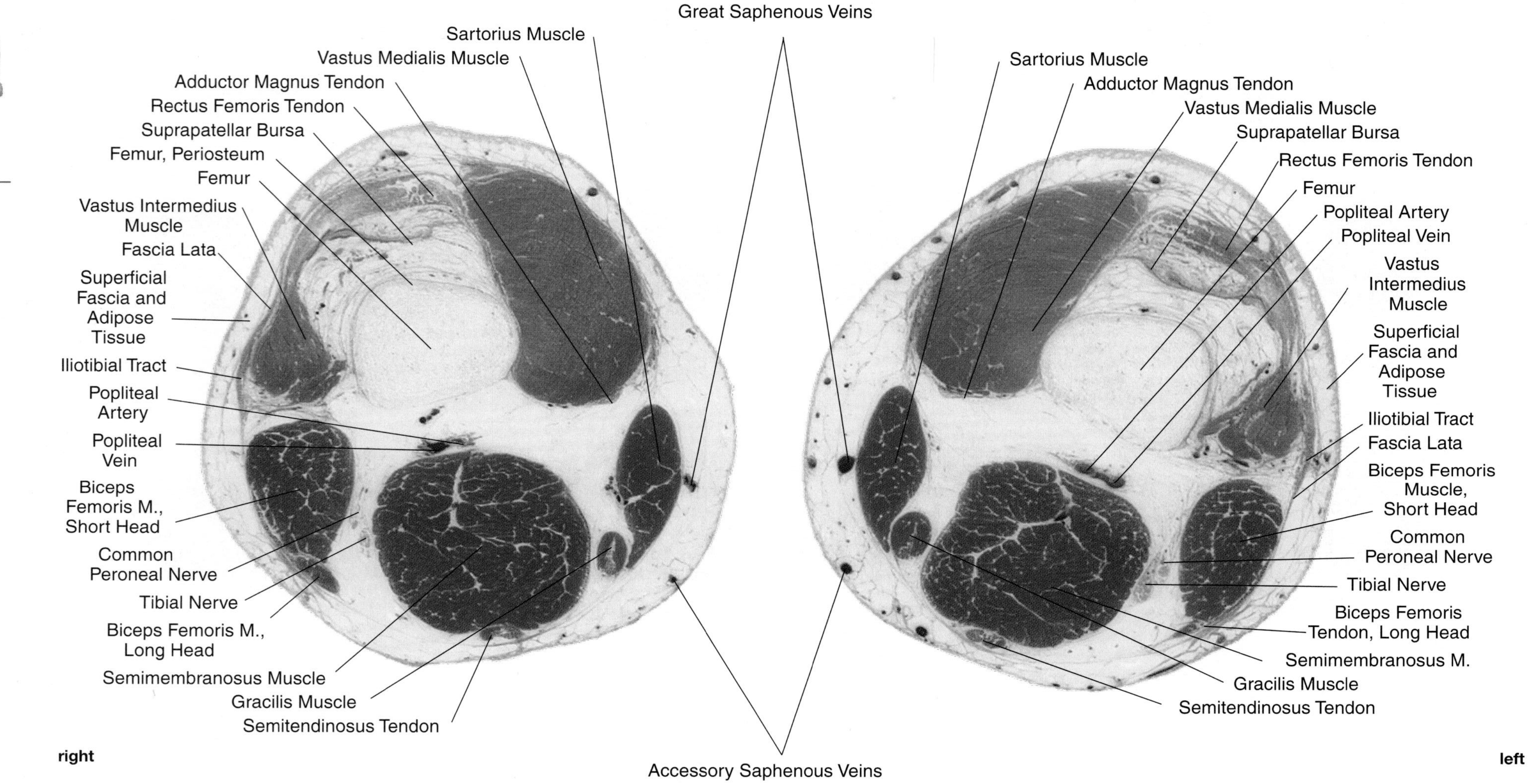
anterior
Great Saphenous Veins
Sartorius Muscle
Vastus Medialis Muscle
Adductor Magnus Tendon
Rectus Femoris Tendon
Suprapatellar Bursa
Femur, Periosteum
Femur
Vastus Intermedius Muscle
Fascia Lata
Superficial Fascia and Adipose Tissue
Iliotibial Tract
Popliteal Artery
Popliteal Vein
Biceps Femoris M., Short Head
Common Peroneal Nerve
Tibial Nerve
Biceps Femoris M., Long Head
Semimembranosus Muscle
Gracilis Muscle
Semitendinosus Tendon
right
Accessory Saphenous Veins
Sartorius Muscle
Adductor Magnus Tendon
Vastus Medialis Muscle
Suprapatellar Bursa
Rectus Femoris Tendon
Femur
Popliteal Artery
Popliteal Vein
Vastus Intermedius Muscle
Superficial Fascia and Adipose Tissue
Iliotibial Tract
Fascia Lata
Biceps Femoris Muscle, Short Head
Common Peroneal Nerve
Tibial Nerve
Biceps Femoris Tendon, Long Head
Semimembranosus M.
Gracilis Muscle
Semitendinosus Tendon
left
posterior

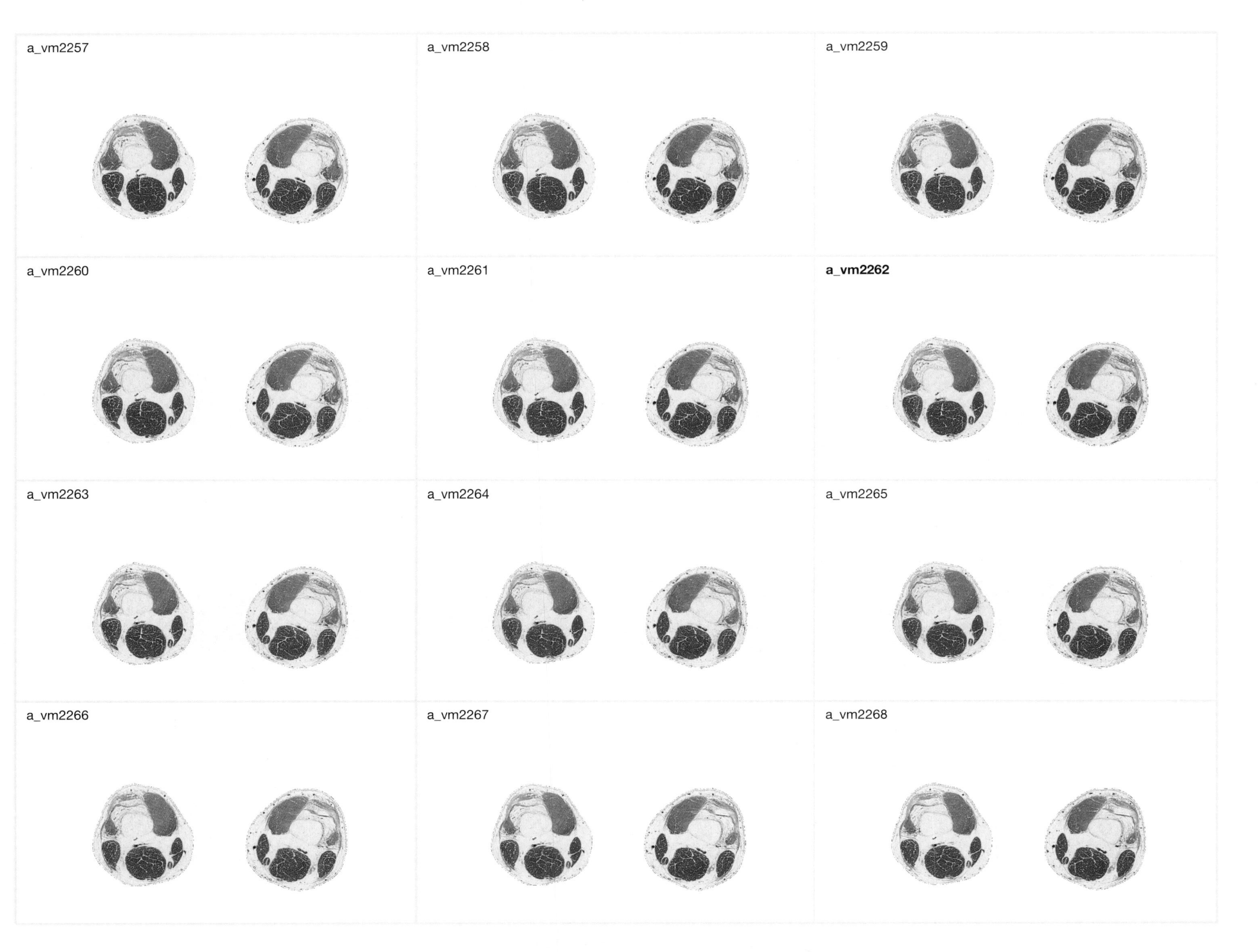
a_vm2257
a_vm2258
a_vm2259
a_vm2260
a_vm2261
a_vm2262
a_vm2263
a_vm2264
a_vm2265
a_vm2266
a_vm2267
a_vm2268

Transverse
a_vm2274

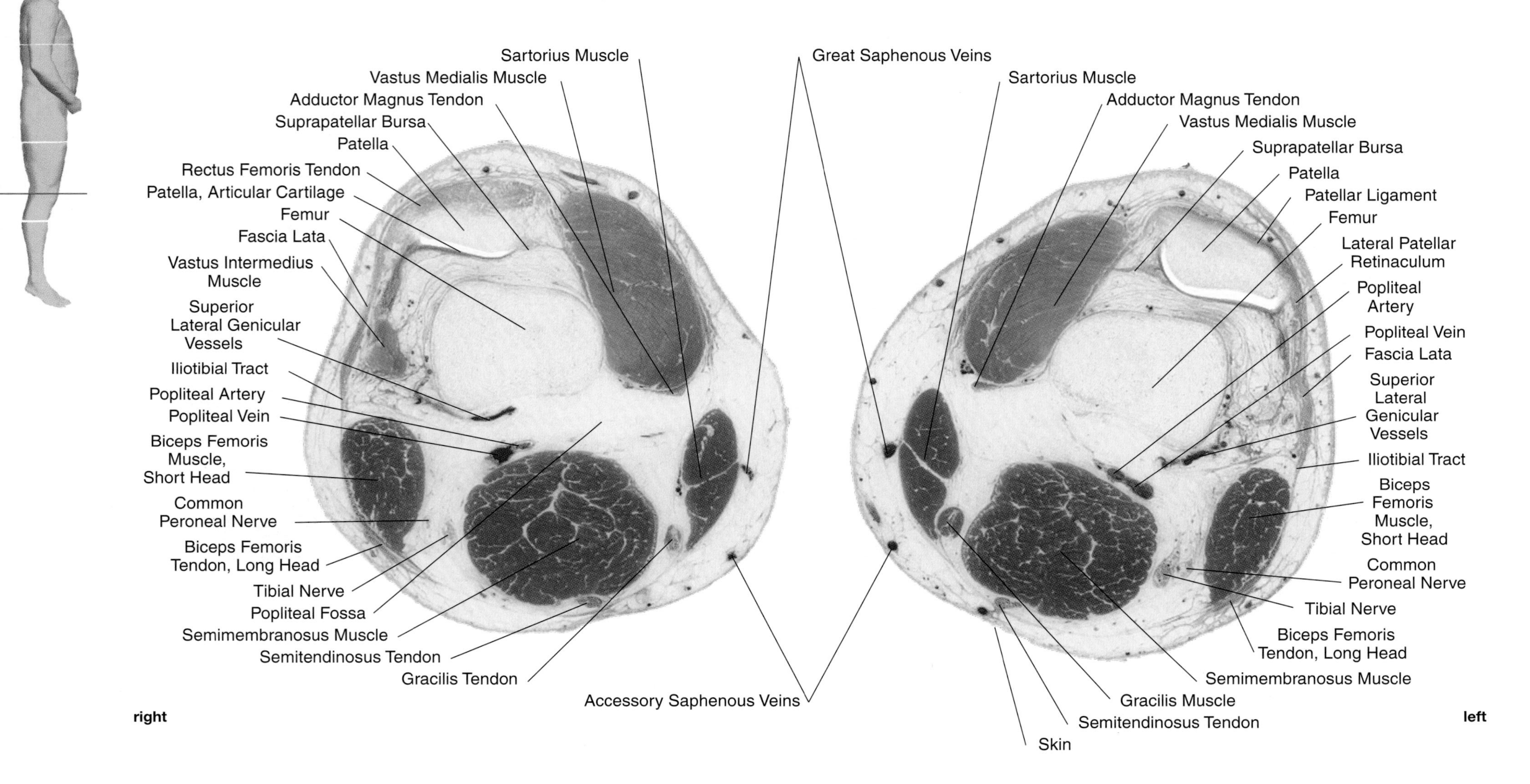

anterior
Sartorius Muscle
Vastus Medialis Muscle
Adductor Magnus Tendon
Suprapatellar Bursa
Patella
Rectus Femoris Tendon
Patella, Articular Cartilage
Femur
Fascia Lata
Vastus Intermedius Muscle
Superior Lateral Genicular Vessels
Iliotibial Tract
Popliteal Artery
Popliteal Vein
Biceps Femoris Muscle, Short Head
Common Peroneal Nerve
Biceps Femoris Tendon, Long Head
Tibial Nerve
Popliteal Fossa
Semimembranosus Muscle
Semitendinosus Tendon
Gracilis Tendon
Great Saphenous Veins
Accessory Saphenous Veins
Sartorius Muscle
Adductor Magnus Tendon
Vastus Medialis Muscle
Suprapatellar Bursa
Patella
Patellar Ligament
Femur
Lateral Patellar Retinaculum
Popliteal Artery
Popliteal Vein
Fascia Lata
Superior Lateral Genicular Vessels
Iliotibial Tract
Biceps Femoris Muscle, Short Head
Common Peroneal Nerve
Tibial Nerve
Biceps Femoris Tendon, Long Head
Semimembranosus Muscle
Gracilis Muscle
Semitendinosus Tendon
Skin
right
left
posterior

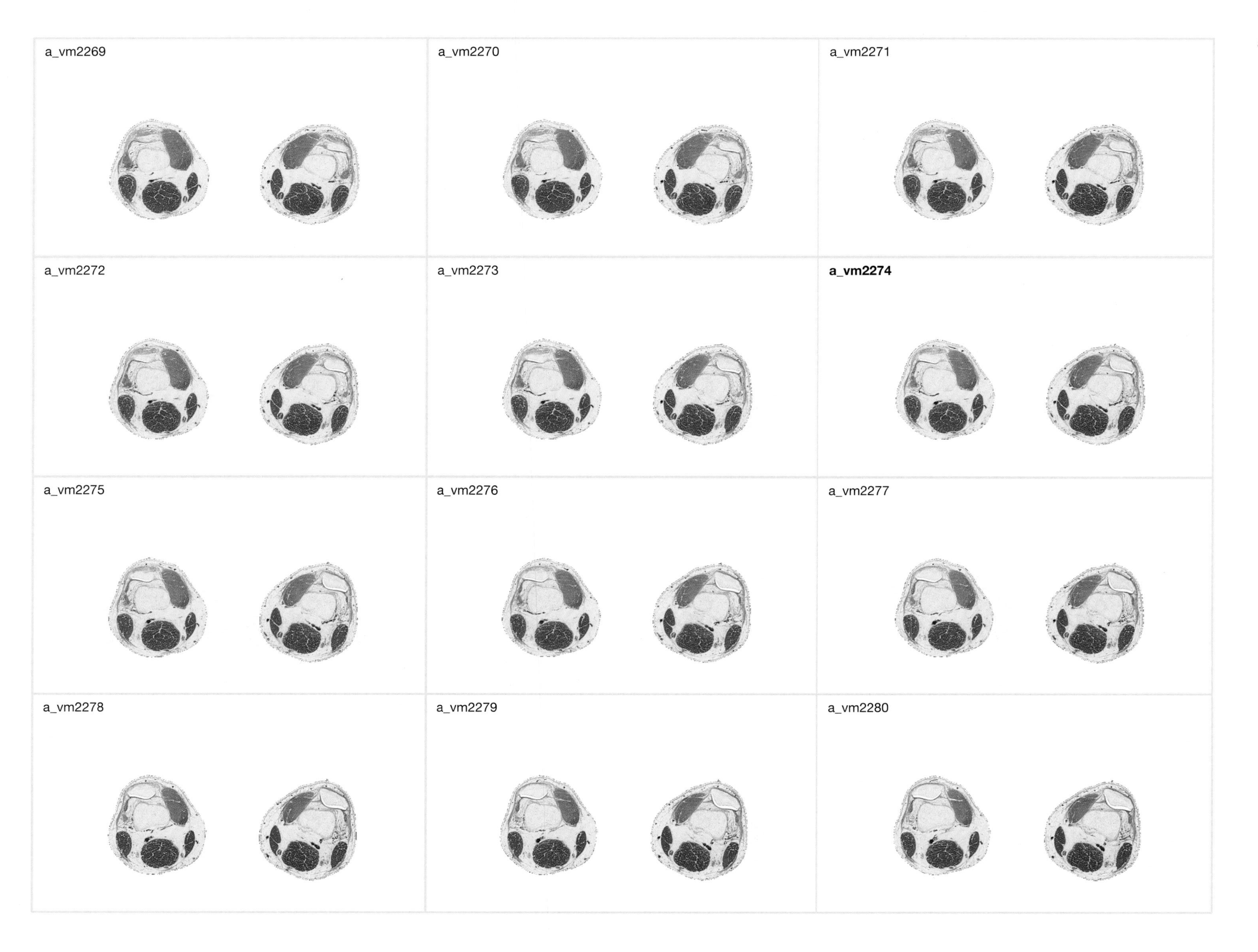
a_vm2269
a_vm2270
a_vm2271
a_vm2272
a_vm2273
a_vm2274
a_vm2275
a_vm2276
a_vm2277
a_vm2278
a_vm2279
a_vm2280

Transverse
a_vm2286

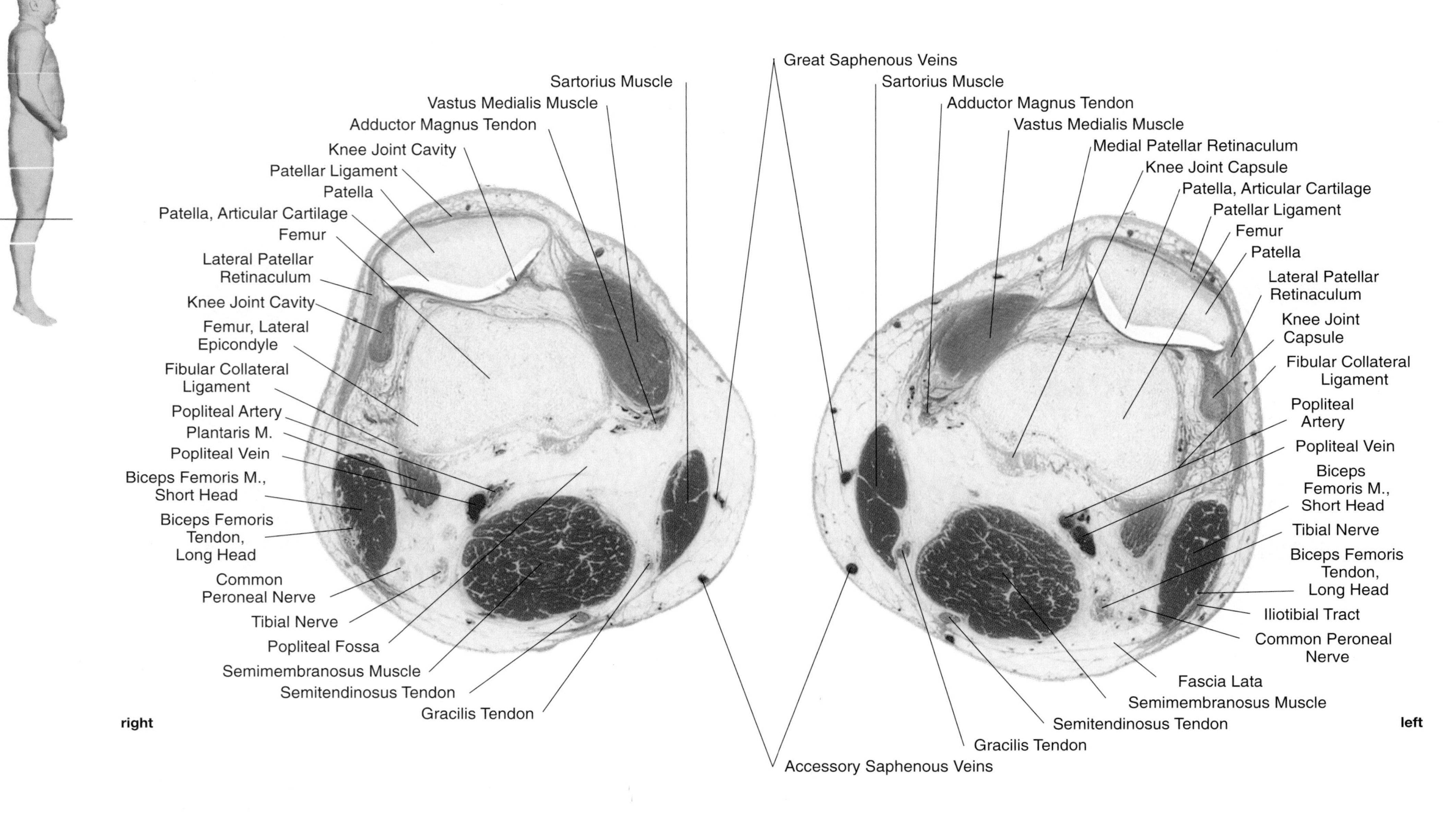

anterior
Great Saphenous Veins
Sartorius Muscle
Vastus Medialis Muscle
Adductor Magnus Tendon
Knee Joint Cavity
Patellar Ligament
Patella
Patella, Articular Cartilage
Femur
Lateral Patellar Retinaculum
Knee Joint Cavity
Femur, Lateral Epicondyle
Fibular Collateral Ligament
Popliteal Artery
Plantaris M.
Popliteal Vein
Biceps Femoris M., Short Head
Biceps Femoris Tendon, Long Head
Common Peroneal Nerve
Tibial Nerve
Popliteal Fossa
Semimembranosus Muscle
Semitendinosus Tendon
Gracilis Tendon
right
Sartorius Muscle
Adductor Magnus Tendon
Vastus Medialis Muscle
Medial Patellar Retinaculum
Knee Joint Capsule
Patella, Articular Cartilage
Patellar Ligament
Femur
Patella
Lateral Patellar Retinaculum
Knee Joint Capsule
Fibular Collateral Ligament
Popliteal Artery
Popliteal Vein
Biceps Femoris M., Short Head
Tibial Nerve
Biceps Femoris Tendon, Long Head
Iliotibial Tract
Common Peroneal Nerve
Fascia Lata
Semimembranosus Muscle
Semitendinosus Tendon
Gracilis Tendon
left
Accessory Saphenous Veins
posterior

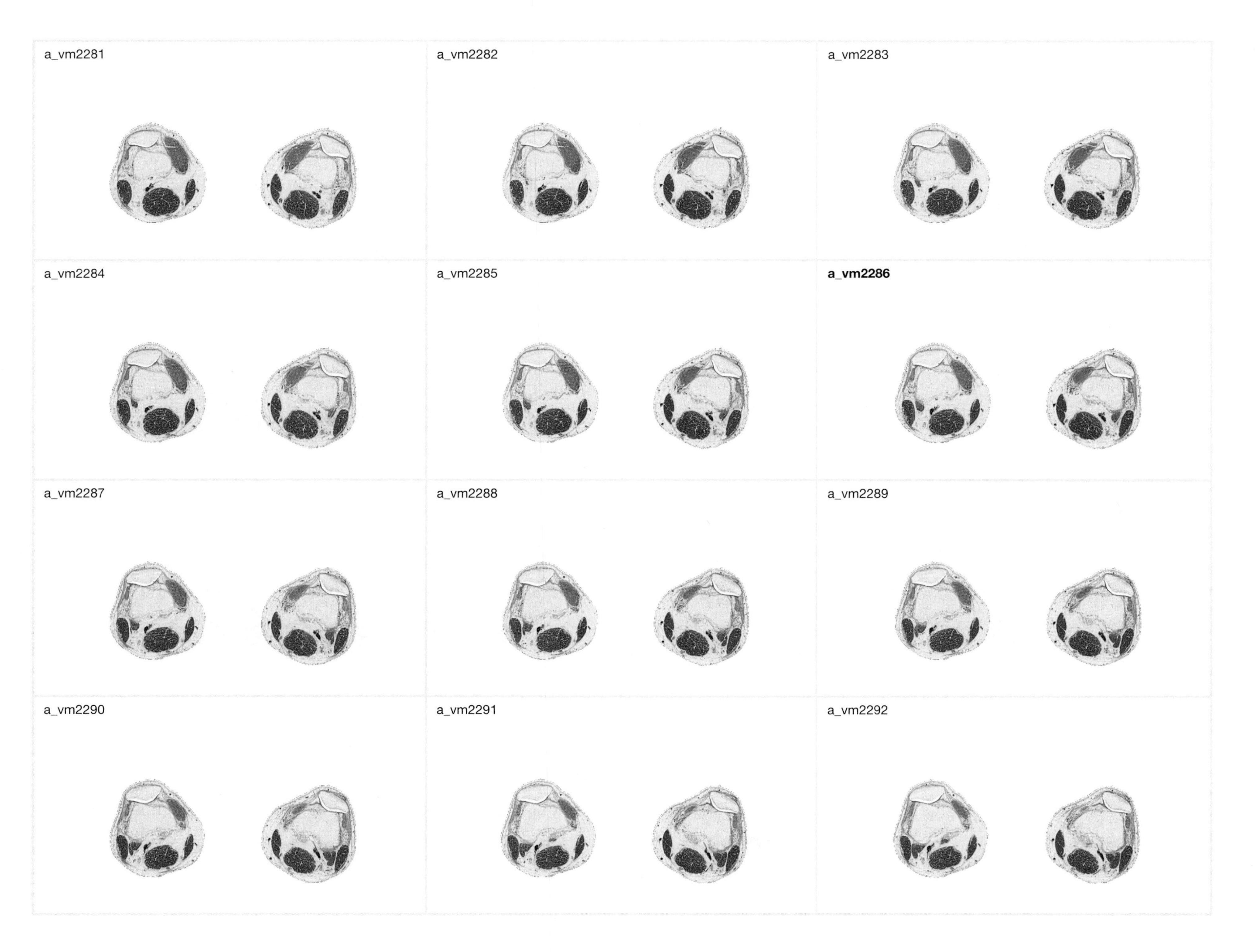
a_vm2281
a_vm2282
a_vm2283
a_vm2284
a_vm2285
a_vm2286
a_vm2287
a_vm2288
a_vm2289
a_vm2290
a_vm2291
a_vm2292

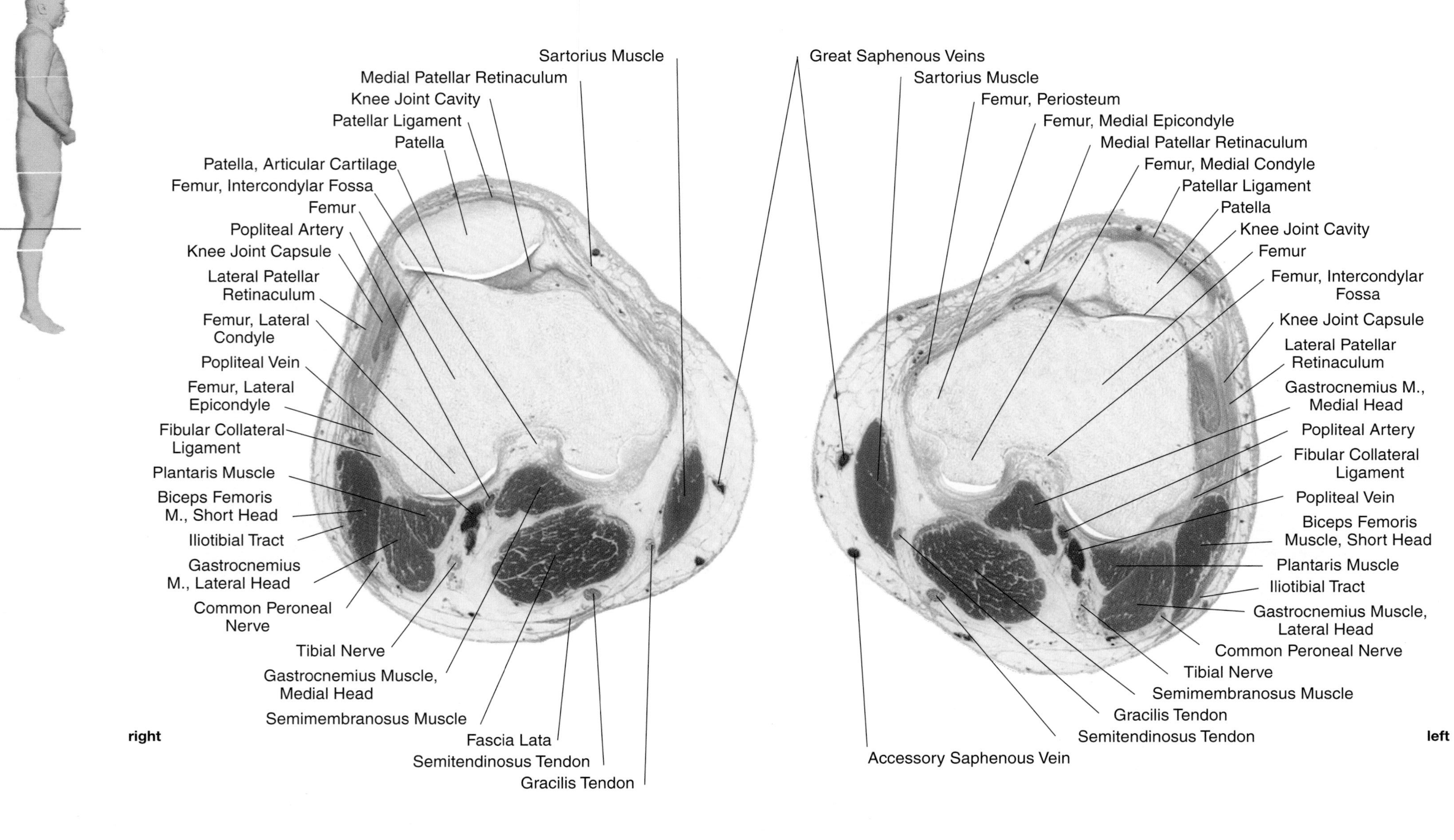
Transverse
a_vm2300
anterior
posterior
right
left
Sartorius Muscle
Medial Patellar Retinaculum
Knee Joint Cavity
Patellar Ligament
Patella
Patella, Articular Cartilage
Femur, Intercondylar Fossa
Femur
Popliteal Artery
Knee Joint Capsule
Lateral Patellar Retinaculum
Femur, Lateral Condyle
Popliteal Vein
Femur, Lateral Epicondyle
Fibular Collateral Ligament
Plantaris Muscle
Biceps Femoris M., Short Head
Iliotibial Tract
Gastrocnemius M., Lateral Head
Common Peroneal Nerve
Tibial Nerve
Gastrocnemius Muscle, Medial Head
Semimembranosus Muscle
Fascia Lata
Semitendinosus Tendon
Gracilis Tendon
Great Saphenous Veins
Sartorius Muscle
Femur, Periosteum
Femur, Medial Epicondyle
Medial Patellar Retinaculum
Femur, Medial Condyle
Patellar Ligament
Patella
Knee Joint Cavity
Femur
Femur, Intercondylar Fossa
Knee Joint Capsule
Lateral Patellar Retinaculum
Gastrocnemius M., Medial Head
Popliteal Artery
Fibular Collateral Ligament
Popliteal Vein
Biceps Femoris Muscle, Short Head
Plantaris Muscle
Iliotibial Tract
Gastrocnemius Muscle, Lateral Head
Common Peroneal Nerve
Tibial Nerve
Semimembranosus Muscle
Gracilis Tendon
Semitendinosus Tendon
Accessory Saphenous Vein

a_vm2293

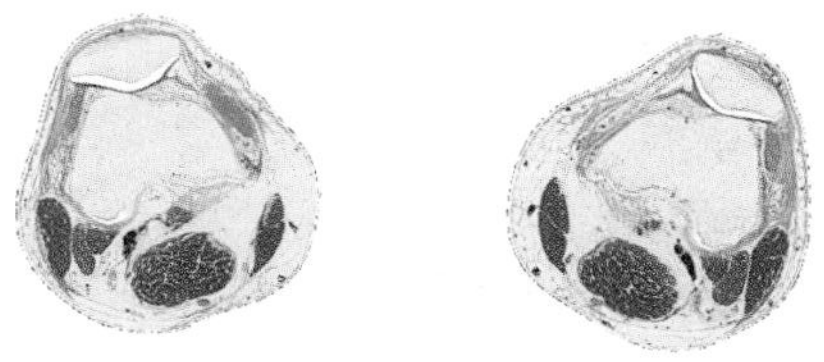

a_vm2294

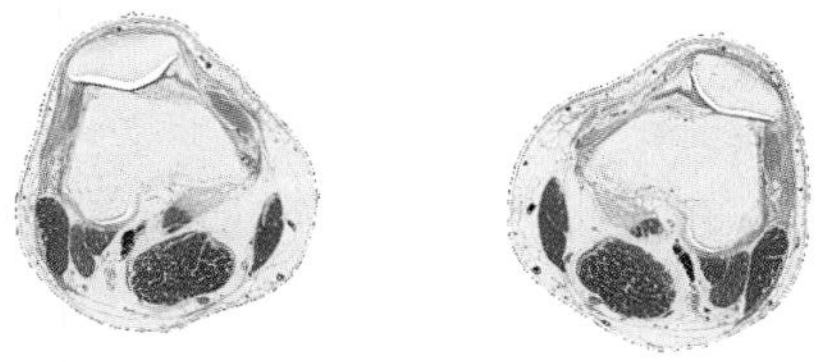

a_vm2295

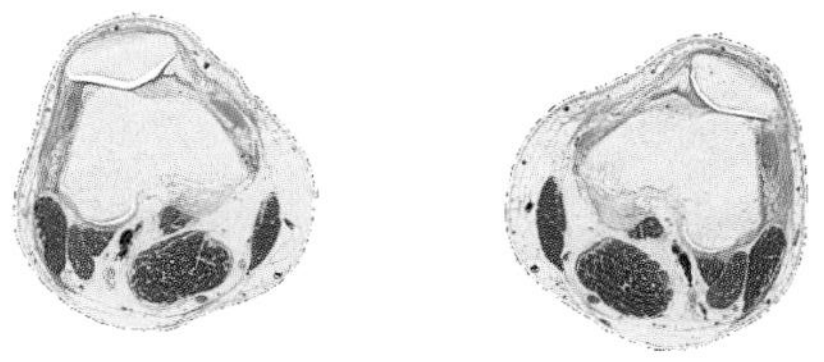

a_vm2296

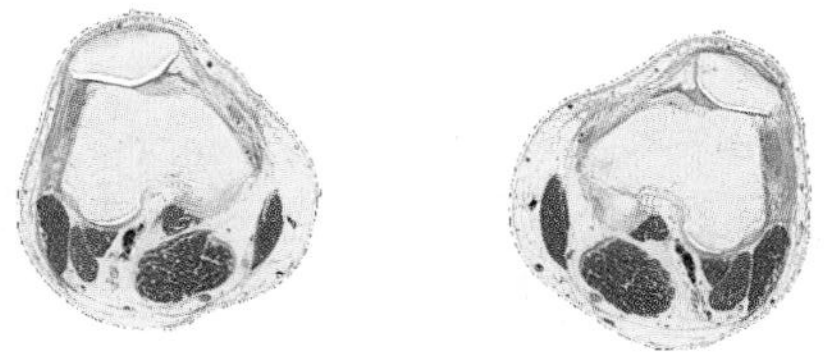

a_vm2297

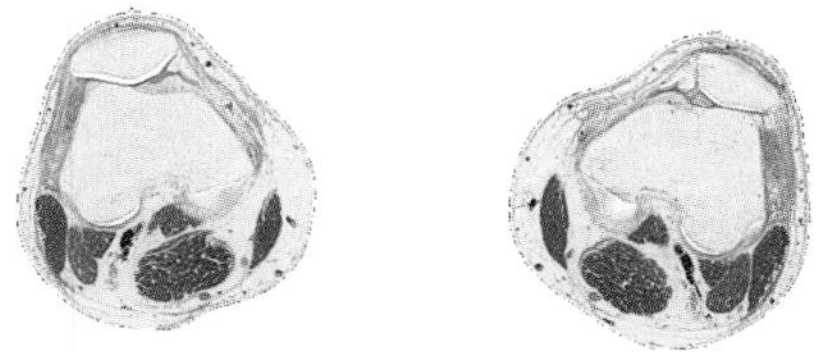

a_vm2298

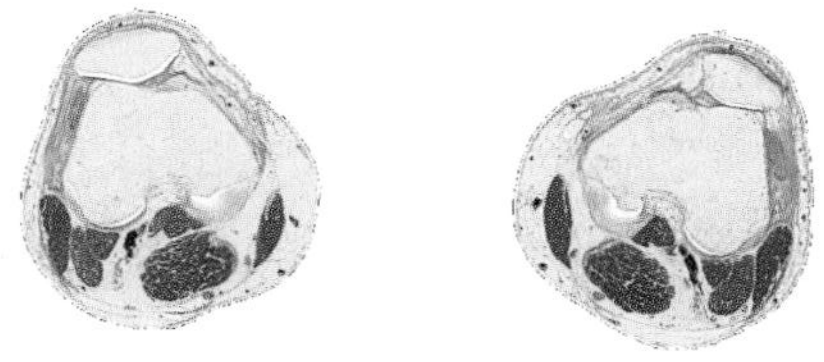

a_vm2299

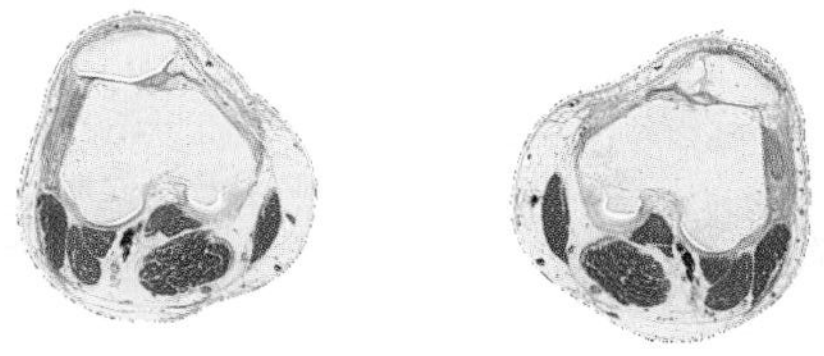

a_vm2300

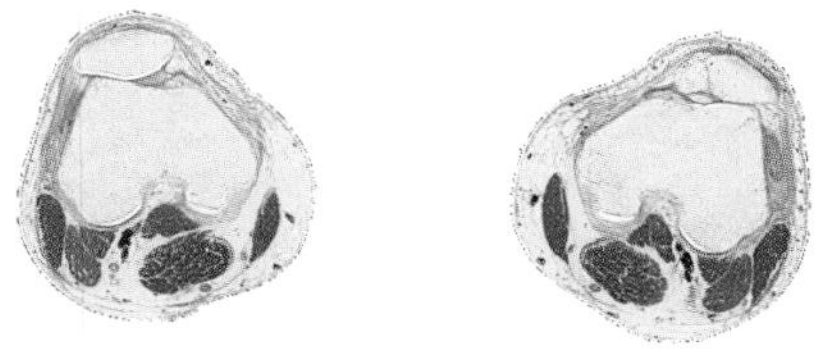

a_vm2301

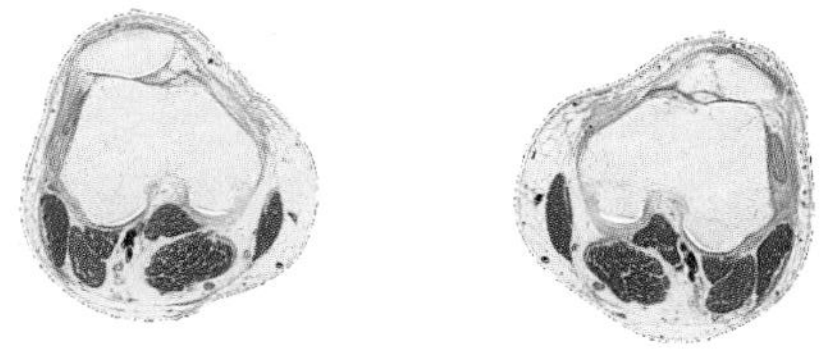

a_vm2302

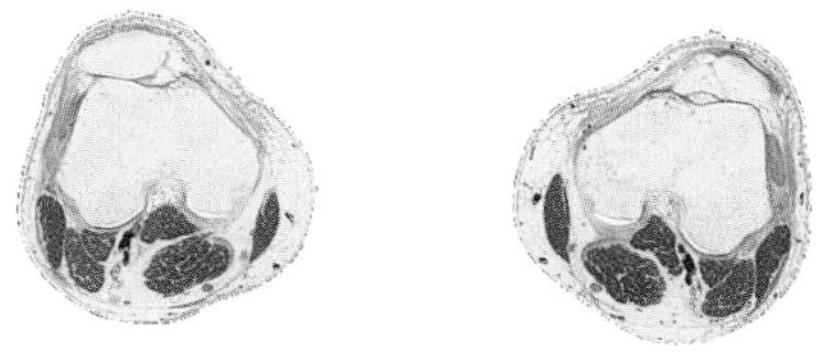

a_vm2303

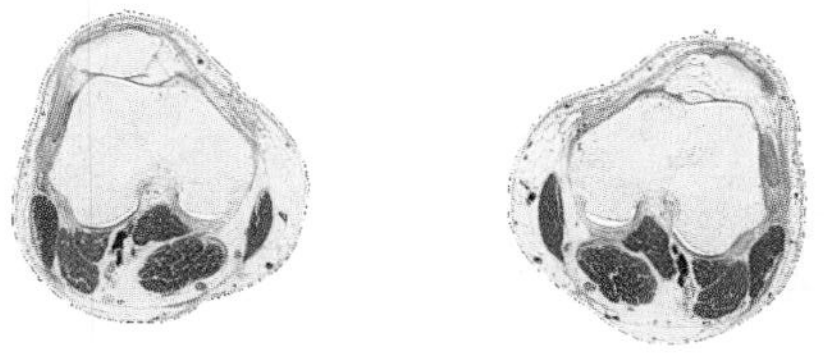

a_vm2304

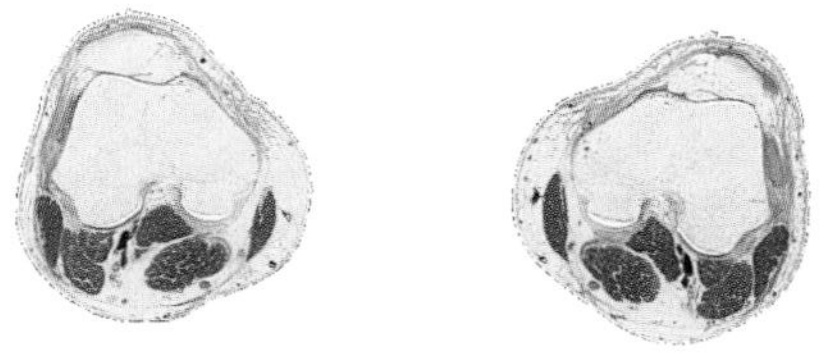

Transverse
a_vm2312

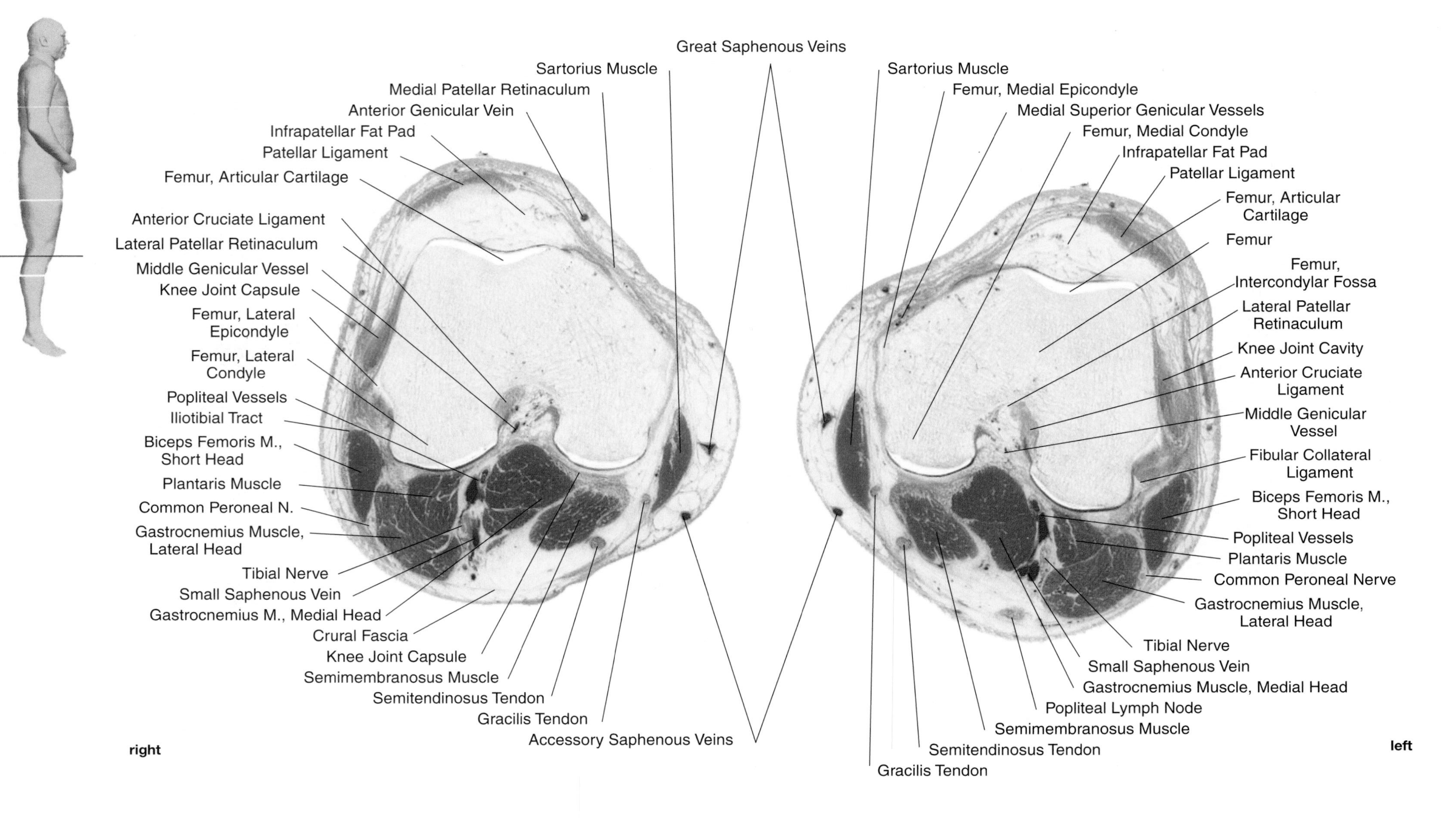

anterior
Great Saphenous Veins
Sartorius Muscle
Medial Patellar Retinaculum
Anterior Genicular Vein
Infrapatellar Fat Pad
Patellar Ligament
Femur, Articular Cartilage
Anterior Cruciate Ligament
Lateral Patellar Retinaculum
Middle Genicular Vessel
Knee Joint Capsule
Femur, Lateral Epicondyle
Femur, Lateral Condyle
Popliteal Vessels
Iliotibial Tract
Biceps Femoris M., Short Head
Plantaris Muscle
Common Peroneal N.
Gastrocnemius Muscle, Lateral Head
Tibial Nerve
Small Saphenous Vein
Gastrocnemius M., Medial Head
Crural Fascia
Knee Joint Capsule
Semimembranosus Muscle
Semitendinosus Tendon
Gracilis Tendon
Accessory Saphenous Veins
right
Sartorius Muscle
Femur, Medial Epicondyle
Medial Superior Genicular Vessels
Femur, Medial Condyle
Infrapatellar Fat Pad
Patellar Ligament
Femur, Articular Cartilage
Femur
Femur, Intercondylar Fossa
Lateral Patellar Retinaculum
Knee Joint Cavity
Anterior Cruciate Ligament
Middle Genicular Vessel
Fibular Collateral Ligament
Biceps Femoris M., Short Head
Popliteal Vessels
Plantaris Muscle
Common Peroneal Nerve
Gastrocnemius Muscle, Lateral Head
Tibial Nerve
Small Saphenous Vein
Gastrocnemius Muscle, Medial Head
Popliteal Lymph Node
Semimembranosus Muscle
Semitendinosus Tendon
Gracilis Tendon
left
posterior

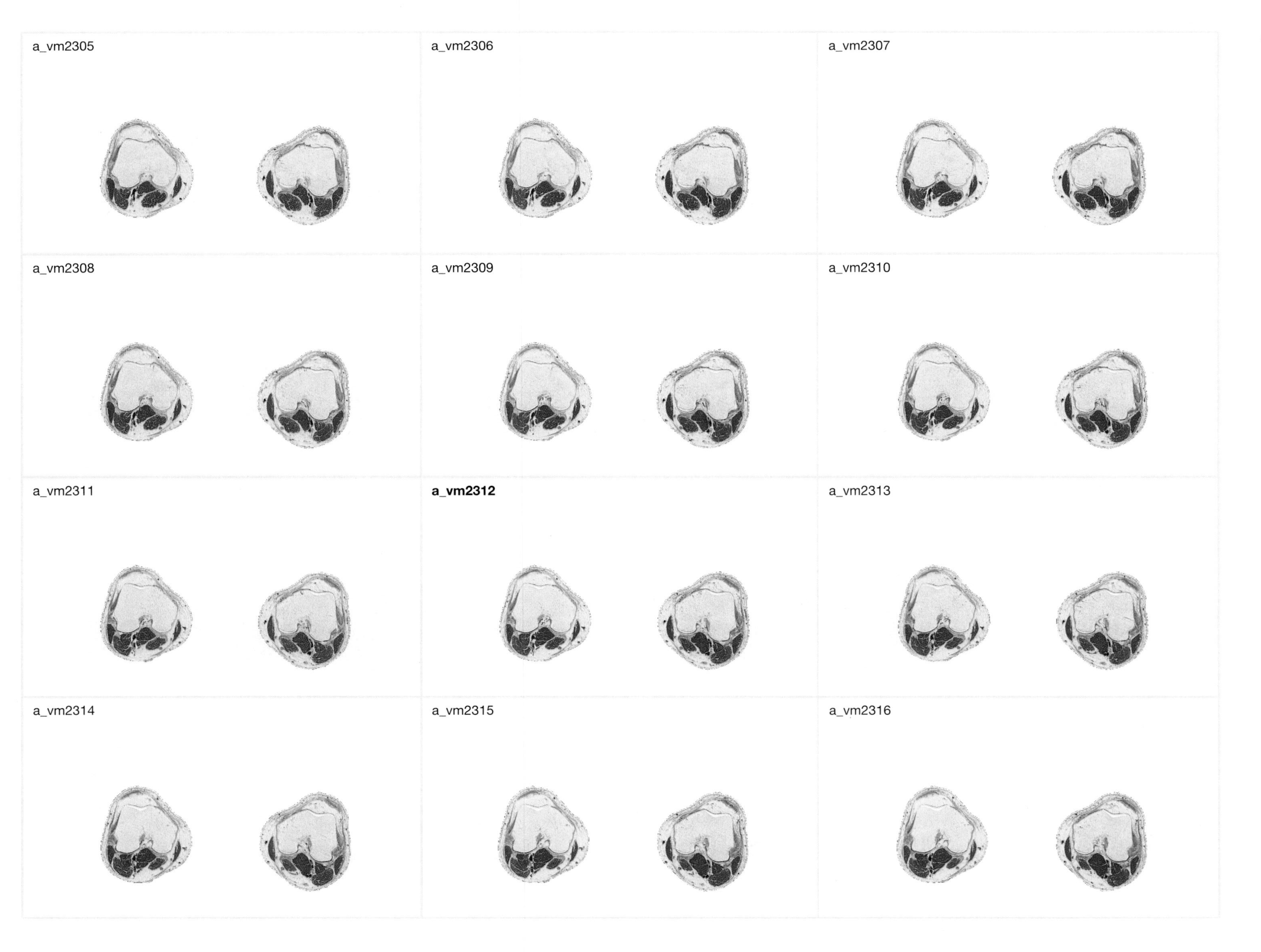
a_vm2305
a_vm2306
a_vm2307
a_vm2308
a_vm2309
a_vm2310
a_vm2311
a_vm2312
a_vm2313
a_vm2314
a_vm2315
a_vm2316

Transverse
a_vm2324

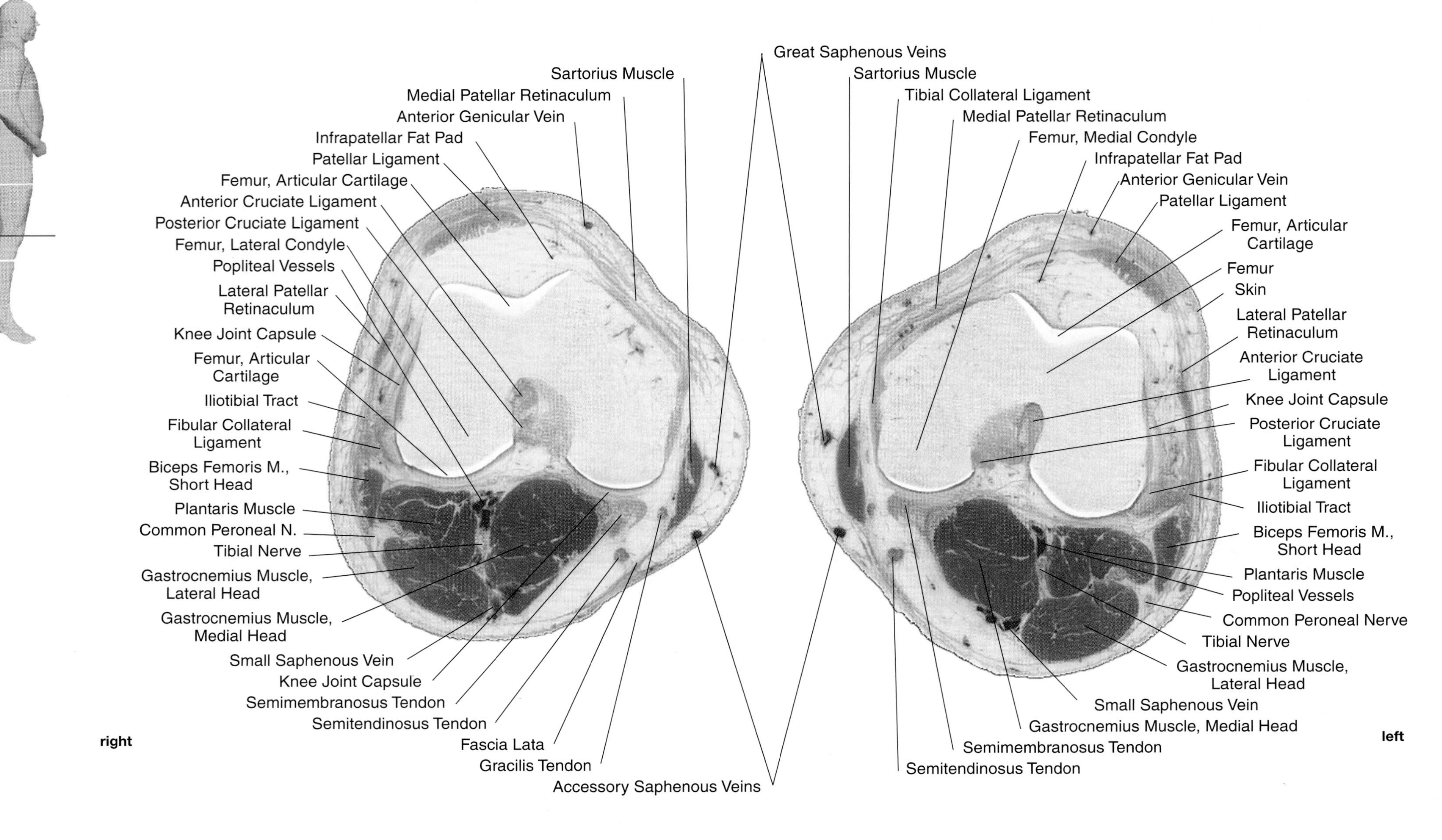
anterior
Sartorius Muscle
Medial Patellar Retinaculum
Anterior Genicular Vein
Infrapatellar Fat Pad
Patellar Ligament
Femur, Articular Cartilage
Anterior Cruciate Ligament
Posterior Cruciate Ligament
Femur, Lateral Condyle
Popliteal Vessels
Lateral Patellar Retinaculum
Knee Joint Capsule
Femur, Articular Cartilage
Iliotibial Tract
Fibular Collateral Ligament
Biceps Femoris M., Short Head
Plantaris Muscle
Common Peroneal N.
Tibial Nerve
Gastrocnemius Muscle, Lateral Head
Gastrocnemius Muscle, Medial Head
Small Saphenous Vein
Knee Joint Capsule
Semimembranosus Tendon
Semitendinosus Tendon
Fascia Lata
Gracilis Tendon
Accessory Saphenous Veins
Great Saphenous Veins
Sartorius Muscle
Tibial Collateral Ligament
Medial Patellar Retinaculum
Femur, Medial Condyle
Infrapatellar Fat Pad
Anterior Genicular Vein
Patellar Ligament
Femur, Articular Cartilage
Femur
Skin
Lateral Patellar Retinaculum
Anterior Cruciate Ligament
Knee Joint Capsule
Posterior Cruciate Ligament
Fibular Collateral Ligament
Iliotibial Tract
Biceps Femoris M., Short Head
Plantaris Muscle
Popliteal Vessels
Common Peroneal Nerve
Tibial Nerve
Gastrocnemius Muscle, Lateral Head
Small Saphenous Vein
Gastrocnemius Muscle, Medial Head
Semimembranosus Tendon
Semitendinosus Tendon
right
left
posterior

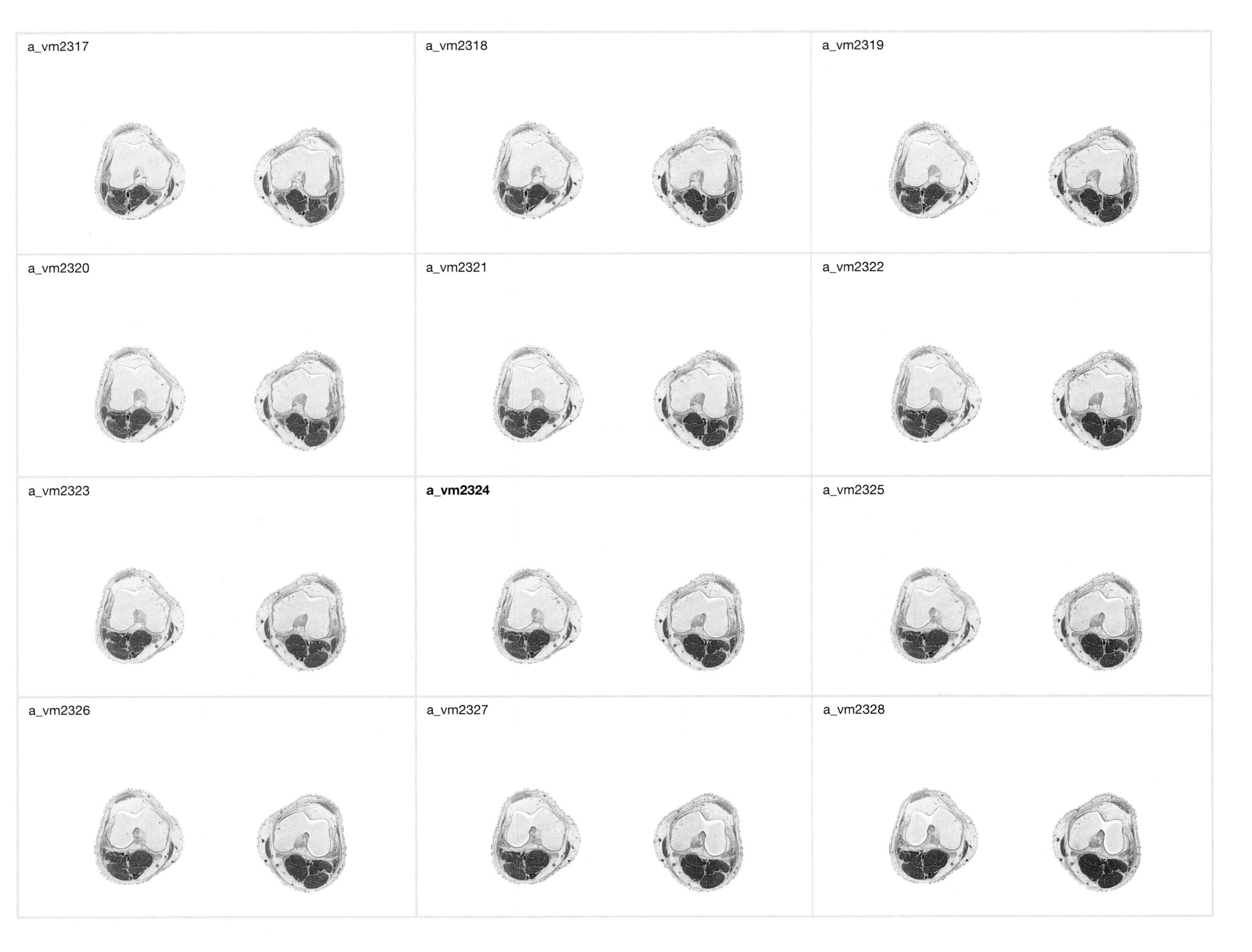
a_vm2317
a_vm2318
a_vm2319
a_vm2320
a_vm2321
a_vm2322
a_vm2323
a_vm2324
a_vm2325
a_vm2326
a_vm2327
a_vm2328

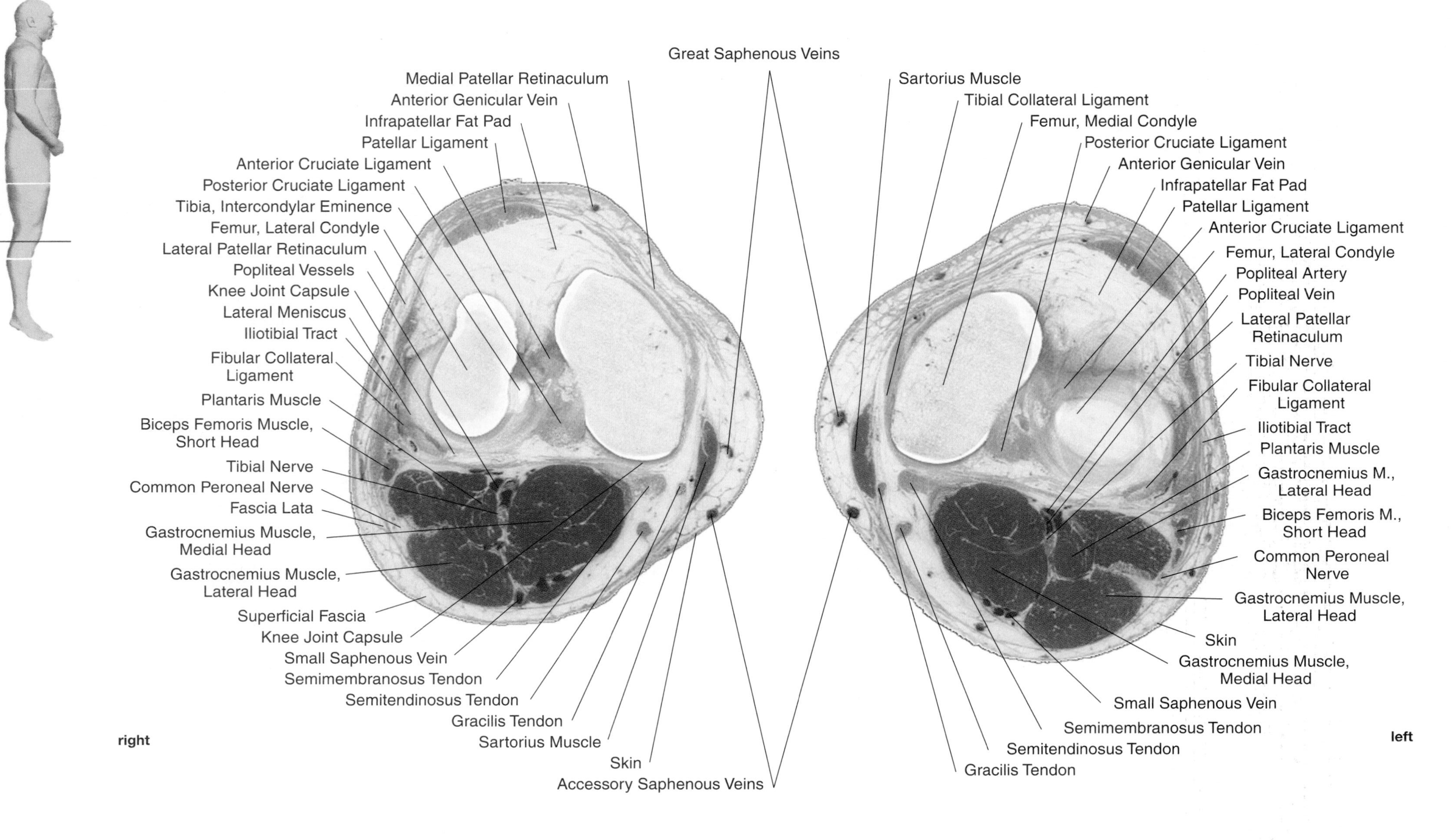
Transverse
a_vm2330
anterior
posterior
right
left
Great Saphenous Veins
Medial Patellar Retinaculum
Anterior Genicular Vein
Infrapatellar Fat Pad
Patellar Ligament
Anterior Cruciate Ligament
Posterior Cruciate Ligament
Tibia, Intercondylar Eminence
Femur, Lateral Condyle
Lateral Patellar Retinaculum
Popliteal Vessels
Knee Joint Capsule
Lateral Meniscus
Iliotibial Tract
Fibular Collateral Ligament
Plantaris Muscle
Biceps Femoris Muscle, Short Head
Tibial Nerve
Common Peroneal Nerve
Fascia Lata
Gastrocnemius Muscle, Medial Head
Gastrocnemius Muscle, Lateral Head
Superficial Fascia
Knee Joint Capsule
Small Saphenous Vein
Semimembranosus Tendon
Semitendinosus Tendon
Gracilis Tendon
Sartorius Muscle
Skin
Accessory Saphenous Veins
Sartorius Muscle
Tibial Collateral Ligament
Femur, Medial Condyle
Posterior Cruciate Ligament
Anterior Genicular Vein
Infrapatellar Fat Pad
Patellar Ligament
Anterior Cruciate Ligament
Femur, Lateral Condyle
Popliteal Artery
Popliteal Vein
Lateral Patellar Retinaculum
Tibial Nerve
Fibular Collateral Ligament
Iliotibial Tract
Plantaris Muscle
Gastrocnemius M., Lateral Head
Biceps Femoris M., Short Head
Common Peroneal Nerve
Gastrocnemius Muscle, Lateral Head
Skin
Gastrocnemius Muscle, Medial Head
Small Saphenous Vein
Semimembranosus Tendon
Semitendinosus Tendon
Gracilis Tendon

a_vm2389 a_vm2390 a_vm2391 a_vm2389-a_vm2400

a_vm2329

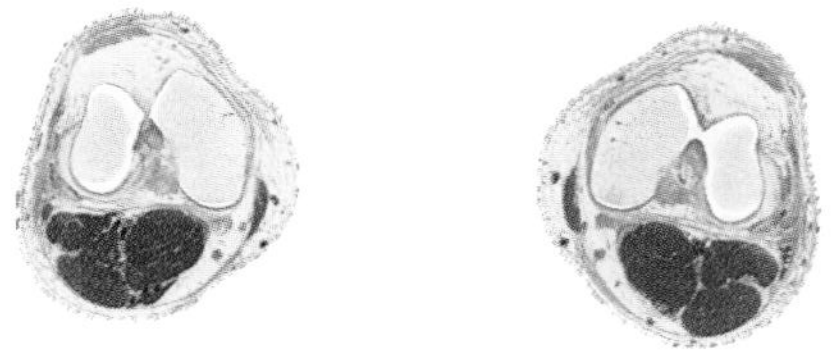

a_vm2330

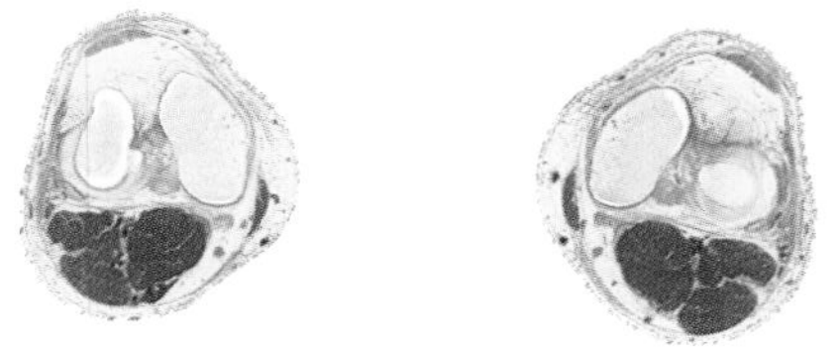

a_vm2331

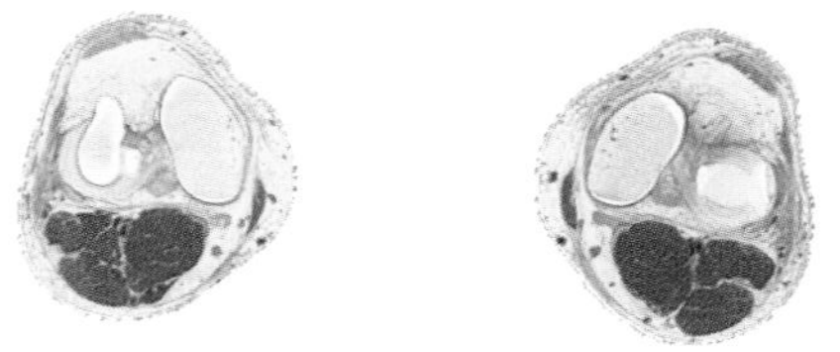

a_vm2332

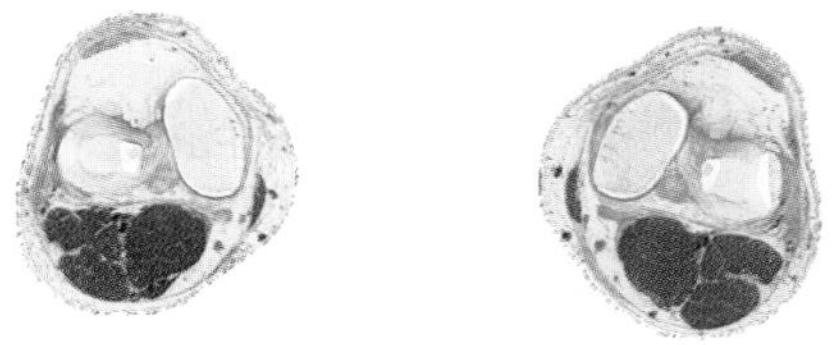

a_vm2333

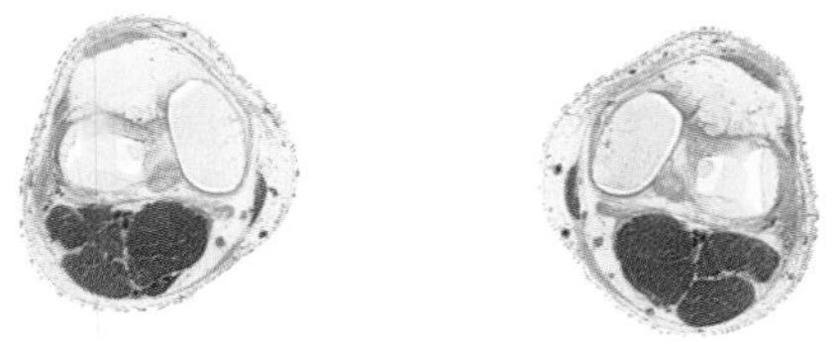

a_vm2334

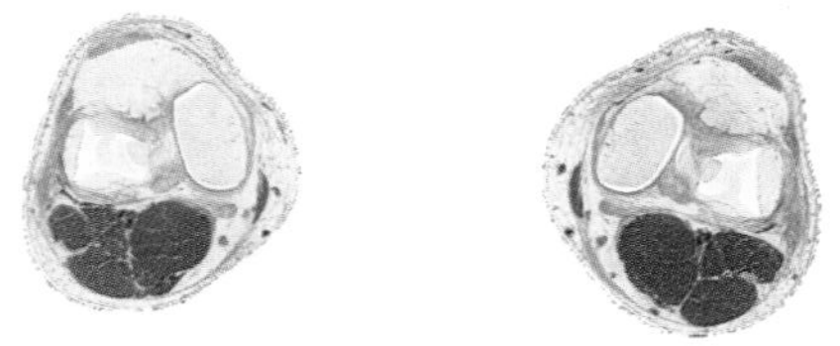

a_vm2335

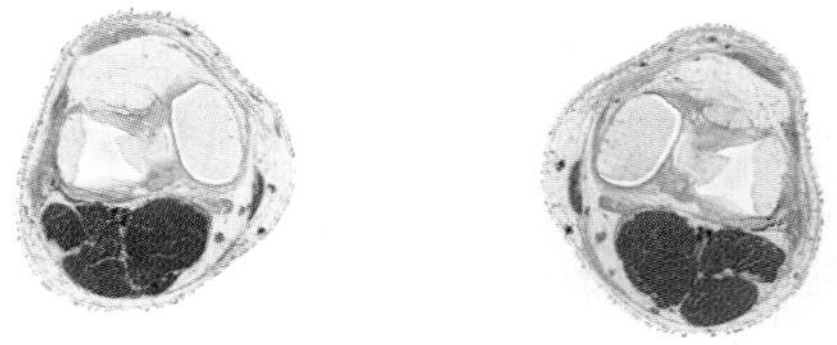

a_vm2336

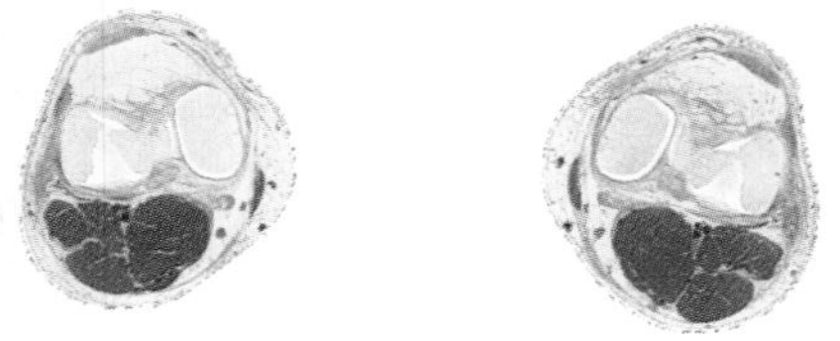

a_vm2337

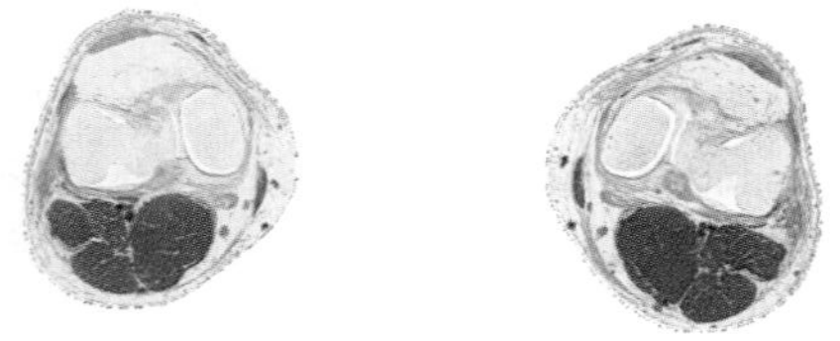

a_vm2338

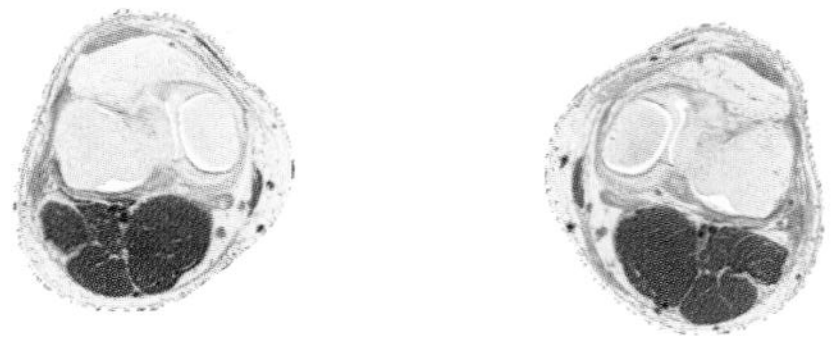

a_vm2339

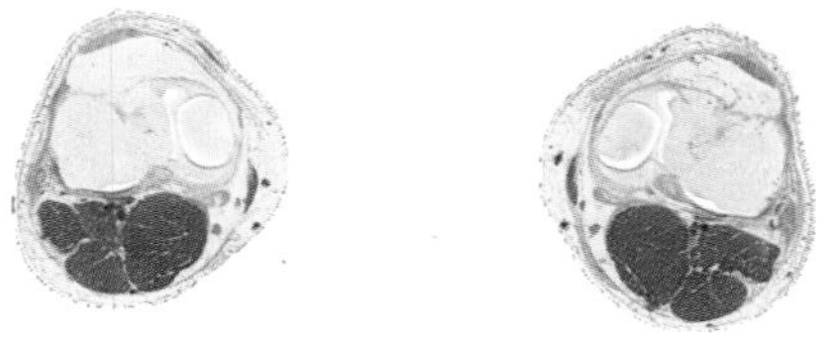

a_vm2340

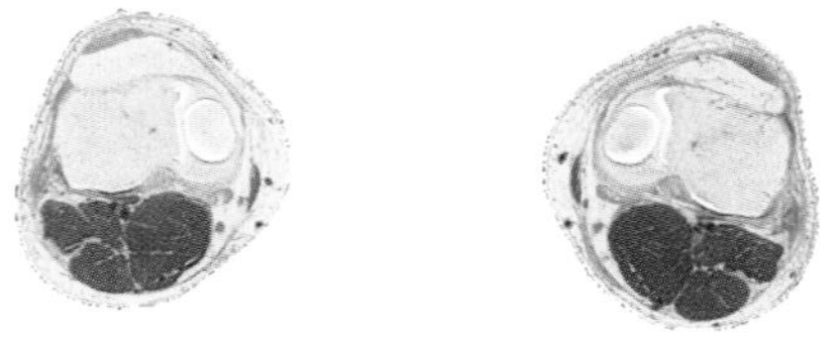

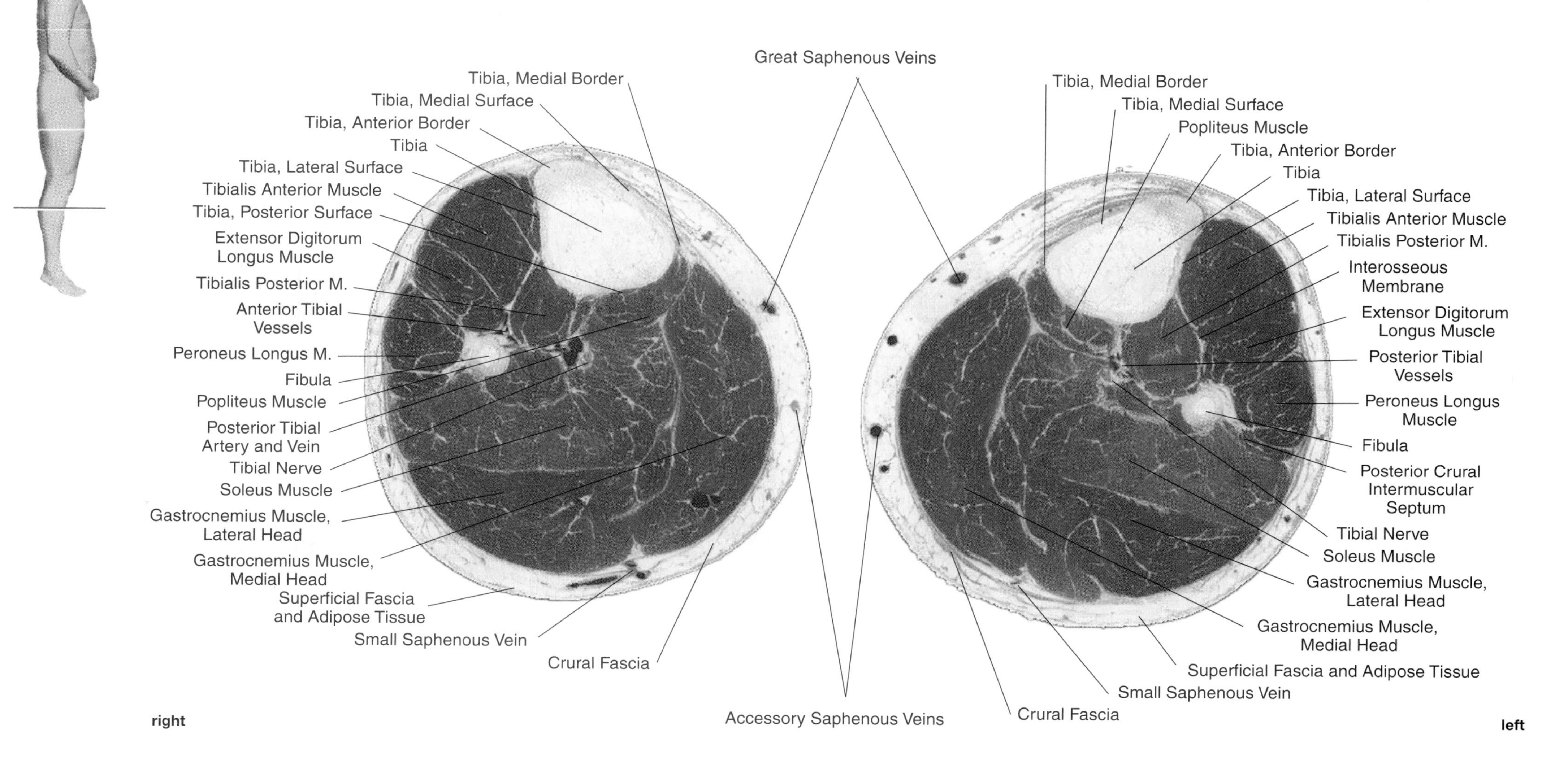
anterior
Great Saphenous Veins
Tibia, Medial Border
Tibia, Medial Surface
Tibia, Anterior Border
Tibia
Tibia, Lateral Surface
Tibialis Anterior Muscle
Tibia, Posterior Surface
Extensor Digitorum Longus Muscle
Tibialis Posterior M.
Anterior Tibial Vessels
Peroneus Longus M.
Fibula
Popliteus Muscle
Posterior Tibial Artery and Vein
Tibial Nerve
Soleus Muscle
Gastrocnemius Muscle, Lateral Head
Gastrocnemius Muscle, Medial Head
Superficial Fascia and Adipose Tissue
Small Saphenous Vein
Crural Fascia
Accessory Saphenous Veins
Tibia, Medial Border
Tibia, Medial Surface
Popliteus Muscle
Tibia, Anterior Border
Tibia
Tibia, Lateral Surface
Tibialis Anterior Muscle
Tibialis Posterior M.
Interosseous Membrane
Extensor Digitorum Longus Muscle
Posterior Tibial Vessels
Peroneus Longus Muscle
Fibula
Posterior Crural Intermuscular Septum
Tibial Nerve
Soleus Muscle
Gastrocnemius Muscle, Lateral Head
Gastrocnemius Muscle, Medial Head
Superficial Fascia and Adipose Tissue
Small Saphenous Vein
Crural Fascia
right
left
posterior

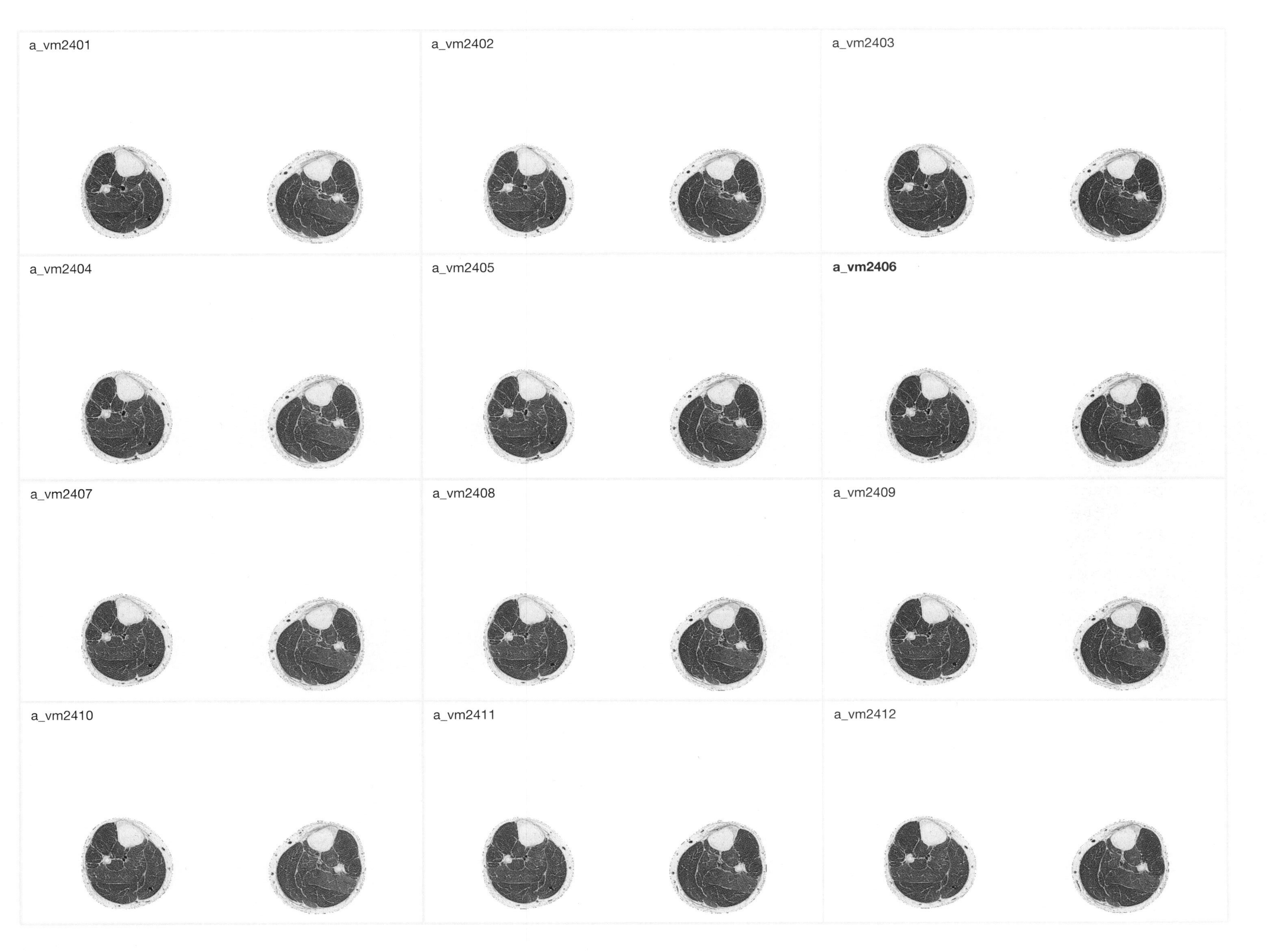
a_vm2401
a_vm2402
a_vm2403
a_vm2404
a_vm2405
a_vm2406
a_vm2407
a_vm2408
a_vm2409
a_vm2410
a_vm2411
a_vm2412

Transverse
a_vm2418

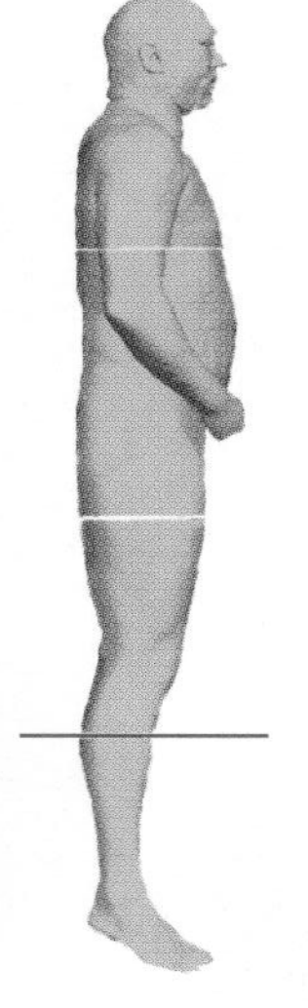

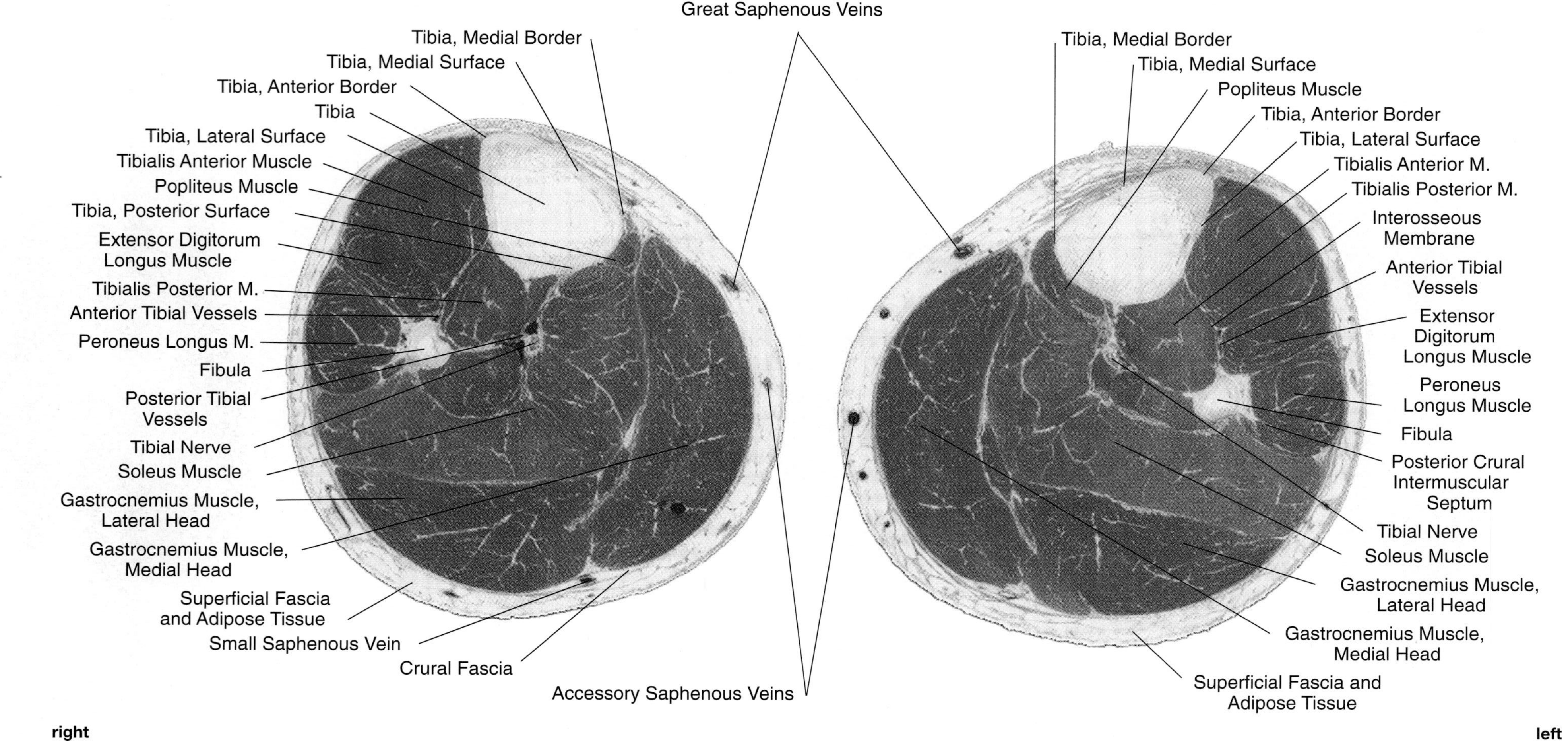

anterior
Great Saphenous Veins
Tibia, Medial Border
Tibia, Medial Surface
Tibia, Anterior Border
Tibia
Tibia, Lateral Surface
Tibialis Anterior Muscle
Popliteus Muscle
Tibia, Posterior Surface
Extensor Digitorum Longus Muscle
Tibialis Posterior M.
Anterior Tibial Vessels
Peroneus Longus M.
Fibula
Posterior Tibial Vessels
Tibial Nerve
Soleus Muscle
Gastrocnemius Muscle, Lateral Head
Gastrocnemius Muscle, Medial Head
Superficial Fascia and Adipose Tissue
Small Saphenous Vein
Crural Fascia
Accessory Saphenous Veins
Tibia, Medial Border
Tibia, Medial Surface
Popliteus Muscle
Tibia, Anterior Border
Tibia, Lateral Surface
Tibialis Anterior M.
Tibialis Posterior M.
Interosseous Membrane
Anterior Tibial Vessels
Extensor Digitorum Longus Muscle
Peroneus Longus Muscle
Fibula
Posterior Crural Intermuscular Septum
Tibial Nerve
Soleus Muscle
Gastrocnemius Muscle, Lateral Head
Gastrocnemius Muscle, Medial Head
Superficial Fascia and Adipose Tissue
right
left
posterior

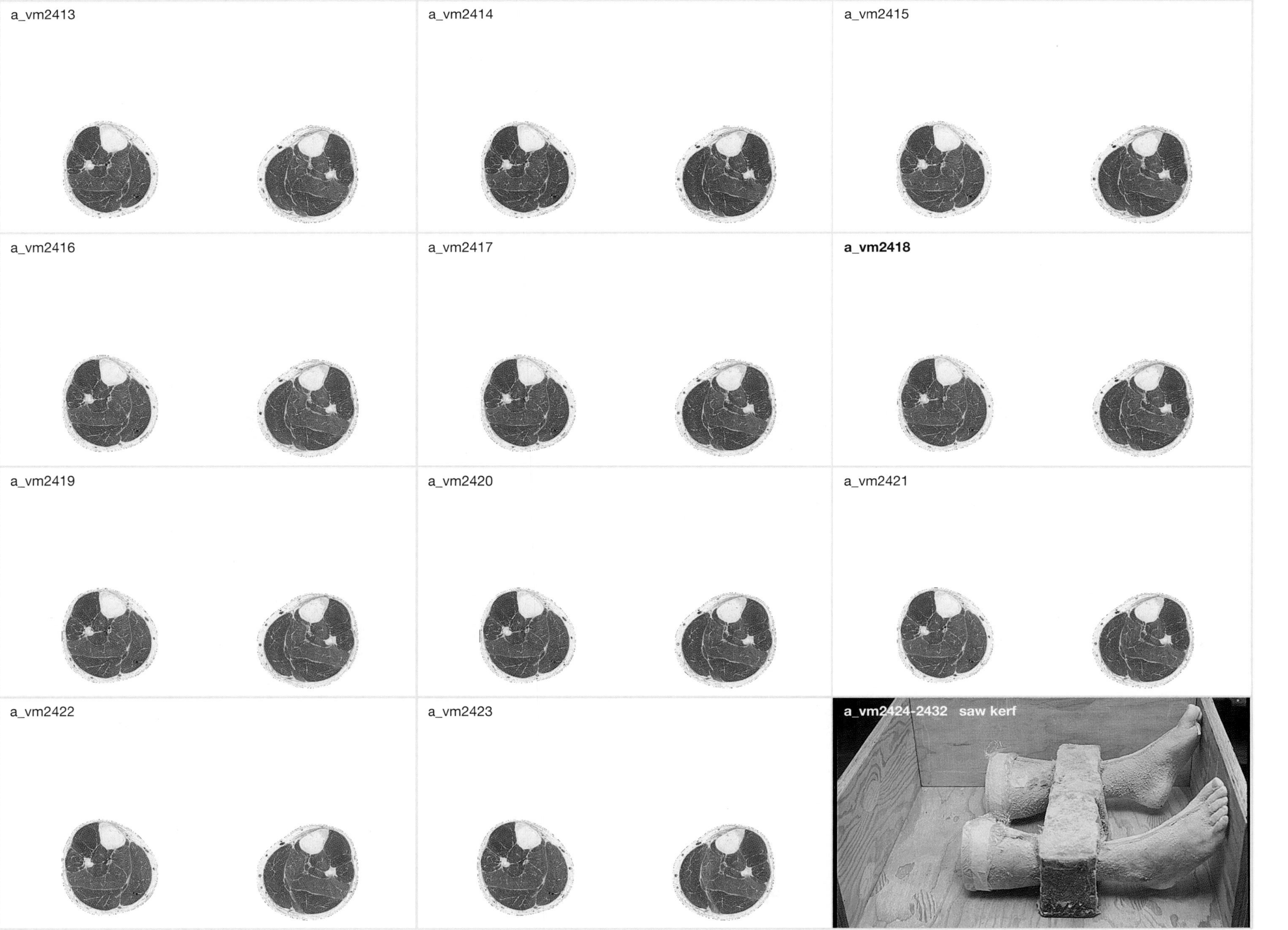
a_vm2413
a_vm2414
a_vm2415
a_vm2416
a_vm2417
a_vm2418
a_vm2419
a_vm2420
a_vm2421
a_vm2422
a_vm2423
a_vm2424-2432 saw kerf

Transverse
a_vm2440

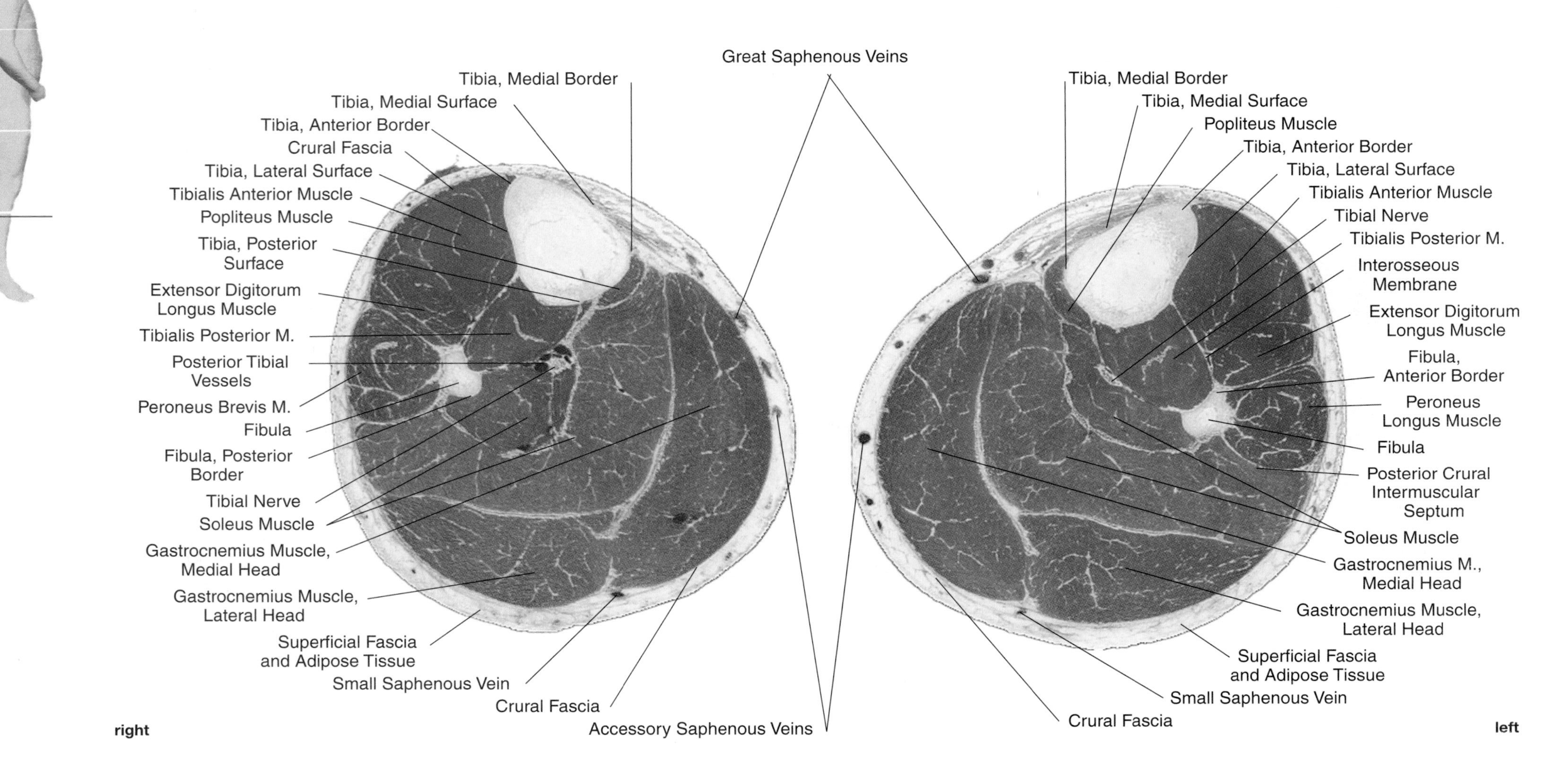

anterior
Great Saphenous Veins
Tibia, Medial Border
Tibia, Medial Surface
Tibia, Anterior Border
Crural Fascia
Tibia, Lateral Surface
Tibialis Anterior Muscle
Popliteus Muscle
Tibia, Posterior Surface
Extensor Digitorum Longus Muscle
Tibialis Posterior M.
Posterior Tibial Vessels
Peroneus Brevis M.
Fibula
Fibula, Posterior Border
Tibial Nerve
Soleus Muscle
Gastrocnemius Muscle, Medial Head
Gastrocnemius Muscle, Lateral Head
Superficial Fascia and Adipose Tissue
Small Saphenous Vein
Crural Fascia
Accessory Saphenous Veins
right
Tibia, Medial Border
Tibia, Medial Surface
Popliteus Muscle
Tibia, Anterior Border
Tibia, Lateral Surface
Tibialis Anterior Muscle
Tibial Nerve
Tibialis Posterior M.
Interosseous Membrane
Extensor Digitorum Longus Muscle
Fibula, Anterior Border
Peroneus Longus Muscle
Fibula
Posterior Crural Intermuscular Septum
Soleus Muscle
Gastrocnemius M., Medial Head
Gastrocnemius Muscle, Lateral Head
Superficial Fascia and Adipose Tissue
Small Saphenous Vein
Crural Fascia
left
posterior

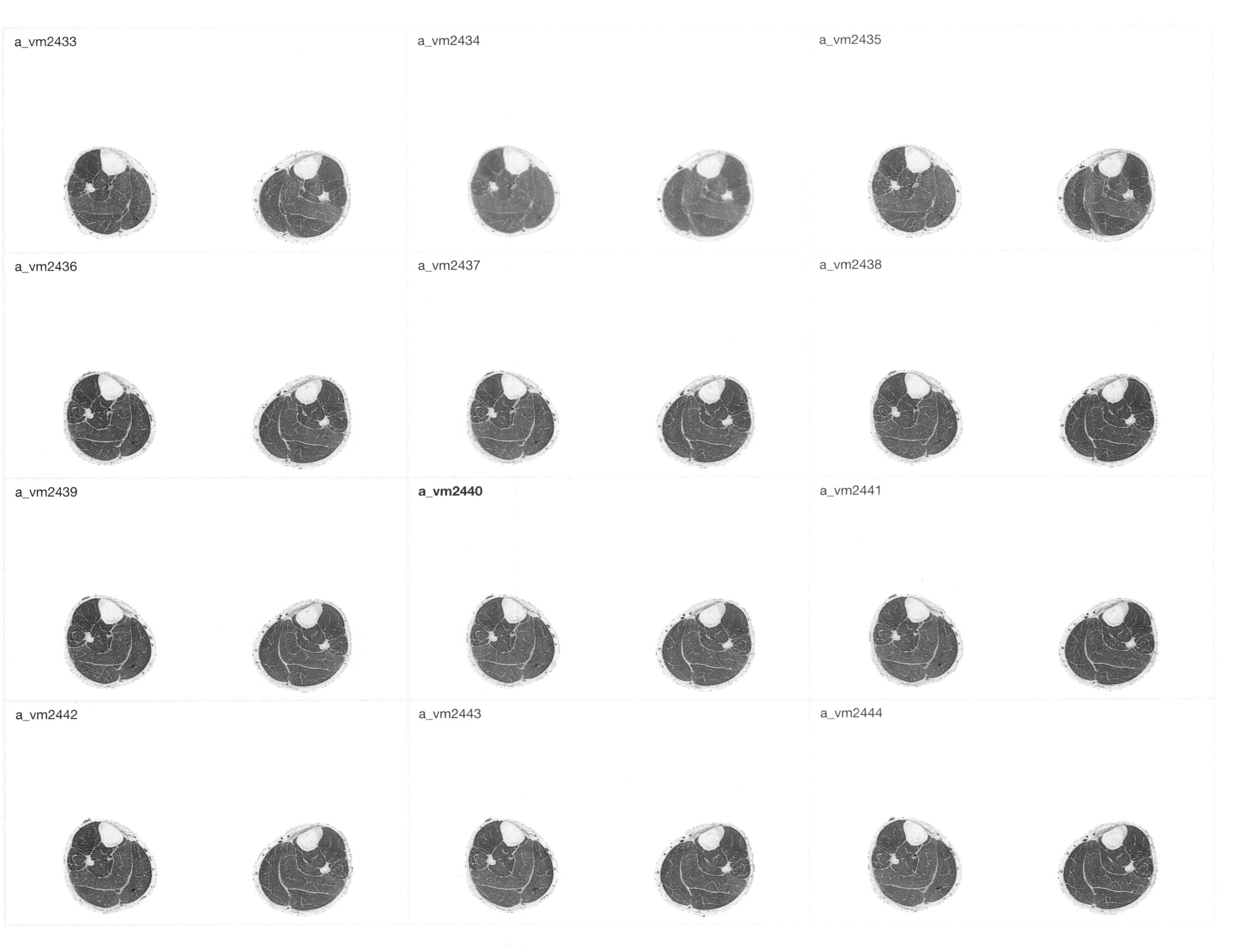
a_vm2433
a_vm2434
a_vm2435
a_vm2436
a_vm2437
a_vm2438
a_vm2439
a_vm2440
a_vm2441
a_vm2442
a_vm2443
a_vm2444

Transverse
a_vm2450

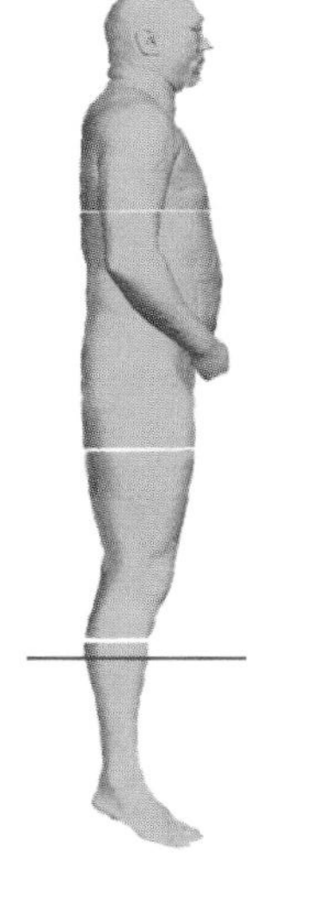

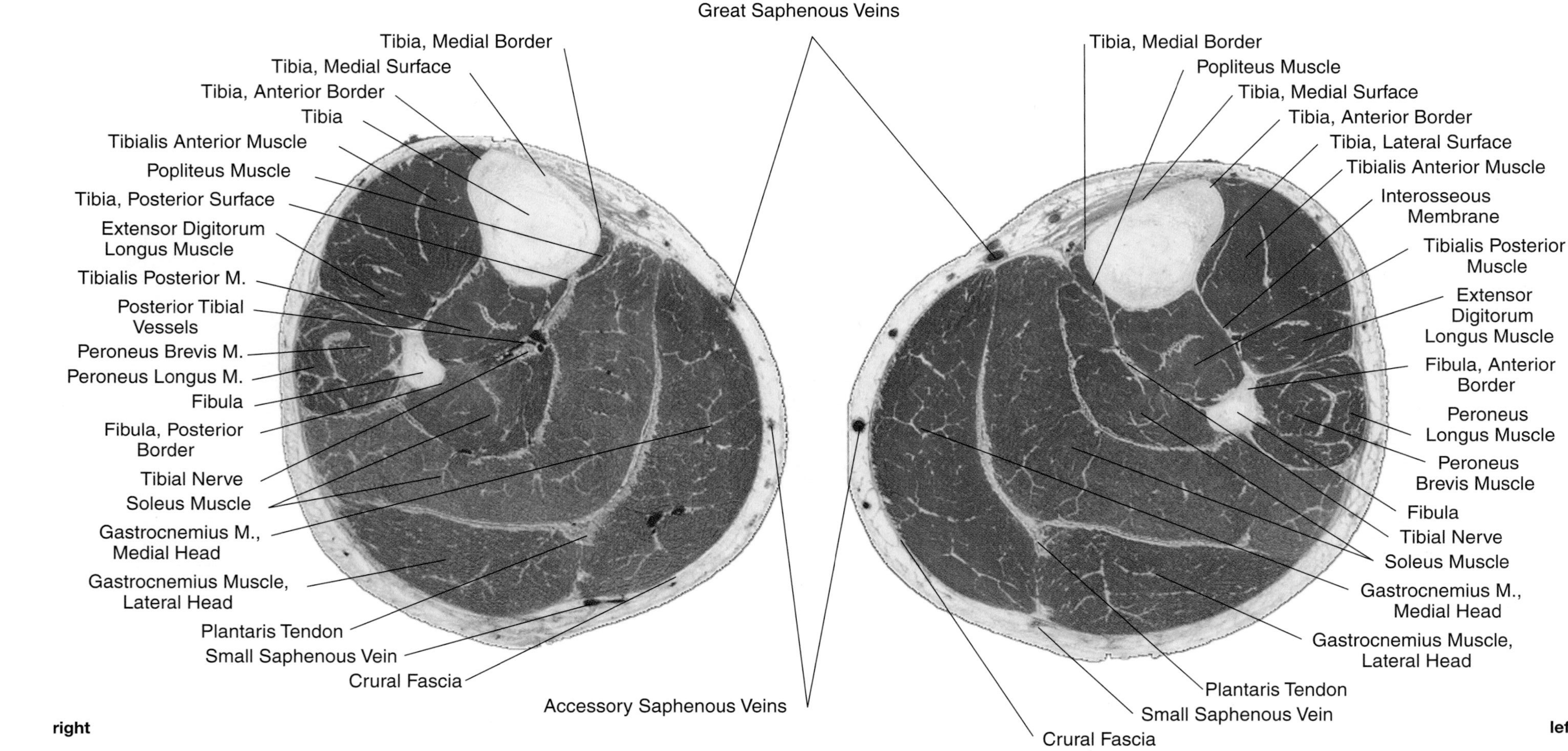
anterior
Great Saphenous Veins
Tibia, Medial Border
Tibia, Medial Surface
Tibia, Anterior Border
Tibia
Tibialis Anterior Muscle
Popliteus Muscle
Tibia, Posterior Surface
Extensor Digitorum Longus Muscle
Tibialis Posterior M.
Posterior Tibial Vessels
Peroneus Brevis M.
Peroneus Longus M.
Fibula
Fibula, Posterior Border
Tibial Nerve
Soleus Muscle
Gastrocnemius M., Medial Head
Gastrocnemius Muscle, Lateral Head
Plantaris Tendon
Small Saphenous Vein
Crural Fascia
Accessory Saphenous Veins
Tibia, Medial Border
Popliteus Muscle
Tibia, Medial Surface
Tibia, Anterior Border
Tibia, Lateral Surface
Tibialis Anterior Muscle
Interosseous Membrane
Tibialis Posterior Muscle
Extensor Digitorum Longus Muscle
Fibula, Anterior Border
Peroneus Longus Muscle
Peroneus Brevis Muscle
Fibula
Tibial Nerve
Soleus Muscle
Gastrocnemius M., Medial Head
Gastrocnemius Muscle, Lateral Head
Plantaris Tendon
Small Saphenous Vein
Crural Fascia
right
left
posterior

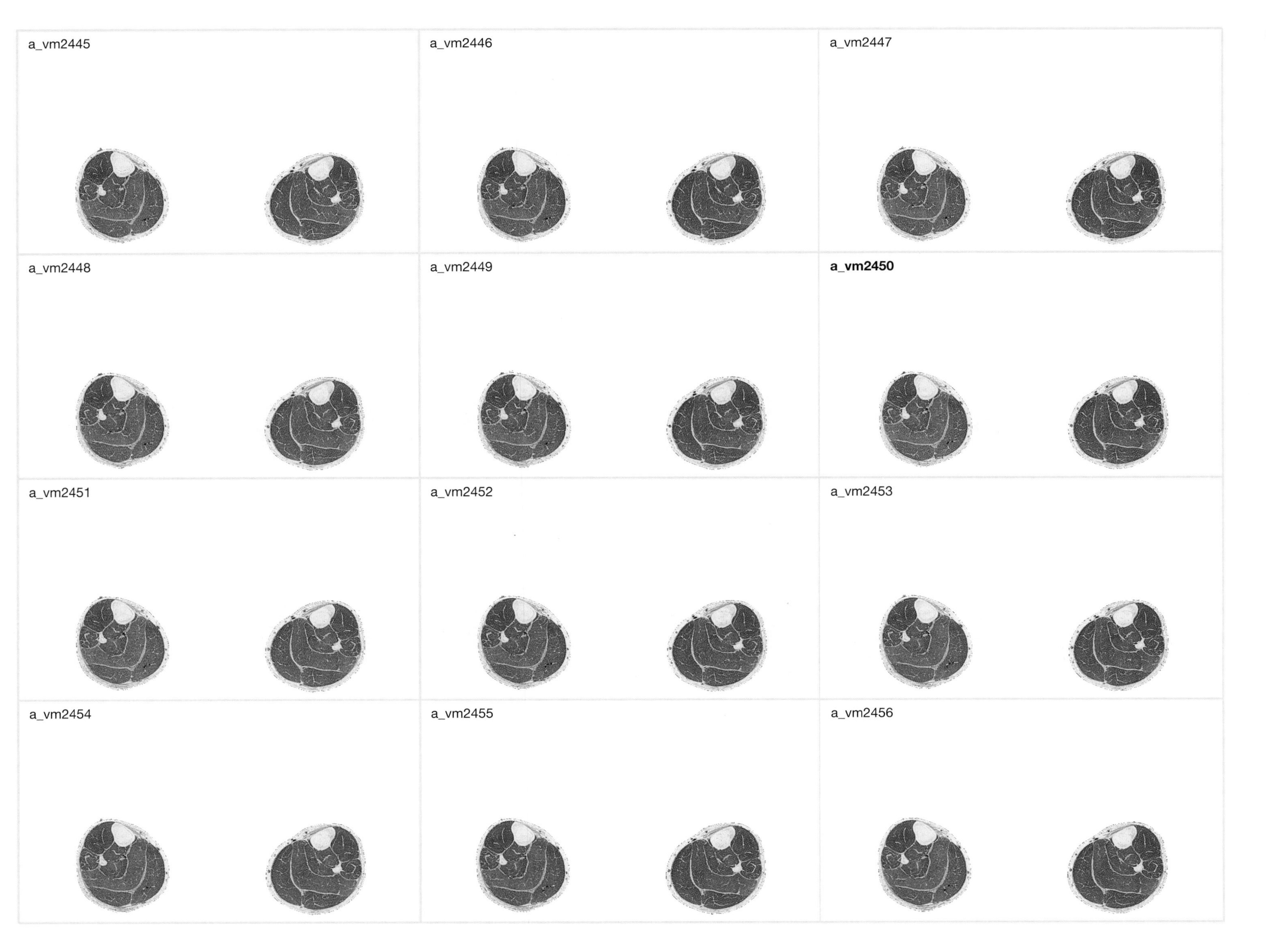
a_vm2445
a_vm2446
a_vm2447
a_vm2448
a_vm2449
a_vm2450
a_vm2451
a_vm2452
a_vm2453
a_vm2454
a_vm2455
a_vm2456

Transverse
a_vm2464

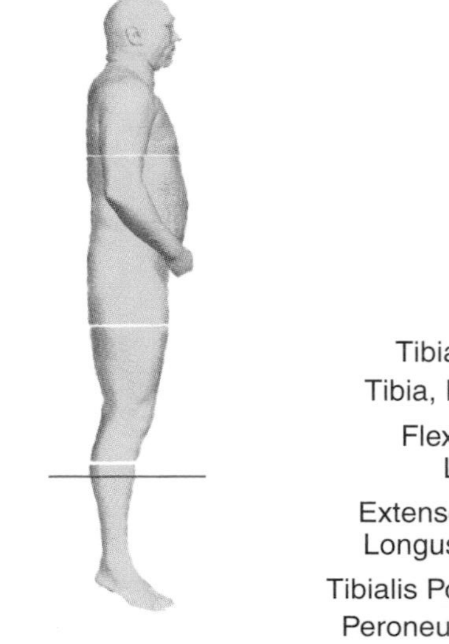

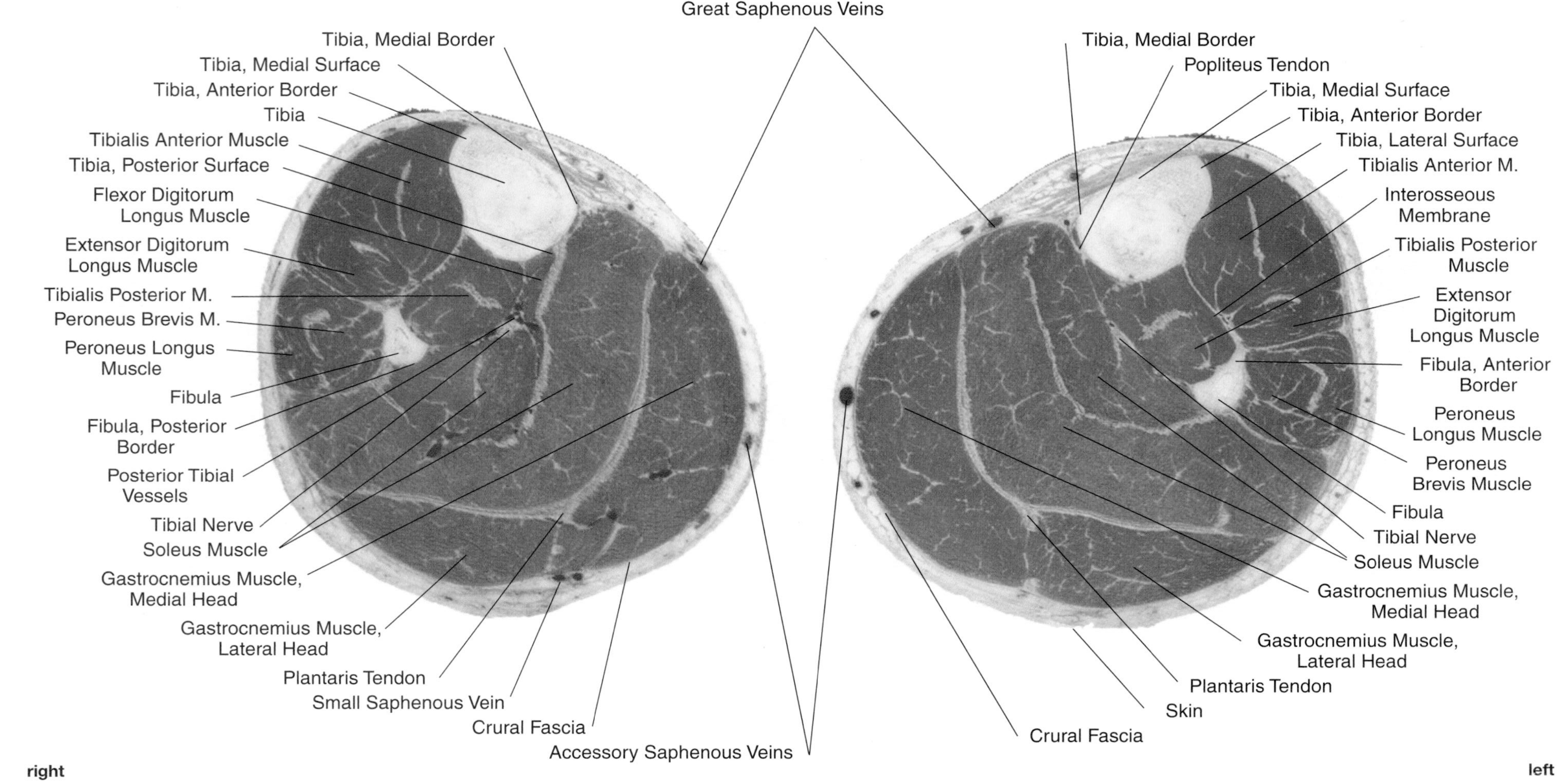

anterior
Great Saphenous Veins
Tibia, Medial Border
Tibia, Medial Surface
Tibia, Anterior Border
Tibia
Tibialis Anterior Muscle
Tibia, Posterior Surface
Flexor Digitorum Longus Muscle
Extensor Digitorum Longus Muscle
Tibialis Posterior M.
Peroneus Brevis M.
Peroneus Longus Muscle
Fibula
Fibula, Posterior Border
Posterior Tibial Vessels
Tibial Nerve
Soleus Muscle
Gastrocnemius Muscle, Medial Head
Gastrocnemius Muscle, Lateral Head
Plantaris Tendon
Small Saphenous Vein
Crural Fascia
Accessory Saphenous Veins
Tibia, Medial Border
Popliteus Tendon
Tibia, Medial Surface
Tibia, Anterior Border
Tibia, Lateral Surface
Tibialis Anterior M.
Interosseous Membrane
Tibialis Posterior Muscle
Extensor Digitorum Longus Muscle
Fibula, Anterior Border
Peroneus Longus Muscle
Peroneus Brevis Muscle
Fibula
Tibial Nerve
Soleus Muscle
Gastrocnemius Muscle, Medial Head
Gastrocnemius Muscle, Lateral Head
Plantaris Tendon
Skin
Crural Fascia
right
left
posterior

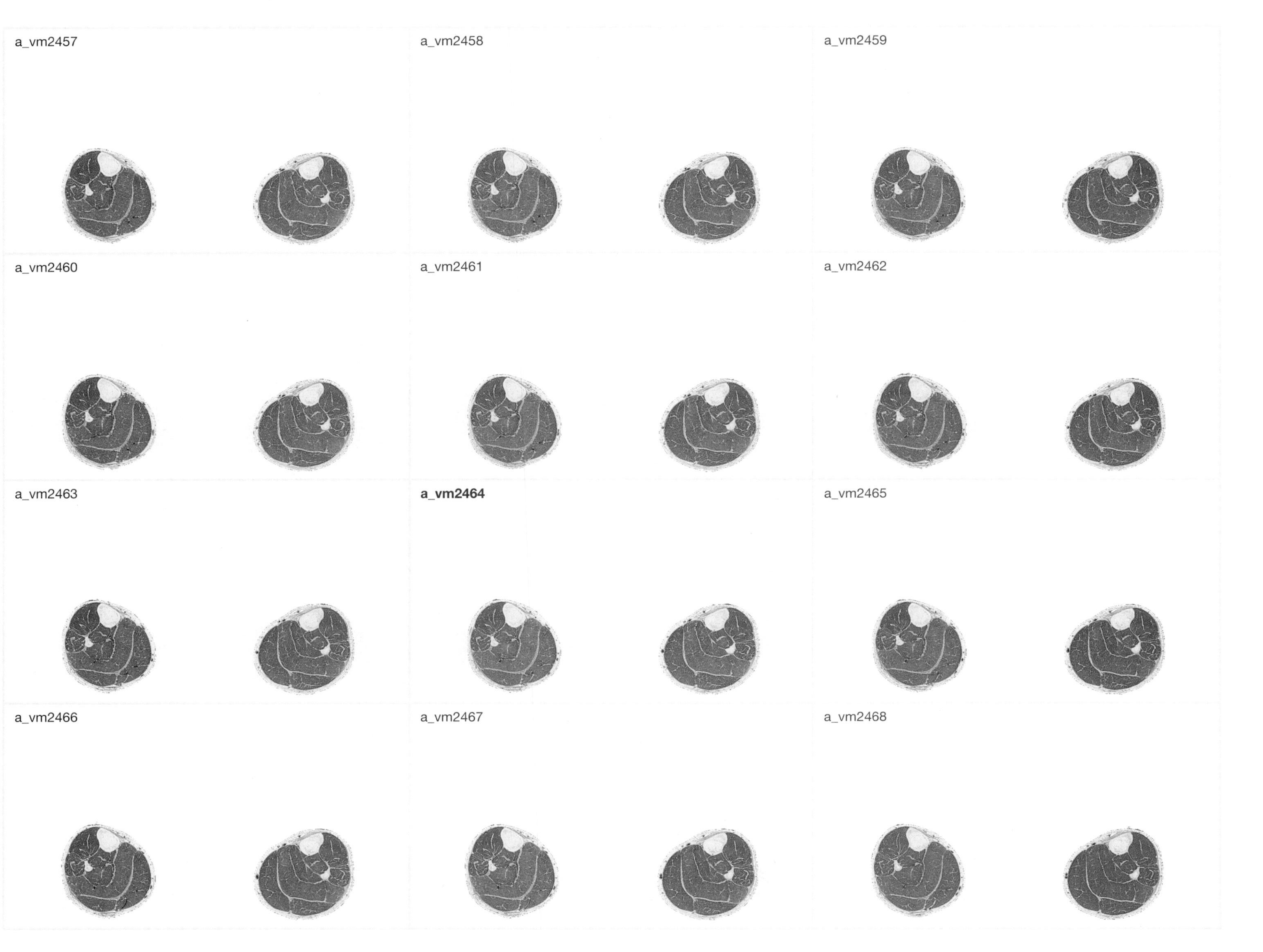
a_vm2457
a_vm2458
a_vm2459
a_vm2460
a_vm2461
a_vm2462
a_vm2463
a_vm2464
a_vm2465
a_vm2466
a_vm2467
a_vm2468

Transverse
a_vm2477

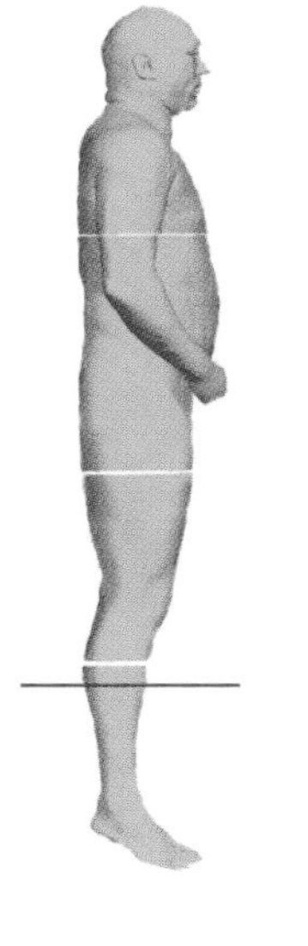

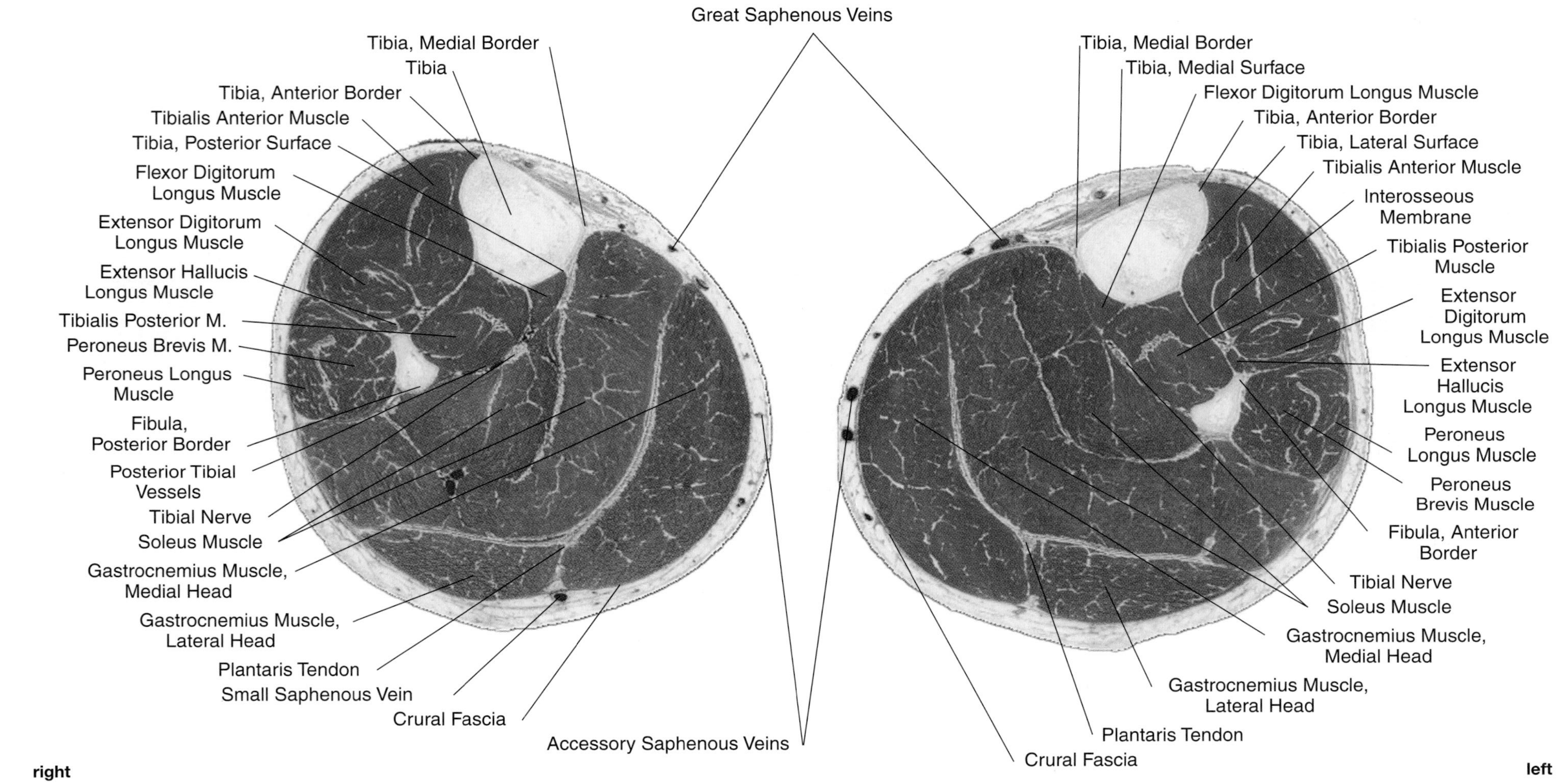

anterior
Great Saphenous Veins
Tibia, Medial Border
Tibia
Tibia, Anterior Border
Tibialis Anterior Muscle
Tibia, Posterior Surface
Flexor Digitorum Longus Muscle
Extensor Digitorum Longus Muscle
Extensor Hallucis Longus Muscle
Tibialis Posterior M.
Peroneus Brevis M.
Peroneus Longus Muscle
Fibula, Posterior Border
Posterior Tibial Vessels
Tibial Nerve
Soleus Muscle
Gastrocnemius Muscle, Medial Head
Gastrocnemius Muscle, Lateral Head
Plantaris Tendon
Small Saphenous Vein
Crural Fascia
Accessory Saphenous Veins
Tibia, Medial Border
Tibia, Medial Surface
Flexor Digitorum Longus Muscle
Tibia, Anterior Border
Tibia, Lateral Surface
Tibialis Anterior Muscle
Interosseous Membrane
Tibialis Posterior Muscle
Extensor Digitorum Longus Muscle
Extensor Hallucis Longus Muscle
Peroneus Longus Muscle
Peroneus Brevis Muscle
Fibula, Anterior Border
Tibial Nerve
Soleus Muscle
Gastrocnemius Muscle, Medial Head
Gastrocnemius Muscle, Lateral Head
Plantaris Tendon
Crural Fascia
right
left
posterior

a_vm2469
a_vm2470
a_vm2471
a_vm2472
a_vm2473
a_vm2474
a_vm2475
a_vm2476
a_vm2477
a_vm2478
a_vm2479
a_vm2480

Transverse
a_vm2487

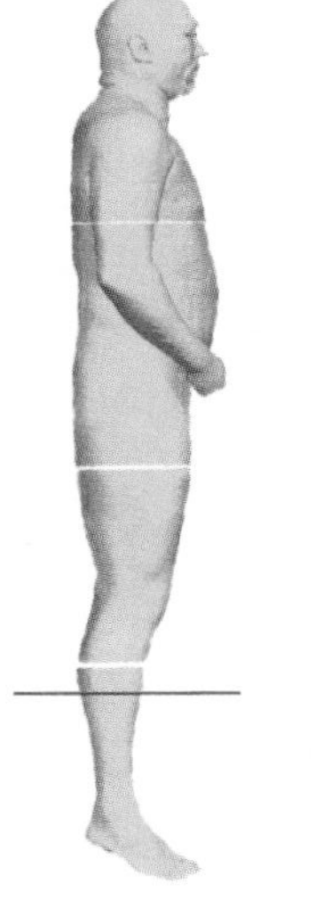

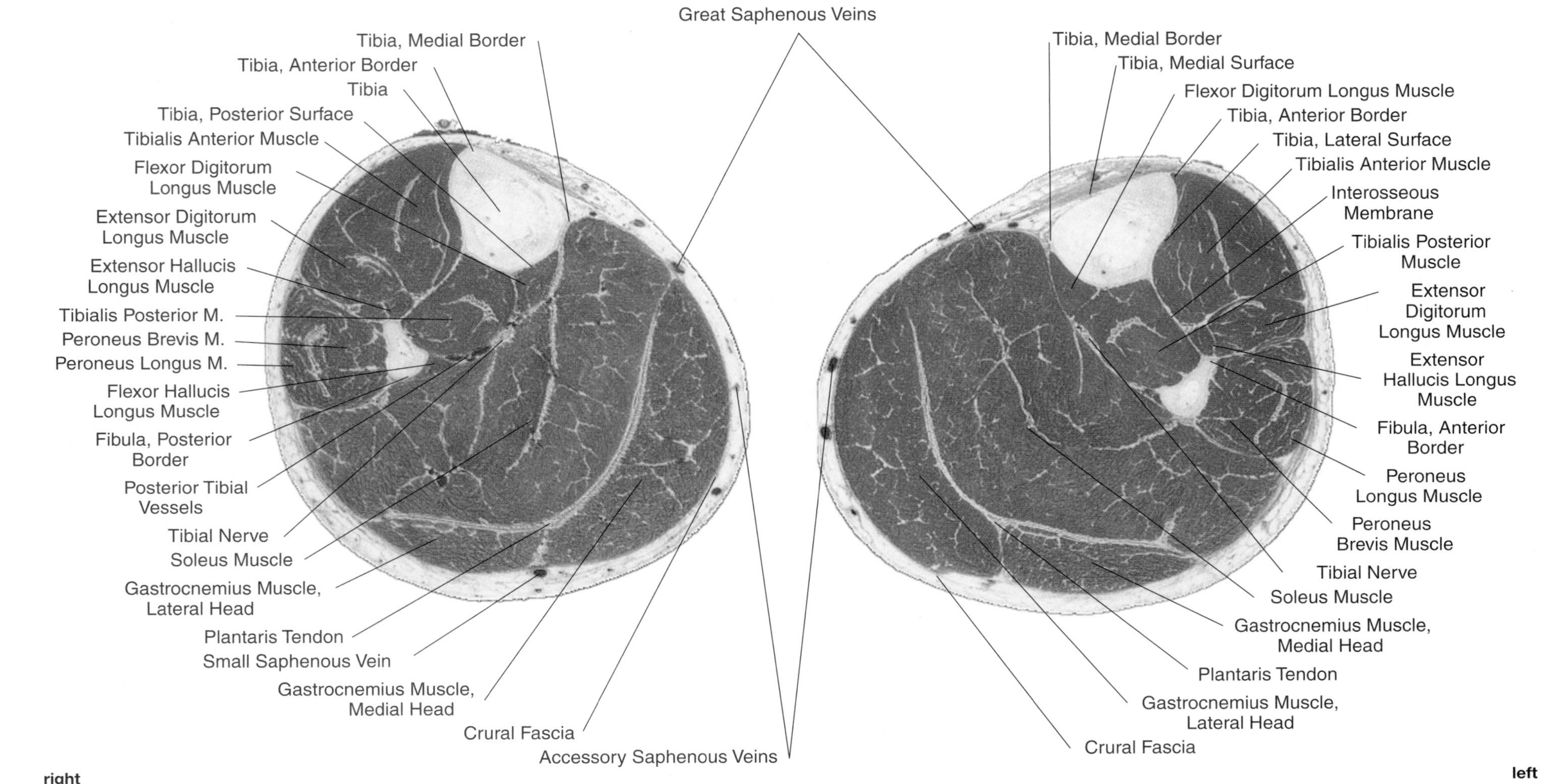

anterior
Great Saphenous Veins
Tibia, Medial Border
Tibia, Anterior Border
Tibia
Tibia, Posterior Surface
Tibialis Anterior Muscle
Flexor Digitorum Longus Muscle
Extensor Digitorum Longus Muscle
Extensor Hallucis Longus Muscle
Tibialis Posterior M.
Peroneus Brevis M.
Peroneus Longus M.
Flexor Hallucis Longus Muscle
Fibula, Posterior Border
Posterior Tibial Vessels
Tibial Nerve
Soleus Muscle
Gastrocnemius Muscle, Lateral Head
Plantaris Tendon
Small Saphenous Vein
Gastrocnemius Muscle, Medial Head
Crural Fascia
Accessory Saphenous Veins
Tibia, Medial Border
Tibia, Medial Surface
Flexor Digitorum Longus Muscle
Tibia, Anterior Border
Tibia, Lateral Surface
Tibialis Anterior Muscle
Interosseous Membrane
Tibialis Posterior Muscle
Extensor Digitorum Longus Muscle
Extensor Hallucis Longus Muscle
Fibula, Anterior Border
Peroneus Longus Muscle
Peroneus Brevis Muscle
Tibial Nerve
Soleus Muscle
Gastrocnemius Muscle, Medial Head
Plantaris Tendon
Gastrocnemius Muscle, Lateral Head
Crural Fascia
right
left
posterior

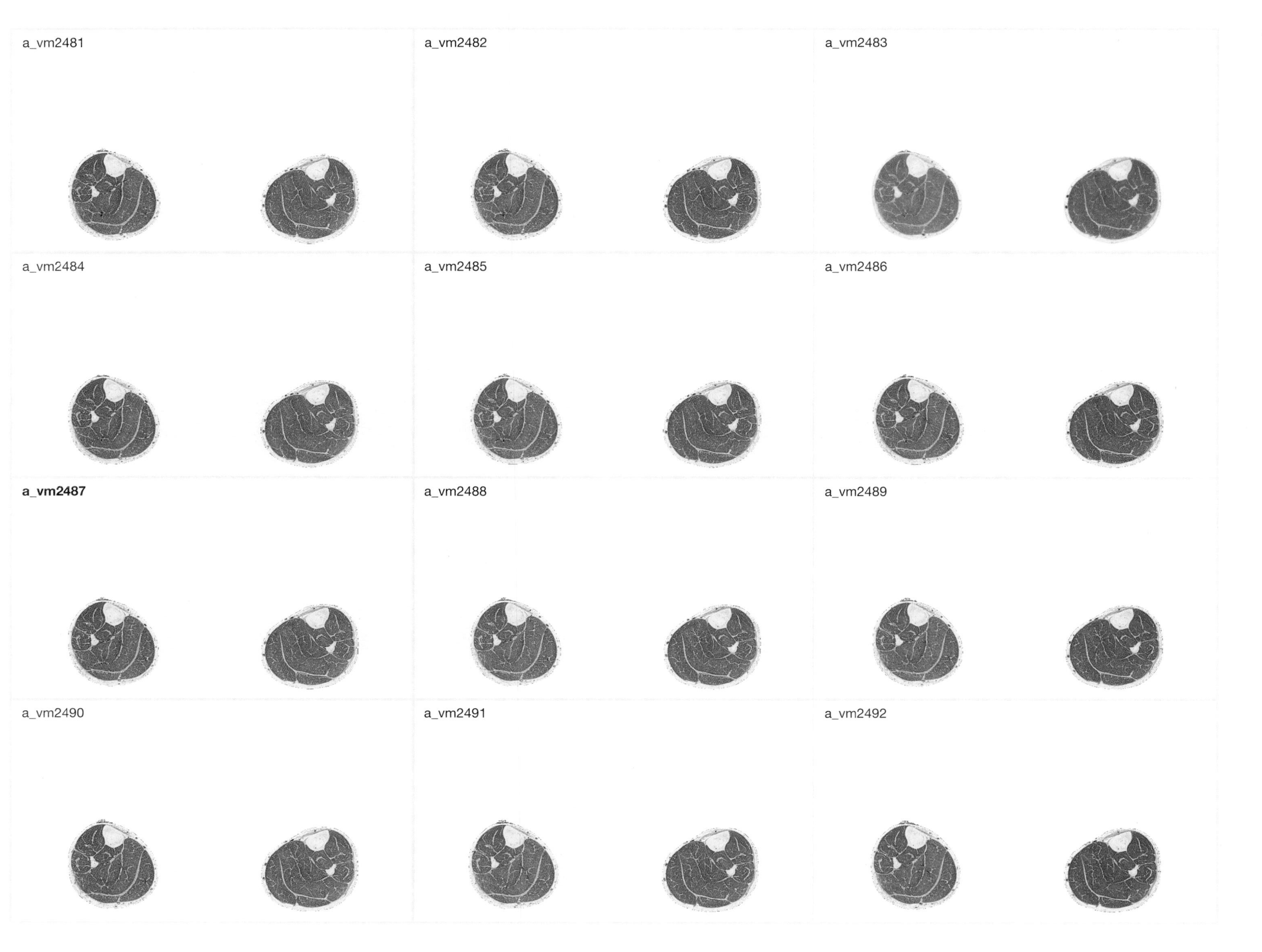
a_vm2481
a_vm2482
a_vm2483
a_vm2484
a_vm2485
a_vm2486
a_vm2487
a_vm2488
a_vm2489
a_vm2490
a_vm2491
a_vm2492

Transverse
a_vm2502

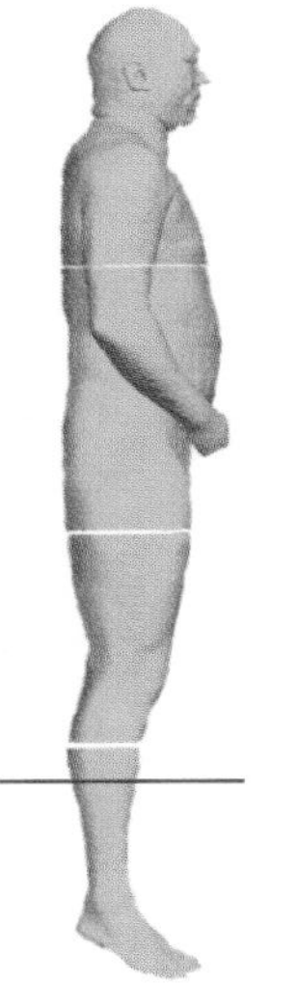

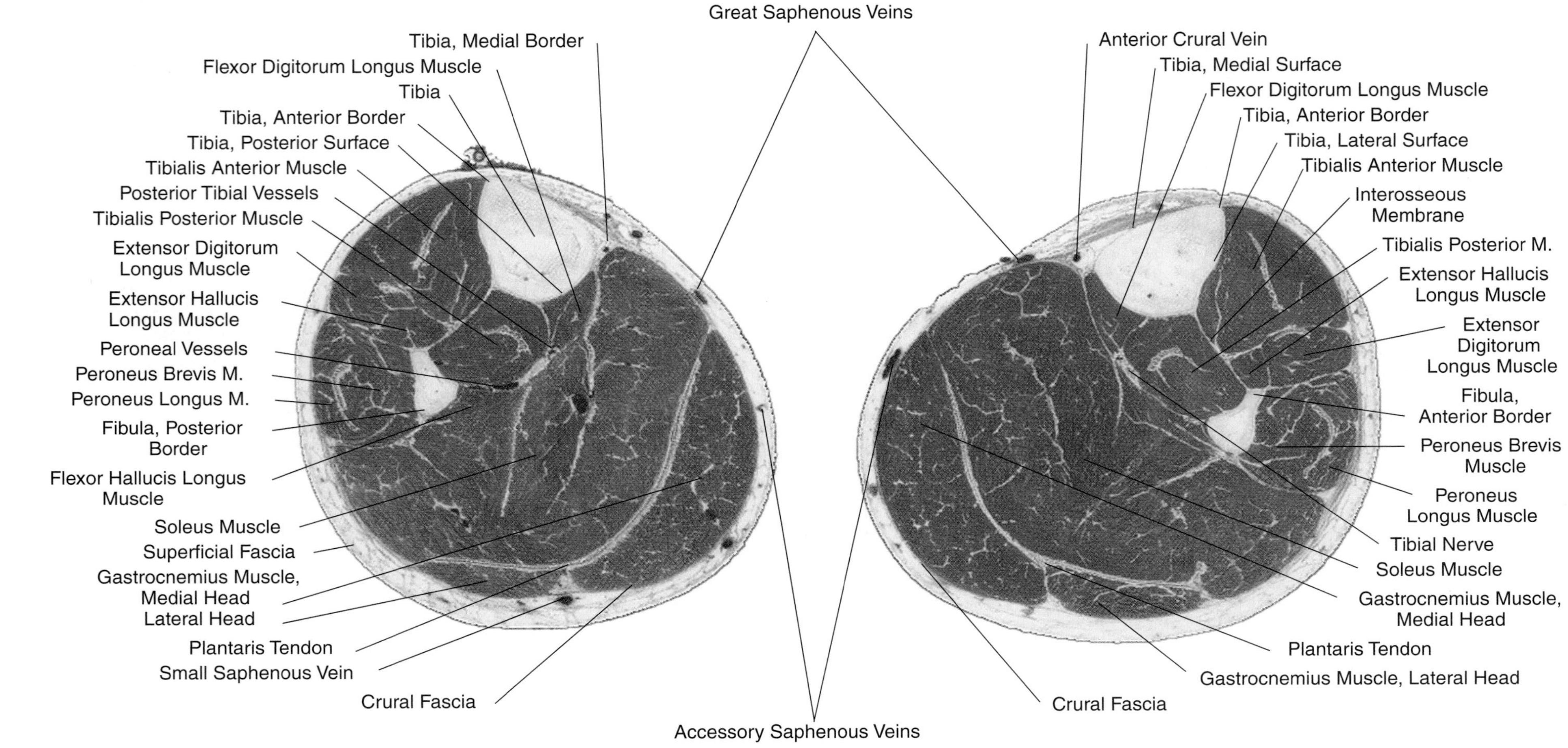

anterior
Great Saphenous Veins
Tibia, Medial Border
Flexor Digitorum Longus Muscle
Tibia
Tibia, Anterior Border
Tibia, Posterior Surface
Tibialis Anterior Muscle
Posterior Tibial Vessels
Tibialis Posterior Muscle
Extensor Digitorum Longus Muscle
Extensor Hallucis Longus Muscle
Peroneal Vessels
Peroneus Brevis M.
Peroneus Longus M.
Fibula, Posterior Border
Flexor Hallucis Longus Muscle
Soleus Muscle
Superficial Fascia
Gastrocnemius Muscle, Medial Head
Lateral Head
Plantaris Tendon
Small Saphenous Vein
Crural Fascia
Accessory Saphenous Veins
Anterior Crural Vein
Tibia, Medial Surface
Flexor Digitorum Longus Muscle
Tibia, Anterior Border
Tibia, Lateral Surface
Tibialis Anterior Muscle
Interosseous Membrane
Tibialis Posterior M.
Extensor Hallucis Longus Muscle
Extensor Digitorum Longus Muscle
Fibula, Anterior Border
Peroneus Brevis Muscle
Peroneus Longus Muscle
Tibial Nerve
Soleus Muscle
Gastrocnemius Muscle, Medial Head
Plantaris Tendon
Gastrocnemius Muscle, Lateral Head
Crural Fascia
right
left
posterior

a_vm2493

a_vm2494

a_vm2495

a_vm2496

a_vm2497

a_vm2498

a_vm2499

a_vm2500

a_vm2501

a_vm2502

a_vm2503

a_vm2504

Transverse
a_vm2510

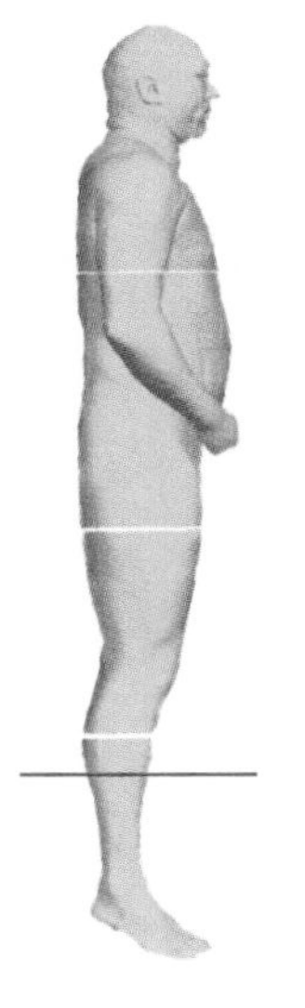

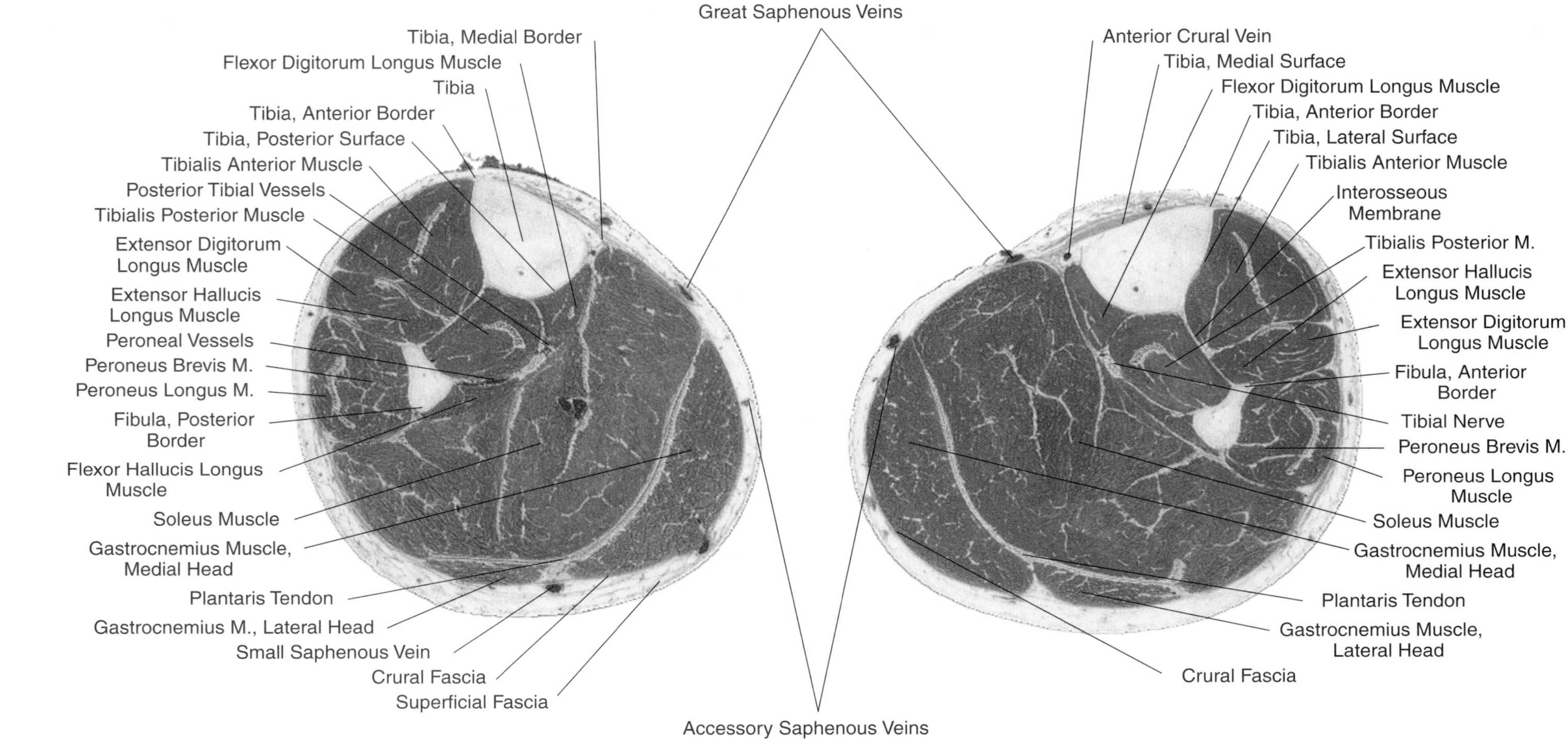

anterior
Great Saphenous Veins
Tibia, Medial Border
Flexor Digitorum Longus Muscle
Tibia
Tibia, Anterior Border
Tibia, Posterior Surface
Tibialis Anterior Muscle
Posterior Tibial Vessels
Tibialis Posterior Muscle
Extensor Digitorum Longus Muscle
Extensor Hallucis Longus Muscle
Peroneal Vessels
Peroneus Brevis M.
Peroneus Longus M.
Fibula, Posterior Border
Flexor Hallucis Longus Muscle
Soleus Muscle
Gastrocnemius Muscle, Medial Head
Plantaris Tendon
Gastrocnemius M., Lateral Head
Small Saphenous Vein
Crural Fascia
Superficial Fascia
Accessory Saphenous Veins
Anterior Crural Vein
Tibia, Medial Surface
Flexor Digitorum Longus Muscle
Tibia, Anterior Border
Tibia, Lateral Surface
Tibialis Anterior Muscle
Interosseous Membrane
Tibialis Posterior M.
Extensor Hallucis Longus Muscle
Extensor Digitorum Longus Muscle
Fibula, Anterior Border
Tibial Nerve
Peroneus Brevis M.
Peroneus Longus Muscle
Soleus Muscle
Gastrocnemius Muscle, Medial Head
Plantaris Tendon
Gastrocnemius Muscle, Lateral Head
Crural Fascia
right
left
posterior

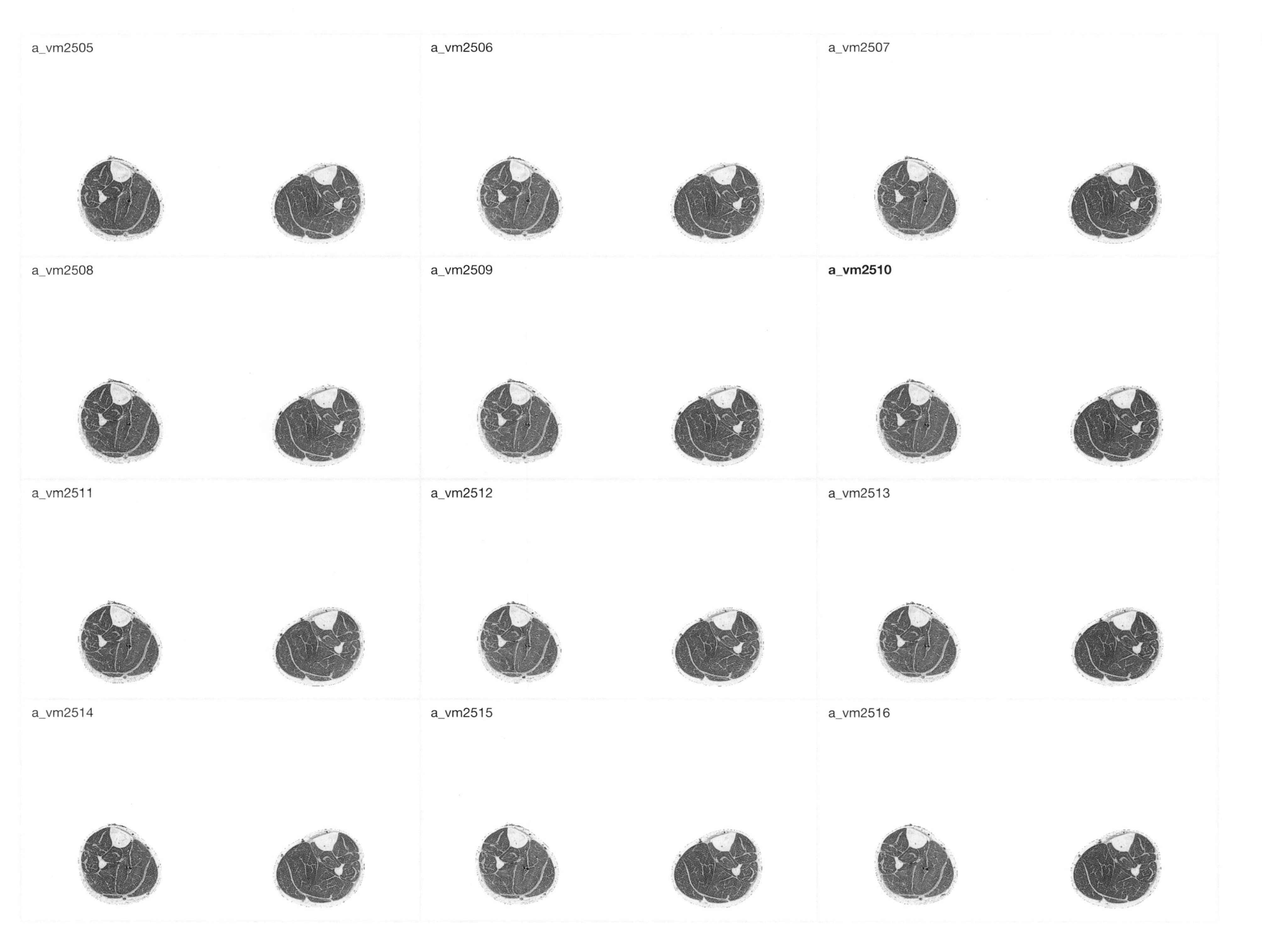
a_vm2505
a_vm2506
a_vm2507
a_vm2508
a_vm2509
a_vm2510
a_vm2511
a_vm2512
a_vm2513
a_vm2514
a_vm2515
a_vm2516

Transverse
a_vm2524

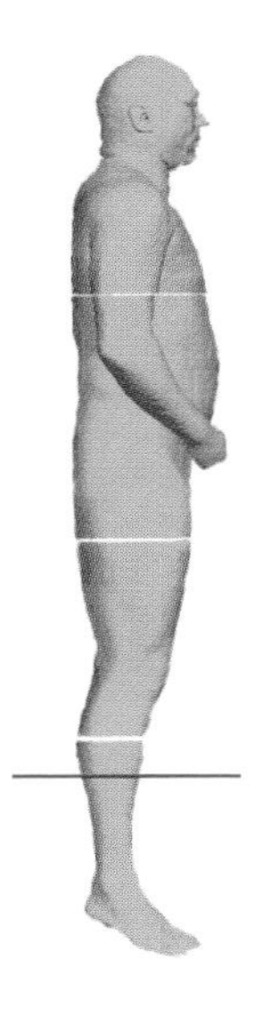

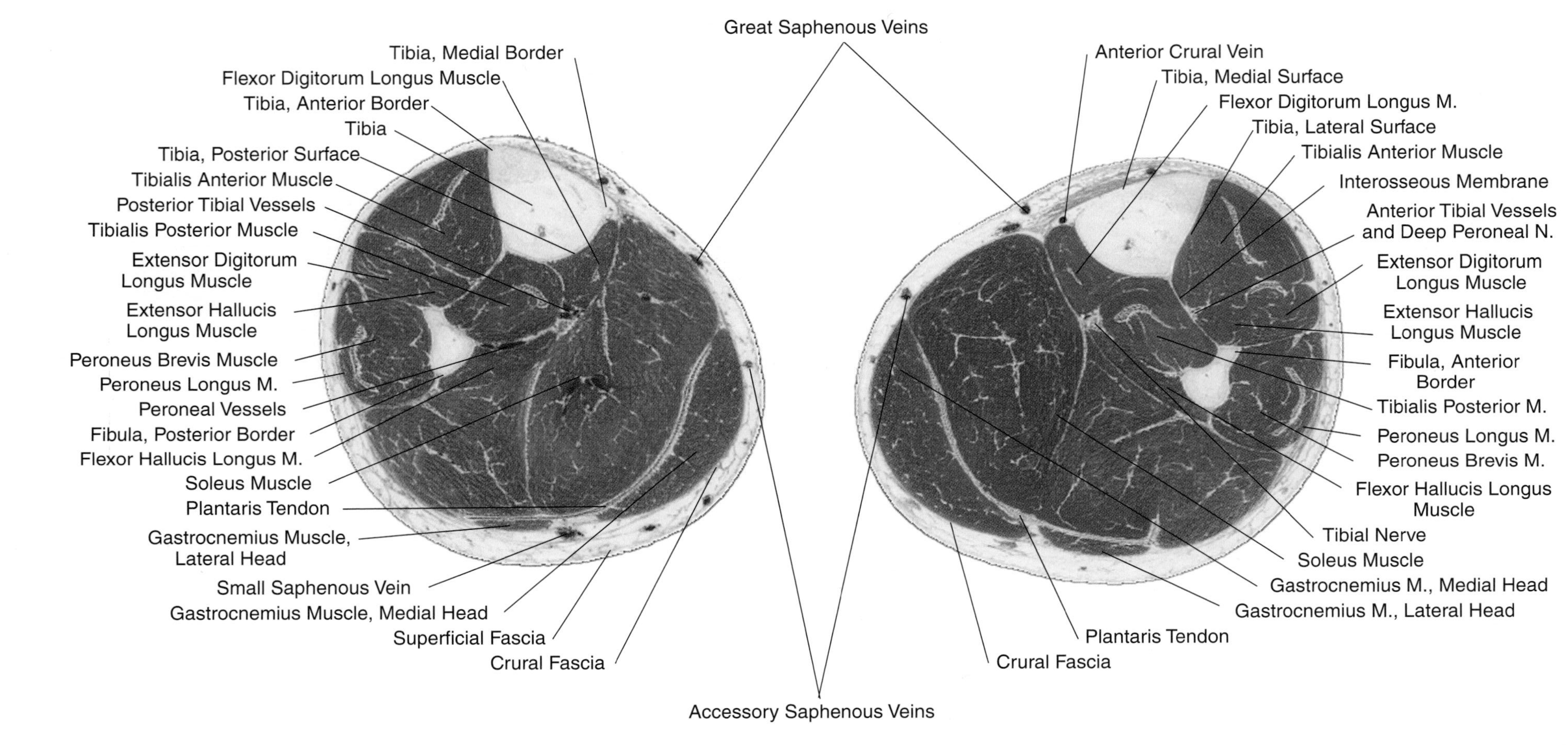
anterior
Great Saphenous Veins
Tibia, Medial Border
Flexor Digitorum Longus Muscle
Tibia, Anterior Border
Tibia
Tibia, Posterior Surface
Tibialis Anterior Muscle
Posterior Tibial Vessels
Tibialis Posterior Muscle
Extensor Digitorum Longus Muscle
Extensor Hallucis Longus Muscle
Peroneus Brevis Muscle
Peroneus Longus M.
Peroneal Vessels
Fibula, Posterior Border
Flexor Hallucis Longus M.
Soleus Muscle
Plantaris Tendon
Gastrocnemius Muscle, Lateral Head
Small Saphenous Vein
Gastrocnemius Muscle, Medial Head
Superficial Fascia
Crural Fascia
Anterior Crural Vein
Tibia, Medial Surface
Flexor Digitorum Longus M.
Tibia, Lateral Surface
Tibialis Anterior Muscle
Interosseous Membrane
Anterior Tibial Vessels and Deep Peroneal N.
Extensor Digitorum Longus Muscle
Extensor Hallucis Longus Muscle
Fibula, Anterior Border
Tibialis Posterior M.
Peroneus Longus M.
Peroneus Brevis M.
Flexor Hallucis Longus Muscle
Tibial Nerve
Soleus Muscle
Gastrocnemius M., Medial Head
Gastrocnemius M., Lateral Head
Plantaris Tendon
Crural Fascia
Accessory Saphenous Veins
right
left
posterior

a_vm2517

a_vm2518

a_vm2519

a_vm2520

a_vm2521

a_vm2522

a_vm2523

a_vm2524

a_vm2525

a_vm2526

a_vm2527

a_vm2528

Transverse
a_vm2632

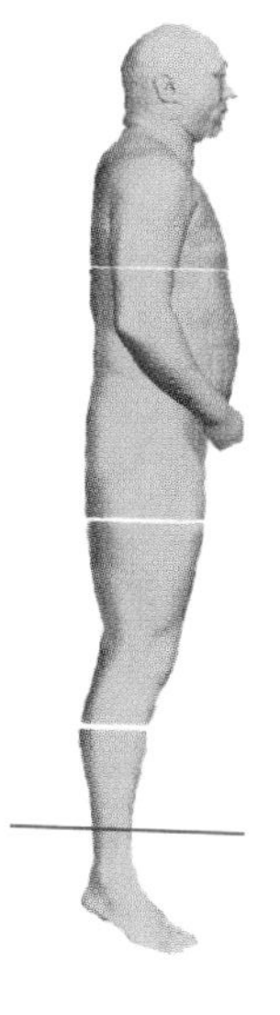

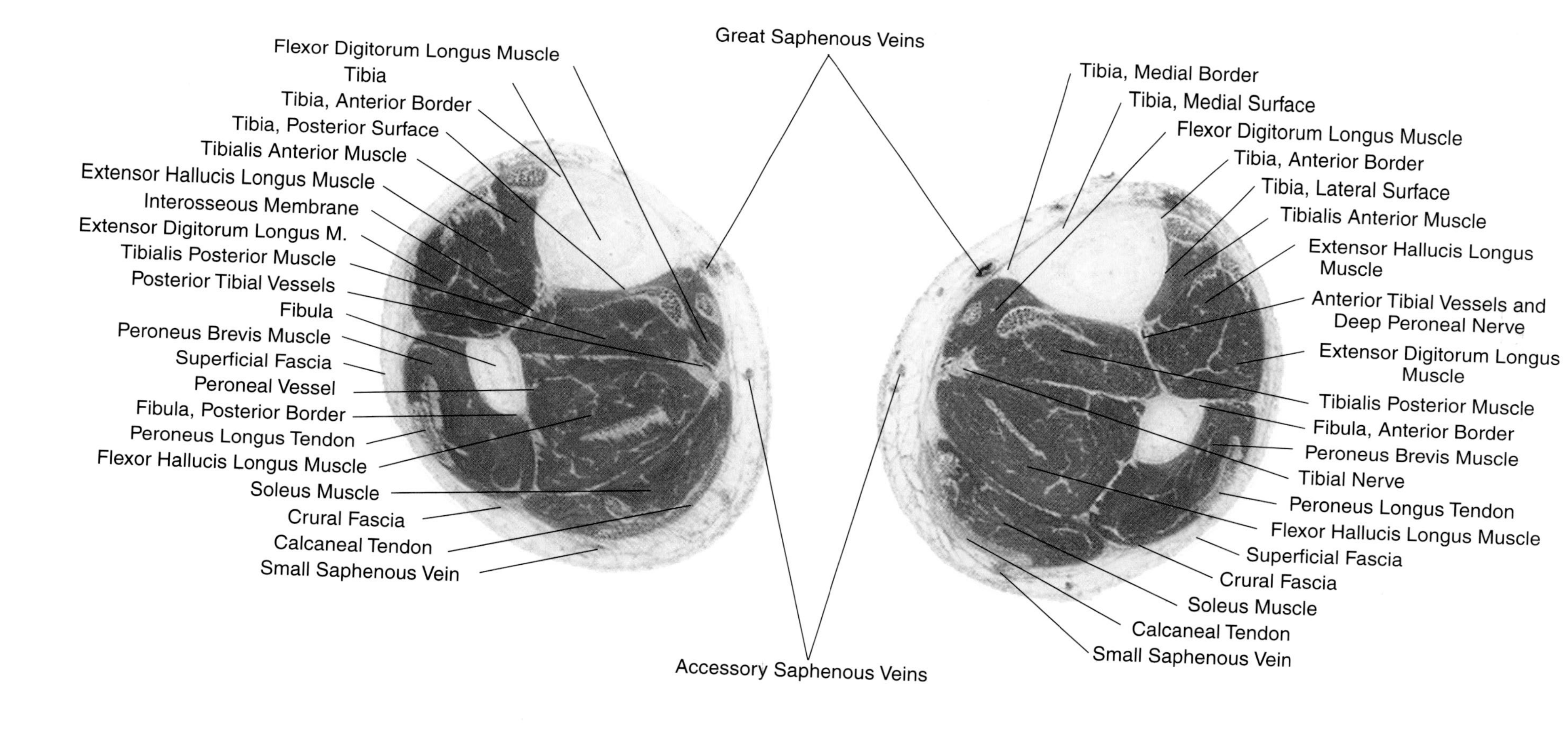
anterior
right
left
posterior
Flexor Digitorum Longus Muscle
Tibia
Tibia, Anterior Border
Tibia, Posterior Surface
Tibialis Anterior Muscle
Extensor Hallucis Longus Muscle
Interosseous Membrane
Extensor Digitorum Longus M.
Tibialis Posterior Muscle
Posterior Tibial Vessels
Fibula
Peroneus Brevis Muscle
Superficial Fascia
Peroneal Vessel
Fibula, Posterior Border
Peroneus Longus Tendon
Flexor Hallucis Longus Muscle
Soleus Muscle
Crural Fascia
Calcaneal Tendon
Small Saphenous Vein
Great Saphenous Veins
Accessory Saphenous Veins
Tibia, Medial Border
Tibia, Medial Surface
Flexor Digitorum Longus Muscle
Tibia, Anterior Border
Tibia, Lateral Surface
Tibialis Anterior Muscle
Extensor Hallucis Longus Muscle
Anterior Tibial Vessels and Deep Peroneal Nerve
Extensor Digitorum Longus Muscle
Tibialis Posterior Muscle
Fibula, Anterior Border
Peroneus Brevis Muscle
Tibial Nerve
Peroneus Longus Tendon
Flexor Hallucis Longus Muscle
Superficial Fascia
Crural Fascia
Soleus Muscle
Calcaneal Tendon
Small Saphenous Vein

a_vm2625

a_vm2626

a_vm2627

a_vm2628

a_vm2629

a_vm2630

a_vm2631

a_vm2632

a_vm2633

a_vm2634

a_vm2635

a_vm2636

Transverse
a_vm2644

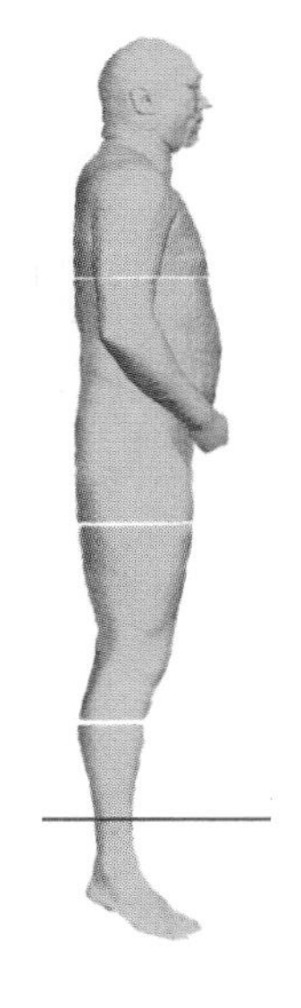

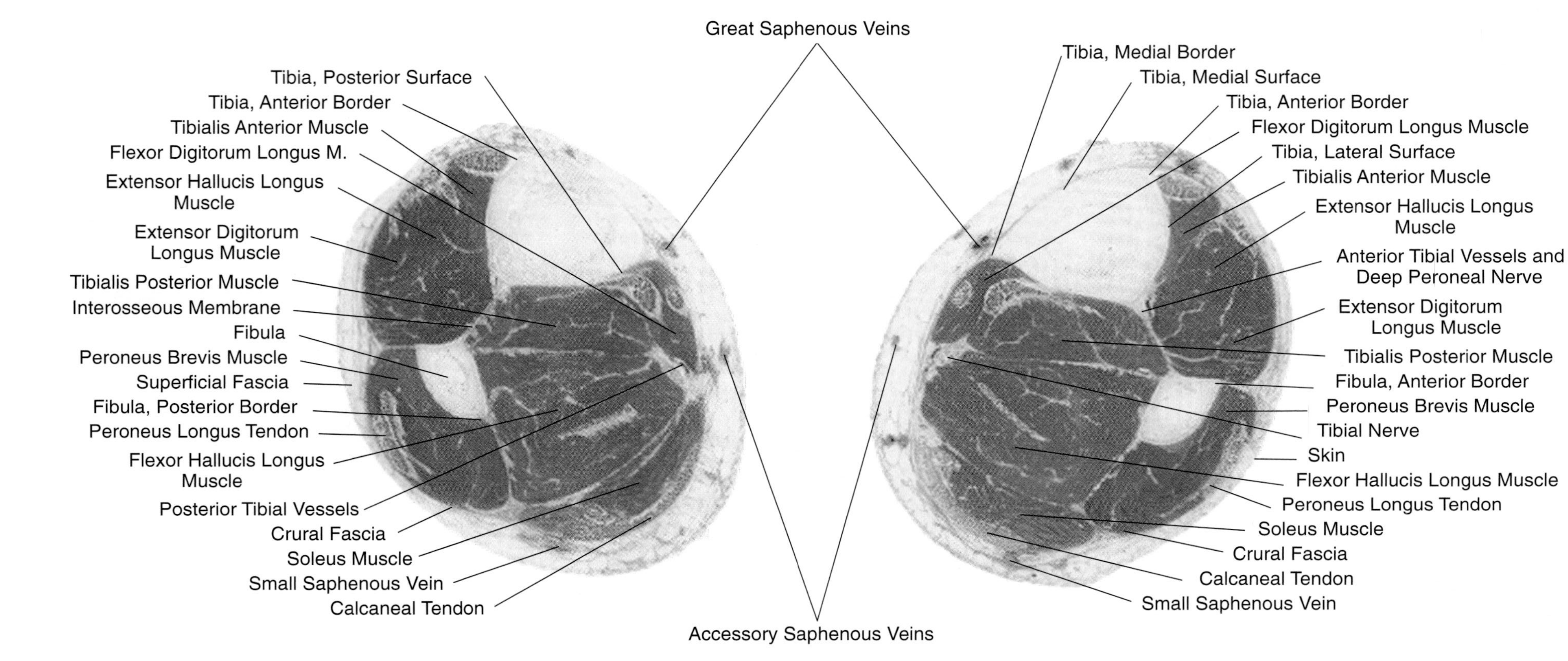

anterior
Great Saphenous Veins
Tibia, Posterior Surface
Tibia, Anterior Border
Tibialis Anterior Muscle
Flexor Digitorum Longus M.
Extensor Hallucis Longus Muscle
Extensor Digitorum Longus Muscle
Tibialis Posterior Muscle
Interosseous Membrane
Fibula
Peroneus Brevis Muscle
Superficial Fascia
Fibula, Posterior Border
Peroneus Longus Tendon
Flexor Hallucis Longus Muscle
Posterior Tibial Vessels
Crural Fascia
Soleus Muscle
Small Saphenous Vein
Calcaneal Tendon
Tibia, Medial Border
Tibia, Medial Surface
Tibia, Anterior Border
Flexor Digitorum Longus Muscle
Tibia, Lateral Surface
Tibialis Anterior Muscle
Extensor Hallucis Longus Muscle
Anterior Tibial Vessels and Deep Peroneal Nerve
Extensor Digitorum Longus Muscle
Tibialis Posterior Muscle
Fibula, Anterior Border
Peroneus Brevis Muscle
Tibial Nerve
Skin
Flexor Hallucis Longus Muscle
Peroneus Longus Tendon
Soleus Muscle
Crural Fascia
Calcaneal Tendon
Small Saphenous Vein
Accessory Saphenous Veins
right
left
posterior

a_vm2637

a_vm2638

a_vm2639

a_vm2640

a_vm2641

a_vm2642

a_vm2643

a_vm2644

a_vm2645

a_vm2646

a_vm2647

a_vm2648

Transverse
a_vm2656

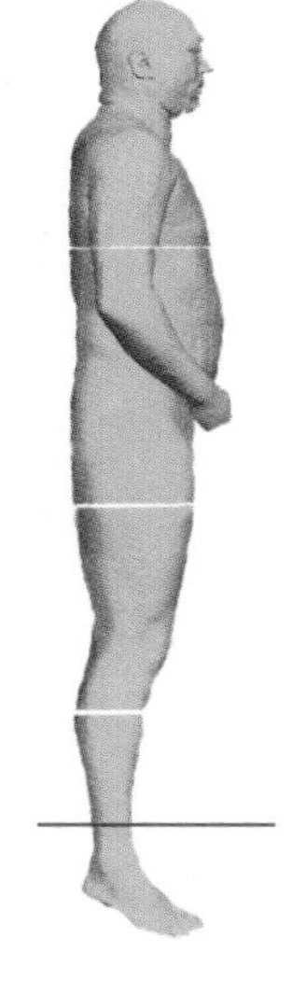

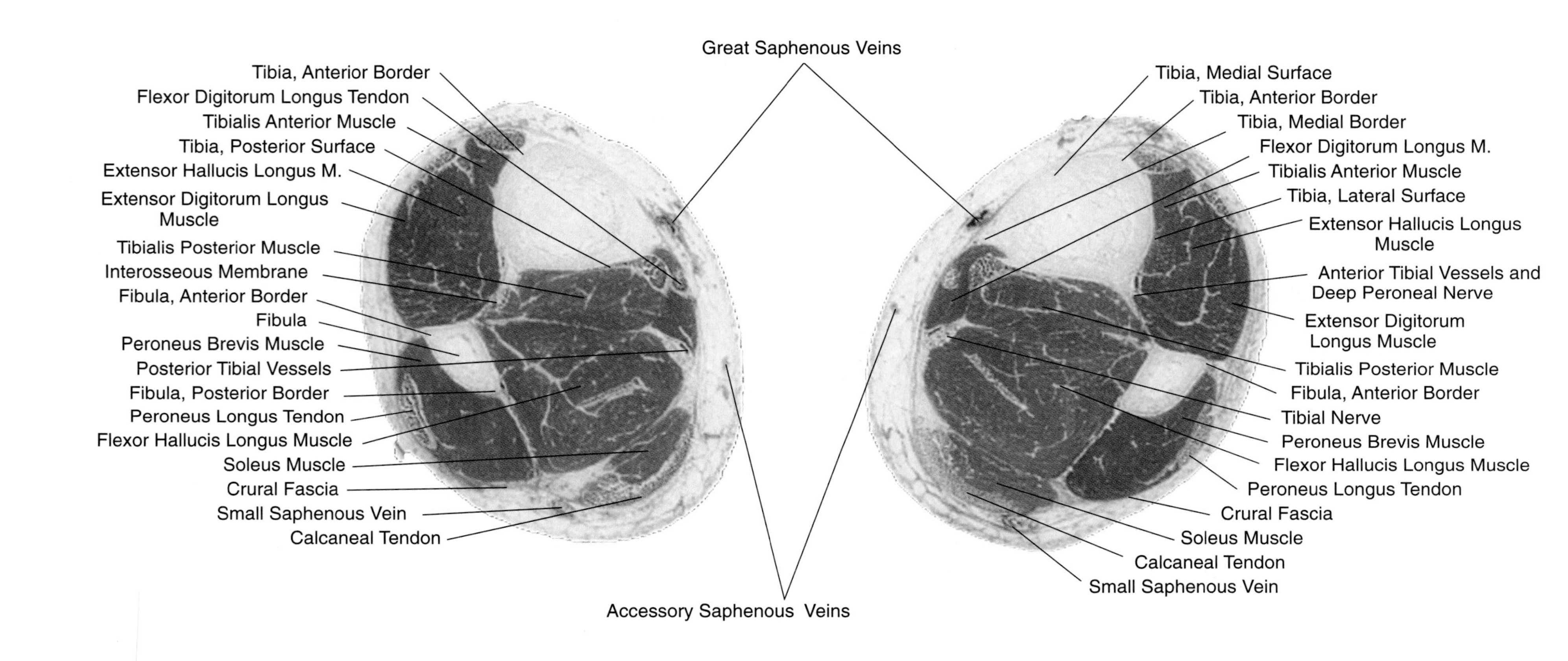

anterior
Great Saphenous Veins
Tibia, Anterior Border
Flexor Digitorum Longus Tendon
Tibialis Anterior Muscle
Tibia, Posterior Surface
Extensor Hallucis Longus M.
Extensor Digitorum Longus Muscle
Tibialis Posterior Muscle
Interosseous Membrane
Fibula, Anterior Border
Fibula
Peroneus Brevis Muscle
Posterior Tibial Vessels
Fibula, Posterior Border
Peroneus Longus Tendon
Flexor Hallucis Longus Muscle
Soleus Muscle
Crural Fascia
Small Saphenous Vein
Calcaneal Tendon
Tibia, Medial Surface
Tibia, Anterior Border
Tibia, Medial Border
Flexor Digitorum Longus M.
Tibialis Anterior Muscle
Tibia, Lateral Surface
Extensor Hallucis Longus Muscle
Anterior Tibial Vessels and Deep Peroneal Nerve
Extensor Digitorum Longus Muscle
Tibialis Posterior Muscle
Fibula, Anterior Border
Tibial Nerve
Peroneus Brevis Muscle
Flexor Hallucis Longus Muscle
Peroneus Longus Tendon
Crural Fascia
Soleus Muscle
Calcaneal Tendon
Small Saphenous Vein
Accessory Saphenous Veins
right
left
posterior

a_vm2649

a_vm2650

a_vm2651

a_vm2652

a_vm2653

a_vm2654

a_vm2655

a_vm2656

a_vm2657

a_vm2658

a_vm2659

a_vm2660

Transverse
a_vm2668

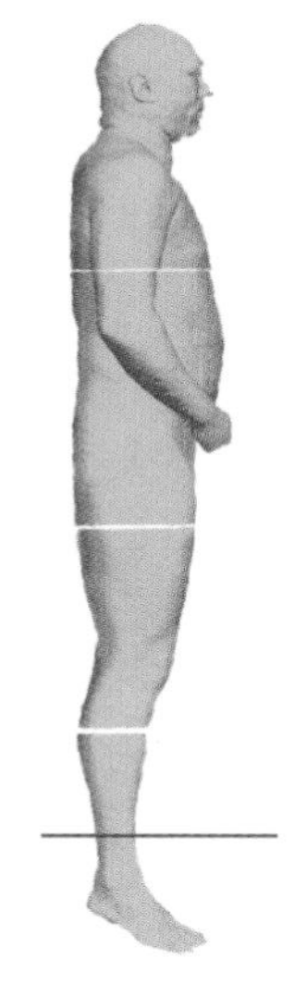

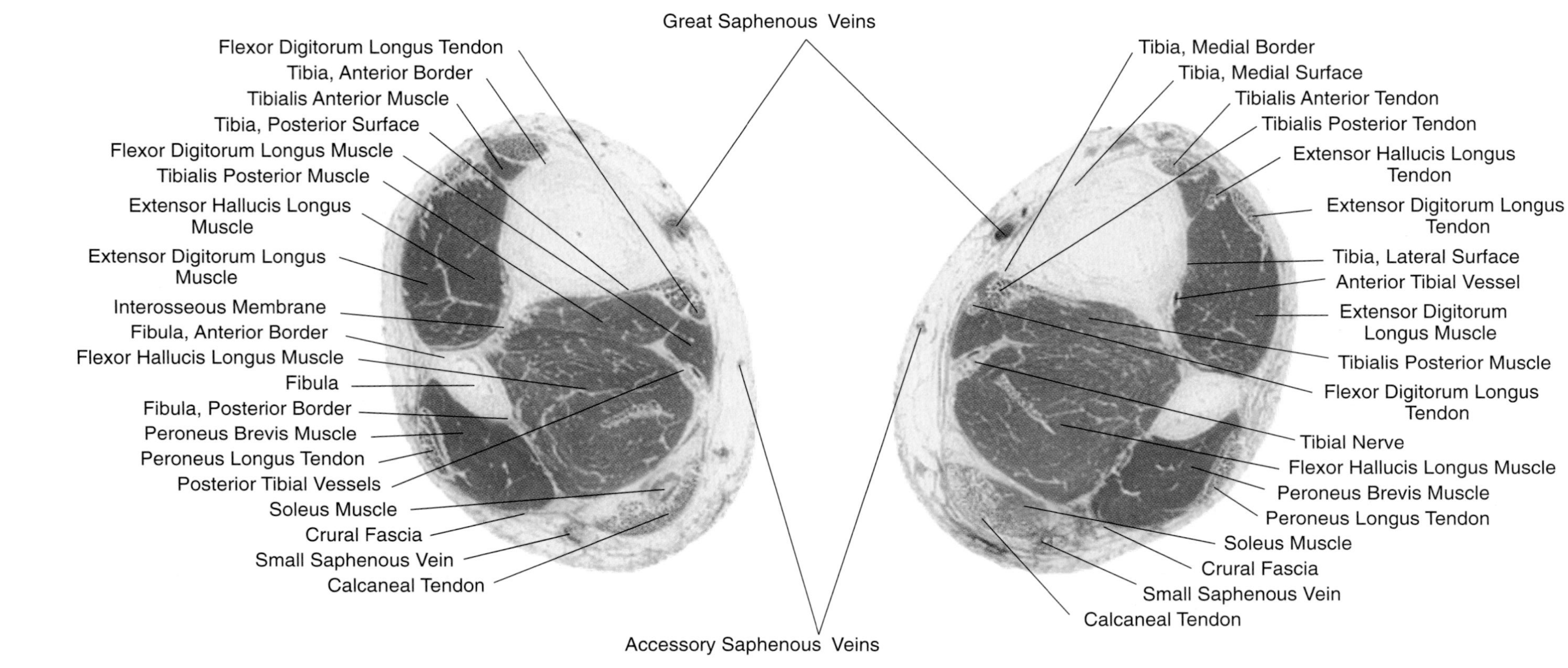
anterior
Great Saphenous Veins
Flexor Digitorum Longus Tendon
Tibia, Anterior Border
Tibialis Anterior Muscle
Tibia, Posterior Surface
Flexor Digitorum Longus Muscle
Tibialis Posterior Muscle
Extensor Hallucis Longus Muscle
Extensor Digitorum Longus Muscle
Interosseous Membrane
Fibula, Anterior Border
Flexor Hallucis Longus Muscle
Fibula
Fibula, Posterior Border
Peroneus Brevis Muscle
Peroneus Longus Tendon
Posterior Tibial Vessels
Soleus Muscle
Crural Fascia
Small Saphenous Vein
Calcaneal Tendon
Tibia, Medial Border
Tibia, Medial Surface
Tibialis Anterior Tendon
Tibialis Posterior Tendon
Extensor Hallucis Longus Tendon
Extensor Digitorum Longus Tendon
Tibia, Lateral Surface
Anterior Tibial Vessel
Extensor Digitorum Longus Muscle
Tibialis Posterior Muscle
Flexor Digitorum Longus Tendon
Tibial Nerve
Flexor Hallucis Longus Muscle
Peroneus Brevis Muscle
Peroneus Longus Tendon
Soleus Muscle
Crural Fascia
Small Saphenous Vein
Calcaneal Tendon
Accessory Saphenous Veins
right
left
posterior

a_vm2661

a_vm2662

a_vm2663

a_vm2664

a_vm2665

a_vm2666

a_vm2667

a_vm2668

a_vm2669

a_vm2670

a_vm2671

a_vm2572

Transverse
a_vm2680

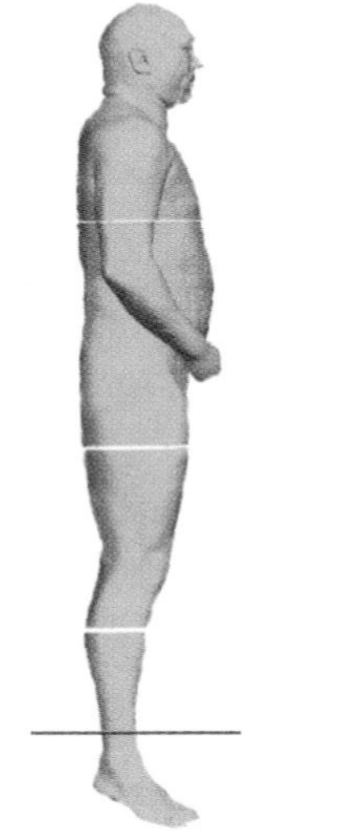

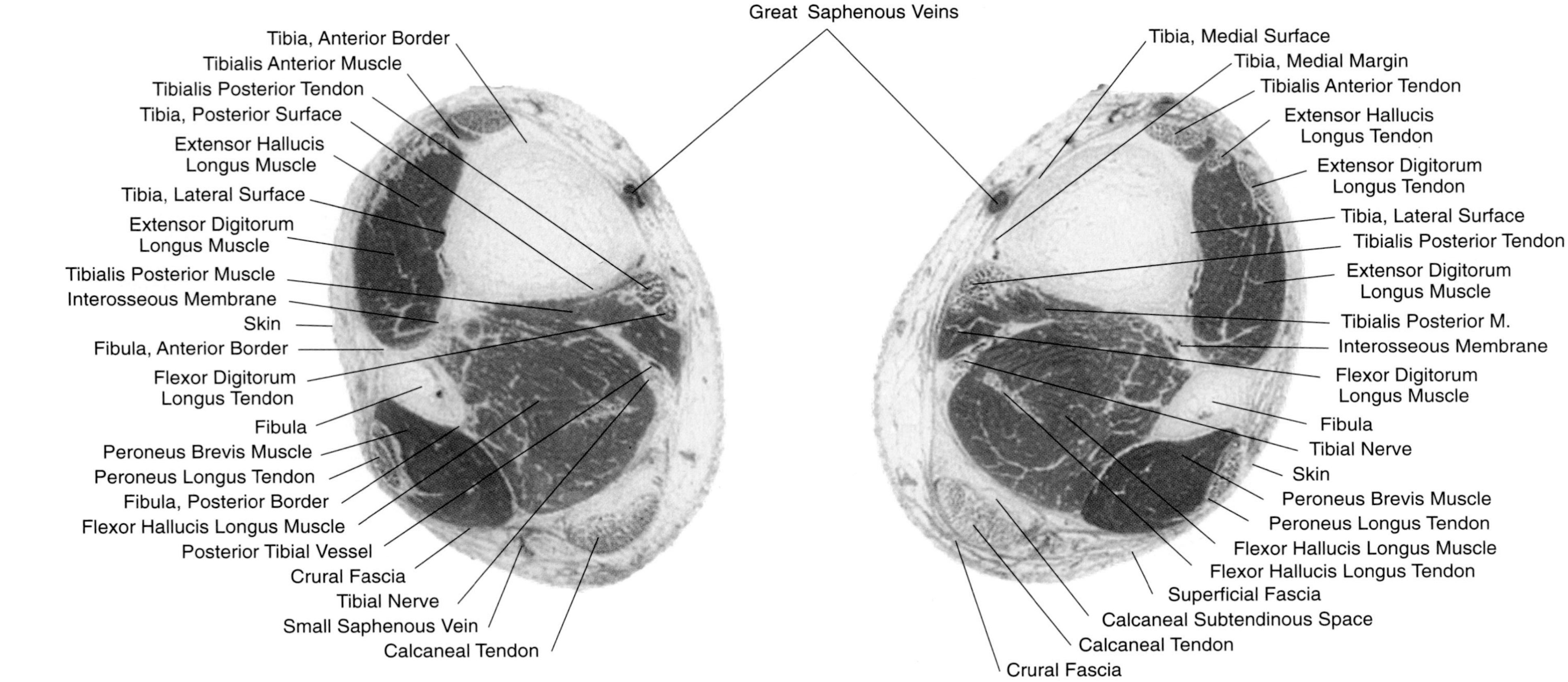
anterior
Great Saphenous Veins
Tibia, Anterior Border
Tibialis Anterior Muscle
Tibialis Posterior Tendon
Tibia, Posterior Surface
Extensor Hallucis Longus Muscle
Tibia, Lateral Surface
Extensor Digitorum Longus Muscle
Tibialis Posterior Muscle
Interosseous Membrane
Skin
Fibula, Anterior Border
Flexor Digitorum Longus Tendon
Fibula
Peroneus Brevis Muscle
Peroneus Longus Tendon
Fibula, Posterior Border
Flexor Hallucis Longus Muscle
Posterior Tibial Vessel
Crural Fascia
Tibial Nerve
Small Saphenous Vein
Calcaneal Tendon
Tibia, Medial Surface
Tibia, Medial Margin
Tibialis Anterior Tendon
Extensor Hallucis Longus Tendon
Extensor Digitorum Longus Tendon
Tibia, Lateral Surface
Tibialis Posterior Tendon
Extensor Digitorum Longus Muscle
Tibialis Posterior M.
Interosseous Membrane
Flexor Digitorum Longus Muscle
Fibula
Tibial Nerve
Skin
Peroneus Brevis Muscle
Peroneus Longus Tendon
Flexor Hallucis Longus Muscle
Flexor Hallucis Longus Tendon
Superficial Fascia
Calcaneal Subtendinous Space
Calcaneal Tendon
Crural Fascia
right
left
posterior

a_vm2673

a_vm2674

a_vm2675

a_vm2676

a_vm2677

a_vm2678

a_vm2679

a_vm2680

a_vm2681

a_vm2682

a_vm2683

a_vm2684

Transverse
a_vm2692

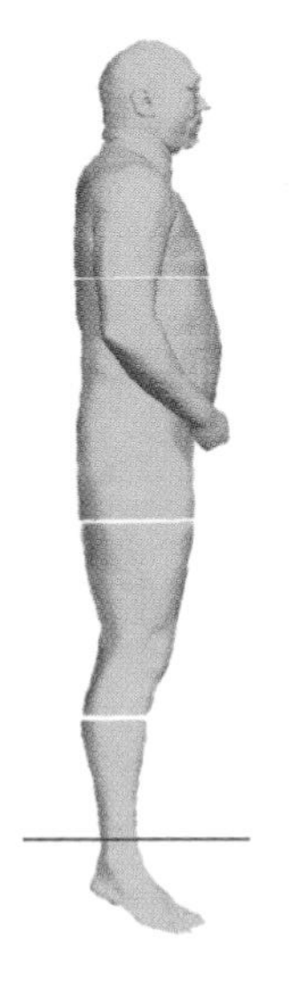

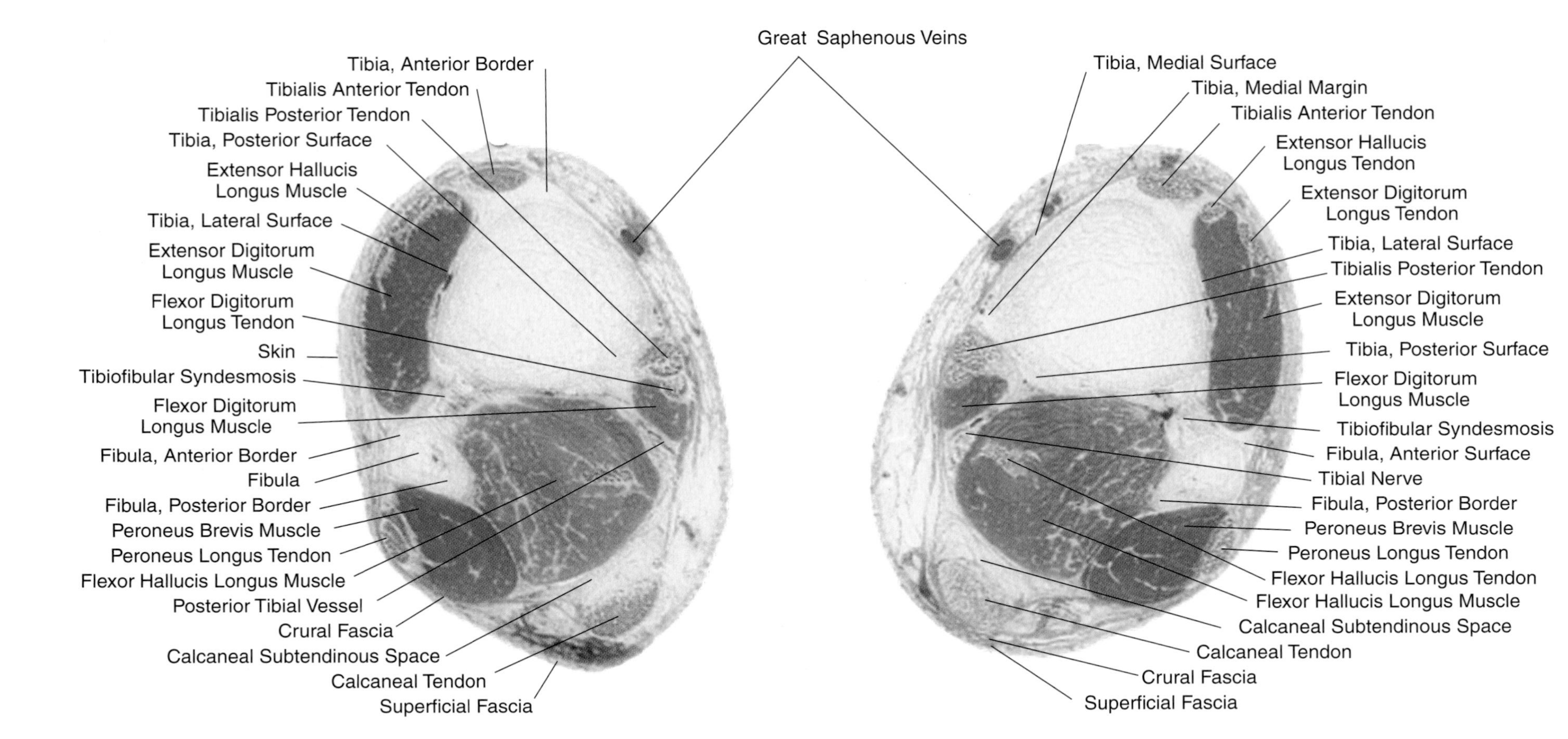
anterior
Great Saphenous Veins
Tibia, Anterior Border
Tibialis Anterior Tendon
Tibialis Posterior Tendon
Tibia, Posterior Surface
Extensor Hallucis Longus Muscle
Tibia, Lateral Surface
Extensor Digitorum Longus Muscle
Flexor Digitorum Longus Tendon
Skin
Tibiofibular Syndesmosis
Flexor Digitorum Longus Muscle
Fibula, Anterior Border
Fibula
Fibula, Posterior Border
Peroneus Brevis Muscle
Peroneus Longus Tendon
Flexor Hallucis Longus Muscle
Posterior Tibial Vessel
Crural Fascia
Calcaneal Subtendinous Space
Calcaneal Tendon
Superficial Fascia
Tibia, Medial Surface
Tibia, Medial Margin
Tibialis Anterior Tendon
Extensor Hallucis Longus Tendon
Extensor Digitorum Longus Tendon
Tibia, Lateral Surface
Tibialis Posterior Tendon
Extensor Digitorum Longus Muscle
Tibia, Posterior Surface
Flexor Digitorum Longus Muscle
Tibiofibular Syndesmosis
Fibula, Anterior Surface
Tibial Nerve
Fibula, Posterior Border
Peroneus Brevis Muscle
Peroneus Longus Tendon
Flexor Hallucis Longus Tendon
Flexor Hallucis Longus Muscle
Calcaneal Subtendinous Space
Calcaneal Tendon
Crural Fascia
Superficial Fascia
right
left
posterior

a_vm2685

a_vm2686

a_vm2687

a_vm2688

a_vm2689

a_vm2690

a_vm2691

a_vm2692

a_vm2693

a_vm2694

a_vm2695

a_vm2696

Transverse
a_vm2704

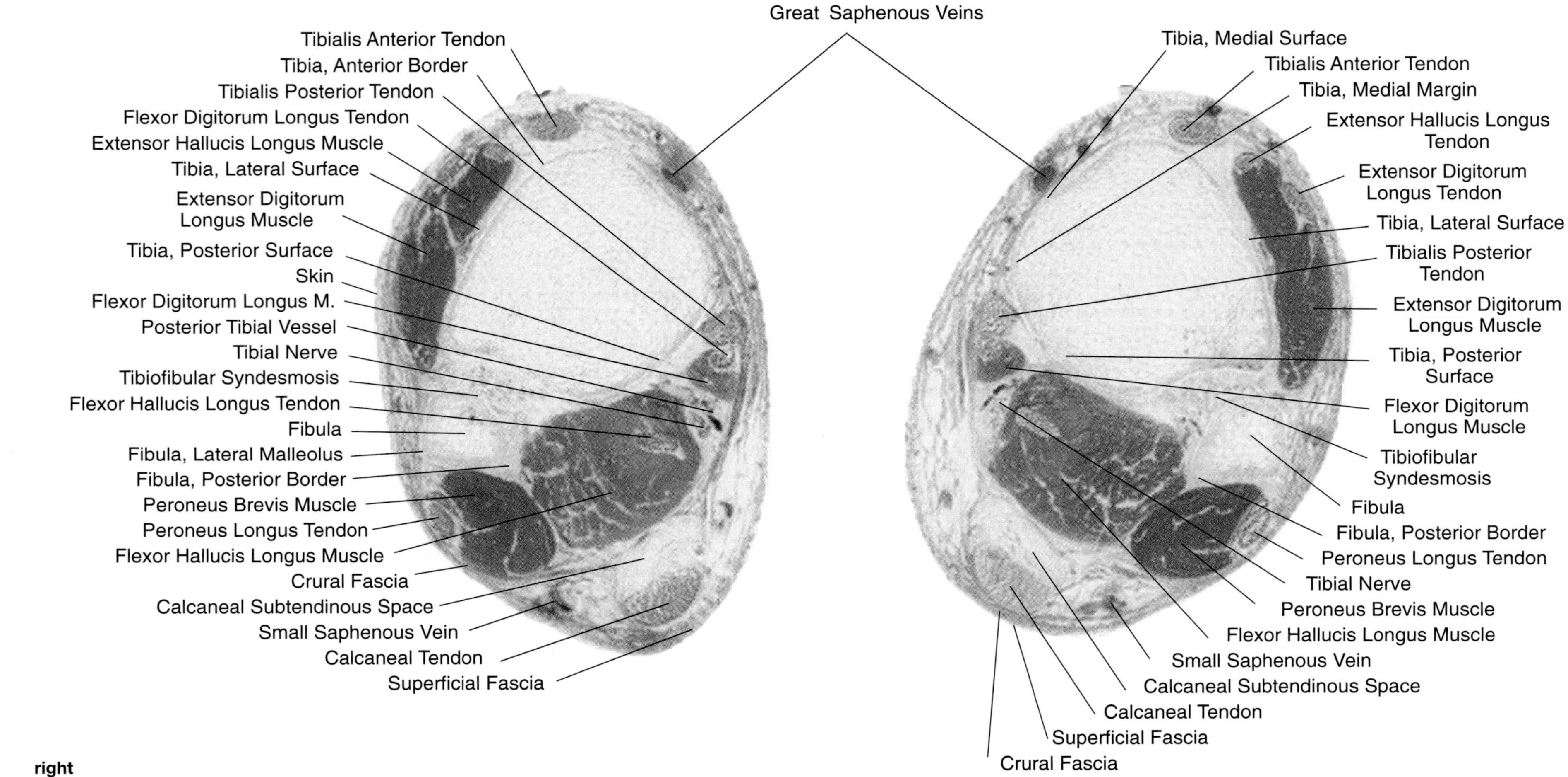

anterior
Great Saphenous Veins
Tibialis Anterior Tendon
Tibia, Anterior Border
Tibialis Posterior Tendon
Flexor Digitorum Longus Tendon
Extensor Hallucis Longus Muscle
Tibia, Lateral Surface
Extensor Digitorum Longus Muscle
Tibia, Posterior Surface
Skin
Flexor Digitorum Longus M.
Posterior Tibial Vessel
Tibial Nerve
Tibiofibular Syndesmosis
Flexor Hallucis Longus Tendon
Fibula
Fibula, Lateral Malleolus
Fibula, Posterior Border
Peroneus Brevis Muscle
Peroneus Longus Tendon
Flexor Hallucis Longus Muscle
Crural Fascia
Calcaneal Subtendinous Space
Small Saphenous Vein
Calcaneal Tendon
Superficial Fascia
Tibia, Medial Surface
Tibialis Anterior Tendon
Tibia, Medial Margin
Extensor Hallucis Longus Tendon
Extensor Digitorum Longus Tendon
Tibia, Lateral Surface
Tibialis Posterior Tendon
Extensor Digitorum Longus Muscle
Tibia, Posterior Surface
Flexor Digitorum Longus Muscle
Tibiofibular Syndesmosis
Fibula
Fibula, Posterior Border
Peroneus Longus Tendon
Tibial Nerve
Peroneus Brevis Muscle
Flexor Hallucis Longus Muscle
Small Saphenous Vein
Calcaneal Subtendinous Space
Calcaneal Tendon
Superficial Fascia
Crural Fascia
right
left
posterior

a_vm2697

a_vm2698

a_vm2699

a_vm2700

a_vm2701

a_vm2702

a_vm2703

a_vm2704

a_vm2705

a_vm2706

a_vm2707

a_vm2708

Transverse
a_vm2757

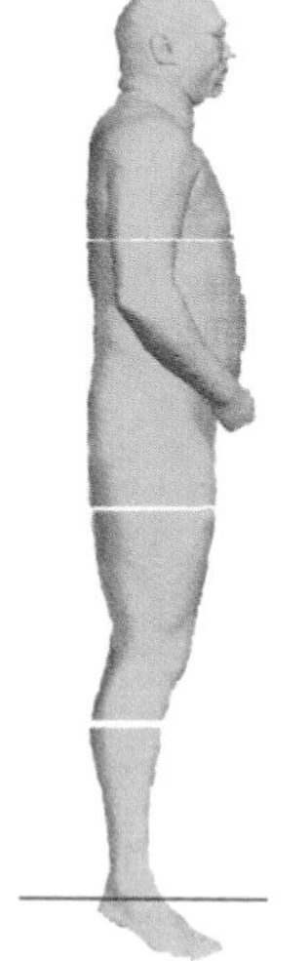

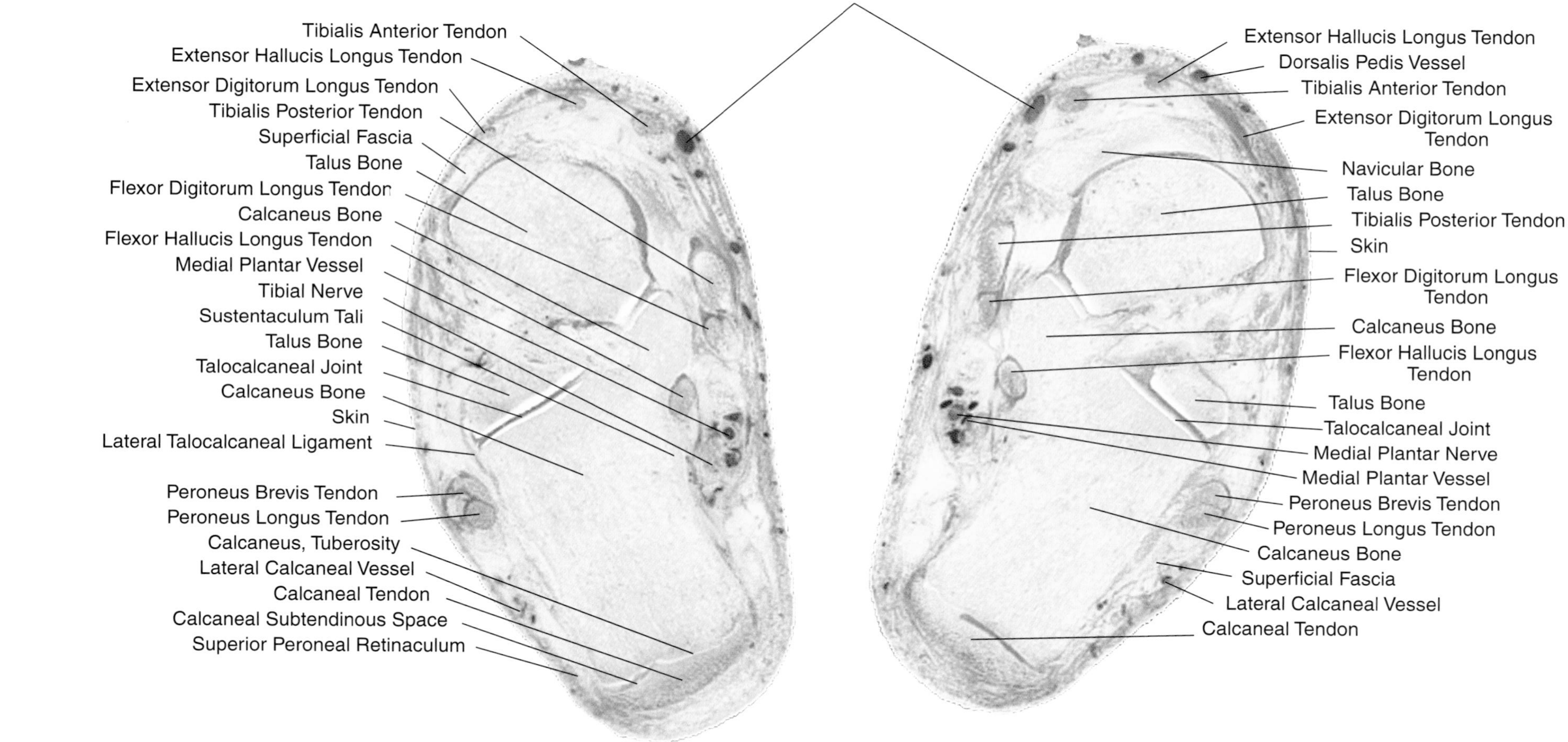
anterior
Great Saphenous Veins
Tibialis Anterior Tendon
Extensor Hallucis Longus Tendon
Extensor Digitorum Longus Tendon
Tibialis Posterior Tendon
Superficial Fascia
Talus Bone
Flexor Digitorum Longus Tendon
Calcaneus Bone
Flexor Hallucis Longus Tendon
Medial Plantar Vessel
Tibial Nerve
Sustentaculum Tali
Talus Bone
Talocalcaneal Joint
Calcaneus Bone
Skin
Lateral Talocalcaneal Ligament
Peroneus Brevis Tendon
Peroneus Longus Tendon
Calcaneus, Tuberosity
Lateral Calcaneal Vessel
Calcaneal Tendon
Calcaneal Subtendinous Space
Superior Peroneal Retinaculum
Extensor Hallucis Longus Tendon
Dorsalis Pedis Vessel
Tibialis Anterior Tendon
Extensor Digitorum Longus Tendon
Navicular Bone
Talus Bone
Tibialis Posterior Tendon
Skin
Flexor Digitorum Longus Tendon
Calcaneus Bone
Flexor Hallucis Longus Tendon
Talus Bone
Talocalcaneal Joint
Medial Plantar Nerve
Medial Plantar Vessel
Peroneus Brevis Tendon
Peroneus Longus Tendon
Calcaneus Bone
Superficial Fascia
Lateral Calcaneal Vessel
Calcaneal Tendon
right
left
posterior

a_vm2757

a_vm2758

a_vm2759

a_vm2760

a_vm2761

a_vm2762

a_vm2763

a_vm2764

a_vm2765

a_vm2766

a_vm2767

a_vm2768

Transverse
a_vm2769

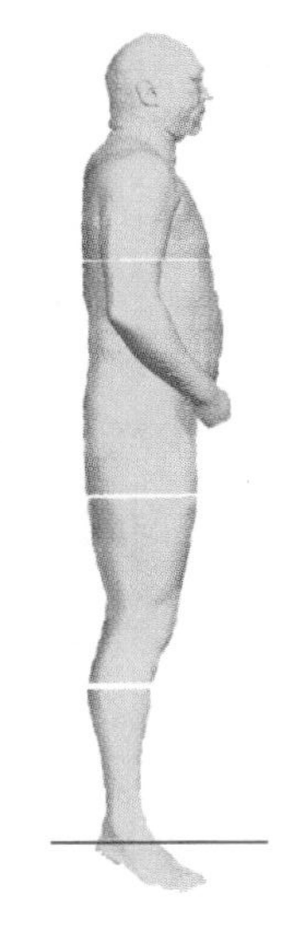

anterior

Extensor Hallucis Longus Tendon
Great Saphenous Vein
Tibialis Anterior Tendon
Extensor Digitorum Longus Tendon
Extensor Hallucis Brevis Muscle
Navicular Bone
Talonavicular Joint
Talus Bone
Tibialis Posterior Tendon
Extensor Digitorum Brevis Muscle
Subtalar Joint
Flexor Digitorum Longus Tendon
Calcaneus Bone
Flexor Hallucis Longus Tendon
Abductor Hallucis Muscle
Peroneus Brevis Tendon
Peroneus Longus Tendon
Medial Plantar Vessel and Nerve
Inferior Peroneal Retinaculum
Calcaneus Bone
Lateral Calcaneal Vessel
Quadratus Plantae Muscle
Skin
Calcaneus, Tuberosity
Calcaneal Tendon
Superficial Fascia

Dorsalis Pedis Vessel
Extensor Hallucis Longus Tendon
Great Saphenous Vein
Tibialis Anterior Tendon
Extensor Hallucis Brevis M.
Extensor Digitorum Longus Tendon
Navicular Bone
Talonavicular Joint
Talus Bone
Tibialis Posterior Tendon
Subtalar Joint
Flexor Digitorum Longus Tendon
Extensor Digitorum Brevis Muscle
Abductor Hallucis Muscle
Flexor Hallucis Longus Tendon
Peroneus Brevis Tendon
Peroneus Longus Tendon
Medial Plantar Vessel and Nerve
Calcaneus Bone
Lateral Calcaneal Vessel
Superficial Fascia
Quadratus Plantae Muscle
Calcaneus, Tuberosity
Calcaneal Tendon

right

left

posterior

a_vm2769

a_vm2770

a_vm2771

a_vm2772

a_vm2773

a_vm2774

a_vm2775

a_vm2776

a_vm2777

a_vm2778

a_vm2779

a_vm2780

Transverse
a_vm2781

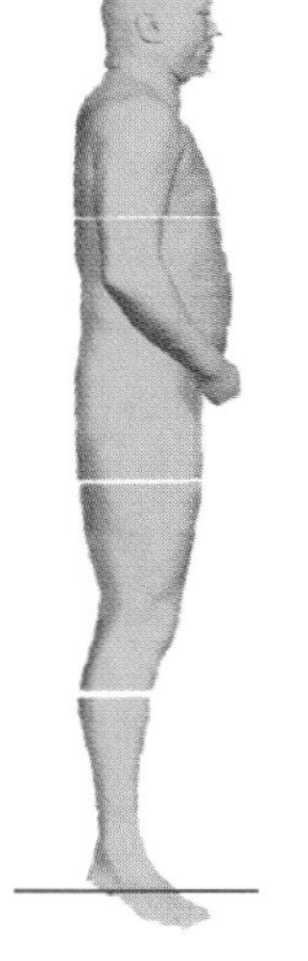

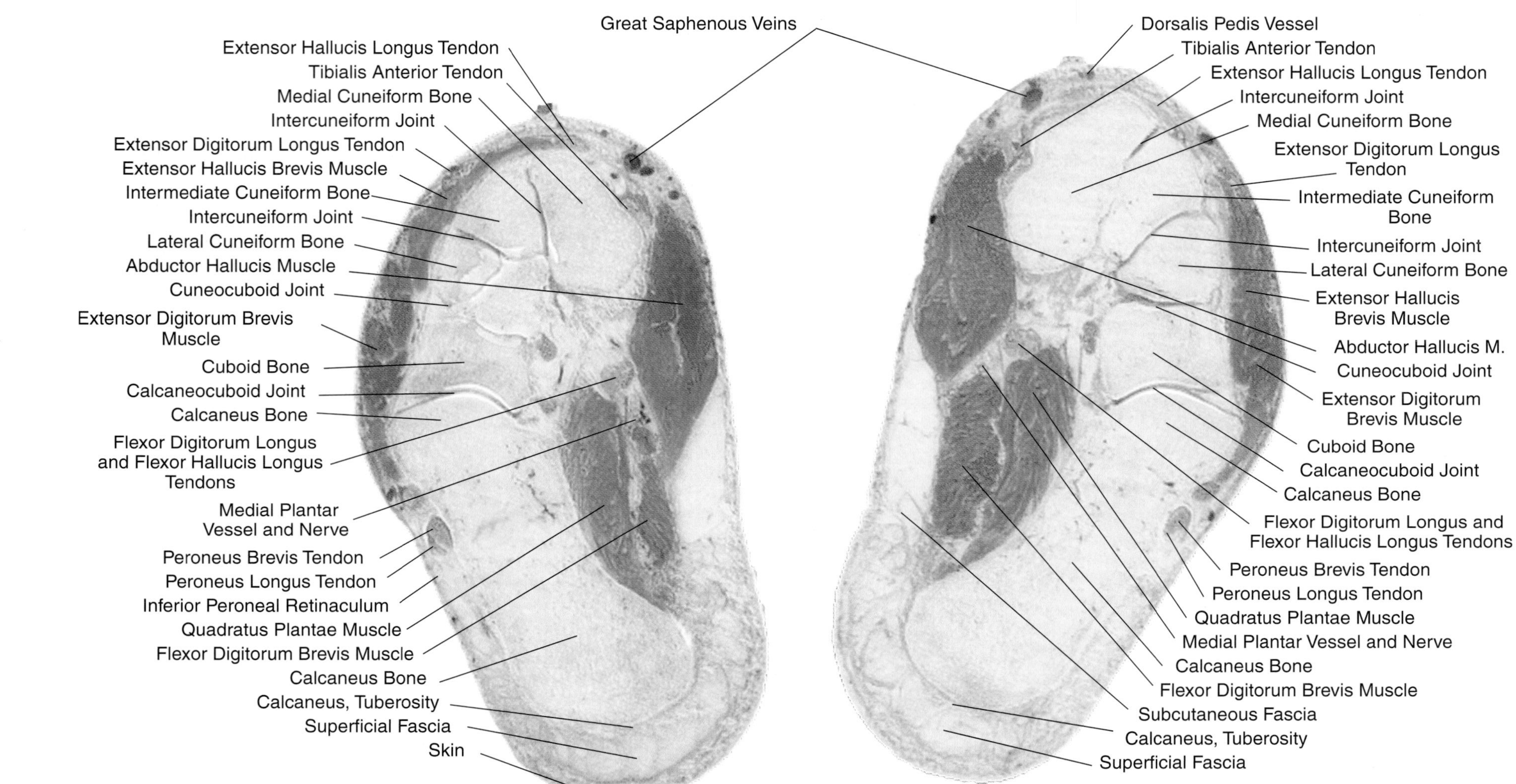
anterior
Great Saphenous Veins
Extensor Hallucis Longus Tendon
Tibialis Anterior Tendon
Medial Cuneiform Bone
Intercuneiform Joint
Extensor Digitorum Longus Tendon
Extensor Hallucis Brevis Muscle
Intermediate Cuneiform Bone
Intercuneiform Joint
Lateral Cuneiform Bone
Abductor Hallucis Muscle
Cuneocuboid Joint
Extensor Digitorum Brevis Muscle
Cuboid Bone
Calcaneocuboid Joint
Calcaneus Bone
Flexor Digitorum Longus and Flexor Hallucis Longus Tendons
Medial Plantar Vessel and Nerve
Peroneus Brevis Tendon
Peroneus Longus Tendon
Inferior Peroneal Retinaculum
Quadratus Plantae Muscle
Flexor Digitorum Brevis Muscle
Calcaneus Bone
Calcaneus, Tuberosity
Superficial Fascia
Skin
Dorsalis Pedis Vessel
Tibialis Anterior Tendon
Extensor Hallucis Longus Tendon
Intercuneiform Joint
Medial Cuneiform Bone
Extensor Digitorum Longus Tendon
Intermediate Cuneiform Bone
Intercuneiform Joint
Lateral Cuneiform Bone
Extensor Hallucis Brevis Muscle
Abductor Hallucis M.
Cuneocuboid Joint
Extensor Digitorum Brevis Muscle
Cuboid Bone
Calcaneocuboid Joint
Calcaneus Bone
Flexor Digitorum Longus and Flexor Hallucis Longus Tendons
Peroneus Brevis Tendon
Peroneus Longus Tendon
Quadratus Plantae Muscle
Medial Plantar Vessel and Nerve
Calcaneus Bone
Flexor Digitorum Brevis Muscle
Subcutaneous Fascia
Calcaneus, Tuberosity
Superficial Fascia
right
left
posterior

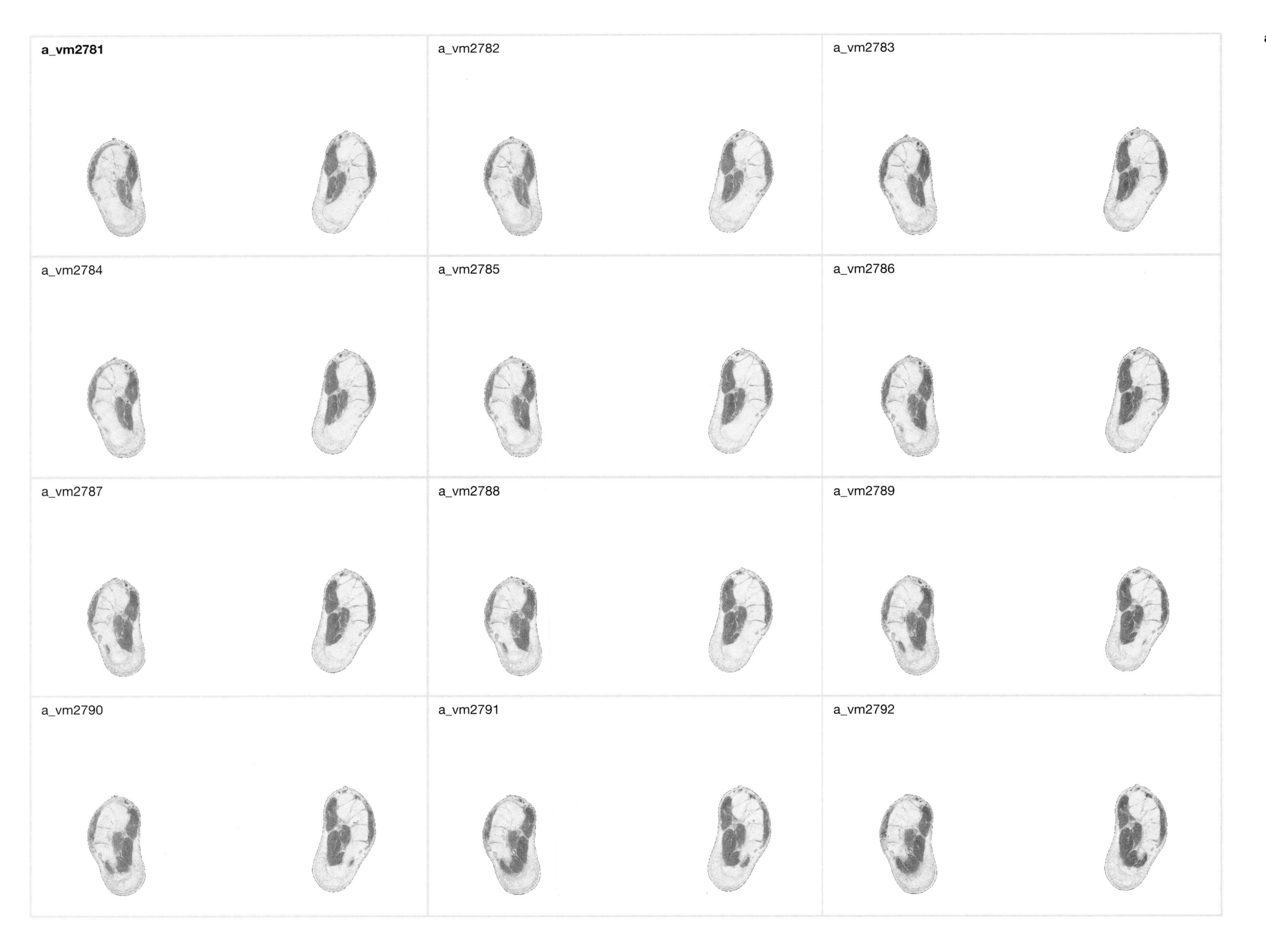
a_vm2781
a_vm2782
a_vm2783
a_vm2784
a_vm2785
a_vm2786
a_vm2787
a_vm2788
a_vm2789
a_vm2790
a_vm2791
a_vm2792

Transverse
a_vm2793

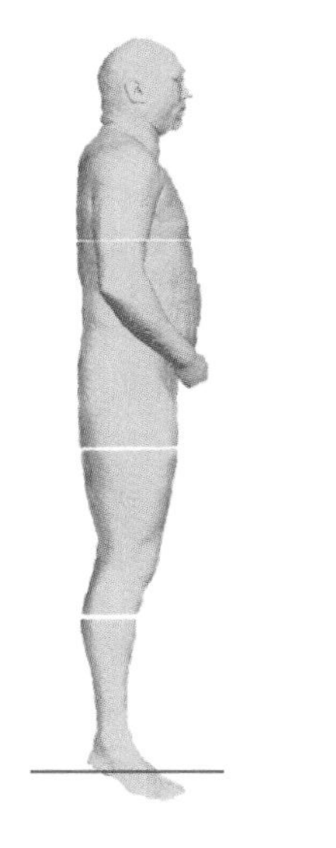

anterior
Great Saphenous Vein
Extensor Hallucis Longus Tendon
Extensor Hallucis Brevis Tendon
First Metatarsal
Second Metatarsal
Medial Cuneiform Bone
Intermediate Cuneiform Bone
Abductor Hallucis Muscle
Lateral Cuneiform Bone
Extensor Digitorum Longus Tendon
Extensor Digitorum Brevis Muscle
Cuboid Bone
Flexor Digitorum Longus and Flexor Hallucis Longus Tendons
Calcaneocuboid Joint
Peroneus Brevis Tendon
Calcaneus Bone
Peroneus Longus Tendon
Quadratus Plantae Muscle
Long Plantar Ligament
Lateral Plantar Vessel and Nerve
Flexor Digitorum Brevis Muscle
Flexor Digiti Minimi Muscle
Skin
Plantar Aponeurosis
Superficial Fascia
right
Great Saphenous Vein
Extensor Hallucis Longus Tendon
Extensor Hallucis Brevis Tendon
First Dorsal Interosseous M.
First Metatarsal
Tarsometatarsal Joint
Second Metatarsal
Medial Cuneiform Bone
Extensor Digitorum Longus Tendon
Intermediate Cuneiform Bone
Lateral Cuneiform Bone
Extensor Digitorum Brevis Muscle
Abductor Hallucis M.
Cuboid Bone
Flexor Hallucis Longus Tendon
Peroneus Brevis Tendon
Flexor Digitorum Longus Tendon
Peroneus Longus Tendon
Quadratus Plantae Muscle
Long Plantar Ligament
Lateral Plantar Vessel and Nerve
Flexor Digitorum Brevis Muscle
Superficial Fascia
Abductor Digiti Minimi Muscle
Plantar Aponeurosis
Skin
Subcutaneous Fascia
left
posterior

a_vm2793

a_vm2794

a_vm2795

a_vm2796

a_vm2797

a_vm2798

a_vm2799

a_vm2800

a_vm2801

a_vm2802

a_vm2803

a_vm2804

Transverse
a_vm2805

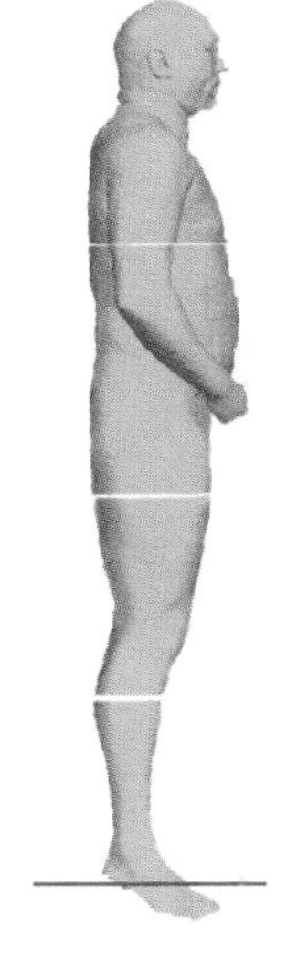

anterior
Great Saphenous Vein
Extensor Hallucis Longus Tendon
Extensor Hallucis Brevis Tendon
First Dorsal Interosseous Muscle
First Metatarsal
Second Metatarsal
Abductor Hallucis Muscle
Third Metatarsal
Extensor Digitorum Brevis M.
Long Plantar Ligament
Lateral Cuneiform Bone
Cuneocuboid Joint
Flexor Hallucis Brevis Muscle
Flexor Hallucis Longus Tendon
Flexor Digitorum Longus Tendon
Cuboid Bone
Quadratus Plantae Muscle
Skin
Peroneus Brevis Tendon
Peroneus Longus Tendon
Flexor Digitorum Brevis Muscle
Lateral Plantar Vessel and Nerve
Abductor Digiti Minimi Muscle
Flexor Digiti Minimi Brevis Muscle
Plantar Aponeurosis
Superficial Fascia
Great Saphenous Vein
Extensor Hallucis Longus Tendon
Extensor Hallucis Brevis Tendon
First Dorsal Interosseous Muscle
First Metatarsal
Abductor Hallucis Muscle
Second Metatarsal
Flexor Hallucis Brevis Muscle
Adductor Hallucis Muscle, Transverse Head
Extensor Digitorum Brevis M.
Third Metatarsal
Long Plantar Ligament
Fourth Metatarsal
Tarsometatarsal Ligament
Flexor Hallucis Longus Tendon
Flexor Digitorum Longus Tendon
Cuboid Bone
Quadratus Plantae Muscle
Flexor Digitorum Brevis Muscle
Peroneus Brevis Tendon
Peroneus Longus Tendon
Lateral Plantar Vessel and Nerve
Skin
Abductor Digiti Minimi Muscle
Flexor Digiti Minimi Brevis Muscle
Plantar Aponeurosis
Superficial Fascia
right
left
posterior

a_vm2805

a_vm2806

a_vm2807

a_vm2808

a_vm2809

a_vm2810

a_vm2811

a_vm2812

a_vm2813

a_vm2814

a_vm2815

a_vm2816

Transverse
a_vm2817

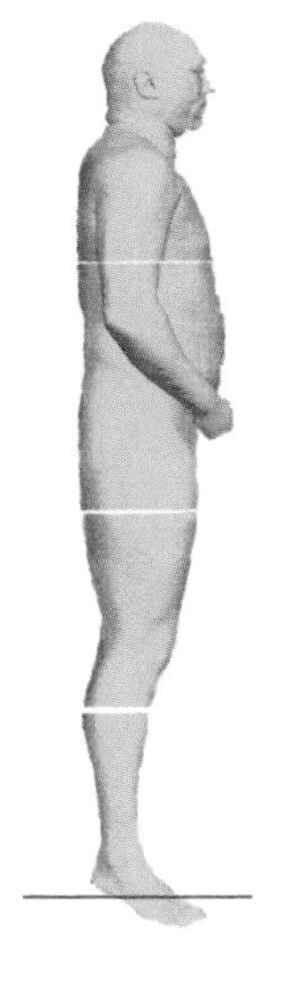

anterior
Great Saphenous Vein
Abductor Hallucis Muscle
Extensor Hallucis Longus Tendon
Extensor Hallucis Brevis Tendon
First Metatarsal
First Dorsal Interosseous Muscle
Second Metatarsal
Adductor Hallucis M., Transverse Head
Second Dorsal Interosseous Muscle
Extensor Digitorum Brevis Muscle
Third Metatarsal
Interosseous Metatarsal Ligament
Adductor Hallucis Muscle, Oblique Head
Fifth Metatarsal
Flexor Hallucis Brevis Muscle
Flexor Hallucis Longus Tendon
Quadratus Plantae Muscle
Fifth Metatarsal
Flexor Digitorum Longus Tendon
Lateral Plantar Vessel and Nerve
Flexor Digitorum Brevis Muscle
Abductor Digiti Minimi Muscle
Flexor Digiti Minimi Brevis Muscle
Plantar Aponeurosis
Skin
Superficial Fascia
Extensor Hallucis Longus Tendon
Extensor Hallucis Brevis Tendon
Abductor Hallucis Muscle
First Metatarsal
Great Saphenous Vein
Flexor Hallucis Brevis Muscle
First Dorsal Interosseous Muscle
Extensor Digitorum Brevis Tendon
Extensor Digitorum Longus Tendon
Second Metatarsal
Second Dorsal Interosseous M.
Adductor Hallucis Muscle, Transverse Head
Extensor Digitorum Brevis Tendon
Third Metatarsal
Third Dorsal Interosseous M.
Extensor Digitorum Longus Tendon
First Plantar Interosseous M.
Fourth Metatarsal
Adductor Hallucis Muscle, Oblique Head
Interosseous Metatarsal Ligament
Fifth Metatarsal
Quadratus Plantae Muscle
Lateral Plantar Vessel and Nerve
Skin
Flexor Digitorum Longus Tendon
Abductor Digiti Minimi Muscle
Flexor Digitorum Brevis Muscle
Superficial Fascia
Flexor Hallucis Longus Tendon
Plantar Aponeurosis
right
left
posterior

a_vm2817

a_vm2818

a_vm2819

a_vm2820

a_vm2821

a_vm2822

a_vm2823

a_vm2824

a_vm2825

a_vm2826

a_vm2827

a_vm2828

Transverse
a_vm2829

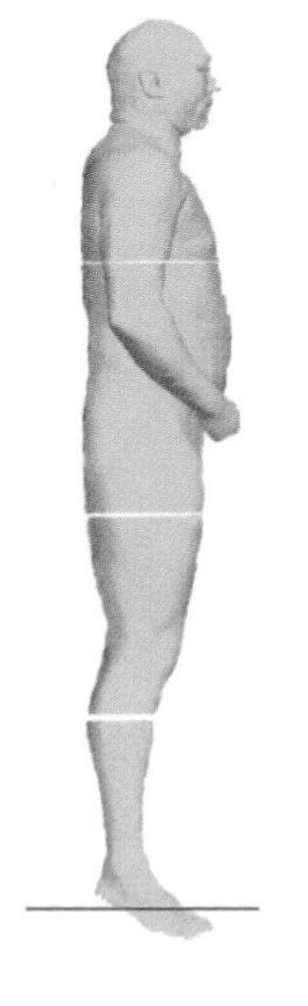

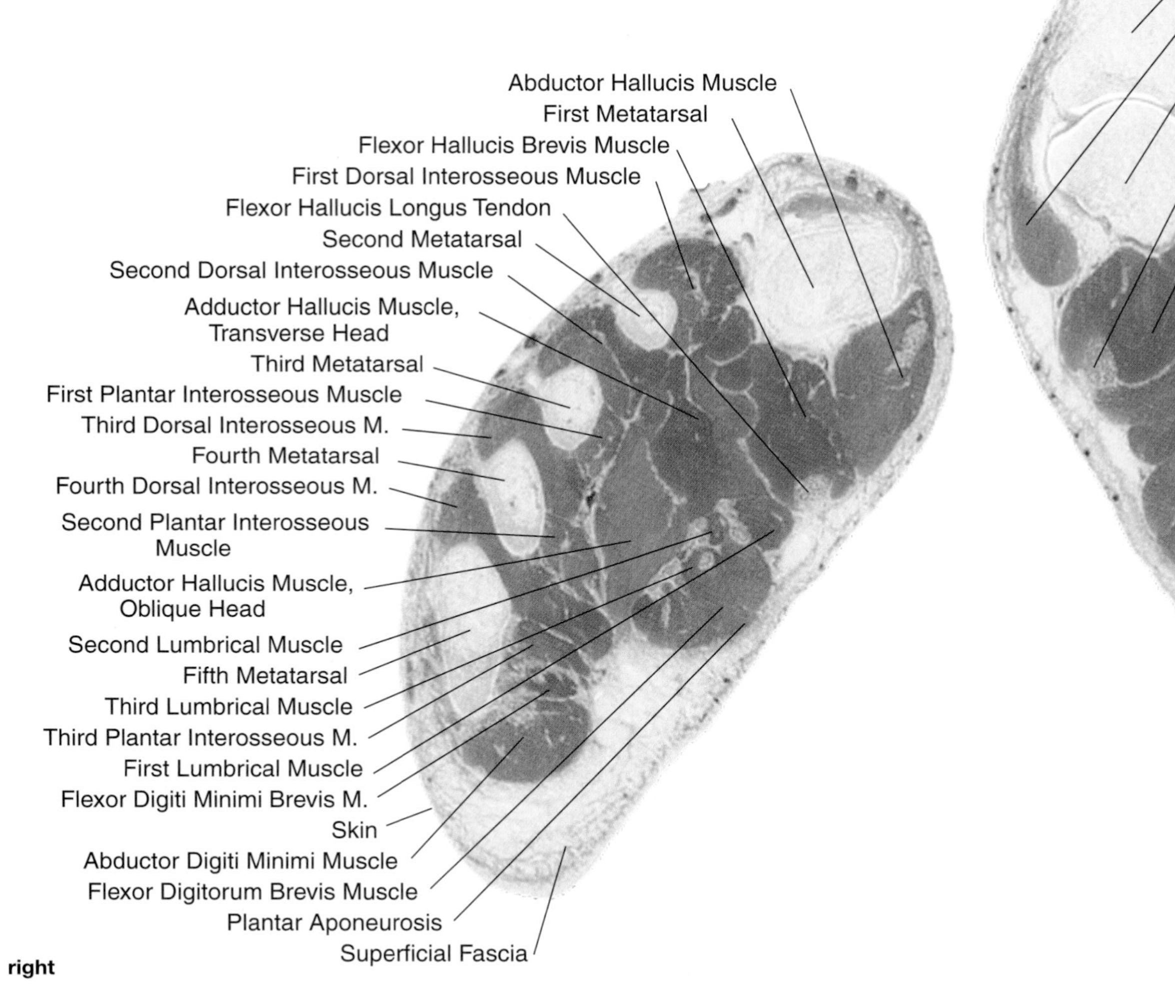

anterior
Abductor Hallucis Muscle
First Metatarsal
Flexor Hallucis Brevis Muscle
First Dorsal Interosseous Muscle
Flexor Hallucis Longus Tendon
Second Metatarsal
Second Dorsal Interosseous Muscle
Adductor Hallucis Muscle, Transverse Head
Third Metatarsal
First Plantar Interosseous Muscle
Third Dorsal Interosseous M.
Fourth Metatarsal
Fourth Dorsal Interosseous M.
Second Plantar Interosseous Muscle
Adductor Hallucis Muscle, Oblique Head
Second Lumbrical Muscle
Fifth Metatarsal
Third Lumbrical Muscle
Third Plantar Interosseous M.
First Lumbrical Muscle
Flexor Digiti Minimi Brevis M.
Skin
Abductor Digiti Minimi Muscle
Flexor Digitorum Brevis Muscle
Plantar Aponeurosis
Superficial Fascia
First Proximal Phalanx
Abductor Hallucis Muscle
First Metatarsal
Flexor Hallucis Longus Tendon
Flexor Hallucis Brevis Muscle
First Dorsal Interosseous Muscle
Adductor Hallucis Muscle, Transverse Head
Second Metatarsal
Second Dorsal Interosseous Muscle
Flexor Digitorum Longus Tendon
Skin
Extensor Digitorum Longus Tendon
Third Metatarsal
First Plantar Interosseous M.
Third Dorsal Interosseous M.
Fourth Metatarsal
Fourth Dorsal Interosseous Muscle
Adductor Hallucis Muscle, Oblique Head
Second Plantar Interosseous Muscle
Fifth Metatarsal
Third Lumbrical Muscle
Fourth Lumbrical Muscle
Third Plantar Interosseous M.
Flexor Digiti Minimi Brevis M.
Flexor Digitorum Brevis M.
Abductor Digiti Minimi Muscle
Superficial Fascia
Plantar Aponeurosis
right
left
posterior

a_vm2829

a_vm2830

a_vm2831

a_vm2832

a_vm2833

a_vm2834

a_vm2835

a_vm2836

a_vm2837

a_vm2838

a_vm2839

a_vm2840

Transverse
a_vm2841

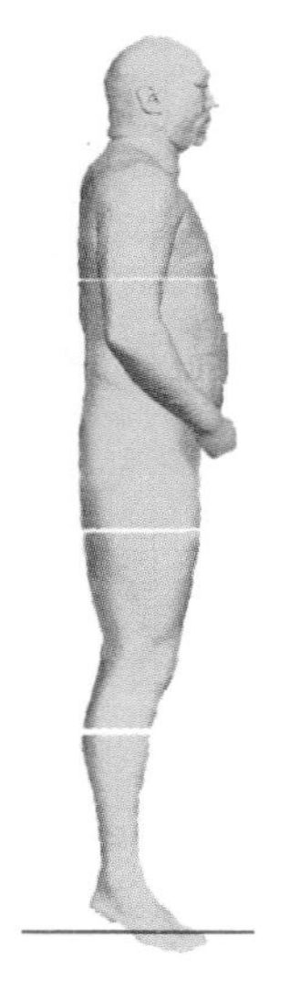

anterior
First Proximal Phalanx
Metatarsal Phalangeal Joint
Abductor Hallucis Muscle
First Metatarsal
Flexor Hallucis Longus Tendon
Flexor Hallucis Brevis Muscle
First Dorsal Interosseous Muscle
Second Metatarsal
Adductor Hallucis Muscle, Transverse Head
Second Dorsal Interosseous Muscle
Third Metatarsal
First Plantar Interosseous Muscle
Third Dorsal Interosseous Muscle
Adductor Hallucis Muscle,
Oblique Head
Fourth Metatarsal
Fourth Dorsal Interosseous M.
First Lumbrical Muscle
Fifth Metatarsal
Second Plantar Interosseous
Muscle
Skin
Second Lumbrical Muscle
Third Plantar Interosseous M.
Third Lumbrical Muscle
Flexor Digiti Minimi Brevis M.
Abductor Digiti Minimi Muscle
Fourth Lumbrical Muscle
Superficial Fascia
Flexor Digitorum Brevis Muscle
Plantar Aponeurosis
right
Abductor Hallucis Tendon
First Distal Phalanx
First Proximal Phalanx
Flexor Hallucis Longus Tendon
First Metatarsal
Second Middle Phalanx
Second Proximal Phalanx
Flexor Hallucis Brevis Tendon
Metatarsal Phalangeal Joint
Second Metatarsal
Third Proximal Phalanx
Flexor Digitorum Brevis Muscle
Second Dorsal Interosseous M.
Adductor Hallucis Muscle,
Transverse Head
Third Metatarsal
First Plantar Interosseous M.
Third Dorsal Interosseous M.
Fourth Metatarsal
Second Plantar
Interosseous Muscle
Fourth Dorsal
Interosseous Muscle
Adductor Hallucis M.,
Oblique Head
Fifth Metatarsal
Third Plantar
Interosseous Muscle
Flexor Digiti Minimi
Brevis Muscle
Abductor Digiti Minimi M.
Skin
Plantar Aponeurosis
Superficial Fascia
left
posterior

a_vm2841

a_vm2842

a_vm2843

a_vm2844

a_vm2845

a_vm2846

a_vm2847

a_vm2848

a_vm2849

a_vm2850

a_vm2851

a_vm2852

Transverse
a_vm2853

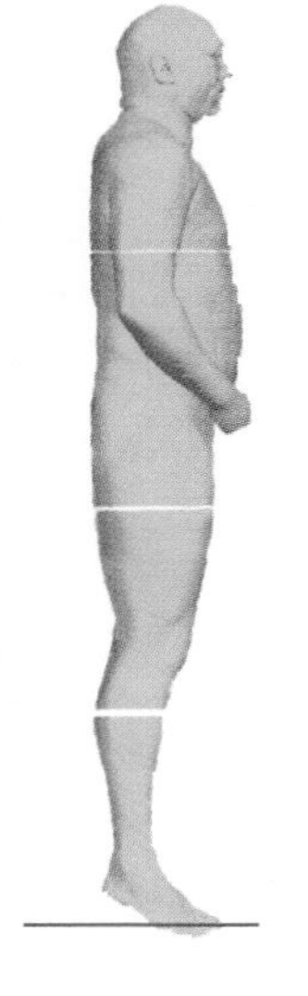

anterior
First Distal Phalanx
First Proximal Phalanx
First Lumbrical Muscle
Second Proximal Phalanx
Second Metatarsal
Second Lumbrical Muscle
Third Proximal Phalanx
Third Metatarsal
First Plantar Interosseous M.
Fourth Proximal Phalanx
Third Lumbrical Muscle
Second Plantar Interosseous Muscle
Fourth Metatarsal
Fourth Lumbrical Muscle
Fourth Dorsal Interosseous Muscle
Fifth Metatarsal
Third Plantar Interosseous Muscle
Flexor Digiti Minimi Brevis Muscle
Abductor Digiti Minimi Muscle
Superficial Fascia
Skin
First Toe
Second Distal Phalanx
Third Distal Phalanx
Third Middle Phalanx
Third Proximal Phalanx
Third Metatarsal
Fourth Middle Phalanx
Third Dorsal Interosseous Tendon
Fourth Proximal Phalanx
Fourth Metatarsal
Fifth Middle Phalanx
Fifth Proximal Phalanx
Third Plantar Interosseous M.
Flexor Digiti Minimi Brevis M.
Fifth Metatarsal
Abductor Digiti Minimi Tendon
Superficial Fascia
Flexor Digitorum Longus and Brevis Tendons
Skin
Plantar Aponeurosis
right
left
posterior

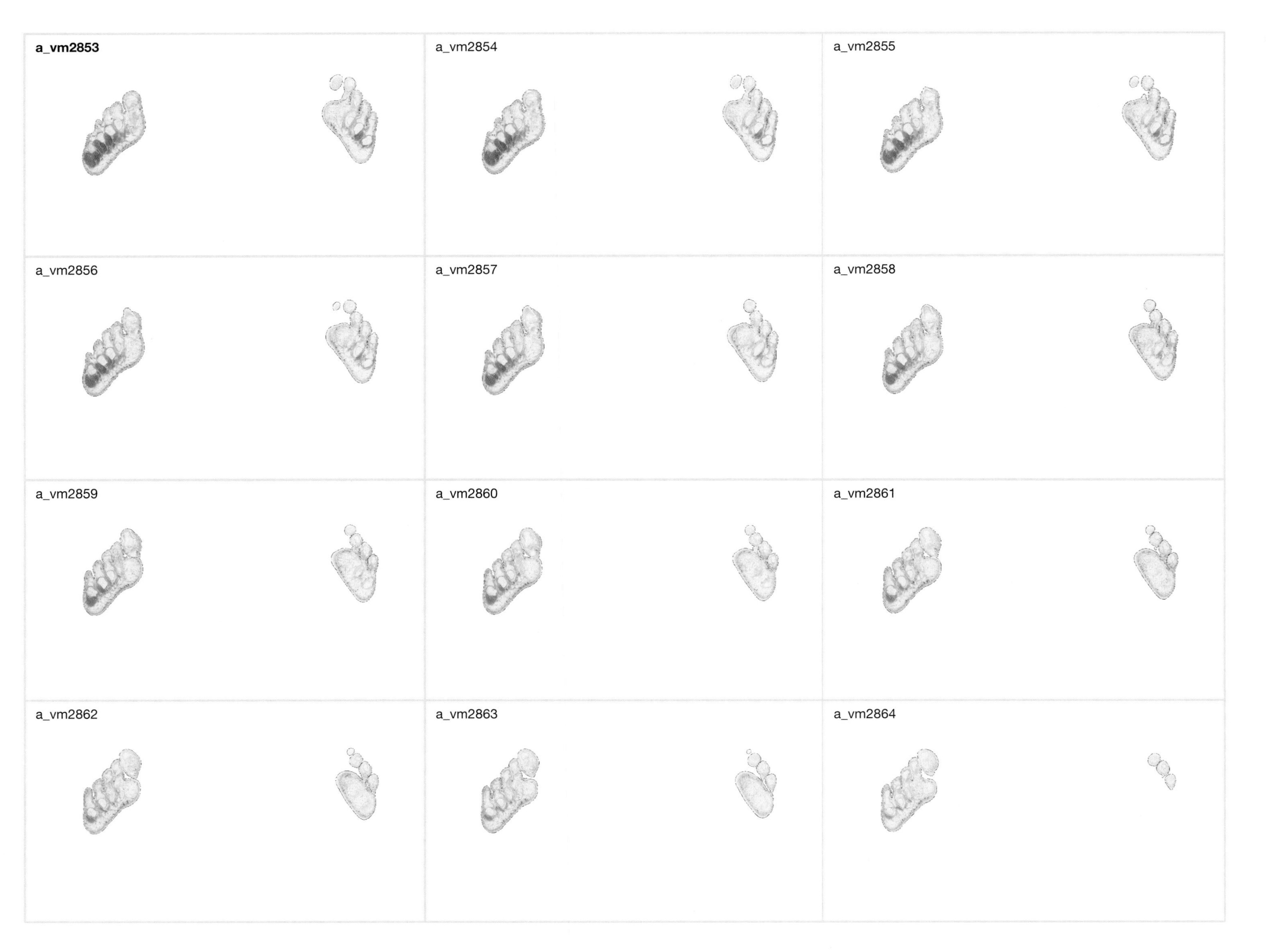
a_vm2853
a_vm2854
a_vm2855
a_vm2856
a_vm2857
a_vm2858
a_vm2859
a_vm2860
a_vm2861
a_vm2862
a_vm2863
a_vm2864

Transverse
a_vm2867

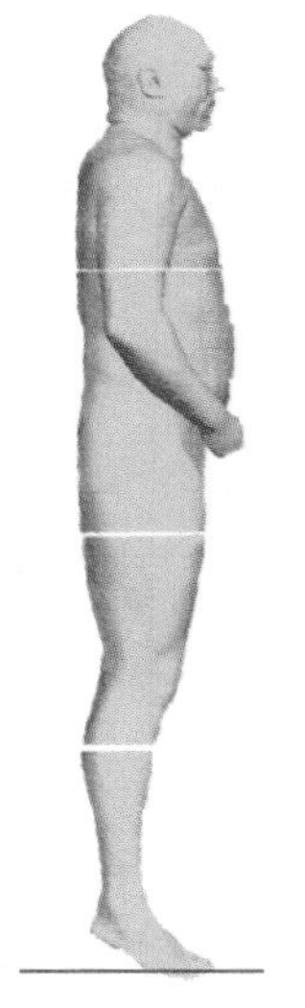

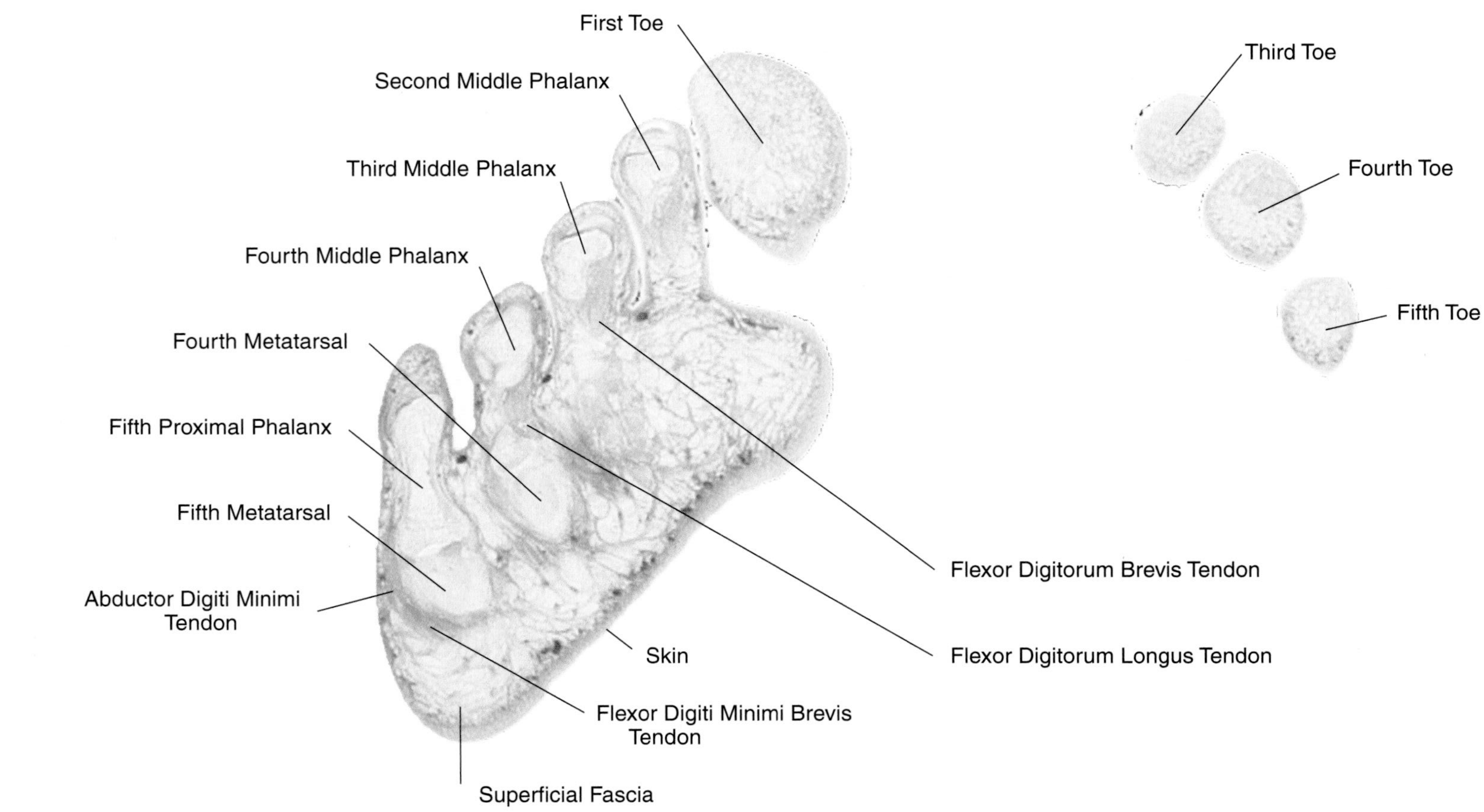
anterior
First Toe
Second Middle Phalanx
Third Middle Phalanx
Fourth Middle Phalanx
Fourth Metatarsal
Fifth Proximal Phalanx
Fifth Metatarsal
Abductor Digiti Minimi Tendon
Skin
Flexor Digiti Minimi Brevis Tendon
Superficial Fascia
Flexor Digitorum Brevis Tendon
Flexor Digitorum Longus Tendon
Third Toe
Fourth Toe
Fifth Toe
right
left
posterior

a_vm2865

a_vm2866

a_vm2867

a_vm2868

a_vm2869

a_vm2870

a_vm2871

a_vm2872

a_vm2873

a_vm2874

a_vm2875

a_vm2876

Transverse
a_vm2877-
a_vm2878

anterior

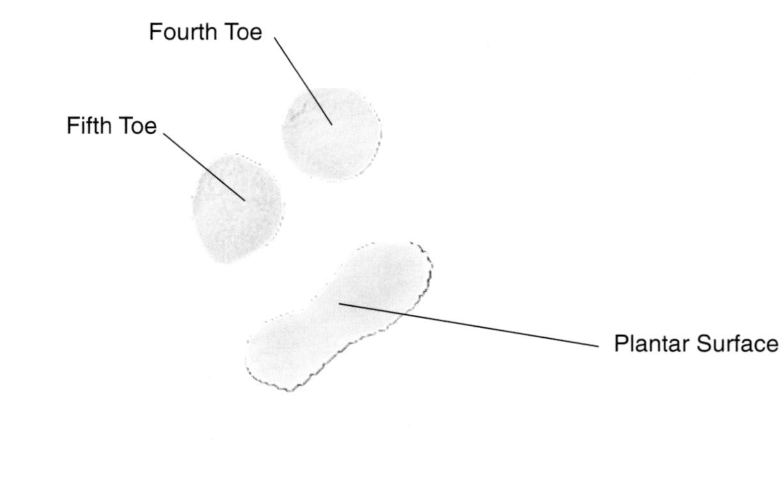
Fourth Toe
Fifth Toe
Plantar Surface

right

left

posterior

a_vm2877
a_vm2878

PART TWO

RECONSTRUCTED IMAGES
Coronal

The coronal images are of vertical planes through the long axis of the body and parallel to or through the coronal suture of the skull. They are presented in one-millimeter increments in this atlas starting at the front. These are standard radiological planes used in direct magnetic resonance and ultrasound imaging. Coronal clinical images are often reconstructed from transverse computed tomography and single photon emission computed tomography images. This plane, like the transverse, is particulary useful for comparison of bilateral anatomical structures.

These images are virtual slices—the Visible Human Male was never physically cut in coronal planes. Each coronal image was computer reconstructed from an edge view of all slices in the transverse image collection as seen from the anterior perspective and each is of a different depth at one-millimeter intervals through that collection.

The maximum resolution of images reconstructed in the coronal plane is determined by the one-millimeter thickness of the original transverse sections. Therefore, the level of detail in the coronal images (one-millimeter pixels) is only one third that seen in the transverse images of the previous part (0.32 millimeter pixels), in each dimension.

The labeling format for coronal images is the same as that used for the transverse images. In the same fashion as the transverse image display, the labeled slice is always repeated from the six coronal images on the same and facing page. The magnification remains the same, however, for all coronal images.

Nose, Tip

Rectus Abdominis Muscle

Umbilicus

First Metacarpal

Thenar Muscles

Second Metacarpal

Third Metacarpal

Palmar Interosseous Muscles

Dorsal Interosseous Muscles

Thumb

Superficial Fascia

First Proximal Phalanx

First Metacarpal

First Dorsal Interosseous Muscle

Second Metacarpal

Second Dorsal Interosseous Muscle

Third Metacarpal

Third Dorsal Interosseous Muscle

Fourth Metacarpal

Fourth Proximal Phalanx

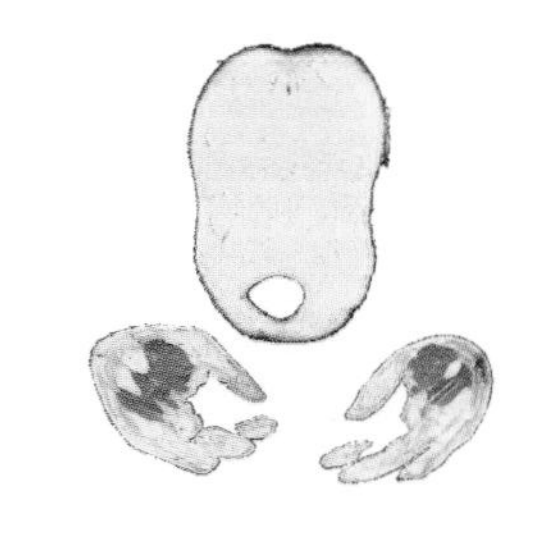

vmc3041

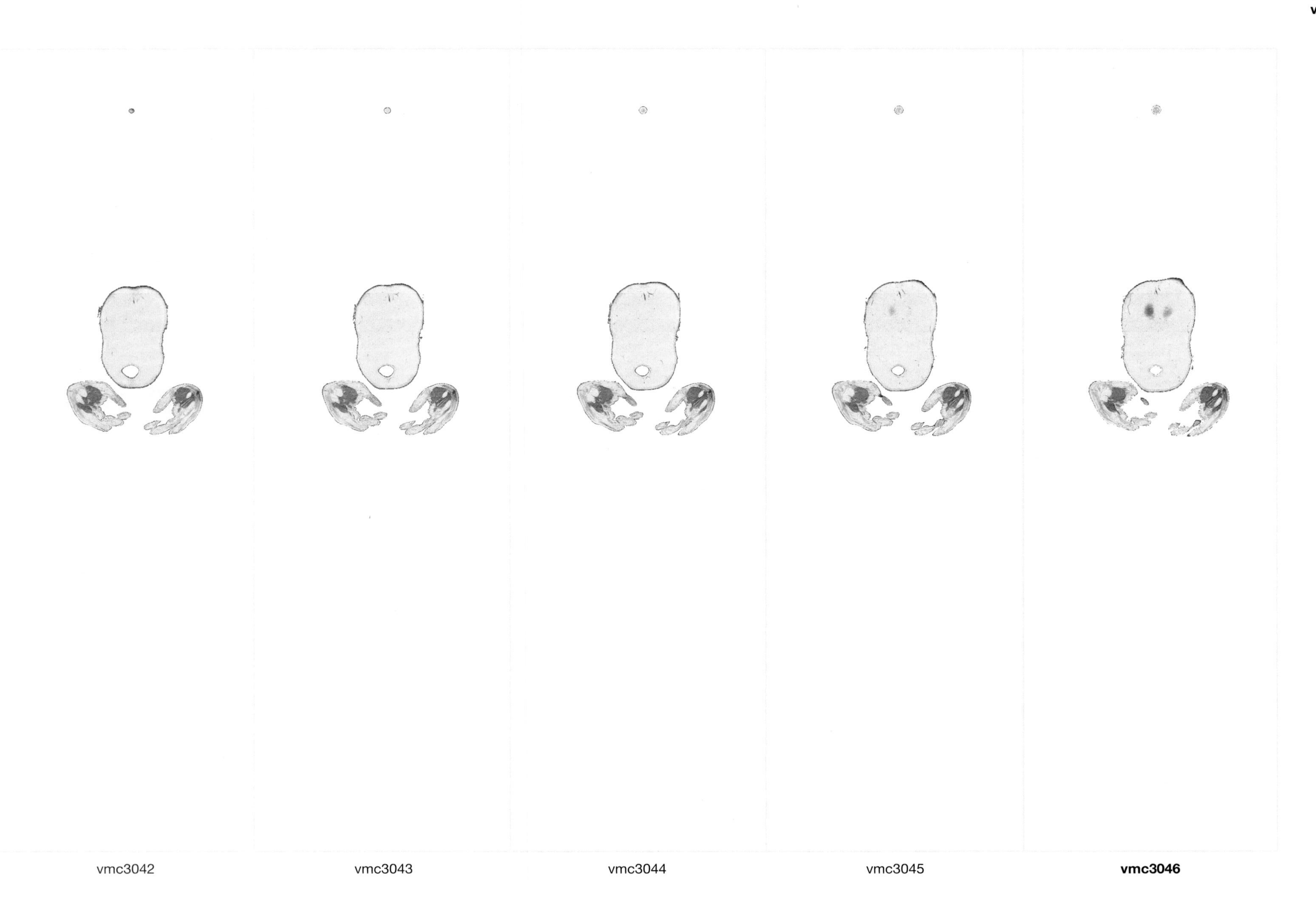
vmc3042
vmc3043
vmc3044
vmc3045
vmc3046

Coronal
vmc3052

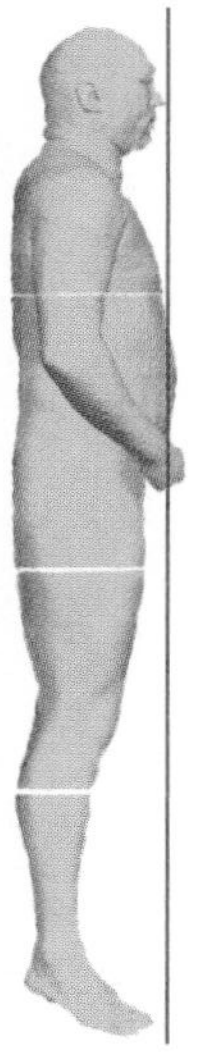

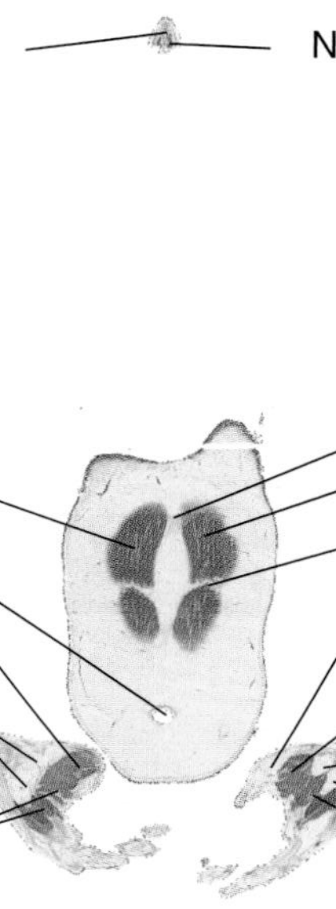

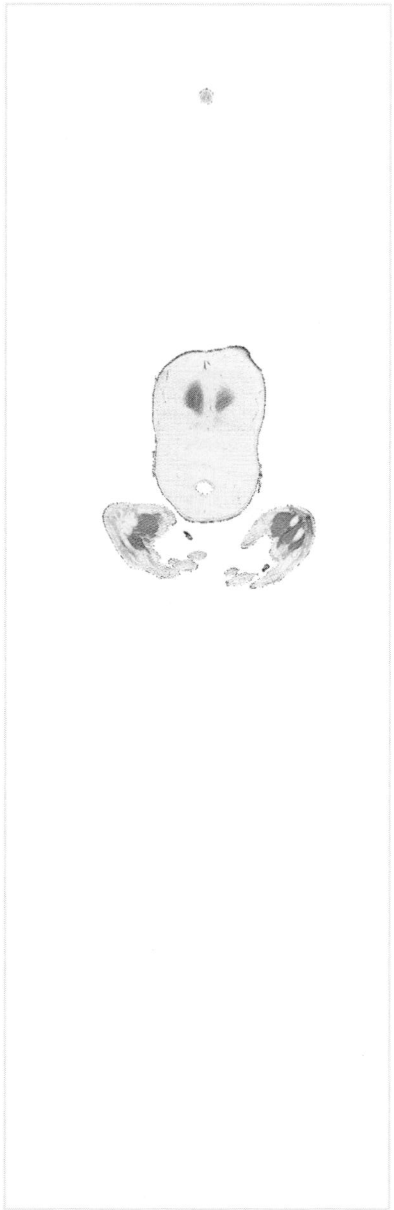

Nose
Naris
Rectus Abdominis Muscle
Umbilicus
Thenar Muscles
Carpal Bones
Palmar Interosseous Muscles
Linea Alba
Rectus Abdominis Muscle
Rectus Abdominis Muscle,
Tendinous Intersection
First Metacarpal
First Dorsal Interosseous Muscle
Second Metacarpal
Third Metacarpal
Palmar Interosseous Muscles

vmc3047

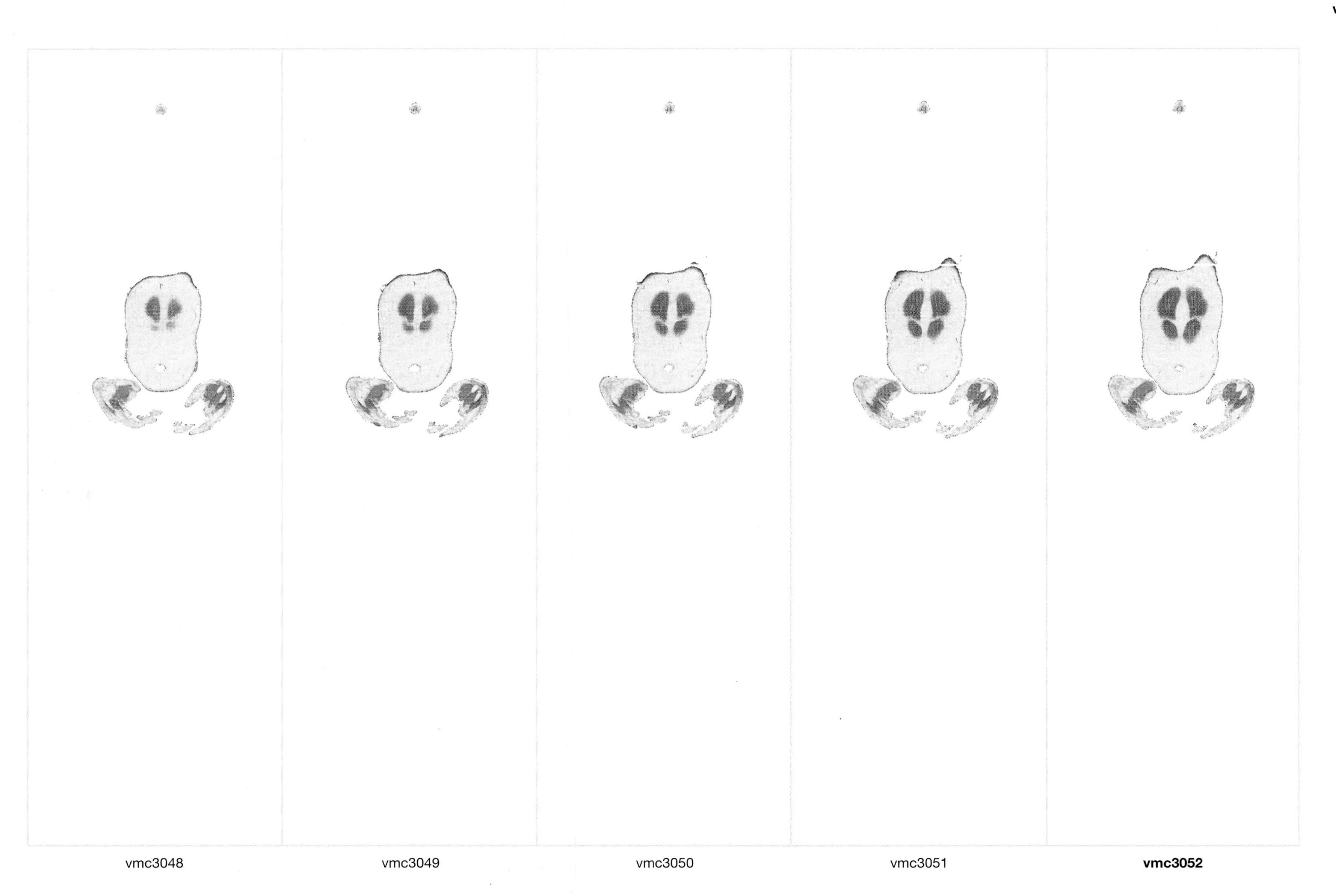
vmc3048
vmc3049
vmc3050
vmc3051
vmc3052

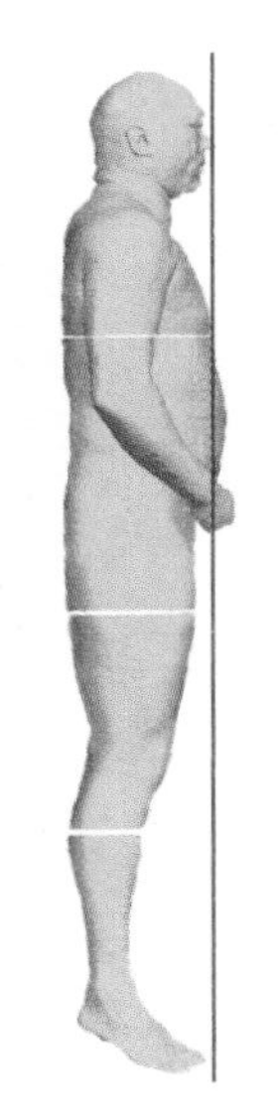

Nose
Upper Lip
Lower Lip
Nasal Septum
Naris

Rectus Abdominis Muscle,
Tendinous Intersections
Rectus Abdominis Muscle
Abductor, Opponens and
Flexor Pollicis Brevis Muscles
Carpal Bones
Palmar Interosseous
Muscle
Linea Alba
Rectus Abdominis Muscle,
Tendinous Intersections
Rectus Abdominis Muscle
Umbilicus
Second Metacarpal
Abductor, Opponens and
Flexor Pollicis Brevis Muscles
Third Metacarpal

First Toe

vmc3053

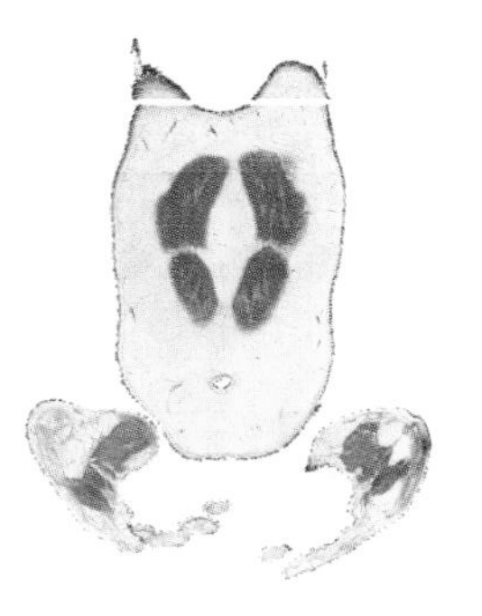

vmc3054

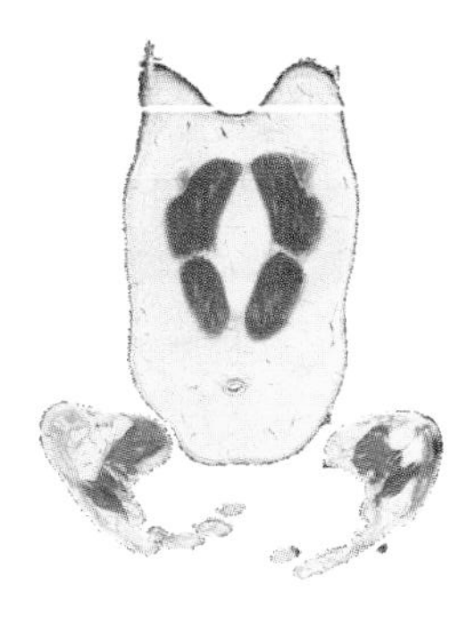

vmc3055

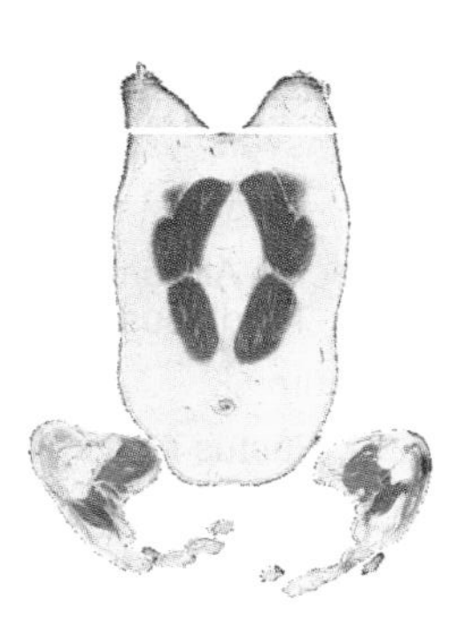

vmc3056

vmc3057

vmc3058

Coronal
vmc3070

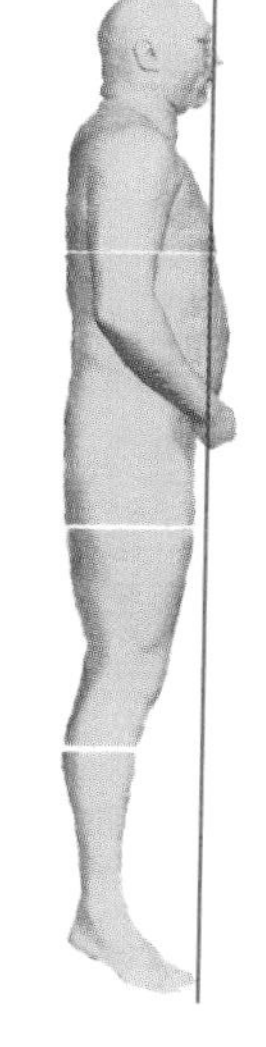

Frontal Bone
Procerus Muscle
Upper Lip
Nasal Septum
Chin
Lower Lip

Superficial Fascia
Pectoralis Major Muscle
Rib 5, Costal Cartilage
Xiphoid Process
External Oblique Muscle
Diaphragm
Diaphragm
Transverse Colon
Ileum
Rectus Abdominis Muscle
Ileum
Mesentery
Rectus Abdominis Muscle, Tendinous Intersection
Radius
Thenar Muscles
Carpal Bones
Carpal Bones
Thenar Muscles
Palmar Interosseous and Lumbrical Muscles with Long Flexor Tendons
Hypothenar Muscles
Fifth Metacarpal

First Distal Phalanx
Second Toe

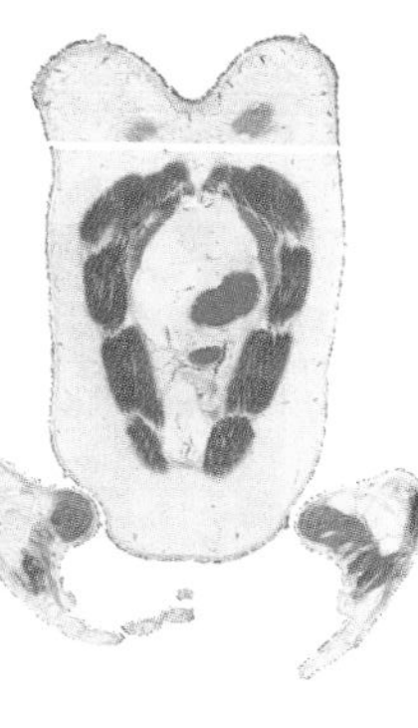

vmc3065

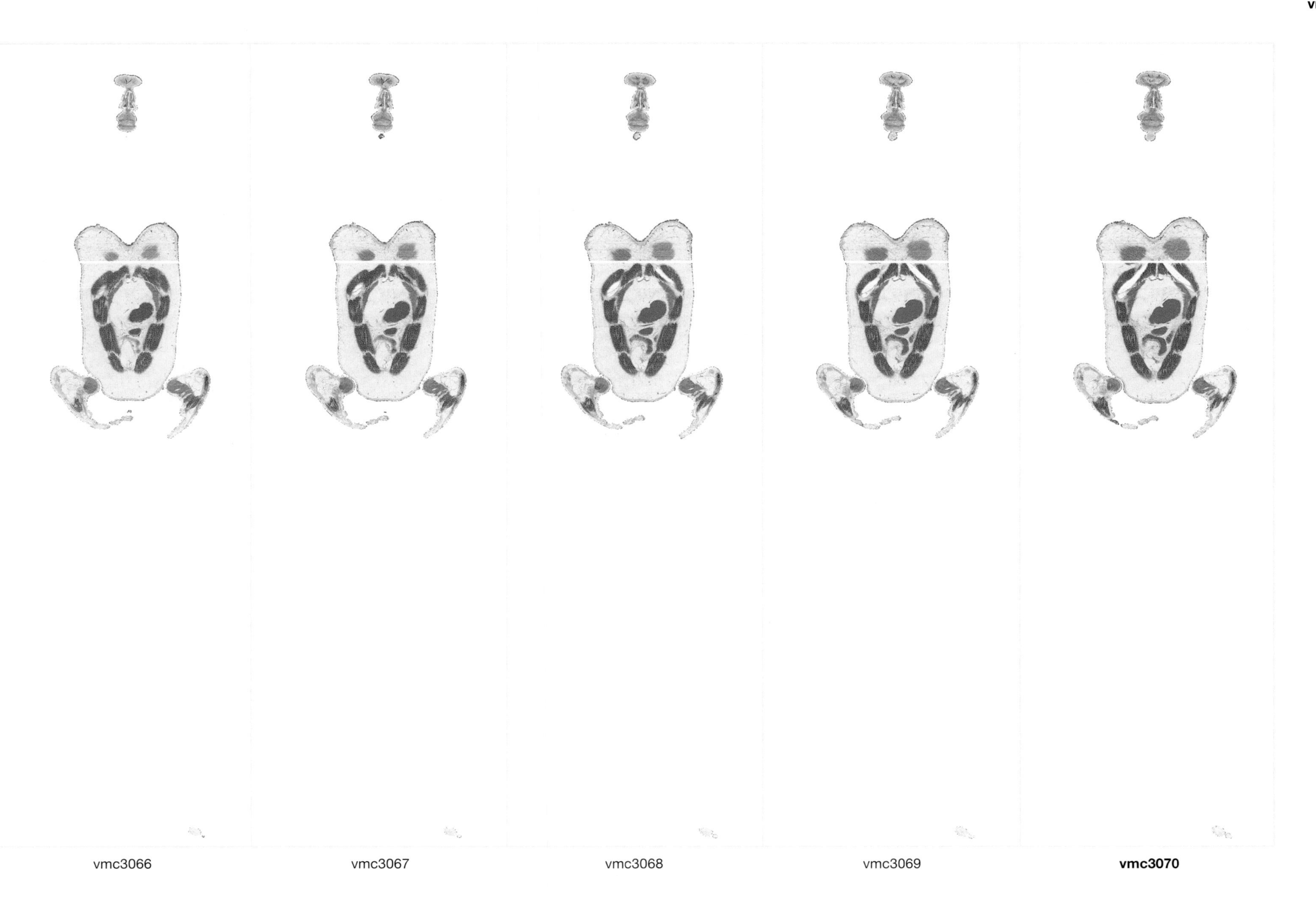

vmc3066 vmc3067 vmc3068 vmc3069 **vmc3070**

Coronal
vmc3076

Frontal Bone
Nasalis Muscle
Nasal Cavity
Orbicularis Oris Muscle
Oral Orifice
Lower Molar Teeth
Superficial Fascia
Sternum
Xiphoid Process
Rib 5, Costal Cartilage
Rib 6, Costal Cartilage
Greater Omentum
Ascending Colon
Diaphragm
Ileum
Mesentery
Rectus Abdominis Muscle, Tendinous Intersection
Radius
Long Extensor Tendons
Ulna
Long Flexor Tendons
Hypothenar Muscles

Frontal Sinus
Nasal Bone
Nasal Septum
Depressor Septi Muscle
Upper Lip
Lower Lip
Chin
Pectoralis Major Muscle
Rib 4, Costal Cartilage
Rib 5, Costal Cartilage
External Oblique Muscle
Stomach
Gastroepiploic Vessel
Diaphragm
Transverse Colon
Jejunum
Rectus Abdominis Muscle
Radius
Carpal Bones
Fifth Metacarpal
Thenar Muscles
Penis

First Toe
First Distal Phalanx
Second Distal Phalanx

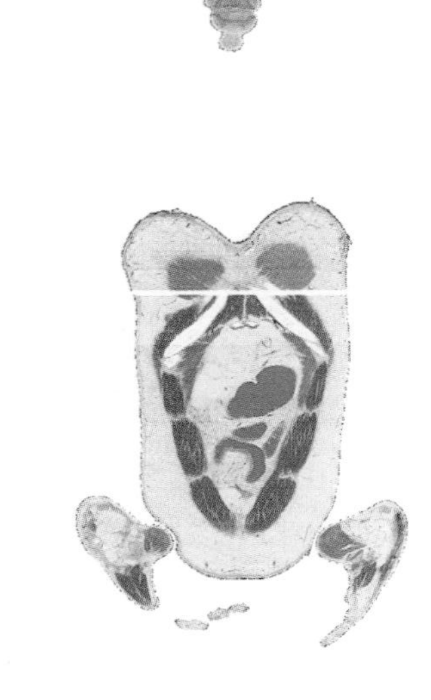

vmc3071

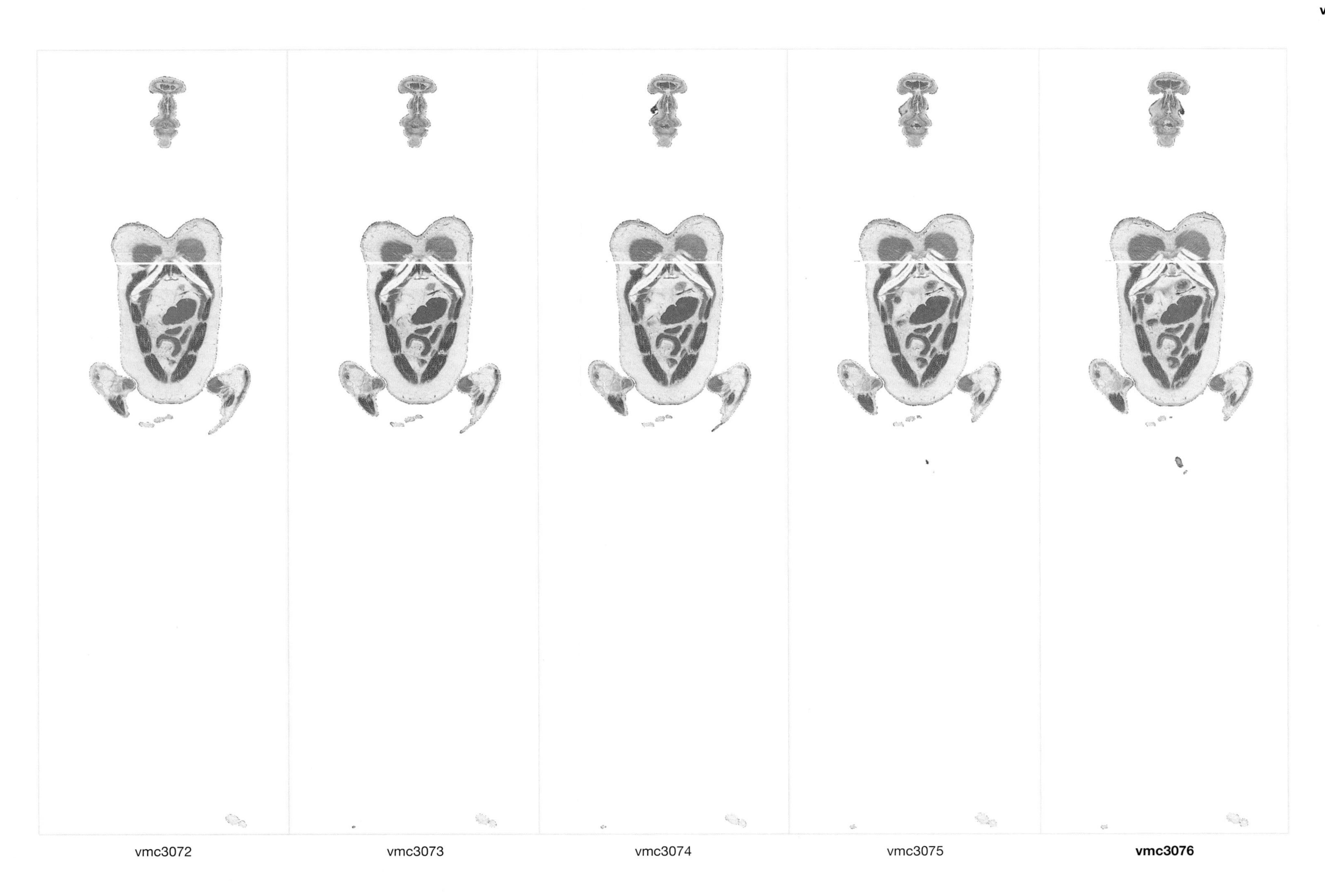
vmc3072
vmc3073
vmc3074
vmc3075
vmc3076

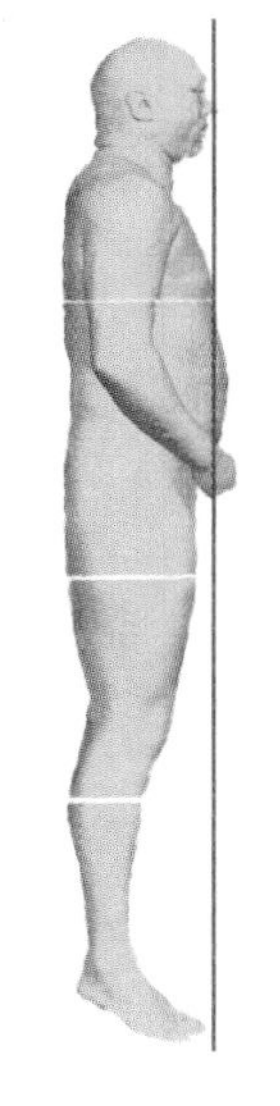

Frontal Bone
Nasal Bone
Nasal Cavity
Facial Vessels
Nasal Septum
Levator Labii Superioris M.
Oral Cavity
Platysma Muscle
Pectoralis Major Muscle
Sternum
Rib 5, Costal Cartilage
Liver, Left Lobe
Gastroepiploic Vessel
Colon, Hepatic Flexure
Transversus Abdominis Muscle
Greater Omentum
Ileum
Mesentery
Ulna
Radius
Long Flexor Tendon
Fifth Metacarpal
Hypothenar Muscles
Rectus Abdominis Muscle
Thigh, Superficial Fascia

Frontal Sinus
Tarsal Plate
Orbicularis Oculi Muscle
Maxillary Bone
Orbicularis Oris Muscle
Lower Incisor Teeth
Mentalis Muscle
Pectoralis Major Muscle
Rib 4, Costal Cartilage
Rib 6, Costal Cartilage
Diaphragm
Stomach
External Oblique Muscle
Transverse Colon
Transversus Abdominis Muscle
Jejunum
Rectus Abdominis Muscle
Radius
Carpal Bones
Thenar Muscles
Hypothenar Muscles

Penis

First Distal Phalanx
Second Toe

First Distal Phalanx
Second Distal Phalanx
Third Toe

vmc3077

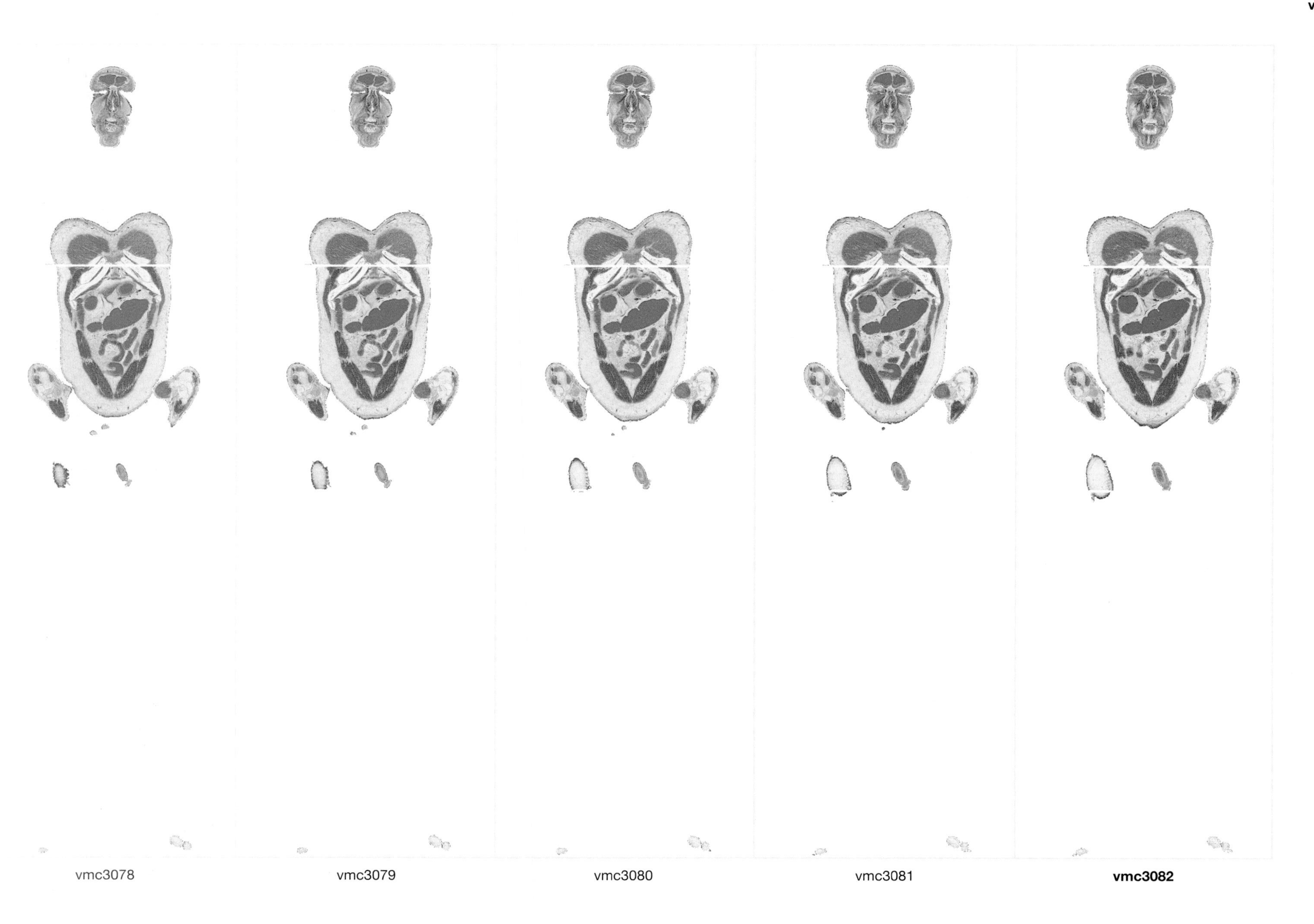

vmc3078 vmc3079 vmc3080 vmc3081 **vmc3082**

Coronal
vmc3088

Frontal Bone
Eye Globe
Lens
Maxillary Bone
Nasal Cavity
Mandible
Platysma Muscle
Superficial Fascia
Sternum
Rib 4, Costal Cartilage
Rib 5, Costal Cartilage
Rib 6, Costal Cartilage
Liver, Right Lobe
Colon, Hepatic Flexure
External Oblique Muscle
Ascending Colon
Ileum
Radius
Palmar Interosseous Muscle
Ulna
Long Flexor Tendons
Rectus Abdominis Muscle
Thigh, Superficial Fascia

Frontal Sinus
Ethmoid Sinuses
Maxillary Sinus
Nasal Septum
Maxillary Bone
Buccinator Muscle
Tongue
Mentalis Muscle
Pectoralis Major Muscle
Rib 4-5, Intercostal Muscles
Anterior Mediastinum
Diaphragm
Stomach
Transverse Colon
Transversus Abdominis Muscle
External Oblique Muscle
Jejunum
Rectus Abdominis Muscle
Radius
Carpal Bones
Thenar Muscles
Hypothenar Muscles
Penis

First Distal Phalanx
Second Distal Phalanx

First Distal Phalanx
Second Distal Phalanx
Third Distal Phalanx

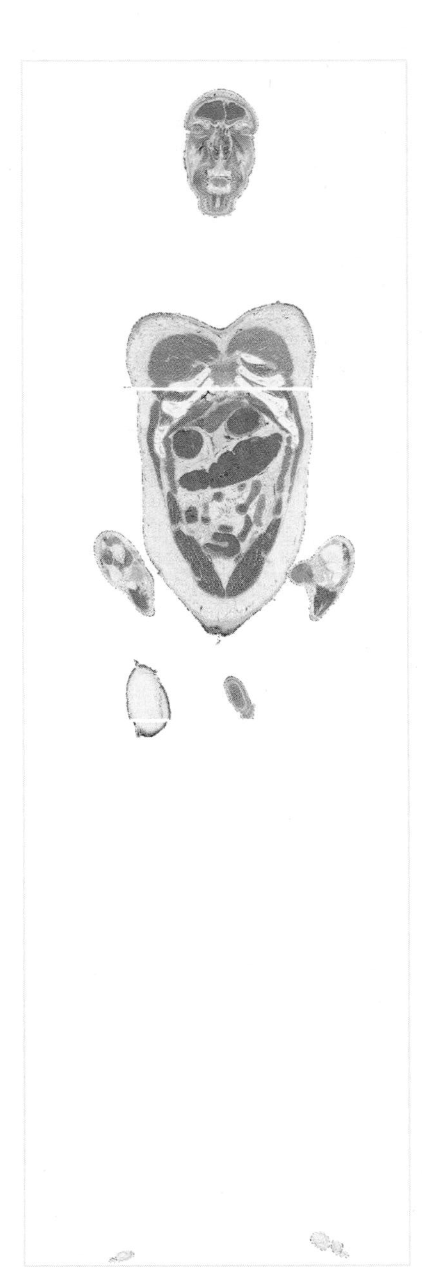

vmc3083

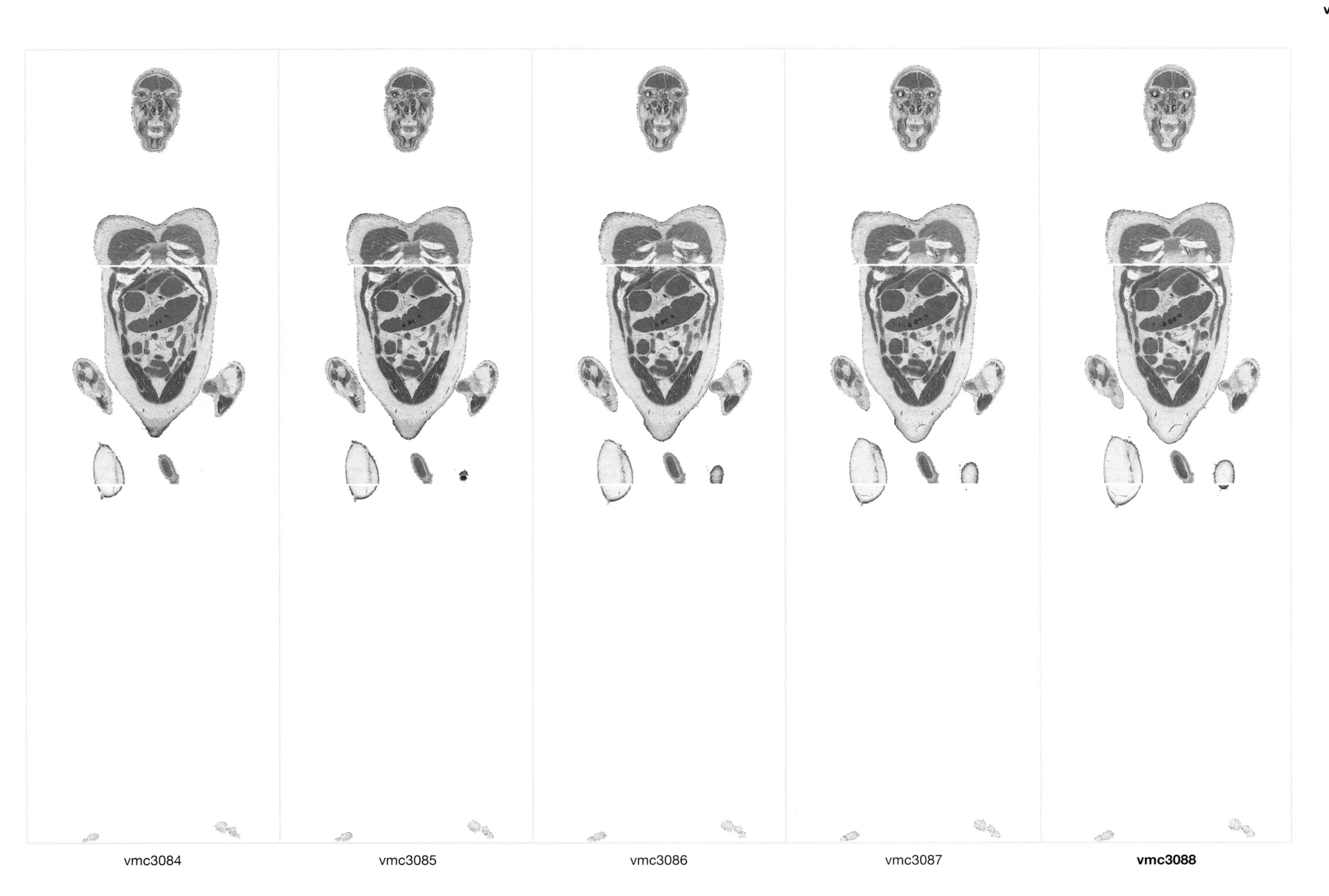
vmc3084
vmc3085
vmc3086
vmc3087
vmc3088

Coronal
vmc3094

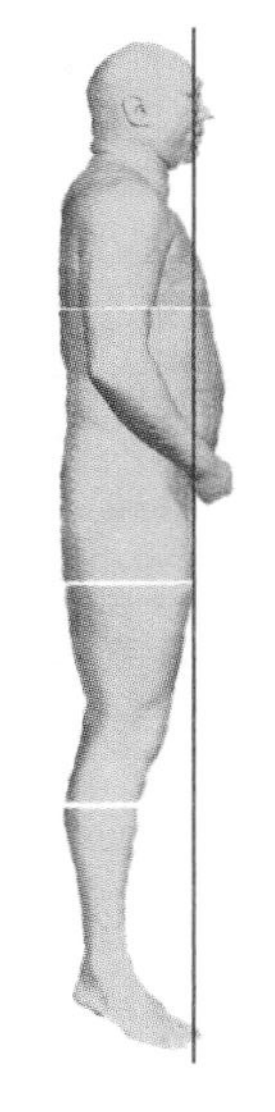

Frontal Bone
Frontal Lobe
Longitudinal Cerebral Fissure
Maxillary Bone
Zygomaticus Major Muscle
Tongue
Genioglossus Muscle
Digastric Muscle, Anterior Belly
Sternum
Rib 4, Costal Cartilage
Pericardium
Liver, Right Lobe
Rib 7, Costal Cartilage
Colon, Hepatic Flexure
Rib 8, Costal Cartilage
Transversus Abdominis Muscle
Ascending Colon
Long Extensor Tendons
Long Flexor Tendons
Ileum
Linea Alba
Artifact
Rectus Femoris Muscle
Thigh, Superficial Fascia

Frontal Sinus
Eye Globe
Nasal Cavity
Maxillary Sinus
Hard Palate
Buccinator Muscle
Mandible
Pectoralis Major Muscle
Rib 3, Costal Cartilage
Rib 3-4, Intercostal Muscles
Rib 5
Liver, Left Lobe
Diaphragm
Stomach
External Oblique Muscle
Transverse Colon
Jejunum
Internal Oblique Muscle
Radius
Ulna
Mesentery
Rectus Abdominis Muscle
Corpus Cavernosum
Penis

First Distal Phalanx
Second Distal Phalanx

First Distal Phalanx
Second Distal Phalanx
Third Distal Phalanx
Fourth Toe

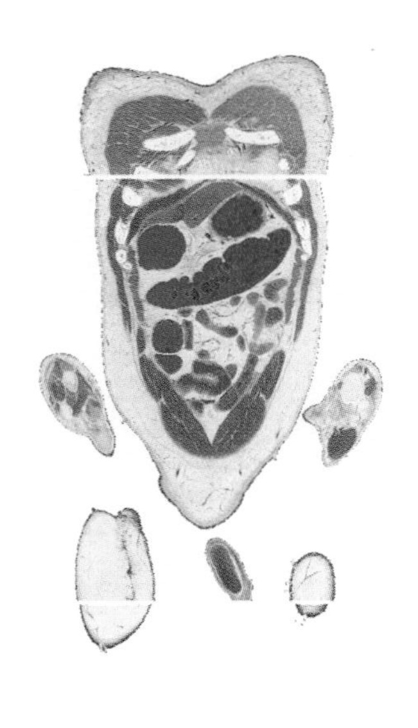

vmc3089

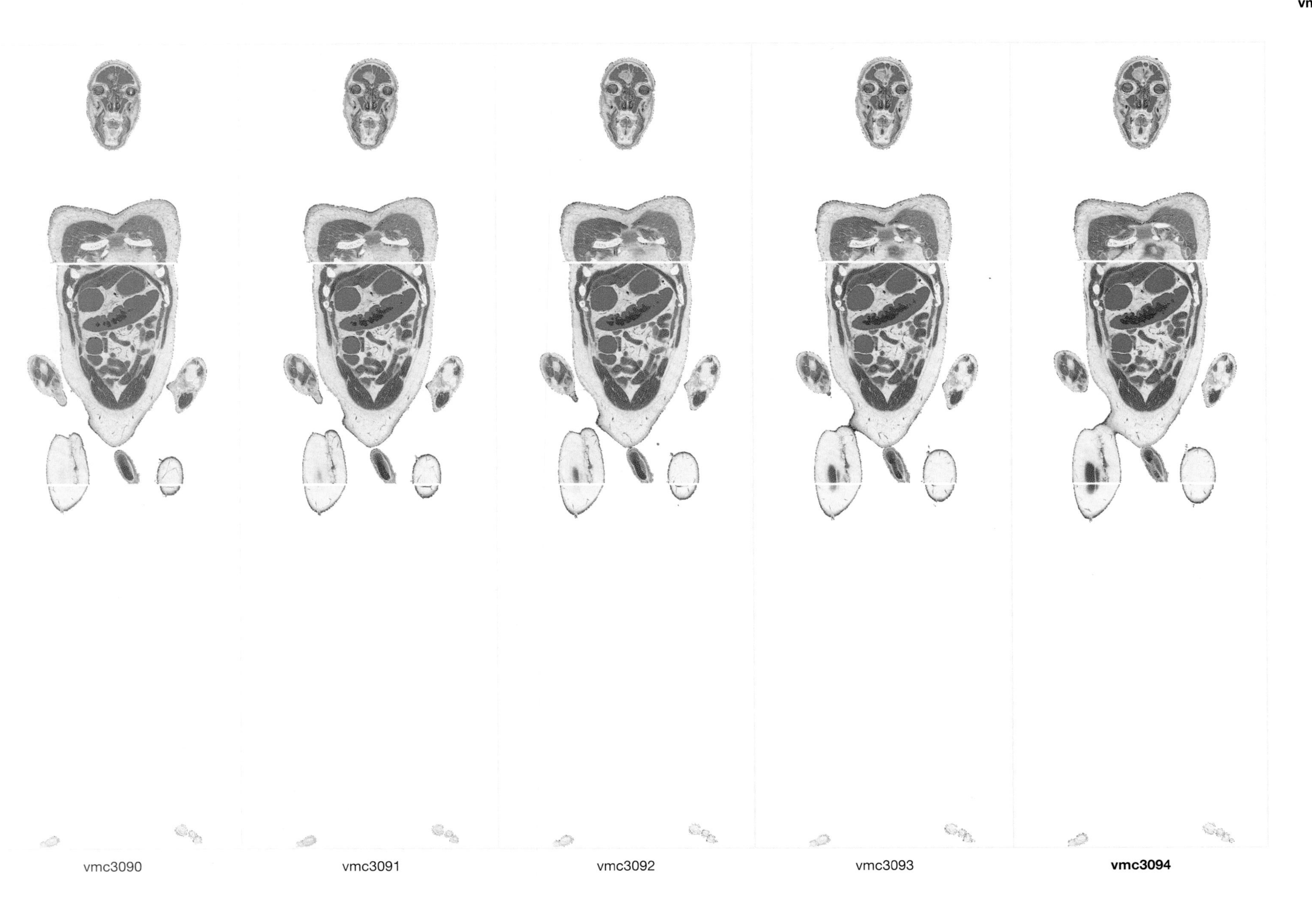
vmc3090
vmc3091
vmc3092
vmc3093
vmc3094

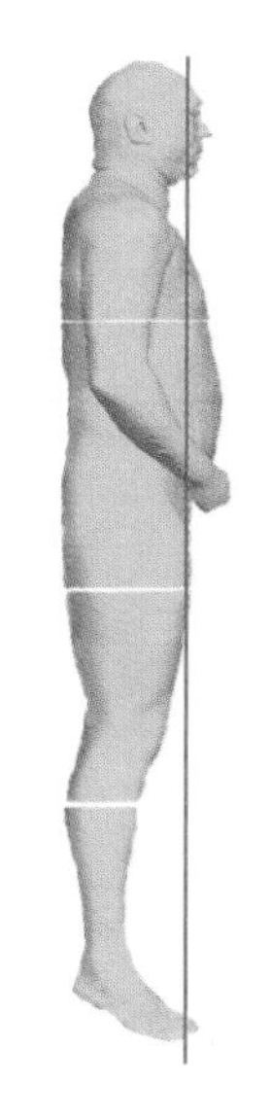

Frontal Lobe
Ethmoid Sinuses
Nasal Cavity
Zygomatic Arch
Hard Palate
Buccinator Muscle
Mandible
Digastric Muscle, Anterior Belly
Sternum
Lung, Upper Lobe
Rib 4
Diaphragm
Liver, Right Lobe
Colon, Hepatic Flexure
Transversus Abdominis Muscle
Ascending Colon
Long Extensor Muscles
Ulna
Long Flexor Muscles
Ileum
Rectus Abdominis Muscle
Sartorius Muscle
Rectus Femoris Muscle
Fascia Lata

Frontal Bone
Superior Sagittal Sinus
Frontal Sinus
Eye Globe
Maxillary Sinus
Tongue
Genioglossus Muscle
Geniohyoid Muscle
Pectoralis Major Muscle
Rib 3, Costal Cartilage
Lung, Upper Lobe
Right Ventricle
Diaphragm
Liver, Left Lobe
Rib 7-8, Intercostal Muscles
Transverse Colon
Jejunum
Brachioradialis Tendon
Radius
Internal Oblique Muscle
External Oblique Muscle
Linea Alba
Superficial Fascia
Corpus Cavernosum
Penis, Glans

First Proximal Phalanx
Second Middle Phalanx
Third Toe

First Proximal Phalanx
Second Middle Phalanx
Third Middle Phalanx
Fourth Distal Phalanx

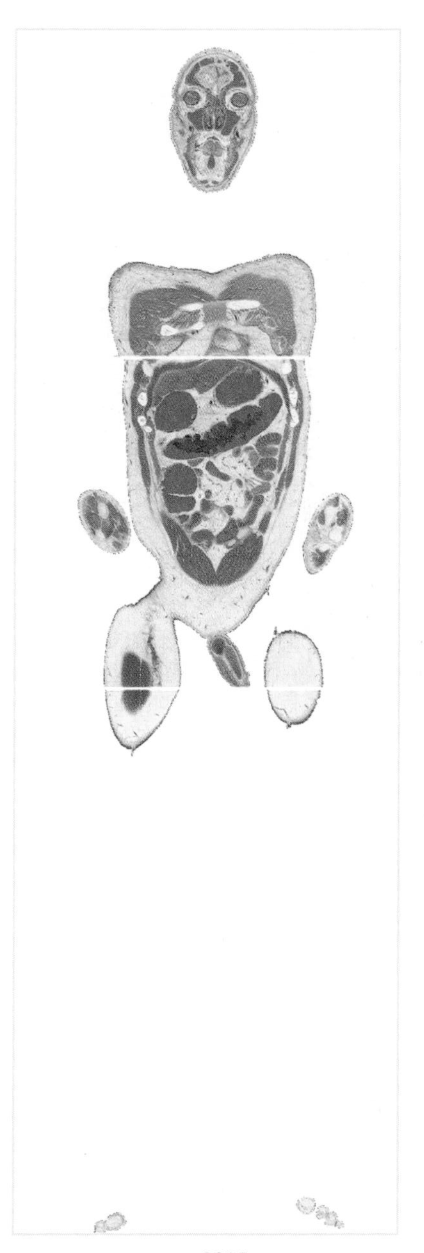

vmc3095

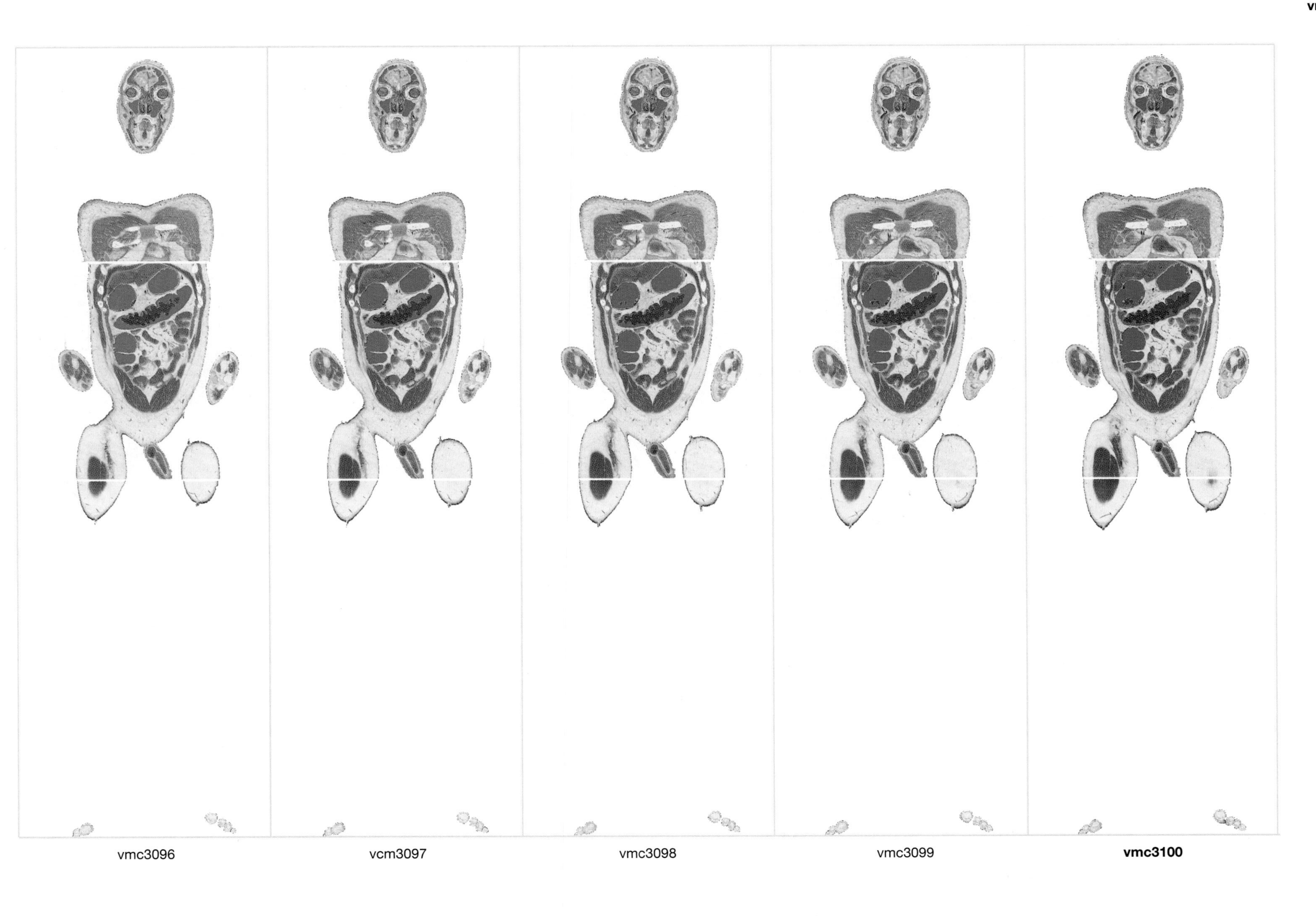

vmc3096 vcm3097 vmc3098 vmc3099 **vmc3100**

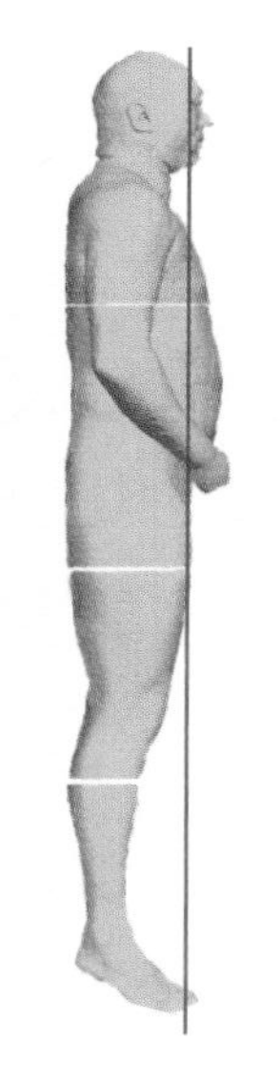

Frontal Lobe
Temporalis Muscle
Optic Nerve
Masseter Muscle
Inferior Nasal Concha
Hard Palate
Digastric Muscle, Anterior Belly
Sternum
Pectoralis Major Muscle
Rib 3, Costal Cartilage
Lung, Upper Lobe
Diaphragm
Liver, Right Lobe
Transversus Abdominis Muscle
Ascending Colon
Internal Oblique Muscle
Long Extensor Muscles
Long Flexor Muscles
External Oblique Muscle
Ileum
Spermatic Cord
Linea Alba
Sartorius Muscle
Vastus Lateralis Muscle
Rectus Femoris Muscle
Vastus Medialis Muscle

Superior Sagittal Sinus
Longitudinal Cerebral Fissure
Orbit
Medial Rectus Muscle
Maxillary Sinus
Tongue
Genioglossus Muscle
Geniohyoid Muscle
Pectoralis Major Muscle
Rib 2-3, Intercostal Muscles
Lung, Upper Lobe
Left Ventricle
Diaphragm
Liver, Left Lobe
Stomach
Transverse Colon
Jejunum
Brachioradialis Tendon
Ulna
Descending Colon
Rectus Abdominis Muscle
Suspensory Ligament
Corpus Cavernosum
Penis
Fascia Lata
Rectus Femoris Muscle
Superficial Fascia

Corpus Spongiosum

First Proximal Phalanx
Second Middle Phalanx
Third Distal Phalanx

First Proximal Phalanx
Second Middle Phalanx
Third Middle Phalanx
Fourth Distal Phalanx

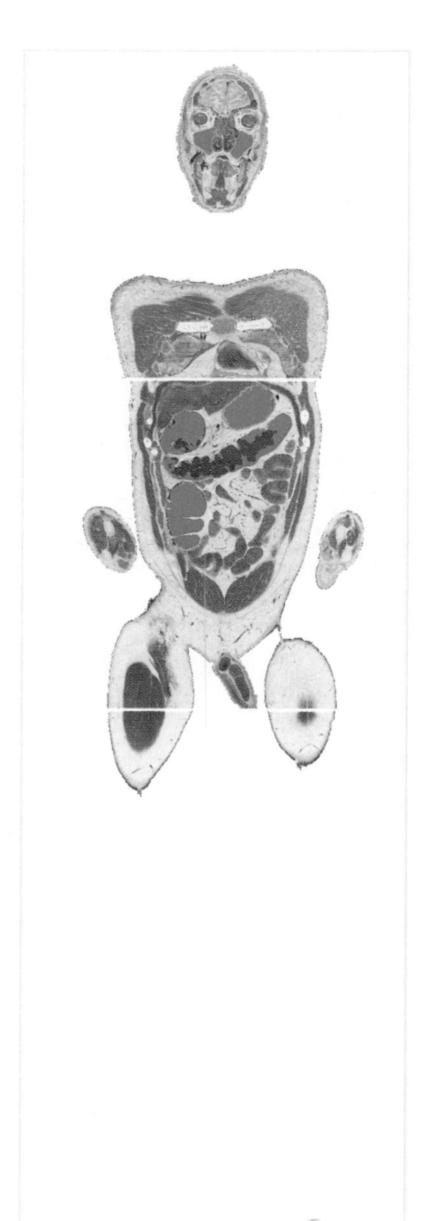

vmc3101

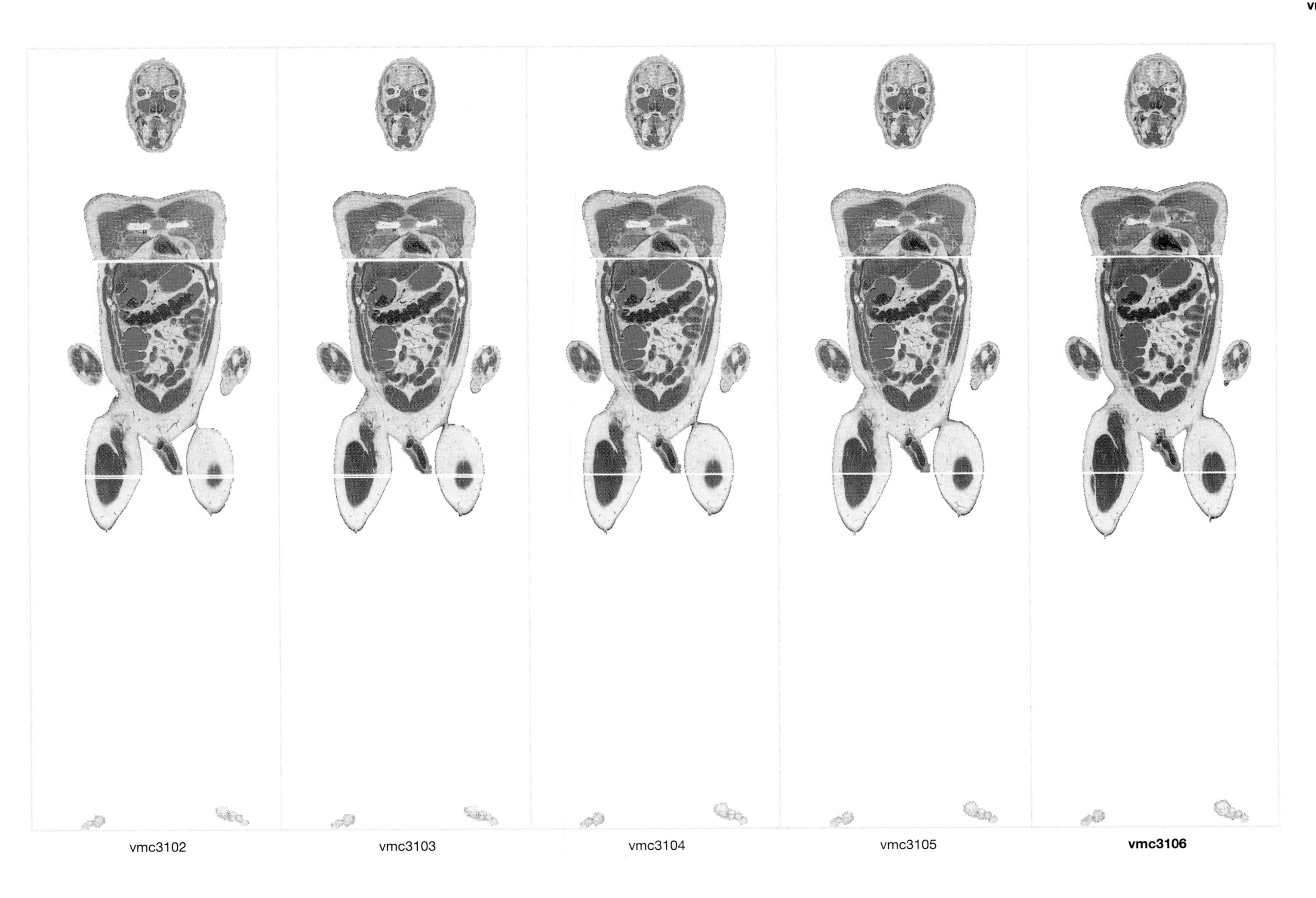
vmc3102
vmc3103
vmc3104
vmc3105
vmc3106

Coronal
vmc3112

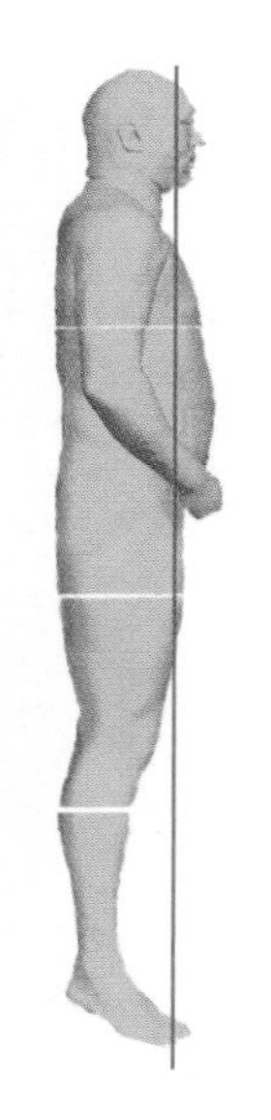

Frontal Bone
Orbit
Temporalis Muscle
Zygomatic Bone
Ethmoid Sinuses
Masseter Muscle
Mylohyoid Muscle
Digastric Muscle, Anterior Belly
Sternum
Rib 2, Costal Cartilage
Internal Thoracic Vessel
Lung, Upper Lobe
Diaphragm
Right Ventricle
Liver, Right Lobe
Gall Bladder
Colon, Hepatic Flexure
Long Extensor Muscles
Ascending Colon
Ulna
Long Flexor Muscles
Ileum
Transversus Abdominis Muscle
Rectus Femoris Muscle
Sartorius Muscle
Vastus Lateralis Muscle
Vastus Medialis Muscle

Superior Sagittal Sinus
Longitudinal Cerebral Fissure
Lateral Rectus Muscle
Nasal Cavity
Maxillary Sinus
Tongue
Mandible
Platysma Muscle
Pectoralis Major Muscle
Rib 3
Atrophic Thymus
Left Ventricle
Interventricular Septum
Liver, Left Lobe
Stomach
External Oblique Muscle
Internal Oblique Muscle
Brachioradialis Tendon
Radius
Ileum
Sigmoid Colon
Rectus Abdominis Muscle
Suspensory Ligament
Sartorius Muscle
Fascia Lata
Rectus Femoris Muscle
Scrotum

Corpus Cavernosum

First Proximal Phalanx
Second Middle Phalanx
Third Middle Phalanx
Fourth Toe

First Proximal Phalanx
Second Proximal Phalanx
Third Proximal Phalanx
Fourth Middle Phalanx
Fifth Toe

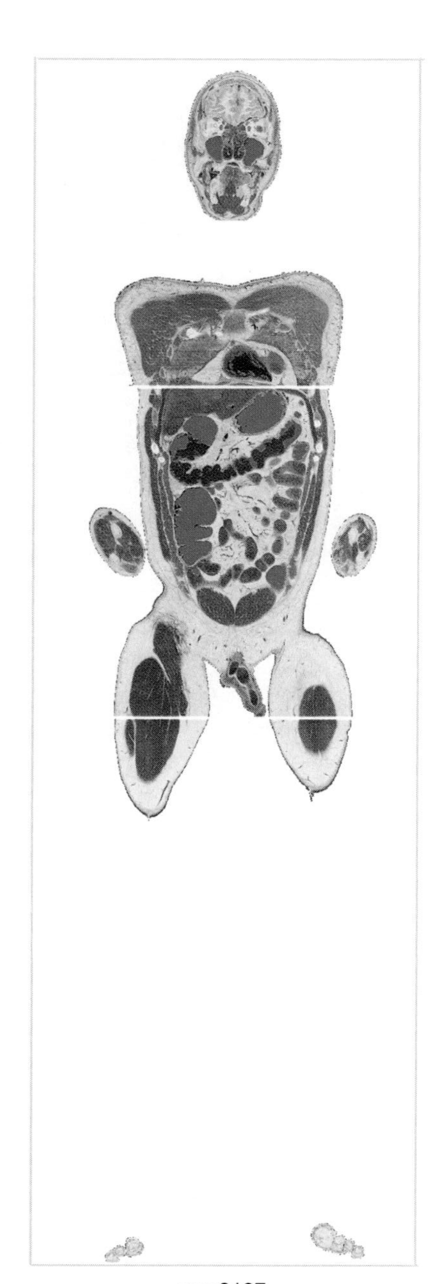

vmc3107

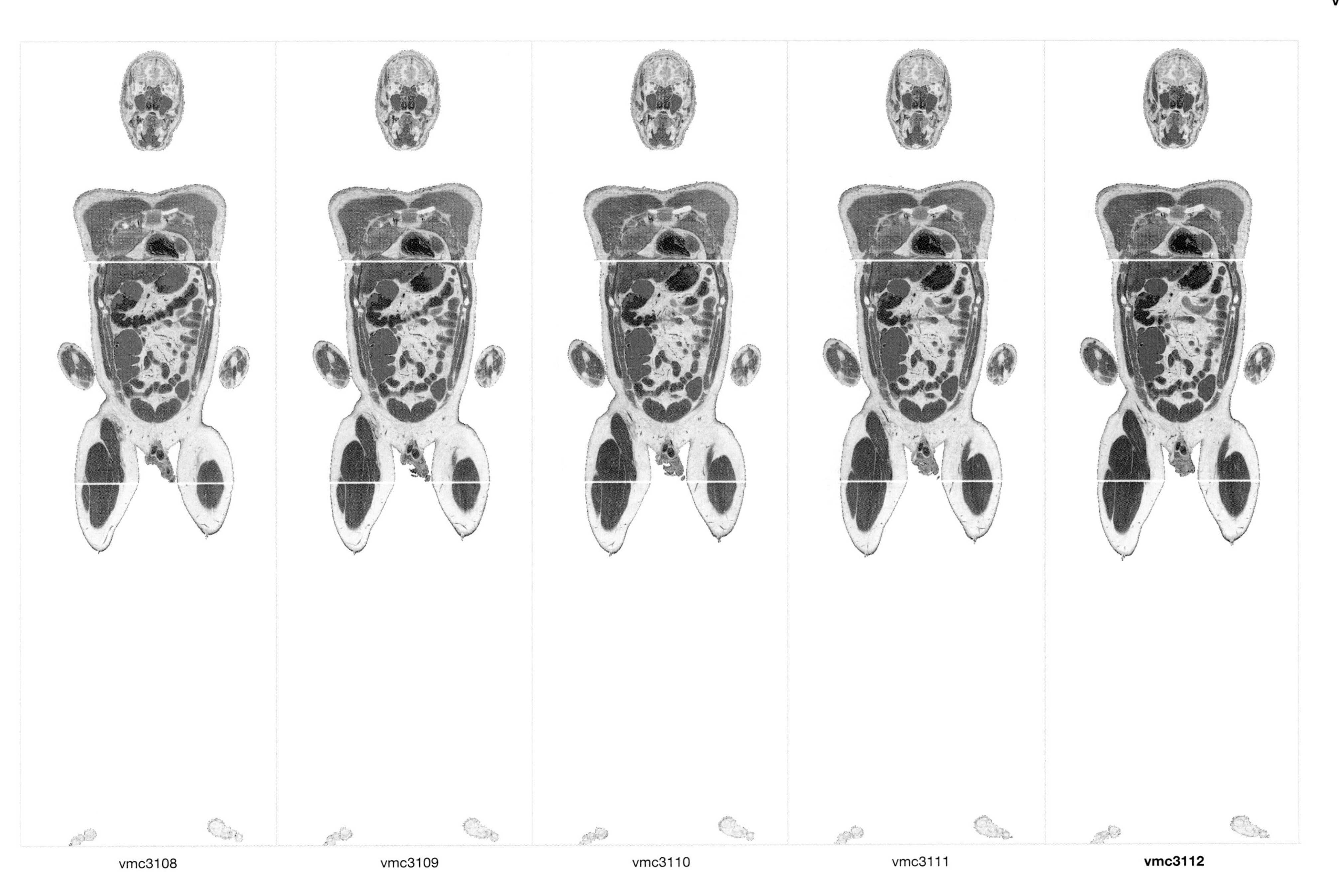

vmc3108 vmc3109 vmc3110 vmc3111 **vmc3112**

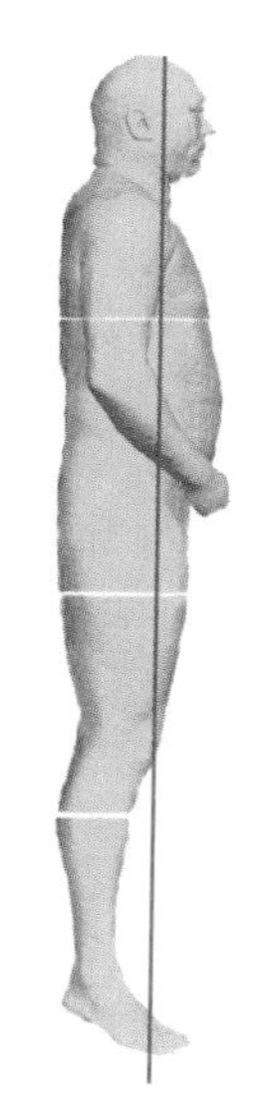

Superior Sagittal Sinus
Lentiform Nucleus
Optic Nerve
Sphenoid Sinus
Hyoglossus Muscle
Epiglottis
Infraglottic Space
Deltoid Muscle
Pectoralis Major Muscle
Pectoralis Minor Muscle
Lung, Horizontal Fissure
Right Atrium
Liver, Right Lobe
Colon, Hepatic Flexure
Pancreas
Portal Vein
Brachioradialis Muscle
Ascending Colon
Ulna
Transversus Abdominis Muscle
Gluteus Medius Muscle
Ileum
Tensor Fascia Lata Muscle
Rectus Femoris Muscle
Bladder
Vastus Intermedius Muscle
Vastus Lateralis Muscle
Vastus Medialis Muscle
Rectus Femoris Tendon
Great Saphenous Vein
Patella

Frontal Bone
Longitudinal Cerebral Fissure
Lateral Fissure
Temporal Lobe
Soft Palate
Masseter Muscle
Tongue
Cricoid Cartilage
Clavicle
Rib 1, Costal Cartilage
Pulmonary Trunk
Diaphragm
Falciform Ligament
Duodenojejunal Junction
Splenic Vein
Radius
Interosseous Membrane
Internal Oblique Muscle
Descending Colon
Sigmoid Colon
Ductus Deferens
Femoral Vein
Pectineus Muscle
Pubic Symphysis
Adductor Longus Muscle
Corpus Cavernosum
Vastus Lateralis Muscle
Scrotum
Rectus Femoris Muscle
Sartorius Muscle
Vastus Medialis Muscle

First Metatarsal
First Dorsal Interosseous Muscle
Second Metatarsal
Second Dorsal Interosseous Muscle
Third Metatarsal
Fourth Metatarsal
Fifth Proximal Phalanx

Abductor Hallucis Muscle
First Metatarsal
Second Metatarsal
Third Metatarsal
Flexor Digitorum Brevis Muscle
Fourth Metatarsal
Fifth Proximal Phalanx

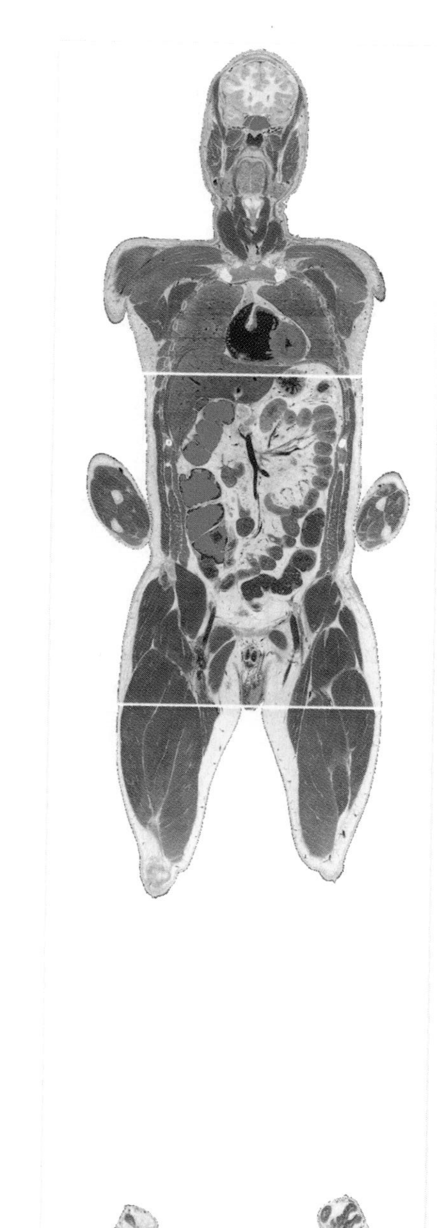

vmc3137

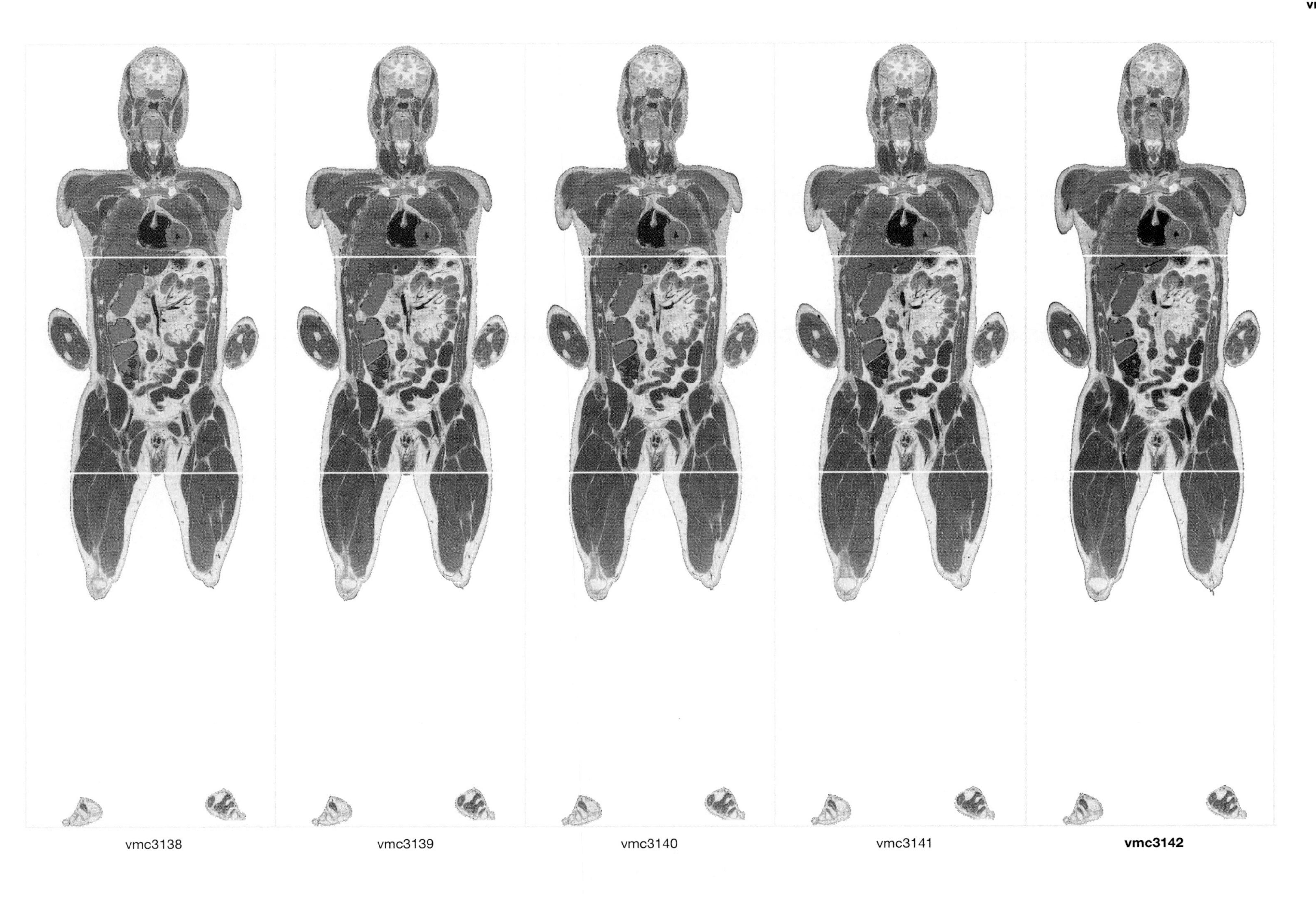

vmc3138 vmc3139 vmc3140 vmc3141 **vmc3142**

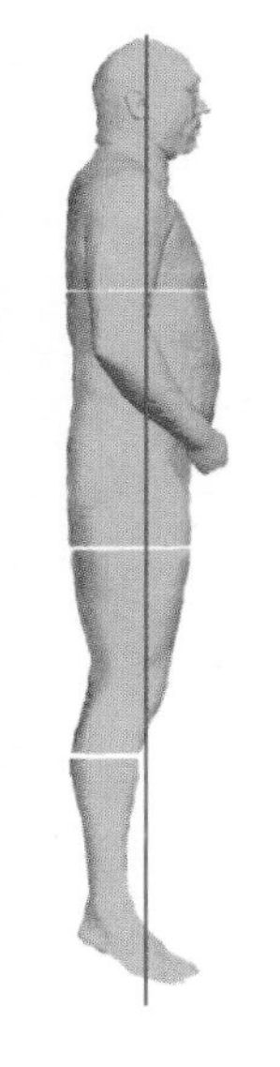

Corpus Callosum
Internal Capsule
Optic Chiasma and Hypophysis
Articular Tubercle
Parotid Gland
Palatine Tonsil
Tongue, Root
Sternocleidomastoid Muscle
Thyroid Gland
Deltoid Muscle
Rib 1, Costal Cartilage
Right Atrium
Liver, Quadrate Lobe
Ascending Colon
Pancreas, Head
Duodenum
Long Extensor Muscles
Long Flexor Muscles
Ileum
Internal Oblique Muscle
Iliac Crest
Gluteus Medius Muscle
Iliacus Muscle
Vastus Intermedius Muscle
External Iliac Vessel
Fascia Lata
Pectineus Muscle
Pubic Symphysis
Corpus Cavernosum
Great Saphenous Vein
Sartorius Muscle
Rectus Femoris Tendon
Patella

Caudate Nucleus
Auditory Tube, Cartilage
Lateral Pterygoid Muscle
Mandible
Medial Pterygoid Muscle
Epiglottis
Vocal Folds
Platysma Muscle
Clavicle
Sternohyoid Muscle
Pectoralis Minor Muscle
Manubrium
Left Ventricle
Stomach
Liver, Left Lobe
Jejunum
Radius
Ulna
External Oblique Muscle
Internal Oblique Muscle
Descending Colon
Mesentery
Rectum
Bladder
Pubis
Vastus Lateralis Muscle
Corpus Spongiosum
Vastus Intermedius Muscle
Sartorius Muscle
Vastus Medialis Muscle
Patella

Abductor Hallucis Muscle

First Metatarsal
First Plantar Interosseous Muscle
First Dorsal Interosseous Muscle
Second Metatarsal
Second Dorsal Interosseous Muscle
Third Metatarsal
Fourth Metatarsal
Fifth Metatarsal

First Metatarsal
First Dorsal Interosseous Muscle
Second Metatarsal
Second Dorsal Interosseous Muscle
Third Metatarsal
Flexor Digitorum Brevis and Longus Tendons
Fourth Metatarsal
Fifth Metatarsal

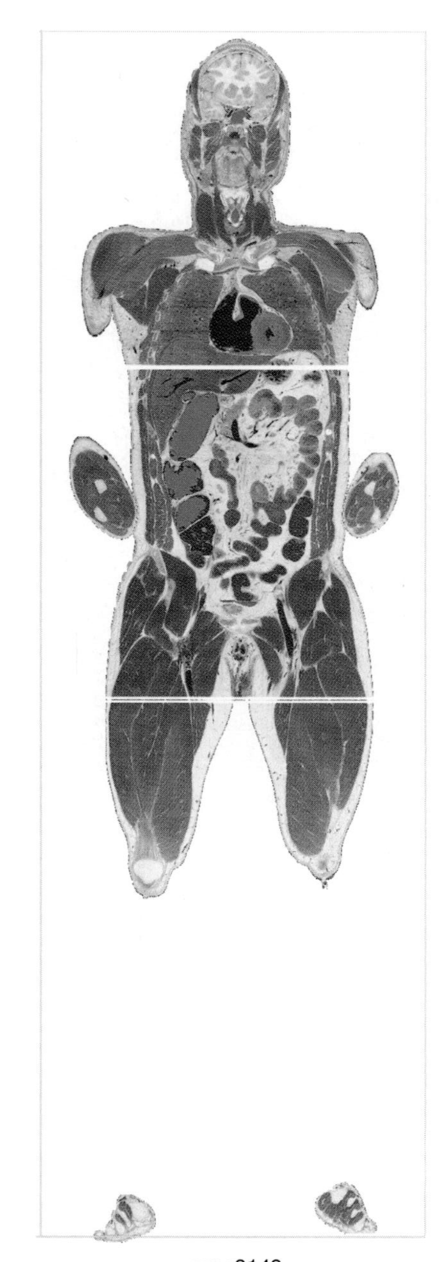

vmc3143

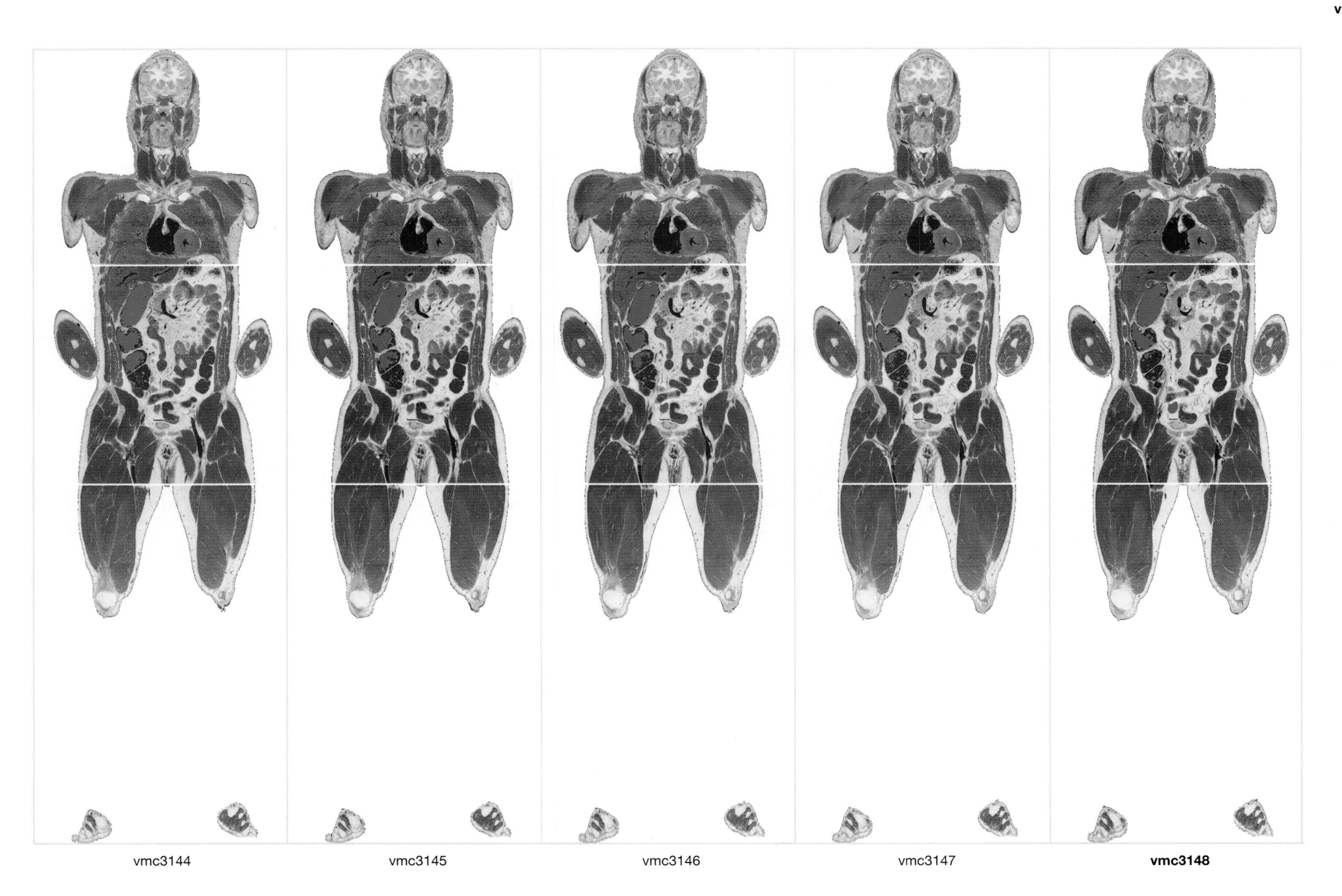

vmc3144 vmc3145 vmc3146 vmc3147 **vmc3148**

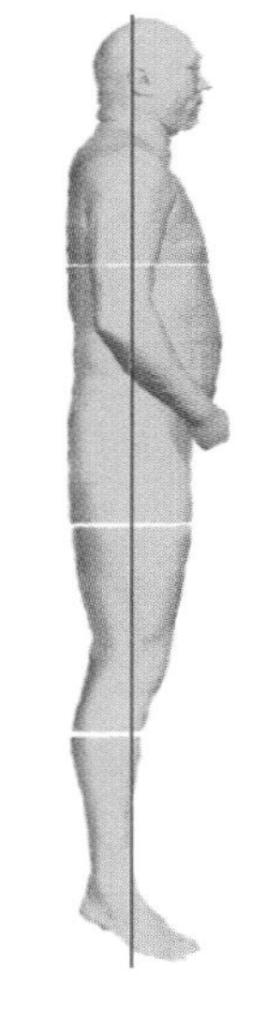

Insula
Temporalis Muscle
Cavernous Sinus
Mandible, Condyle
Longus Capitis Muscle
Parotid Gland
Sternocleidomastoid Muscle
Internal Jugular Vein
Clavicle
Pectoralis Minor Muscle
Pectoralis Major Muscle
Serratus Anterior Muscle
Diaphragm
Liver, Right Lobe
Cystic Duct
Brachioradialis Muscle
Ascending Colon
Long Extensor Muscles
Long Flexor Muscles
External Oblique Muscle
Internal Oblique Muscle
Ilium
Psoas Major Muscle
Gluteus Medius Muscle
Iliacus Muscle
Vastus Intermedius Muscle
Bladder
Adductor Longus Muscle
Sartorius Muscle
Vastus Lateralis Muscle
Suprapatellar Bursa
Patellar Tendon
Great Saphenous Vein

Superior Sagittal Sinus
Frontal Lobe
Caudate Nucleus
Temporal Lobe
Lateral Pterygoid Muscle
Medial Pterygoid Muscle
Palatine Tonsil
Epiglottis
Arytenoid Cartilage
Cricoid Cartilage
Trachea
Rib 1, Costal Cartilage
Lung, Upper Lobe
Left Ventricle
Stomach
Portal Vein
Duodenojejunal Junction
Pancreas, Head
Jejunum
Transversus Abdominis Muscle
Iliac Crest
Descending Colon
Sigmoid Colon
Pubic Symphysis
Penis, Crus
Femoral Vessels
Sartorius Muscle
Vastus Intermedius Muscle
Vastus Lateralis Muscle
Vastus Medialis Muscle
Rectus Femoris Tendon
Patella

Abductor Hallucis Muscle

Adductor Hallucis Muscle
First Metatarsal
Second Metatarsal
Second Plantar Interosseous Muscle
Third Metatarsal
Fourth Metatarsal
Plantar Aponeurosis

First Metatarsal
Medial Cuneiform Bone
Flexor Digitorum Longus Tendon
Second Dorsal Interosseous Muscle
Third Metatarsal
Fourth Metatarsal
Flexor Digitorum Brevis Muscle

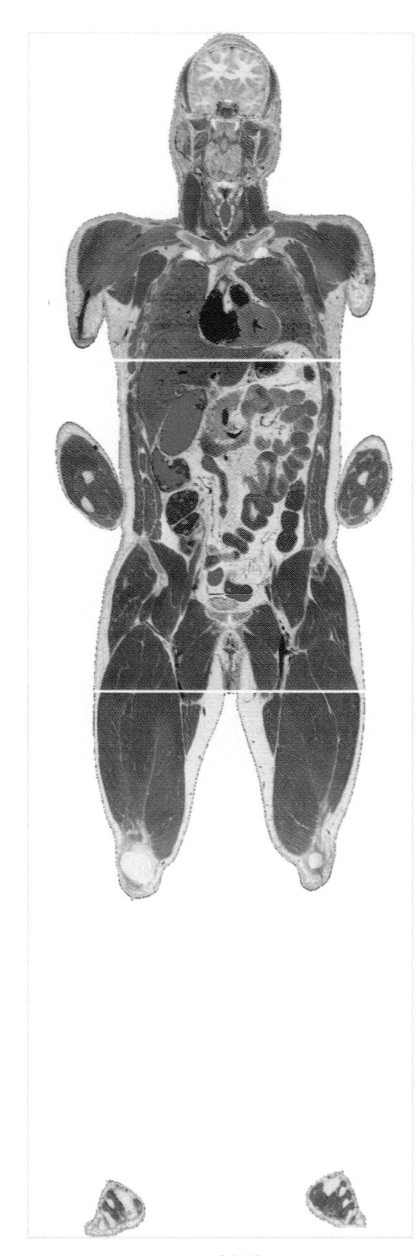

vmc3149

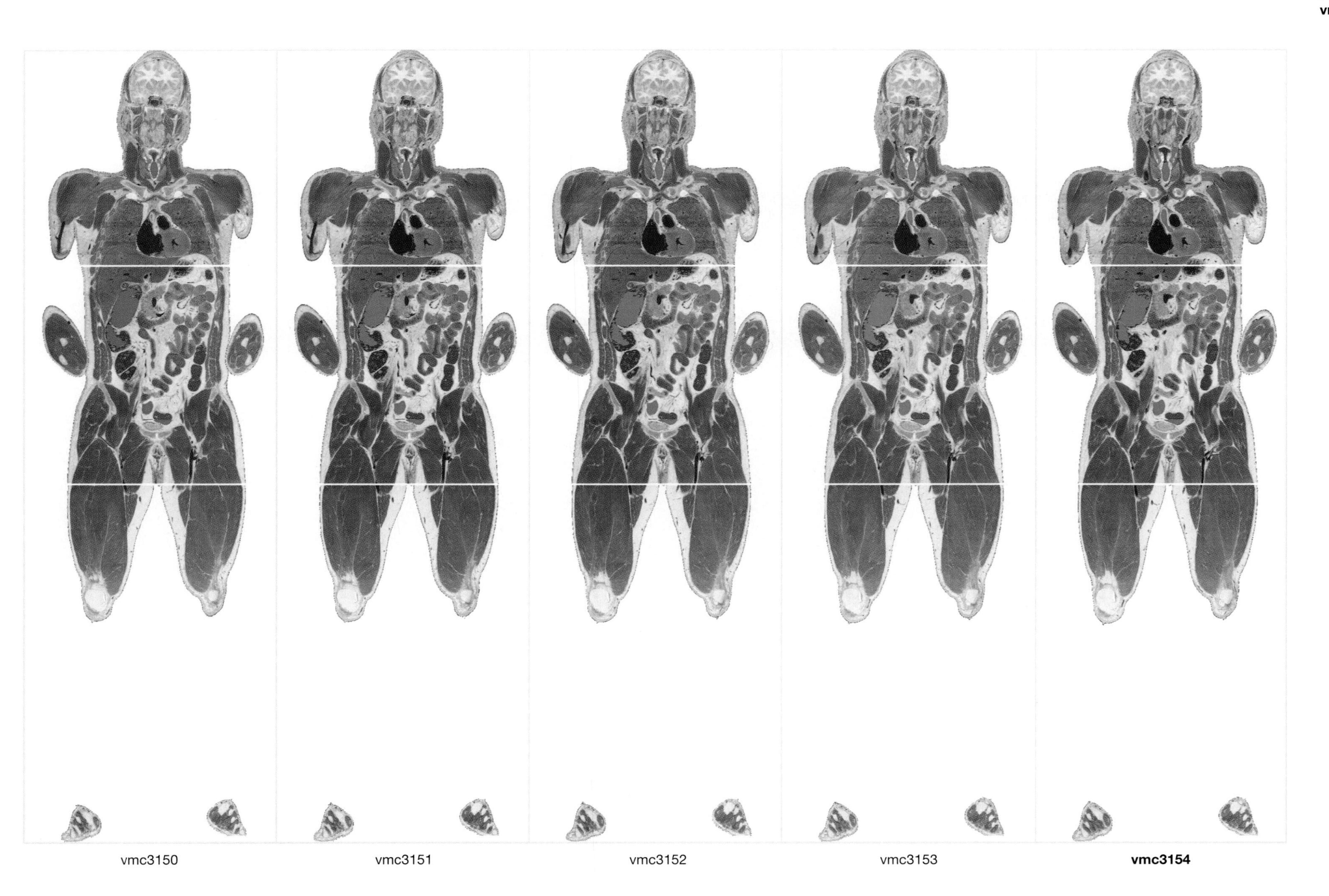

vmc3150 vmc3151 vmc3152 vmc3153 **vmc3154**

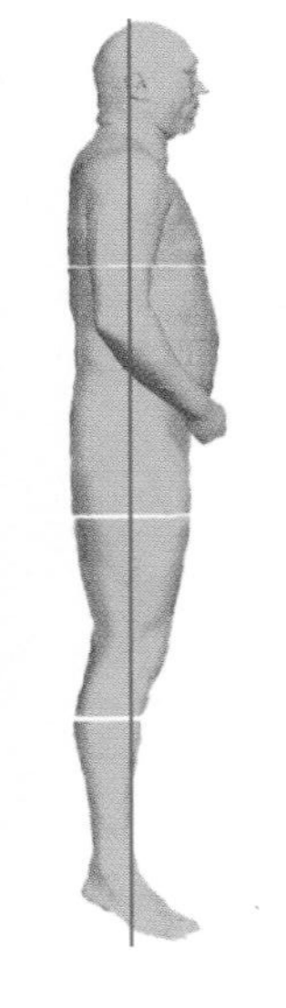

Longitudinal Cerebral Fissure
Corpus Callosum
Lentiform Nucleus
Amygdala
Tensor Veli Palatini Muscle
Longus Capitis Muscle
Digastric Muscle, Posterior Belly
Laryngopharynx
Humerus, Head
Internal Jugular Vein
Superior Vena Cava
Biceps Brachii Muscle
Ascending Aorta
Right Atrium
Coronary Sinus
Liver, Right Lobe
Colon, Hepatic Flexure
Pancreas, Head
Long Flexor Muscles
Ascending Colon
Internal Oblique Muscle
Psoas Major Muscle
Gluteus Medius Muscle
Iliacus Muscle
Rectus Femoris Muscle
Pubis, Superior Ramus
Obturator Externus Muscle
Adductor Longus Muscle
Pubic Symphysis
Vastus Intermedius Muscle
Sartorius Muscle
Suprapatellar Bursa
Great Saphenous Vein
Patellar Tendon
Adductor Hallucis Muscle
First Metatarsal
Medial Cuneiform Bone
Second Metatarsal
Second Dorsal Interosseous Muscle
Third Dorsal Interosseous Muscle
Plantar Aponeurosis
Fourth Dorsal Interosseous Muscle

Frontal Bone
Lateral Ventricle
Temporalis Muscle
Pons
Mandible, Condyle
Retromandibular Vein
Pharyngeal Constrictor Muscle
Deltoid Muscle
Cricoid Cartilage
Pectoralis Minor Muscle
Cephalic Vein
Brachiocephalic Vein
Trachea
Pulmonary Trunk
Left Ventricle
Portal Vein
Radius
Ulna
Jejunum
External Oblique Muscle
Sigmoid Colon
Inferior Mesenteric Vessel
Rectus Femoris Muscle
Iliopsoas Muscle
Femoral Vessel
Pectineus Muscle
Bladder
Adductor Longus Muscle
Penis, Crus
Adductor Canal
Articularis Genu Muscle
Rectus Femoris Tendon
Patella
Vastus Medialis Muscle
Abductor Hallucis Muscle
First Metatarsal
Medial Cuneiform Bone
Second Metatarsal
Third Metatarsal
Flexor Digitorum Brevis and Longus Muscles
Fourth Metatarsal
Fifth Metatarsal

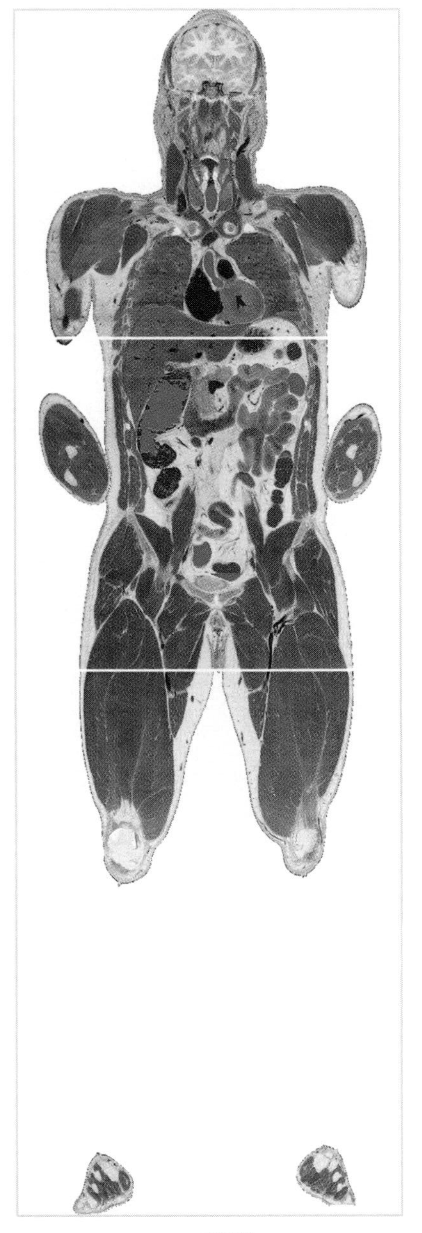

vmc3155

vmc3156

vmc3157

vmc3158

vmc3159

vmc3160

Coronal
vmc3184

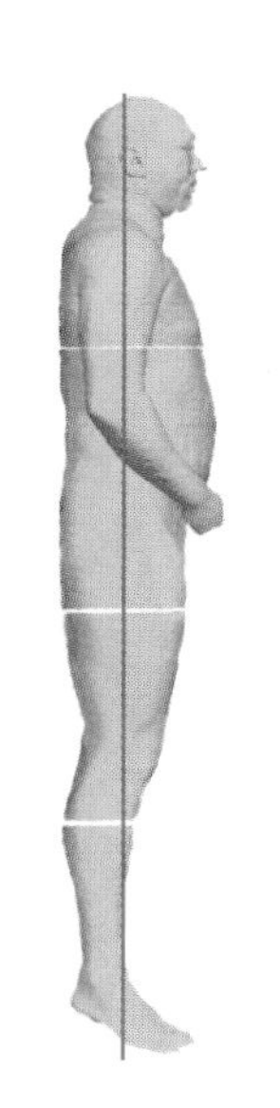

Cerebellum
Middle Cerebellar Peduncle
Obliquus Capitis Inferior Muscle
Sternocleidomastoid Muscle
Levator Scapulae Muscle
Clavicle
Scapula, Acromion
Coracoid Process
Aorta, Arch
Pulmonary Artery
Serratus Anterior Muscle
Right Ventricle
Inferior Vena Cava
Liver, Right Lobe
Ascending Colon
Inferior Vena Cava
Diaphragm, Crus
Long Flexor Muscles
Kidney
Transversus Abdominis Muscle
Psoas Major Muscle
Iliofemoral Ligament
Femur, Neck
Obturator Internus Muscle
Obturator Externus Muscle
Gracilis Muscle
Vastus Lateralis Muscle
Adductor Canal
Sartorius Muscle
Femur, Medial Epicondyle
Femur, Lateral Condyle
Knee Joint
Tibia
Tibialis Anterior Muscle
Tibia, Shaft
Medial Cuneiform Bone
Navicular Bone
Intermediate Cuneiform Bone
Lateral Cuneiform Bone
Flexor Digitorum Brevis Muscle
Fifth Metatarsal
Plantar Aponeurosis

Superior Sagittal Sinus
Parietal Lobe
Lateral Fissure
Medulla Oblongata
Mastoid Air Cells
Digastric Muscle, Posterior Belly
Brachial Plexus, C5 Root
Humerus
Longus Cervicis Muscle
Deltoid Muscle
Coracobrachialis Muscle
Esophagus
Biceps Brachii Muscle
Trachea
Lung, Upper Lobe
Left Ventricle
Stomach
Ulna
Descending Colon
Iliacus Muscle
Ilium
Gluteus Medius Muscle
Gluteus Minimus Muscle
Tensor Fascia Lata Muscle
Femur, Head
Sigmoid Colon
Bladder
Adductor Longus Muscle
Penis, Crus
Vastus Intermedius Muscle
Femur, Shaft
Vastus Medialis Muscle
Patellar Ligament
Tibial Tuberosity
Femur, Lateral Condyle
Abductor Hallucis Muscle
Navicular Bone
Lateral Cuneiform Bone
Quadratus Plantae Muscle
Fourth Metatarsal
Fifth Metatarsal
Abductor Digiti Minimi Muscle

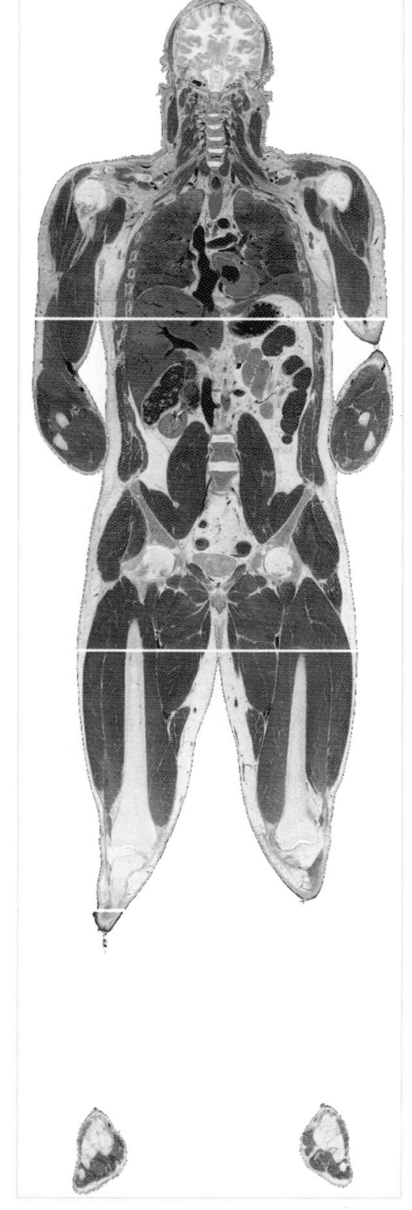

vmc3179

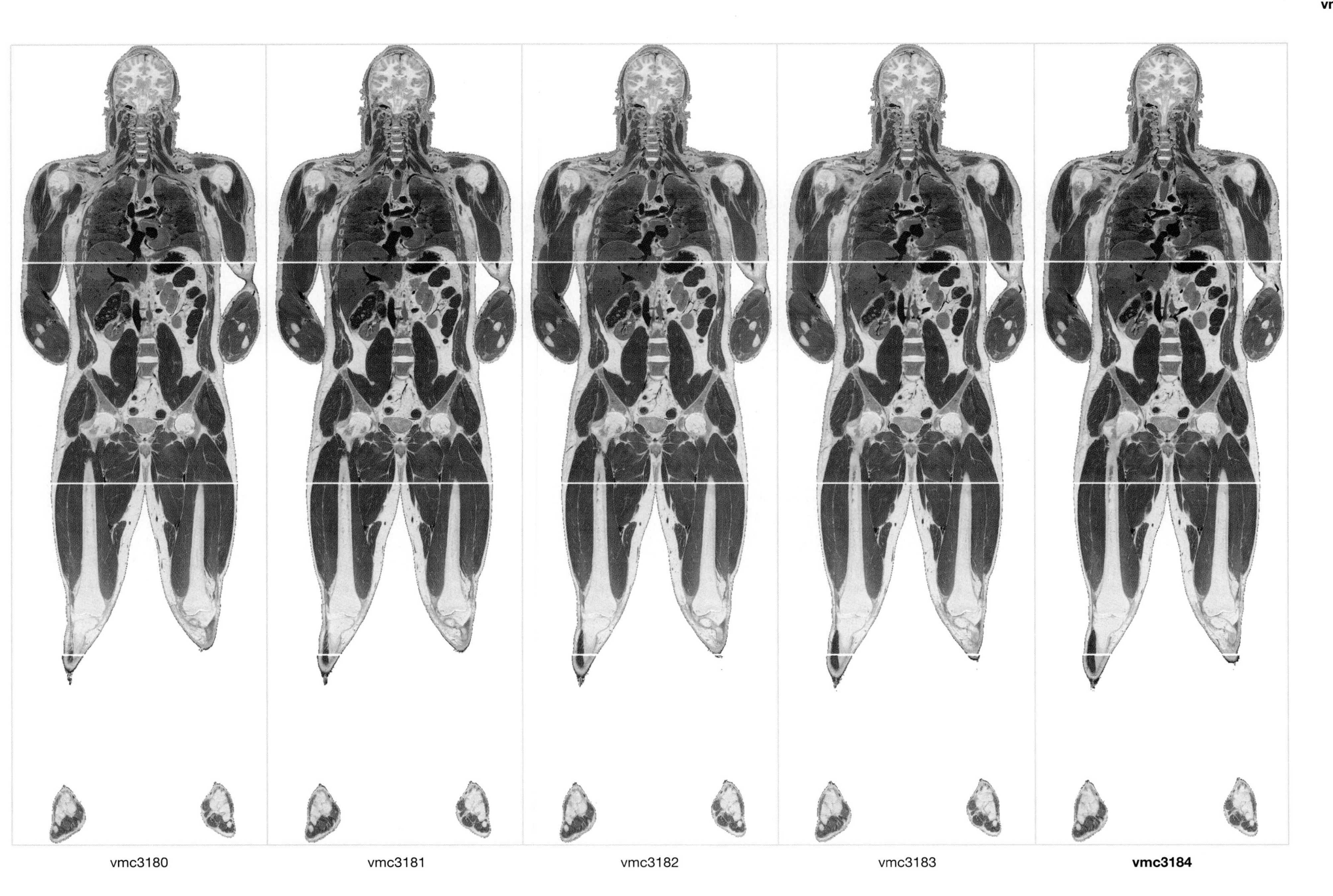
vmc3180
vmc3181
vmc3182
vmc3183
vmc3184

Coronal
vmc3190

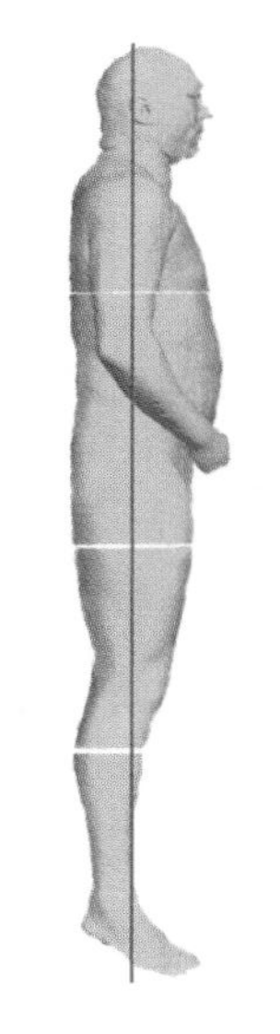

Splenium
Lateral Ventricle
Cerebellum
Sigmoid Sinus
Fourth Ventricle
Medulla Oblongata
Levator Scapulae Muscle
Shoulder Joint
Lung, Upper Lobe
Subscapularis Muscle
Pulmonary Artery
Right Ventricle
Triceps Brachii Muscle
Inferior Vena Cava
Diaphragm, Crus
Radius
Kidney
Long Flexor Muscles
L4-5, Intervertebral Disc
Femur, Head
Femur, Greater Trochanter
Obturator Internus Muscle
Iliopsoas Tendon
Obturator Externus Muscle
Bladder
Pubis, Inferior Ramus
Adductor Brevis Muscle
Gracilis Muscle
Adductor Magnus Muscle
Adductor Canal
Sartorius Muscle
Vastus Medialis Muscle
Tibial Collateral Ligament
Crural Fascia
Tibialis Anterior Muscle
Tibia, Shaft
Tibialis Anterior Tendon
Navicular Bone
Medial Cuneiform Bone
Quadratus Plantae Muscle
Fifth Metatarsal
Abductor Digiti Minimi Muscle

Parietal Bone
Superior Sagittal Sinus
Temporalis Muscle
Mastoid Air Cells
Obliquus Capitis Inferior Muscle
Sternocleidomastoid Muscle
Spinal Cord
Humerus, Head
Deltoid Muscle
Esophagus
Aorta, Arch
Lung, Oblique Fissure
Lung, Lower Lobe
Stomach
Spleen
Pancreas
Descending Colon
Renal Vein
Internal Oblique Muscle
Psoas Major Muscle
Iliac Crest
Gluteus Minimus Muscle
Gluteus Medius Muscle
Sigmoid Colon
Penis, Crus
Gracilis Muscle
Vastus Lateralis Muscle
Femoral Vessel
Sartorius Muscle
Vastus Intermedius Muscle
Femur, Shaft
Knee Joint Cavity
Knee Joint
Tibia, Condyles
Tibialis Anterior Muscle
Abductor Hallucis Muscle
Talus Bone
Lateral Cuneiform Bone
Extensor Digitorum Brevis Muscle
Cuboid Bone
Fourth Metatarsal
Fifth Metatarsal

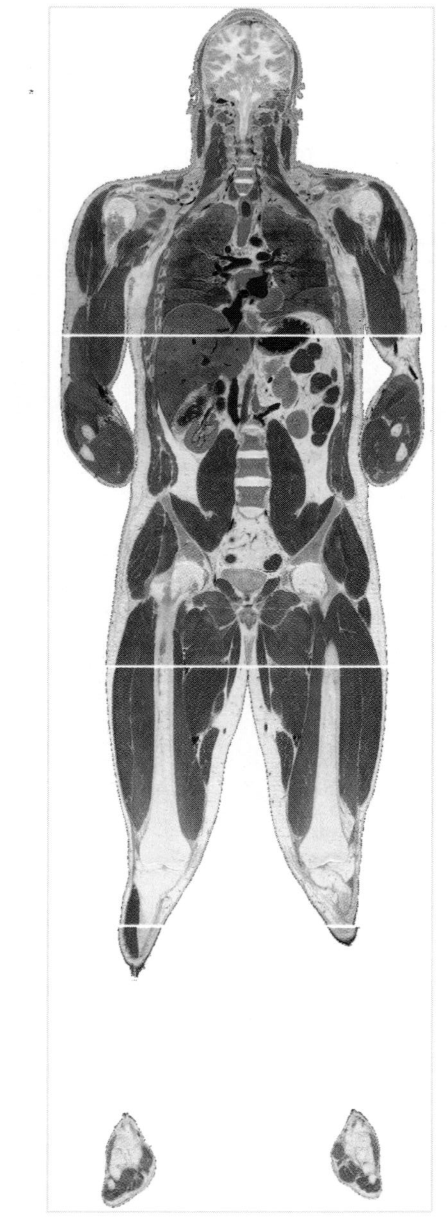

vmc3185

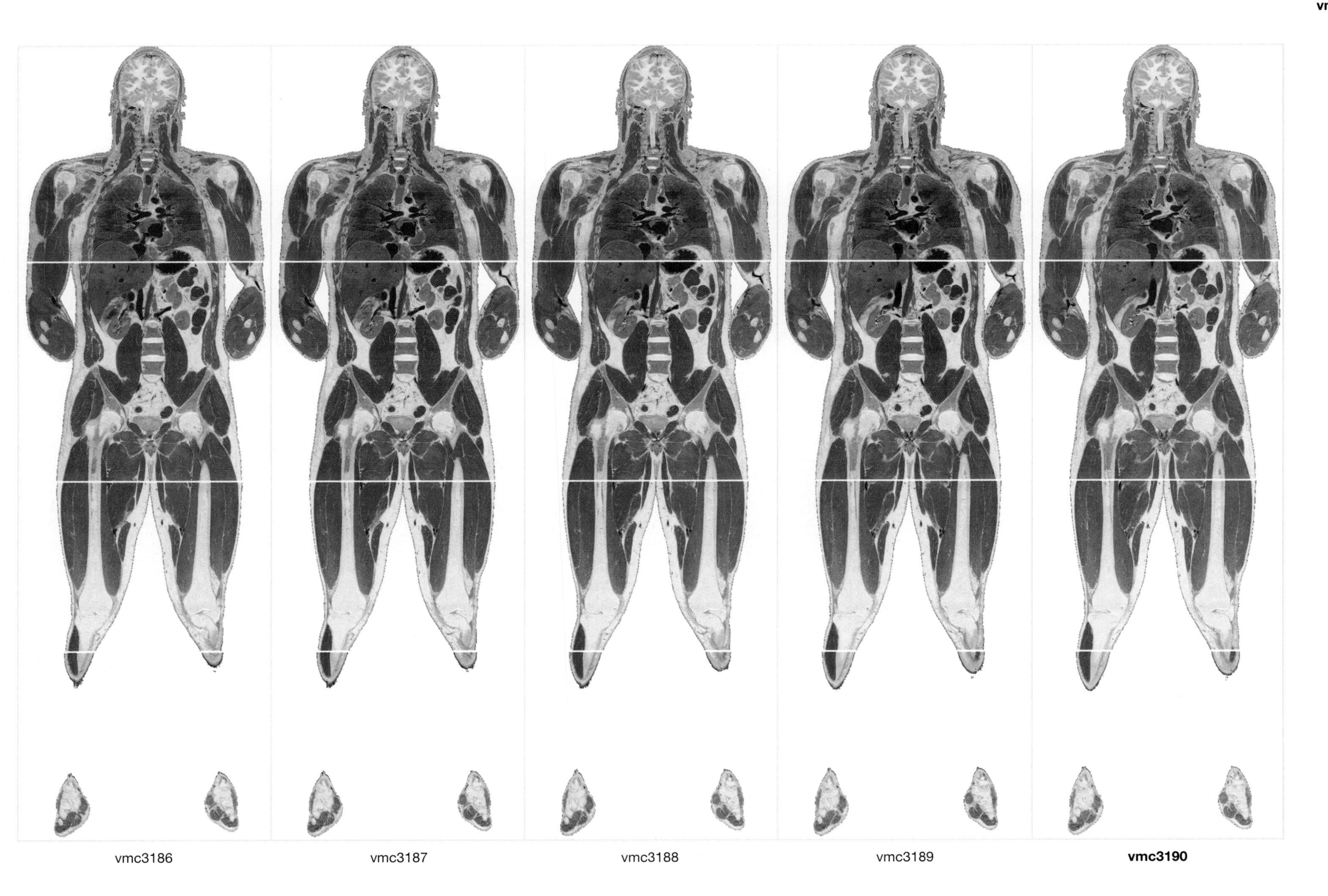

vmc3186 vmc3187 vmc3188 vmc3189 **vmc3190**

Coronal
vmc3196

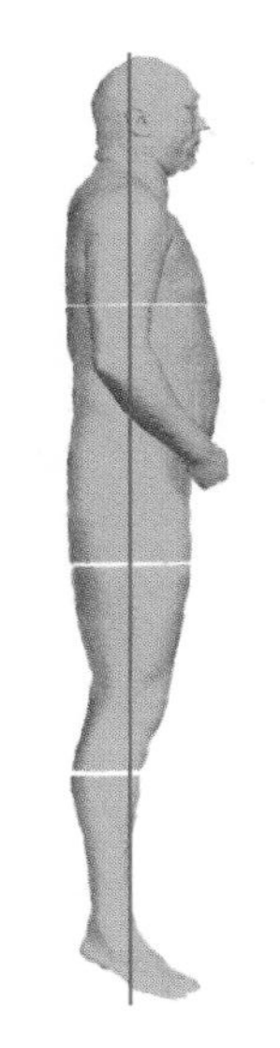

Splenium
Fornix
Temporal Lobe
Tentorium
Transverse Sinus
Trapezius Muscle
Clavicle
Scapula, Acromion
Supraspinatus Muscle
Deltoid Muscle
Subscapularis Muscle
Esophagus
Triceps Brachii Muscle
Liver, Right Lobe
Esophagus
Inferior Vena Cava
Diaphragm, Crus
Kidney
Psoas Major Muscle
Ilium
Gluteus Maximus Muscle
Gluteus Minimus Muscle
Bladder
Puborectalis Muscle
Prostate Gland
Bulbospongiosus Muscle
Femur, Linea Aspera
Sartorius Muscle
Vastus Intermedius Muscle
Femur, Lateral Condyle
Anterior Cruciate Ligament
Tibia, Lateral Condyle
Tibialis Anterior Muscle
Crural Fascia
Tibia, Shaft
Tibialis Anterior Tendon
Extensor Hallucis Longus Tendon
Talus Bone
Navicular Bone
Extensor Digitorum Brevis Muscle
Abductor Hallucis Muscle
Flexor Digiti Minimi Muscle

Superior Sagittal Sinus
Falx Cerebri
Lateral Ventricle
Temporalis Muscle
Cerebellum
Sternocleidomastoid Muscle
Levator Scapulae Muscle
Spinal Cord
Humerus
Trachea, Carina
Primary Bronchus
Lung, Oblique Fissure
Stomach
Spleen
Pancreas
Radius
Ulna
Transversus Abdominis Muscle
L4-5, Intervertebral Disc
Iliacus Muscle
Femur, Greater Trochanter
Femur, Neck
Obturator Internus Muscle
Obturator Externus Muscle
Obturator Membrane
Urogenital Diaphragm
Ischiocavernosus Muscle
Vastus Lateralis Muscle
Gracilis Muscle
Adductor Magnus Muscle
Vastus Medialis Muscle
Lateral Meniscus
Femur, Medial Condyle
Tibia, Medial Condyle
Tibia, Shaft
Crural Fascia
Tibia
Tibia, Medial Malleolus
Talus Bone
Extensor Digitorum Brevis Muscle
Quadratus Plantae Muscle
Flexor Digitorum Brevis Muscle

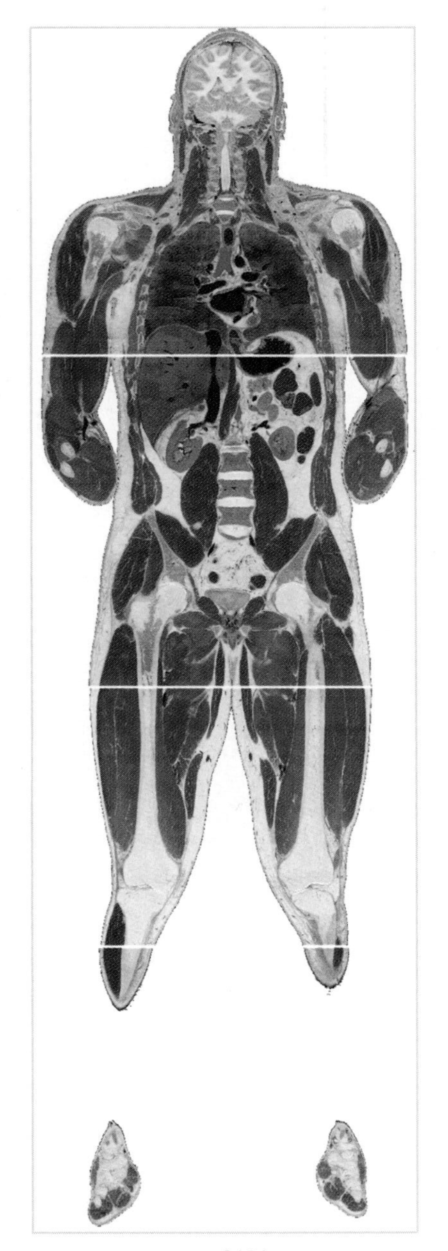

vmc3191

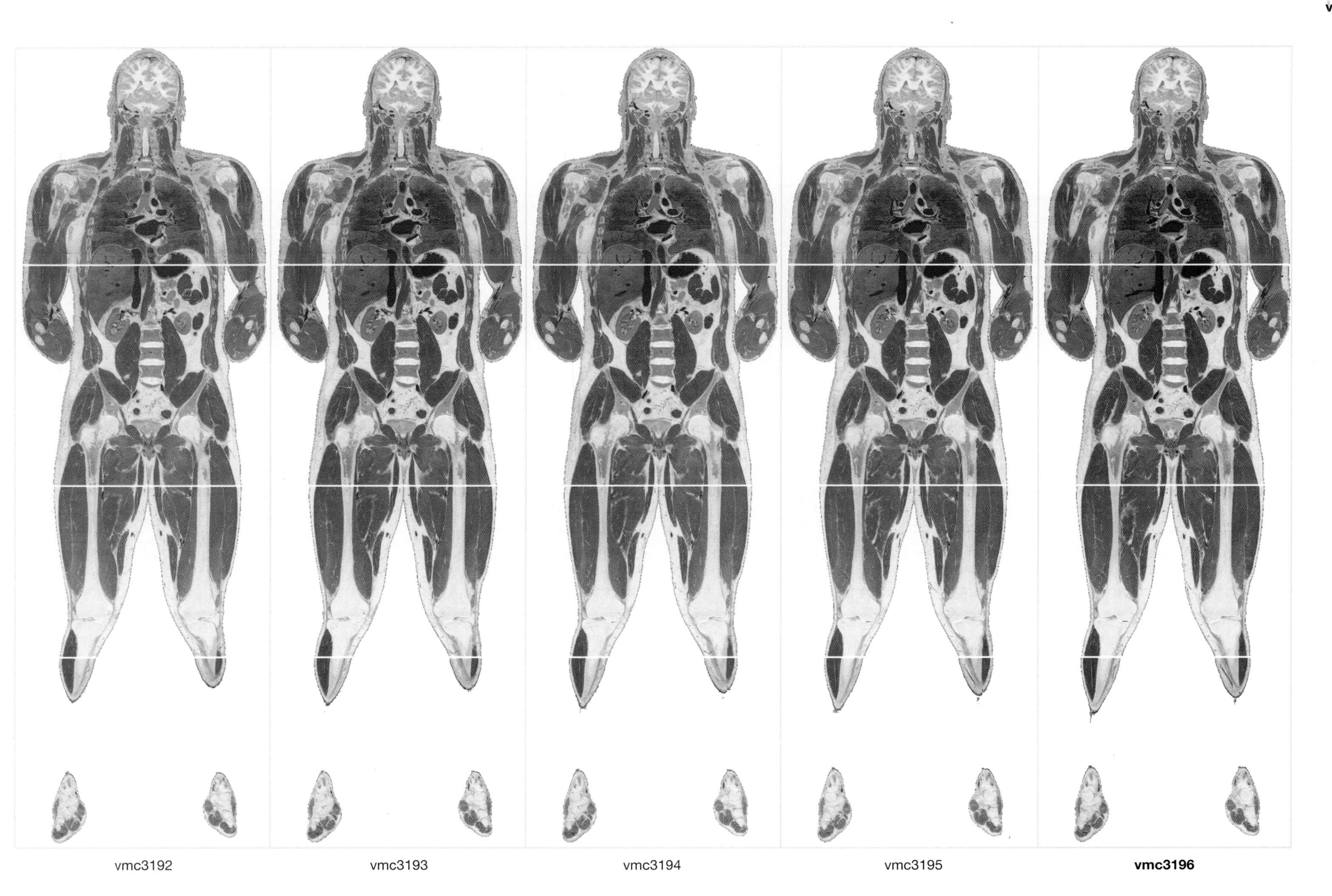

vmc3192 vmc3193 vmc3194 vmc3195 **vmc3196**

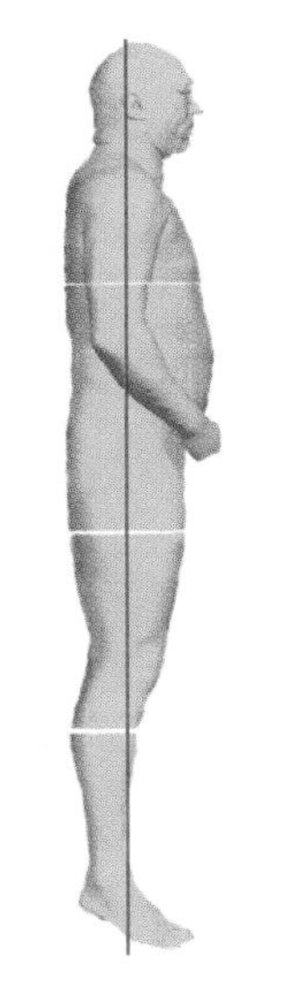

Parietal Bone
Parietal Lobe
Transverse Sinus
Rectus Capitis Posterior Major Muscle
Obliquus Capitis Inferior Muscle
Trapezius Muscle
Levator Scapulae Muscle
Subscapularis Muscle
Teres Major Muscle
Lung, Upper Lobe
Triceps Brachii Muscle
Liver, Right Lobe
Latissimus Dorsi Muscle
Kidney
Psoas Major Muscle
Flexor Carpi Ulnaris Muscle
Ilium
Gluteus Minimus Muscle
Gluteus Medius Muscle
Ischium
Prostate Gland
Obturator Membrane
Bulbospongiosus Muscle
Fascia Lata
Adductor Magnus Muscle
Sartorius Muscle
Biceps Femoris Muscle, Short Head
Femur, Medial Condyle
Anterior Cruciate Ligament
Tibia, Medial Condyle
Tibialis Anterior Muscle
Extensor Digitorum Longus Muscle
Tibia, Shaft
Crural Fascia
Abductor Hallucis Muscle
Quadratus Plantae Muscle
Extensor Hallucis Brevis Muscle
Extensor Digitorum Brevis Muscle
Flexor Digitorum Brevis Muscle
Abductor Digiti Minimi Muscle

Superior Sagittal Sinus
Longitudinal Cerebral Fissure
Tentorium
Longissimus Capitis Muscle
Sternocleidomastoid Muscle
Semispinalis Cervicis Muscle
Deltoid Muscle
Spinal Cord
Humerus
Rib 4-5, Intercostal Muscles
Biceps Brachii Muscle
Stomach
Spleen
Colon, Splenic Flexure
Radius
Ulna
Internal Oblique Muscle
Iliacus Muscle
Sigmoid Colon
Femur, Head
Femur, Greater Trochanter
Obturator Internus Muscle
Obturator Externus Muscle
Adductor Brevis Muscle
Vastus Lateralis Muscle
Gracilis Muscle
Femur, Shaft
Vastus Intermedius Muscle
Femur, Medial Epicondyle
Femur, Lateral Condyle
Lateral Meniscus
Tibia, Lateral Condyle
Medial Meniscus
Tibia, Shaft
Tibia, Medial Malleolus
Peroneus Tertius Muscle
Flexor Digitorum Longus Tendon
Flexor Hallucis Longus Tendon
Cuboid Bone
Extensor Digitorum Brevis Muscle
Fifth Metatarsal

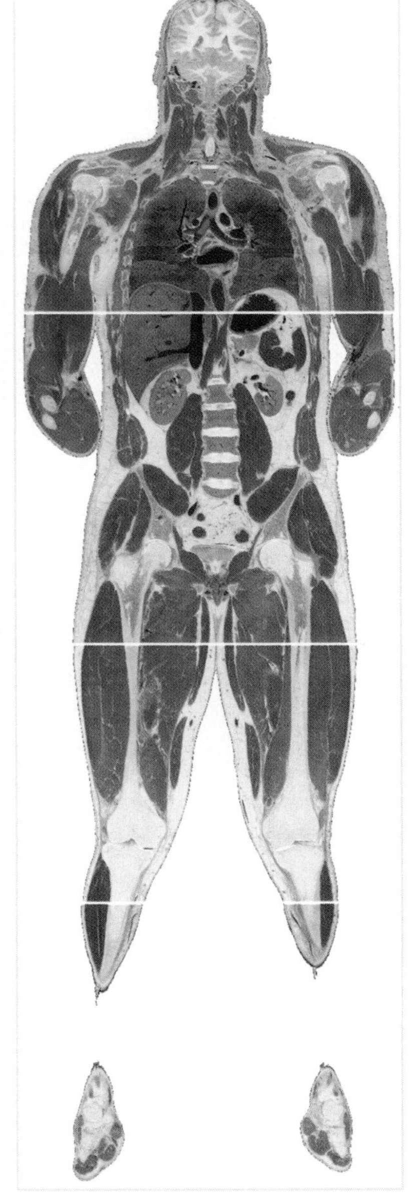

vmc3197

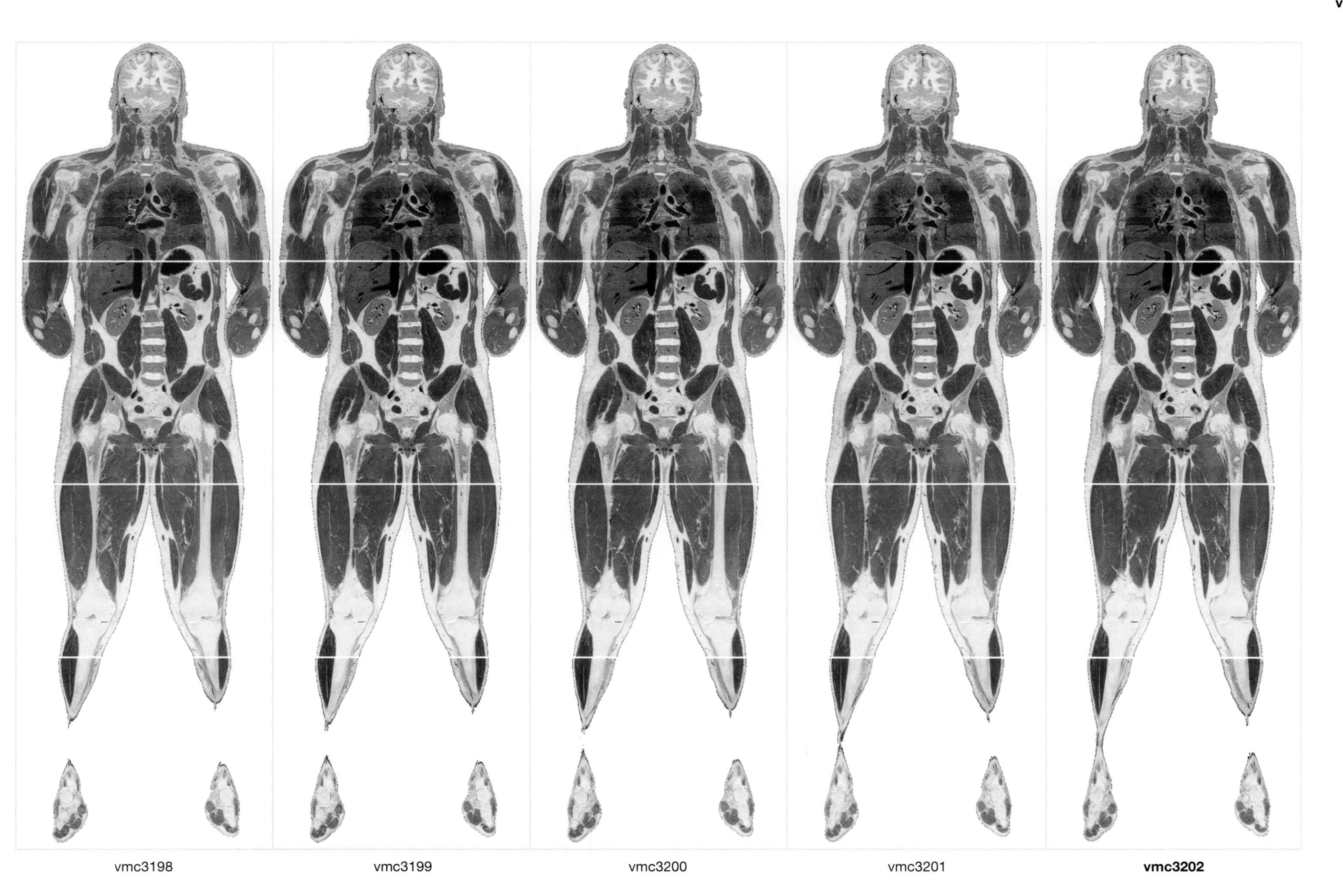
vmc3198
vmc3199
vmc3200
vmc3201
vmc3202

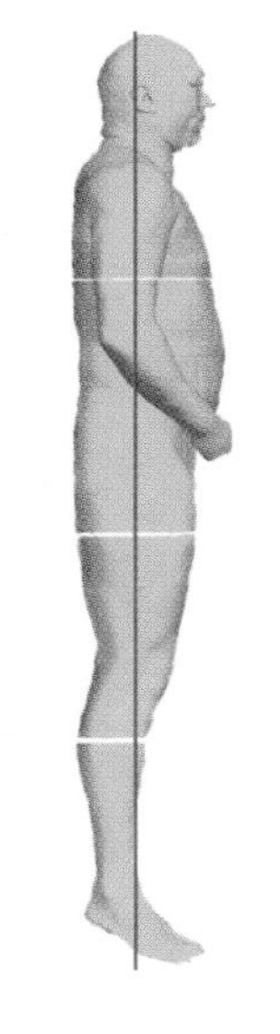

Parietal Bone
Central Sulcus
Temporal Lobe
Transverse Sinus
Cerebellum, Vermis
Splenius Capitis Muscle
Trapezius Muscle
Levator Scapulae Muscle
Subscapularis Muscle
Teres Major Muscle
Thoracic Aorta
Latissimus Dorsi Muscle
Liver, Right Lobe
Diaphragm, Crus
Kidney
L2-3, Intervertebral Disc
Flexor Carpi Ulnaris Muscle
Psoas Major Muscle
Common Iliac Vein
Gluteus Minimus Muscle
Gluteus Maximus Muscle
Prostate Gland
Pubococcygeus Muscle
Bulbospongiosus Muscle
Adductor Magnus Muscle
Sartorius Muscle
Popliteal Vessel
Biceps Femoris Muscle
Femur, Medial Condyle
Anterior Cruciate Ligament
Tibia, Medial Condyle
Tibialis Anterior Muscle
Soleus Muscle
Tibia
Tibialis Anterior Tendon
Abductor Hallucis Muscle
Peroneus Tertius Tendon
Quadratus Plantae Muscle
Extensors Hallucis and Digitorum Brevis Muscles
Flexor Digitorum Brevis Muscle
Abductor Digiti Minimi Muscle

Superior Sagittal Sinus
Falx Cerebri
Longitudinal Cerebral Fissure
Longissimus Capitis Muscle
Sternocleidomastoid Muscle
Semispinalis Cervicis Muscle
Supraspinatus Muscle
Spinal Cord
Pulmonary Artery
Humerus
Bronchus
Biceps Brachii Muscle
Stomach
Pancreas, Tail
Spleen
Radius
Ulna
Transversus Abdominis Muscle
Iliacus Muscle
Ilium
Sigmoid Colon
Femur, Head
Obturator Externus Muscle
Adductor Brevis Muscle
Ischiocavernosus Muscle
Vastus Lateralis Muscle
Gracilis Muscle
Adductor Magnus Muscle
Vastus Intermedius Muscle
Femur, Lateral Condyle
Lateral Meniscus
Medial Meniscus
Gastrocnemius Muscle
Great Saphenous Vein
Extensor Digitorum Longus Muscle
Tibia, Medial Malleolus
Peroneus Tertius Muscle
Flexor Digitorum Longus Tendon
Talus Bone
Extensor Digitorum Brevis Muscle
Cuboid Bone
Fifth Metatarsal

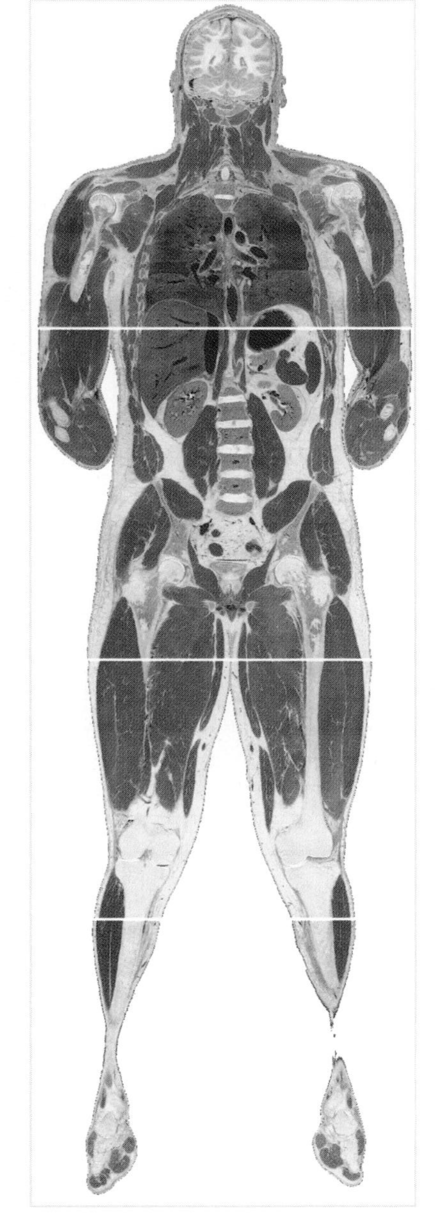

vmc3203

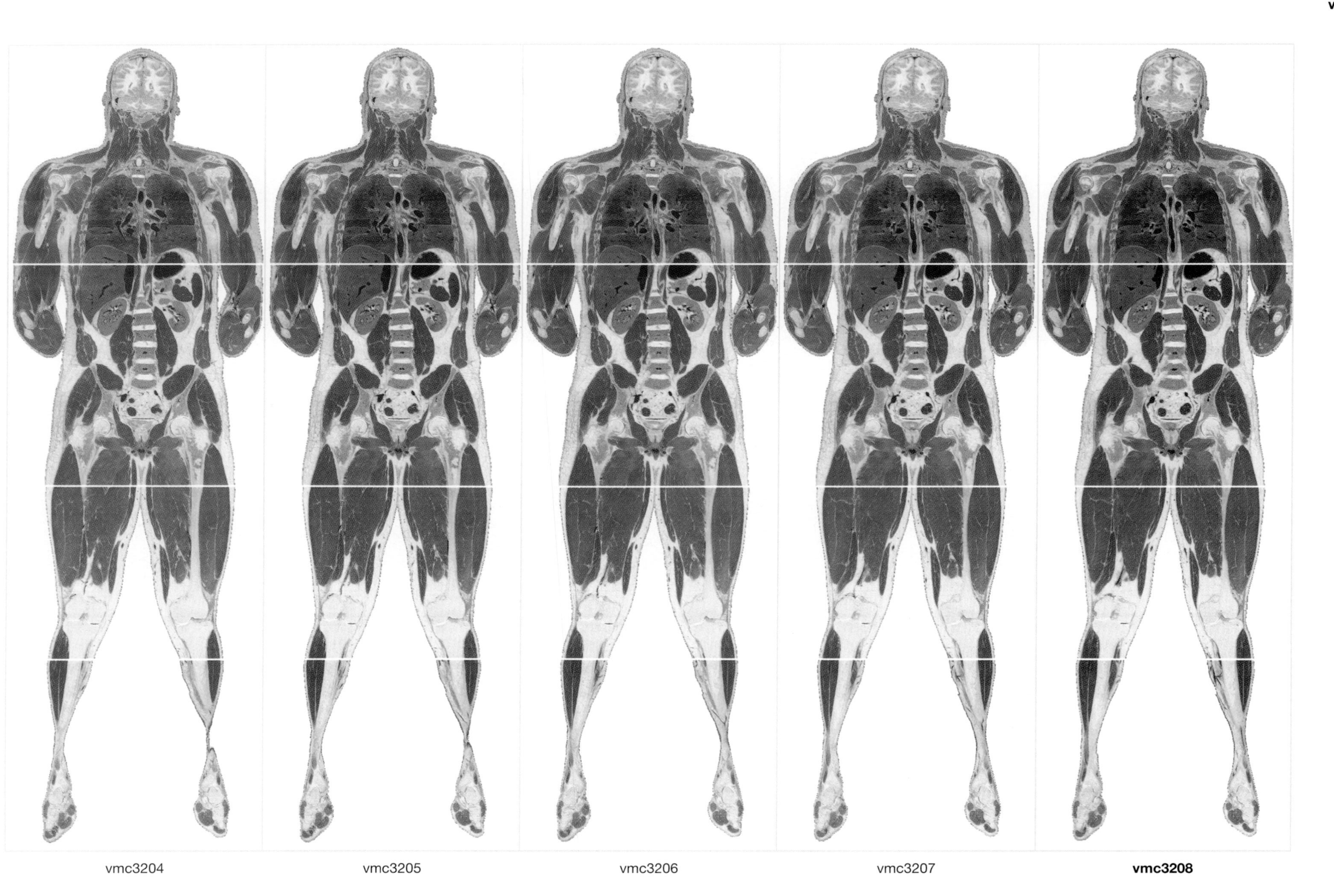
vmc3204
vmc3205
vmc3206
vmc3207
vmc3208

Parietal Bone
Central Sulcus
Temporal Bone
Rectus Capitis Posterior Major Muscle
Obliquus Capitis Inferior Muscle
Semispinalis Capitis Muscle
Trapezius Muscle
Levator Scapulae Muscle
Scapula
Teres Minor Muscle
Serratus Anterior Muscle
Esophagus
Azygos Vein
Brachioradialis Muscle
T12, Vertebral Body
Kidney, Adipose Capsule
Quadratus Lumborum Muscle
Subarachnoid Space
Sacrum
Gluteus Minimus Muscle
Gluteus Maximus Muscle
Seminal Vesicle
Prostate Gland
Bulbospongiosus Muscle
Adductor Magnus Muscle
Semimembranosus Muscle
Sartorius Muscle
Biceps Femoris Muscle
Femur, Medial Condyle
Anterior Cruciate Ligament
Tibia, Medial Condyle
Tibialis Posterior Muscle
Gastrocnemius Muscle
Extensor Hallucis Longus Muscle
Tibialis Anterior Muscle
Abductor Hallucis Muscle
Quadratus Plantae Muscle
Fibula, Lateral Malleolus
Extensors Hallucis and Digitorum Brevis Muscles
Flexor Digitorum Brevis Muscle
Abductor Digiti Minimi Muscle

Superior Sagittal Sinus
Falx Cerebri
Tentorium
Cerebellum, Lateral Hemisphere
Longissimus Capitis Muscle
Sternocleidomastoid Muscle
Semispinalis Cervicis Muscle
T1, Transverse Process
Lung, Upper Lobe
Rib 5-6, Intercostal Muscles
Pulmonary Vein
Biceps Brachii Muscle
Stomach
Pancreas, Tail
Spleen
Radius
Ulna
External Oblique Muscle
Sacroiliac Joint
Ilium
Obturator Internus Muscle
Femur, Greater Trochanter
Obturator Externus Muscle
Ischiocavernosus Muscle
Vastus Lateralis Muscle
Gracilis Muscle
Adductor Magnus Muscle
Vastus Intermedius Muscle
Femur, Lateral Condyle
Lateral Meniscus
Medial Meniscus
Great Saphenous Vein
Popliteus Muscle
Extensor Digitorum Longus Muscle
Peroneus Longus Muscle
Tibia
Peroneus Tertius Muscle
Flexor Digitorum Longus Tendon
Flexor Hallucis Longus Tendon
Cuboid Bone
Fifth Metatarsal

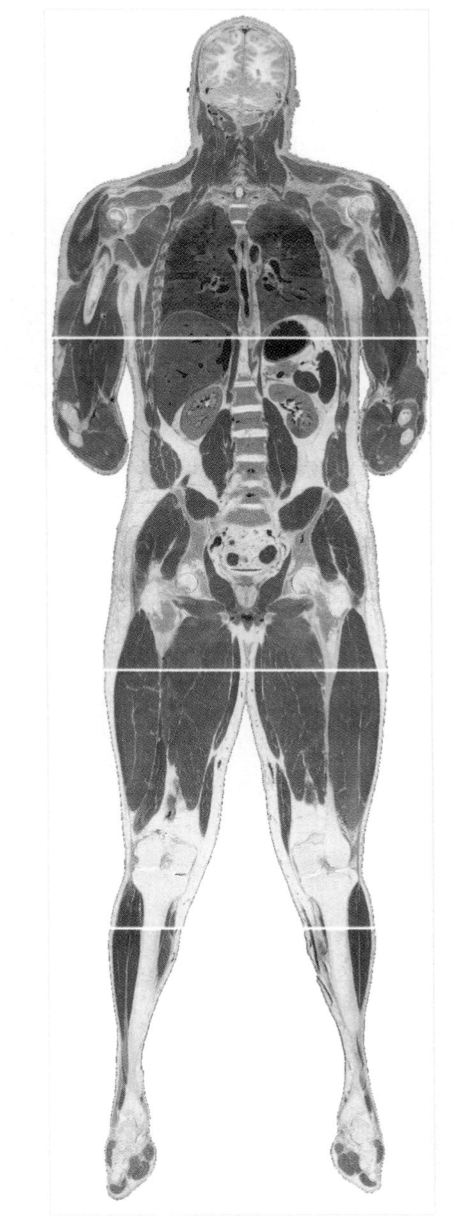

vmc3209

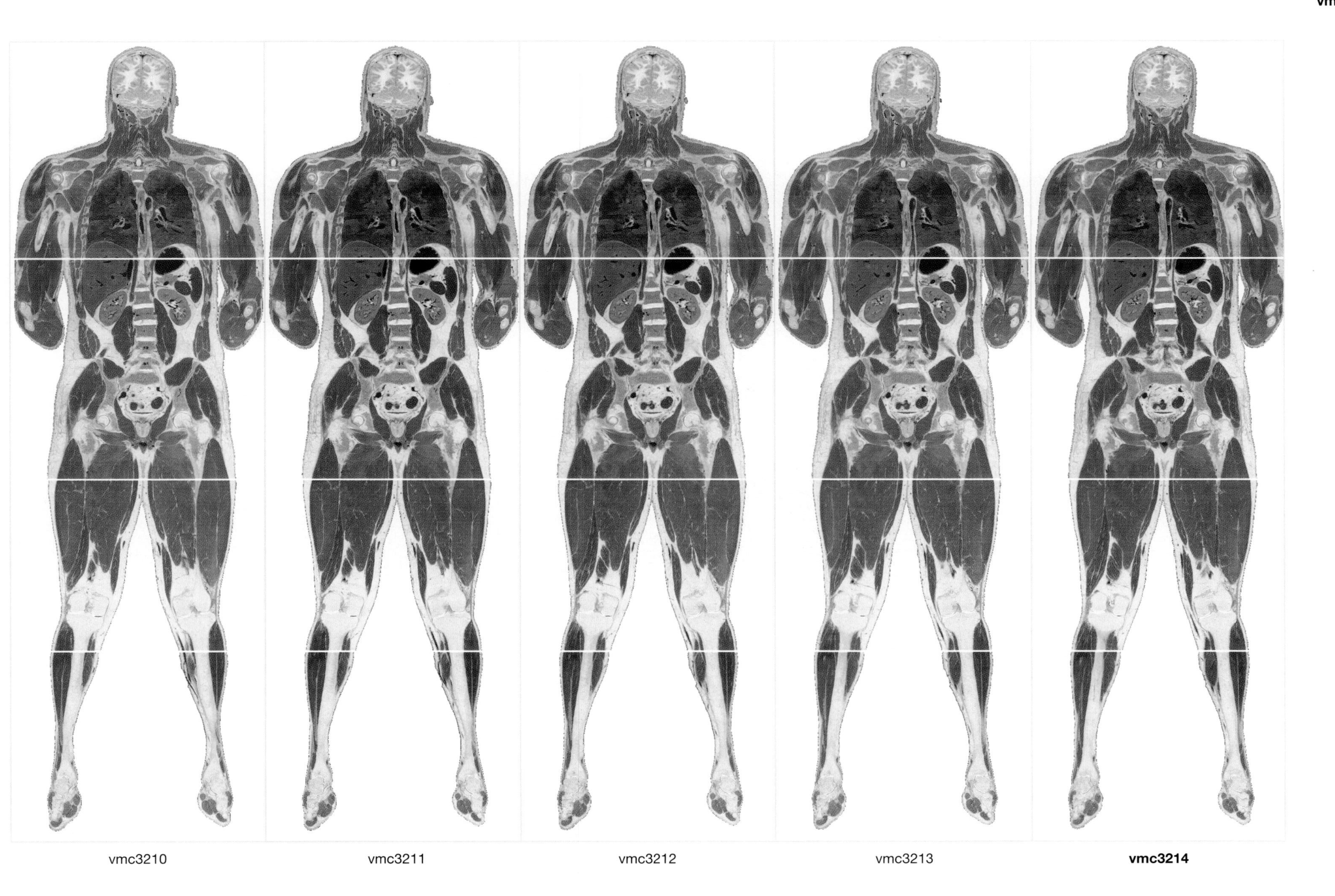

vmc3210 vmc3211 vmc3212 vmc3213 **vmc3214**

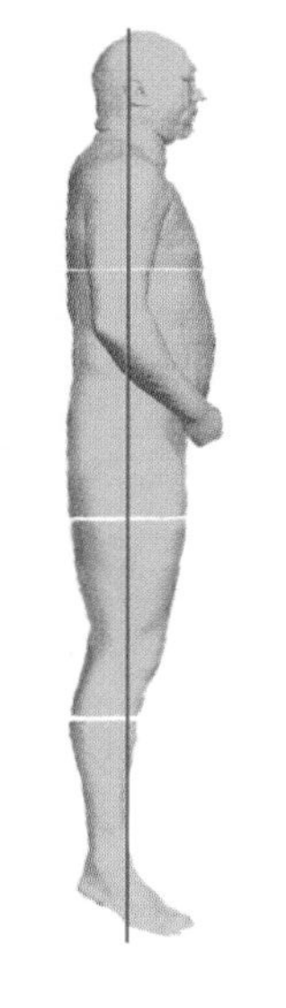

Temporal Bone
Calcarine Sulcus
Tentorium
Transverse Sinus
Rectus Capitis Posterior Major M.
Trapezius Muscle
Supraspinatus Muscle
Infraspinatus Muscle
Deltoid Muscle
Lung, Upper Lobe
Serratus Anterior Muscle
Latissimus Dorsi Muscle
Diaphragm
Liver, Right Lobe
Brachialis Muscle
Humerus, Capitulum
Radius
L1-2, Intervertebral Disc
Psoas Major Muscle
Quadratus Lumborum Muscle
Sacrum, Lateral Wing
Sacroiliac Joint
Internal Iliac Vein
Sigmoid Colon
Seminal Vesicle
Prostate Gland
Gracilis Muscle
Vastus Lateralis Muscle
Adductor Magnus Muscle
Semimembranosus Muscle
Biceps Femoris Muscle
Gastrocnemius Muscle, Lateral Head
Femur, Lateral Condyle
Tibialis Posterior Muscle
Soleus Muscle
Gastrocnemius Muscle, Medial Head
Peroneus Brevis Muscle
Tibia, Shaft
Peroneus Tertius Muscle
Talus Bone
Abductor Hallucis Muscle
Flexor Digiti Minimi Muscle

Superior Sagittal Sinus
Longitudinal Cerebral Fissure
Occipital Lobe
Cerebellum
Rectus Capitis Posterior Minor M.
Semispinalis Cervicis Muscle
Trapezius Muscle
Spinal Cord
Scapula
Teres Major Muscle
Humerus
Thoracic Aorta
Stomach
Spleen
Kidney
Ulna
Internal Oblique Muscle
Cauda Equina
Gluteus Medius Muscle
Ilium, Body
Gluteus Minimus Muscle
Femur, Greater Trochanter
Pubis
Obturator Externus Muscle
Ischiocavernosus Muscle
Bulbospongiosus Muscle
Vastus Lateralis Muscle
Biceps Femoris Muscle, Short Head
Adductor Magnus Muscle
Gracilis Muscle
Popliteal Vessel
Femur, Lateral Condyle
Posterior Cruciate Ligament
Sartorius Tendon
Peroneus Longus Muscle
Peroneus Brevis Muscle
Tibia, Shaft
Tibia, Medial Malleolus
Ankle Joint
Extensor Digitorum Brevis Muscle
Cuboid Bone
Flexor Digitorum Brevis Muscle

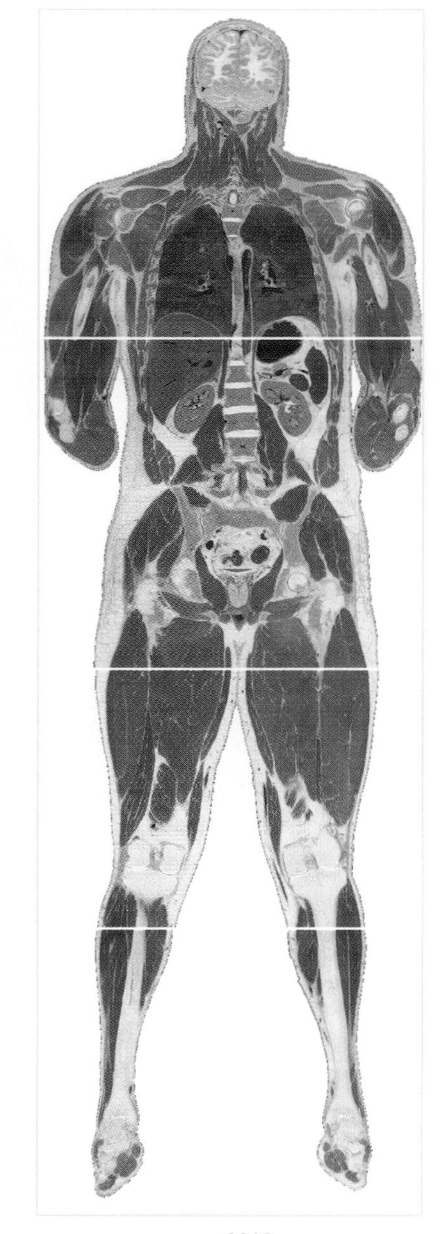

vmc3215

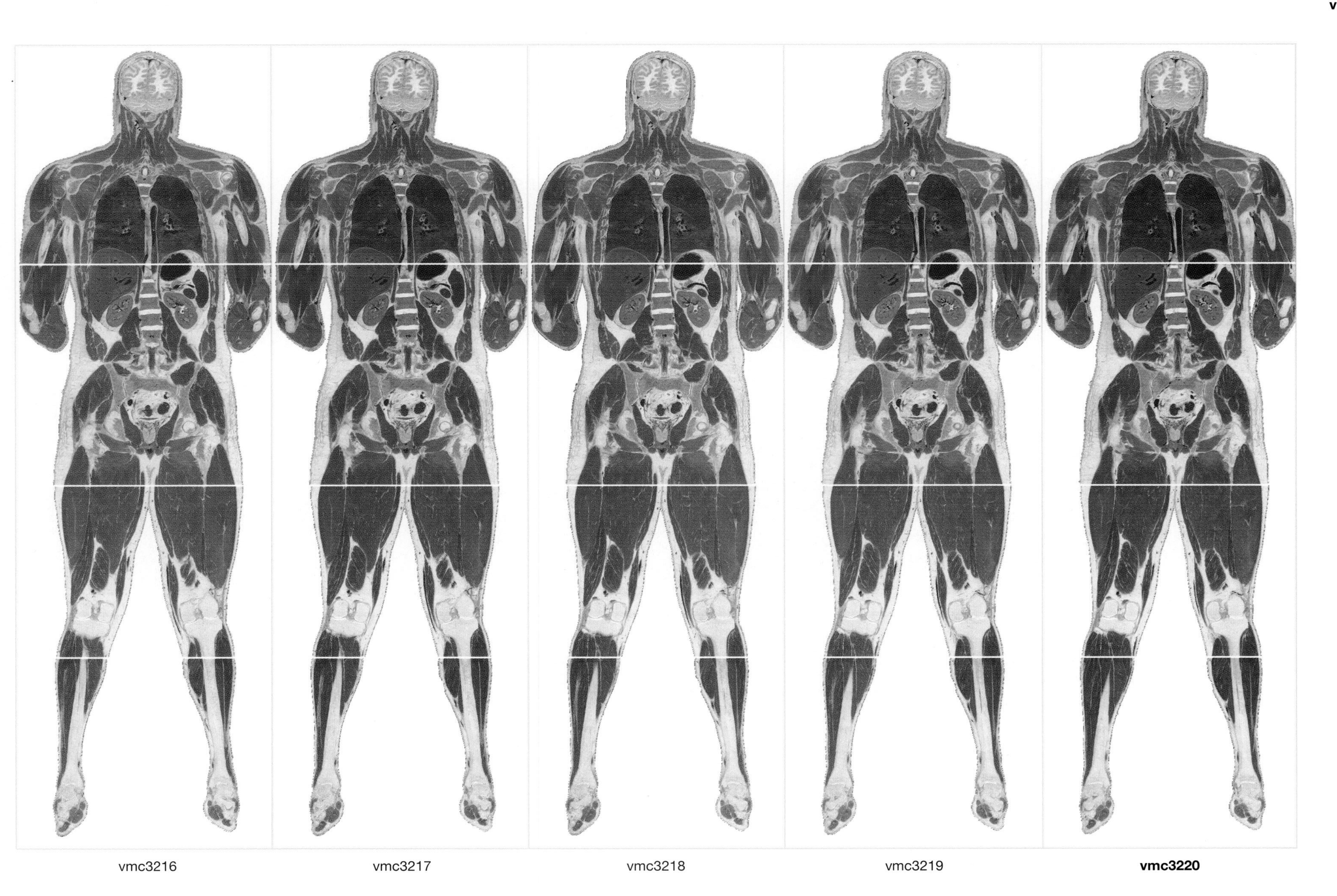

vmc3216 vmc3217 vmc3218 vmc3219 **vmc3220**

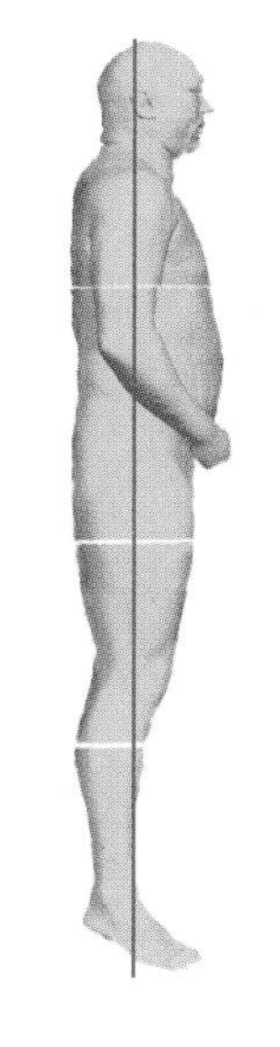

Calcarine Sulcus
Tentorium
Transverse Sinus
Cerebellum
Rectus Capitis Posterior Minor M.
Splenius Capitis Muscle
Semispinalis Capitis Muscle
Levator Scapulae Muscle
Infraspinatus Muscle
Subscapularis Muscle
Teres Major Muscle
Latissimus Dorsi Muscle
Serratus Anterior Muscle
Humerus
Diaphragm
Liver, Right Lobe
Renal Fascia
Psoas Major Muscle
L1-2, Intervertebral Disc
Quadratus Lumborum Muscle
Ilium
Sacroiliac Joint
Internal Iliac Vein
Pubis
Levator Ani Muscle
Ischiocavernosus Muscle
Iliotibial Tract
Vastus Lateralis Muscle
Sciatic Nerve
Biceps Femoris Muscle
Semimembranosus Muscle
Lateral Meniscus
Fibula, Head
Popliteus Muscle
Great Saphenous Vein
Tibialis Posterior Muscle
Peroneus Longus Muscle
Tibia, Shaft
Tibia, Medial Malleolus
Ankle Joint
Calcaneus Bone
Plantar Aponeurosis

Superior Sagittal Sinus
Parietal Lobe
Occipital Lobe
Rectus Capitis Posterior Major M.
Trapezius Muscle
Supraspinatus Muscle
Spinal Cord
Deltoid Muscle
Scapula
Subscapularis Muscle
Lung, Upper Lobe
Lung, Oblique Fissure
Thoracic Aorta
Stomach
Spleen
Radius
Ulna
Adrenal Gland
Kidney
External Oblique Muscle
Iliacus Muscle
Gluteus Medius Muscle
Gluteus Minimus Muscle
S1-2, Intervertebral Disc
Sigmoid Colon
Obturator Externus Muscle
Seminal Vesicle
Vastus Lateralis Muscle
Obturator Internus Muscle
Prostatic Urethra
Adductor Magnus Muscle
Gracilis Muscle
Semimembranosus Muscle
Femur, Lateral Condyle
Posterior Cruciate Ligament
Sartorius Muscle
Gastrocnemius Muscle, Medial Head
Talus Bone
Quadratus Plantae Muscle
Extensor Digitorum Brevis Muscle
Flexor Digitorum Brevis Muscle
Flexor Digiti Minimi Muscle

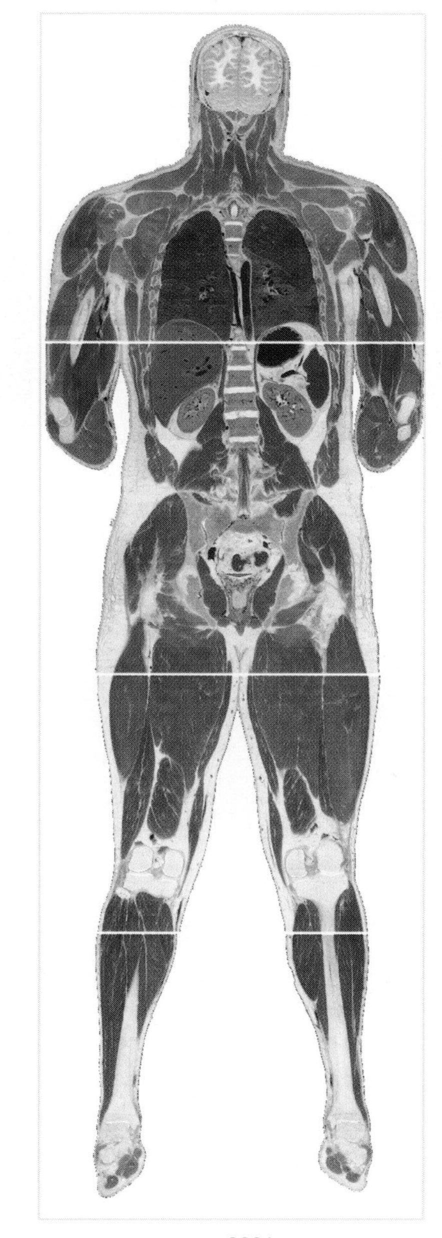

vmc3221

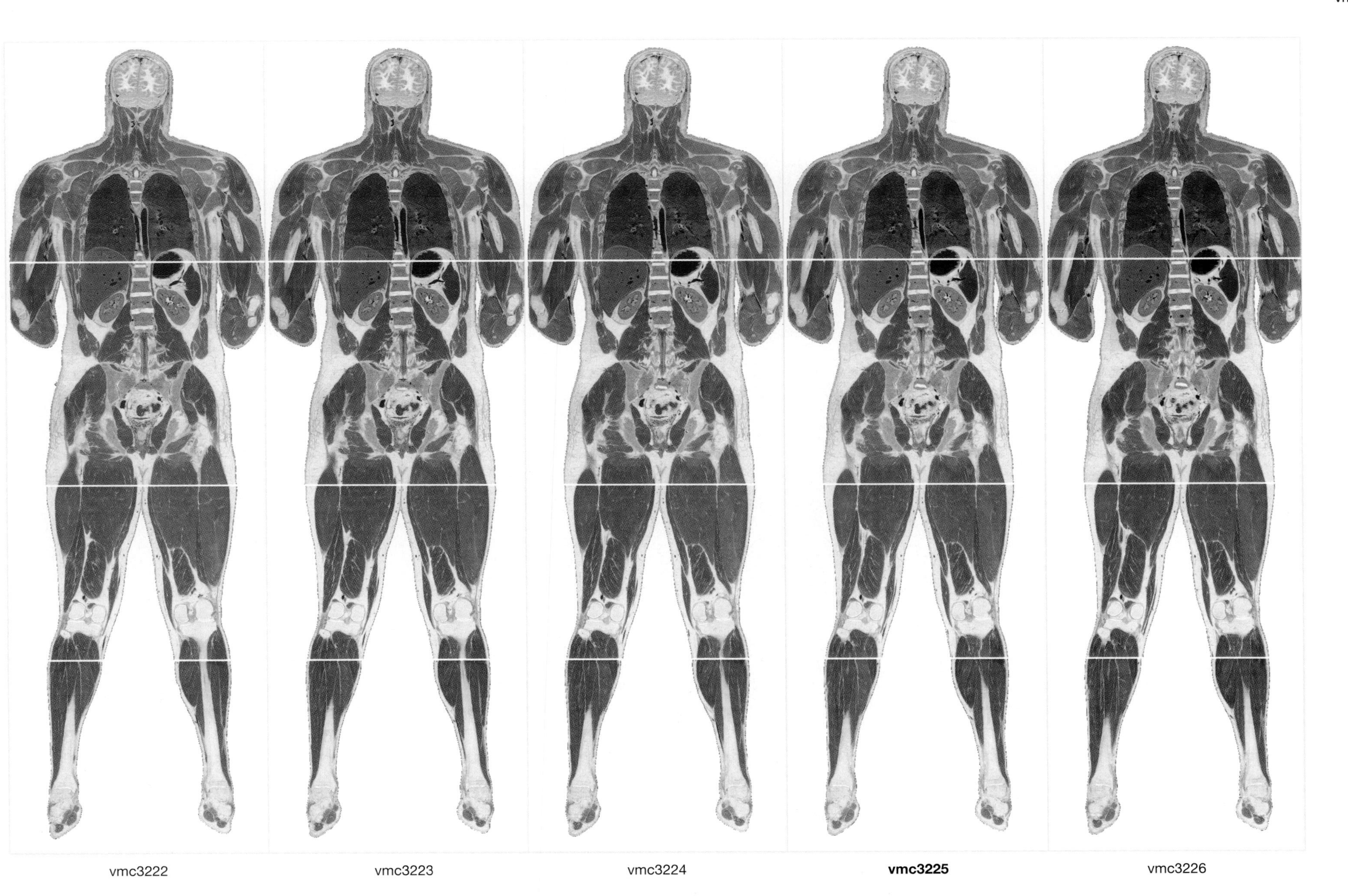
vmc3222
vmc3223
vmc3224
vmc3225
vmc3226

Coronal
vmc3231

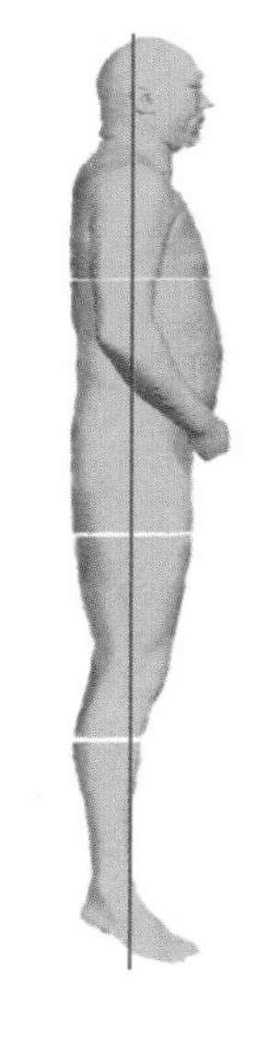

Parietal Bone
Calcarine Sulcus
Transverse Sinus
Occipital Bone
Splenius Capitis Muscle
Trapezius Muscle
Supraspinatus Muscle
Deltoid Muscle
Infraspinatus Muscle
Lung, Lower Lobe
Diaphragm
Humerus
Liver, Right Lobe
Brachialis Muscle
Humerus, Trochlea
Kidney
Vertebral Canal
Sacrum, Lateral Wing
Gluteus Medius Muscle
Levator Ani Muscle
Quadratus Femoris Muscle
Adductor Magnus Muscle
Gracilis Muscle
Semimembranosus Muscle
Sciatic Nerve
Biceps Femoris Muscle
Sartorius Muscle
Femur, Medial Condyle
Popliteus Muscle
Fibula, Head
Gastrocnemius Muscle, Lateral Head
Gastrocnemius Muscle, Medial Head
Soleus Muscle
Peroneus Longus Muscle
Flexor Hallucis Longus Muscle
Peroneus Brevis Muscle
Tibia
Talus Bone
Quadratus Plantae Muscle
Calcaneus Bone
Abductor Digiti Minimi Muscle
Plantar Aponeurosis

Superior Sagittal Sinus
Parietal Lobe
Occipital Lobe
Tentorium
Cerebellum
Semispinalis Capitis Muscle
Trapezius Muscle
Scapula, Spine
Subscapularis Muscle
Spinal Cord
Teres Major Muscle
Lung, Oblique Fissure
Lung, Lower Lobe
Serratus Anterior Muscle
Brachioradialis Muscle
Stomach
Spleen
External Oblique Muscle
Kidney
Internal Oblique Muscle
T12-L1, Intervertebral Disc
Subarachnoid Space
Ilium
Gluteus Maximus Muscle
Inferior Gemellus Muscle
Obturator Internus Muscle
Ischiorectal Fossa
Anus
Urogenital Diaphragm
Vastus Lateralis Muscle
Adductor Magnus Muscle
Semimembranosus Muscle
Popliteal Vessel
Tibia, Lateral Condyle
Popliteus Muscle
Gastrocnemius Muscle, Medial Head
Soleus Muscle
Talus Bone
Quadratus Plantae Muscle
Calcaneus Bone
Flexor Digitorum Brevis Muscle
Abductor Digiti Minimi Muscle

vmc3227

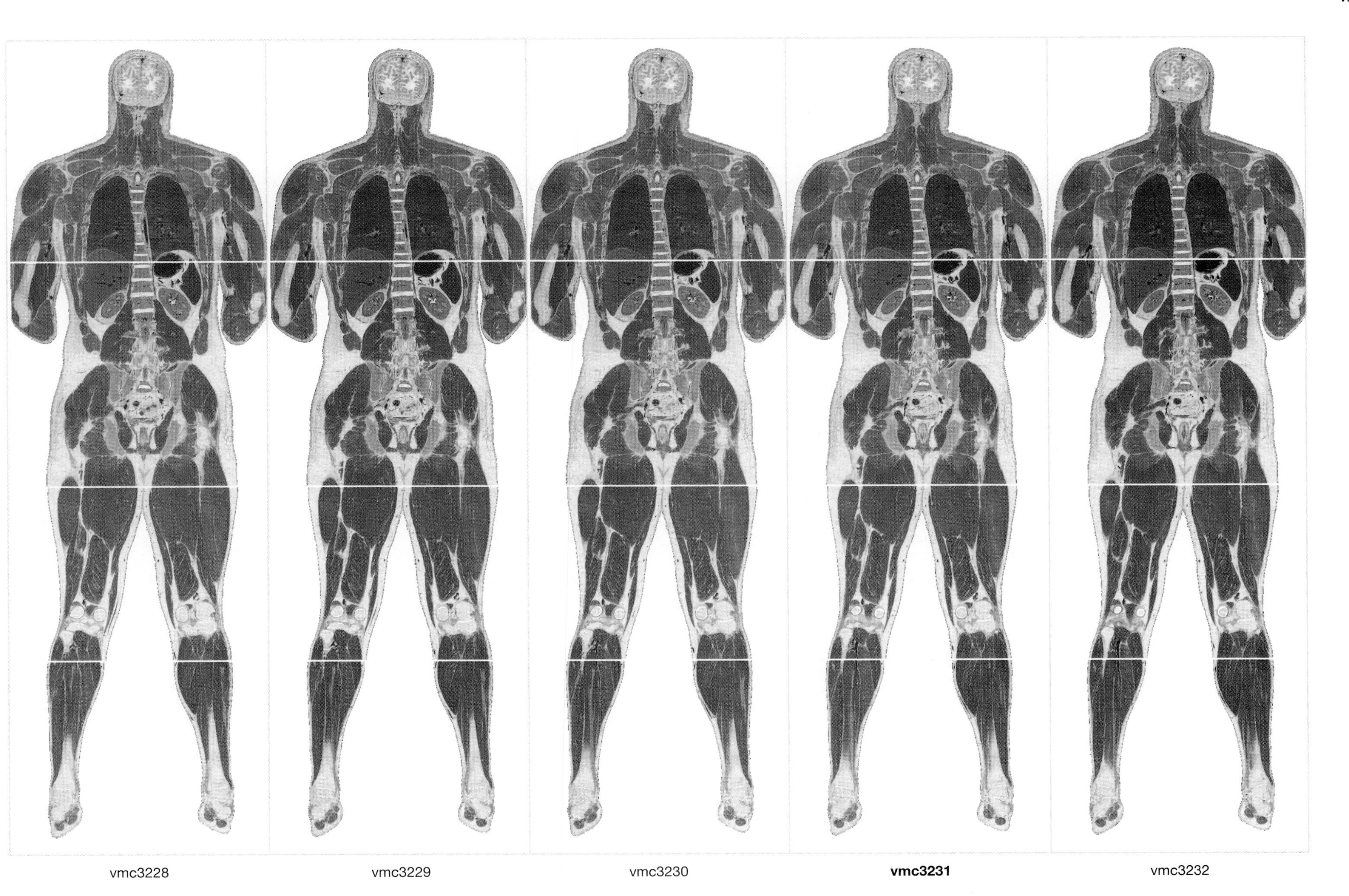

vmc3228 vmc3229 vmc3230 **vmc3231** vmc3232

Coronal
vmc3237

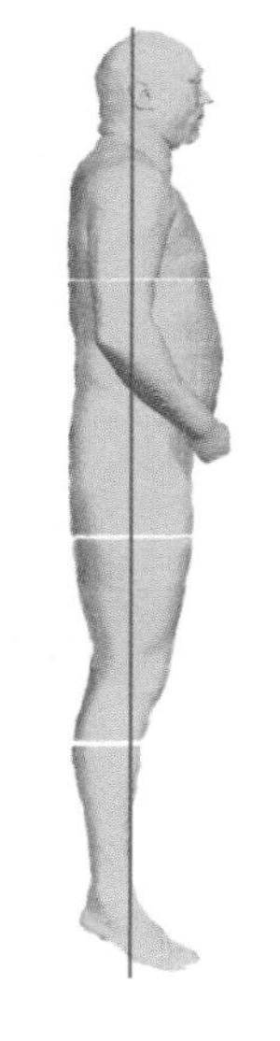

Longitudinal Cerebral Fissure
Calcarine Sulcus
Straight Sinus
Transverse Sinus
Cerebellum
Splenius Capitis Muscle
Trapezius Muscle
Supraspinatus Muscle
Infraspinatus Muscle
Spinal Cord
Serratus Anterior Muscle
Lung, Lower Lobe
Diaphragm
Liver, Right Lobe
External Oblique Muscle
Humerus, Trochlea
Ulna, Coronoid Process
Kidney
Psoas Major Muscle
Subarachnoid Space
Quadratus Lumborum Muscle
Spinalis Muscle
Gluteus Medius Muscle
Sacrum
Piriformis Muscle
Obturator Internus Tendon
Obturator Internus Muscle
Vastus Lateralis Muscle
Ischium
Ischiorectal Fossa
Biceps Femoris Muscle, Long Head
Gracilis Muscle
Sartorius Muscle
Small Saphenous Vein
Popliteal Vessels
Gastrocnemius Muscle, Medial Head
Soleus Muscle
Tibia
Fibula, Lateral Malleolus
Talus Bone
Calcaneus Bone
Plantar Aponeurosis

Superior Sagittal Sinus
Parietal Bone
Parietal Lobe
Tentorium
Occipitofrontalis Muscle, Occipital Belly
Semispinalis Capitis Muscle
Scapula, Spine
Deltoid Muscle
Subscapularis Muscle
Teres Minor Muscle
Teres Major Muscle
Triceps Brachii Muscle
Lung, Oblique Fissure
Humerus
Latissimus Dorsi Muscle
Spleen
Stomach
Internal Oblique Muscle
Multifidus Muscle
Ilium
Sacral Foramen
Levator Ani Muscle
Anus
Quadratus Femoris Muscle
Central Tendinous Point
Adductor Magnus Muscle
Semitendinosus Muscle
Semimembranosus Muscle
Biceps Femoris Muscle
Gracilis Muscle
Femur, Lateral Condyle
Tibia, Lateral Condyle
Fibula, Head
Popliteus Muscle
Gastrocnemius Muscle, Lateral Head
Peroneus Longus Muscle
Tibia
Long Saphenous Vein
Fibula, Lateral Malleolus
Quadratus Plantae Muscle
Abductor Digiti Minimi Muscle
Flexor Digitorum Brevis Muscle

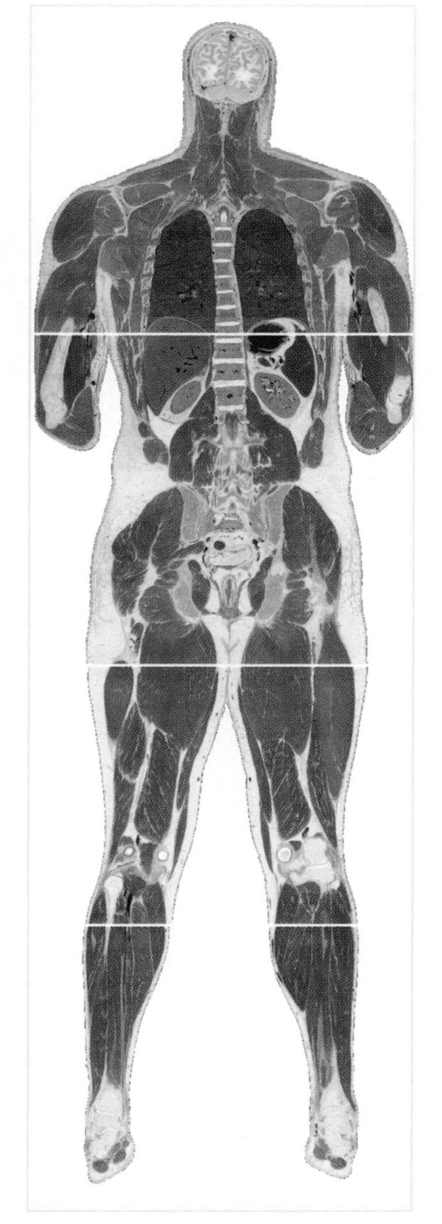

vmc3233

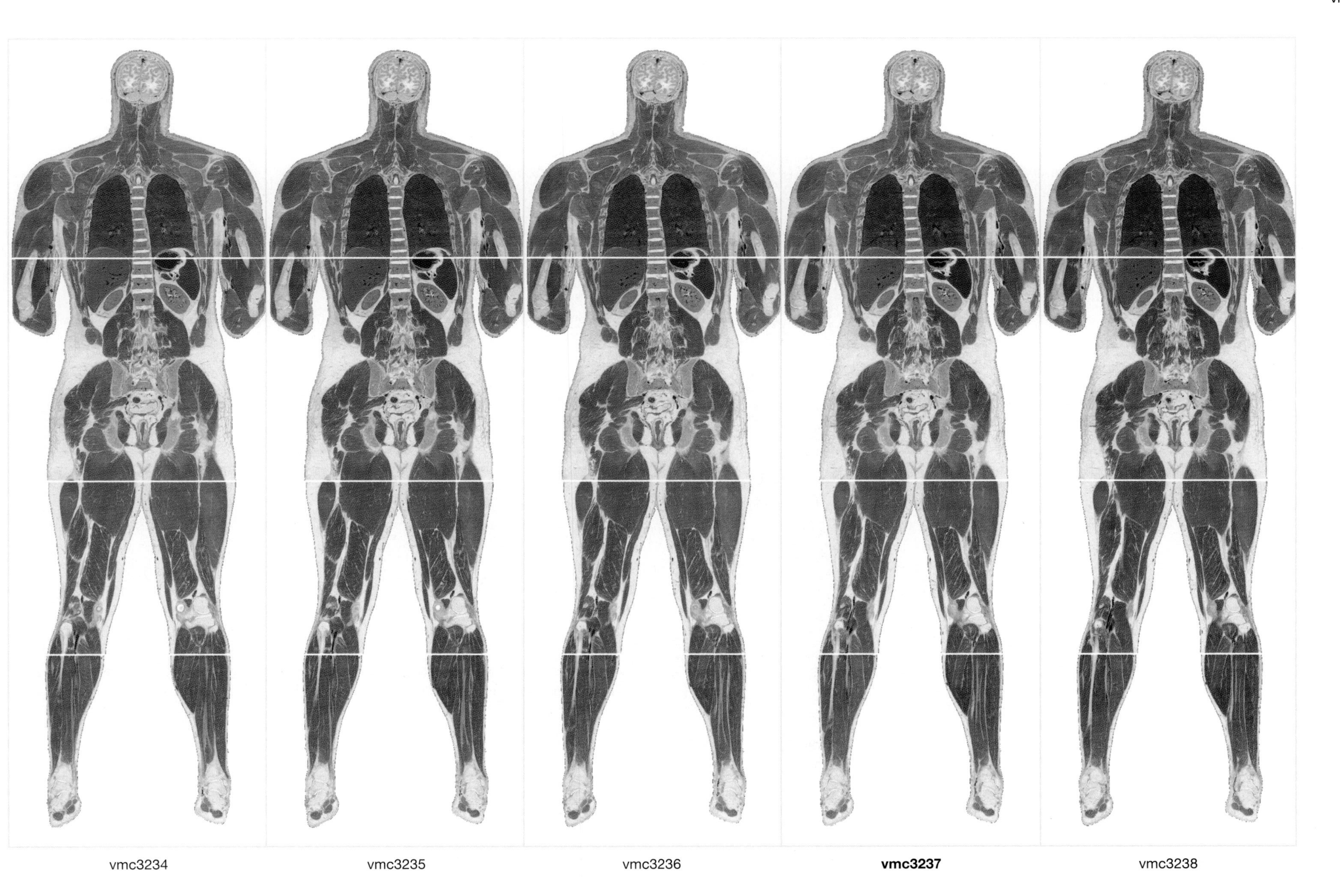

vmc3234 vmc3235 vmc3236 **vmc3237** vmc3238

Coronal
vmc3242

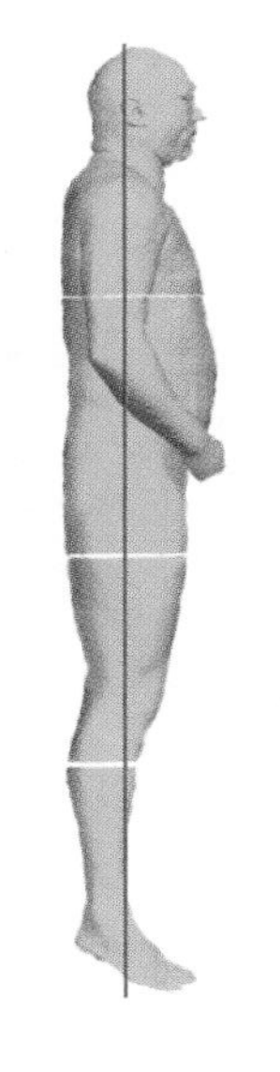

Superior Sagittal Sinus
Calcarine Sulcus
Occipitofrontalis Muscle, Occipital Belly
Straight Sinus
Splenius Capitis Muscle
Levator Scapulae Muscle
Subscapularis Muscle
Infraspinatus Muscle
Deltoid Muscle
Teres Major Muscle
Latissimus Dorsi Muscle
Triceps Brachii Muscle
Serratus Anterior Muscle
Liver, Right Lobe
Olecranon Process
Humerus, Trochlea
External Oblique Muscle
Quadratus Lumborum Muscle
L2, Transverse Process
Longissimus Muscle
Ilium
Gluteus Maximus Muscle
Sacrum
Sigmoid Colon
Obturator Internus Muscle
Ischiorectal Fossa
Vastus Lateralis Muscle
Biceps Femoris Muscle, Short Head
Biceps Femoris Muscle, Long Head
Semitendinosus Muscle
Gracilis Muscle
Small Saphenous Vein
Sartorius Muscle
Gastrocnemius Muscle, Medial Head
Soleus Muscle
Peroneus Longus Muscle
Fibula, Shaft
Flexor Hallucis Longus Muscle
Peroneus Brevis Muscle
Fibula, Lateral Malleolus
Ankle Joint
Quadratus Plantae Muscle

Occipital Bone
Longitudinal Cerebral Fissure
Occipital Lobe
Splenius Capitis Muscle
Semispinalis Capitis Muscle
Trapezius Muscle
Supraspinatus Muscle
Spinal Cord
Lung, Oblique Fissure
Lung, Lower Lobe
Latissimus Dorsi Muscle
Serratus Anterior Muscle
Diaphragm
Spleen
Kidney
Subarachnoid Space
L3, Spinous Process
Quadratus Lumborum Muscle
Multifidus Muscle
Gluteus Medius Muscle
Anus
Gluteus Maximus Muscle
Levator Ani Muscle
Adductor Magnus Muscle
Vastus Lateralis Muscle
Semitendinosus Muscle
Semimembranosus Muscle
Biceps Femoris Muscle
Femur, Lateral Condyle
Fibula, Head
Gastrocnemius Muscle, Lateral Head
Gastrocnemius Muscle, Medial Head
Soleus Muscle
Flexor Hallucis Longus Muscle
Tibia
Ankle Joint
Talus Bone
Fibula, Lateral Malleolus
Calcaneus Bone
Abductor Hallucis Muscle
Abductor Digiti Minimi Muscle
Quadratus Plantae Muscle

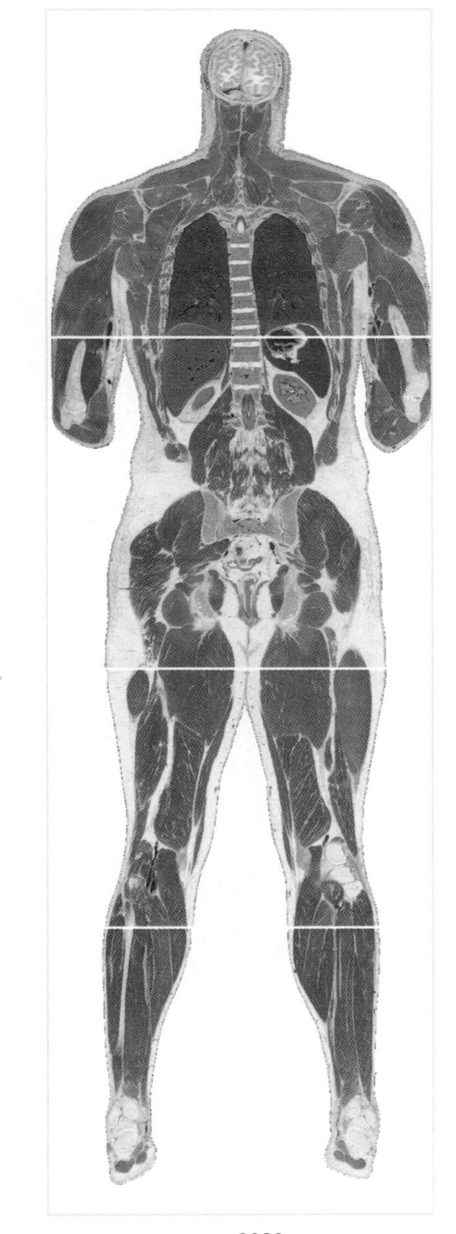

vmc3239

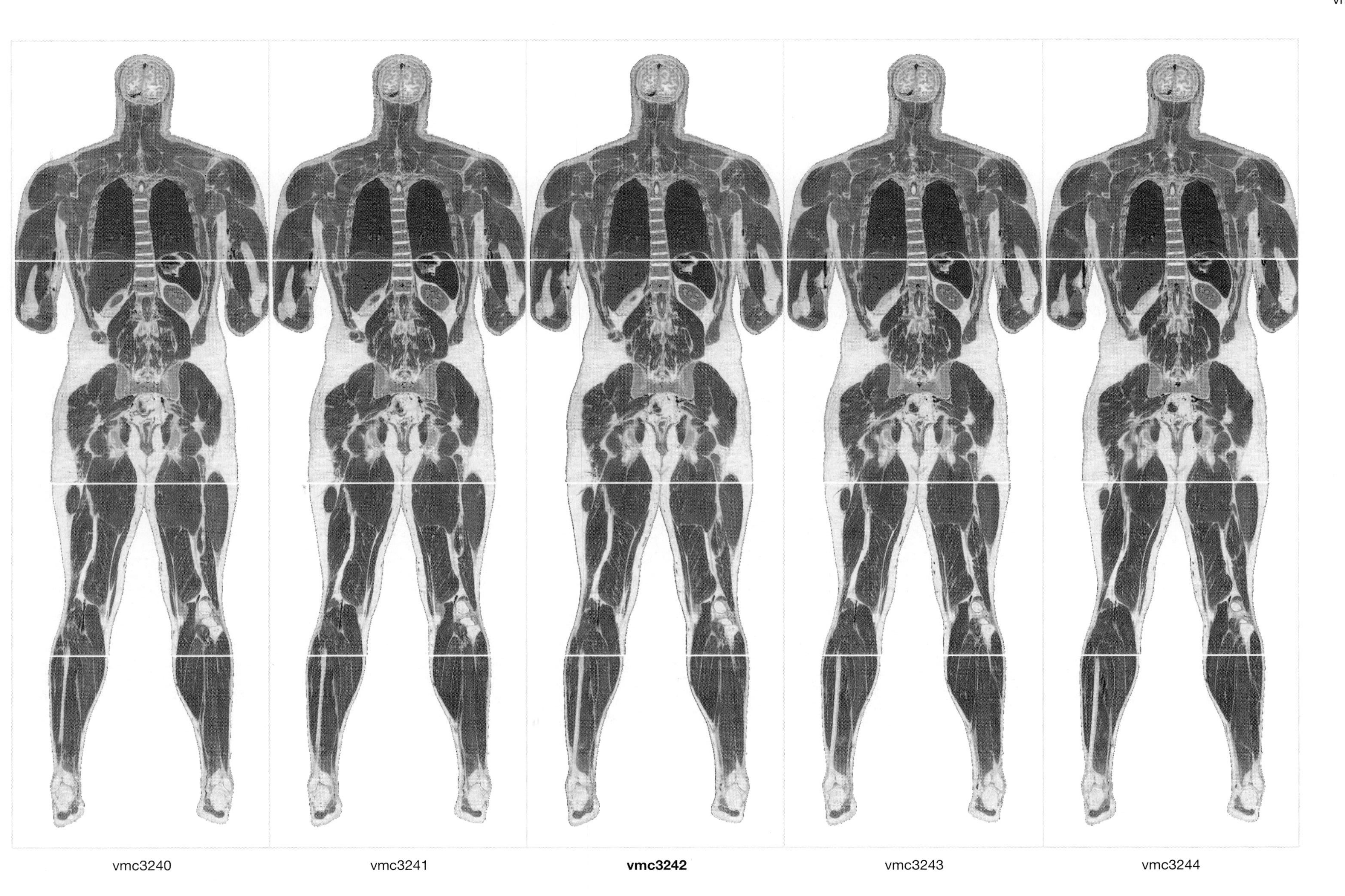

vmc3240 vmc3241 **vmc3242** vmc3243 vmc3244

Coronal
vmc3248

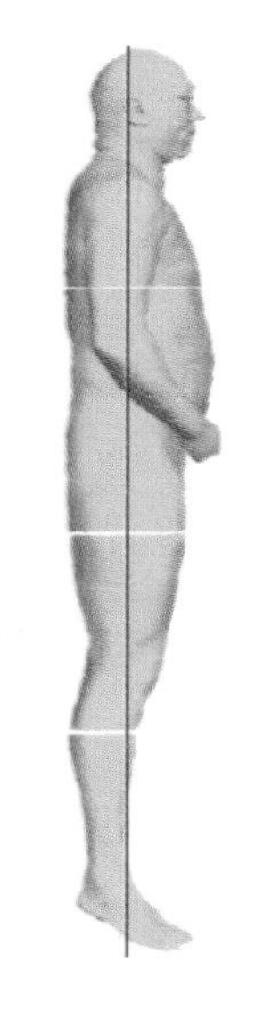

Superior Sagittal Sinus
Occipital Bone
Occipital Vessel
Trapezius Muscle
Levator Scapulae Muscle
Multifidus Muscle
Deltoid Muscle
T6-7, Intervertebral Disc
T7, Vertebra
Lung, Lower Lobe
Serratus Anterior Muscle
Triceps Brachii Muscle
Latissimus Dorsi Muscle
Liver, Right Lobe
Humerus, Medial Condyle
Ulna
T12, Vertebra
Spinal Cord
Iliocostalis Muscle
Longissimus Muscle
Gluteus Medius Muscle
Gluteus Maximus Muscle
Sigmoid Colon
Ischial Tuberosity
Ischiorectal Fossa
Anus
Adductor Magnus Muscle
Biceps Femoris Muscle
Gracilis Muscle
Biceps Femoris Tendon
Great Saphenous Vein
Gastrocnemius Muscle, Medial Head
Soleus Muscle
Peroneus Longus Muscle
Peroneus Brevis Muscle
Fibula, Shaft
Flexor Hallucis Longus Muscle
Crural Fascia
Tibia, Medial Malleolus
Ankle Joint
Abductor Digiti Minimi Muscle
Abductor Hallucis Muscle

Sagittal Suture
Occipital Lobe
Semispinalis Capitis Muscle
Trapezius Muscle
Supraspinatus Muscle
Scapula, Spine
Infraspinatus Muscle
Subscapularis Muscle
Spinal Cord
Teres Major Muscle
Rib 6
Rib 6-7, Intercostal Muscles
Diaphragm
Humerus
Basilic Vein
Common Flexor Tendon
Common Extensor Tendon
Kidney
Quadratus Lumborum Muscle
Longissimus Muscle
Ilium
L5, Spinous Process
Piriformis Muscle
Sacrum
Gemelli Muscles
Obturator Internus Muscle
Vastus Lateralis Muscle
Levator Ani Muscle
Semitendinosus Muscle
Semimembranosus Muscle
Small Saphenous Vein
Fibula, Head
Great Saphenous Vein
Gastrocnemius Muscle, Medial Head
Soleus Muscle
Peroneus Longus Muscle
Crural Fascia
Fibula, Shaft
Fibula, Lateral Malleolus
Calcaneus Bone
Abductor Hallucis Muscle
Abductor Digiti Minimi Muscle

vmc3245

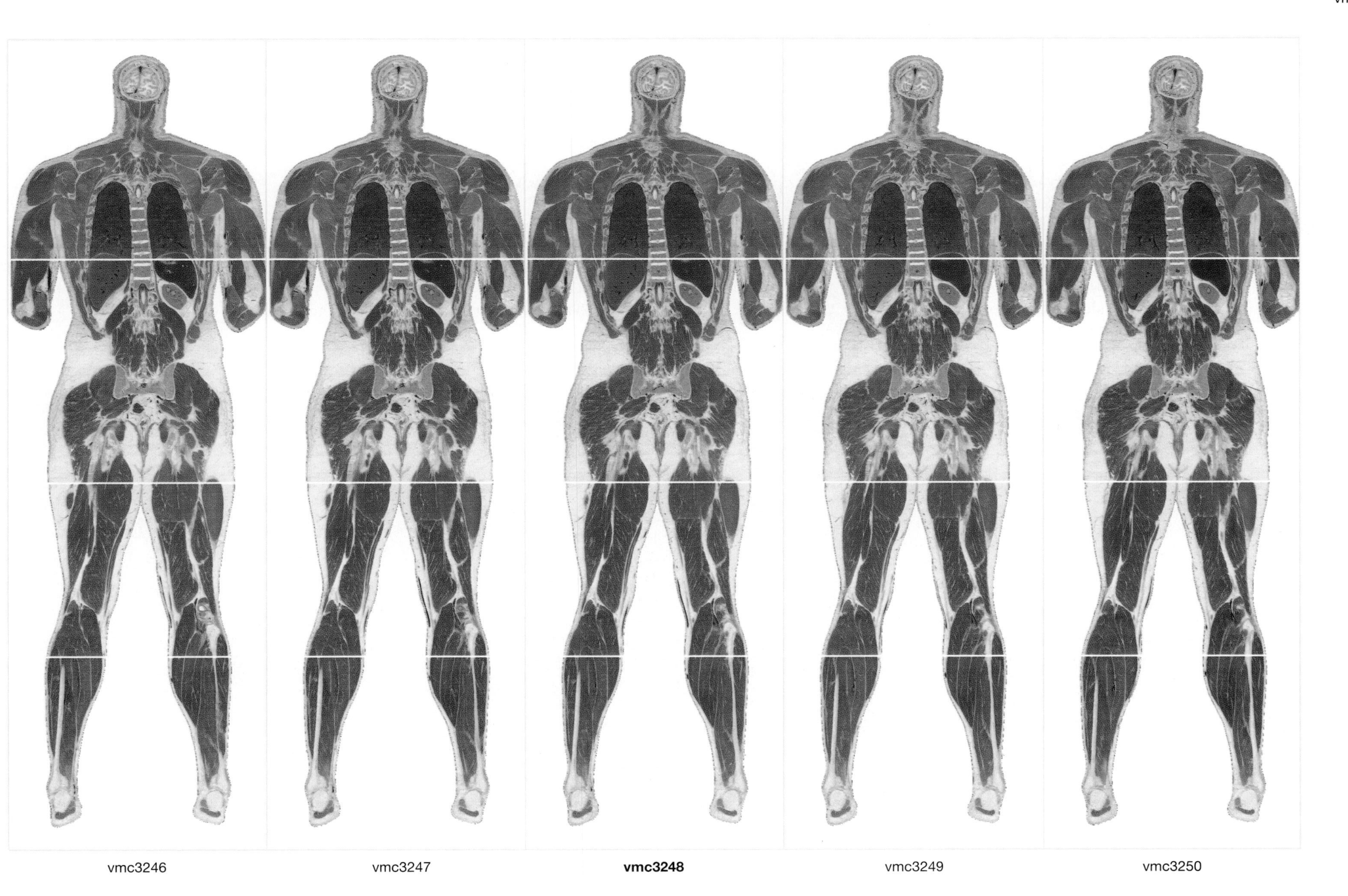

vmc3246 vmc3247 **vmc3248** vmc3249 vmc3250

Coronal
vmc3254

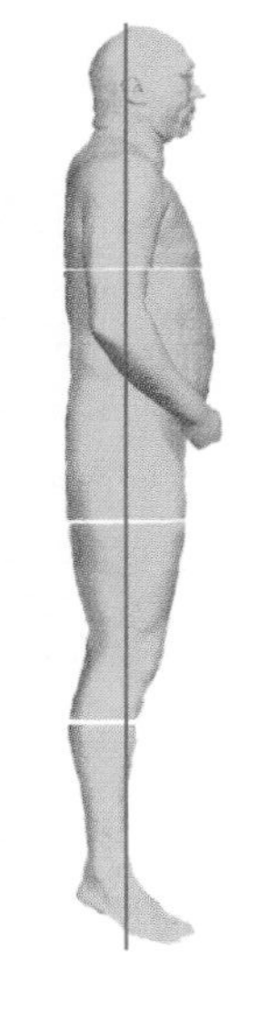

Longitudinal Cerebral Fissure
Internal Occipital Protuberance
Occipital Vessel
Trapezius Muscle
Levator Scapulae Muscle
Rhomboid Major Muscle
Scapula, Spine
Multifidus Muscle
Deltoid Muscle
Subscapularis Muscle
Teres Major Muscle
Triceps Brachii Muscle, Long Head
Serratus Anterior Muscle
Latissimus Dorsi Muscle
Diaphragm
Ulna
Humerus, Medial Epicondyle
Liver, Right Lobe
Spinal Cord
L3, Superior Articular Process
Gluteus Medius Muscle
L5, Spinous Process
Ilium
Sacrotuberous Ligament
Ischiorectal Fossa
Anus
Adductor Magnus Muscle
Biceps Femoris Muscle, Long Head
Biceps Femoris Muscle, Short Head
Semimembranosus Muscle
Plantaris Muscle
Gastrocnemius Muscle, Lateral Head
Gastrocnemius Muscle, Medial Head
Soleus Muscle
Crural Fascia
Peroneus Brevis Muscle
Flexor Hallucis Longus Muscle
Fibula, Lateral Malleolus
Calcaneus Bone
Quadratus Plantae Muscle
Abductor Hallucis Muscle

Occipital Bone
Occipital Lobe
Trapezius Muscle
Supraspinatus Muscle
Subscapularis Muscle
Infraspinatus Muscle
Deltoid Muscle
Teres Minor Muscle
Rib 4
Lung, Oblique Fissure
Spinal Cord
Latissimus Dorsi Muscle
Lung, Lower Lobe
Diaphragm
Kidney
Quadratus Lumborum Muscle
External Oblique Muscle
Iliocostalis Muscle
Longissimus Muscle
Sacroiliac Joint
Sacrum
Gluteus Maximus Muscle
Piriformis Muscle
Sigmoid Colon
Ischial Tuberosity
Levator Ani Muscle
Vastus Lateralis Muscle
External Anal Sphincter
Semitendinosus Muscle
Semimembranosus Muscle
Plantaris Muscle
Small Saphenous Vein
Gastrocnemius Muscle, Lateral Head
Great Saphenous Vein
Peroneus Longus Muscle
Soleus Muscle
Fibula, Shaft
Flexor Hallucis Longus Muscle
Fibula, Lateral Malleolus
Calcaneus Bone
Plantar Aponeurosis

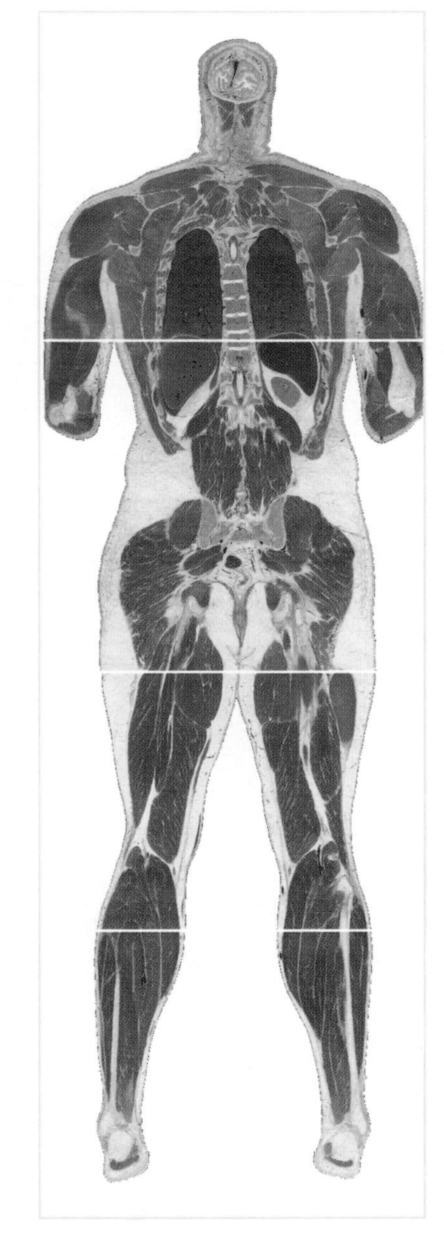

vmc3251

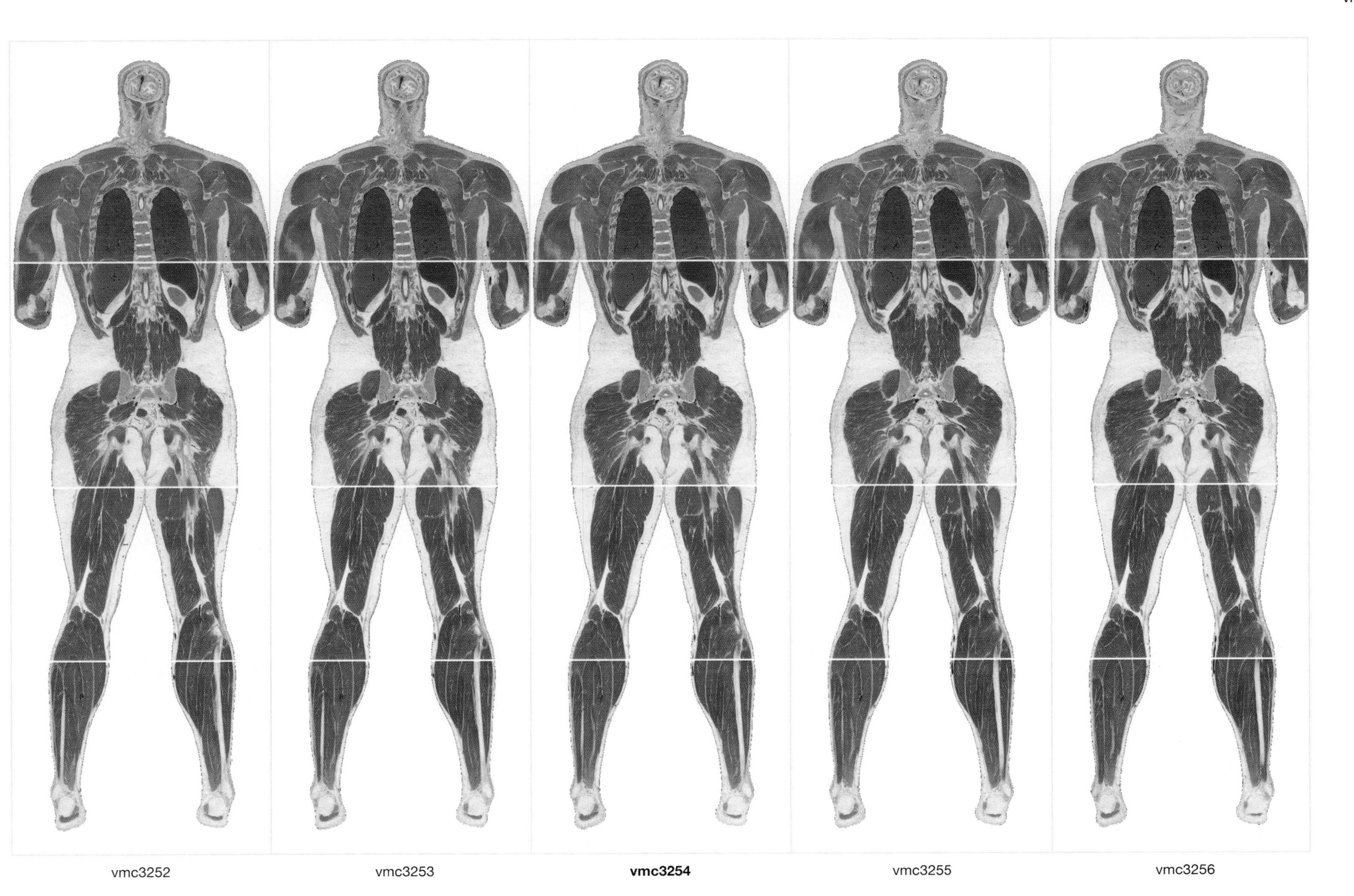
vmc3252
vmc3253
vmc3254
vmc3255
vmc3256

Coronal
vmc3259

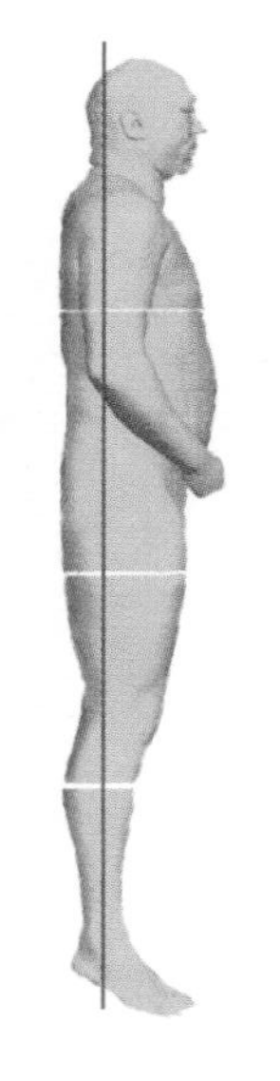

Diploe
Outer Table
Occipital Bone
Diploic Vessel
Multifidus Muscle
Supraspinatus Muscle
Deltoid Muscle
Rib 4-5, Intercostal Muscles
Triceps Brachii Muscle, Long Head
Subscapularis Muscle
Spinal Cord
Rib 7
Serratus Anterior Muscle
Lung, Lower Lobe
Triceps Brachii Tendon
Diaphragm
Liver, Right Lobe
Latissimus Dorsi Muscle
Superficial Fascia and Adipose Tissue
Spinalis Muscle
Multifidus Muscle
Ilium
Sacrum
Piriformis Muscle
Ischiorectal Fossa
External Anal Sphincter
Adductor Magnus Muscle
Biceps Femoris Muscle
Fascia Lata
Semitendinosus Muscle
Semimembranosus Muscle
Great Saphenous Vein
Plantaris Muscle
Gastrocnemius Muscle, Medial Head
Gastrocnemius Muscle, Lateral Head
Soleus Muscle
Tibialis Posterior Muscle
Flexor Hallucis Longus Muscle
Peroneus Brevis Muscle
Calcaneus Bone
Quadratus Plantae Muscle
Flexor Digitorum Brevis Muscle

Inner Table
Galea Aponeurotica
Occipital Lobe
Occipitofrontalis Muscle, Occipital Belly
Trapezius Muscle
Supraspinatus Muscle
Scapula, Spine
Infraspinatus Muscle
Deltoid Muscle
Subscapularis Muscle
Scapula
Teres Major Muscle
Triceps Brachii Muscle, Long Head
Lung, Lower Lobe
Diaphragm
Spleen
Quadratus Lumborum Muscle
Longissimus Muscle
Iliocostalis Muscle
Gluteus Medius Muscle
Sacroiliac Joint
Gluteus Maximus Muscle
Internal Iliac Vein
Levator Ani Muscle
Sigmoid Colon
Anus
Iliotibial Tract
Biceps Femoris Muscle, Long Head
Adductor Magnus Muscle
Semitendinosus Muscle
Biceps Femoris Tendon
Semimembranosus Muscle
Small Saphenous Vein
Gastrocnemius Muscle, Lateral Head
Soleus Muscle
Fibula, Shaft
Peroneus Brevis Muscle
Flexor Hallucis Longus Muscle
Fibula, Lateral Malleolus
Peroneus Longus Tendon
Calcaneus Bone
Plantar Aponeurosis

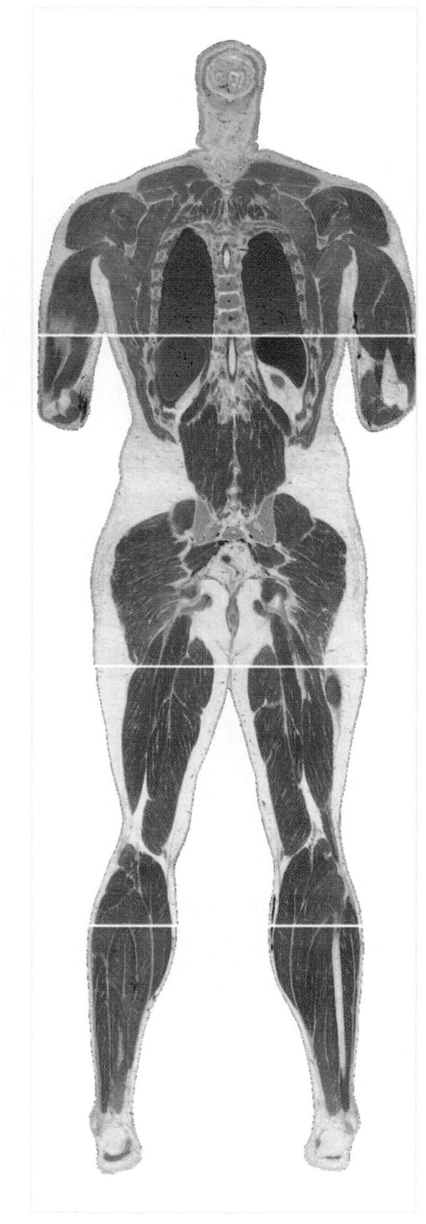

vmc3257

vmc3258

vmc3259

vmc3260

vmc3261

vmc3262

Coronal
vmc3271

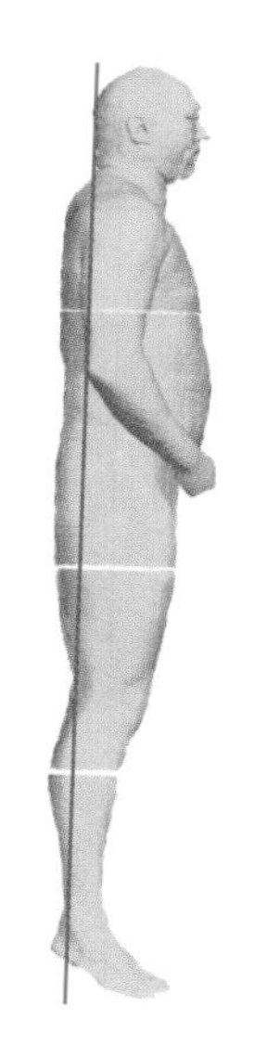

Occipital Vein
Scalp
Trapezius Muscle
Trapezius Muscle
Levator Scapulae Muscle
Rhomboid Major Muscle
Rhomboid Major Muscle
Scapula, Spine
Infraspinatus Muscle
Deltoid Muscle
Deltoid Muscle
Infraspinatus Muscle
Multifidus Muscle
Teres Minor Muscle
Subscapularis Muscle
Teres Major Muscle
Teres Major Muscle
Lung, Lower Lobe
Scapula
Triceps Brachii Muscle
Triceps Brachii Muscle
Serratus Anterior Muscle
Vertebral Canal
Radius
Diaphragm
Ulna
Liver, Right Lobe
Rib 11-12, Intercostal Muscle
Spinalis Muscle
Latissimus Dorsi Muscle
Longissimus Muscle
Iliocostalis Muscle
Superficial Fascia and Adipose Tissue
Ilium
Dorsal Sacroiliac Ligament
Sacroiliac Joint
Middle Rectal Vessel
Sacrum
Gluteus Maximus Muscle
Piriformis Muscle
Anococcygeal Ligament
Coccygeus Muscle
Ischiorectal Fossa
Gluteus Maximus Muscle
External Anal Sphincter
Anus
Intergluteal Crease
Adductor Magnus Muscle
Adductor Magnus Muscle
Fascia Lata
Fascia Lata
Semimembranosus Muscle
Semitendinosus Muscle
Biceps Femoris Muscle
Biceps Femoris Muscle
Semitendinosus Muscle
Semimembranosus Muscle
Plantaris Muscle
Plantaris Muscle
Gastrocnemius Muscle, Medial Head
Soleus Muscle
Gastrocnemius Muscle, Medial Head
Soleus Muscle
Gastrocnemius Muscle, Lateral Head
Crural Fascia
Crural Fascia
Soleus Muscle
Gastrocnemius Muscle, Lateral Head
Calcaneal Tendon
Calcaneal Tendon
Lateral Malleolar Vessel
Calcaneus Bone
Calcaneus Bone

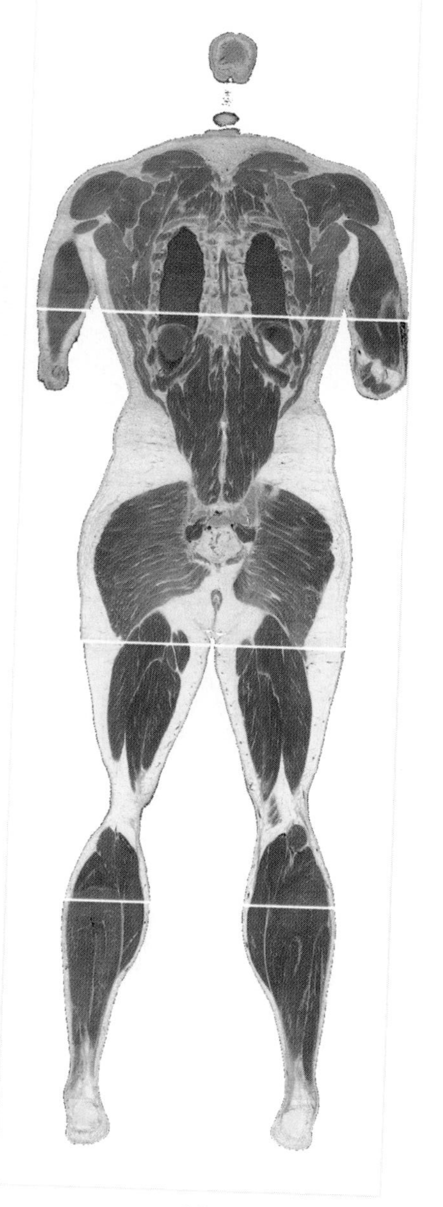

vmc3269

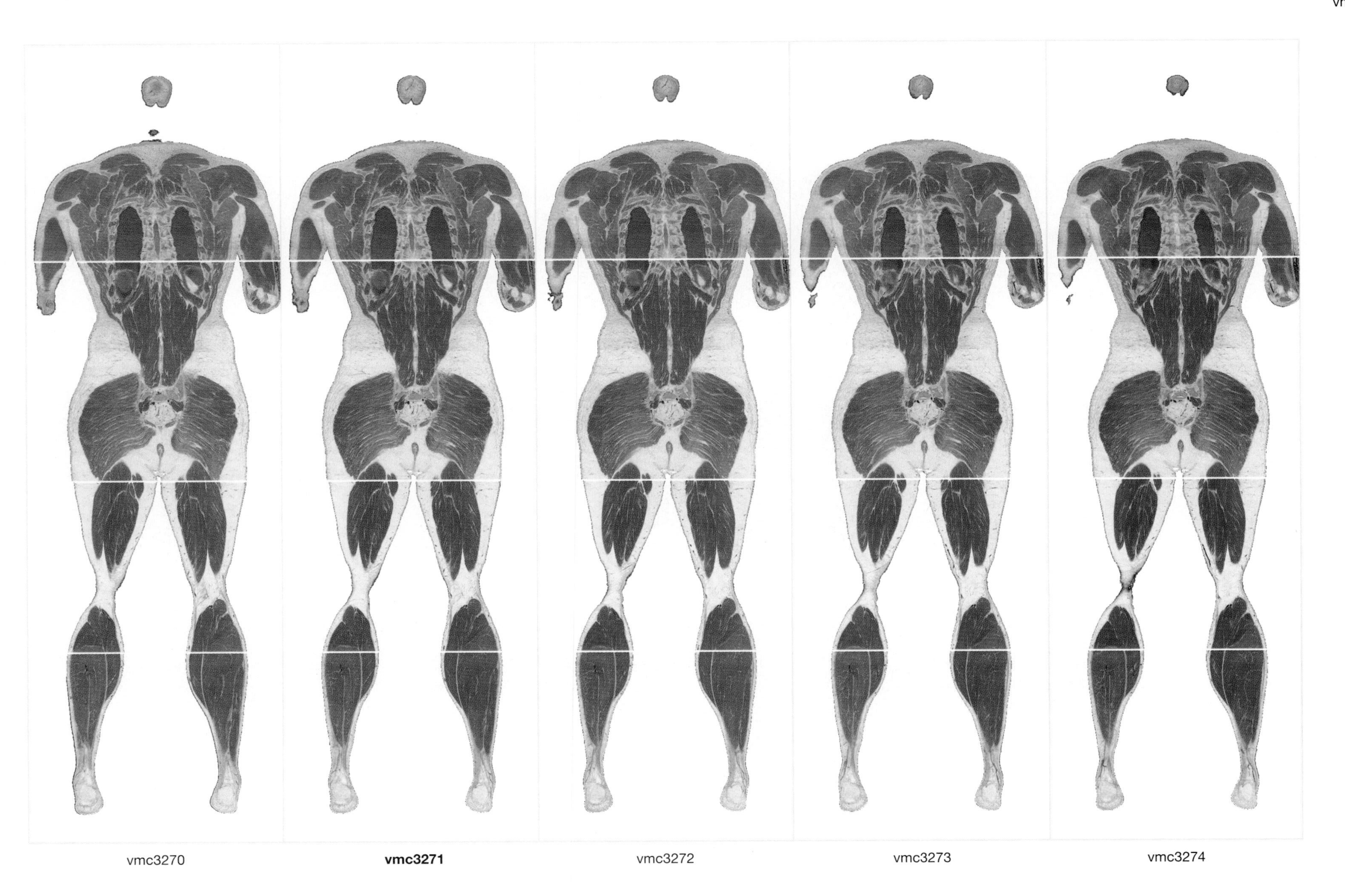

vmc3270 **vmc3271** vmc3272 vmc3273 vmc3274

Coronal
vmc3276

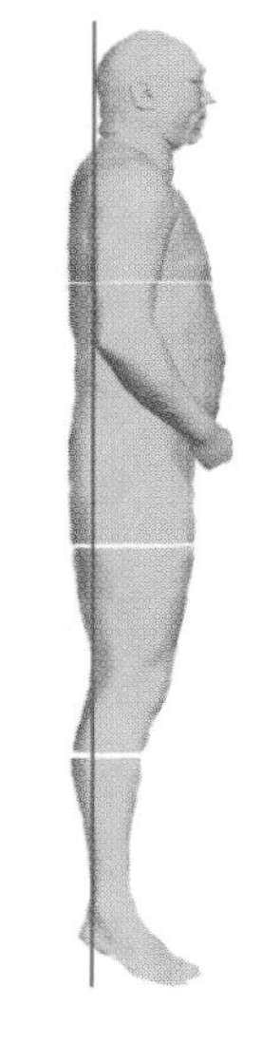

Rhomboid Minor Muscle
Trapezius Muscle
Levator Scapulae Muscle
Scapula
Semispinalis Cervicis Muscle
Deltoid Muscle
Axillary Vessel
Multifidus Muscle
Teres Major Muscle
Lung, Lower Lobe
T8, Spinous Process
Rib 10
Latissimus Dorsi Muscle
Lumbodorsal Fascia
Superficial Fascia
Sacrum
Median Sacral Vessel
Gluteus Maximus Muscle
Anococcygeal Ligament
Ischiorectal Fossa
Biceps Femoris Muscle
Fascia Lata
Semitendinosus Muscle
Semimembranosus Muscle
Superficial Fascia
Plantaris Muscle
Gastrocnemius Muscle, Medial Head
Soleus Muscle
Crural Fascia
Calcaneal Tendon
Calcaneus Bone
Plantar Aponeurosis

Rhomboid Major Muscle
Trapezius Muscle
Scapula, Spine
Infraspinatus Muscle
Deltoid Muscle
Subscapularis Muscle
Rib 4-5, Intercostal Muscles
Scapula
Teres Major Muscle
Triceps Brachii Muscle
Ligamentum Flavum
Triceps Brachii Tendon
Latissimus Dorsi Muscle
Rib 10-11, Intercostal Muscles
Rib 11
Iliocostalis Muscle
Longissimus Muscle
Spinalis Muscle
Coccygeus Muscle
Coccyx
External Anal Sphincter
Intergluteal Crease
Biceps Femoris Muscle
Semitendinosus Muscle
Semimembranosus Muscle
Fascia Lata
Superficial Fascia
Plantaris Muscle
Gastrocnemius Muscle, Medial Head
Soleus Muscle
Crural Fascia
Calcaneal Tendon
Calcaneus Bone

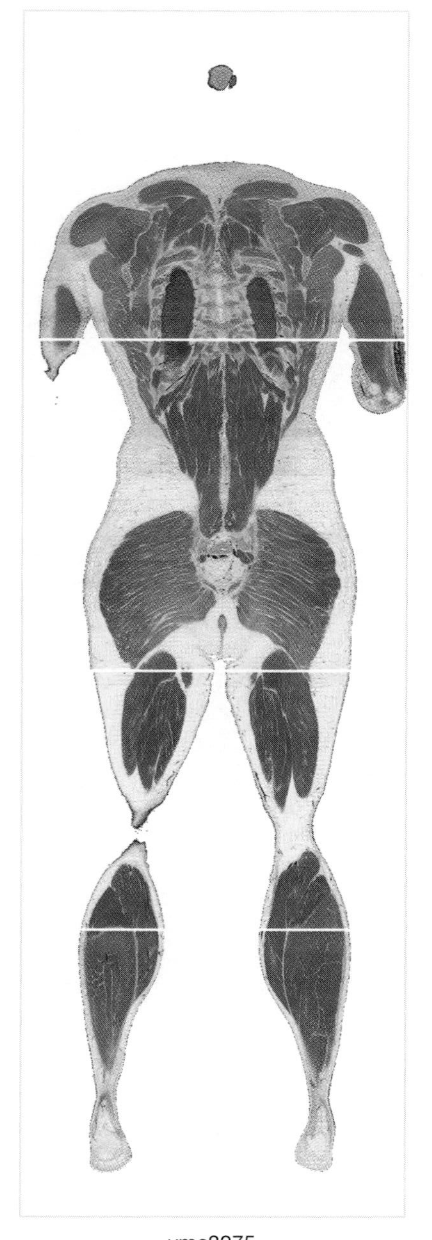

vmc3275

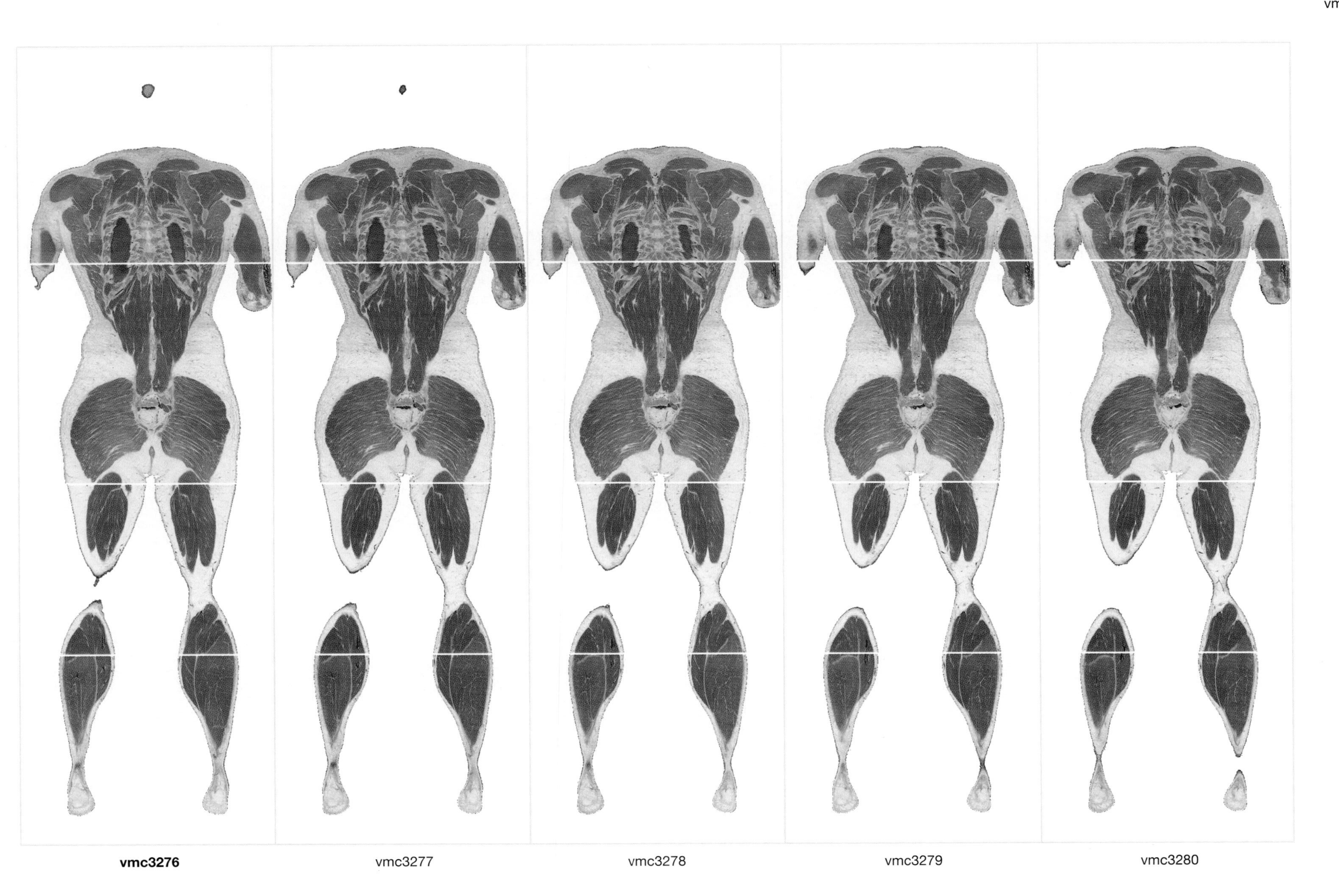

vmc3276 vmc3277 vmc3278 vmc3279 vmc3280

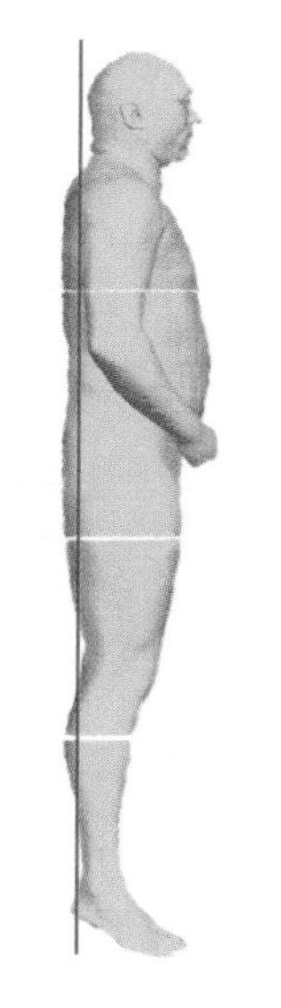

Trapezius Muscle
Rhomboid Minor Muscle
Rhomboid Major Muscle
Longissimus Thoracis Muscle
Infraspinatus Muscle
Subscapularis Muscle
Rib 6-7, Intercostal Muscles
Teres Major Muscle
Rib 7
Lung, Lower Lobe
Latissimus Dorsi Muscle
Spinalis Muscle
Longissimus Muscle
Iliocostalis Muscle
Supraspinous Ligament
Superficial Fascia and Adipose Tissue
Longissimus Muscle
Gluteus Maximus Muscle
Intergluteal Crease
Biceps Femoris Muscle
Semimembranosus Muscle
Semitendinosus Muscle
Fascia Lata
Plantaris Muscle
Gastrocnemius Muscle, Lateral Head
Gastrocnemius Muscle, Medial Head
Soleus Muscle
Crural Fascia
Small Saphenous Vein
Calcaneal Tendon
Calcaneus Bone

Trapezius Muscle
Rhomboid Minor Muscle
Rhomboid Major Muscle
Subscapularis Muscle
Deltoid Muscle
Infraspinatus Muscle
Scapula
Teres Major Muscle
Serratus Anterior Muscle
Triceps Brachii Muscle
Latissimus Dorsi Muscle
Rib 10
Spinalis Muscle
Iliocostalis Muscle
Longissimus Muscle
Lumbodorsal Fascia
Superficial Fascia and Adipose Tissue
Longissimus Muscle
Gluteus Maximus Muscle
Sacrum
Internal Iliac Vessel
Semimembranosus Muscle
Semitendinosus Muscle
Biceps Femoris Muscle, Long Head
Fascia Lata
Small Saphenous Vein
Plantaris Muscle
Gastrocnemius Muscle, Medial Head
Gastrocnemius Muscle, Lateral Head
Crural Fascia
Soleus Muscle
Calcaneal Tendon
Calcaneus Bone

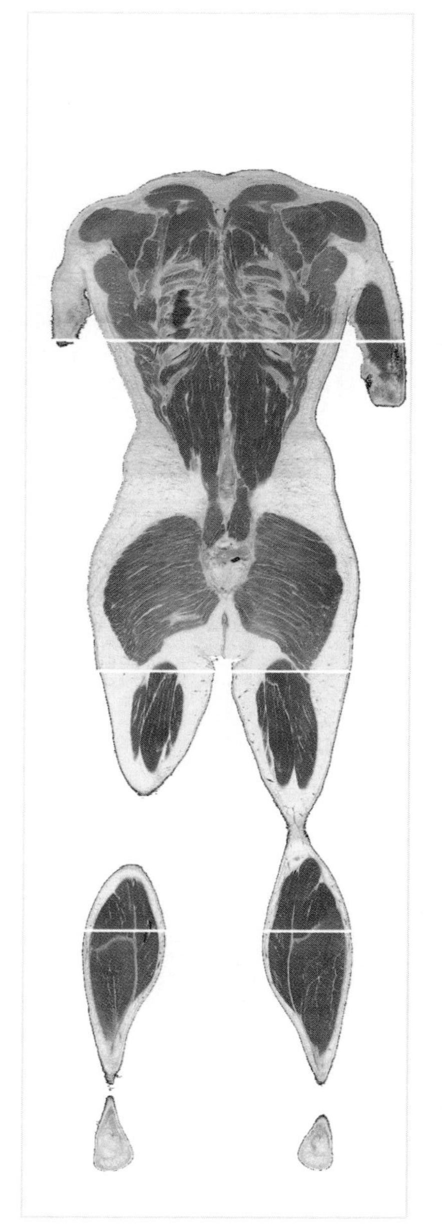

vmc3281

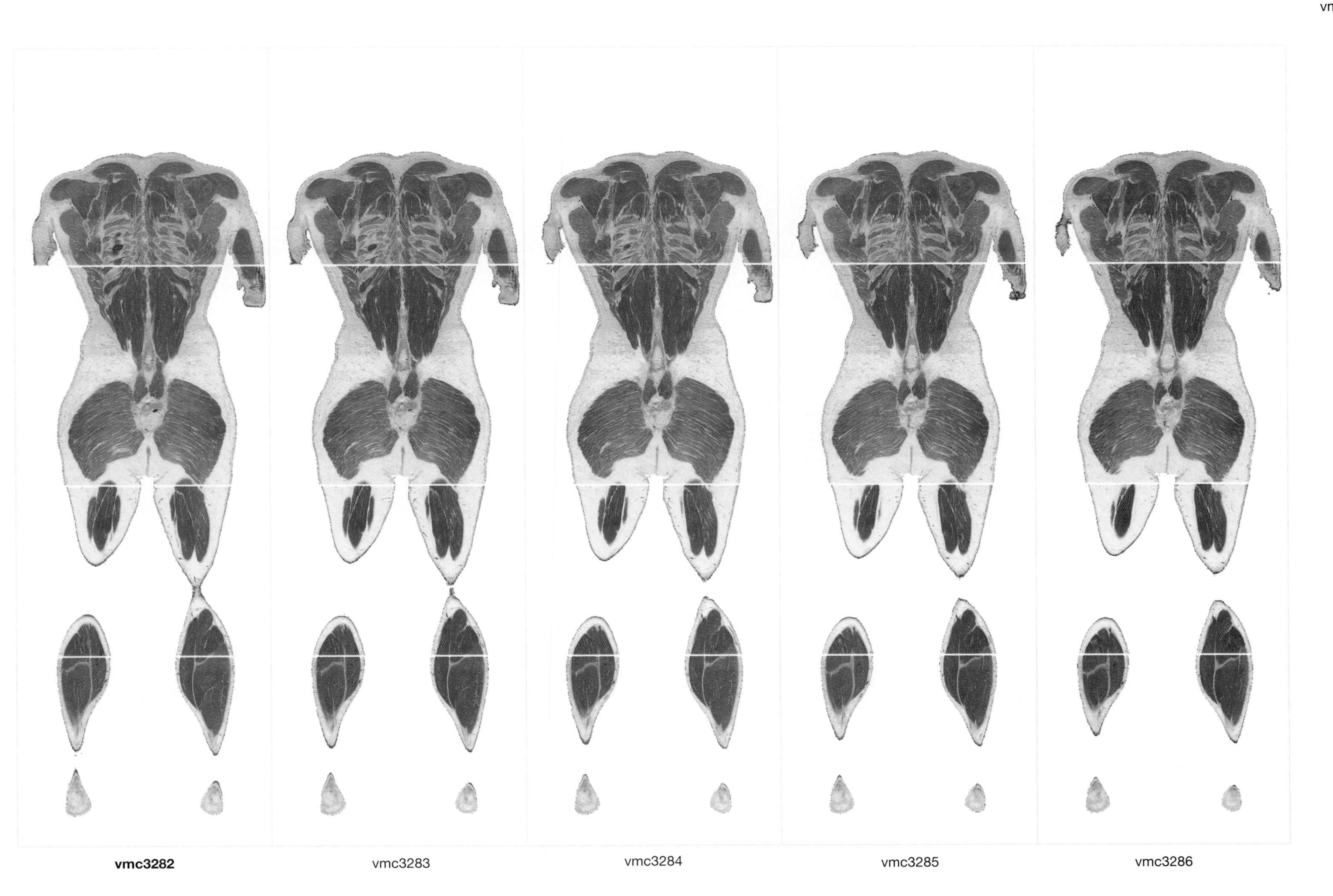

vmc3282 vmc3283 vmc3284 vmc3285 vmc3286

Coronal
vmc3288

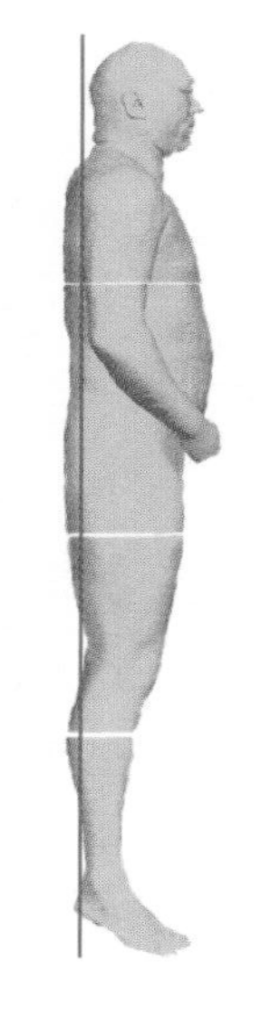

Ligamentum Nuchae
Trapezius Muscle
Multifidus Muscle
Rhomboid Major Muscle
Deltoid Muscle
Longissimus Thoracis Muscle
Rib 6-7, External Intercostal Muscles
Teres Major Muscle
Scapula
Rib 8
Spinalis Muscle
Longissimus Muscle
Latissimus Dorsi Muscle
Supraspinous Ligament
Superficial Fascia
Medial Sacral Crest
Median Sacral Vessels
Coccyx
Intergluteal Crease
Fascia Lata
Semitendinosus Muscle
Superficial Fascia
Small Saphenous Vein
Plantaris Muscle
Gastrocnemius Muscle, Lateral Head
Gastrocnemius Muscle, Medial Head
Soleus Muscle
Crural Fascia
Superficial Fascia
Skin
Calcaneal Tendon
Calcaneus Bone

Trapezius Muscle
Rhomboid Major Muscle
Subscapularis Muscle
Deltoid Muscle
Infraspinatus Muscle
Scapula
Teres Major Muscle
Serratus Anterior Muscle
Triceps Brachii Muscle
Rib 9
Latissimus Dorsi Muscle
Spinalis Muscle
Iliocostalis Muscle
Longissimus Muscle
Lumbar Fascia
Longissimus Muscle
Sacrum
Gluteus Maximus Muscle
Anococcygeal Ligament
Superficial Fascia
Semitendinosus Muscle
Biceps Femoris Muscle, Long Head
Small Saphenous Vein
Plantaris Muscle
Gastrocnemius Muscle, Medial Head
Gastrocnemius Muscle, Lateral Head
Crural Fascia
Soleus Muscle
Superficial Fascia
Skin
Heel

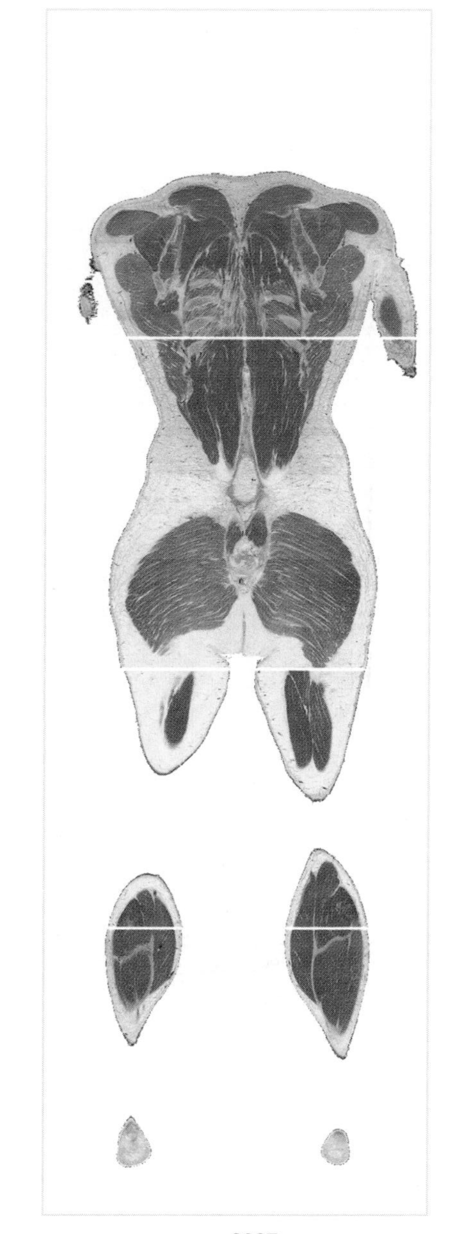

vmc3287

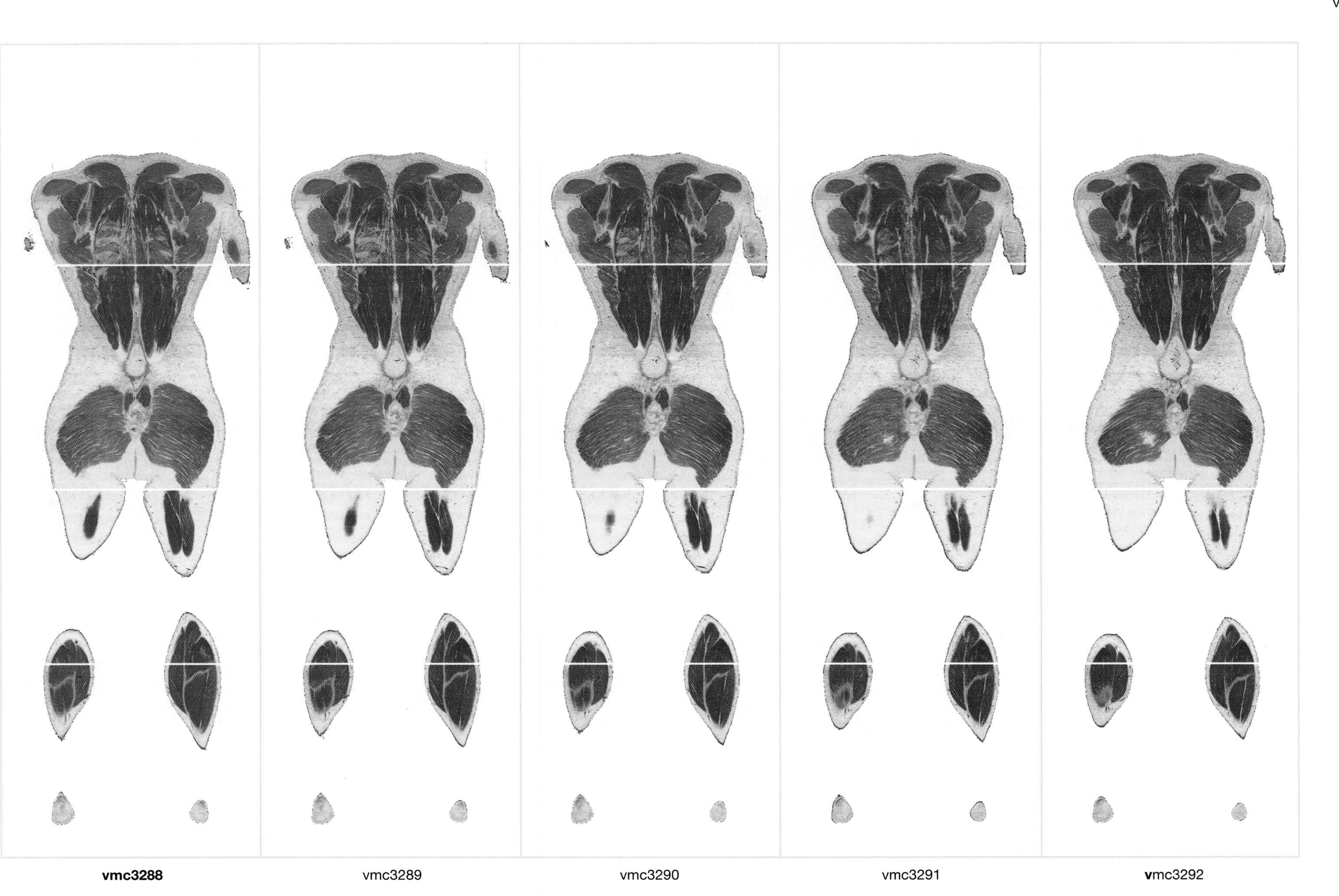
vmc3288
vmc3289
vmc3290
vmc3291
vmc3292

Coronal
vmc3293

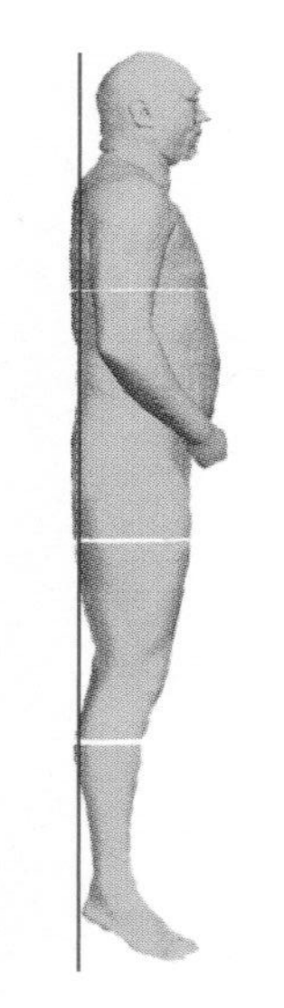

Trapezius Muscle
Rhomboid Major Muscle
Infraspinatus Muscle
Deltoid Muscle
Multifidus Muscle
Longissimus Thoracis Muscle
Scapula
Latissimus Dorsi Muscle
Spinalis Muscle
Longissimus Muscle
Latissimus Dorsi Muscle
Supraspinous Ligament
Superficial Fascia and Adipose Tissue
Gluteus Maximus Muscle
Coccyx
Anococcygeal Ligament

Trapezius Muscle
Rhomboid Major Muscle
Deltoid Muscle
Infraspinatus Muscle
Subscapularis Muscle
Scapula
Teres Major Muscle
Longissimus Muscle
Serratus Anterior Muscle
Latissimus Dorsi Muscle
Spinalis Muscle
Iliocostalis Muscle
Longissimus Muscle
Lumbar Fascia
Longissimus Muscle
Sacrum
Gluteus Maximus Muscle
Intergluteal Crease
Fascia Lata
Semitendinosus Muscle
Biceps Femoris Muscle, Long Head

Crural Fascia
Gastrocnemius Muscle, Medial Head
Gastrocnemius Muscle, Lateral Head
Soleus Muscle
Small Saphenous Vein

Plantaris Muscle
Gastrocnemius Muscle, Medial Head
Crural Fascia
Gastrocnemius Muscle, Lateral Head
Soleus Muscle
Small Saphenous Vein

Calcaneal Tendon
Calcaneus Bone

Heel

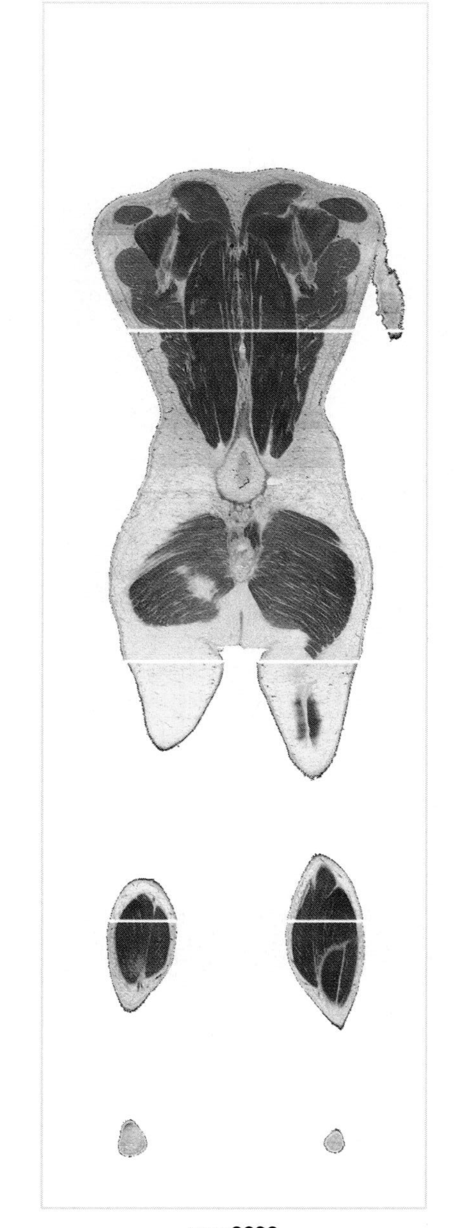

vmc3293

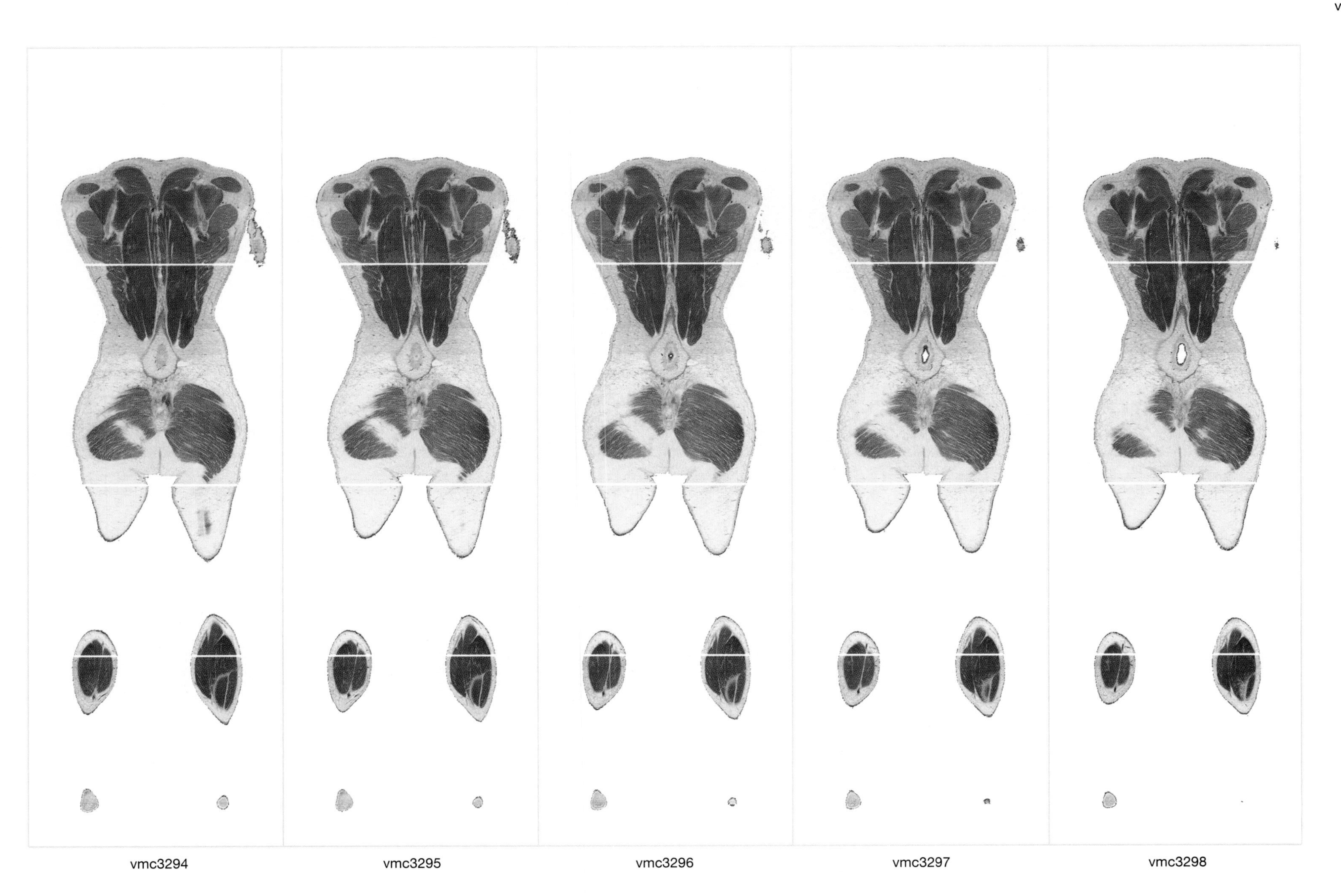

vmc3294 vmc3295 vmc3296 vmc3297 vmc3298

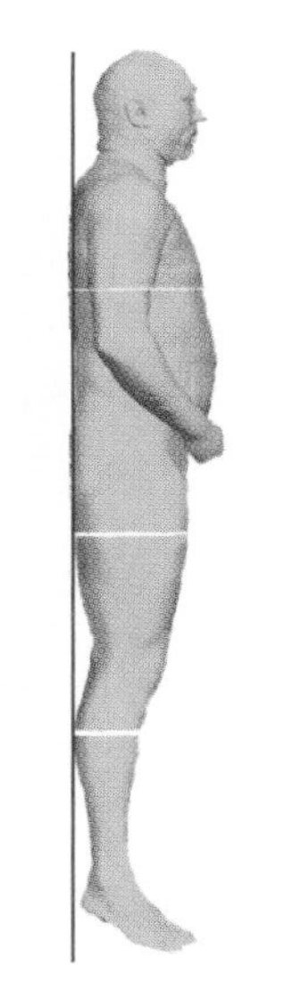

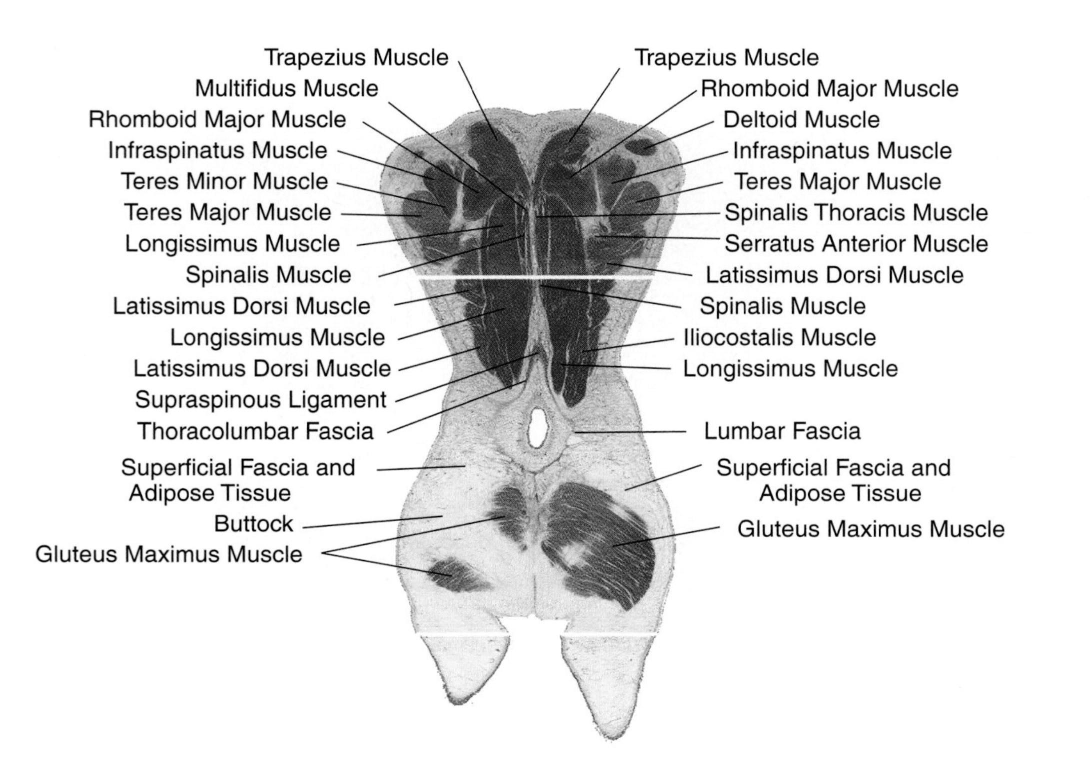
Trapezius Muscle
Multifidus Muscle
Rhomboid Major Muscle
Infraspinatus Muscle
Teres Minor Muscle
Teres Major Muscle
Longissimus Muscle
Spinalis Muscle
Latissimus Dorsi Muscle
Longissimus Muscle
Latissimus Dorsi Muscle
Supraspinous Ligament
Thoracolumbar Fascia
Superficial Fascia and Adipose Tissue
Buttock
Gluteus Maximus Muscle
Trapezius Muscle
Rhomboid Major Muscle
Deltoid Muscle
Infraspinatus Muscle
Teres Major Muscle
Spinalis Thoracis Muscle
Serratus Anterior Muscle
Latissimus Dorsi Muscle
Spinalis Muscle
Iliocostalis Muscle
Longissimus Muscle
Lumbar Fascia
Superficial Fascia and Adipose Tissue
Gluteus Maximus Muscle

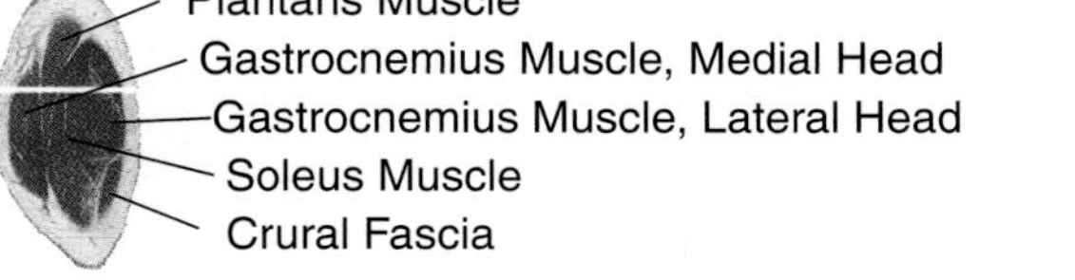
Gastrocnemius Muscle, Medial Head
Gastrocnemius Muscle, Lateral Head
Crural Fascia
Small Saphenous Vein
Plantaris Muscle
Gastrocnemius Muscle, Medial Head
Gastrocnemius Muscle, Lateral Head
Soleus Muscle
Crural Fascia
Heel

vmc3299

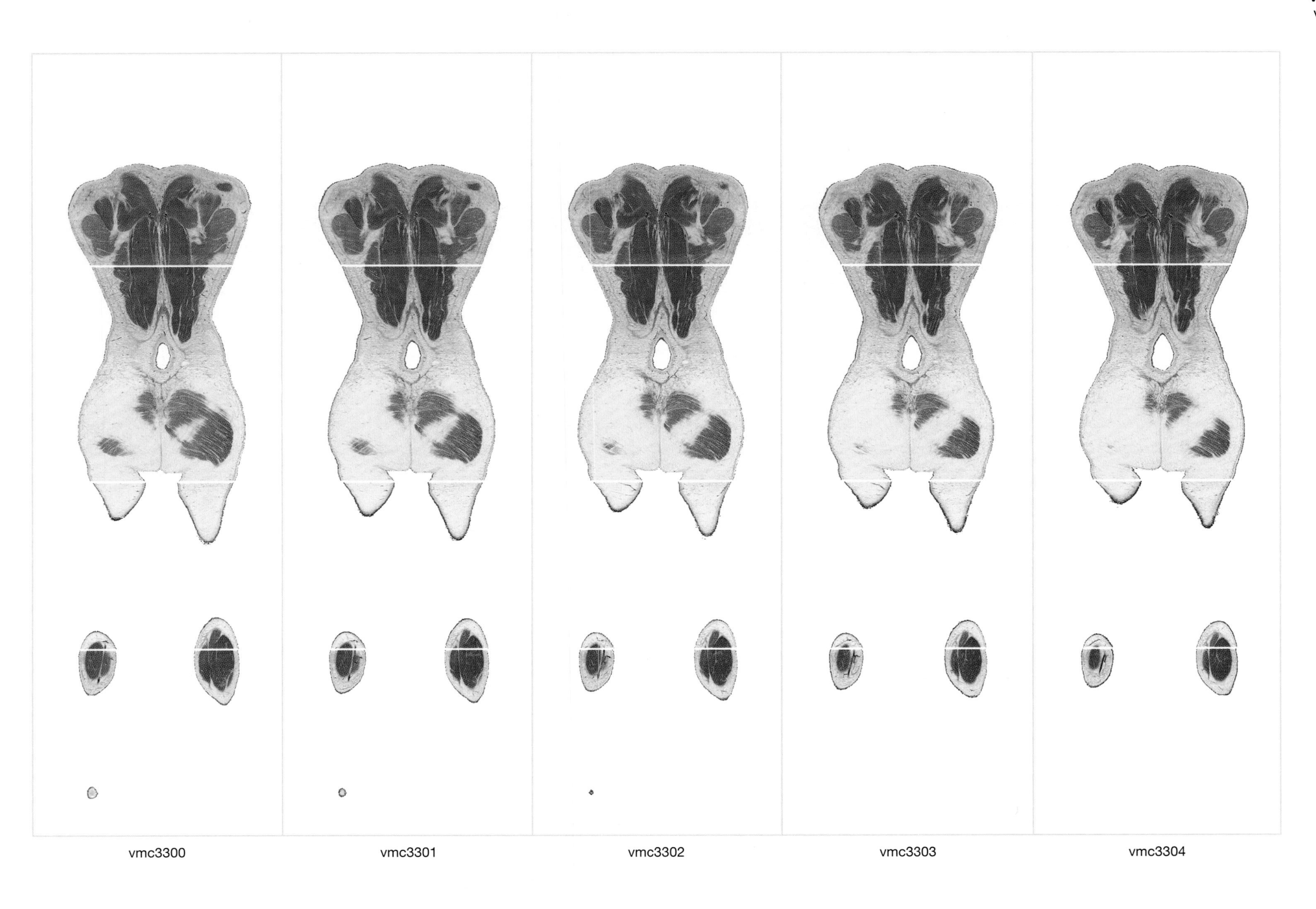
vmc3300
vmc3301
vmc3302
vmc3303
vmc3304

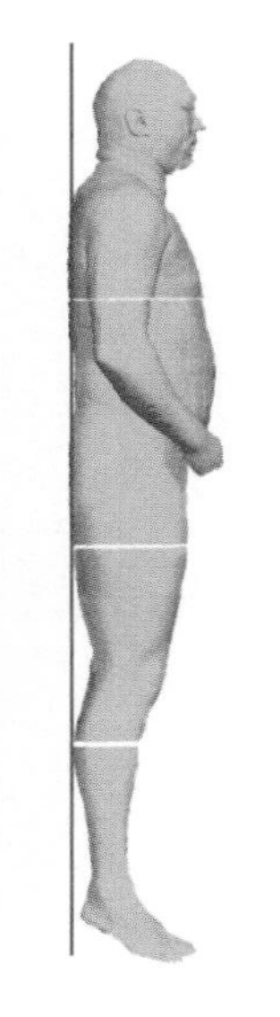

Trapezius Muscle
Spinalis Muscle
Infraspinatus Muscle
Teres Major Muscle
Longissimus Muscle
Latissimus Dorsi Muscle
Thoracolumbar Fascia
Lumbar Fascia
Gluteus Maximus Muscle
Buttock
Intergluteal Crease
Trapezius Muscle
Infraspinatus Muscle
Spinalis Thoracis Muscle
Teres Major Muscle
Latissimus Dorsi Muscle
Longissimus Thoracis Muscle
Spinalis Muscle
Longissimus Muscle
Latissimus Dorsi Muscle
Iliocostalis Muscle
Longissimus Muscle
Gluteus Maximus Muscle

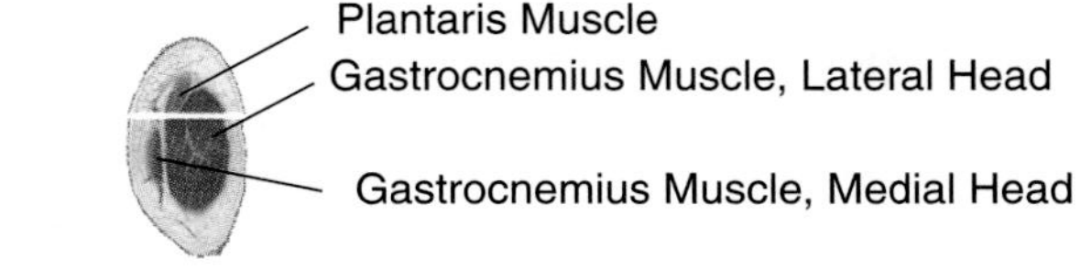

Gastrocnemius Muscle, Lateral Head
Gastrocnemius Muscle, Medial Head
Plantaris Muscle
Gastrocnemius Muscle, Lateral Head
Gastrocnemius Muscle, Medial Head

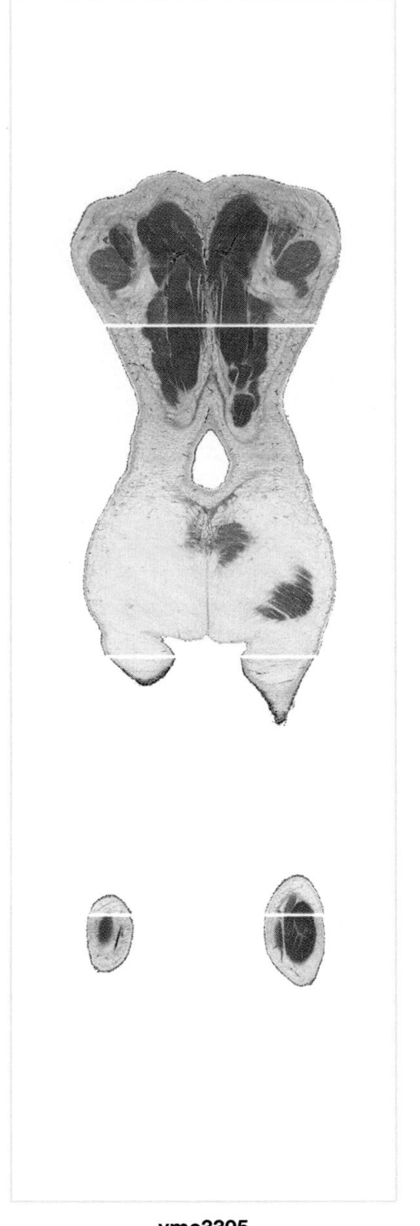

vmc3305

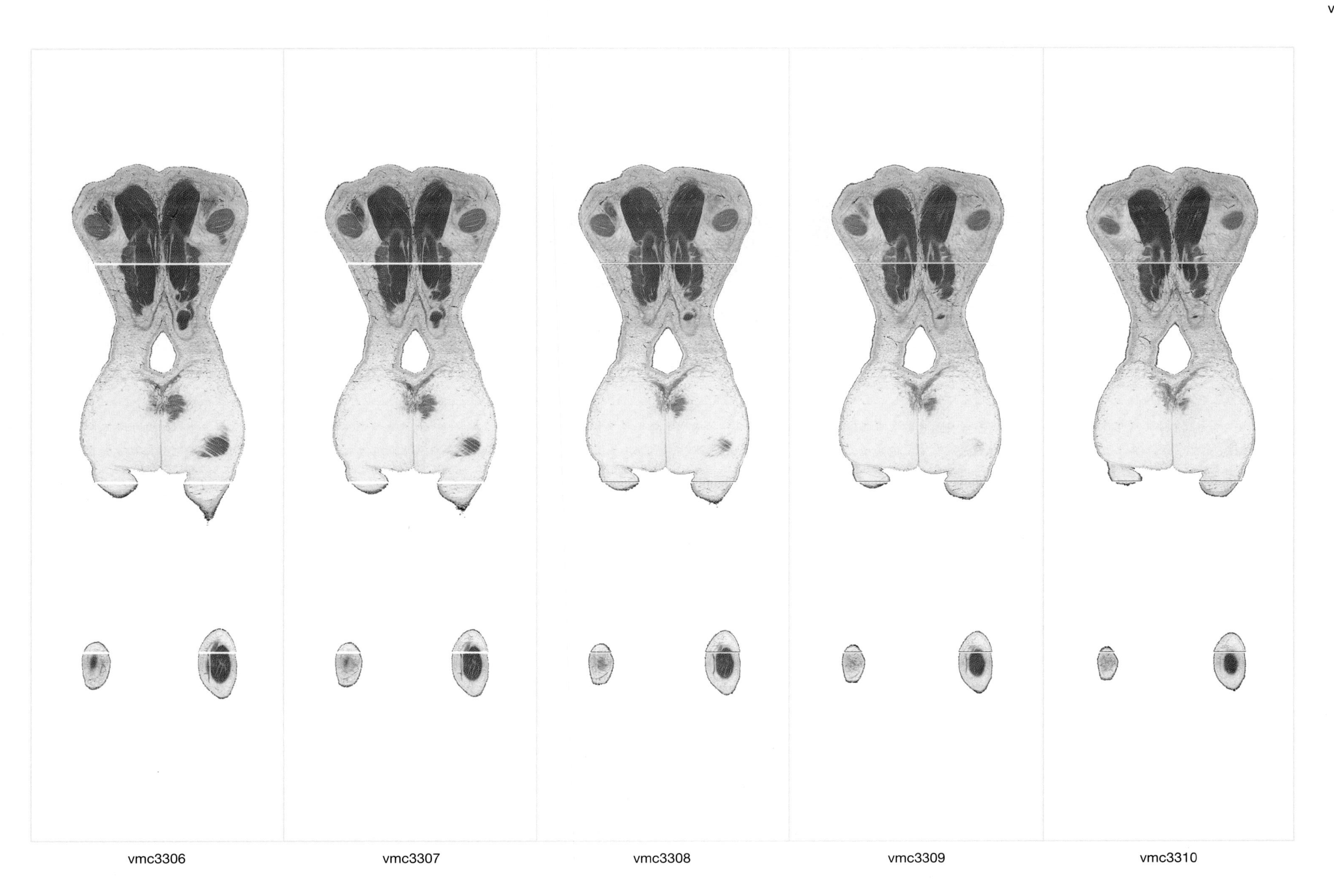
vmc3306
vmc3307
vmc3308
vmc3309
vmc3310

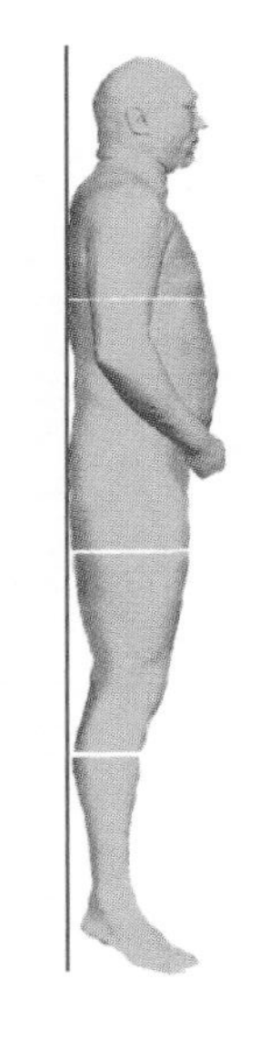

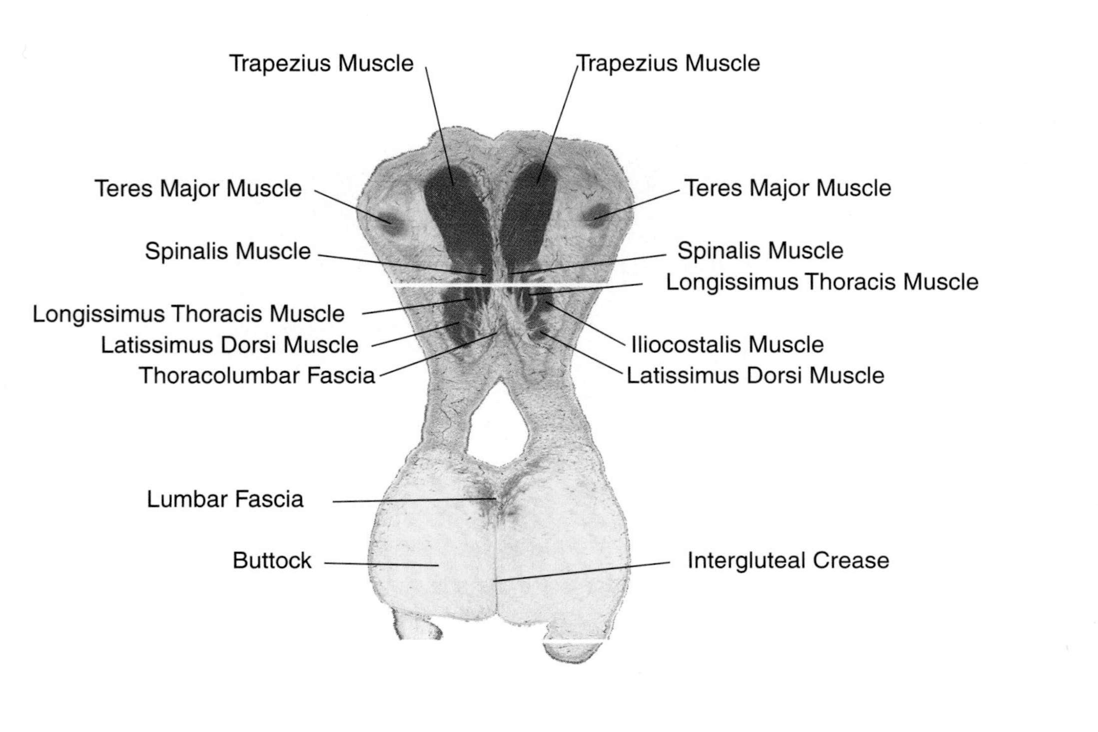
Trapezius Muscle
Trapezius Muscle
Teres Major Muscle
Teres Major Muscle
Spinalis Muscle
Spinalis Muscle
Longissimus Thoracis Muscle
Longissimus Thoracis Muscle
Latissimus Dorsi Muscle
Iliocostalis Muscle
Thoracolumbar Fascia
Latissimus Dorsi Muscle
Lumbar Fascia
Buttock
Intergluteal Crease

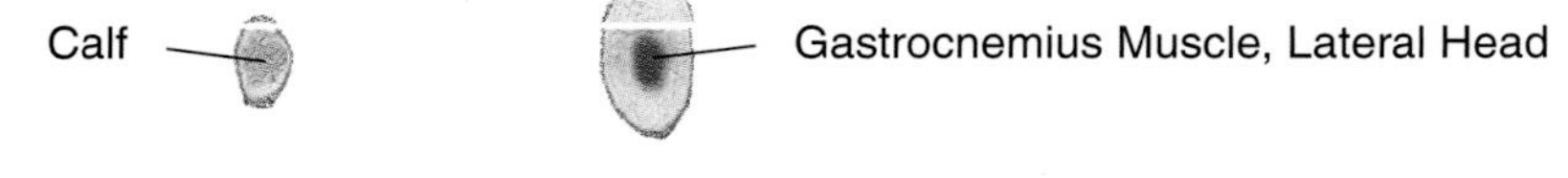
Calf
Gastrocnemius Muscle, Lateral Head

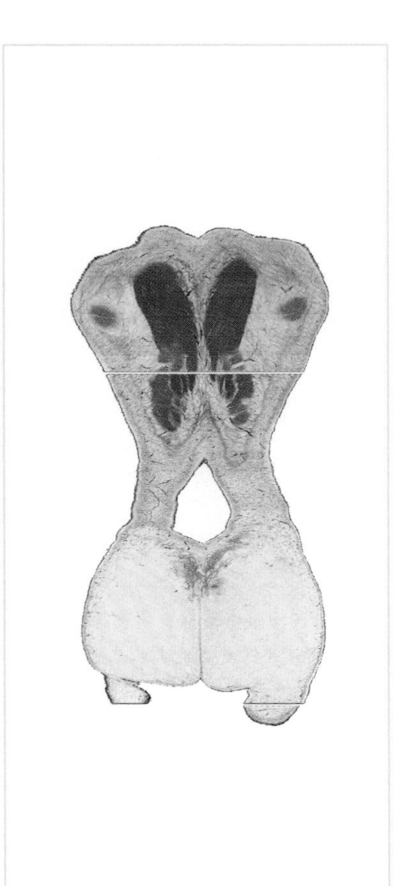

vmc3311

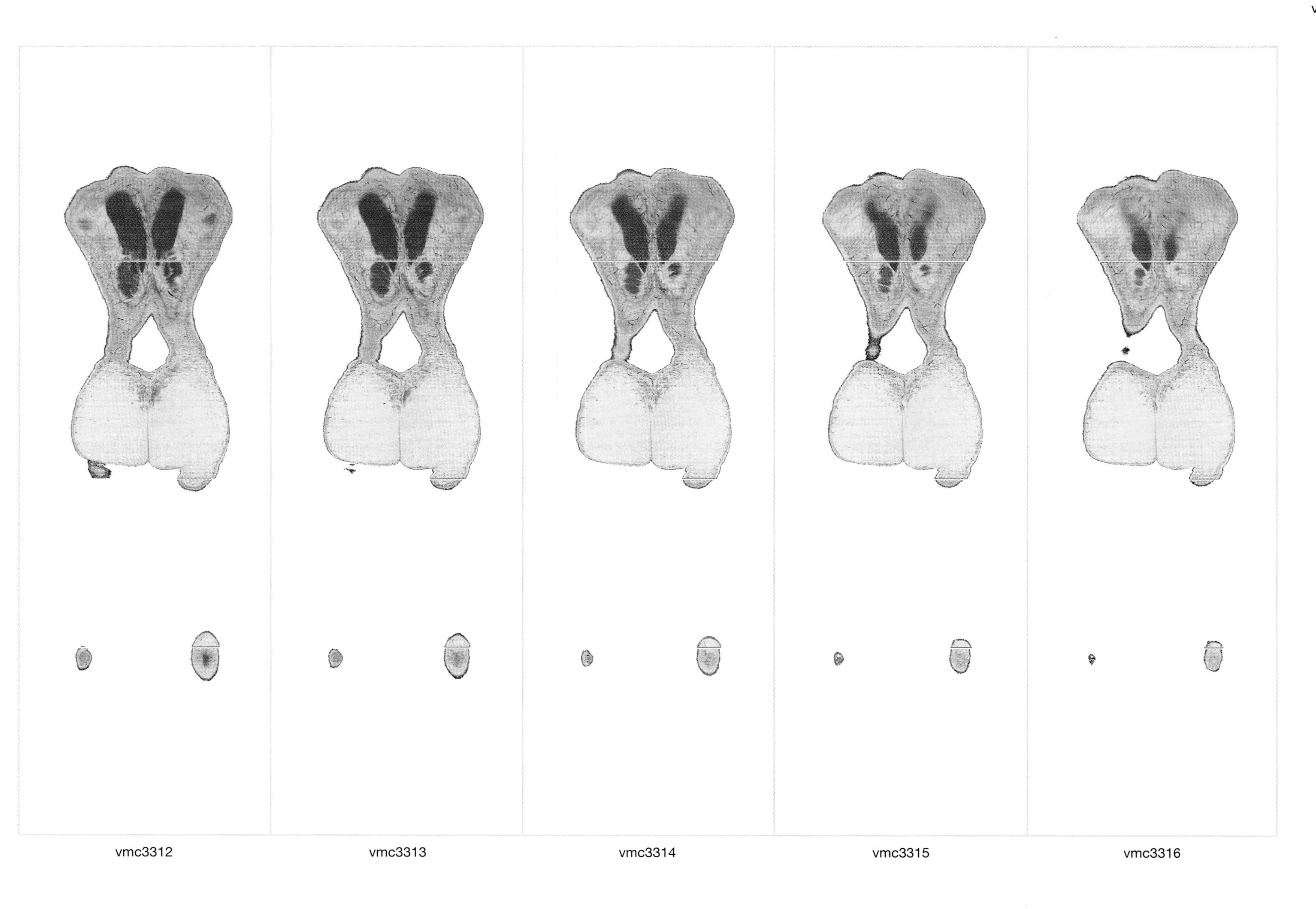

vmc3312 vmc3313 vmc3314 vmc3315 vmc3316

Coronal
vmc3317

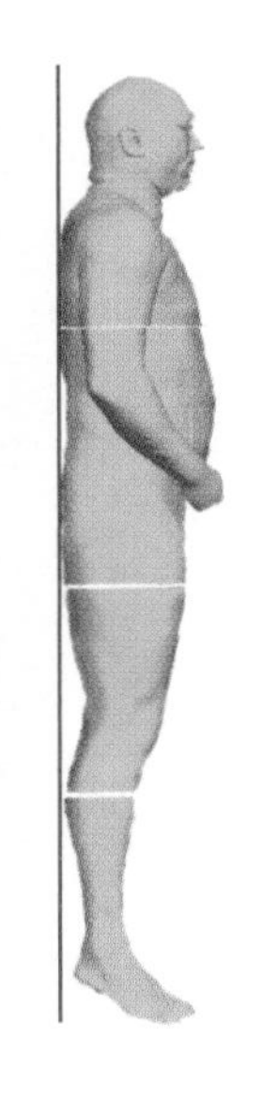

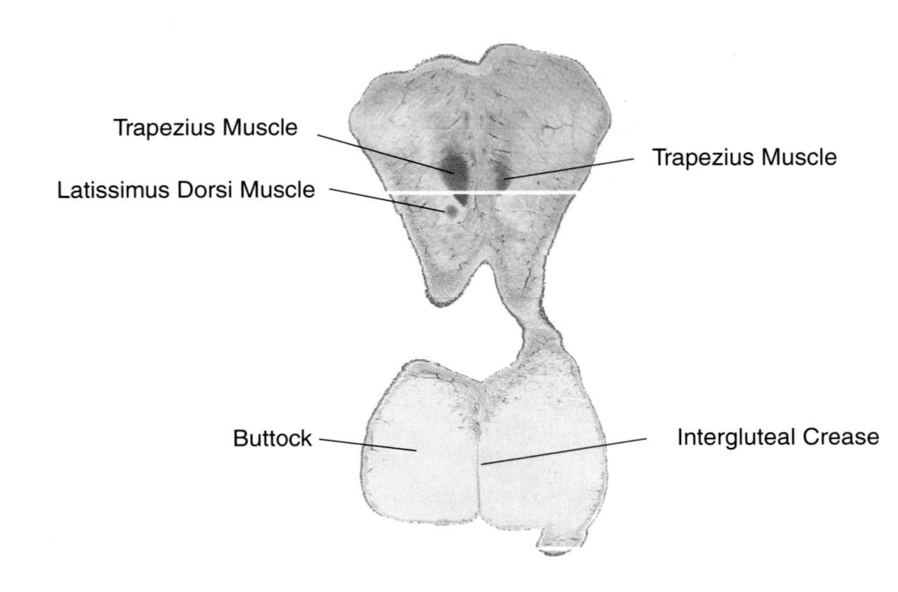

Trapezius Muscle
Latissimus Dorsi Muscle
Trapezius Muscle
Buttock
Intergluteal Crease

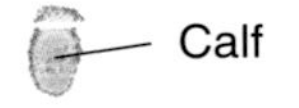

Calf

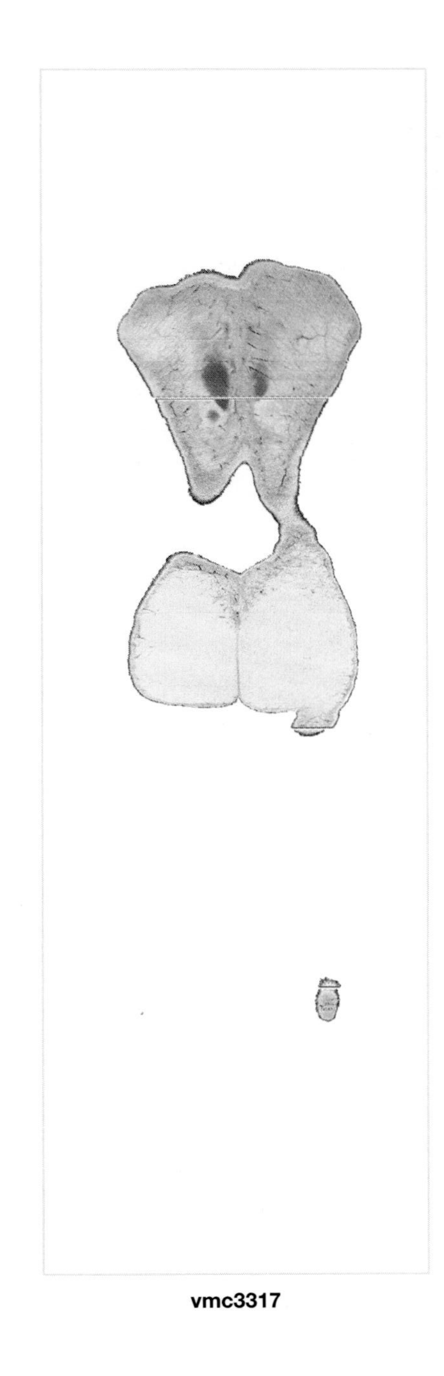

vmc3317

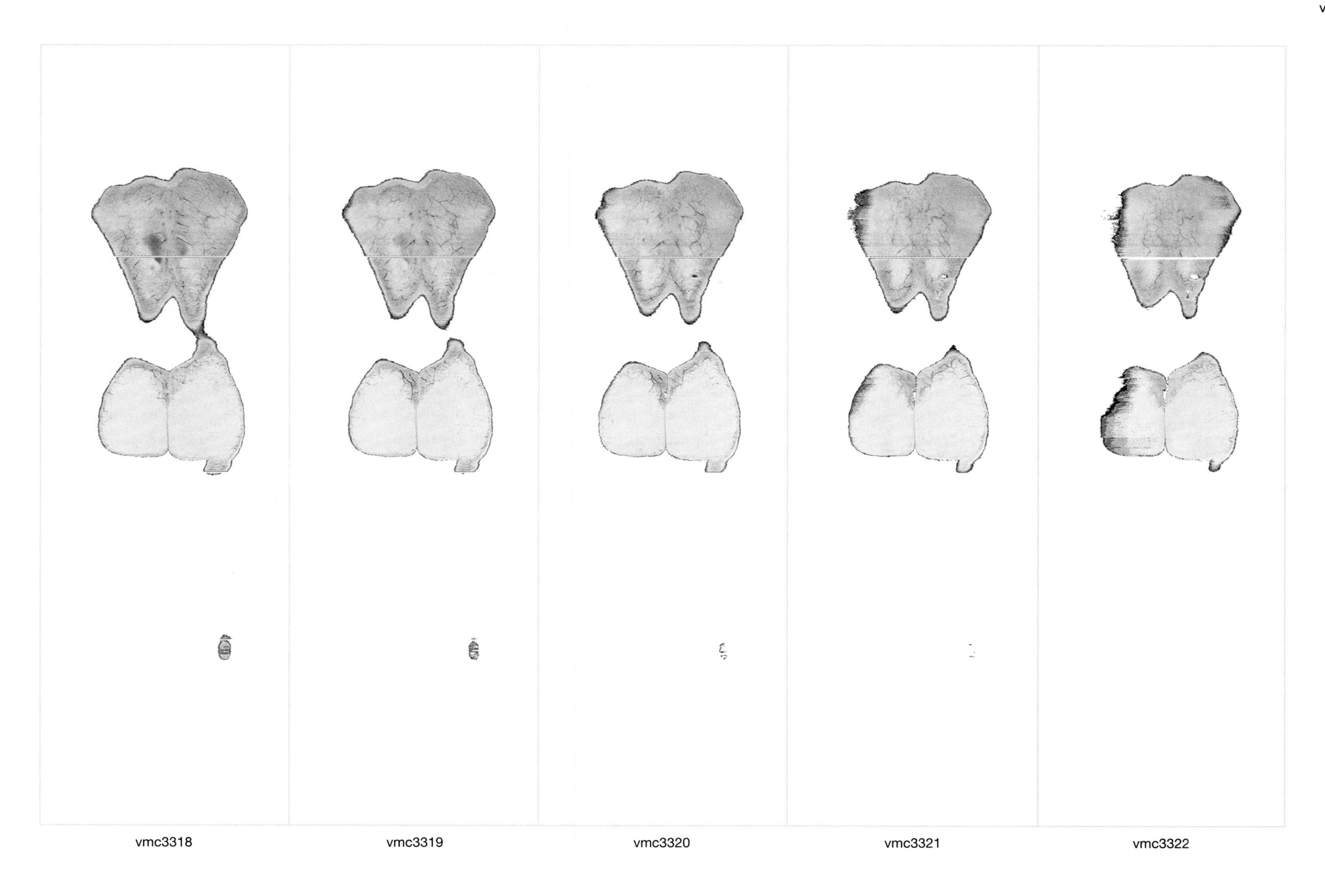
vmc3318
vmc3319
vmc3320
vmc3321
vmc3322

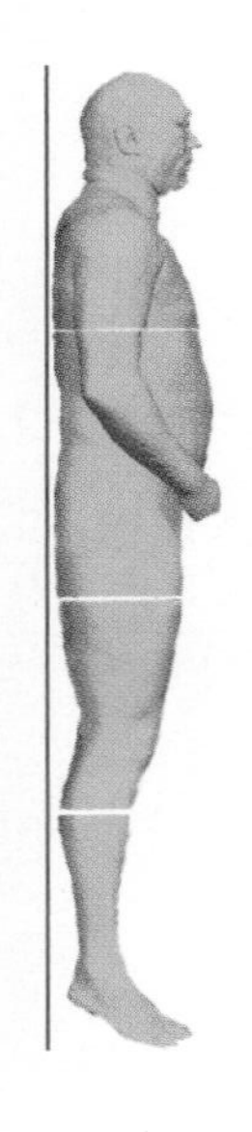

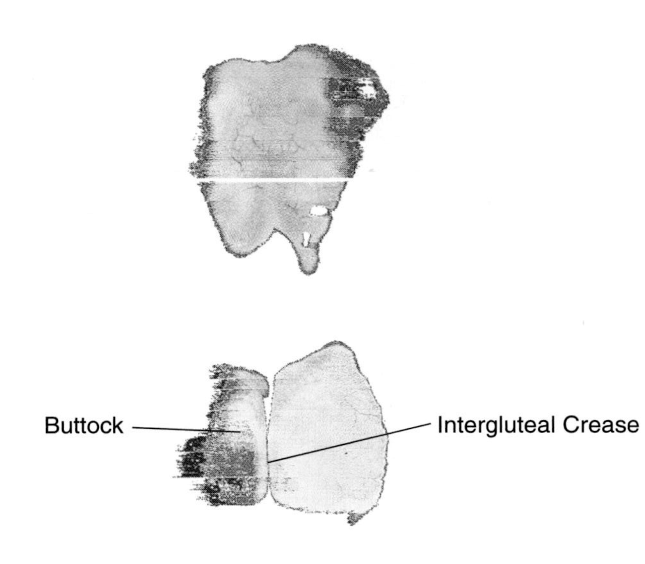
Buttock
Intergluteal Crease

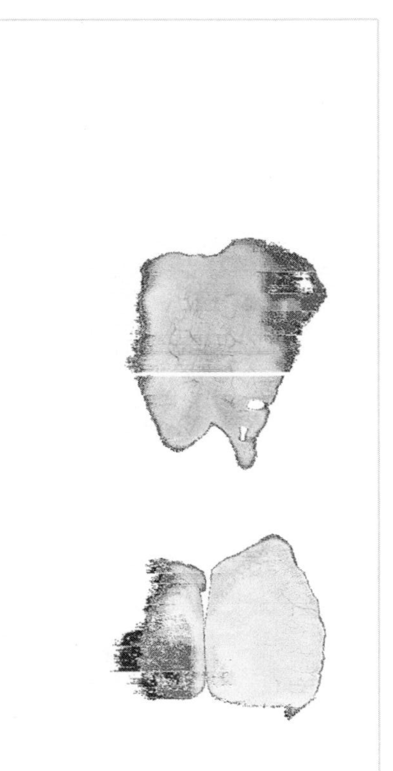

vmc3323

PART THREE

RECONSTRUCTED IMAGES

Sagittal

The sagittal images are of vertical planes through the long axis of the body parallel to or through that of the skull's sagittal suture. They are standard radiological planes used for direct magnetic resonance and ultrasound imaging. Sagittal clinical images are often reconstructed from transverse computed tomography and single photon emission computed tomography images. A sagittal image best demonstrates the arrangement of our anatomy from front to back. A special sagittal plane, the median sagittal plane, biscets many upaired anatomical structures (nose, mouth, spinal cord, etc.) and is the plane of symmetry for most bilateral anatomical structures (upper- and lower-extremity structures, eyes, kidneys, etc.).

These images, like their coronal counterparts, are virtual slices—the Visible Human Male was never physically cut in sagittal planes. Each sagittal image was computer reconstructed from an edge view of all slices in the transverse image collection as seen from the left lateral perspective and each is of a different depth at one-millimeter intervals through that collection.

The maximum resolution of images reconstructed in the sagittal plane is equivalent to the coronal plane reconstructions and is determined by the one-millimeter thickness of the original transverse sections.

The labeling format for sagittal images is the same as that used for the transverse and coronal images. In the same fashion as the coronal image display, the labeled slice is always repeated from the ten sagittal images on the same and facing pages.

Unlike transverse images, which include all slices, these reconstructed sagittal images begin at a full section through the juntion of the left shoulder and the chest wall and continue through most of the right arm. The scale remains the same for all sagittal images.

Sagittal
vms5158

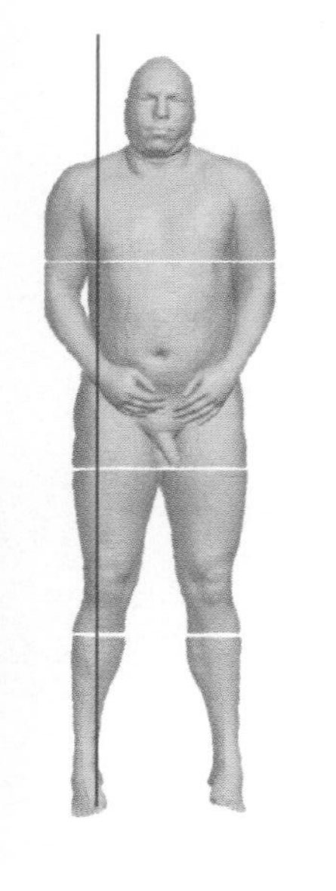

Clavicle
Coracoclavicular Ligament
Deltoid Muscle
Coracobrachialis Muscle
Cephalic Vein
Musculocutaneous Nerve
Pectoralis Major Muscle
Axillary Vessel
Brachial Plexus
Axilla
Serratus Anterior Muscle
External Oblique Muscle
Rib 8
Rib 9
Rib 9-10, Intercostal M.
Rib 10
Radius
Pronator Quadratus M.
Flexor Digitorum Superficialis Muscle
Flexor Carpi Ulnaris Muscle
Tensor Fascia Lata Muscle
Rectus Femoris Muscle
Vastus Intermedius Muscle
Fascia Lata
Articularis Genu
Rectus Femoris Tendon
Patella
Patellar Ligament
Anterior Cruciate Ligament
Tibial Tuberosity
Tibia, Shaft
Tibialis Posterior Muscle
Small Saphenous Vein
Fifth Metatarsal
Fifth Proximal Phalanx
Fifth Middle Phalanx

Coracoid Process
Trapezius Muscle
Supraspinatus Muscle
Scapula, Spine
Infraspinatus Muscle
Subscapularis Muscle
Scapula, Lateral Border
Teres Major Muscle
Latissimus Dorsi Muscle
External Oblique Muscle
Superficial Fascia and Adipose Tissue
Gluteus Medius Muscle
Gluteus Minimus Muscle
Femur, Greater Trochanter
Gluteus Maximus Muscle
Femur, Shaft
Vastus Lateralis Muscle
Biceps Femoris Muscle, Long Head
Short Head
Popliteal Fossa
Knee Joint, Cavity
Femur, Lateral Condyle
Knee Joint
Tibia, Lateral Condyle
Popliteal Vessel
Popliteus Muscle
Soleus Muscle
Crural Fascia
Superficial Fascia
Abductor Digiti Minimi Muscle
Heel
Flexor Digiti Minimi Brevis M.
Long Flexor Tendon

vms5153

vms5154

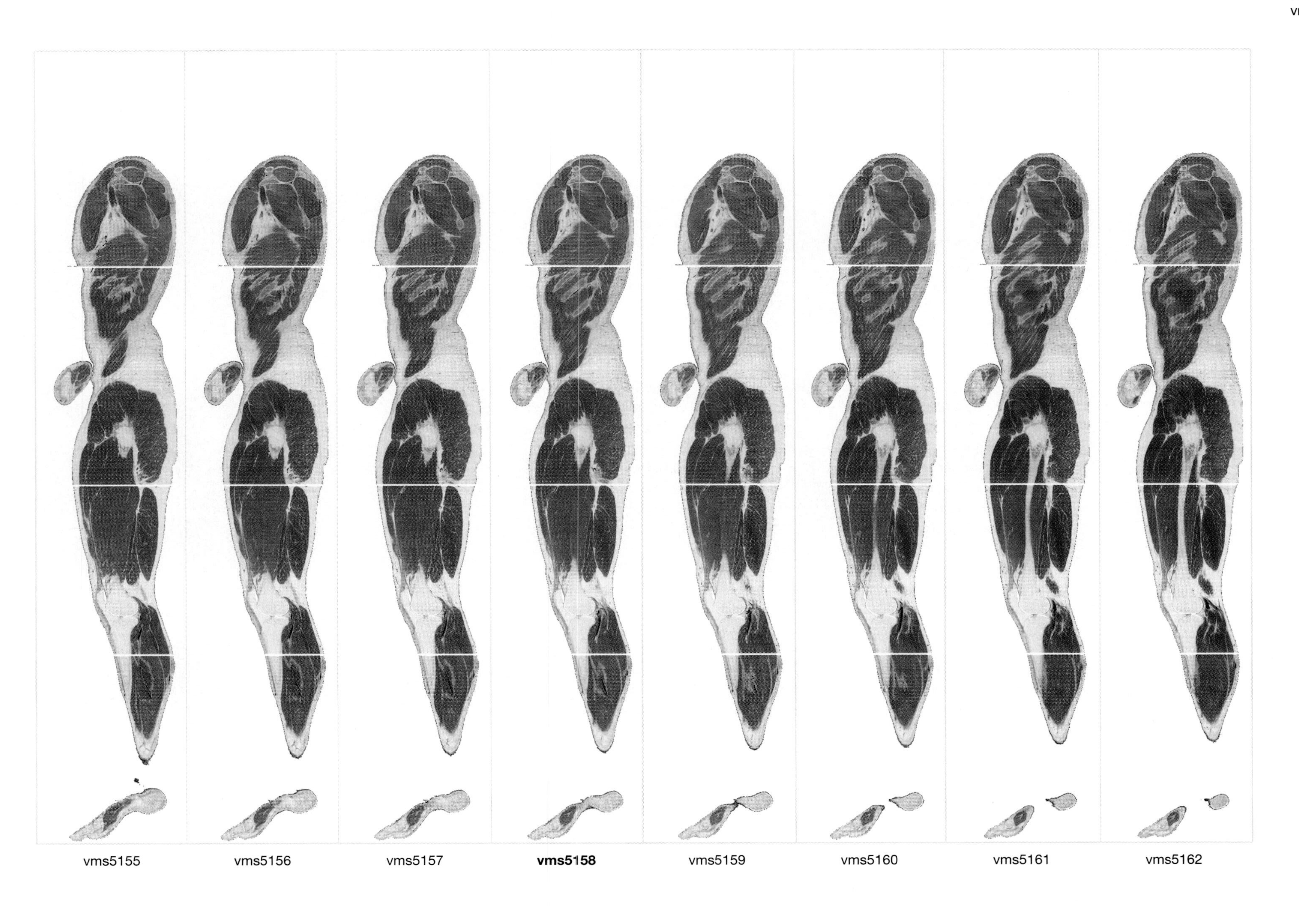
vms5155
vms5156
vms5157
vms5158
vms5159
vms5160
vms5161
vms5162

Sagittal
vms5168

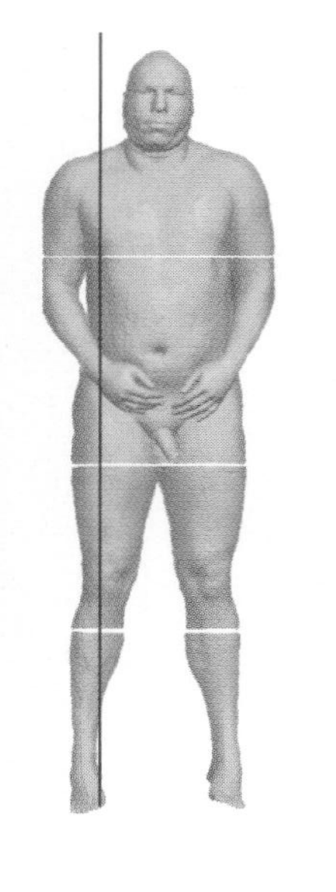

Clavicle
Brachial Plexus
Deltoid Muscle
Cephalic Vein
Pectoralis Minor Muscle
Pectoralis Major Muscle
Axilla
Serratus Anterior Muscle
Rib 6-7, Intercostal M.
Liver, Right Lobe
Diaphragm
External Oblique Muscle
Rib 11-12, Intercostal M.
External Oblique Muscle
Radius
Carpal Bones
Long Flexor Tendons
Extensor Digitorum Tendon
Hypothenar Muscles
Ilium, Crest
Tensor Fascia Lata Muscle
Iliofemoral Ligament
Adductor Magnus Muscle
Rectus Femoris Muscle
Rectus Femoris Tendon
Suprapatellar Bursa
Patella
Patellar Ligament
Anterior Cruciate Ligament
Popliteus Muscle
Tibialis Posterior Muscle

Trapezius Muscle
Supraspinatus Muscle
Scapula, Spine
Scapula
Subscapularis Muscle
Infraspinatus Muscle
Scapula, Lateral Border
Teres Major Muscle
Rib 5
Serratus Anterior Muscle
Diaphragm
Latissimus Dorsi Muscle
Rib 12, Costal Cartilage
Gluteus Medius Muscle
Gluteus Maximus Muscle
Femur, Greater Trochanter
Femur, Neck
Quadratus Femoris M.
Femur, Shaft
Fascia Lata
Biceps Femoris Muscle,
Long Head
Short Head
Vastus Intermedius Muscle
Semimembranosus Muscle
Popliteal Vessel and Nerve
Femur, Lateral Condyle
Gastrocnemius Muscle,
Medial Head
Gastrocnemius Muscle,
Lateral Head
Crural Fascia

Abductor Digiti Minimi Muscle
Flexor Digiti Minimi Brevis Muscle

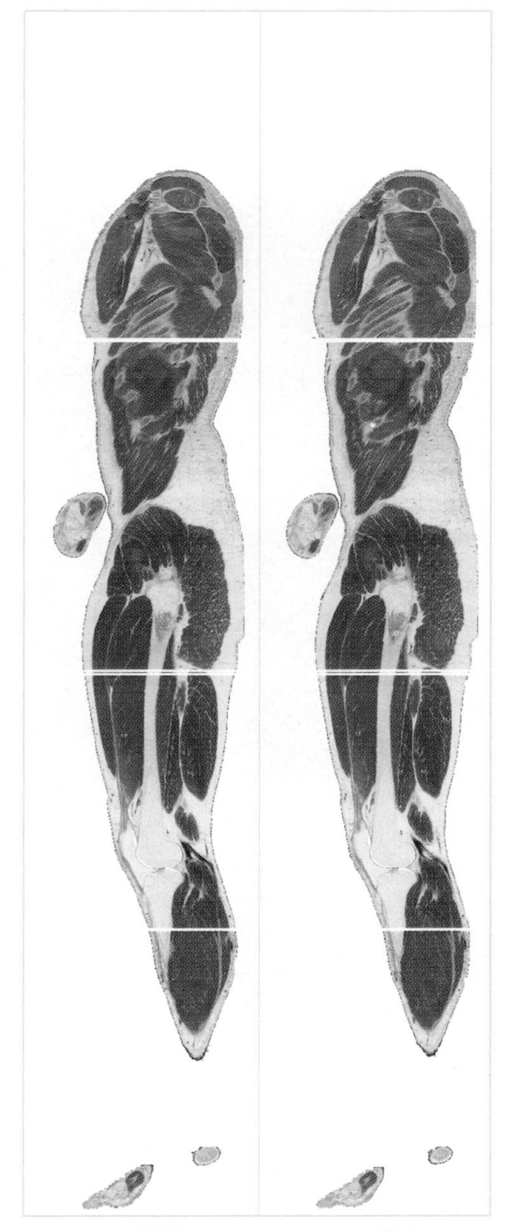
vms5163
vms5164

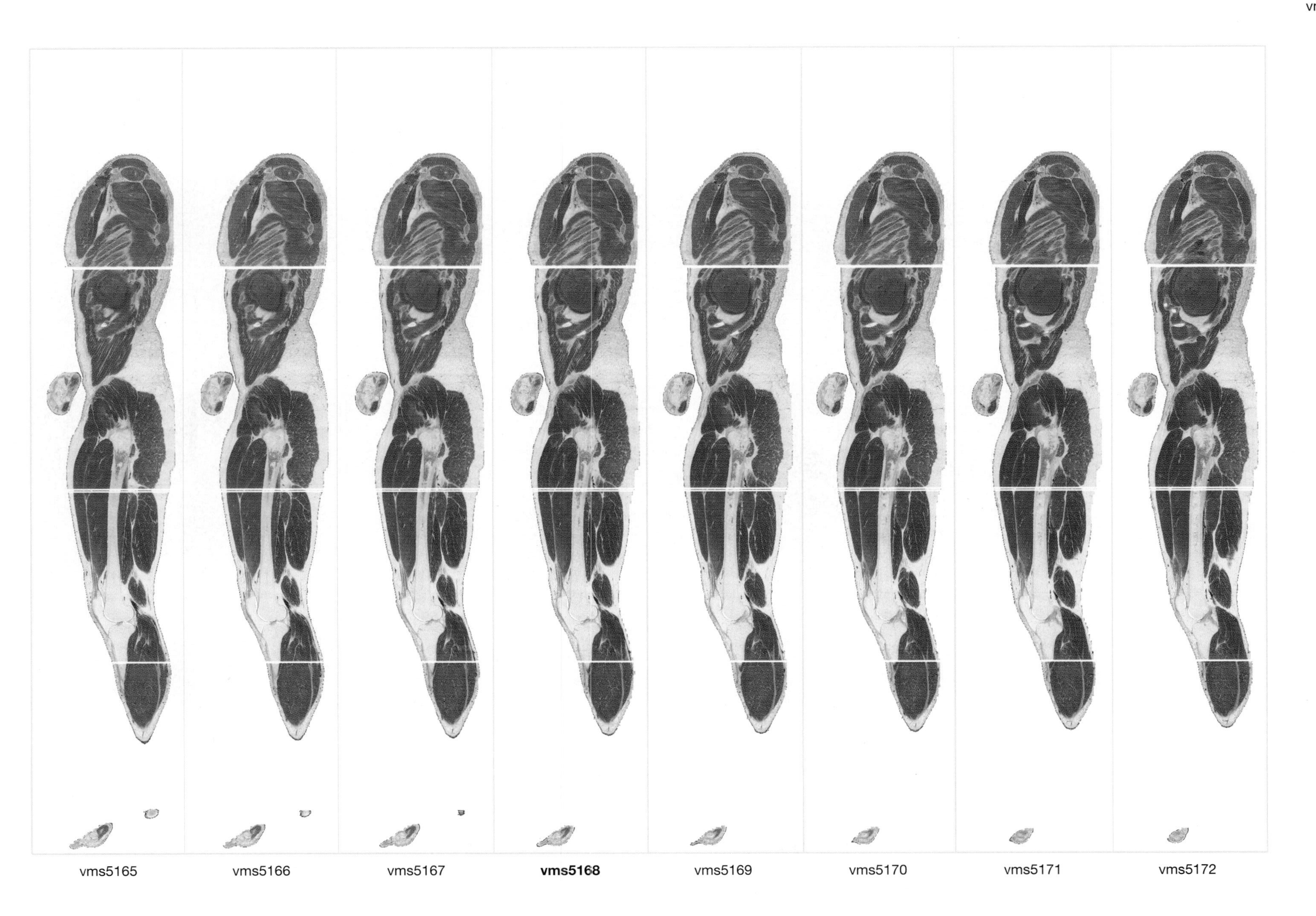
vms5165
vms5166
vms5167
vms5168
vms5169
vms5170
vms5171
vms5172

Sagittal
vms5178

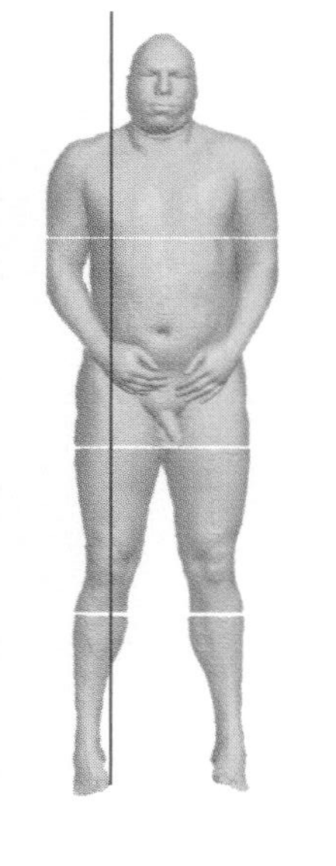

Suprascapular Vessel
Clavicle
Subclavius Muscle
Deltoid Muscle
Cephalic Vein
Axillary Vessel
Pectoralis Major Muscle
Brachial Plexus
Pectoralis Minor Muscle
Axilla
Rib 9
Liver, Right Lobe
Intra-abdominal Fat
Transversus Abdominis M.
External Oblique Muscle
Long Flexor Tendons
Carpal Bones
Fourth Metacarpal
Fourth Dorsal Interosseous M.
Fifth Metacarpal
Hypothenar Muscles
Sartorius Muscle
Iliofemoral Ligament
Femur, Neck
Femur, Intertrochanteric Crest
Rectus Femoris Muscle
Vastus Intermedius Muscle
Femur, Shaft
Vastus Medialis Muscle
Rectus Femoris Tendon
Patella
Infrapatellar Fat Pad
Femur, Lateral Condyle
Patellar Ligament
Tibia, Shaft
Popliteus Muscle

Supraspinatus Muscle
Trapezius Muscle
Scapula, Spine
Infraspinatus Muscle
Subscapularis Muscle
Rib 5
Serratus Anterior Muscle
Teres Minor Muscle
Scapula, Inferior Angle
Rib 8
Diaphragm
Rib 11-12, Intercostal M.
Rib 12
Latissimus Dorsi Muscle
External Oblique Muscle
Internal Oblique Muscle
Transversus Abdominis M.
Gluteus Medius Muscle
Ilium, Anterior Inferior Spine
Obturator Internus Tendon
Gluteus Minimus Muscle
Quadratus Femoris Muscle
Adductor Magnus Muscle
Biceps Femoris Muscle, Long Head
Short Head
Semimembranosus M.
Posterior Cruciate Ligament
Knee Joint
Tibia
Gastrocnemius Muscle, Medial Head
Soleus Muscle
Small Saphenous Vein

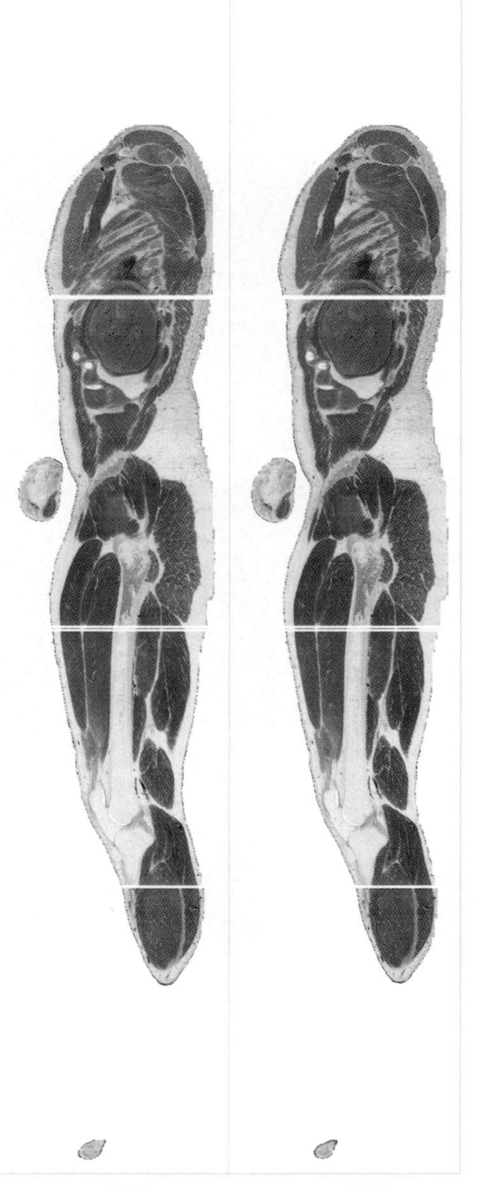

vms5173 vms5174

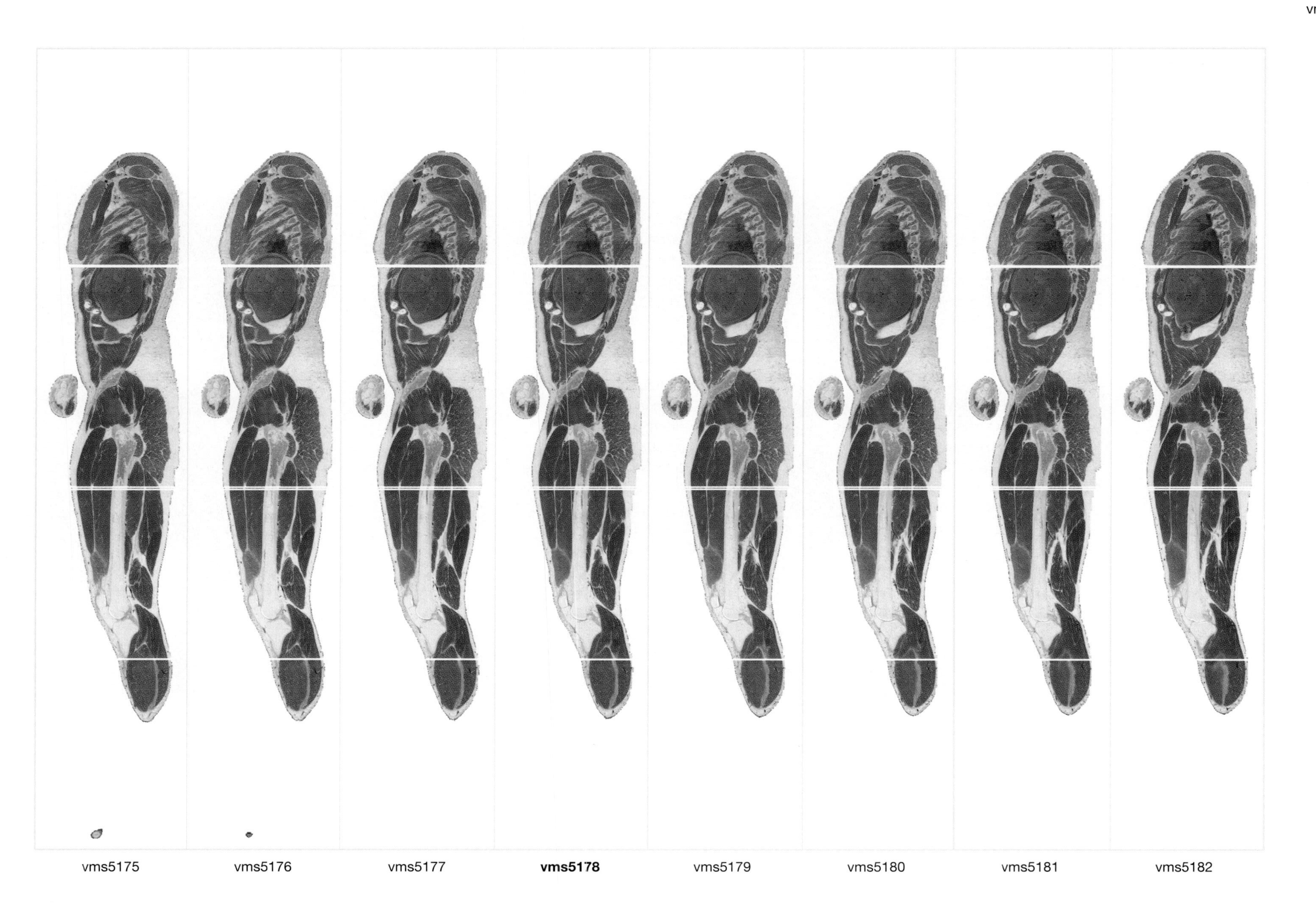

vms5175 vms5176 vms5177 **vms5178** vms5179 vms5180 vms5181 vms5182

Sagtital
vms5188

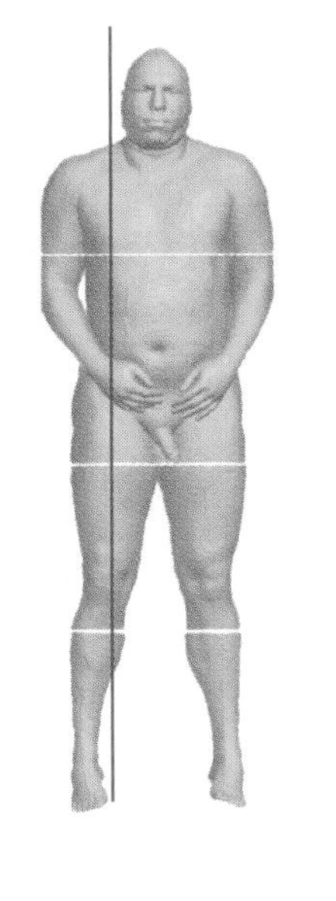

Clavicle
Subclavius Muscle
Cephalic Vein
Axilla
Axillary Vessel
Brachial Plexus
Pectoralis Minor Muscle
Pectoralis Major Muscle
Rib 4
Lung, Middle Lobe
External Oblique Muscle
Liver, Right Lobe
Rib 8, Costal Cartilage
External Oblique Muscle
Internal Oblique Muscle
Transversus Abdominis M.
Carpal Bones
Third Dorsal Interosseous
Long Flexor Tendons
Fourth Dorsal Interosseous M.
Fifth Metacarpal
Hypothenar Muscles
Sartorius Muscle
Iliofemoral Ligament
Iliopsoas Muscle
Femur, Neck
Rectus Femoris Muscle
Vastus Intermedius Muscle
Vastus Medialis Muscle
Medial Patellar Retinaculum
Knee Joint
Tibia, Periosteum
Crural Fascia
Soleus Muscle

Trapezius Muscle
Serratus Anterior Muscle
Supraspinatus Muscle
Scapula, Spine
Infraspinatus Muscle
Subscapularis Muscle
Rib 5
Serratus Anterior Muscle
Lung, Upper Lobe
Lung, Oblique Fissure
Lung, Lower Lobe
Diaphragm
Pararenal Fat
Renal Fascia
Ascending Colon
Iliacus Muscle
Ilium
Gluteus Medius Muscle
Gluteus Minimus Muscle
Gluteus Maximus Muscle
Gemelli and Obturator Internus Tendons
Quadratus Femoris M.
Femur, Intertrochanteric Crest
Biceps Femoris Muscle, Long Head
Adductor Magnus Muscle
Semitendinosus Muscle
Femur, Shaft
Semimembranosus M.
Popliteal Vessel
Gastrocnemius Muscle, Medial Head
Femur, Medial Condyle

vms5183

vms5184

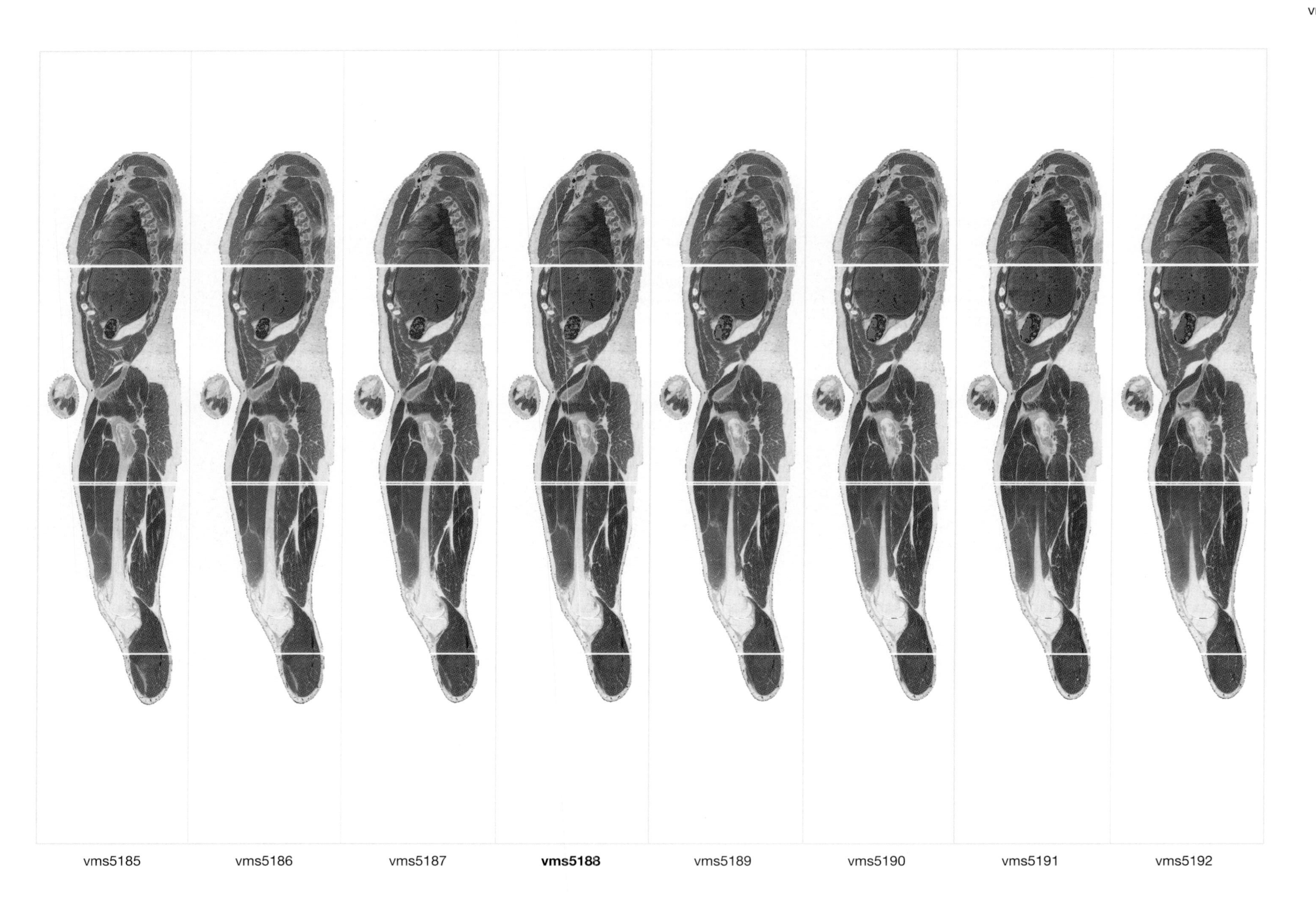

vms5185 vms5186 vms5187 **vms5188** vms5189 vms5190 vms5191 vms5192

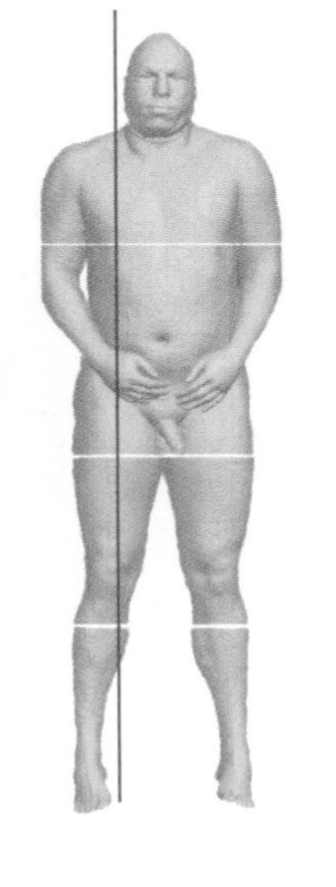

Serratus Anterior Muscle
Clavicle
Subclavius Muscle
Cephalic Vein
Axillary Vessel & Brachial Plexus
Pectoralis Major Muscle
Pectoralis Minor Muscle
Rib 2-3, Intercostal Muscles
Lung, Horizontal Fissure
Lung, Middle Lobe
Rib 4
Liver, Right Lobe
External Oblique Muscle
Transversus Abdominis M.
Ascending Colon
Internal Oblique Muscle
Trapezium Bone
First Dorsal Interosseous M.
Second Metacarpal
Third Metacarpal
Third Dorsal Interosseous M.
Long Flexor Tendons
Fifth Metacarpal
Abductor Digiti Minimi M.
Sartorius Muscle
Rectus Femoris Tendon
Rectus Femoris Muscle
Vastus Intermedius Muscle
Adductor Magnus Muscle
Adductor Canal, Opening
Knee Joint Cavity
Gastrocnemius Muscle, Medial Head

Omohyoid Muscle, Inferior Belly
Trapezius Muscle
Supraspinatus Muscle
Rib 2
Scapula
Subscapularis Muscle
Trapezius Muscle
Lung, Upper Lobe
Lung, Oblique Fissure
Lung, Lower Lobe
Diaphragm
Iliocostalis Thoracis M.
Rib 10-11, Intercostal M.
Perirenal Fat
Renal Fascia
Latissimus Dorsi Muscle
Internal Oblique Muscle
Transversus Abdominis M.
Ilium, Crest
Iliacus Muscle
Gluteus Medius Muscle
Gluteus Maximus Muscle
Ilium, Body
Femur, Head
Quadratus Femoris Muscle
Femur, Lesser Trochanter
Biceps Femoris Muscle, Long Head
Semitendinosus Muscle
Semimembranosus Muscle
Femur, Medial Condyle
Knee Joint
Tibia, Medial Condyle

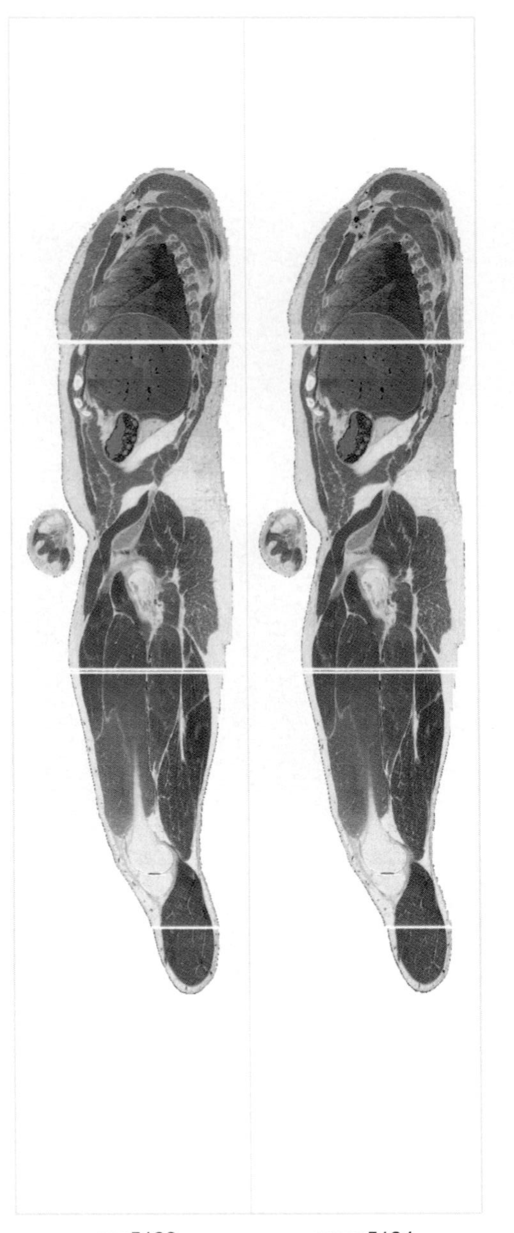

vms5193 vmsv5194

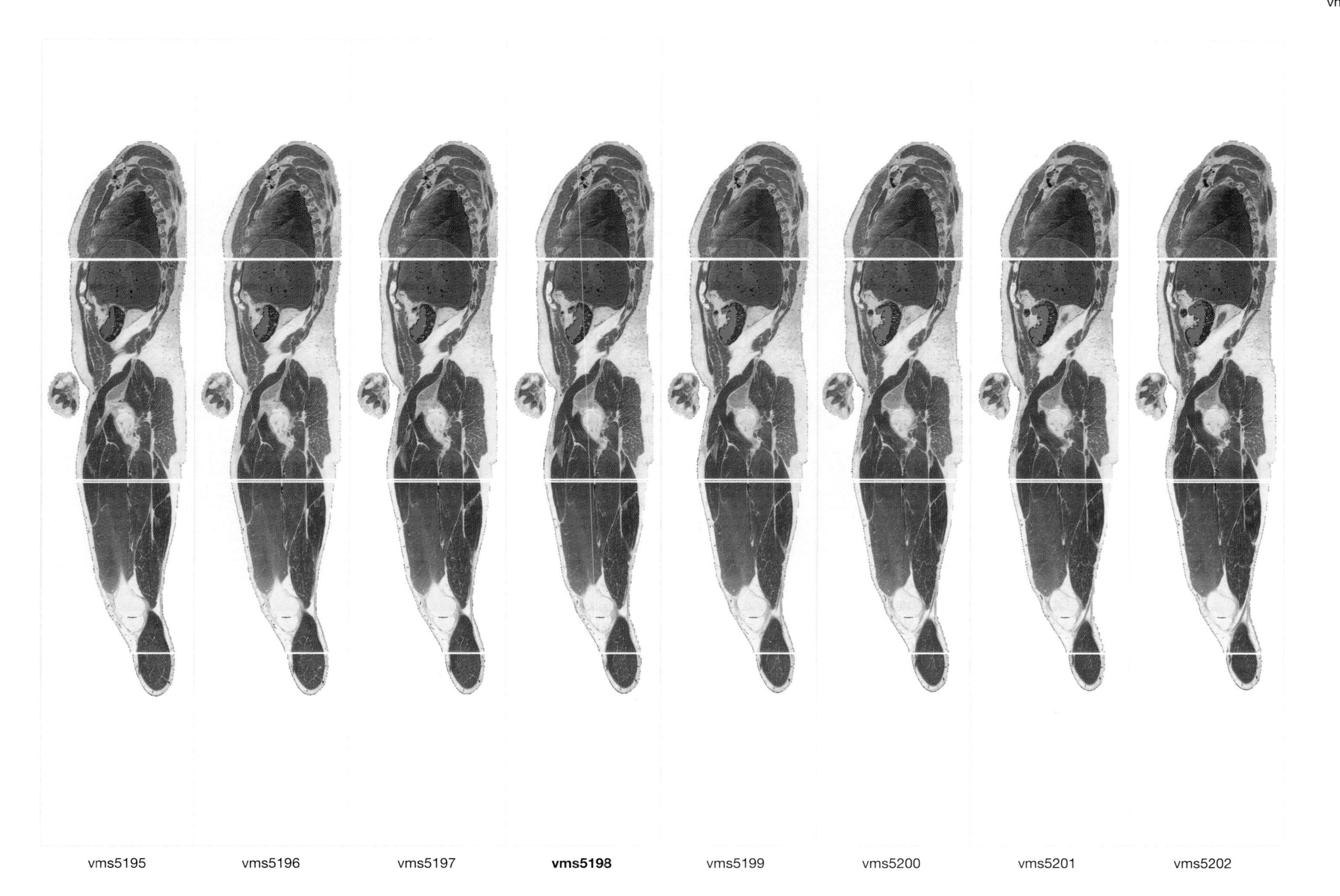

vms5195 vms5196 vms5197 **vms5198** vms5199 vms5200 vms5201 vms5202

Sagittal
vms5238

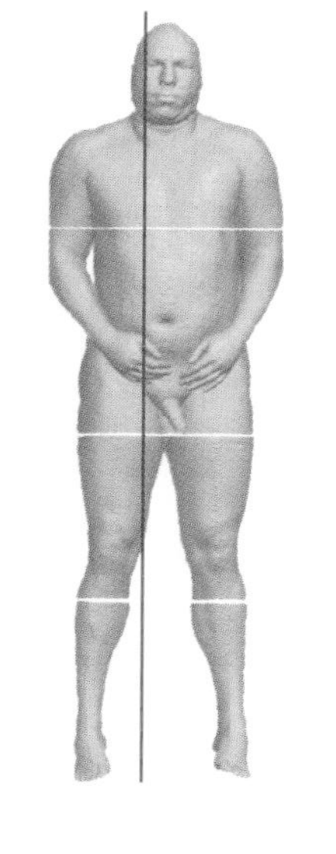

Frontal Bone
Temporalis Muscle
Orbicularis Oculi Muscle
Superficial Temporal Vein
Masseter Muscle
Facial Vein
Platysma Muscle
Sternocleidomastoid Muscle
Communicating Vein
Scalenus Medius Muscle
Brachial Plexus
Subclavian Vein
Pectoralis Major Muscle
Diaphragm
Colon, Hepatic Flexure
Portal Vein
Ascending Colon
Tendinous Intersection
Rectus Abdominis Muscle
First Metacarpal, Head
Second Metacarpal
Thenar Muscles
External Oblique Aponeurosis
Internal Oblique Muscle
Femoral Vein
Iliacus Muscle
Pubis
Pectineus Muscle
Great Saphenous Vein
Adductor Longus Muscle
Sartorius Muscle
Vastus Medialis Muscle
Great Saphenous Vein

Parietal Bone
Lateral Fissure
Occipitofrontalis Muscle, Occipital Belly
Temporal Lobe
External Acoustic Meatus
Auricular Cartilage
Splenius Capitis Muscle
Trapezius Muscle
Levator Scapulae Muscle
Rhomboid Major Muscle
Trapezius Muscle
Lung, Upper Lobe
Iliocostalis Thoracis M.
Liver, Caudate Lobe
Kidney
Latissimus Dorsi Muscle
Iliocostalis Muscle
Quadratus Lumborum M.
Gluteus Medius Muscle
Psoas Major Muscle
Ilium
Piriformis Muscle
Gluteus Maximus Muscle
Obturator Internus Muscle
Ischium
Semitendinosus Muscle
Obturator Externus Muscle
Adductor Brevis Muscle
Semimembranosus Muscle
Adductor Magnus Muscle
Gracilis Muscle

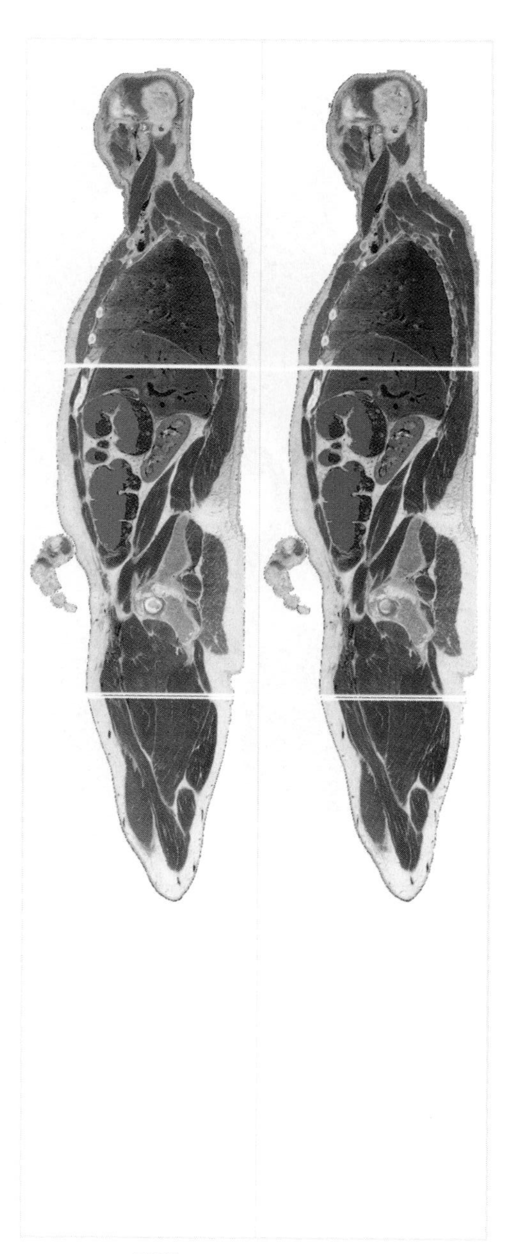

vms5233 vms5234

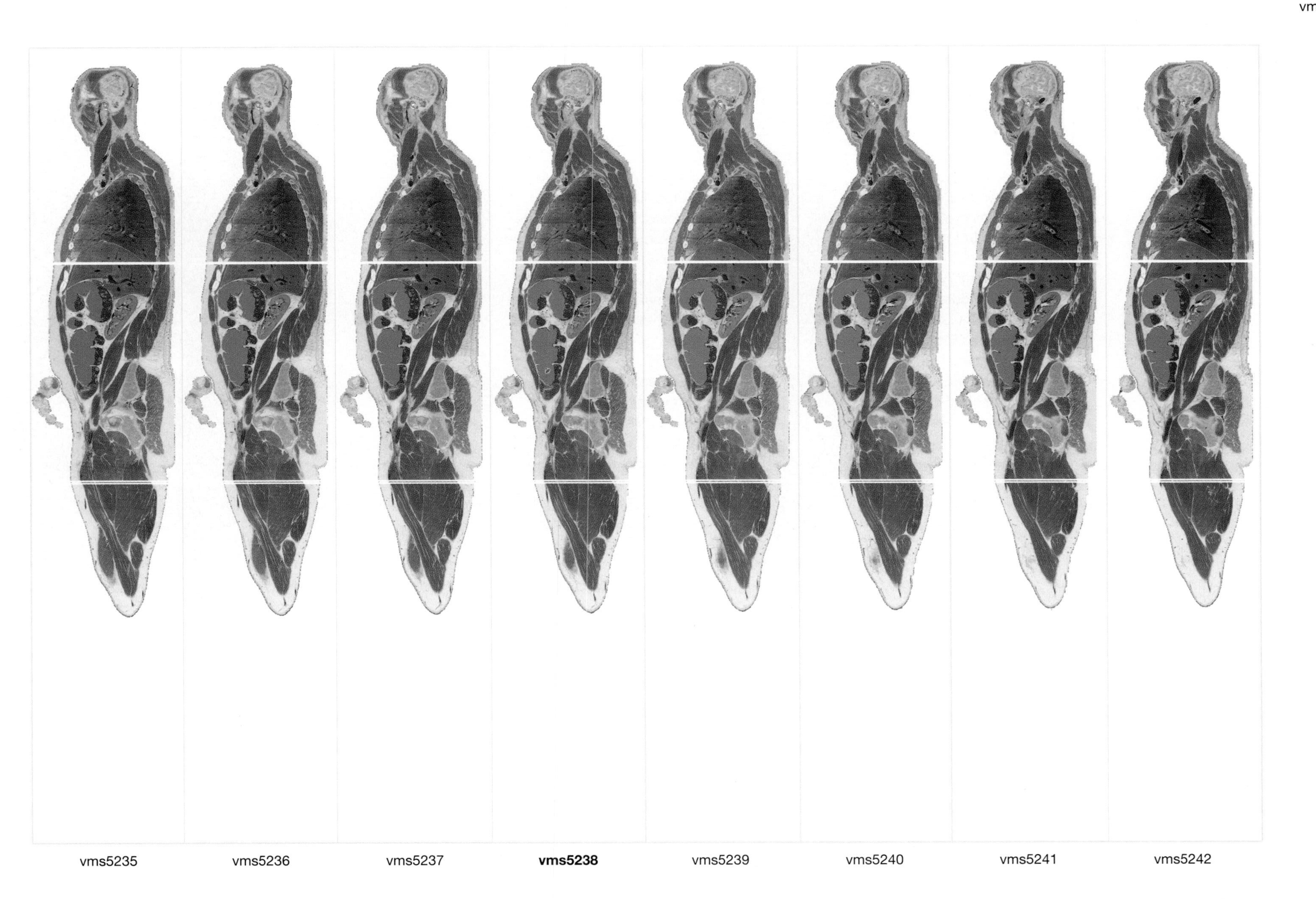
vms5235
vms5236
vms5237
vms5238
vms5239
vms5240
vms5241
vms5242

Sagittal
vms5248

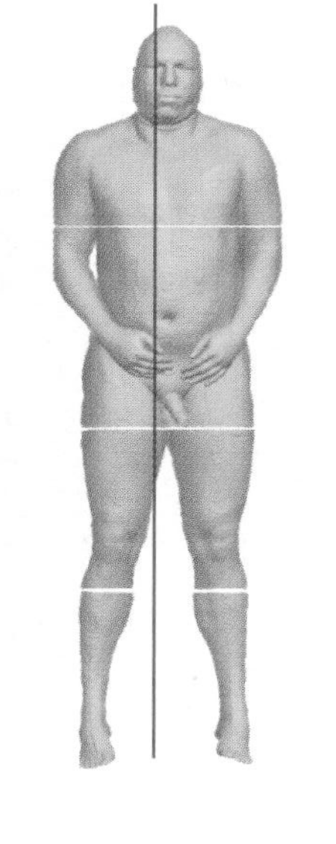

Frontal Bone
Frontal Lobe
Temporalis Muscle
Lateral Pterygoid Muscle
Masseter Muscle
Facial Vein
Digastric M., Posterior Belly
Internal Jugular Vein
Sternocleidomastoid M.
Scalenus Anterior Muscle
Clavicle
Pectoralis Major Muscle
Lung, Middle Lobe
Liver, Right Lobe
Colon, Hepatic Flexure
Ascending Colon
Tendinous Intersection
Rectus Abdominis Muscle
Cecum
Inguinal Canal, Internal Ring
Spermatic Cord
Femoral Vein
Femoral Vessel
Pubis
Pectineus Muscle
Adductor Brevis Muscle
Adductor Longus Muscle
Sartorius Muscle
Superficial Fascia
Great Saphenous Vein

Galea Aponeurotica
Parietal Bone
Parietal Lobe
Lateral Fissure
Transverse Sinus
Cerebellum
Semispinalis Capitis M.
Trapezius Muscle
Levator Scapulae Muscle
Scalenus Posterior Muscle
Brachial Plexus
Lung, Upper Lobe
Pulmonary Artery
Diaphragm
Adrenal Gland
Kidney
Duodenum
Longissimus Muscle
Psoas Major Muscle
Ilium
Sacrum, Lateral Wing
Piriformis Muscle
Obturator Vessel
Gluteus Maximus Muscle
Obturator Internus Muscle
Ischium
Obturator Externus Muscle
Semimembranosus Muscle
Adductor Magnus Muscle
Gracilis Muscle

vms5243

vms5244

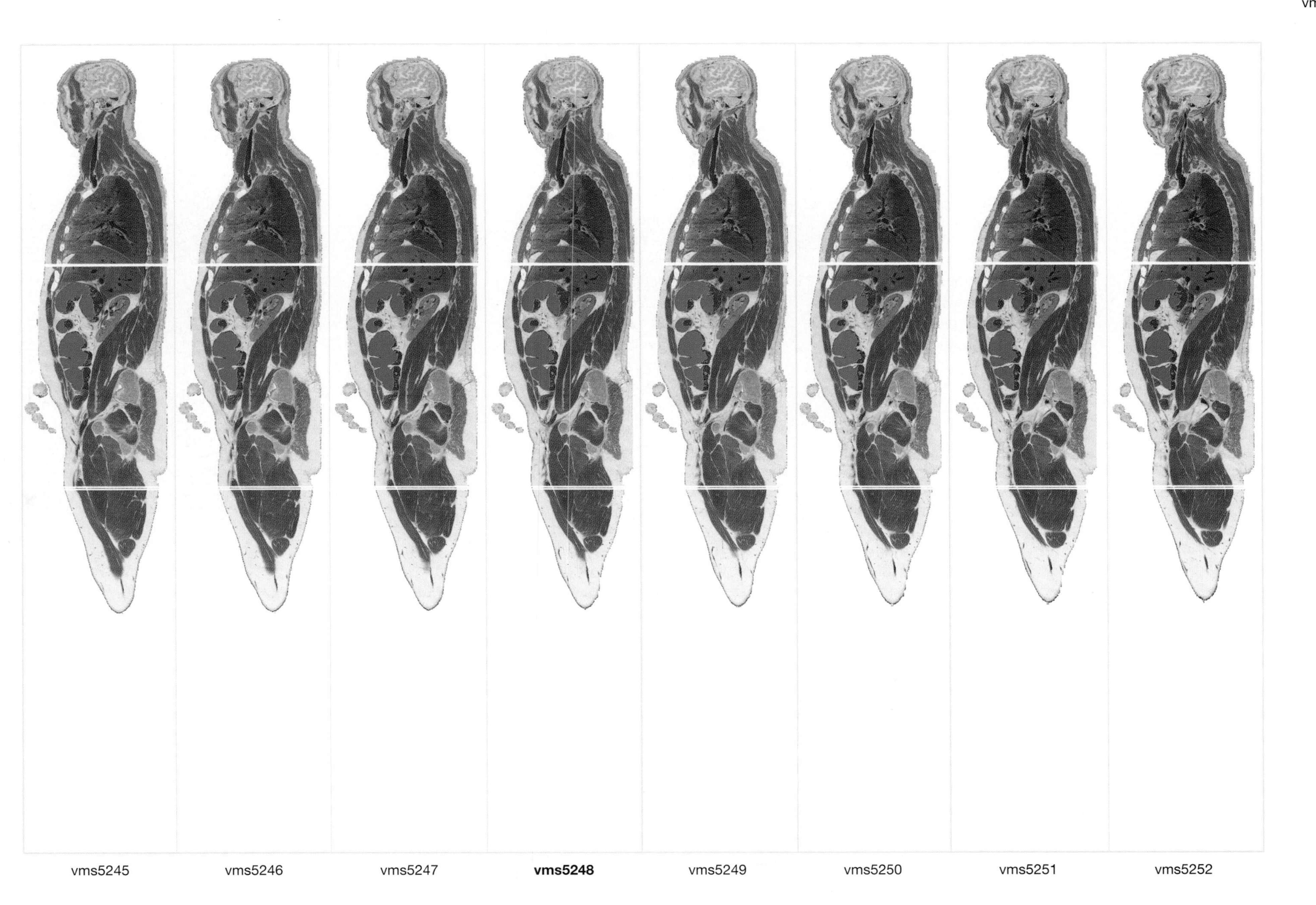

vms5245 vms5246 vms5247 **vms5248** vms5249 vms5250 vms5251 vms5252

Sagittal
vms5258

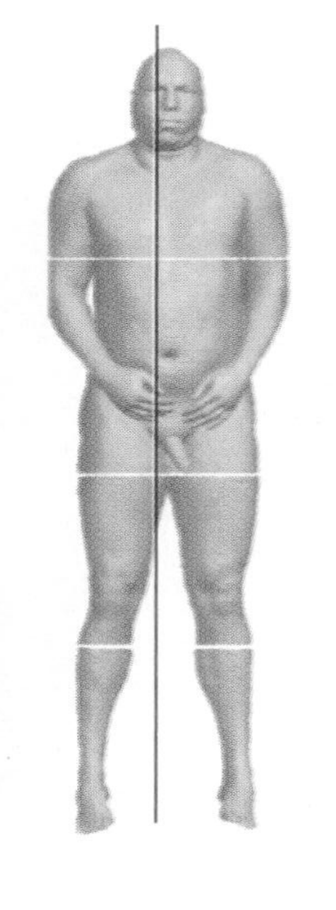

Frontal Bone
Frontal Lobe
Frontal Sinus
Eye Globe
Orbicularis Oculi Muscle
Maxillary Sinus
Temporal Lobe
Orbicularis Oris Muscle
Buccinator Muscle
Mandible
Medial Pterygoid Muscle
Common Carotid Artery
Sternocleidomastoid M.
Brachiocephalic Vein
Lung, Middle Lobe
Right Atrium
Secondary Bronchus
Liver, Right Lobe
Transverse Colon
Ascending Colon
Ileum
Rectus Abdominis Muscle
Spermatic Cord
Pubis, Superior Ramus
Pectineus Muscle
Adductor Longus Muscle
Adductor Brevis Muscle
Vastus Medialis Muscle
Great Saphenous Vein
Superficial Fascia

Galea Aponeurotica
Parietal Bone
Parietal Lobe
Lateral Fissure
Transverse Sinus
Tentorium
Cerebellum
Sigmoid Sinus
Obliquus Capitis Inferior M.
Trapezius Muscle
Lateral Pterygoid Muscle
Styloglossus Muscle
Levator Scapulae Muscle
Lung, Upper Lobe
Pulmonary Artery
Diaphragm
Adrenal Gland
Kidney
Duodenum
Iliocostalis Muscle
Psoas Major Muscle
Ilium
Sacroiliac Joint
Sacrum, Lateral Wing
Gluteus Maximus Muscle
Obturator Internus Muscle
Ischium
Obturator Externus Muscle
Adductor Magnus Muscle
Gracilis Muscle

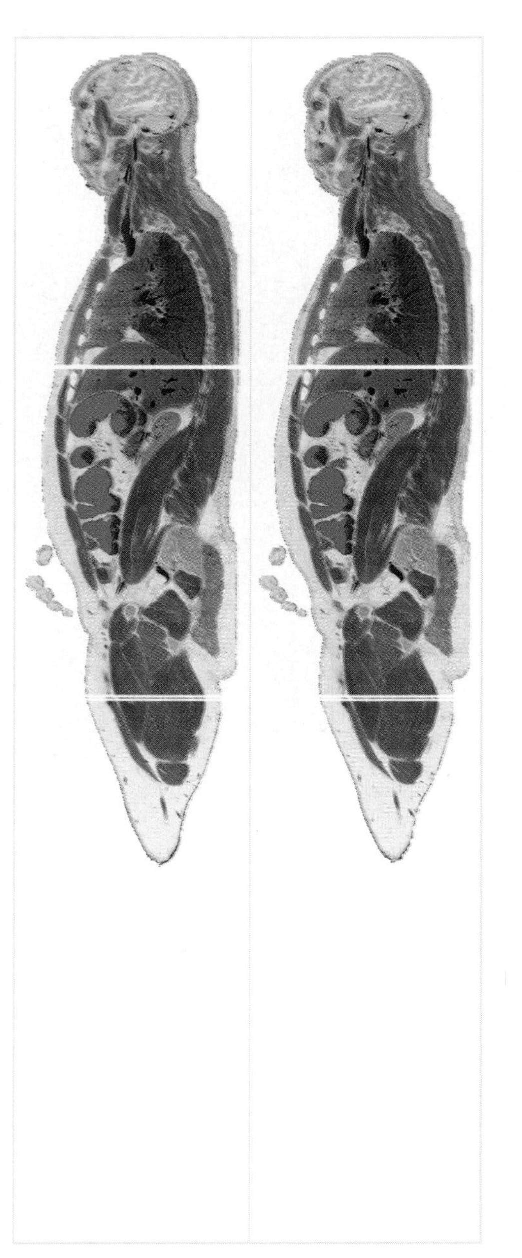

vms5253 vms5254

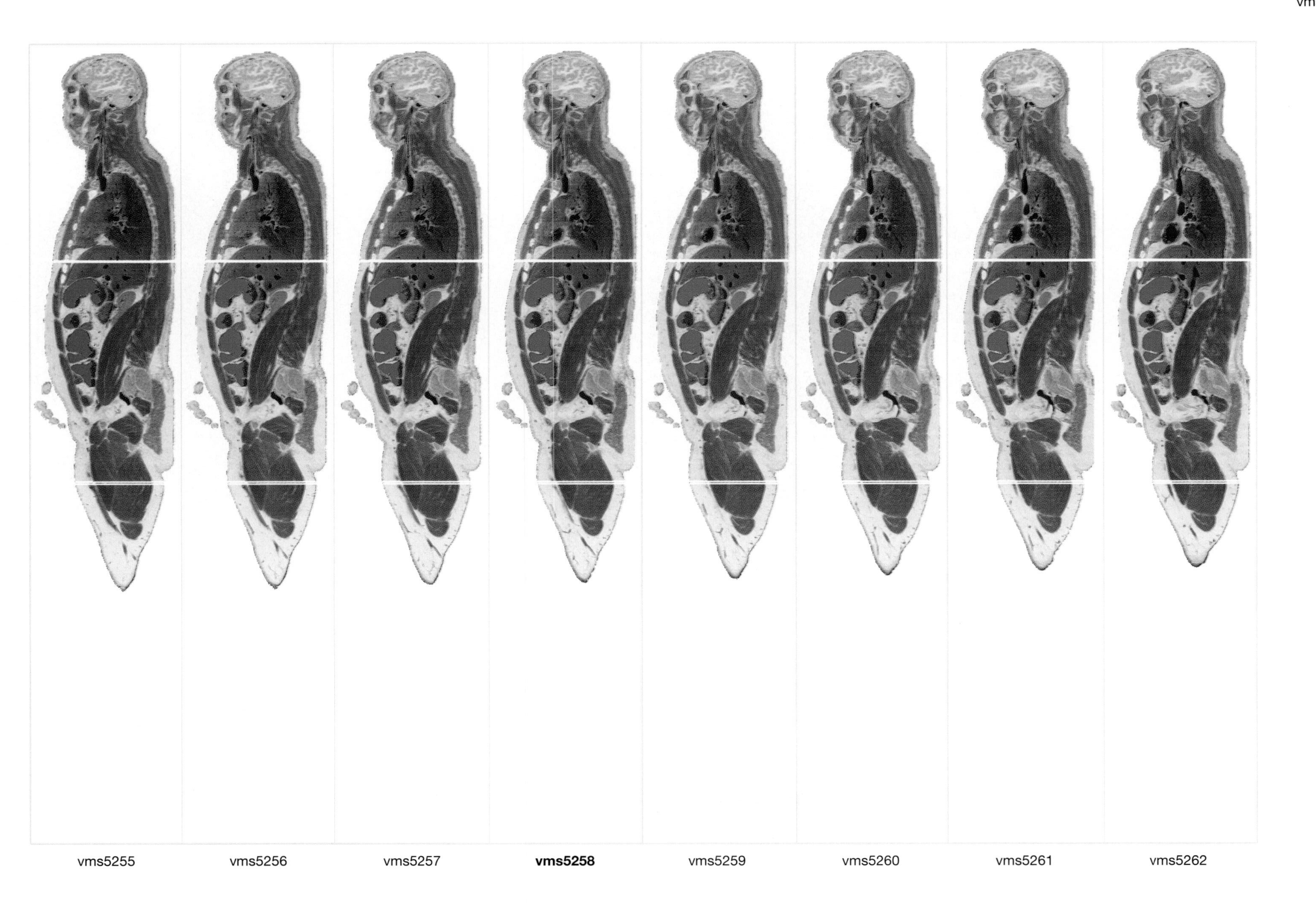

vms5255 vms5256 vms5257 **vms5258** vms5259 vms5260 vms5261 vms5262

Sagittal
vms5268

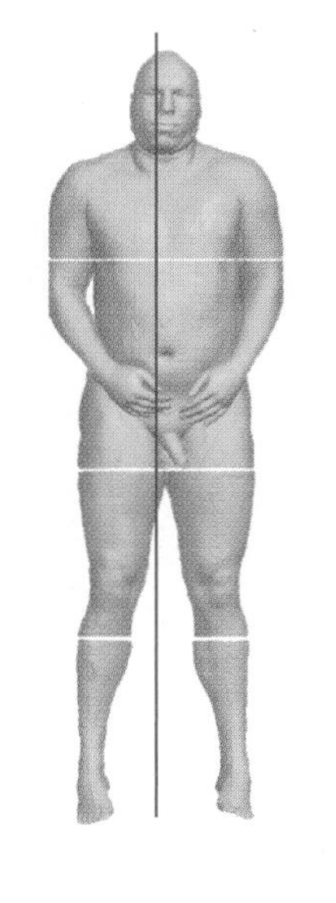

Frontal Bone
Frontal Lobe
Frontal Sinus
Eye Globe
Orbicularis Oculi Muscle
Maxillary Sinus
Orbicularis Oris Muscle
Mandible
Buccinator Muscle
Medial Pterygoid Muscle
Common Carotid Artery
Sternocleidomastoid M.
Superior Vena Cava
Right Atrium
External Oblique Muscle
Liver, Quadrate Lobe
Colon, Hepatic Flexure
Ascending Colon
Ileum
Rectus Abdominis Muscle
Spermatic Cord
Pubis, Superior Ramus
Pectineus Muscle
Adductor Brevis Muscle
Adductor Longus Muscle
Fascia Lata
Superficial Fascia
Gracilis Muscle
Galea Aponeurotica
Insula
Lateral Fissure
Transverse Sinus
Tentorium
Cerebellum
Sigmoid Sinus
Obliquus Capitis Inferior M.
Trapezius Muscle
Semispinalis Cervicis M.
Thyroid Gland
Lung, Upper Lobe
Primary Bronchus
Pulmonary Artery
Diaphragm
Adrenal Gland
Portal Vein
Inferior Vena Cava
Longissimus Muscle
Duodenum
Ilium
Psoas Major Muscle
Piriformis Muscle
Internal Iliac Vein
Gluteus Maximus Muscle
Bladder
Adductor Magnus Muscle

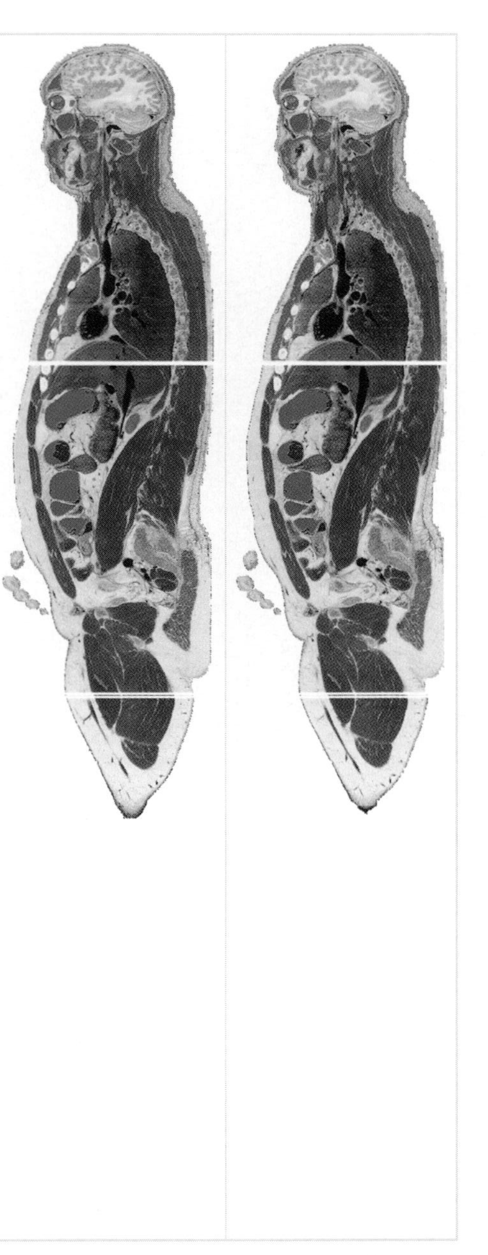

vms5263 vms5264

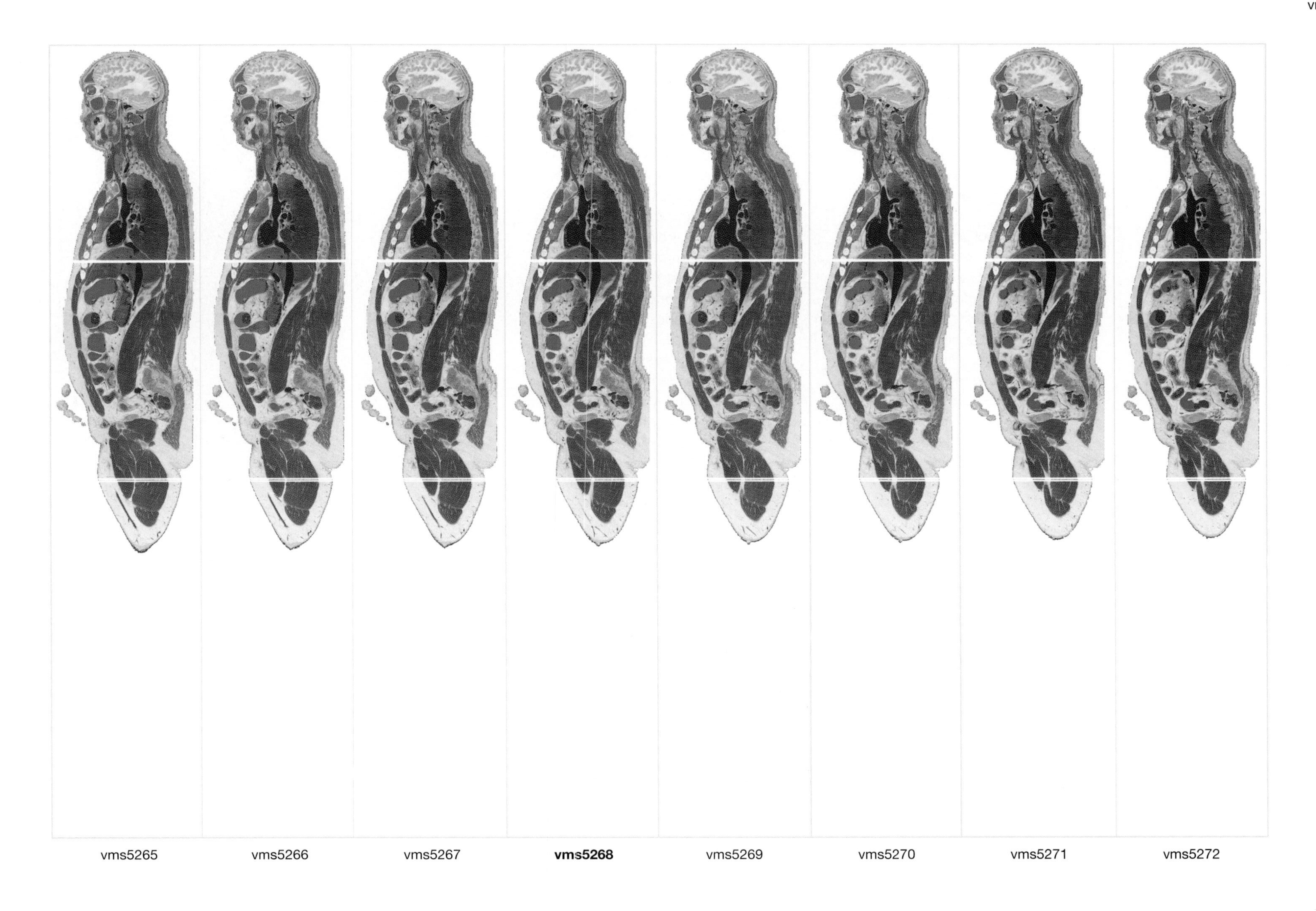
vms5265
vms5266
vms5267
vms5268
vms5269
vms5270
vms5271
vms5272

Sagittal
vms5278

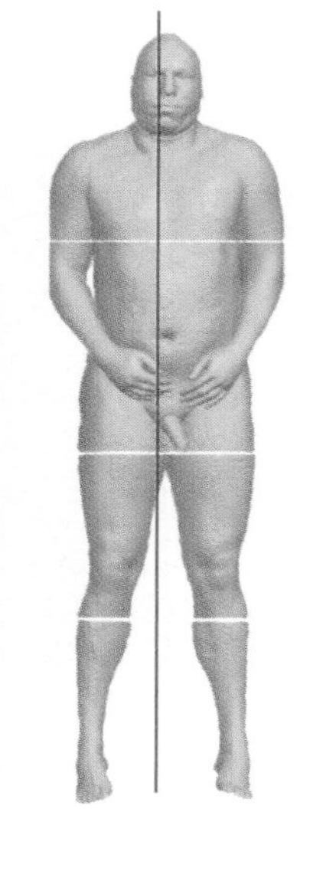

Frontal Bone
Lentiform Nucleus
Frontal Sinus
Lateral Rectus Muscle
Temporal Lobe
Orbicularis Oris Muscle
Hard Palate
Palatine Tonsil
Tongue
Sternohyoid Muscle
Manubrium
Lung, Upper Lobe
Superior Vena Cava
Right Atrium
Liver, Quadrate Lobe
Rectus Abdominis M.
Transverse Colon
Umbilicus
Ileum
Bladder
Pubis
Obturator Externus Muscle
Adductor Brevis Muscle
Adductor Longus Muscle
Pubis, Inferior Ramus
Gracilis Muscle
Superficial Fascia

Parietal Bone
Parietal Lobe
Lateral Ventricle
Transverse Sinus
Hippocampus
Cerebellum
Longus Capitis Muscle
Longus Cervicis Muscle
Trapezius Muscle
Thyroid Gland
Primary Bronchus
Superior Vena Cava
Pulmonary Vein
Inferior Vena Cava
Portal Vein
Pancreas
Duodenum
Longissimus Muscle
Jejunum
Psoas Major Muscle
Sacrum, Lateral Wing
Gluteus Maximus Muscle
Piriformis Muscle
Gluteus Maximus Muscle
Seminal Vesicle
Obturator Internus Muscle
Ischiocavernosus Muscle
Fascia Lata

vms5273

vms5274

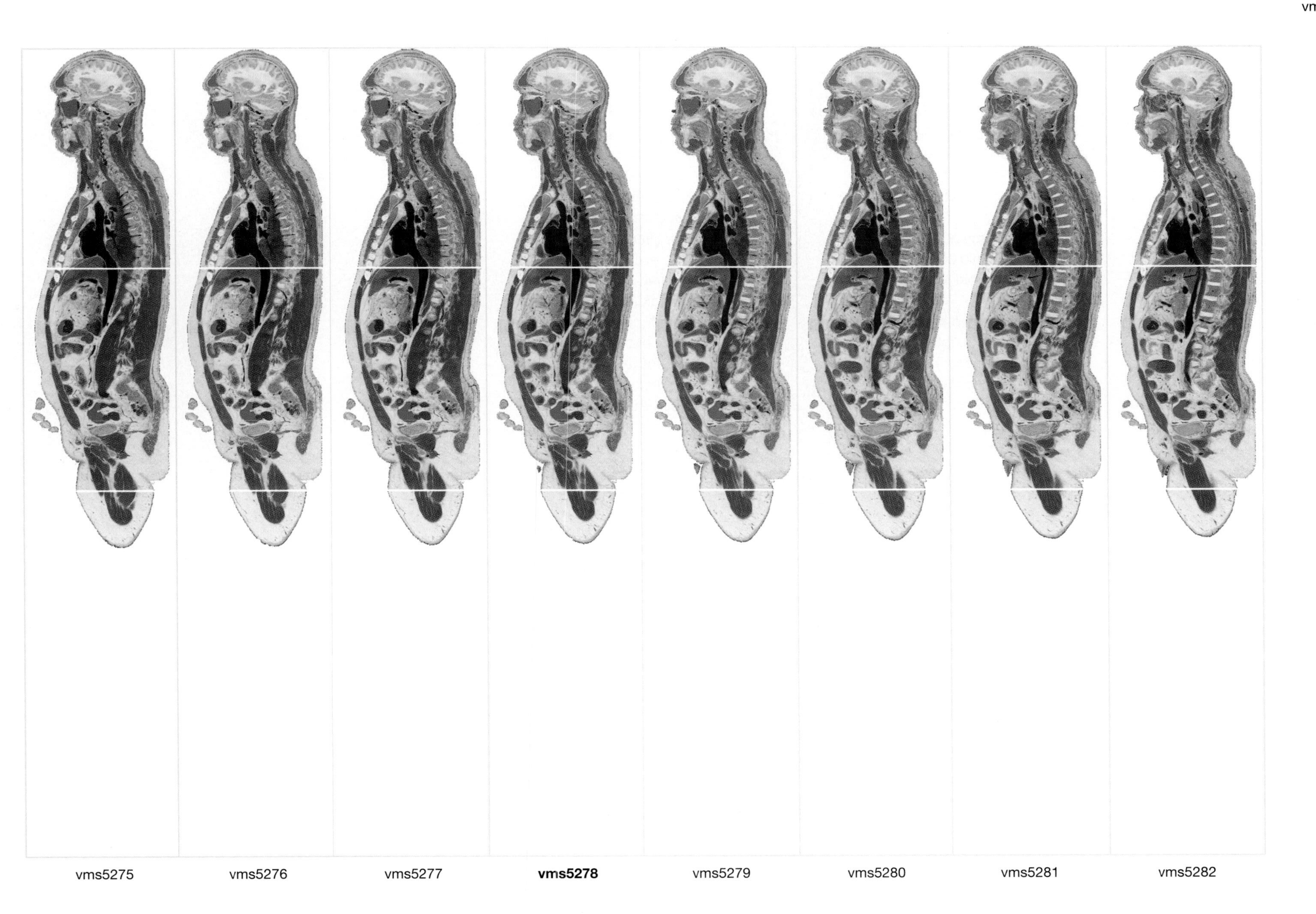

vms5275 vms5276 vms5277 **vms5278** vms5279 vms5280 vms5281 vms5282

Sagittal
vms5288

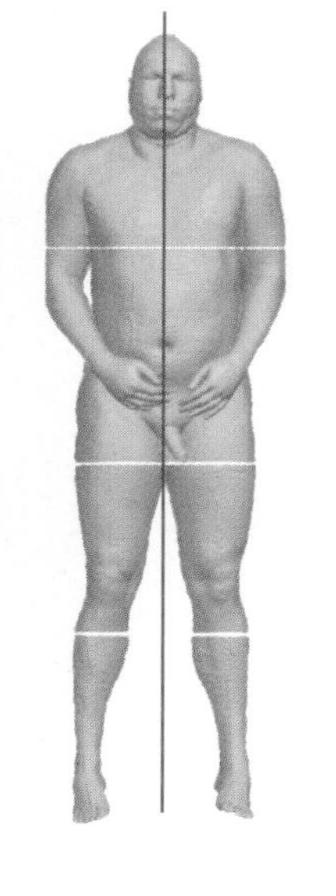

Frontal Bone
Frontal Lobe
Frontal Sinus
Caudate Nucleus
Ethmoid Sinuses
Lentiform Nucleus
Inferior Concha
Hard Palate
Tongue
Genioglossus Muscle
Geniohyoid Muscle
Mylohyoid Muscle
Pons
Sternohyoid Muscle
Thyroid Gland, Isthmus
Manubrium
Brachiocephalic Vein
Right Atrium
Liver, Quadrate Lobe
Rectus Abdominis M.
Transverse Colon
Umbilicus
Ileum
Bladder
Pubic Symphysis
Adductor Longus Muscle
Prostate Gland
Corpus Cavernosum
Ischiocavernosus Muscle
Urogenital Diaphragm

Parietal Bone
Parietal Lobe
Parietooccipital Sulcus
Dorsal Thalamus
Confluence of Sinuses
Tentorium
Cerebellum
Cerebellum, Dentate Nuc.
Medulla Oblongata
Multifidus Muscle
C2, Axis, Body
Spinal Cord
Epiglottis
Arytenoid Cartilage
Cricoid Cartilage
Esophagus
Pulmonary Artery
Left Atrium
Aorta, Arch
Lung, Lower Lobe
Inferior Vena Cava
Diaphragm
Splenic Vein
Duodenum, Superior Part
Pancreas
Coccygeus Muscle
Sacrum
Levator Ani Muscle
Seminal Vesicle
Gracilis Muscle

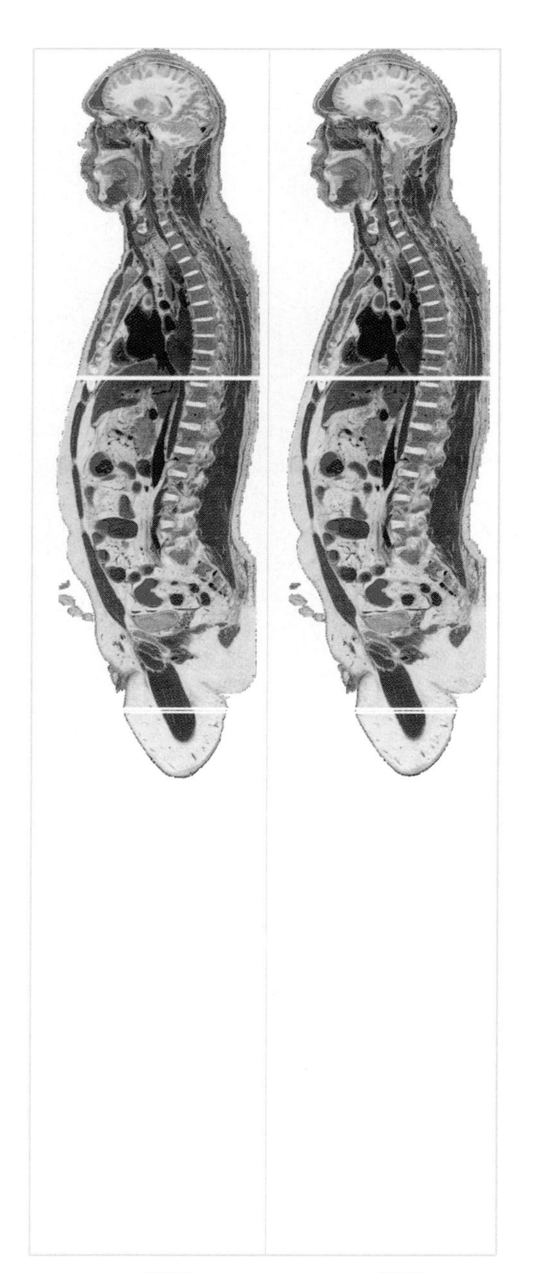

vms5283 vms5284

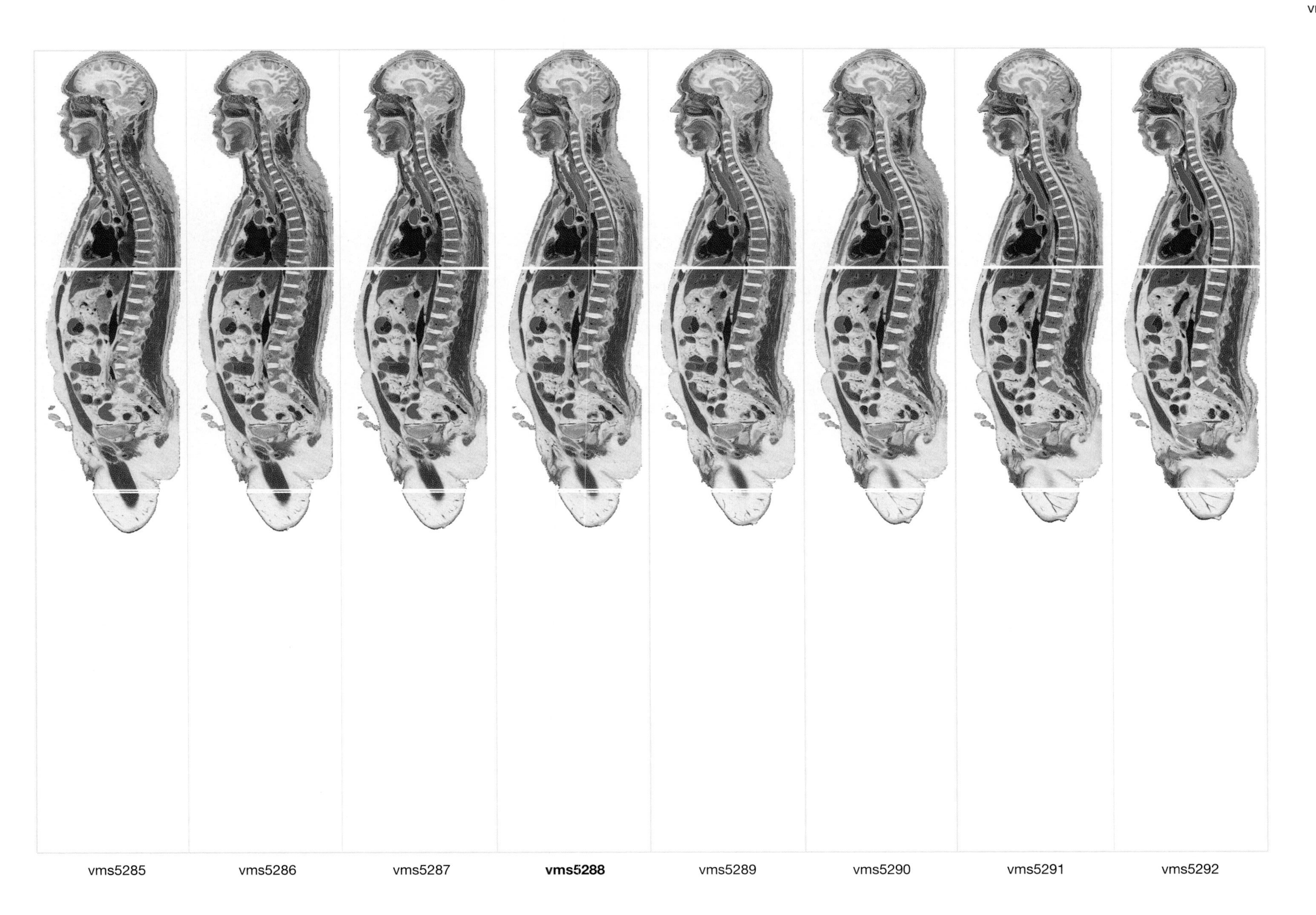

vms5285 vms5286 vms5287 **vms5288** vms5289 vms5290 vms5291 vms5292

Sagittal
vms5298

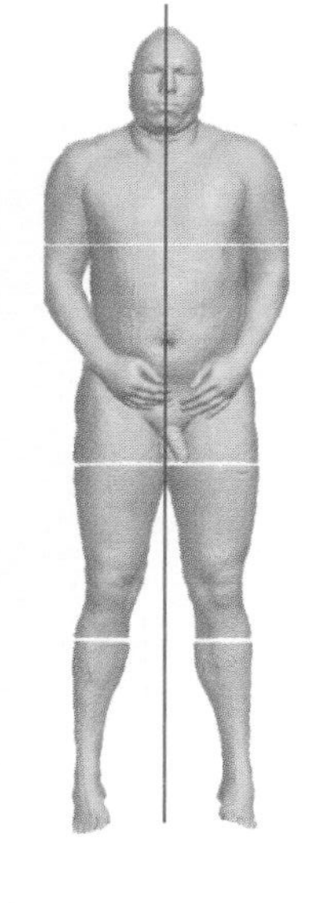

Corpus Callosum
Frontal Lobe
Frontal Sinus
Sphenoid Sinus
Nasal Cavity
Hard Palate
Orbicularis Oris Muscle
Mandible
Genioglossus Muscle
Epiglottis
Cricoid Cartilage
Trachea
Brachiocephalic Vein
Right Ventricle
Rectus Abdominis M.
Pylorus
Linea Alba
Transverse Colon
Umbilicus
Ileum
Bladder
Pubic Symphysis
Corpus Cavernosum
Penis
Testicle
Scrotum
Penis, Crus
Urethra

Superior Sagittal Sinus
Dorsal Thalamus
Corpus Callosum, Splenium
Occipital Lobe
Tentorium
Semispinalis Capitis M.
Rectus Capitis Posterior Minor Muscle
Medulla Oblongata
Multifidus Muscle
Trapezius Muscle
Spinal Cord
Esophagus
Left Atrium
Ascending Aorta
Pancreas
Diaphragm, Crus
Superior Mesenteric Vein
Duodenum
Inferior Vena Cava
Common Iliac Artery
L5-S1, Intervertebral Disc
Sigmoid Colon
Coccygeus Muscle
Rectum
External Anal Sphincter
Prostate Gland
Urogenital Diaphragm

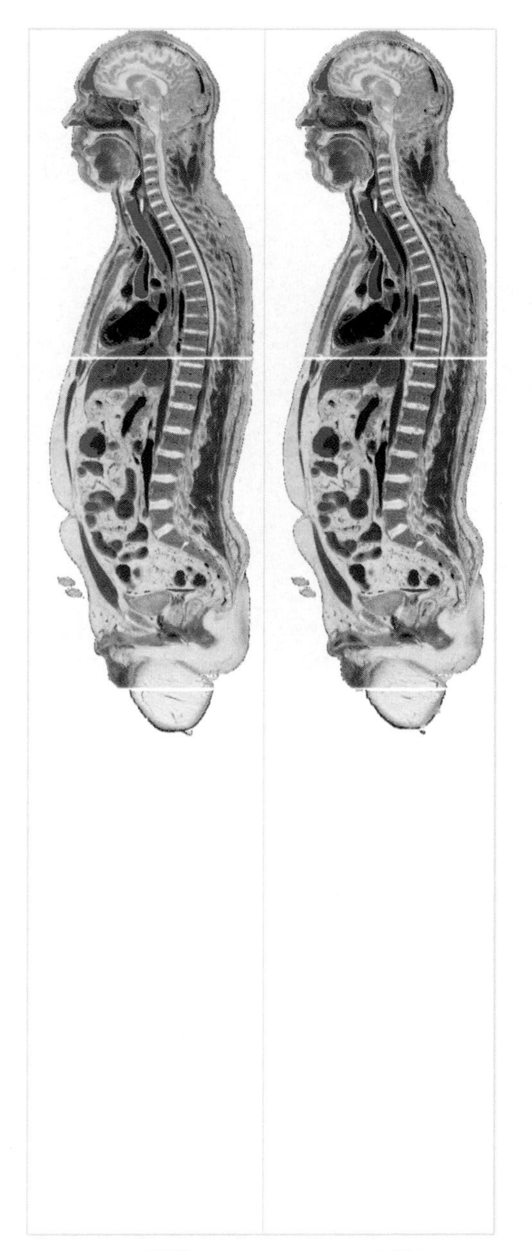

vms5293 vms5294

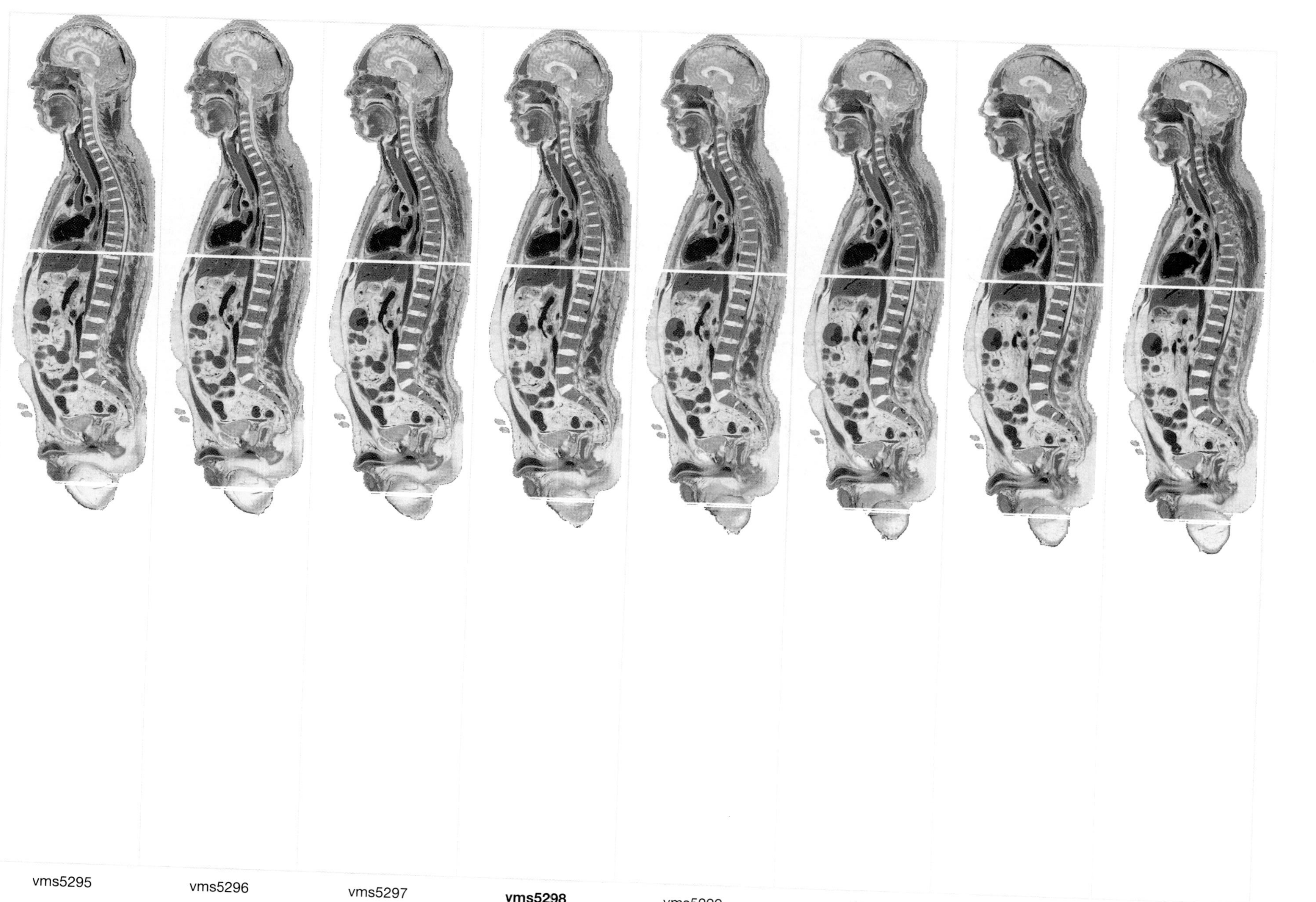
vms5295
vms5296
vms5297
vms5298
vms5299
vms5300
vms5301
vms5302

Sagittal
vms5408

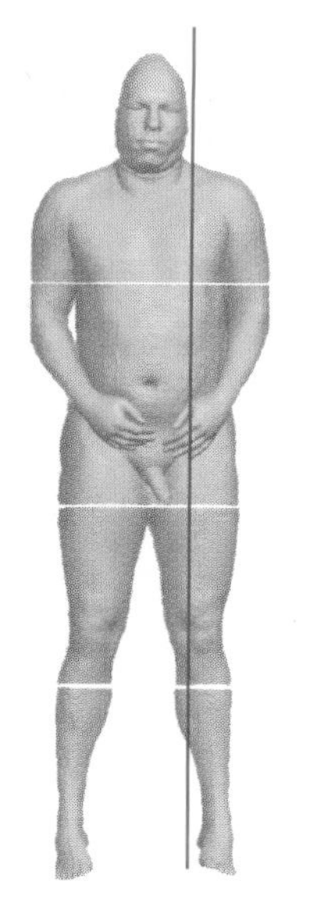

Clavicle
Subclavius Muscle
Axilla
Cephalic Vein
Serratus Anterior Muscle
Axillary Vessel
Pectoralis Major Muscle
Pectoralis Minor Muscle
Lung, Upper Lobe
Rib 4
Rib 4-5, Intercostal Muscles
Lung, Lingula
Diaphragm
Transverse Colon
Jejunum
Transversus Abdominis M.
External Oblique Muscle
Internal Oblique Muscle
First Metacarpal
First Dorsal Interosseous
Abductor Pollicis Muscle
Third Metacarpal
Flexor Pollicis Brevis Muscle
Sartorius Muscle
Iliofemoral Ligament
Femur, Head
Rectus Femoris Muscle
Iliopsoas Muscle
Femur, Shaft
Vastus Medialis Muscle
Vastus Medialis Muscle
Femur, Shaft
Medial Patellar Retinaculum
Knee Joint Cavity
Femur, Medial Condyle
Knee Joint
Tibia, Shaft
Popliteus Muscle
Posterior Cruciate Ligament
Trapezius Muscle
Supraspinatus Muscle
Infraspinatus Muscle
Scapula, Spine
Subscapularis Muscle
Scapula
Rib 5
Lung, Oblique Fissure
Scapula, Inferior Angle
Teres Major Muscle
Serratus Anterior Muscle
Lung, Lower Lobe
Latissimus Dorsi Muscle
Spleen
Colon, Splenic Flexure
Renal Fascia
Kidney
Serratus Post. Inferior M.
Descending Colon
Quadratus Lumborum M.
Iliacus Muscle
Gluteus Medius Muscle
Piriformis Muscle
Gluteus Maximus Muscle
Superior Gemellus Muscle
Inferior Gemellus Muscle
Quadratus Femoris Muscle
Femur, Intertrochanteric Crest
Femur, Neck
Biceps Femoris Muscle
Adductor Magnus Muscle
Sciatic Nerve
Semimembranosus M.
Popliteal Vein
Gastrocnemius Muscle, Lateral Head
Gastrocnemius Muscle, Medial Head
Soleus Muscle

vms5403

vms5404

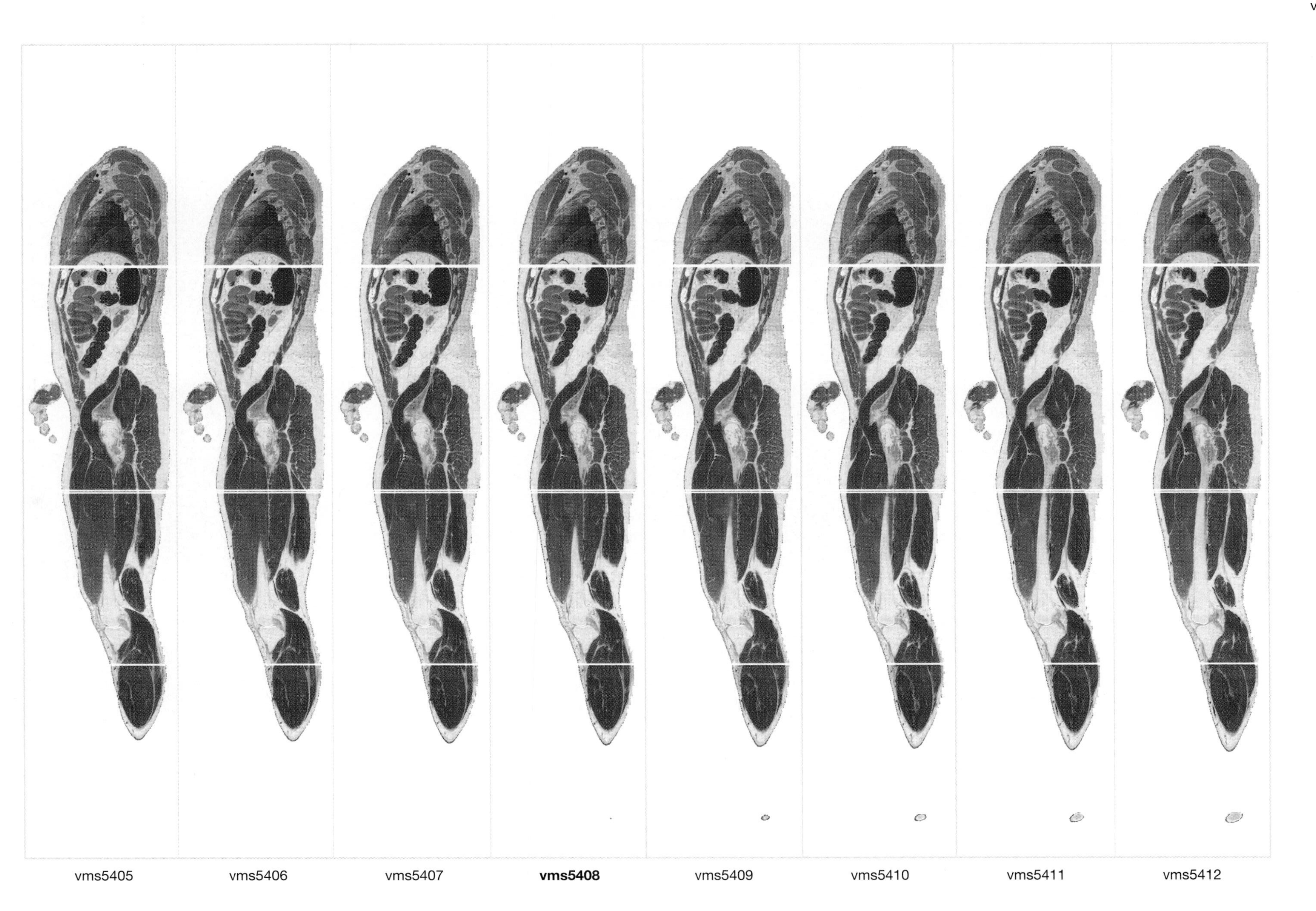

vms5405 vms5406 vms5407 **vms5408** vms5409 vms5410 vms5411 vms5412

Sagittal
vms5418

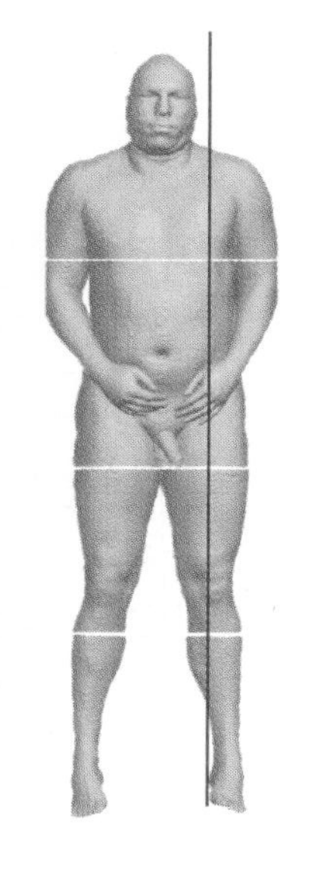

Clavicle
Deltoid Muscle
Axillary Vessels and Nerve
Cephalic Vein
Pectoralis Minor Muscle
Serratus Anterior Muscle
Pectoralis Major Muscle
Rib 4
Lung, Upper Lobe
Lung, Oblique Fissure
Transverse Colon
Jejunum
External Oblique Muscle
Internal Oblique Muscle
Transversus Abdominis M.
First Metacarpal
First Dorsal Interosseous
Opponens Pollicis M.
Second Metacarpal
Second Dorsal Interosseous
Third Metacarpal
Sartorius Muscle
Rectus Femoris Tendon
Iliopsoas Muscle
Iliofemoral Ligament
Femoral Vessel
Vastus Intermedius Muscle
Rectus Femoris Muscle
Femur, Neck
Vastus Medialis Muscle
Rectus Femoris Tendon
Suprapatellar Bursa
Femur, Lateral Condyle
Tibia, Lateral Condyle
Anterior Cruciate Ligament
Tibia, Shaft
Soleus Muscle
Tibialis Posterior Muscle
Feet, Superficial Fascia

Trapezius Muscle
Supraspinatus Muscle
Scapula, Spine
Infraspinatus Muscle
Subscapularis Muscle
Serratus Anterior Muscle
Scapula, Inferior Angle
Teres Major Muscle
Lung, Lower Lobe
Diaphragm
Latissimus Dorsi Muscle
Spleen
Rib 11
Rib 11-12, Intercostal M.
External Oblique Muscle
Internal Oblique Muscle
Iliacus Muscle
Ilium
Gluteus Medius Muscle
Gluteus Maximus Muscle
Gluteus Minimus Muscle
Piriformis Tendon
Obturator Internus Tendon
Quadratus Femoris M.
Femur, Intertrochanteric Crest
Femur, Shaft
Biceps Femoris Muscle, Long Head
Adductor Magnus Muscle
Biceps Femoris Muscle, Short Head
Popliteal Nerve
Semimembranosus Muscle
Popliteal Vein
Medial Meniscus
Popliteus Muscle
Soleus Tendon
Great Saphenous Vein

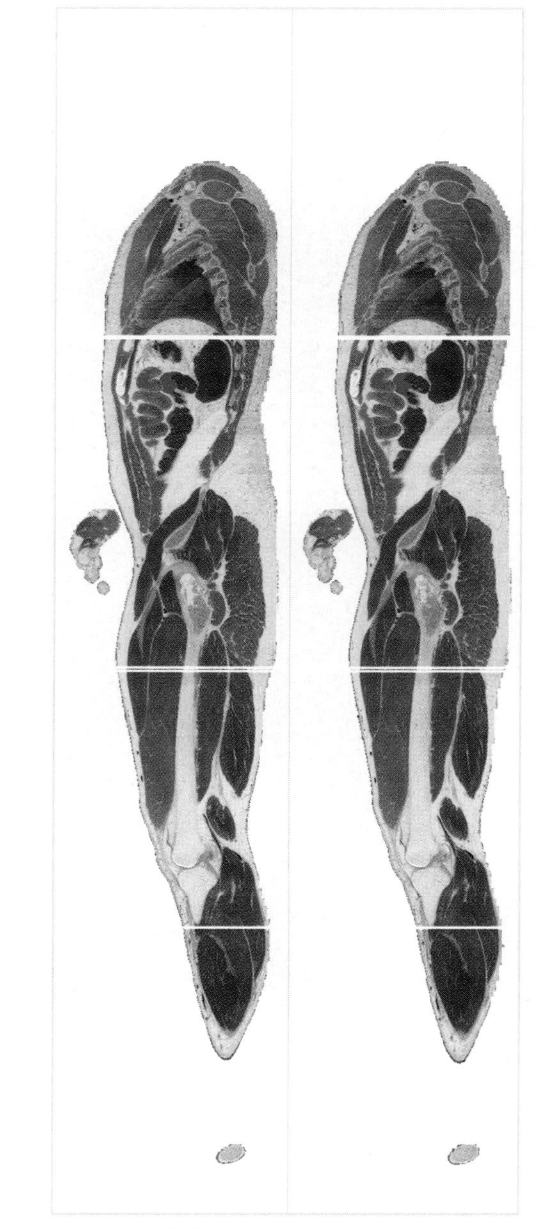

vms5413
vms5414

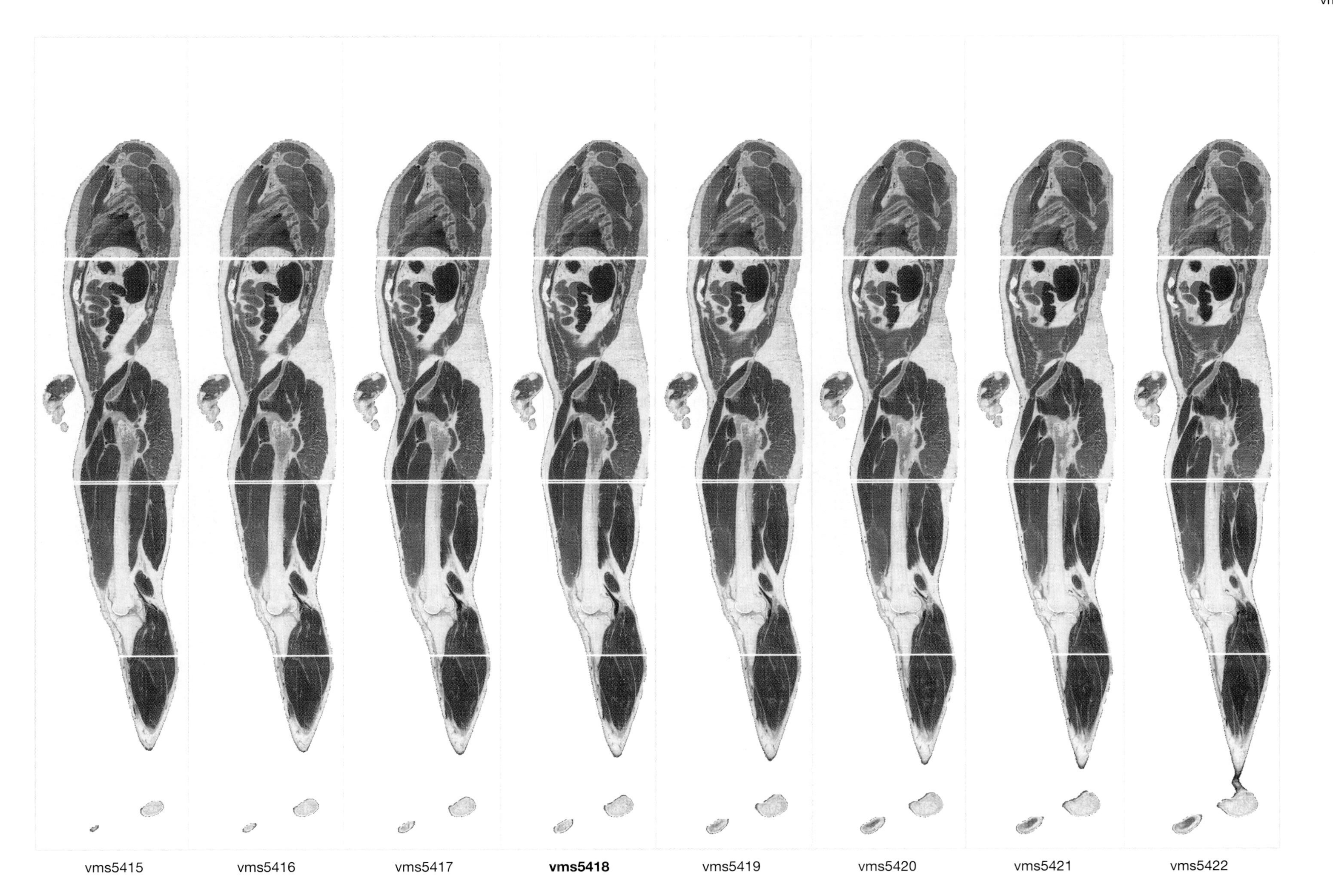

vms5415 vms5416 vms5417 **vms5418** vms5419 vms5420 vms5421 vms5422

Sagittal
vms5448

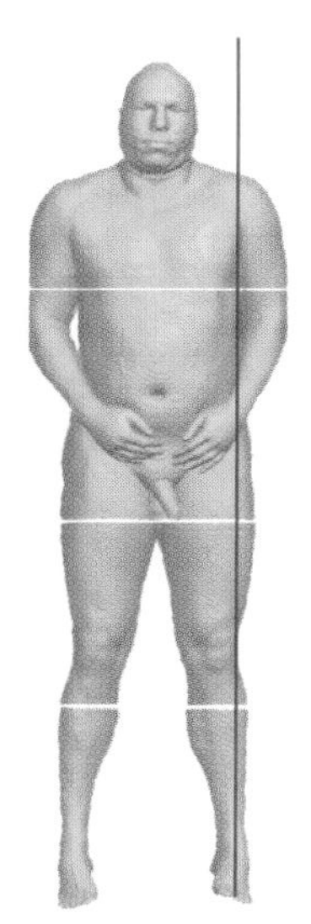

Clavicle
Shoulder Joint
Deltoid Muscle
Glenohumeral Ligament
Deltoid Muscle
Scapula
Coracobrachialis Muscle
Pectoralis Major Muscle
Axillary Vessel and Brachial Plexus
Thoracolumbar Vein
Rib 9-10, Intercostal Muscles
Rib 10
Internal Oblique Muscle
External Oblique Muscle
Radius
Long Flexor Tendons
Carpal Bones
Fourth Metacarpal
Fourth Dorsal Interosseous M.
Fifth Metacarpal
Hypothenar Muscles
Tensor Fascia Lata Muscle
Vastus Lateralis Muscle
Rectus Femoris Muscle
Rectus Femoris Tendon
Articularis Genu Muscle
Suprapatellar Bursa
Patella
Patellar Ligament
Tibia, Periosteum
Tibia, Shaft
Crural Fascia
Tibia, Medial Malleolus
Medial Cuneiform Bone
First Metatarsal
Adductor Hallucis Muscle
Flexor Digitorum Longus Tendon

Trapezius Muscle
Supraspinatus Muscle
Scapula, Spine
Deltoid Muscle
Infraspinatus Muscle
Scapula, Neck
Teres Minor Muscle
Teres Major Muscle
Subscapularis Muscle
Latissimus Dorsi Muscle
Rib 11
External Oblique Muscle
Superficial Fascia
Gluteus Medius Muscle
Gluteus Minimus Muscle
Femur, Greater Trochanter
Gluteus Maximus Muscle
Buttock
Vastus Intermedius M.
Fascia Lata
Biceps Femoris Muscle, Long Head
Short Head
Femur, Lateral Condyle
Knee Joint
Lateral Meniscus
Tibia, Lateral Condyle
Gastrocnemius Muscle, Lateral Head
Tibialis Posterior Muscle
Flexor Digitorum Longus Muscle
Flexor Hallucis Longus M.
Talus Bone
Quadratus Plantae Muscle
Calcaneus Bone
Flexor Digiti Minimi Brevis M.
Abductor Digiti Minimi M.
Plantar Aponeurosis

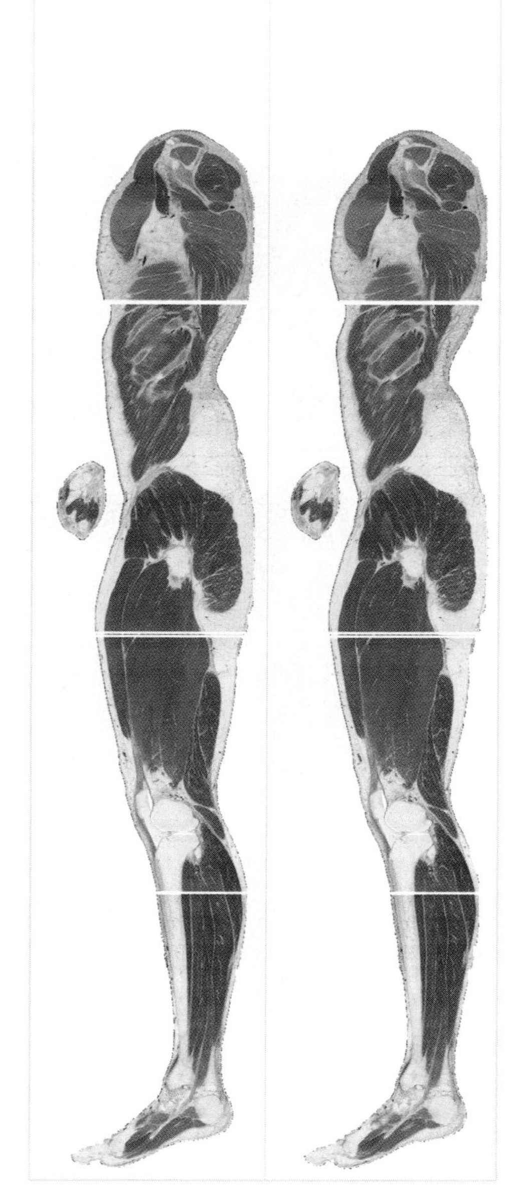
vms5443
vms5444

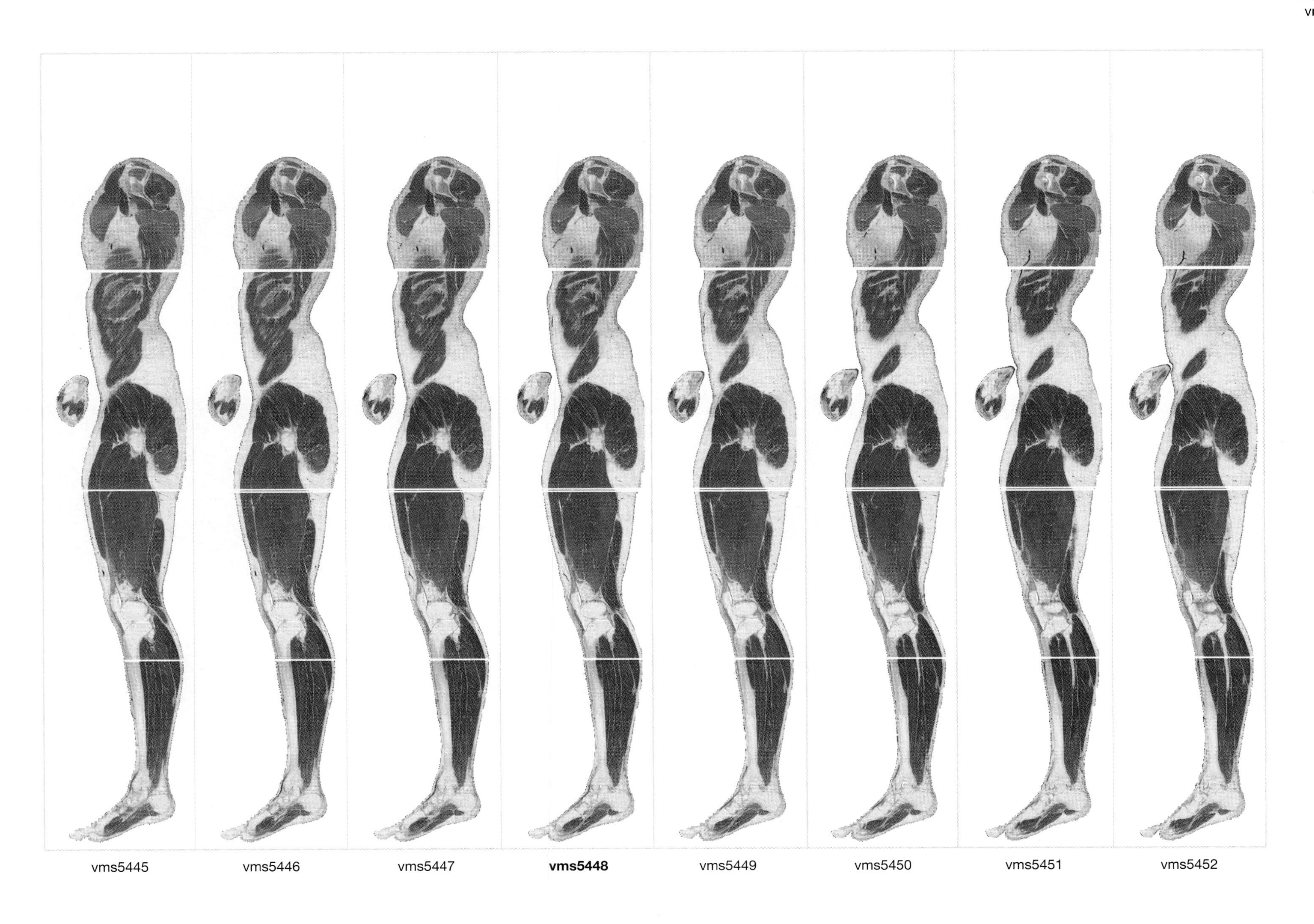

vms5445 vms5446 vms5447 **vms5448** vms5449 vms5450 vms5451 vms5452

Sagittal
vms5458

Humerus, Head
Humerus, Lesser Tubercle
Shoulder Joint, Capsule
Deltoid Muscle
Coracobrachialis Muscle
Biceps Brachii Muscle
Pectoralis Major Muscle
Long Flexor Tendons
Pronator Quadratus Muscle
Radius
Carpal Bones
Long Extensor Tendons
Hypothenar Muscles
Tensor Fascia Lata Muscle
Rectus Femoris Muscle
Vastus Lateralis Muscle
Fascia Lata
Knee Joint Cavity
Patella
Patellar Ligament
Lateral Meniscus
Tibia, Lateral Condyle
Tibialis Anterior Muscle
Extensor Digitorum Longus M.
Fibula, Shaft
Tibia
Navicular Bone
Medial Cuneiform Bone
Plantar Aponeurosis
Adductor Hallucis Muscle
First Dorsal Interosseous M.
First Metatarsal

Clavicle
Trapezius Muscle
Scapula, Acromion
Supraspinatus Muscle
Deltoid Muscle
Infraspinatus Muscle
Teres Minor Muscle
Subscapularis Tendon
Teres Major Muscle
Latissimus Dorsi Muscle
External Oblique Muscle
Superficial Fascia and Adipose Tissue
External Oblique Muscle
Gluteus Medius Muscle
Gluteus Maximus Muscle
Buttock
Vastus Intermedius Muscle
Biceps Femoris Muscle, Short Head
Fibular Collateral Ligament
Biceps Femoris Tendon, Long Head
Fibula, Head
Crural Fascia
Gastrocnemius Muscle, Lateral Head
Fibula, Shaft
Peroneus Brevis Muscle
Ankle Joint
Talus Bone
Quadratus Plantae Muscle
Calcaneus Bone
Abductor Digiti Minimi M.
Flexor Digitorum Brevis M.

vms5453

vms5454

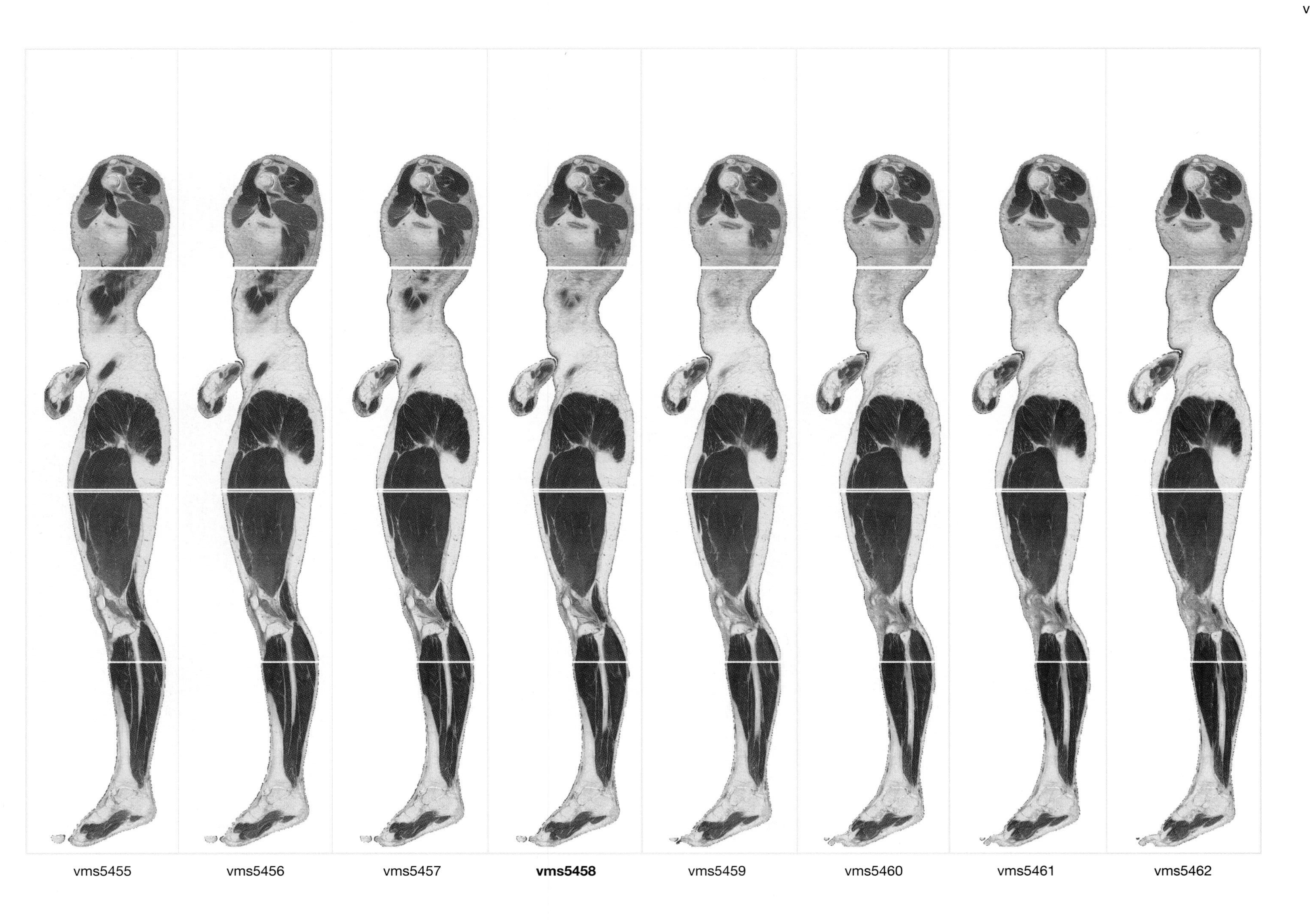
vms5455
vms5456
vms5457
vms5458
vms5459
vms5460
vms5461
vms5462

Acromioclavicular Ligament
Humerus, Head
Deltoid Muscle
Glenohumeral Ligament
Humerus, Neck
Biceps Brachii Muscle, Long Head
Anterior Axillary Fold
Superficial Fascia and Adipose Tissue
Flexor Digitorum Profundus Muscle
Radius
Scaphoid Bone
Extensor Carpi Radialis Brevis Tendon
Carpal Bones
Tensor Fascia Lata Muscle
Vastus Lateralis Muscle
Rectus Femoris Muscle
Vastus Intermedius Muscle
Lateral Patellar Retinaculum
Tibialis Anterior Muscle
Extensor Hallucis Longus Muscle
Tibialis Anterior Tendon
Tibia
Talus Bone
Navicular Bone
Quadratus Plantae Muscle
Second Metatarsal
First Dorsal Interosseous Muscle
Adductor Hallucis Muscle
Plantar Aponeurosis
Second Proximal Phalanx
Scapula, Acromion
Supraspinatus Muscle
Deltoid Muscle
Infraspinatus Muscle
Articular Capsule
Teres Major Muscle
Posterior Humeral Circumflex Vessel
Posterior Axillary Fold
Flexor Digitorum Superficialis Muscle
Flexor Carpi Ulnaris Muscle
Hypothenar Muscles
Gluteus Maximus Muscle
Gluteus Medius Muscle
Superficial Fascia and Adipose Tissue
Fascia Lata
Lateral Genicular Vessel
Biceps Femoris Tendon
Fibula, Head
Gastrocnemius Muscle, Lateral Head
Extensor Digitorum Longus Muscle
Crural Fascia
Fibula, Shaft
Peroneus Brevis Muscle
Lateral Cuneiform Bone
Calcaneus Bone
Cuboid Bone
Abductor Digiti Minimi M.
Peroneus Longus Tendon
Flexor Digitorum Brevis M.

vms5463

vms5464

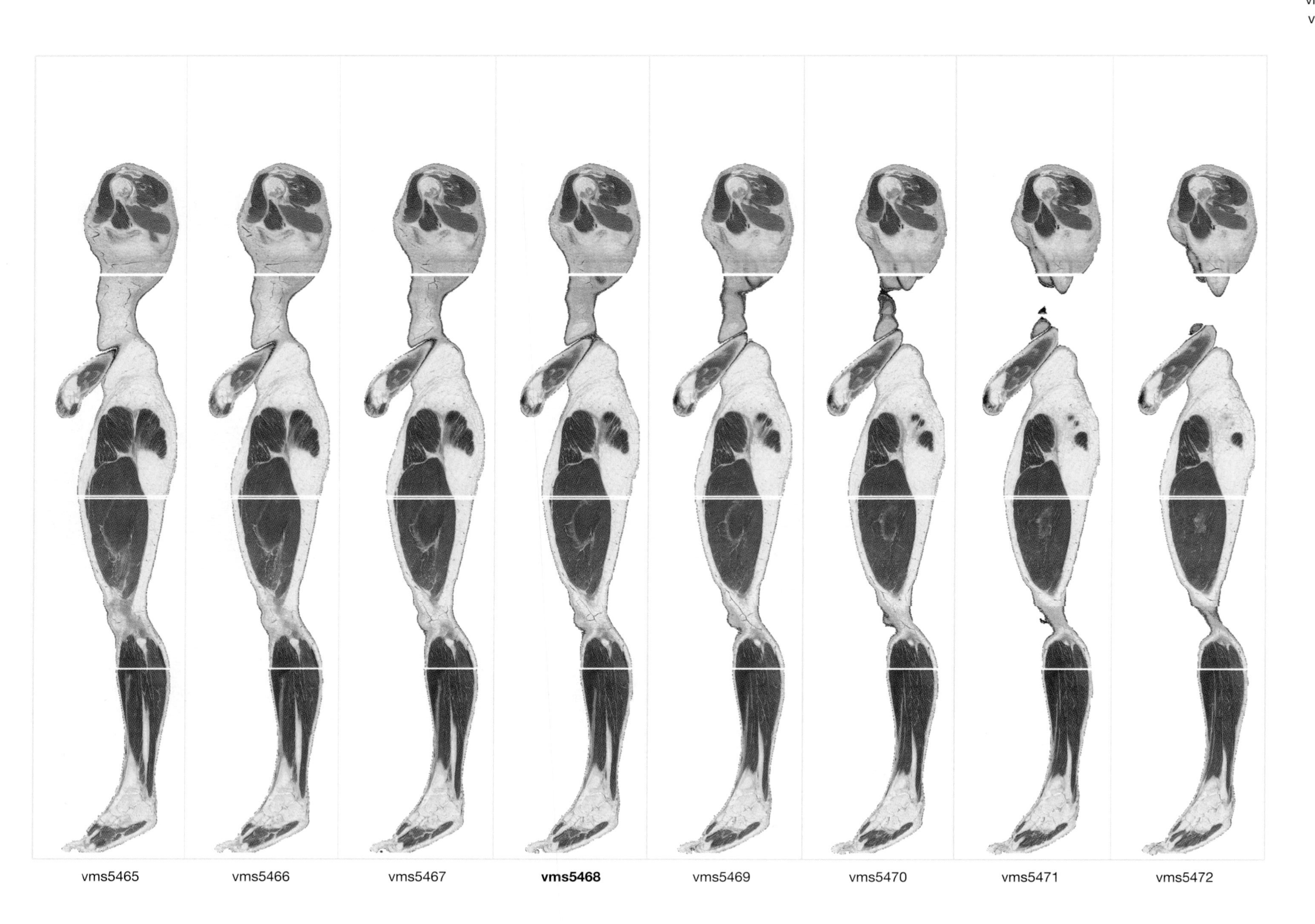
vms5465
vms5466
vms5467
vms5468
vms5469
vms5470
vms5471
vms5472

Sagittal
vms5478

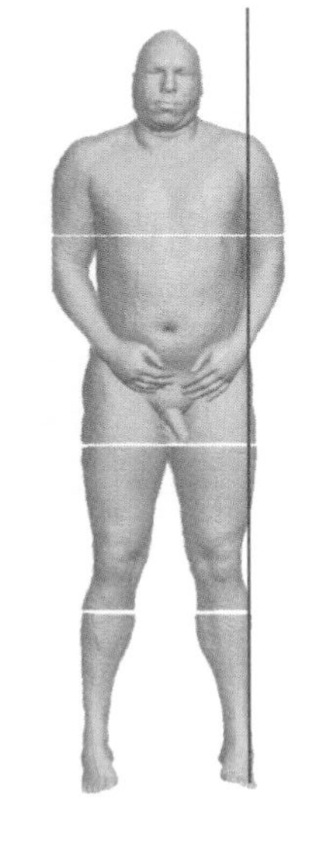

Scapula, Acromion
Humerus, Head
Humerus, Neck
Humerus, Shaft
Deltoid Muscle
Coracobrachialis Muscle
Brachial Vessels and Nerves
Biceps Brachii Muscle
Brachialis Muscle
Cephalic Vein
Brachioradialis Tendon
Radius
Carpal Bones
Ulna
Superficial Fascia
Tensor Fascia Lata Muscle
Iliotibial Tract
Vastus Lateralis Muscle
Fascia Lata

Shoulder Joint
Infraspinatus Muscle
Deltoid Muscle
Teres Minor Muscle
Triceps Brachii Muscle, Long Head
Teres Major Muscle
Superficial Fascia and Adipose Tissue
Extensor Carpi Radialis Brevis Muscle
Extensor Carpi Ulnaris Muscle
Flexor Digitorum Longus Muscle
Pronator Quadratus Muscle
Gluteus Medius Muscle
Buttock
Superficial Fascia and Adipose Tissue
Vastus Lateralis Muscle

Tibialis Anterior Muscle
Extensor Digitorum Longus Muscle
Extensor Hallucis Longus Muscle
Tibialis Anterior Tendon
Intermediate Cuneiform Bone
Third Metatarsal
Second Dorsal Interosseous Muscle
Third Dorsal Interosseous M.
Plantar Aponeurosis

Gastrocnemius Muscle, Lateral Head
Peroneus Brevis Muscle
Crural Fascia
Talus Bone
Fibula
Peroneus Longus Tendon
Fibula, Lateral Malleolus
Cuboid Bone
Calcaneus Bone
Fourth Metatarsal
Flexor Digitorum Brevis M.

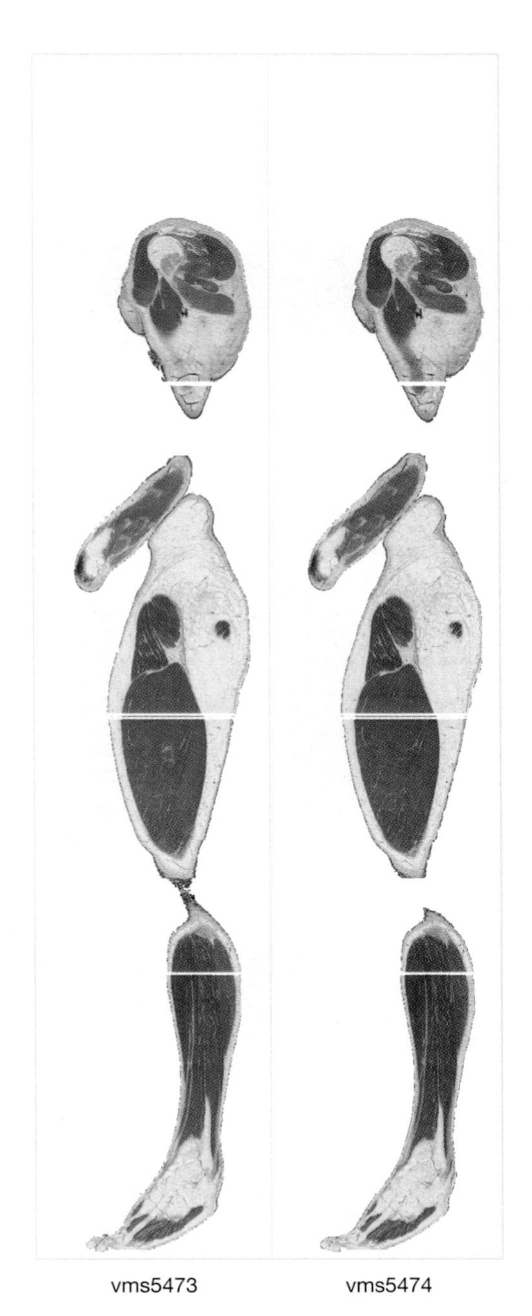

vms5473 vms5474

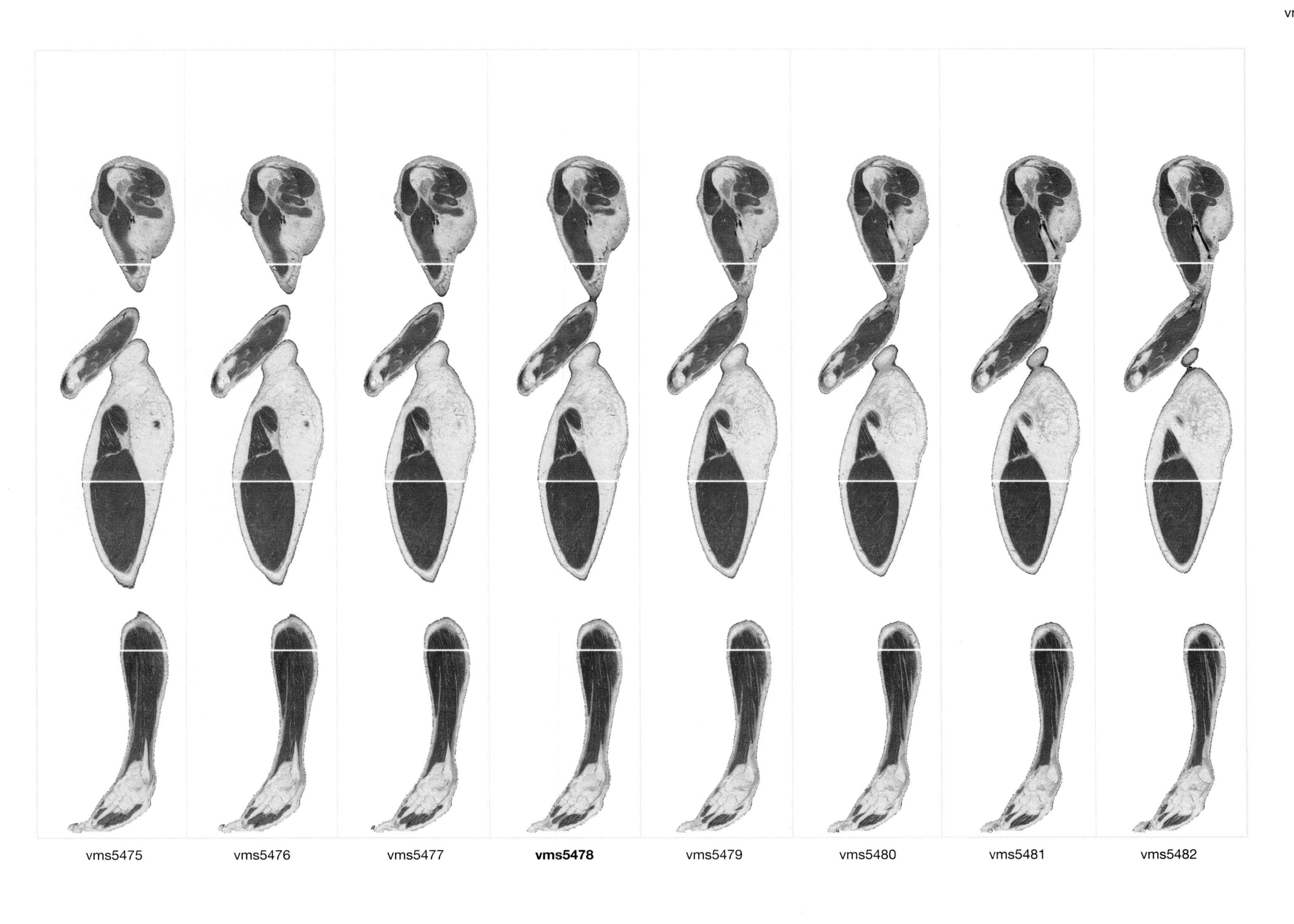
vms5475
vms5476
vms5477
vms5478
vms5479
vms5480
vms5481
vms5482

Sagittal
vms5488

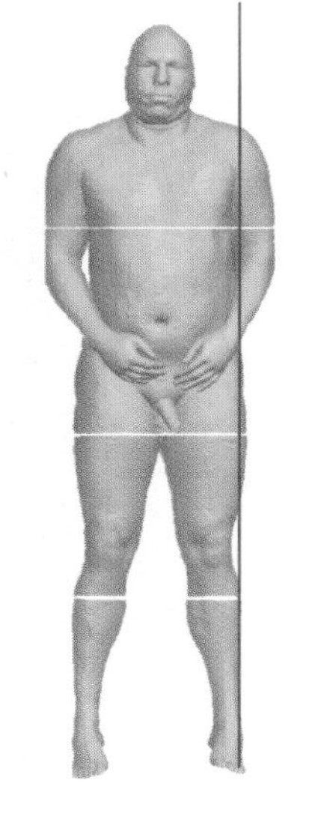

Rotator Cuff
Humerus, Head
Humerus, Neck
Humerus, Shaft
Deltoid Muscle
Coracobrachialis Muscle
Median and Ulnar Nerves
Biceps Brachii Muscle
Extensor Carpi Radialis Longus Muscle
Extensor Carpi Radialis Brevis Muscle
Brachioradialis Muscle
Radius
Extensor Pollicis Longus Muscle
Extensor Digitorum Muscle
Ulna
Vastus Lateralis Muscle
Skin

Scapula, Acromion
Infraspinatus Muscle
Deltoid Muscle
Teres Minor Muscle
Triceps Brachii Muscle,
Long Head
Medial Head
Brachial Vessel
Brachial Vein
Flexor Digitorum Superficialis Muscle
Flexor Digitorum Profundus Muscle
Flexor Carpi Ulnaris Muscle
Buttock
Superficial Fascia and Adipose Tissue
Fascia Lata

Extensor Digitorum Longus Muscle
Anterior Crural Intermuscular Septum
Lateral Cutaneous Vein
Extensor Hallucis Longus Muscle
Extensor Digitorum Brevis Muscle
Intermediate Cuneiform Bone
Third Metatarsal
Third Dorsal Interosseous Muscle
Fourth Metatarsal
Fourth Proximal Phalanx

Peroneus Longus Muscle
Crural Fascia
Fifth Metatarsal
Fibula
Fibula, Lateral Malleolus
Calcaneus Bone
Cuboid Bone
Plantar Aponeurosis
Fourth Dorsal Interosseous M.
Flexor Digiti Minimi Brevis M.

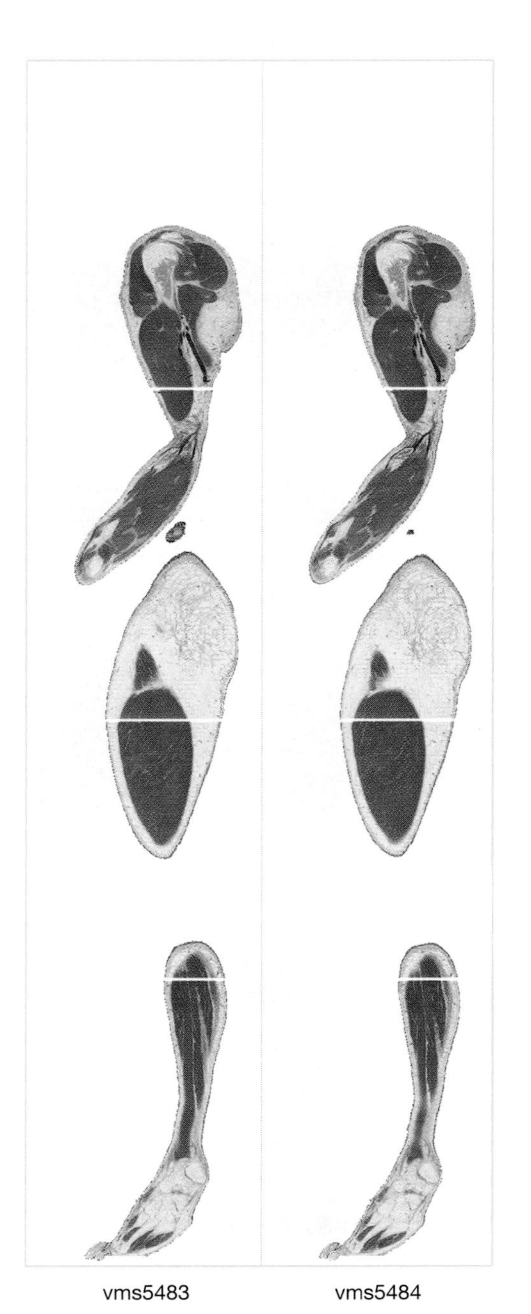
vms5483
vms5484

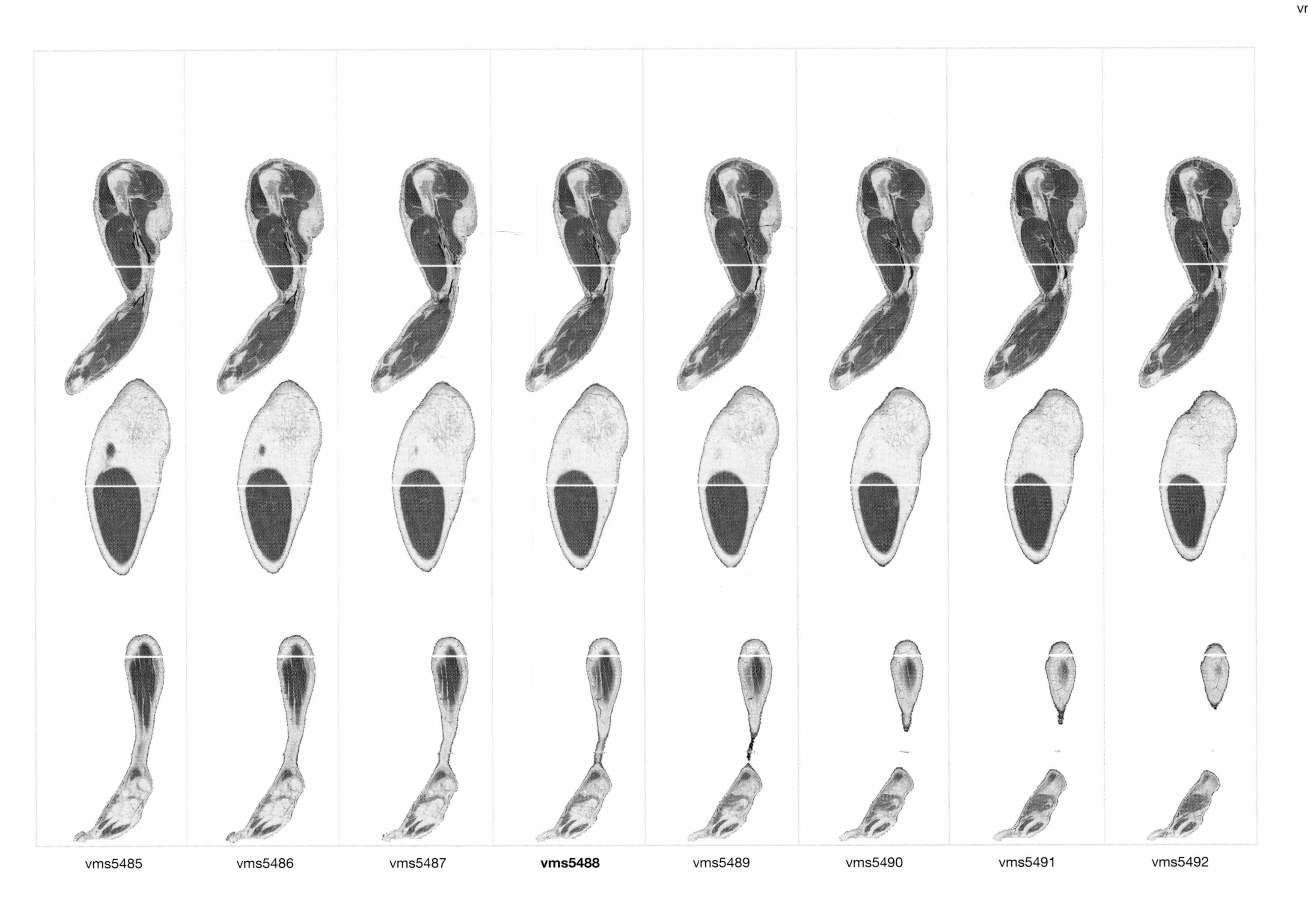
vms5485
vms5486
vms5487
vms5488
vms5489
vms5490
vms5491
vms5492

Sagittal
vms5498

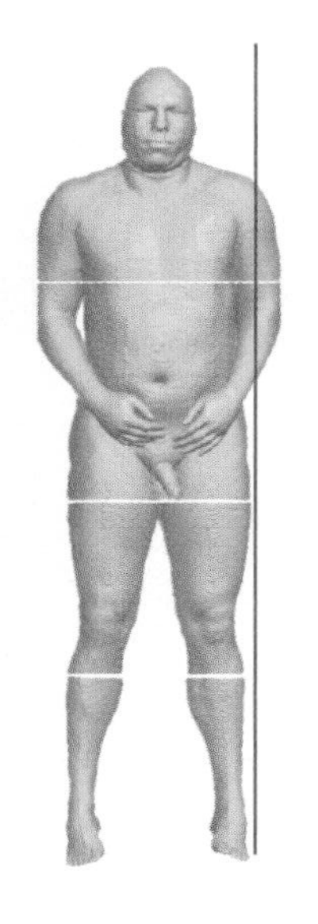

Humerus, Greater Tubercle
Deltoid Muscle
Cephalic Vein
Biceps Brachii Muscle
Coracobrachialis Muscle
Brachialis Muscle
Brachioradialis Muscle
Extensor Carpi Radialis Brevis Muscle
Extensor Carpi Radialis Longus Muscle
Radius
Extensor Digitorum Muscle
Extensor Carpi Ulnaris Muscle
Ulna
Flexor Digitorum Profundus Muscle

Deltoid Muscle
Infraspinatus Tendon
Teres Minor Muscle
Deltoid Muscle
Triceps Brachii Muscle, Lateral Head
Humerus, Shaft
Triceps Brachii Muscle, Long Head
Brachial Vessel and Nerves
Brachialis Muscle
Flexor Digitorum Superficialis Muscle
Flexor Carpi Ulnaris Muscle

Superficial Fascia and Adipose Tissue
Vastus Lateralis Muscle
Fascia Lata

Fifth Middle Phalanx
Fifth Distal Phalanx
Extensor Digitorum Brevis Muscle
Abductor Digiti Minimi Muscle
Flexor Digiti Minimi Brevis M.

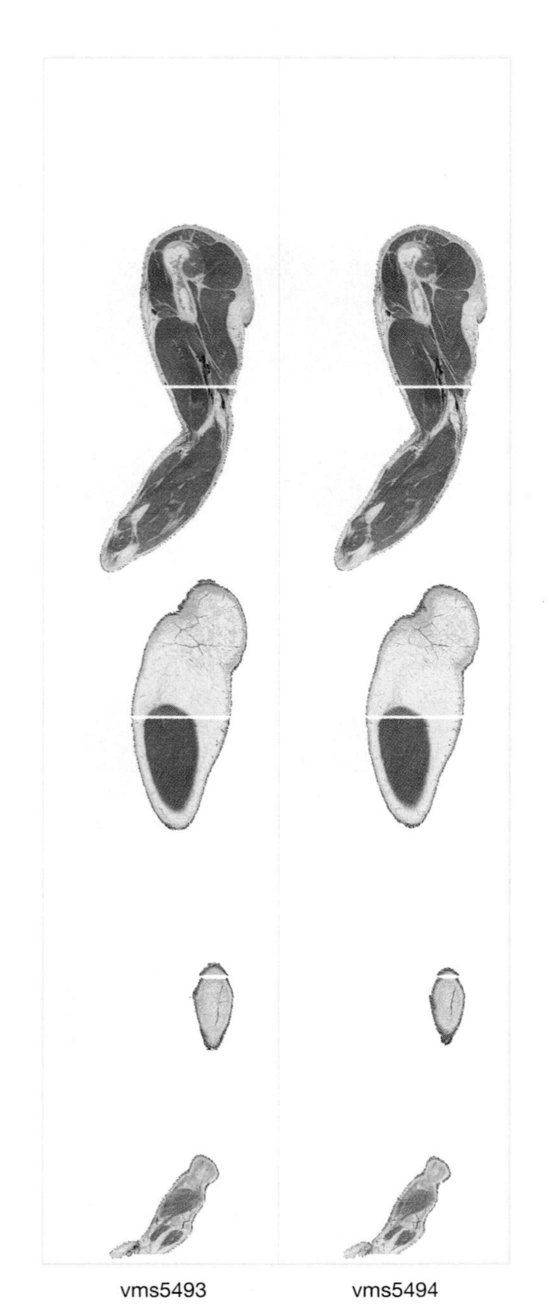

vms5493 vms5494

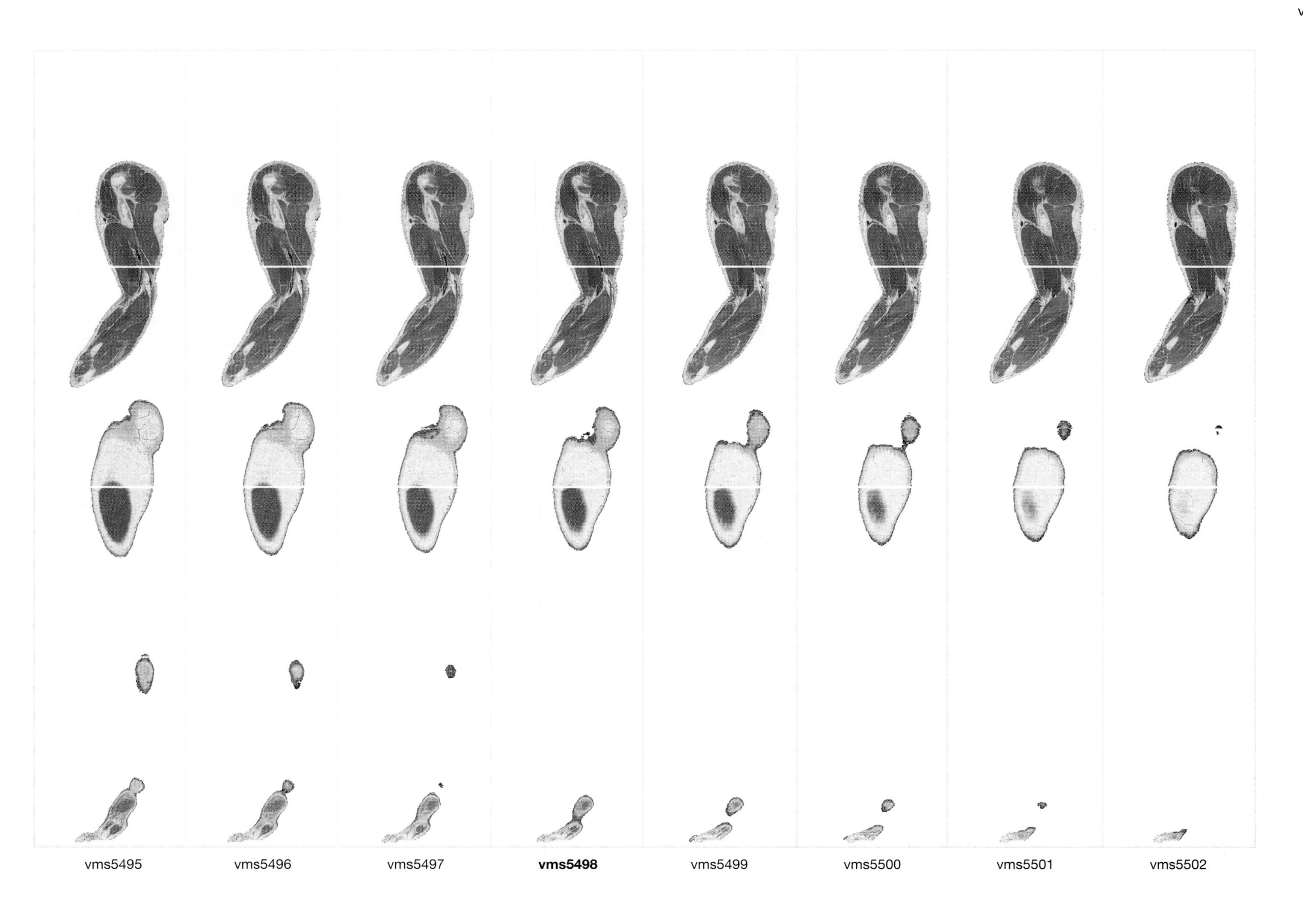
vms5495
vms5496
vms5497
vms5498
vms5499
vms5500
vms5501
vms5502

Sagittal
vms5508

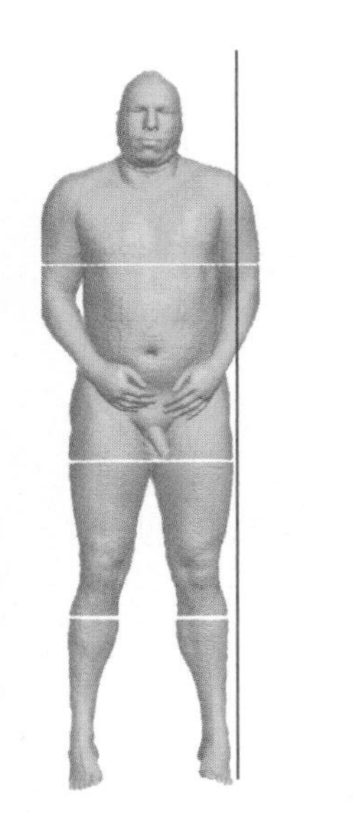

Deltoid Muscle
Cephalic Vein
Biceps Brachii Muscle
Brachialis Muscle
Brachioradialis Muscle
Extensor Carpi Radialis Longus Muscle
Radius
Extensor Digitorum Muscle
Ulna
Extensor Carpi Ulnaris Muscle
Triceps Brachii Muscle, Lateral Head
Humerus, Shaft
Triceps Brachii Muscle, Long Head
Pronator Teres Muscle
Flexor Digitorum Superficialis Muscle
Flexor Digitorum Profundus Muscle
Flexor Carpi Ulnaris Muscle

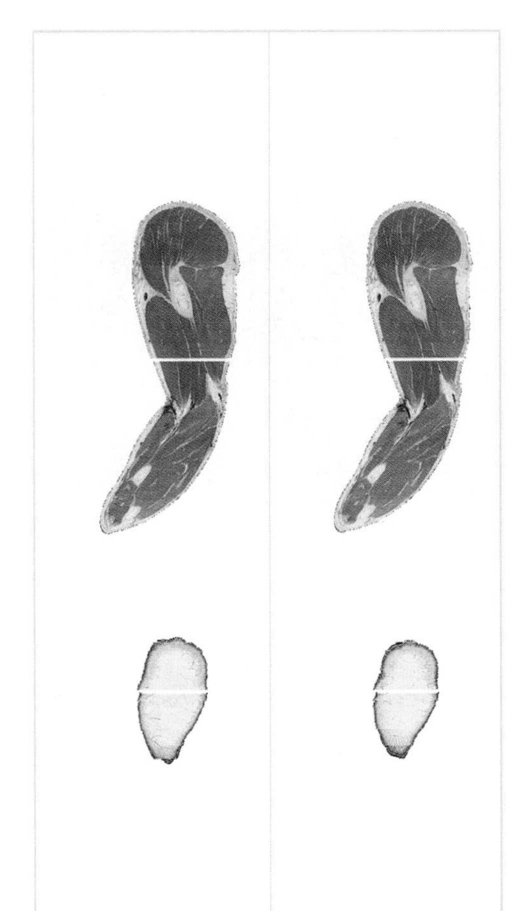

vms5503 vms5504

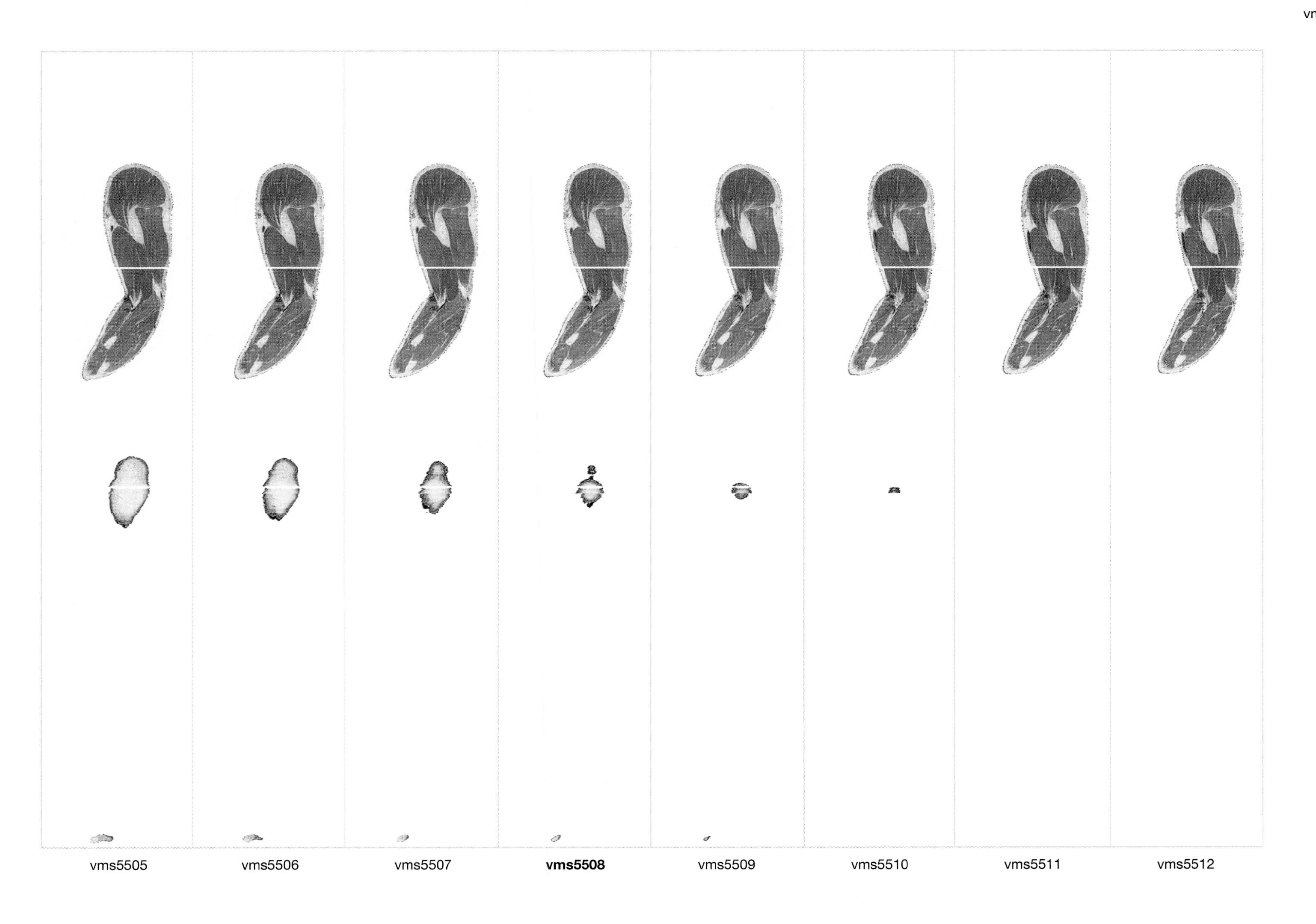
vms5505
vms5506
vms5507
vms5508
vms5509
vms5510
vms5511
vms5512

Sagittal
vms5518

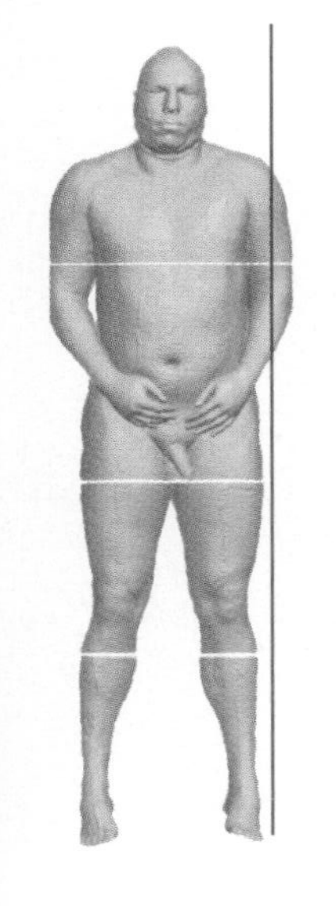

Deltoid Muscle
Triceps Brachii Muscle, Lateral Head
Humerus, Deltoid Tuberosity
Biceps Brachii Muscle
Humerus, Shaft
Brachialis Muscle
Triceps Brachii Muscle, Long Head
Brachioradialis Muscle
Pronator Teres Muscle
Extensor Carpi Radialis Longus Muscle
Flexor Digitorum Superficialis Muscle
Radius
Extensor Digitorum Muscle
Flexor Digitorum Profundus Muscle
Ulna
Extensor Carpi Ulnaris Muscle

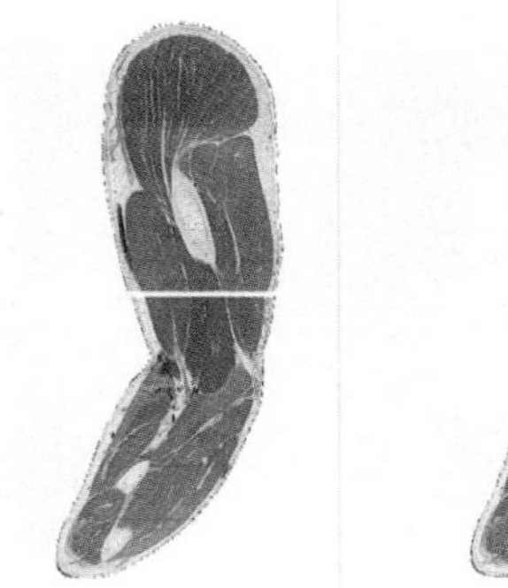

vms5513

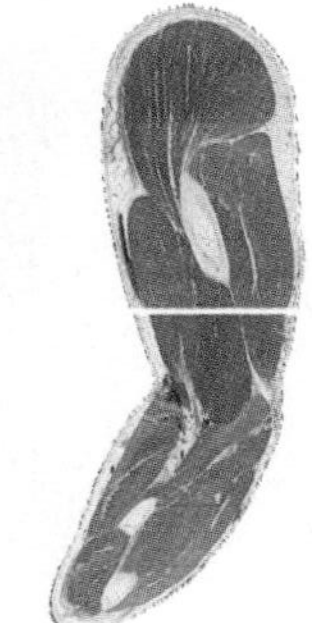

vms5514

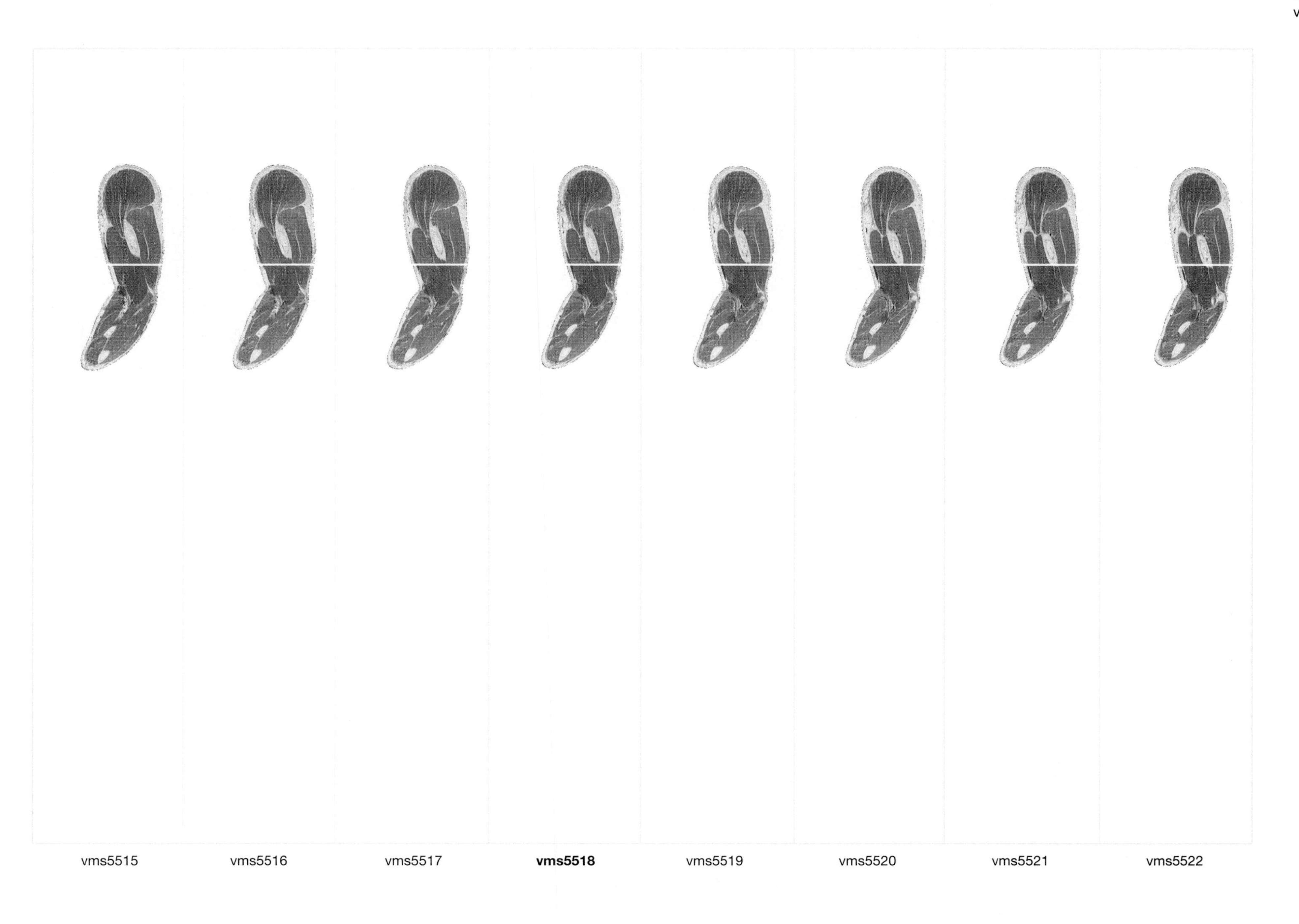
vms5515
vms5516
vms5517
vms5518
vms5519
vms5520
vms5521
vms5522

Sagittal
vms5528

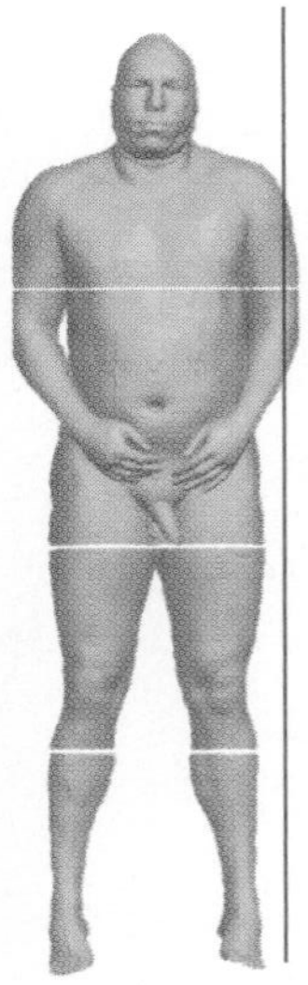

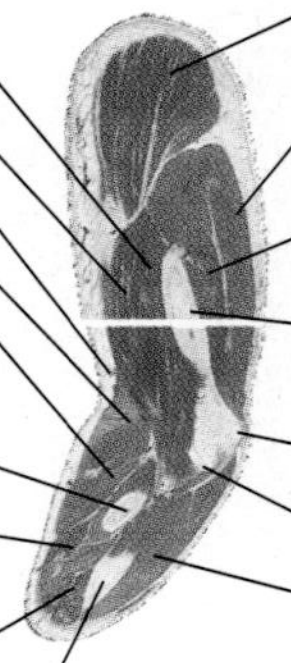

Brachialis Muscle
Biceps Brachii Muscle
Cephalic Vein
Brachioradialis Muscle
Extensor Carpi Radialis Longus Muscle
Radius
Extensor Carpi Radialis Brevis Muscle
Extensor Digitorum Longus Muscle
Ulna
Deltoid Muscle
Triceps Brachii Muscle, Long Head
Triceps Brachii Muscle, Lateral Head
Humerus, Shaft
Humerus, Medial Epicondyle
Humerus, Trochlea
Flexor Digitorum Superficialis Muscle

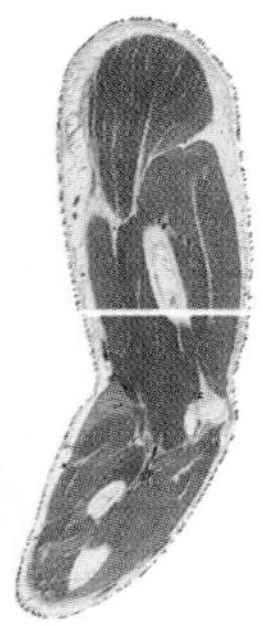

vms5523

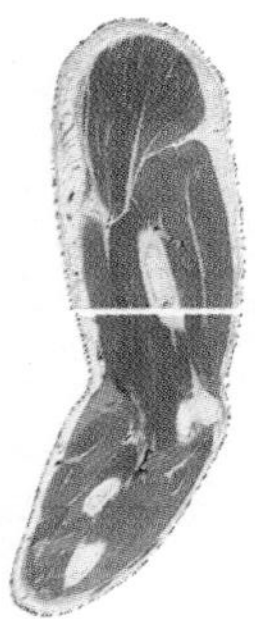

vms5524

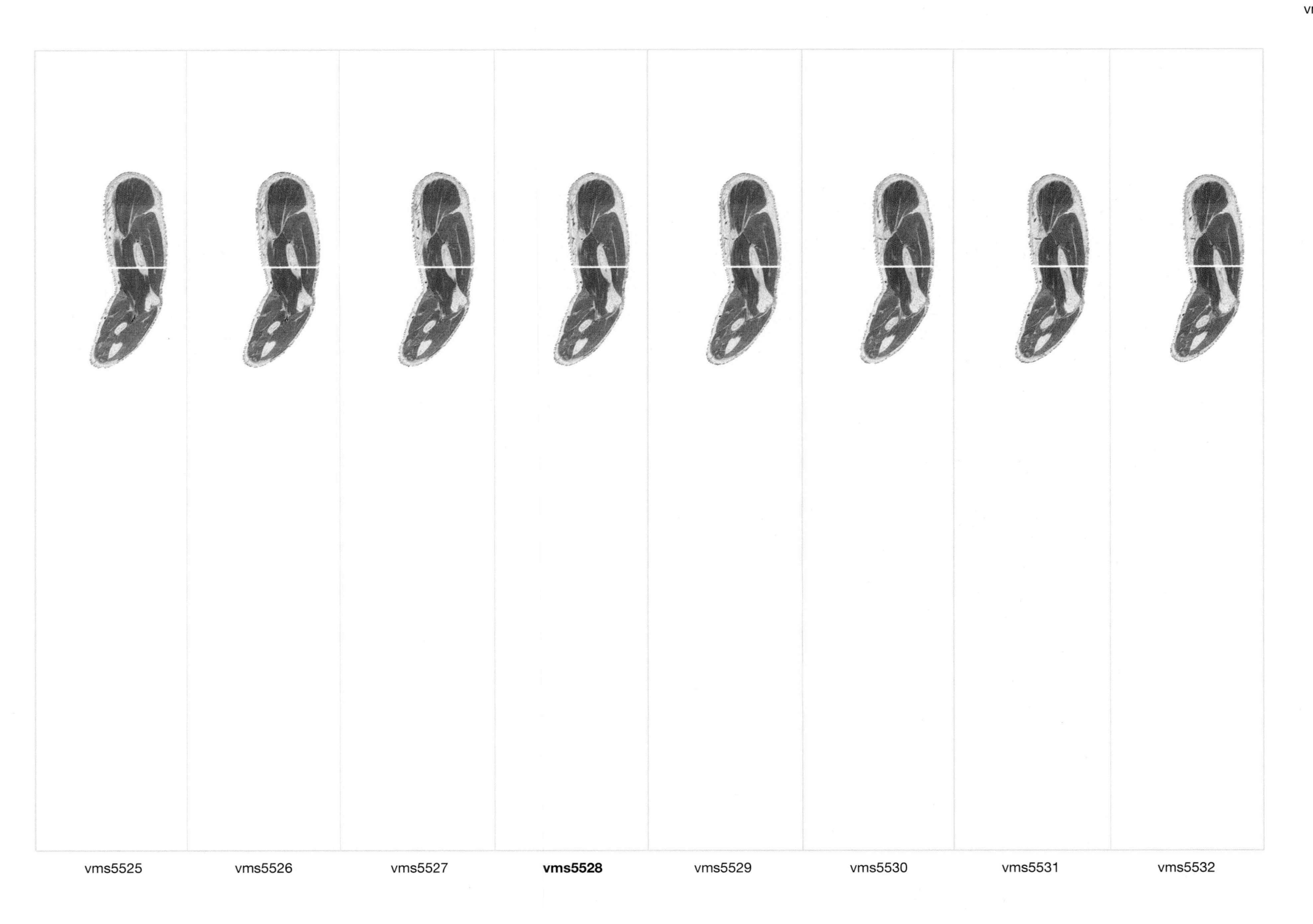
vms5525
vms5526
vms5527
vms5528
vms5529
vms5530
vms5531
vms5532

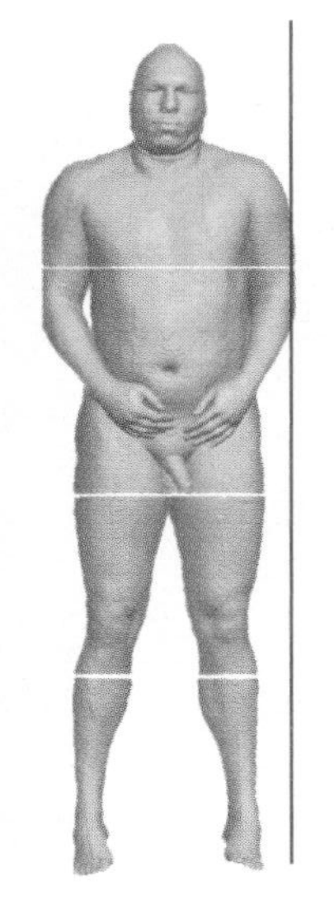

Deltoid Muscle
Triceps Brachii Muscle, Long Head
Brachial Fascia
Biceps Brachii Muscle
Triceps Brachii Muscle, Lateral Head
Cephalic Vein
Brachioradialis Muscle
Humerus, Shaft
Radius
Brachialis Muscle
Supinator Muscle
Humerus, Trochlea
Extensor Carpi Radialis Longus Muscle
Elbow Joint
Ulna
Extensor Carpi Radialis Brevis Muscle
Ulna, Shaft
Flexor Digitorum Profundus Muscle

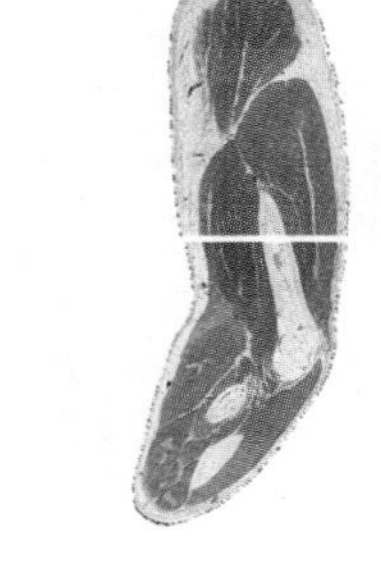

cms5533

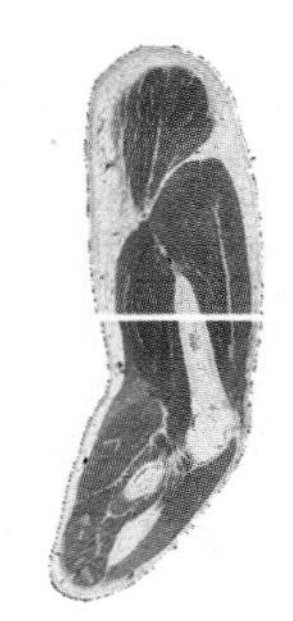

vms5534

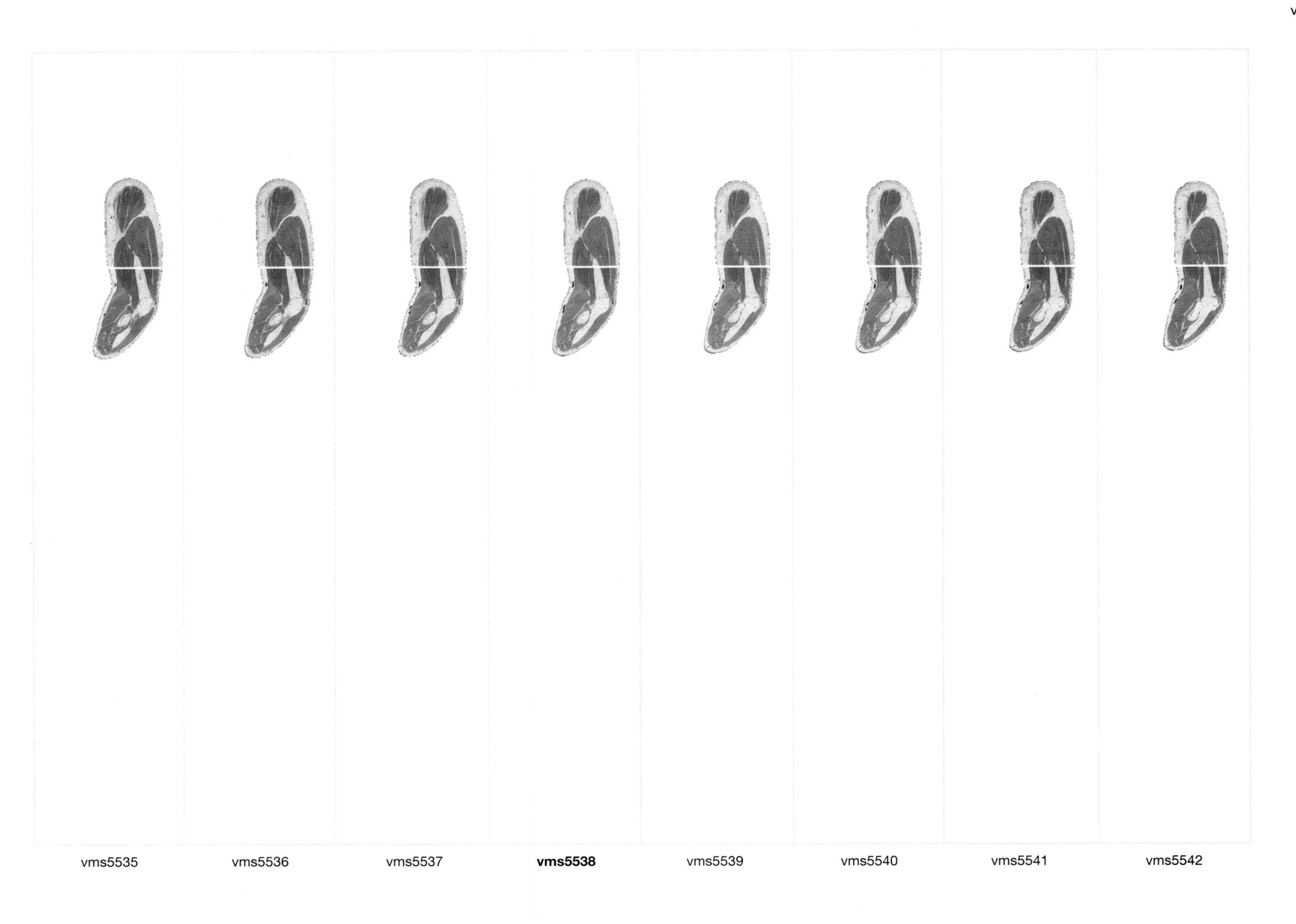
vms5535
vms5536
vms5537
vms5538
vms5539
vms5540
vms5541
vms5542

PART FOUR

Image Gallery

The following twenty pages contain a potpourri of some exciting views and applications of the Visible Human Male dataset. The first display is of images from the transverse, coronal, and sagittal parts. These images are particularly interesting slices because they show collections of anatomical structures not easily visualized in conventional displays (e.g., a transverse section through the optic chiasma where both optic nerves are visualized in the same slice, the median sagittal plane, and a coronal image through the entire length of the body). These images are displayed here without labels and leader lines and at maximum size. Next, there are three "slices" that are not of any of the standard planes. The first of these is cut in the Frankel imaging plane, which is widely used in clinical computed tomography to examine the base of the skull. The other two images are not planes at all but rather curved surfaces taken through the center of the vertebral column and through the center of the spinal cord. These reconstructions show how a volume of computed image data can be manipulated to demonstrate entire lengths of anatomical structures (i.e., the spinal cord or vertebral column).

This part also contains ray-traced renderings of the classified data of the Visible Human Project. The 3D effect has been achieved by illuminating the surfaces of the visualized structures and analyzing the reflected light. The classified data used to create these pictures is courtesy of Gold Standard Multimedia, Inc. These pictures are unique in that their color and/or texture are attribuites of the original cross-sectional data. Some of the images demonstrate that each structure can be uniquely identified by assignment of different colors to distinct structures.

The part concludes with three pages displaying developments at the University of Colorado Center for Human Simulation. These images demonstrate how the classified data of volume-based image data is being utilized for simulating medical porocedures. In addition to anatomical tissue type classification, each voxel has also been designated with physical characteristics such that the user can now "feel" the data as well as see the cut planes and 3D renderings.

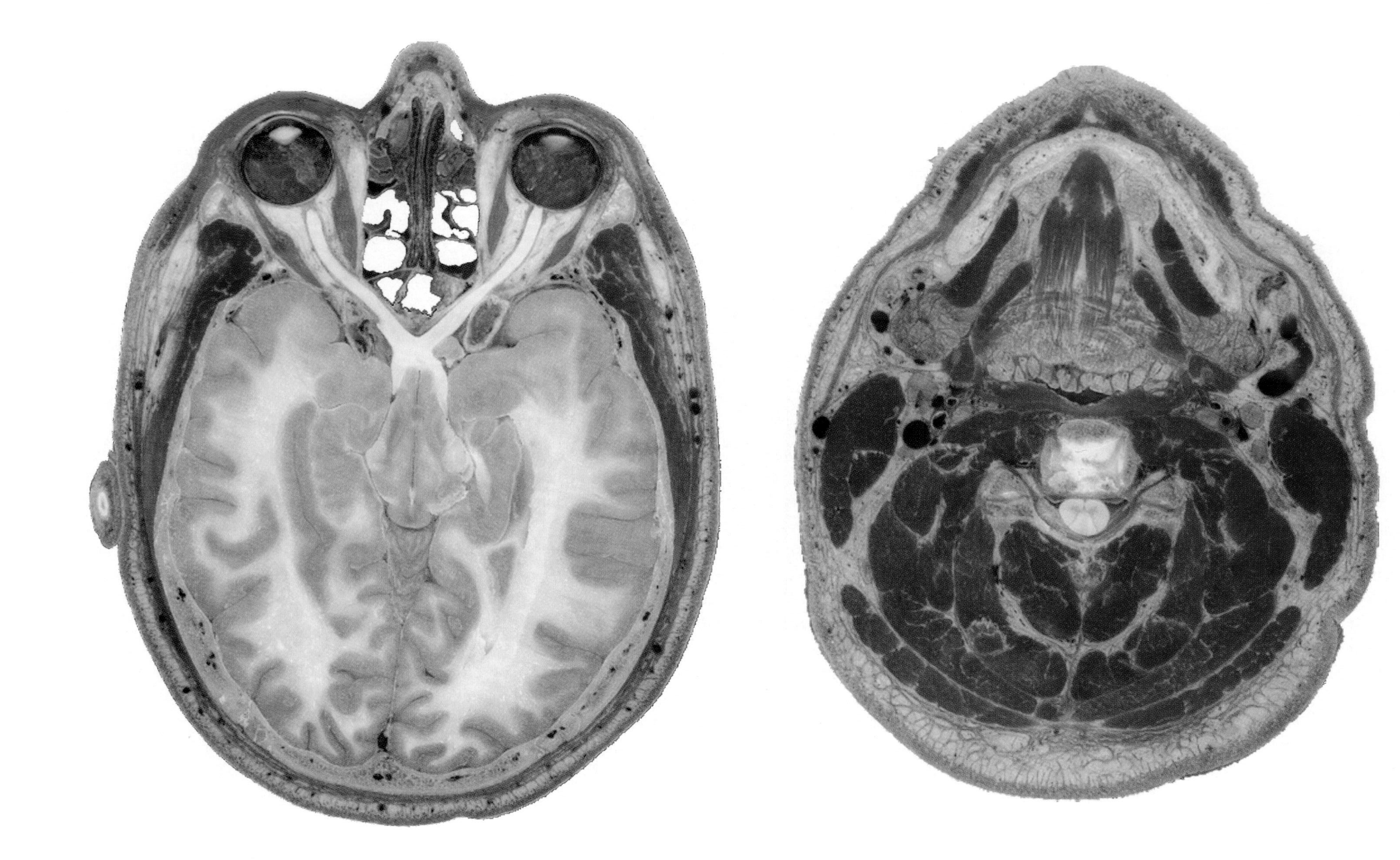

a_vm1107

a_vm1212

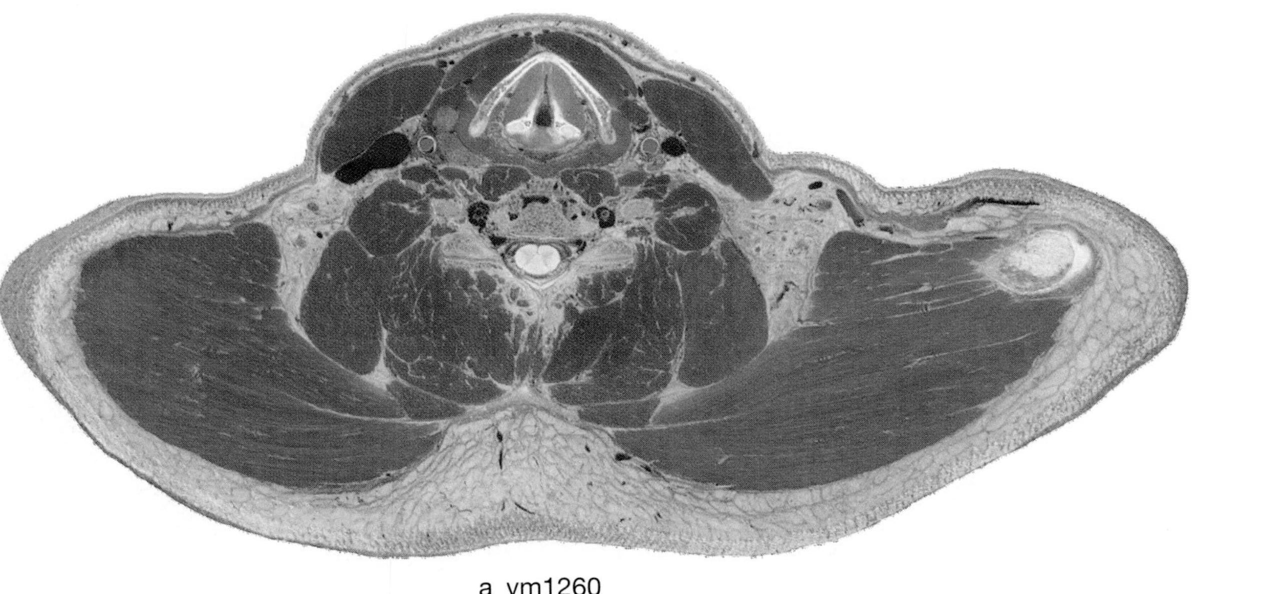

a_vm1260

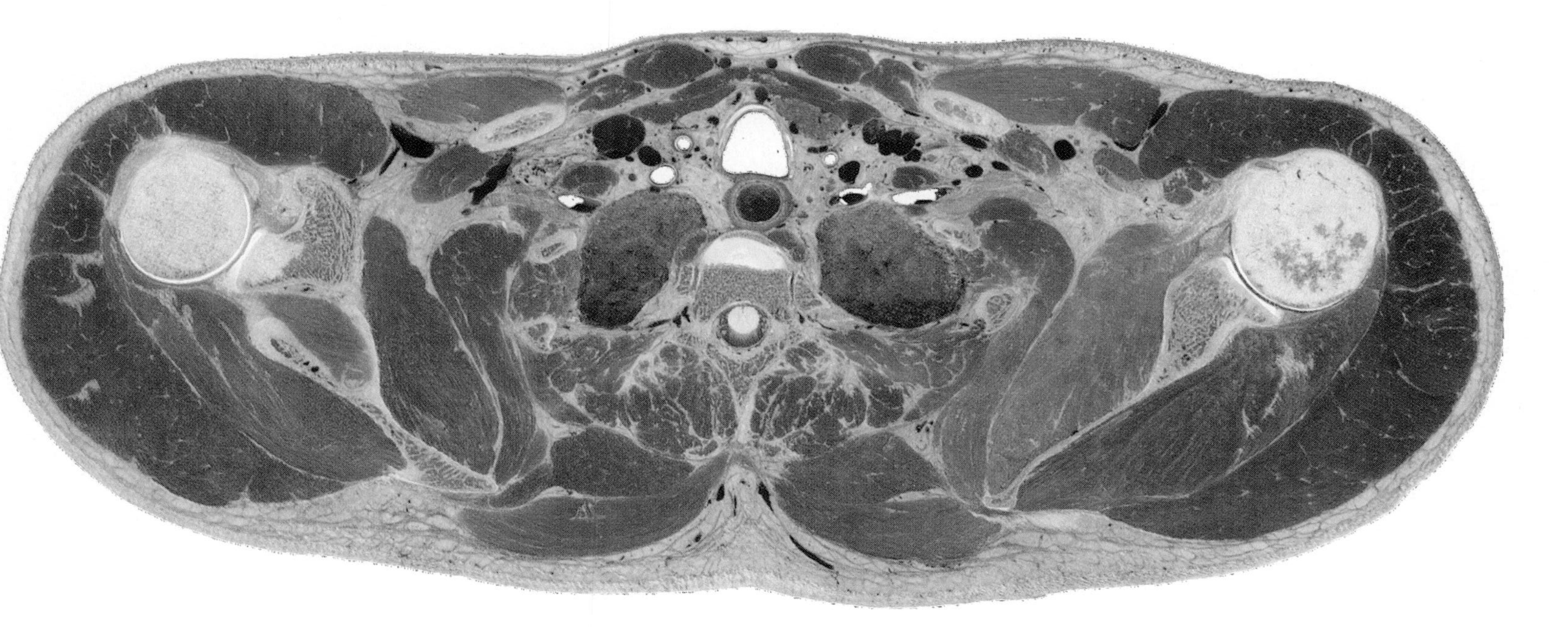

a_vm1307

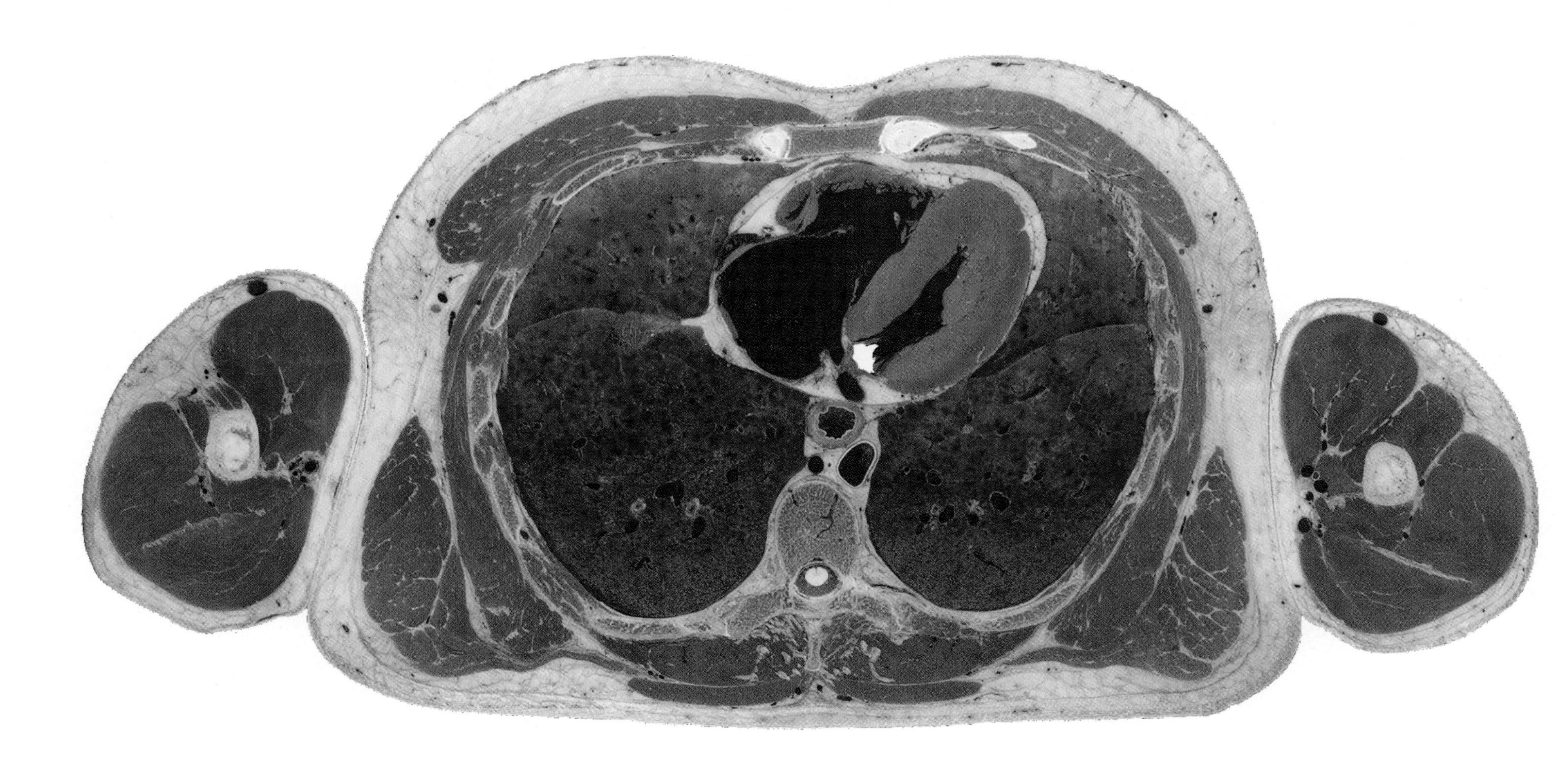

a_vm1452

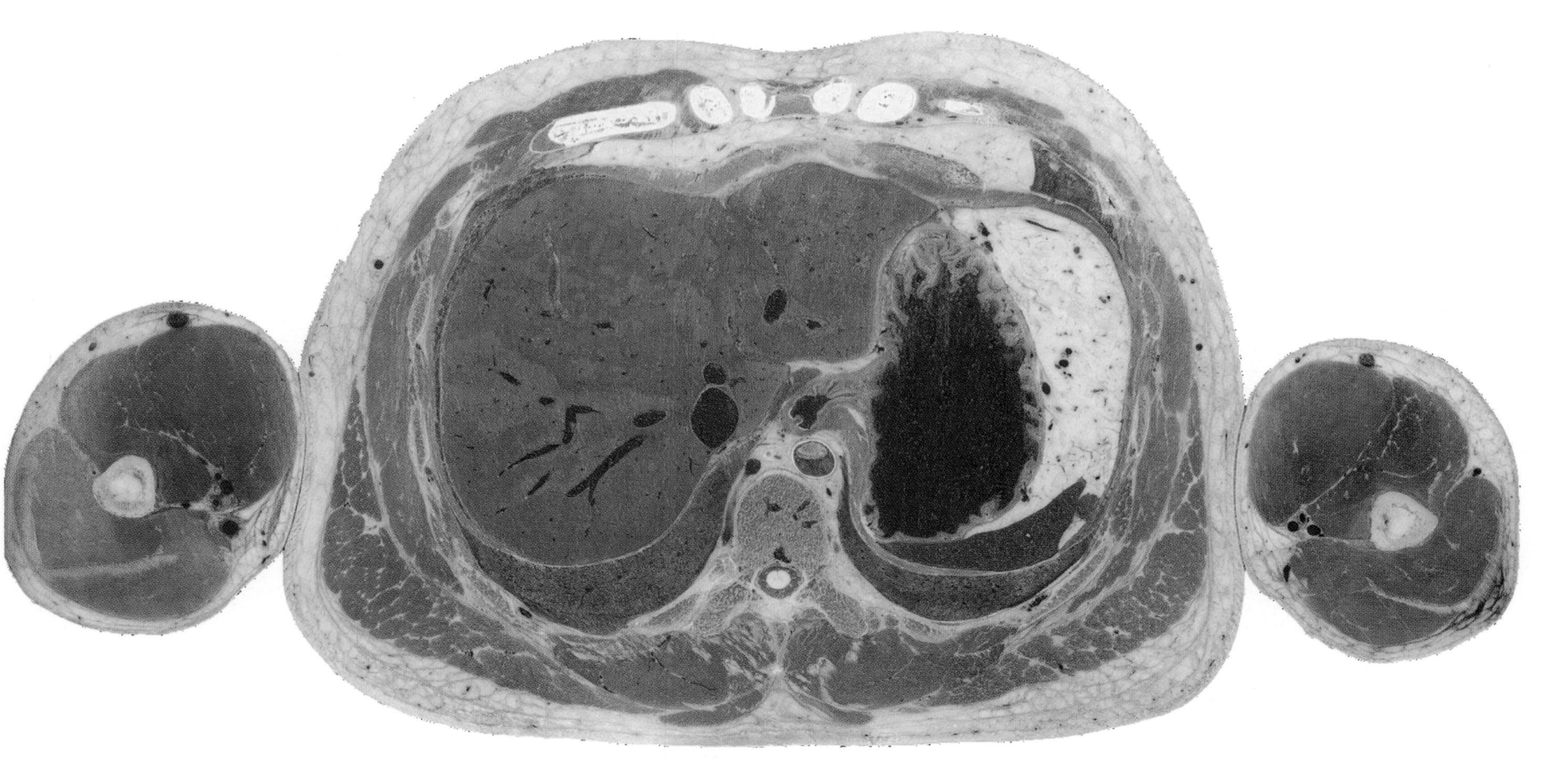

a_vm1510

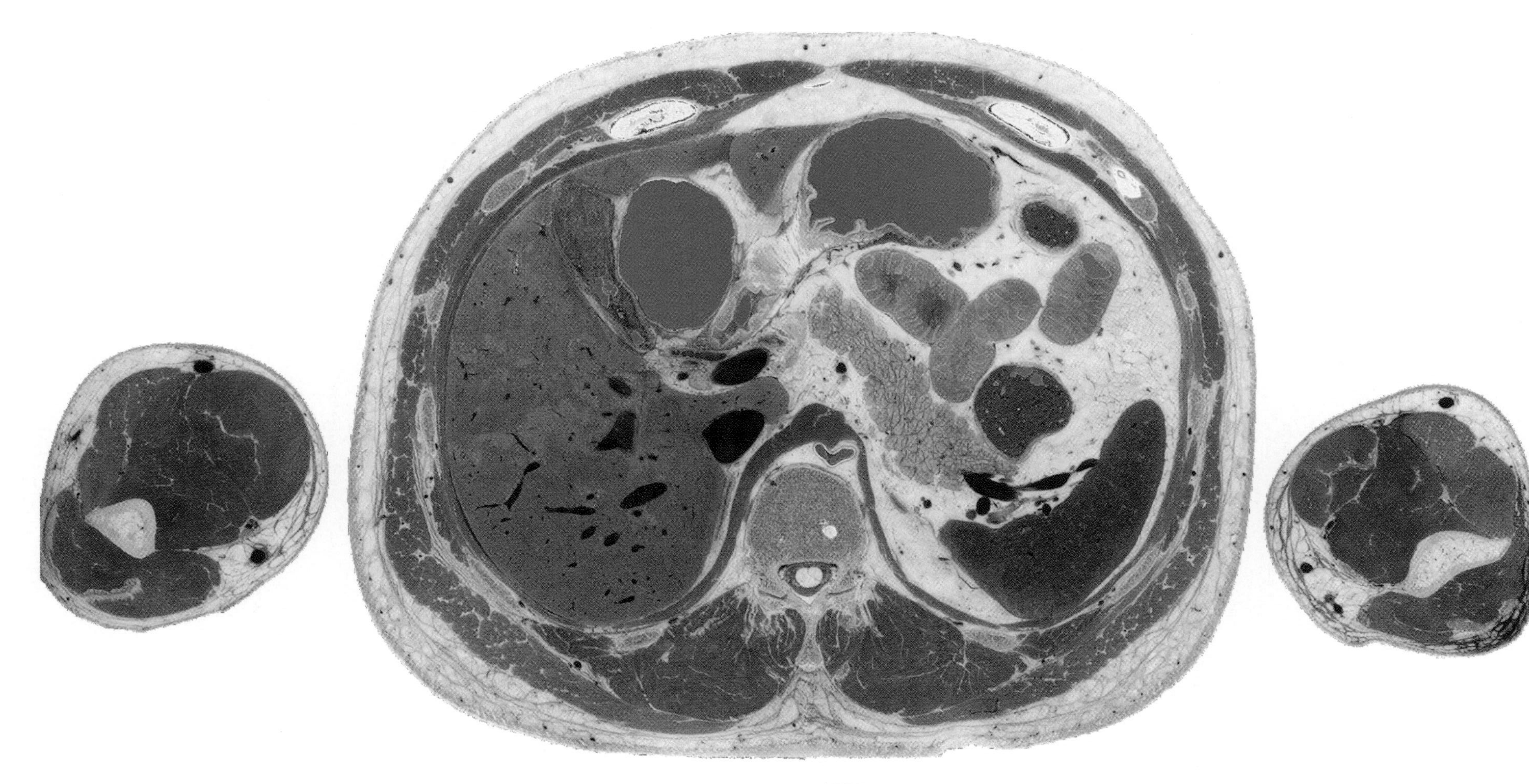

a_vm1555

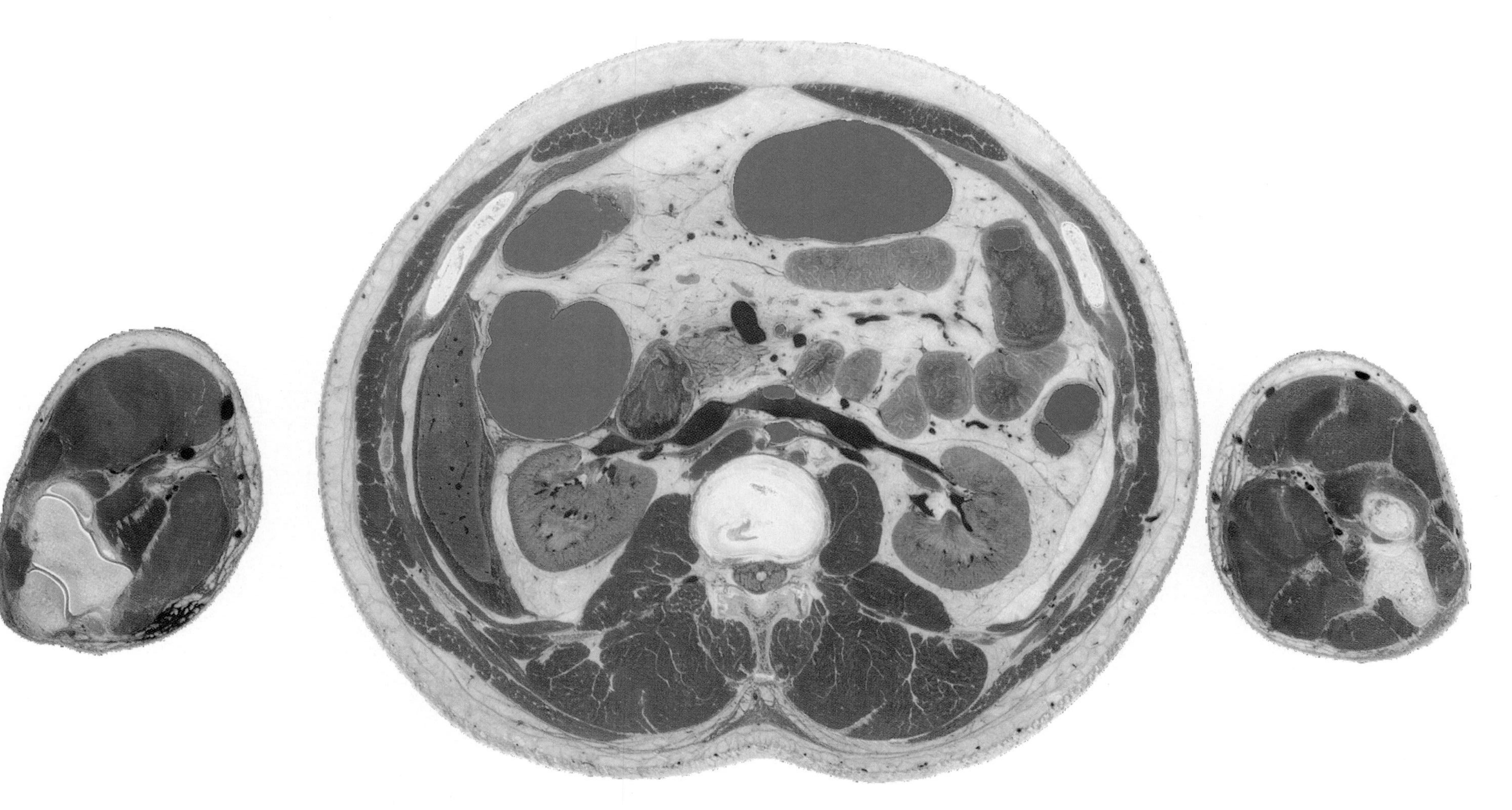

a_vm1616

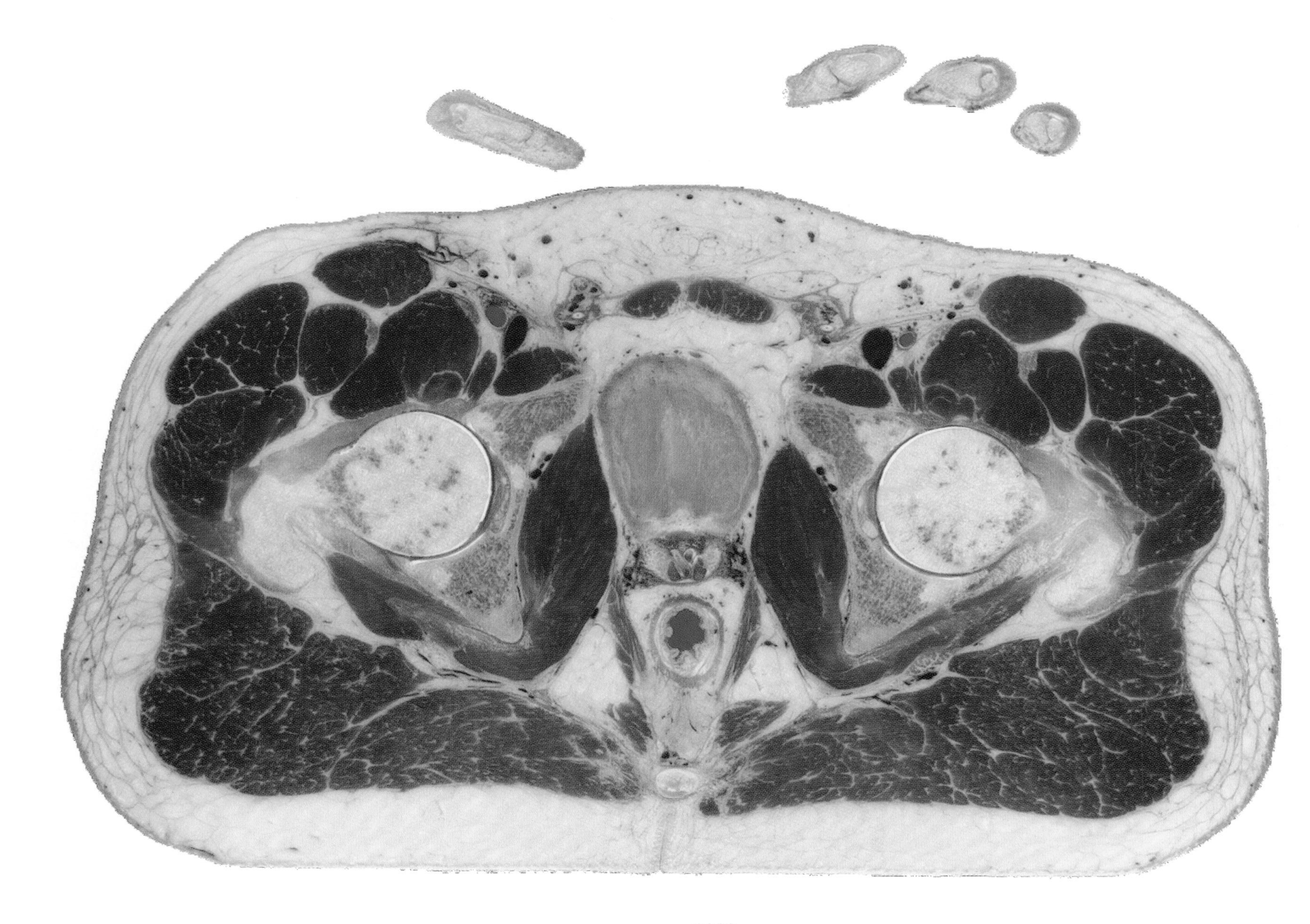

a_vm1886

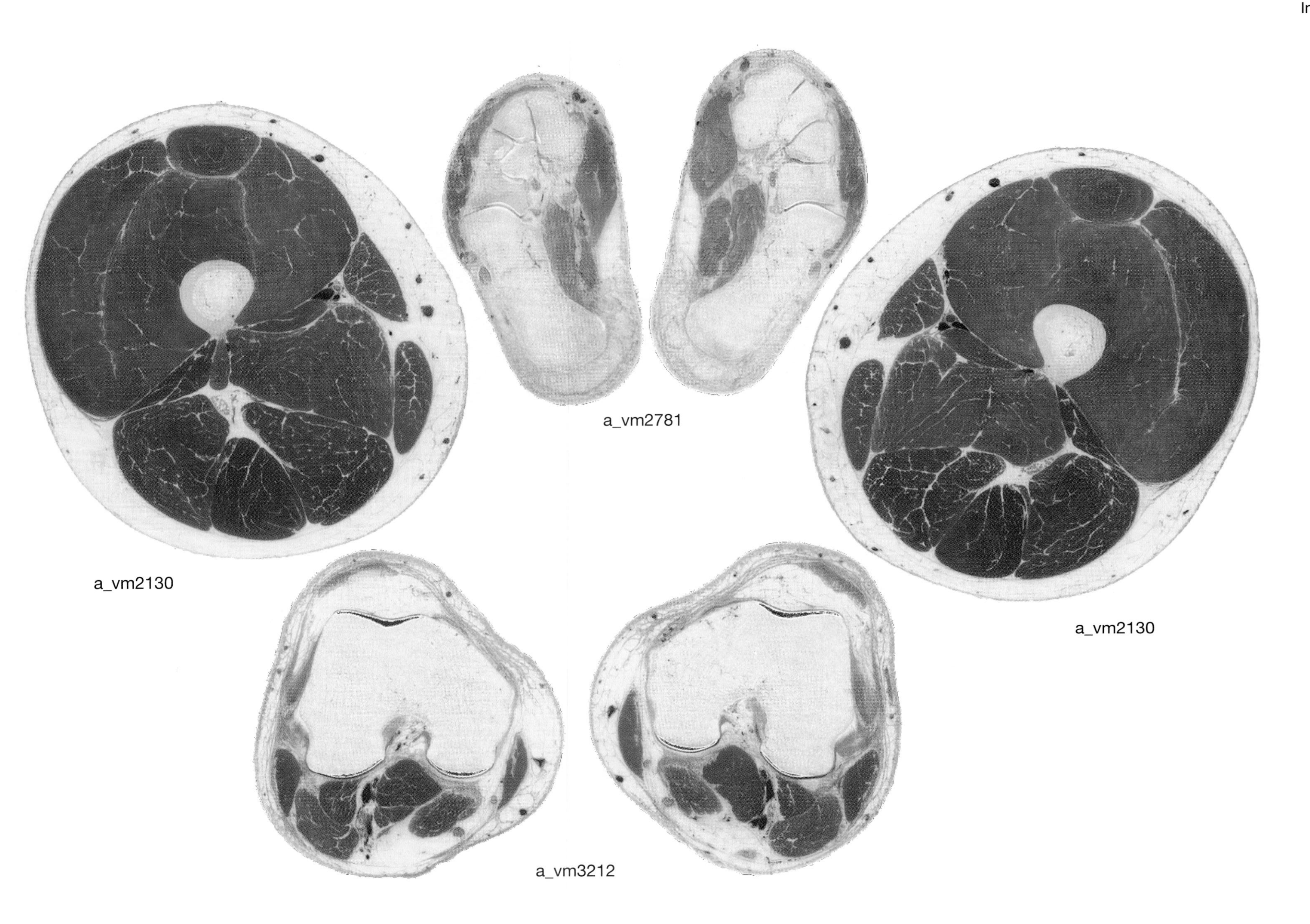
a_vm2781
a_vm2130
a_vm2130
a_vm3212

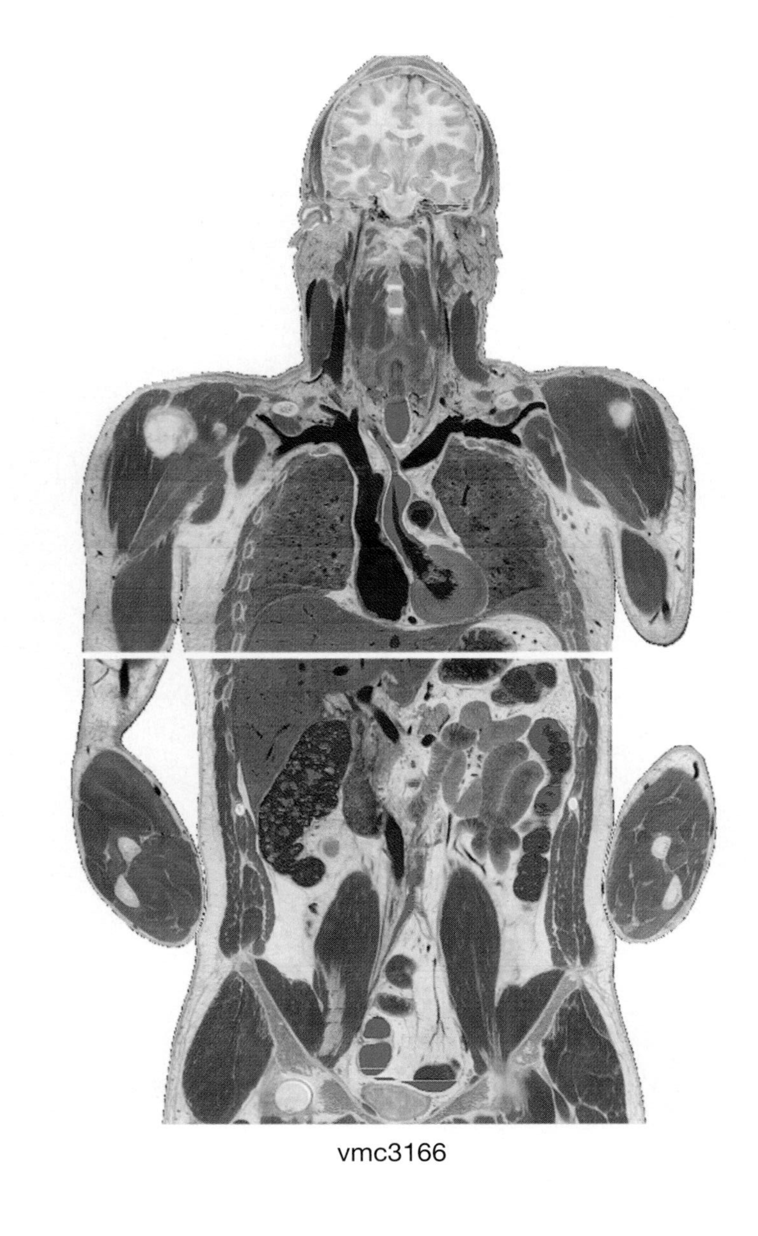

vmc3166

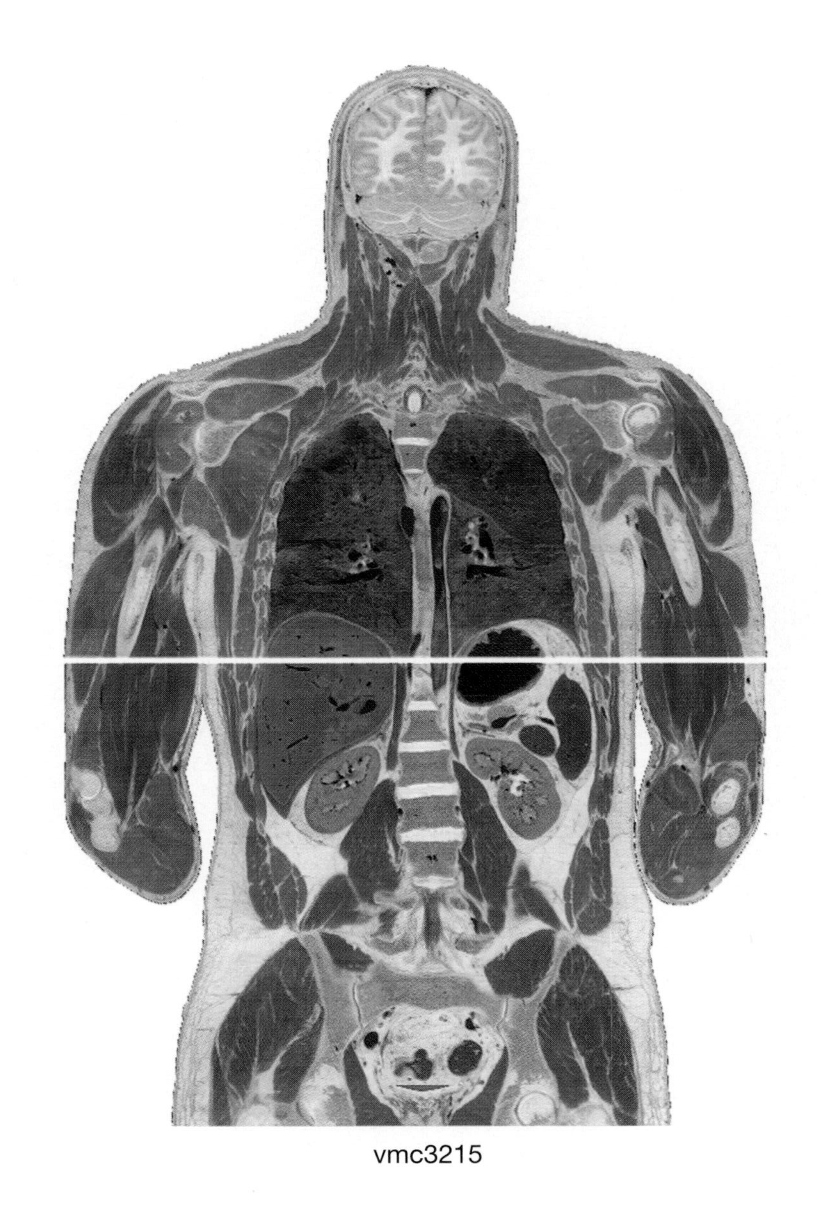

vmc3215

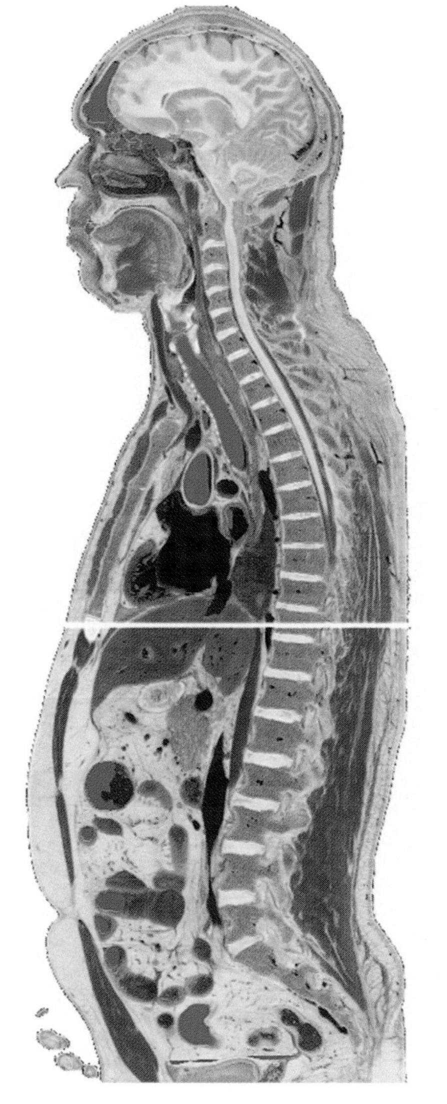

vms5288

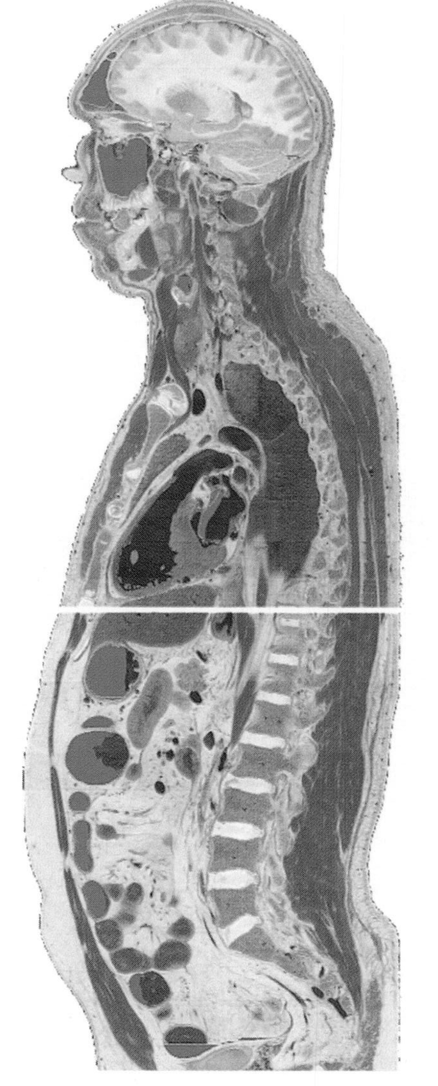

vms5318

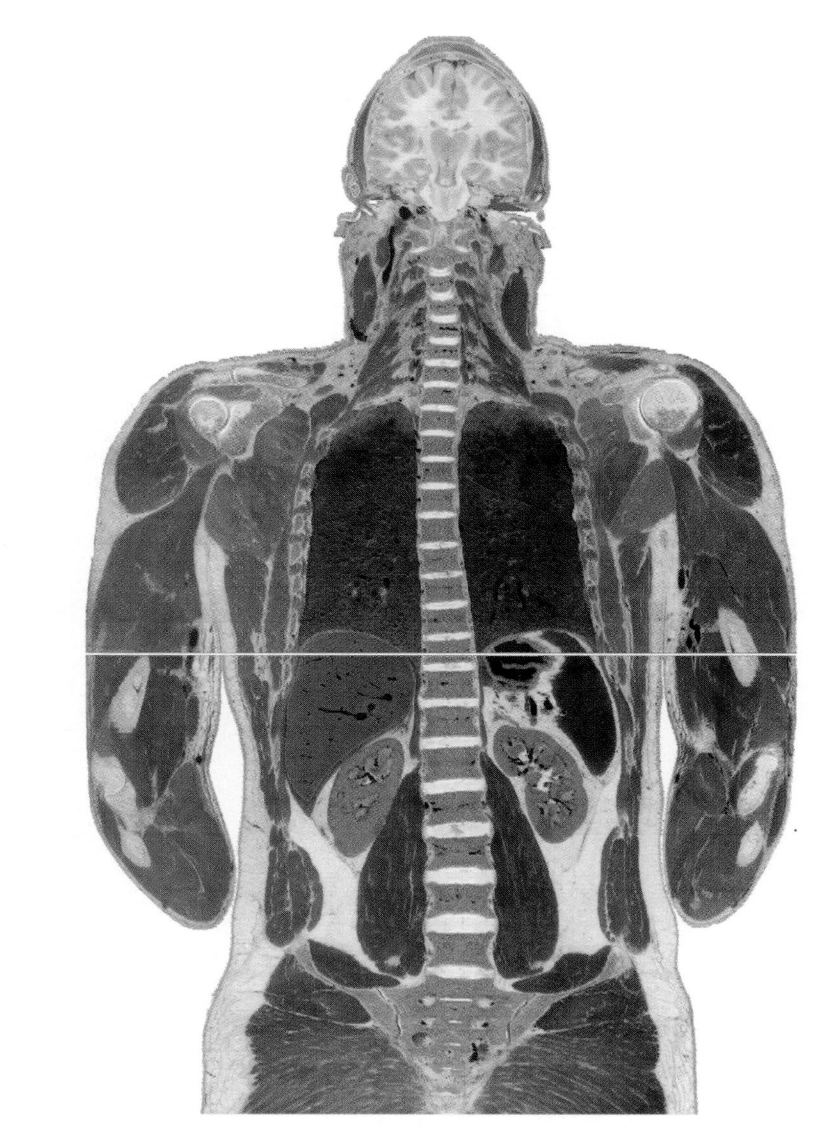

Curved surface through center of spinal cord

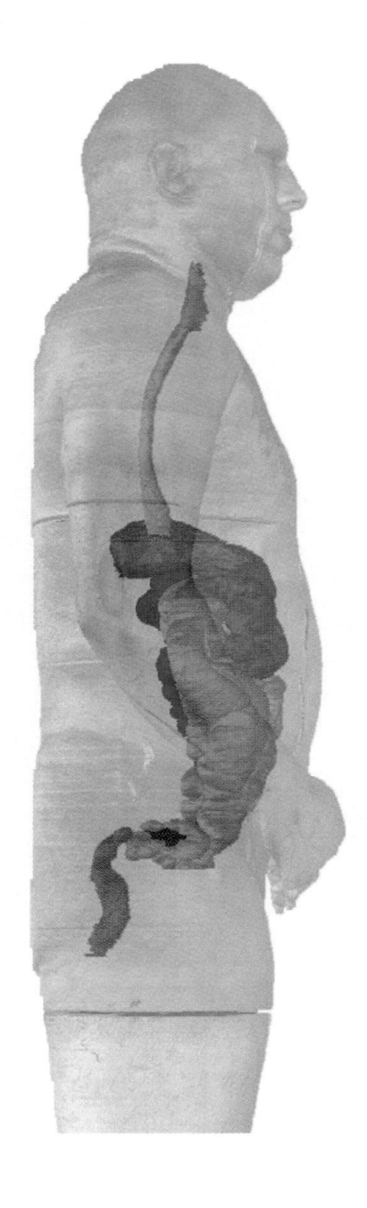

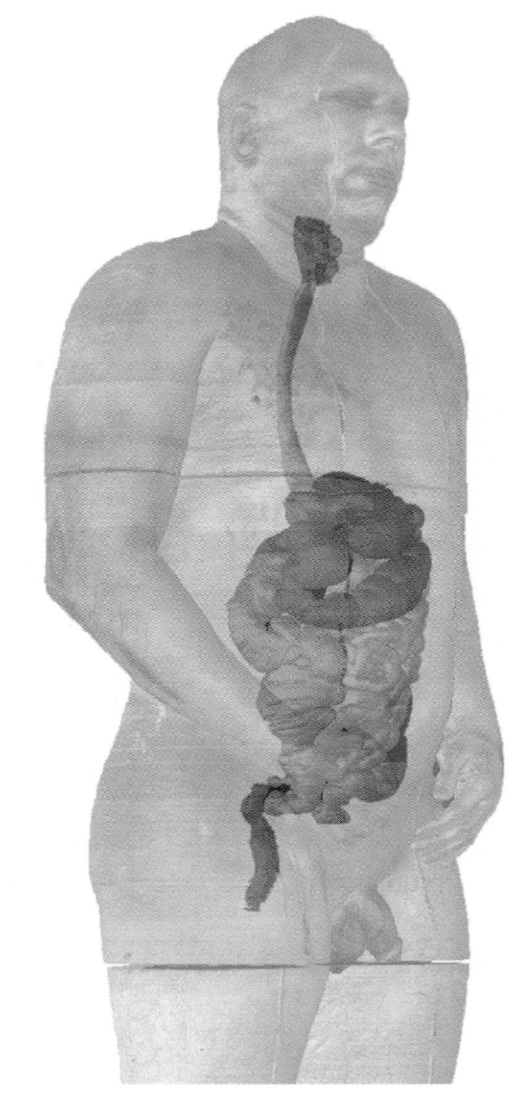

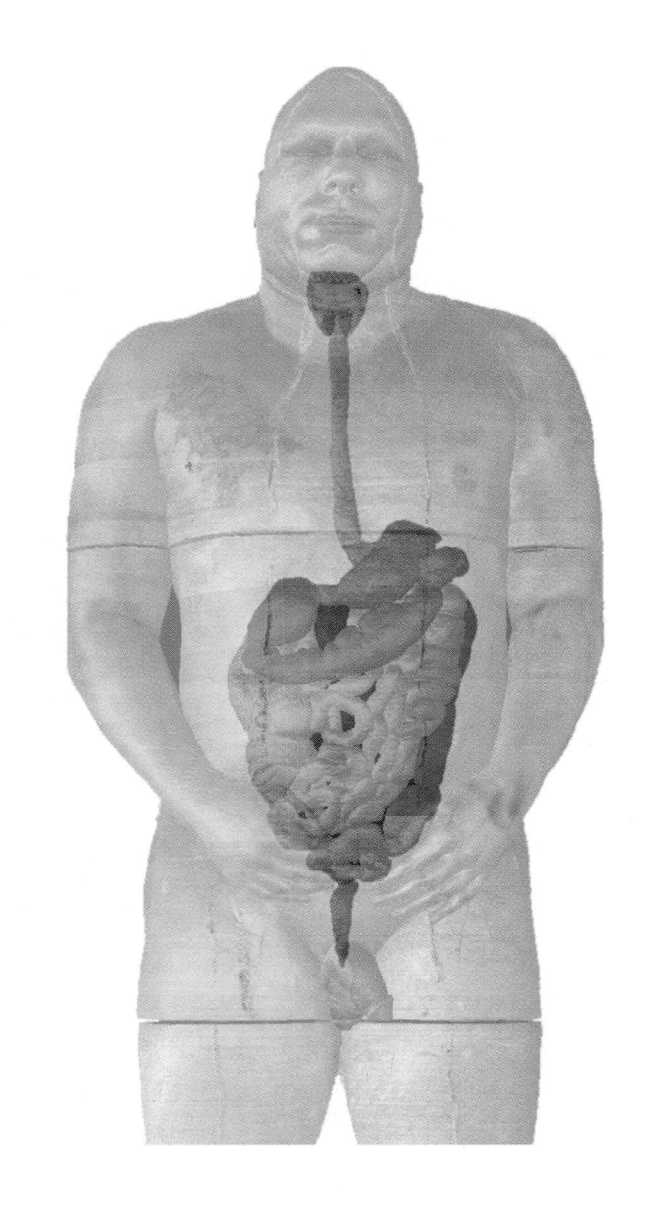

Digestive System

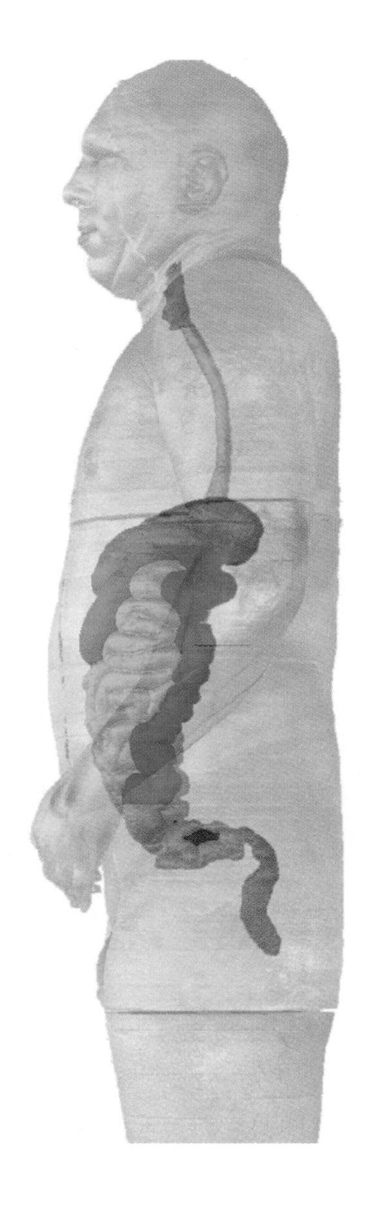

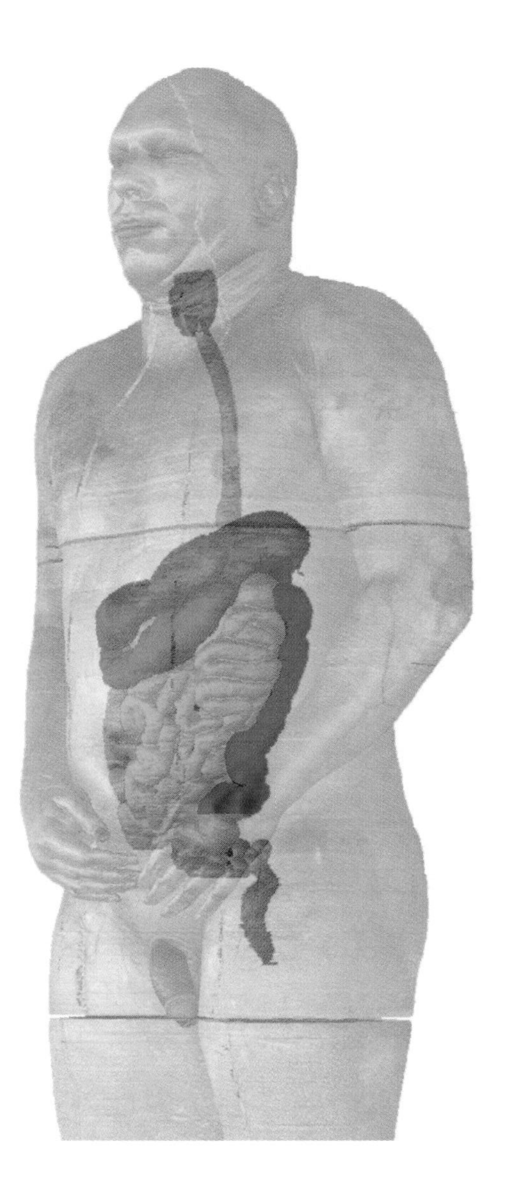

Digestive System

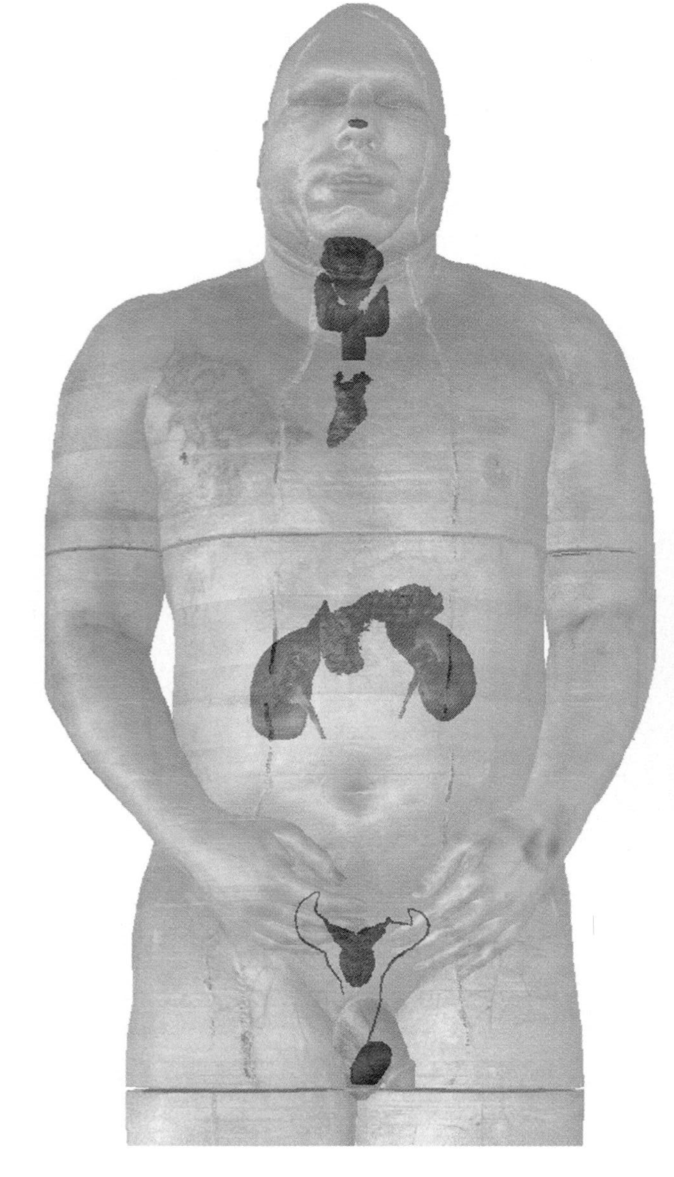

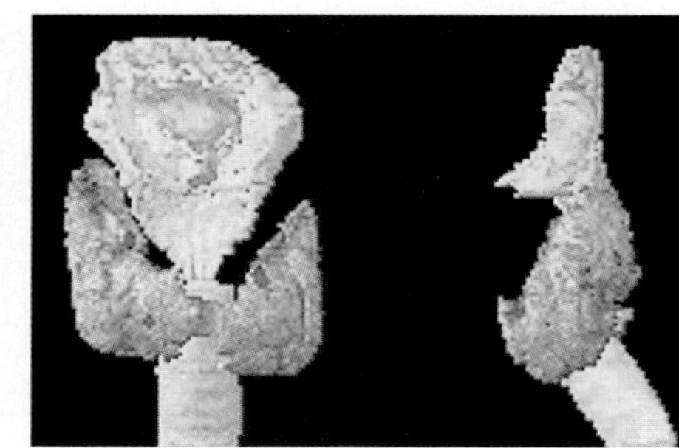

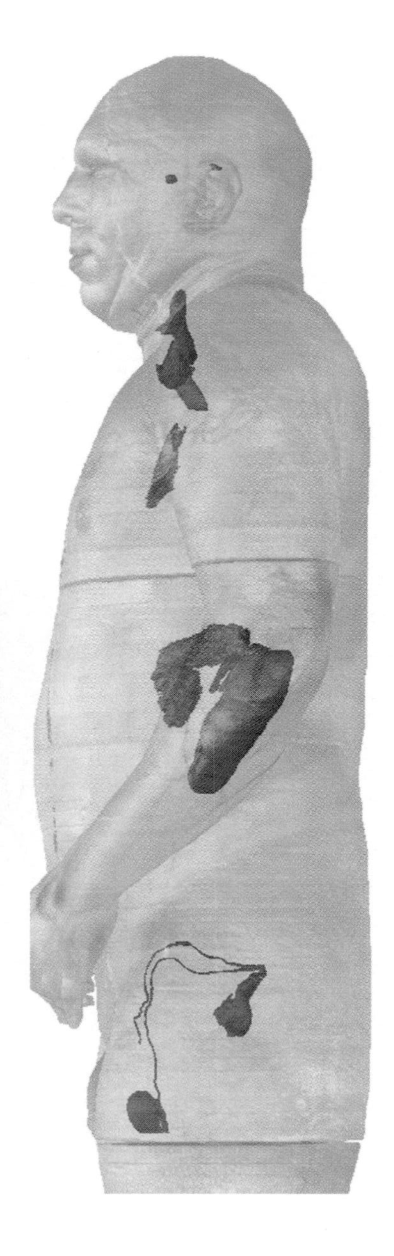

Endocrine System

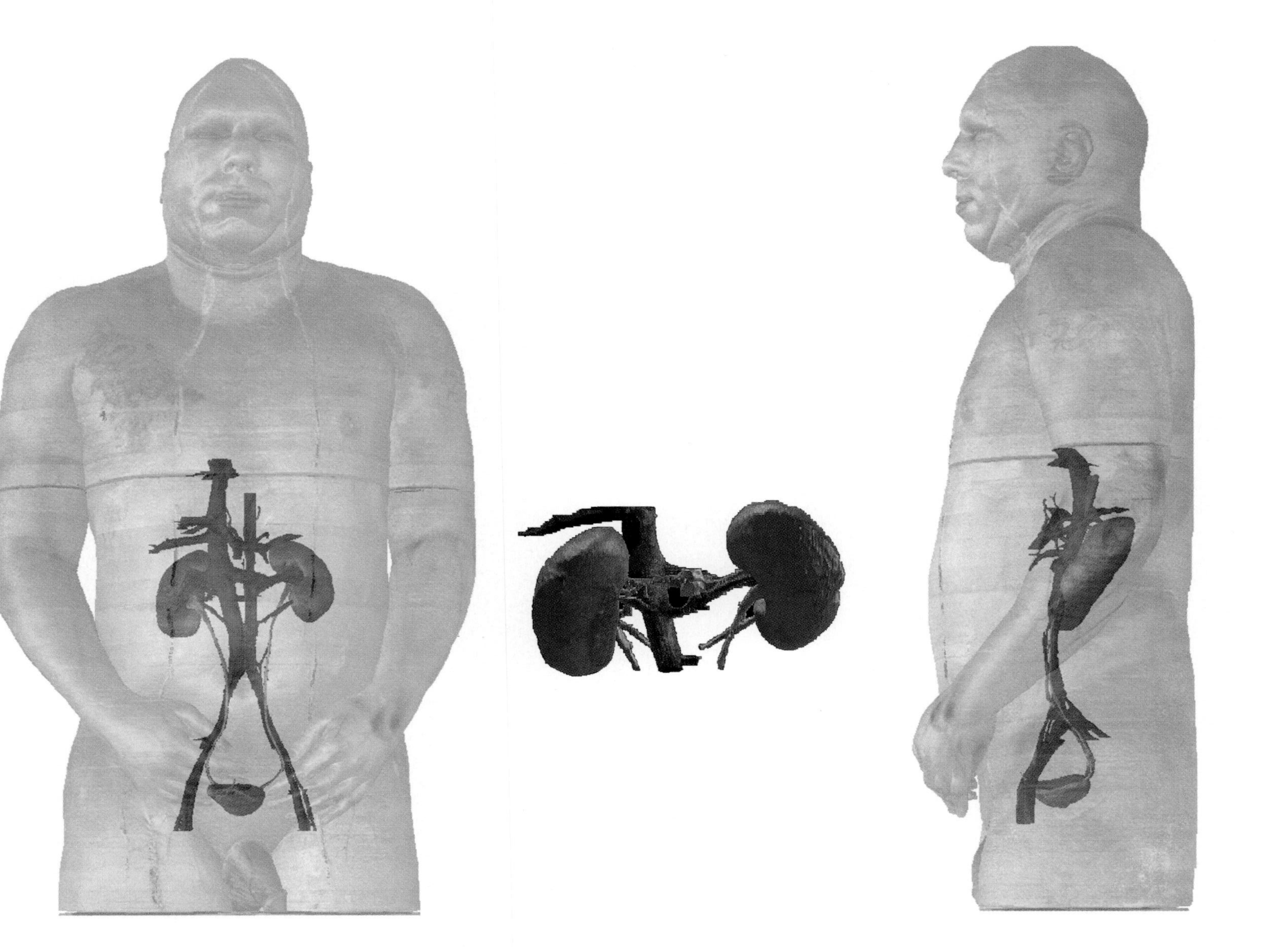

Urinary System

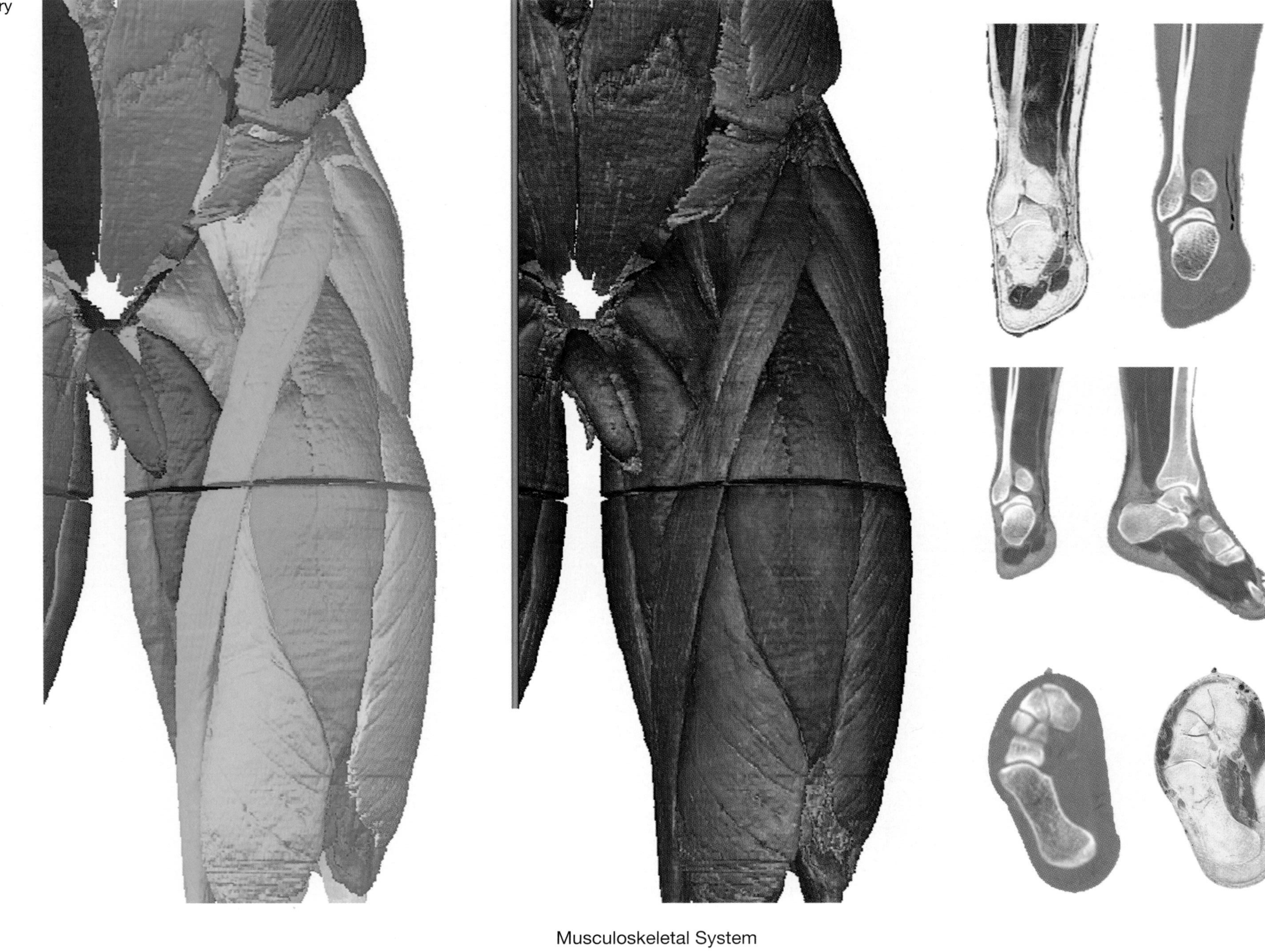

Musculoskeletal System

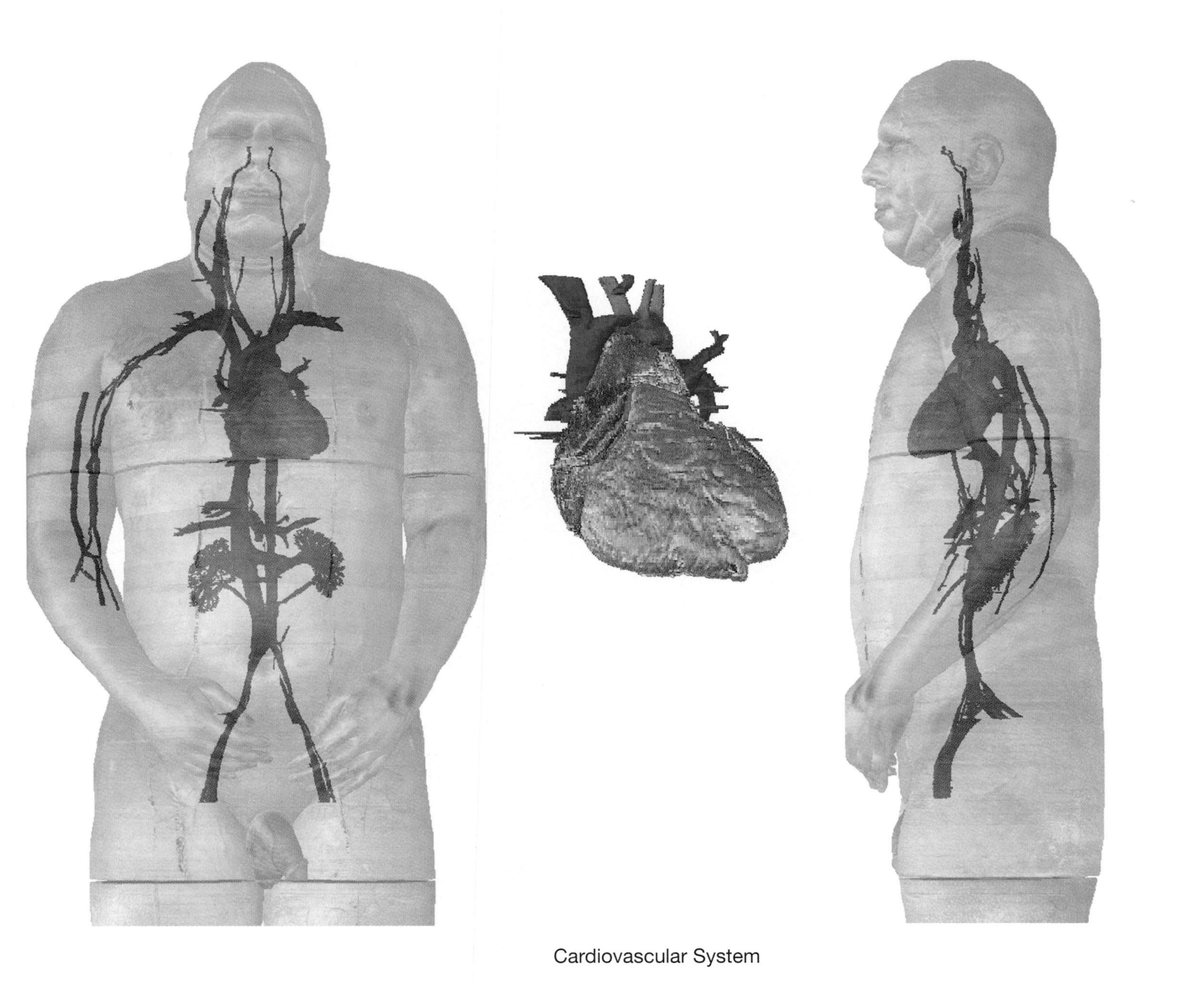

Cardiovascular System

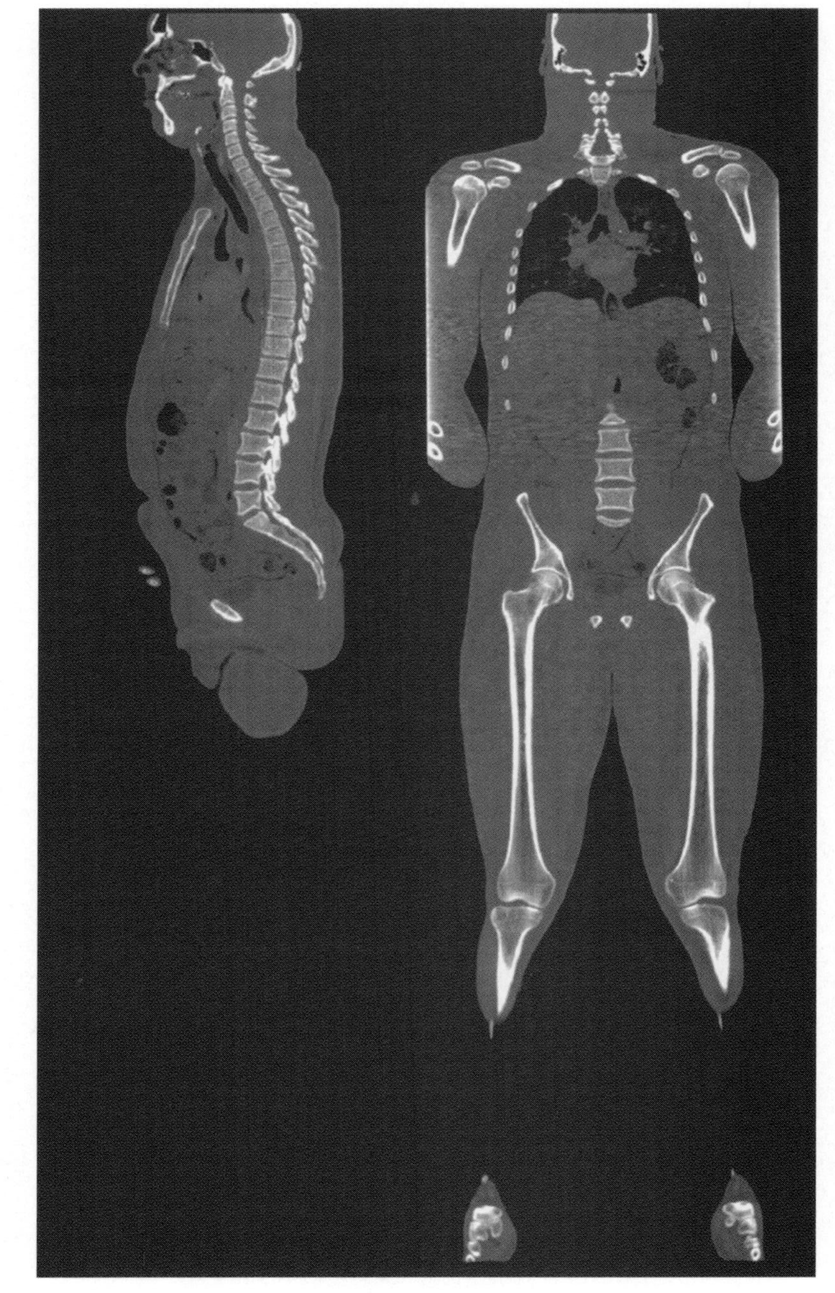

CT Skeletal Images

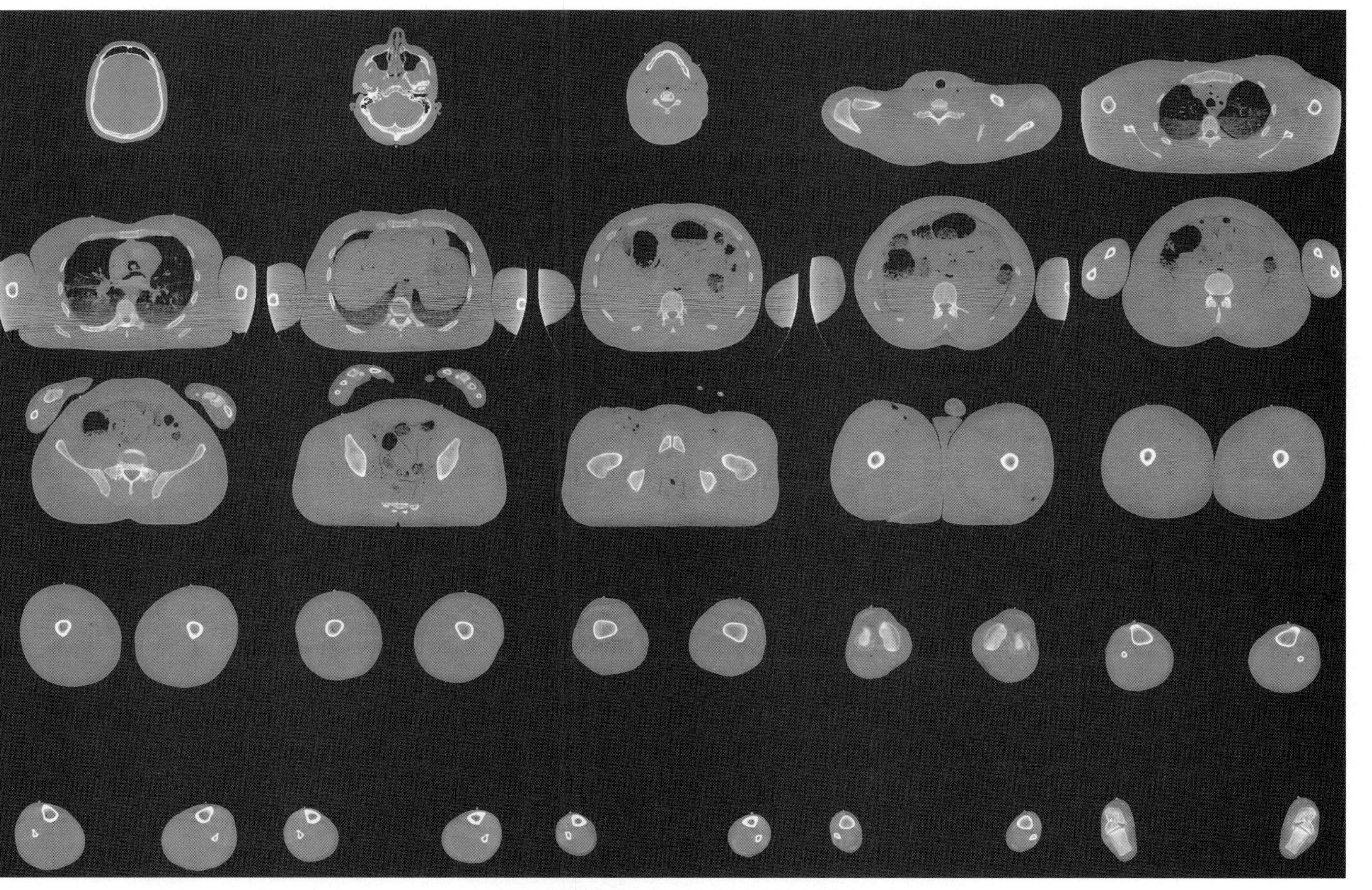

The 1871 Transverse CT Cross-Sections

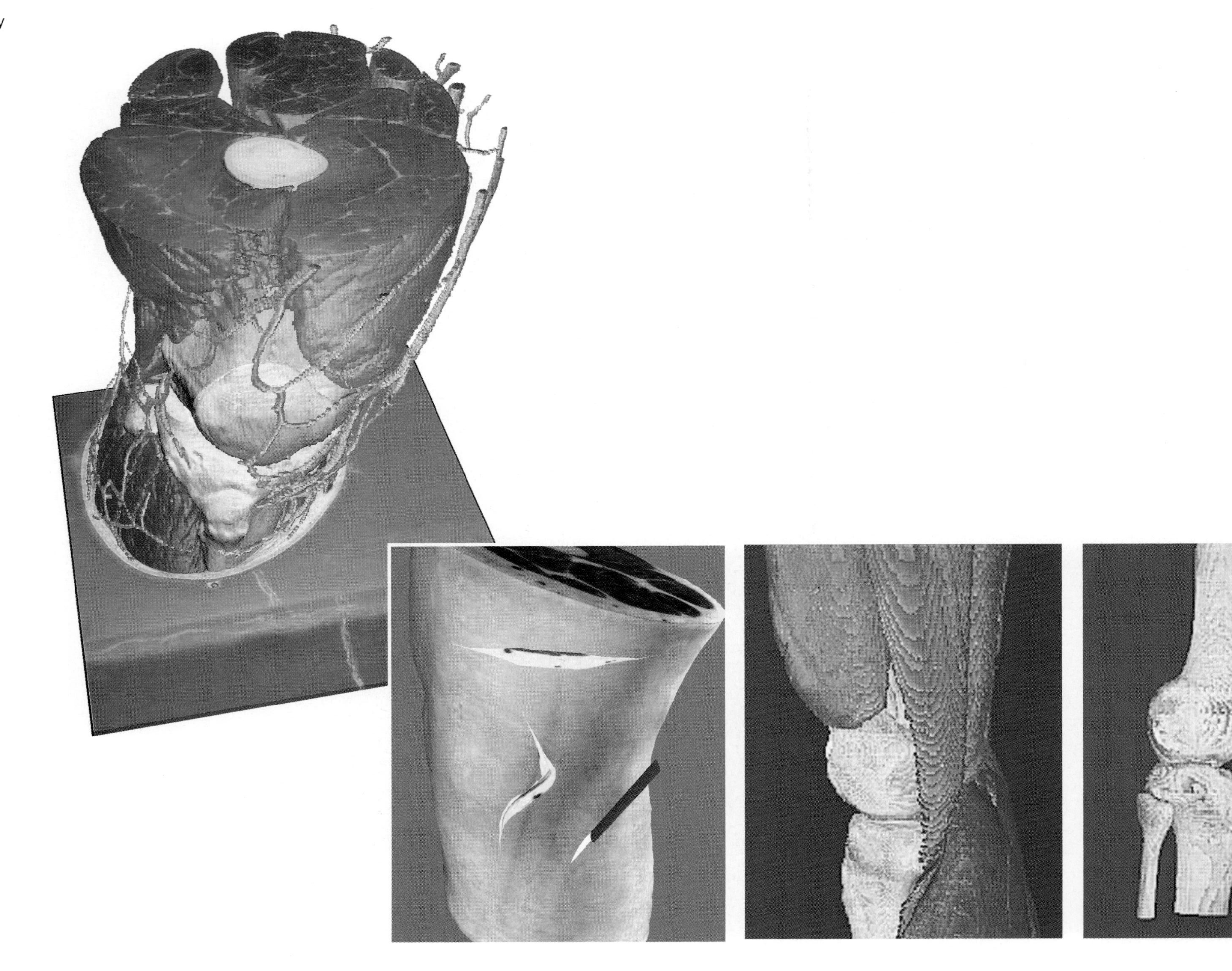

INDEX

Abductor
  digiti minimi muscle, 1829–1861, 2793–2853, 3178–3248, 5168–5198, 5448–5498
    tendon, 2853–2867
  hallucis muscle, 2767–2841, 3136–3254, 5428–5438
    tendon, 2841
  pollicis brevis muscle, 3058–3064, 5388–5408
  pollicis longus muscle, 1740–1750
Accessory saphenous veins, 2026–2668
Acetabular fossa, 1886
Acetabulum, 1861–1897, 5398
Acromioclavicular ligament, 5468
Adductor
  brevis muscle, 1922–2058, 3178–3208, 5238–5278, 5358
  canal, 2026–2226, 3160–3190
    opening, 2214–2226, 5198
  hallucis muscle, 3154–3172, 5428–5468
    oblique head, 2817–2841
    transverse head, 2805–2841
  longus muscle, 1933–2154, 3136–3184, 5208–5398
  magnus muscle, 1933–2238, 3178–3271, 5168–5418
    tendon, 2238–2286
  pollicis muscle, 1789–1800
Adenohypophysis, 1112
Adipose tissue, 1420–2440, 3259–3299, 5158, 5428–5498
Adrenal gland, 1555–1595, 3225, 5248–5268
Amygdala, 1107–1112, 3160

Anconeus muscle, 1583–1610
Ankle joint, 3220–3248, 5458
Anococcygeal ligament, 1897–1933, 3265–3293
Anterior
  axillary fold, 5468
  commissure, 1095
  cruciate ligament, 2312–2330, 3196–3214, 5168, 5418
  crural intermuscular septum, 5488
  crural vein, 2502–2536
  ethmoid sinus, 1112
  femoral vein, 1940–2154, 2312–2330
  interventricular vessel, 1428–1473, 3130
  jugular vein, 1307–1317, 3130–3136
  longitudinal ligament, 1352–1750
  mediastinum, 3088
  nasal spine, 1160
  talofibular ligament, 2725
  tibial vessels, 2406–2745
Annular tracheal ligament, 1285
Anus, 1951–1980, 3231–3271
Aorta,
  abdominal, 1510–1719, 3166
  arch, 1362, 3172–3190, 5288–5318
  ascending, 1381–1428, 3160–3166, 5298
  thoracic, 1381–1504, 3208–3225, 5308
Arteries, see specific artery or vessel by name
Articularis genu muscle, 2190–2238, 3160–3166, 5438–5448
Artifact, 1134–1178, 1521–1570, 1778, 1873, 3094

Aryepiglottic fold, 1243
Arytenoid cartilage, 1260, 3154, 5288
Atrium, left 1405–1436, 5288–5318
Atrium, right, 1405–1473, 3130–3178, 5258–5288
Auditory tube, cartilage, 3148
Auricle, left 1405, 3172
Auricular cartilage, 1107–1160, 5218–5238, 5368–5378
Axilla, 5168–5188, 5398–5438
Axillary
  artery, 1326
  nerve, 5418
  vein, 1326, 5368–5378
  vessel, 1381, 3276, 5178–5208, 5368–5448
Azygos vein, 1381–1521, 3214

Basilic vein, 1405–1583, 3248
Biceps brachii muscle, 1473–1595, 3160–3214, 5458–5538
  long head, 1381–1521, 5468
  short head, 1340–1521
Biceps brachii tendon, 1610
  long head, 1317–1340
  short head, 1326
    origin, 1317
Biceps femoris muscle, 3208–3282, 5378–5408
  long head, 1951–2262, 3237–3293, 5158–5208, 5368–5448

Biceps femoris muscle (*continued*)
  short head, 2082–2330, 3202–3254, 5158–5178, 5418–5458
Biceps femoris tendon, 2341, 3258, 5468
  long head, 2262–2286, 3248, 5458
Bicuspid valve, 5328
Bladder, 1873–1897, 3142–3196, 5268–5318
Brachial
  artery, 1452–1490
  fascia, 1473–1510, 5538
  nerves, 5478–5498
  plexus, 1272–1388, 3172, 5168–5448
    C4 root, 3178
    C5 root, 3184
  vein, 1405–1521, 5488
  vessels, 1504–1616, 5478–5498
Brachialis muscle, 1452–1616, 3220–3231, 5478–5538
  tendon, 1629–1640
Brachiocephalic artery, 1317–1352
Brachiocephalic vein, 1326–1362, 3160–3172, 5258–5328
Brachioradialis muscle, 1510–1682, 3130–3231, 5488–5538
  tendon, 1692–1762, 3100–3112, 5478
Bregma, 1012–1018
Bronchial lymph node, 1405
Bronchopulmonary lymph node, 1428
Bronchus, 3208, 5338
Buccal fat pad, 1178–1197
Buccinator muscle, 1178–1197, 3088–3124, 5258–5328
Bulbospongiosus muscle, 1940–1951, 3196–3220
Buttock, 1922, 3299–3323, 5438–5488

C1, atlas, 1160, 3172
  anterior arch, 1160
  posterior arch, 5318
  superior articular facet, 1160
C2, axis, 3166
  body, 1178–1188, 5288
  dens, 1160, 3172
  dorsal root ganglion, 1178
  posterior arch, 1188
  spinous process, 1188–1197
  superior articular facet, 1178
  transverse foramen, 1188
  transverse process, 1188
C2–3, articular capsule, 1197
C2–3, intervertebral disc, 1197
C3
  body, 1197–1212
  posterior arch, 1212
  posterior articular facet, 1212
  spinal nerve root, 1212
  spinous process, 1212
  superior articular facet, 1197
C3–4, intervertebral disc, 1212
C4
  body, 1218, 3178
  pedicle, 1218
  posterior arch, 1218
  spinal nerve, 1218, 1243
  spinal nerve roots, 1218
  superior articular facet, 1212
  transverse foramen, 1218
C4–5, intervertebral disc, 1235
C5
  body, 1235–1243, 3172
  cervical spinal nerve, 1235
  dorsal root ganglion, 1235
  inferior articular process, 1243
  pedicle, 1235
  posterior arch, 1235–1243
  spinal nerve, 1260–1272
  spinal nerve root, 1235
  spinous process (bifid) , 1243
  transverse foramen, 1235–1243
  ventral rootlets, 1243
C6
  body, 1260
  posterior arch, 1260
  spinal root, 1260
  spinous process, 1260
  superior articular process, 1243
  transverse foramen, 1260
C6–7, intervertebral disc, 1272
C7
  body, 1272–1285
  inferior articular process, 1272
  posterior arch, 1272
  spinous process, 1272
C7–T1, intervertebral disc, 1285
Calcaneal subtendinous space, 2680–2757, 3265
Calcaneal tendon, 2536–2767, 3271–3293, 5428–5438
Calcaneocuboid joint, 2781–2793
Calcaneus bone, 2745–2793, 3225–3293, 5428–5488
Calcaneus tuberosity, 2757–2781
Calcarine sulcus, 1095–1112, 3220–3242
Calf, 3311–3317
Capitate bone, 1778–1800
Cardiac vein, middle 1461–1473
Carotid canal, 1134
Carpal bones, 3052–3088, 5168–5188, 5448–5478
Cauda equina, 1640–1762, 3220
Caudate nucleus, 1072, 3148–3154, 5288
  head, 1095
  tail, 1082–1095
Cavernous sinus, 3154
Cecum, 1750–1800, 3118–3130, 5228–5368
Central
  sulcus, 1048–1060, 3208–3214
  tegmental tract, 1112–1150
  tendinous point, 1933, 3237
Cephalic vein, 1317–1740, 3124–3166, 5168–5538
Cerebellar cortex, 1107–1150
Cerebellar peduncle, middle 1134, 3178–3184
Cerebellum, 3184–3237, 5248–5338
  dentate nucleus, 5288
  emboliform nucleus, 1134
  fastigial nucleus, 1134
  globose nucleus, 1134
  lateral hemisphere, 1150, 3214
  vermis, 1134–1150, 3208
Cerebral
  aqueduct, 1112
  cortex, 1107–1122
  peduncle, 1107–1112, 3172
Chin, 3070–3076
Cingulate gyrus, 1060–1095, 5308
Cingulate sulcus, 1072–1082
Cisterna chyli, 1570
Cisterna magna, 1160–1178
Claustrum, 1082–1095
Clavicle, 1260–1340, 3142–3196, 5168–5458
Clivus, 1122
Coccygeus muscle, 1886, 3271–3276, 5288–5298
Coccyx, 1861–1897, 3276–3293, 5308–5318
Cochlea, 1134, 5328
Colic
  lymph node, 1692–1719
  vein, 1610
    middle 1583–1595
  vessels, middle 1539–1629
Collateral sulcus, 1107–1122, 5328
Colon
  ascending, 1583–1740, 3076–3184, 5188–5268
  descending, 1539–1812, 3106–3190, 5338–5438
  hepatic flexure, 1549–1570, 3082–3172, 5238–5268
  sigmoid, 1812–1861, 3112–3259, 5298–5348
  splenic flexure, 1521, 3166, 3202, 5348–5408
  transverse, 1539–1719, 3064–3106, 5208–5418
Common
  bile duct, 1570–1595, 3166
  carotid artery, 1243–1352, 5258–5328
  extensor tendon, 3248
  flexor tendon, 3248
  hepatic bile duct, 1549
  iliac artery, 1730–1778, 5298
  iliac vein, 1750–1778, 3178–3208
  peroneal nerve, 2178–2382
Communicating vein, 5238, 5358
Confluence of sinuses, 5288
Coracoacromial ligament, 5438

Coracobrachialis muscle, 1307–1461, 3178–3184, 5428–5498
Coracoid process, 1293–1307, 3178–3184, 5158, 5438
Corona radiata, 1072
Coronal suture, 1012–1072
Coronary
  artery, left 1405
  artery, right, 1452–1473
  sinus, 1436, 3160
    valve, 1452
  vessel, right, 1428–1436
Corpus callosum, 1072, 3130–3172, 5298
  genu, 1082, 5308
  splenium, 1082–1095, 5298
Corpus cavernosum, 1933–1992, 3094–3148, 5288–5338
  fibrous capsule, 1951
Corpus spongiosum, 1951–2010, 3106–3148
Corticospinal fibers, 1122–1134
Costophrenic space, 1521–1539
Costovertebral joint, 1317
Cricoid cartilage, 1272–1285, 3142–3160, 5288–5298
Cricothyroid ligament, 1272
Cricothyroid muscle, 1272
Crista terminalis, 1428–1461
Crural fascia, 2312–2745, 3190–3299, 5168–5208, 5368–5488
Cuboid bone, 2781–2805, 3190–3220, 5468–5488
Cuneate fasciculus, 1160
Cuneiform cartilage, 1243
Cuneocuboid joint, 2781–2805
Cystic duct, 1549–1555, 3154

Deep
  brachial vessel, 1461–1521
  cerebral vein, 1188
  cervical fascia, 1178–1272
  cervical lymph node, 1212–1260
  cervical veins, 1160–1212
  femoral artery, 1908–1922
  femoral vein, 1922–1992, 2046–2190
  femoral vessel, 1992–2036, 2154–2166, 5398
  peroneal nerve, 2524–2745
Deltoid ligament, 2740
Deltoid muscle, 1272–1436, 3136–3299, 5168–5178, 5418–5538
Dense connective tissue, 1028, 1134–1307
Dense subcutaneous tissue, 1012–1082, 1212
Dental restoration, 1178
Depressor
  anguli oris muscle, 1178–1218
  labii inferioris muscle, 1212–1218
  septi muscle, 1160, 3076
  supercilii muscle, 1095
Dermis, 1012–1028

Diaphragm, 1461–1640, 3064–3271, 5168–5438
  crus, 1521–1640, 3184–3208, 5298–5318
  esophogeal hiatus, 1510
Digastric muscle
  anterior belly, 1235, 3094–3124, 5308–5318
  posterior belly, 1150–1218, 3160–3184, 5248–5338
Diploe, 1018–1082, 3259
Diploic vessel, 1018–1150, 3259
Dorsal
  interosseous muscles, 3046
  interosseous muscles, first
    of the foot, 2793–2841, 3124–3148, 5408–5468
    of the hand, 1789–1829, 3046–3052, 5198, 5218–5228
  sacroiliac ligament, 3271
  thalamus, 1082, 3166–3178, 5288–5308
Dorsalis pedis vessel, 2757–2781
Dorsolateral nucleus, 1095
Ductus deferens, 1835–2010, 3142
  ampulla, 1886
Duodenojejunal flexure, 1640
Duodenojejunal junction, 3142–3154
Duodenum, 1555–1657, 3148–3178, 5248–5328
  superior part, 5288
Dura mater, 1028–1095

Ejaculatory duct, 1897–1908
Elbow joint, 5538
Epidermis, 1012–1028
Epididymis, 1980–1992, 2010
Epiglottis, 1218–1243, 3130–3154, 5288–5298
Esophagus, 1285–1504, 3178–3214, 5288–5298
Ethmoid bone, 1107–1122
  crista galli, 1095
  middle nasal concha, 1134
  superior nasal concha, 1122
Ethmoid sinuses, 1122, 3088–3124, 5288
  middle 1107
Extensor carpi radialis
  brevis muscle, 1629–1750, 5478–5538
    tendon, 5468
  longus muscle, 1539–1730, 3118, 5488–5538
    tendon, 1750–1762
Extensor carpi ulnaris muscle, 5478–5518
    tendon, 1762
Extensor digitorum
  brevis muscle, 2767–2817, 3190–3225, 5488–5498
    tendon, 2817
  longus muscle, 2353–2725, 3202–3214, 5458–5528
    tendon, 2668–2829
  muscle, 1657–1740, 5488–5518
    tendon, 1750–1800, 5168
Extensor hallucis
  brevis muscle, 2767–2781, 3202–3214
    tendon, 2793–2817
  longus muscle, 2477–2740, 3214, 5468–5488
    tendon, 2668–2817, 3196
Extensor pollicis
  brevis muscle, 1750–1762
  longus muscle, 5488
External
  acoustic meatus, 1134, 3172, 5218–5368
  anal sphincter, 1922–1980, 3254–3276, 5298–5308
  capsule, 1082–1095
  carotid artery, 1212–1235
  ear, 1122–1134
  iliac artery, 1778–1886
  iliac vein, 1789–1886
  iliac vessel, 3148
  intercostal muscle, 1629
  jugular vein, 1243–1285, 3136–3172, 5228
  oblique muscle, 1521–1800, 3064–3254, 5158–5458
    aponeurosis, 1812–1886, 5238
  occipital protuberance, 1134
  perineal fascia, 1964
  pudendal vessel, 1922–1933
Extreme capsule, 1082–1095
Eye globe, 3088–3100, 5258–5328
Eyelid, tarsus, 1095

Facial vein, 1122–1218, 5238–5248
Facial vessels, 3082
Falciform ligament, 3124–3142
Falx cerebri, 1028–1095, 3196–3214
Fascia
  between adductor and hamstring muscles, 2010–2226
  lata, 1992–2330, 3100–3293, 5168–5498
  see also specific full name
Fauces, isthmus, 1188
Femoral
  artery, 1897–2166
  hemorrhage, 1940–1951
  nerve, 1719–2118
  vein, 1897–2214, 3142–3166, 5238–5388
  vessels, 1964, 2166–2202, 3124–3190, 5208–5418
Femur, 1933–2324
  articular cartilage, 2312–2324
  greater trochanter, 1886–1908, 3190–3220, 5158–5168, 5428–5448
  head, 1861–1897, 3166–3208, 5198–5228, 5368–5408
    fovea of head, 1873
    ligament of head, 1873
  intercondylar fossa, 2300–2312
  intertrochanteric crest, 1908–1933, 5178–5188, 5408–5418
  lateral condyle, 2300–2330, 3166–3242, 5168–5178, 5418–5448
  lateral epicondyle, 2286–2312

Femur (*continued*)
lesser trochanter, 1940–1951, 5198
linea aspera, 2010–2178, 3196
medial condyle, 2300–2341, 3166–3231, 5188–5408
medial epicondyle, 2300–2312, 3184–3202, 5388
neck, 1886–1922, 3184–3196, 5168–5188, 5408–5428
periosteum, 2166–2300
shaft, 3172–3202, 5158–5188, 5398–5438
Fibula, 2382–2745, 3178, 5428–5488
anterior border, 2440–2692
anterior surface, 2692
apex, 2353–2368, 3225–3248, 5458
head, 5468
lateral malleolus, 2704–2745, 3214–3259, 5478–5488
posterior border, 2440–2704
shaft, 3242–3259, 5458–5468
Fibular collateral ligament, 2286–2353, 5458
Fimbria, 1095–1107
Flexor carpi
radialis muscle, 1657–1682, 1719–1740
tendon, 1750–1762
ulnaris muscle, 1610–1616, 1657–1762, 3202–3208, 5468–5508
Flexor digiti minimi
brevis muscle, of the foot, 2793–2853, 3196–3225, 5168, 5438–5498
tendon, 2867
brevis muscle, of the hand, 1835–1850
longus muscle, of the hand, 1829
Flexor digitorum
brevis muscle, 2781–3265, 5428–5478
tendon, 2853–2867, 3148
longus muscle, 2464–2704, 3160, 5428–5478
tendon, 2656–3214, 5448
profundus muscle, 1629–1778, 5468–5538
superficialis muscle, 1629–1778, 5468–5528
Flexor hallucis
brevis muscle, 2805–2841, 5438
tendon, 2841
longus muscle, 2487–2725, 3231–3265, 5438–5448
tendon, 2680–2841, 3202–3214, 5438
Flexor pollicis
brevis muscle, 1789–1829, 3058–3064, 5408
longus muscle, 1682–1750
Flexor retinaculum, 2716
Forehead, 3064
Fornix, 1082–1107, 3196
Frontal
bone, 1012–1095, 3070–3160, 5238–5288, 5338–5358
gyrus, middle 1048–1082
lobe, 1028–1095, 3094–3154, 5248–5328
pole, 1048
sinus, 1048–1095, 3076–3100, 5258–5338
vein, 5228

Galea aponeurotica, 1012–1060, 3259–5268, 5348–5368
Gall bladder, 1549–1570, 3112–3130, 5228
Gastric vein, left 1539–1555
Gastric vessel, left 1504–1521
Gastrocnemius muscle, 3208–3214
lateral head, 2300–2536, 3220–3311, 5168, 5398–5478
medial head, 2300–2560, 3220–3305, 5168–5208, 5368–5408
Gastroepiglottic vein, right, 1570
Gastroepiploic vessel, 3076–3082
Gemelli muscles, 3248
tendons, 5188
Genicular vessel, middle 2312
Genioglossus muscle, 1188–1212, 3094–3118, 5288–5298
Geniohyoid muscle, 1218, 3100–3106, 5288
Glenohumeral ligament, 1307, 5448–5468
Globus pallidus, 1095
external segment, 1095
internal segment, 1095
Gluteus
maximus muscle, 1778–2046, 3196–3305, 5158–5468
medius muscle, 1740–1908, 3142–3265, 5158–5478
tendon, 1861–1922
minimus muscle, 1778–1873, 3172–3225, 5158–5208, 5368–5448
tendon, 1922
Gracilis
fasciculus, 1160
muscles, 1951–2274, 3178–3248, 5228–5358
tendon, 2274–2368, 5218, 5368
nucleus, 1160
Gray matter, 1028–1072
Great cardiac vein, 1428–1436
Great saphenous veins, 1933–2817, 3118–3259, 5208–5438
Greater omentum, 1595–1616, 3076–3082
Greater sciatic notch, 1829–1861
Gyrus rectus, 1095

Habenulopeduncular tract, 1107–1112
Hamate bone, 1778–1800
hamulus, 1800
Hamstring tendon, 1922–1951
Hard palate, 1160, 3094–3124, 5278–5308
Heel, 3288–3299
Hepatic artery, 1539–1570
Hepatic vein, 1490–1555
Hip joint, capsule, 1897
Hippocampal sulcus, 1107
Hippocampus, 1095–1112, 3172–3178, 5278, 5318
Humerus, 1317–1616, 3184–3248
capitulum, 1583–1616, 3220
deltoid tuberosity, 1388–1428, 5518, 5498
head, 1285–1326, 3160–3190, 5458–5488
lateral epicondyle, 1570–1583
lateral supracondylar ridge, 1539–1555
lesser tubercle, 3172, 5458
medial condyle, 3248
medial epicondyle, 1570–1583, 3254–3265, 5528
medial supracondylar ridge, 1549–1555
neck, 5468–5488
olecranon fossa, 1570
shaft, 1452, 5478–5538
surgical neck, 1326
trochlea, 1583–1610, 3231–3265, 5528–5538
Hyoglossus muscle, 1197–1218, 3130–3142
Hyoid bone, 1235, 3130–3136, 5308
Hypophysis, 3148
Hypothalamus, 1107
Hypothenar muscles, 1778–1812, 3070–3088, 5168–5188, 5448–5468

IIliofemoral ligament, 1922
Ileocecal junction, 1740, 3124
Ileocolic vein, 1657–1762
Ileum, 1657–1835, 3064–3148, 5258–5388
Iliacus muscle, 1740–1812, 3136–3225, 5188–5238, 5378, 5408–5428
Iliocolic vein, 1640
Iliocostalis lumborum muscle, 1521–1762
Iliocostalis muscle, 3248–3311, 5218–5258, 5378–5388
Iliocostalis thoracis muscle, 5198, 5238
Iliofemoral ligament, 1873–1922, 3172–3184, 5168–5208, 5388–5428
Iliolumbar vessels, 1740–1951, 3160–3178, 5188–5418
Iliopsoas muscle, 1800–1951, 3160–3178, 5188–5208, 5398, 5408–5418
tendon, 1922–1964, 3166–3190, 5388
Iliotibial tract, 1861–2353, 3225–3259, 5478
Ilium, 1740–1886, 3154–3271, 5188–5438
anterior inferior spine, 3136, 5178
body, 1850–1861, 3220, 5198, 5368–5398
crest, 1812, 3148–3190, 5168–5198, 5378–5398
Indusium griseum, 1082
Inferior
alveolar vessel, 1218
cerebellar peduncle, 1150
colliculus, 1112
brachium, 1112
concha, 5288
epigastric vessel, 1812–1850
frontal gyrus, 1072
frontal sulcus, 1072
gemellus muscle, 1886–1908, 3231, 5208, 5368–5408
gluteal vessel, 1861–1933
labial vein, 1235–1243
mesenteric vein, 1595–1800, 5328

mesenteric vessel, 3160
nasal concha, 1150, 3106–3118
oblique muscle, 1122–1150
peroneal retinaculum, 2767–2781
rectus muscle, 1112
temporal gyrus, 1107–1122
temporal sulcus, 1107–1122
thyroid vein, 1285–1326
vena cava, 1473–1740, 3166–3196, 5268–5298
valve, 1473
Infraglottic space, 1272, 3142
Infrapatellar fat pad, 2312–2353, 5178
Infraspinatus muscle, 1293–1405, 3220–3305, 5158–5188, 5388–5498
tendon, 5498
Inguinal
canal, 3124, 5228–5368
internal ring, 3130
external ring, 3136, 5248, 5378
ligament, 3124–3136
lymph node, 1940–1964
Inner table, 1028–1082, 3259
Insula, 1082–1095, 3154–3166, 5268, 5328
Interbronchial lymph node, 1388
Intercostal vein, 1381–1510
Intercostal vessel, 1452–1510
Intercuneiform joint, 2781
Intergluteal crease, 1850, 3265–3323
Intermediate cuneiform bone, 2781–2793, 3172–3184, 5478–5488
Intermediate mass, 1095
Internal
anal sphincter, 1951–1964
arcuate fibers, 1150
capsule, 3148
anterior limb, 1082
genu, 1082
posterior limb, 1082–1095
carotid artery, 1122–1235, 3166
iliac artery, 1778–1850
iliac vein, 1789–1850, 3220–3259, 5268, 5328–5338
iliac vessel, 3282
intercostal muscle, 1657
jugular vein, 1160–1307, 3154–3172, 5248
oblique muscle, 1640–1850, 3094–3237, 5178–5448
occipital protuberance, 1112–1134, 3254
pudendal vessel, 1908–1933
thoracic vein, 1405
thoracic vessel, 1420–1510, 3112–3130
Interosseous
membrane, 1671–1762, 2368–2680, 3124–3178
ligament
metatarsal, 2817
sacroiliac, 1778–1800

Interosseous muscle
dorsal
first, of the foot, 2793–2841, 3124–3148
first, of the hand, 1789–1829, 3046–3052, 5198, 5218–5228, 5408–5468
fourth, of the foot, 2829–2853, 3160, 5448–5488
fourth, of the hand, 5178–5188
second, of the foot, 2817–2841,3136–3160, 5478
second, of the hand, 1800–1829, 3046, 5418–5438
tendon, third, of the foot, 2853
third, of the foot, 2817–2841, 3160, 5478–5488
third, of the hand, 1829–1835, 3046, 5188–5198
palmar, 3046–3088
first, 1800–1812, 5208, 5428–5438
second, 1829–1835, 5428
tendons, 3070
third, 1829–1850, 2829–2853
plantar
first, 2817–2853, 3148
second, 2829–2853, 3154
Interparietal sulcus, 1060
Interspinalis muscle, 1235
Interspinous ligament, 1235
Interventricular septum, 1436–1461, 3112
Intestinal
lymph node, 1762
vein, 3130
vessel, 5368
Intraabdominal adipose tissue, 1473–1521
Intraabdominal fat, 5178
Ischial tuberosity, 1922–1940, 3248–3254, 5218–5358
Ischiocavernosus muscle, 1933–1940, 3196–3225, 5278–5328
Ischiorectal fossa, 1886–1964, 3231–3276, 5308–5338
Ischium, 1873–1940, 3202–3237, 5238–5368
spine, 1873

Jejunal vein, 1629–1657
Jejunal vessels, 1595–1671
Jejunum, 1549–1762, 3076–3178, 5278–5418
Jugular bulb, 1150
Jugular canal, 1150

Kidney, 1570–1682, 3172–3254, 5208–5408
adipose capsule, 3214
hilus, 5378
renal pelvis, 1610
renal sinus fat, 1595–1610
Knee joint, 3178–3190, 5178–5198, 5388–5448
articular cartilage, 5398
capsule, 2286–2330, 5378
cavity, 2286–2312, 3178–3190, 5198, 5398–5458

L1
body, 1595–1610
spinous process, 1610–1616
L1–2, intervertebral disc, 1616–3225
L2
body, 1629–1640
spinous process, 1657
transverse process, 1640, 3242
L2–3, intervertebral disc, 3208
L3
body, 1657–1682, 5318
inferior articular process, 1682–1703
lamina, 1692
superior articular process, 1682–1703, 3242–3254
transverse process, 1671
L3–4, intervertebral disc, 1657–1692
L4
body, 1703–1719
inferior articular process, 1719–1730
lamina, 1719–1730
mammillary process, 1703
spinous process, 1719–1740
transverse process, 1703
L4–5, intervertebral disc, 1730, 3190–3196
L5
body, 1740–1762, 3178, 5328
inferior articular process, 1750
lamina, 1750
mammillary process, 1740
nerve, 1812
spinous process, 1750–1778, 3248–3254
transverse process, 1740–1750
L5–S1, intervertebral disc, 1762–1778, 5298–5318
Lacrimal bone, 1107–1112
Lacrimal lake, 1028
Lambdoidal suture, 1082–1150
Laryngeal orifice, 1218
Laryngeal ventricle, 1243
Laryngopharynx, 1235–1243, 3160
Larynx, inlet, 1243
Lateral
calcaneal vessel, 2757–2767
cerebellar hemisphere, 1134
cricoarytenoid muscle, 1260
cuneiform bone, 2781–2805, 3172–3190, 5468
cutaneous vein, 5488
femoral circumflex vessels, 1908–1964
femoral intermuscular septum, 1940–2238
fissure, 1048–1082, 3130–3184, 5238–5358
genicular vessel, 5468
intermuscular septum, 1504–1510
lemniscus, 1122–1134
malleolar vessel, 3271
meniscus, 2330, 3196–3225, 5448–5458

Lateral (*continued*)
  occipital gyrus, 1072–1122
  patellar retinaculum, 2274–2353, 3166, 5468
  plantar nerve, 2793–2817
  plantar vessel, 2793–2817
  pterygoid muscle, 1134–1160, 3136–3154, 5248–5258, 5328
  recess, 1150
  rectus muscle, 1107–1112, 3112–3118, 5278
  talocalcaneal ligament, 2745–2757
  ventricle, 1072–1082, 3136–3196, 5278, 5318
    anterior horn, 1082
    choroid plexus, 1095–1107
    inferior horn, 1107
Latissimus dorsi muscle, 1405–1692, 3202–3317, 5158–5458
Lens, 1112, 3088
Lentiform nucleus, 3142–3160, 5278–5318
Levator
  anguli oris muscle, 1160
  ani muscle, 1886–1933, 3225–3265, 5288–5318
  costae muscle, 1521–1583
  costalis muscle, 1405–1510
  labii superioris muscle, 1134–1160, 3082, 5318
  palpebrae superioris muscle, 1095
  scapulae muscle, 1178–1293, 3184–3276, 5218–5358
  veli palatini muscle, 1150–1160
Ligament(s), see full name of ligament
Ligamentum flavum, 1197, 1243–1405, 1730, 3276
Ligamentum nuchae, 1188–1362, 3288
Ligamentum venosum, 1521
Linea alba, 1570–1829, 3052–3106, 5298
Linea semilunaris, 1583–1812, 5388–5398
Lingual gyrus, 1107–1218
Lip, lower, 1188–1197, 3058–3076
Lip, upper, 1178, 3058–3076
Liver
  caudate lobe, 1510–1549, 5238
  left lobe, 1490–1570, 3082–3148, 5308–5318
  quadrate lobe, 3148, 5268–5288
  right lobe, 1461–1629, 3088–3271, 5168–5258
Locus ceruleus 1134
Long
  extensor muscles,of the hand 3100–3172
    tendons, 3076–3094, 5458
  flexor muscles, of the hand 3100–3190
    tendons, 1778–1873, 3064–3094, 5168–5198, 5438–5458
  plantar ligament, 2793–2805
  saphenous vein, 3237
Longissimus capitis muscle, 1160–1260, 3202–3214
Longissimus muscle, 1583–1829, 3242–3305, 5248–5278
Longissimus thoracis muscle, 1293–1570, 3282–3311
Longitudinal cerebral fissure, 1028–1122, 3094–3254
Longus capitis muscle, 1150–1243, 3154–3166, 5278
Longus cervicis muscle, 1188–1317, 3172–3184, 5278–5318
Loose connective tissue, 1012–1028, 1134
Lumbar
  artery, 1671
  fascia, 3288–3311
  lymph node, 1616–1682
  vein, 1657–1682
    ascending, 1703
  vessel, 1719
Lumbodorsal fascia, 1682–1829, 3276–3282
Lumbrical muscles, 1812, 3064
  first, of the foot, 2829–2853
  fourth, of the foot, 2829–2853
  second of the foot, 2829–2853
  tendons, of the hand, 3070
  third, of the foot, 2829–2853
Lunate bone, 1778
Lung
  horizontal fissure, 1405–1420, 3142, 5198–5208
  lingula, 5348–5408
  lower lobe, 1362–1521, 3172–3282, 5188–5418
  middle lobe, 1405–1473, 5188–5258
  oblique fissure, 1362–1473, 3190–3254, 5188–5418
  upper lobe, 1307–1504, 3100–3225, 5188–5418
Lymph node(s), *see* specific full name

Mammillary body, 1112
Mammillothalamic fasciculus, 1107–1112
Mandible, 1188–1235, 3088–3148, 5258–5328
  condyle, 1134–1150, 3154–3160
  coronoid process, 1134–1150
  ramus, 1150–1197
Mandibular canal, 1188–1197
Manubrium, 3148–5328
  suprasternal notch, 3136
Masseter muscle, 1122–1197, 3106–3142, 5228–5358
Mastoid air cells, 1122–1160, 3184–3190, 5338–5358
Maxillary bone, 1107–1178, 3082–3094
Maxillary sinus, 1122–1160, 3088–3124, 5258–5338
Medial
  cuneiform bone, 2781–2793, 3154–3190, 5448–5458
  femoral intermuscular septum, 1980–2214
  frontal gyrus, 1072–1082
  genicular vessels, 2353
  lemniscus, 1107–1134
  lingual raphe, 1178
  longitudinal fasciculus, 1112–1150
  meniscus, 2341, 3202–3214, 5418
  patellar retinaculum, 2286–2368, 5188–5218, 5368, 5408
  plantar nerve, 2757–2781
  pterygoid muscle, 1150–1188, 3136–3154, 5258–5268
  rectus muscle, 1107–1112, 3106
  sacral crest, 3288
  superior genicular vessels, 2312
Median
  nerve, 1420–1570, 5488
  raphe, 1150
  sacral vessels, 3276–3288
Medulla oblongata, 3184–3190, 5288–5298
  raphe, 1160
Membranous interventricular septum, 1452
Mentalis muscle, 1212–1235, 3082–3088
Metacarpal
  fifth, 1812–1861, 3070–3082, 5178–5198, 5428–5448
  first, 1778–1789, 3046–3064, 5208, 5218–5228, 5388–
    head, 5238–5428
  fourth, 1812–1850, 3046, 5178, 5428–5448
  second, 1789–1835, 3046–3058, 5198–5238, 5398–5438
  third, 1789–1850, 3046–3058, 5198, 5218, 5398–5428
Mesencephalic tegmentum, 1112–1122
Mesentery, 1610–1800, 3070–3148, 5378–5388
Metatarsal
  fifth, 2817–2867, 3148–3214, 5488
  first, 2793–2841, 3118–3172, 5438–5458
  fourth, 2805–2867, 3130–3190, 5478–5488
  phalangeal joint, 2841
  second, 2793–2853, 3118–3172, 5468
  third, 2805–2853, 3124–3178, 5478–5488
Midbrain, 3178
Mouth, 5318
Multifidus muscle, 1197–1778, 3237–3299, 5288–5348
Muscles, see full name of muscle
Muscular vessel, 1980, 2536–2548
Mylohyoid muscle, 1188–1235, 3112–3130, 5288–5318
  raphe, 1235

Naris, 1150, 3052–3064
Nasal bone, 1107–1112, 3076–3082
Nasal cavity, 1107–1134, 3076–3124, 5298–5308
  vestibule, 1150
Nasal septal cartilage, 1122–1150
Nasal septum, 3058–3088
Nasalis muscle, 1150, 3076
Nasolacrimal canal, 1122–1134
Nasopharynx, 1178–1188, 3136
Navicular bone, 2757–2767, 3178–3196, 5458–5468
Neurohypophysis, 1112
Nose, 3052–3058
  tip, 3046

Obliquus capitis inferior muscle, 1178–1188, 3184–3214, 5258–5328
Obliquus capitis superior muscle, 1150–1160
Obturator
  externus muscle, 1908–1940, 3160–3225, 5208–5388
    tendon, 1897

foramen, 1897–1922
internus muscle, 1829–1933, 3172–3248, 5238–5358
tendon, 1886, 3237, 5178–5218, 5398–5418
membrane, 1908–1922, 3196–3202
nerve, 1886–1897
vessel, 1886–1908, 5248
Occipital
bone, 1082–1150, 3231–3265, 5318
basal portion, 1134
internal occipital crest, 1150
outer table, 1134
condyle, 1160, 3172
lobe, 1060–1122, 3220–3259, 5298–5328
vein, 3271
vessel, 1060–1178, 3248–3265
Occipitofrontalis muscle
frontal belly, 1048–1082
occipital belly, 1072–1122, 3237–3259, 5228–5238, 5358
Olfactory sulcus, 1095
Omohyoid muscle
inferior belly, 1285–1293, 3172, 5198–5388
superior belly, 1235–1285
Opponens digiti minimi muscle, 1829–1850
Opponens pollicis muscle, 1778, 3058–3064, 5418
Optic
chiasma, 1107
nerve, 1107, 3106–3142
radiation, 1095–1112
tract, 1107
Oral cavity, 1178, 3082–3124
Oral crifice, 3076
Orbicularis oculi muscle, 1095–1134, 3082, 5238–5268
Orbicularis oris muscle, 1178–1197, 3064–3082, 5258–5318
Orbit, 3106–3124, 5318–5348
Orbital cavity, 1095–1122
Orbital gyrus, 1082–1095
Oropharynx, 1197–1218
Outer table, 1018–1082, 3259

Palatine bone, 1150–1160
Palatine tonsil, 1178–1197, 3148–3154
Palatoglossus muscle, 1178–1197
Palatopharyngeus muscle, 1178
Pancreas, 3142–3196, 5278–5358
head, 1570–1616, 3148–3160
body, 1549–1616, 3148–3160
tail, 1549–1555, 3208–3214, 5368
Papillary muscle, 1461
Paracentral lobule, 1048–1060
Parahippocampal gyrus, 1107–1122
Pararenal fat, 1616–1682, 5188–5208
Parathyroid gland, 1285
Parietal
bone, 1012–1112, 3166–3237, 5238–5358
lobe, 1048–1082, 3166–3237, 5248–5348
scalp vessel, 5368
Parietooccipital sulcus, 1060–1082, 5288
Parietotemporopontine fibers, 1107–1122
Parotid
duct, 1188
gland, 1150–1197, 3148–3178, 5218, 5348–5368
lymph node, 1212
Patella, 2274–2300, 3136–3166, 5168–5178, 5428–5458
articular cartilage, 2274–2300
Patellar ligament, 2274–2382, 3172–3184, 5168–5178, 5428–5458
Patellar tendon, 3154–3166
Pectinate muscle, 1461
Pectineus muscle, 1861–1992, 3136–3166, 5238–5398
Pectoralis major muscle, 1285–1504, 3064–3154, 5168–5458
Pectoralis minor muscle, 1317–1436, 3118–3172, 5168–5208, 5378–5428
Penis, 1951–1992, 3076–3106, 5298
bulb, 1933–1940
crus, 1933–1940, 3154–3190, 5298
dorsal vessel, 1933
dorsum, 1922
glans, 1992, 2010, 3100
Perfusion artifact, 1951–1980
Pericardiophrenic vessel, 1428–1473
Pericardium, 1381–1504, 3094–3178, 5368–5378
Perineal vessels, 1940
Periosteum, 3265, 5368
Perirenal fat, 1570–1682, 5198
Peritoneal cavity, 1490, 1610–1640
Peritoneum, 1850–1861
Peroneal vessels, 2502–2632
Peroneus
brevis muscle, 2440–2740, 3220–3265, 5458–5478
tendon, 2745–2805,
longus muscle, 2368–2620, 3214–3265, 5488
tendon, 2536–2805, 3259, 5468–5478
tertius muscle, 3202–3220
tendon, 2740, 3208
Phalanx
fifth distal, of the foot, 3124–3130, 5498
fifth distal, of the hand, 1908
fifth middle, of the foot, 2853, 5498
fifth middle, of the hand, 1873–1897, 3118–3136
fifth proximal, of the foot, 2853–2867, 3130–3142
fifth proximal, of the hand, 1861–1886
first distal, of the foot, 2841–2853, 3064–3094
first distal, of the hand, 1800–1835
first proximal, of the foot, 2829–2853, 3100–3124, 5428–5438
first proximal, of the hand, 1800–1812, 3046, 5378
fourth distal, of the foot, 3100–3118
fourth distal, of the hand, 1897–1908
fourth middle, of the foot, 2853–2867, 3112–3130
fourth middle, of the hand, 1861–1897
fourth proximal, of the foot, 2853, 3118–3136, 5488
fourth proximal, of the hand, 1850–1886
second distal, of the foot, 2853, 3076–3094
second distal, of the hand, 1850–1873
second middle, of the foot, 2841–2867, 3100–3112
second middle, of the hand, 1850–1873, 5378
second proximal, of the foot, 2841–2853, 3112–3124, 5468
second proximal, of the hand, 1829–1861
third distal, of the foot, 2853, 3088–3106
third distal, of the hand, 1886–1897
third middle, of the foot, 2853–2867, 3100–3124
third middle, of the hand, 1873–1886, 5378
third proximal, of the foot, 2841–2853, 3112–3136
third proximal, of the hand, 1850–1873
Pharyngeal constrictor muscle, 1178–1272, 3160
Phrenic nerve, 1293
Piriform recess, 1235–1243, 5318
Piriformis muscle, 1829–1873, 3237–3271, 5218–5408
tendon, 1873–1886, 5208, 5418
Pisiform bone, 1789–1800
Plantar aponeurosis, 2793–2853, 3154–3276, 5428–5488
Plantar surface, 2877
Plantaris muscle, 2286–2368, 3254–3305
tendon, 2368–2536
Platysma muscle, 1197–1293, 3082–3148, 5228–5368
Pleural cavity, 1473
Pons, 3160–3166, 5288
basilar sulcus, 1122–1134
tegmentum, 1134
Pontine nuclei, 1134
Pontine raphe, 1122–1134
Pontocerebellar fibers, 1122–1134
Popliteal
artery, 2238–2368
fossa, 2274–2286, 5158–5208, 5368–5438
lymph node, 2312
nerve, 5168, 5418
vein, 2226–2368, 5408–5418
vessels, 2226, 2312–2353, 3208–3237, 5168–5188, 5398
Popliteus muscle, 2341–2450, 3214–3237, 5168–5178, 5408–5418
tendon, 2464
Portal vein, 1510–1570, 3142–3178, 5238–5278
Postcentral gyrus, 1048
Postcentral sulcus, 1048
Posterior
atlantooccipital membrane, 1160
auricular vein, 1150–1188, 5368
axillary fold, 5468
cricoarytenoid muscle, 1272
cruciate ligament, 2324–2341, 3220–3225, 5178, 5408

Posterior (*continued*)
crural intermuscular septum, 2382–2440
femoral vein, 2010–2058
femoral vessel, 1992
humeral circumflex vessel, 5468
longitudinal ligament, 1235–1762
talocalcaneal ligament, 2740
talofibular ligament, 2740
tibial artery, 2394–2406
tibial vein, 2394–2406
tibial vessels, 2382–2692, 2745
tibiofibular ligament, 2725
Precentral gyrus, 1028–1060
Precentral sulcus, 1028–1048
Precommissural septum, 1095
Precuneus, 1048–1072
Pretracheal lymph node, 1362
Prevesicle space, 1897–1908
Primary bronchus, 1381–1405, 3196, 5268–5278
Procerus muscle, 1095, 3064–3070
Pronator quadratus muscle, 1762–1789, 5458–5478
Pronator teres muscle, 1570–1616, 5508–5518
Prostate gland, 1886, 3196–3220, 5288–5298
lateral lobe, 1897–1908
middle lobe, 1897–1908
Prostatic urethra, 3225
Prostatic vein, 1897
Prostatic vessel, 1908–1922
Psoas major mscle, 1583–1812, 3154–3237, 5228–5378
Pubic symphysis, 1897–1922, 3142–3172, 5288–5298
Pubis, 1873–1908, 3148–3225, 5238–5278
inferior ramus, 1922–1940, 3190, 5278, 5308–5328
superior ramus, 1886–1897, 3160, 5258–5358
Pubococcygeus muscle, 3208
Pubofemoral ligament, 1908, 3178
Puborectalis muscle, 3196
Pulmonary artery, 1381–1405, 3166–3208, 5248–5338
Pulmonary trunk, 1388–1405, 3142–3160, 5308–5318
Pulmonary vein, 1405–1436, 3214, 5278
Putamen, 1082–1095
Pylorus, 1539–1570, 3118–3130, 5298–5368
Pyramidal tract, 1150–1160
Pyramids, decussation, 1160

Quadratus femoris muscle, 1897–1951, 3231–3237, 5168–5208, 5378–5418
Quadratus lumborum muscle, 1583–1730, 3214–3259, 5208–5408
Quadratus plantae muscle, 2767–2817, 3178–3259, 5428–5468

Radial nerve, 1436–1521
Radial tuberosity, 1640
Radial vessel, 1629–1657
Radius, 1616–1762, 3064–3271, 5168, 5448–5538
head, 1595–1610
interosseous crest, 1657–1719
neck, 1616
radial tuberosity, 1629
shaft, 1640–1750
styloid process, 1762
Rectal vessels, 1812
middle, 1850–1873, 3265–3271
Rectocolon junction, 1835
Rectovesicle pouch, 1861
Rectum, 1850–1940, 3148, 5298–5308
longitudinal muscle, 1933–1940
Rectus abdominis muscle, 1510–1897, 3046–3130, 5218–5388
tendinous intersection, 1629–1778, 3052–3076, 5218–5388
Rectus capitis
anterior muscle, 1150–1160
posterior major muscle, 1160–1178, 3202–3225
posterior minor muscle, 1150–1160, 3225, 5298
Rectus femoris muscle, 1873–2190, 3094–3166, 5168–5208, 5398–5468
tendon, 1850–2274, 3142–3160, 5168–5198, 5418–5448
Red nucleus, 1112
Renal
artery, 1610–1640
fascia, 1583–1682, 3225, 5188–5218, 5388–5408
medulla, 1570
pelvis, 1595–1657
vein, 1610–1640, 3190, 5308–5368
Reticular formation, 1107–1160
Retromandibular vein, 1134–1218, 3160, 5348
Retropharyngeal space, 1293
Rhinal sulcus, 1107–1112
Rhomboid major muscle, 1272–1420, 3254–3299, 5208–5378
Rhomboid minor muscle, 1243–1293, 3276–3282, 5208
Rib 1, 1317–1340
costal cartilage, 1340–1362, 3136–3154
head, 1293
neck, 1293
tubercle, 1285
Rib 1–2, intercostal muscles, 1293–1362, 5368
Rib 2, 1307–1388, 5198, 5388
costal cartilage, 1381–1388, 3112–3118
Rib 2–3, intercostal muscles, 1307–1405, 3106, 5198, 5378
Rib 3, 1307–1420, 3112–3124, 5218–5388
costal cartilage, 1420–1428, 3094–3106
head, 1326
Rib 3–4, intercostal muscles, 1326–1452, 3094
Rib 4, 1340–1452, 3100, 3254, 5188–5198, 5408–5418
costal cartilage, 1452–1473, 3076–3094
head, 1340
Rib 4–5, intercostal muscles, 1340–1490, 3088, 3259, 3202, 3276, 5408
Rib 5, 1381–1473, 3094, 5168–5188, 5408–5438
costal cartilage, 1490–1504, 3070–3088
head, 1362
Rib 5–6, intercostal muscles, 1381–1510, 3214, 5428
Rib 6, 1405–1490, 3248
costal cartilage, 1504–1555, 3076–3088, 5378
head, 1388
neck, 1381
Rib 6–7
external intercostal muscles, 3288
intercostal muscles, 1405–1549, 3248, 3282, 5168
internal intercostal muscles, 1555
Rib 7, 1420–1549, 3259–3282, 5428
costal cartilage, 1504–1595, 3094
Rib 7–8
intercostal muscles, 1428–1570, 3100, 3265
internal intercostal muscles, 1555
Rib 8, 1452–1583, 3288, 5178
costal cartilage, 1595–1610, 3094, 5188, 5398
Rib 8–9, intercostal muscles, 1473–1610
Rib 9, 1473–1595, 3288, 5178
costal cartilage, 1610–1616, 3130, 5428
head, 1461
Rib 9–10, intercostal muscles, 1504–1640, 5448
Rib 10, 1510–1616, 3276–3282, 5448
costal cartilage, 5438
head, 1504
Rib 10–11
intercostal muscles, 1539–1640, 3276, 5198
internal intercostal muscle, 1629
Rib 11, 1539–1616, 3276, 5418–5448
costal cartilage, 1629–1640
head, 1521
Rib 11–12
intercostal muscles, 1555–1640, 3265–3271, 5168–5178, 5418
internal intercostal muscle, 1629
Rib 12, 1570–1640, 5178, 5428
costal cartilage, 1657, 5168, 5438
Rima glottidis, 1260
Rotator cuff, 5488

S1, nerve, 1812
S1–2, intervertebral disc, 1789, 3225
Sacral
canal, 1789–1835
foramen, 1789, 3237
first, 1800, 5328
second, 1800
nerve, 1850
vessels, middle, 1789–1861

Sacroiliac joint, 1762–1812, 3214–3271, 5258
  articular surface, 1789–1812
Sacrotuberous ligament, 1850–1873, 3254, 5338
Sacrum, 1829–1850, 3214–3293, 5288–5358
  alae, 1762–1800
  body, 1778–1812
  lateral wing, 1778, 3220–3231, 5248–5278
Sagittal suture, 1012–1072, 3248
Sartorius muscle, 1829, 2046–2353, 3100–3242, 5178–5428
  tendon, 2353–2368, 3220, 5218, 5368
Scalenus
  anterior muscle, 1260–1317, 3172–3178, 5248, 5328–5348
  medius muscle, 1243–1317, 3178, 5228–5358
  posterior muscle, 1218–1293, 5248, 5338–5358
Scalp, 3271, 5378
Scalp vessel, 1012–1018
Scaphoid bone, 1778, 5468
Scapula, 1272–1436, 3214–3293, 5168–5198, 5368–5448
  acromion, 1272, 3184–3196, 5458–5488
  inferior angle, 5178–5418
  lateral border, 5158–5168
  medial border, 1307–1340
  neck, 5448
  spine, 1285–1307, 3231–3276, 5158–5188, 5388–5448
  vertebral border, 1352
Scapuloclavicular joint, 1285
Sciatic nerve, 1829–2178, 3225–3231, 5408
Scrotum, 1964–2026, 3112–3142, 5298–5328
Secondary bronchus, 1381–1436, 5258
Semicircular canal, 1122
Semilunar fold, 1629–1750
Semimembranosus muscle, 2026–2312, 3214–3282, 5168–5418
  tendon, 1964–2010, 2324–2353, 5208, 5378–5388
Seminal vesicle, 1873–1886, 3214–3225, 5278–5288, 5328
Semispinalis capitis muscle, 1134–1235, 3214–3248, 5248–5348
  lateral part, 1150–1197
  medial part, 1150–1197
Semispinalis cervicis muscle, 1197–1272, 3202–3276, 5268
Semitendinosus muscle, 1933–2250, 3237–3293, 5188–5398
  tendon, 2262–2368
Septum pellucidum, 1082
Serratus anterior muscle, 1285–1539, 3154–3299, 5168–5218, 5368–5438
Serratus posterior inferior muscle, 1583–1616, 5408
Short gastric vein, 1521–1539
Short gastric vessel, 1490–1510
Shoulder joint, 1307–1326, 3190, 5448–5478
  articular capsule, 5468
Sigmoid
  mesocolon, 1829–1861
  sinus, 3178–3190, 5258–5268
  vessels, 1800–1835
Skin, 1060–2867, 3288, 5488
Skull, periosteum, 5228
Small intestinal vessels, 1657–1778
Small saphenous vein, 2312, 2608–2745, 3237–3299, 5178, 5218, 5398, 5438
Soft palate, 3142
Soleus muscle, 2353–2668, 3208–3299, 5178–5188, 5388–5438
  tendon, 5418
Spermatic cord, 1861–1964, 3106–3136, 5248–5358
Sphenoid bone, 1095–1150
  foramen spinosum, 1134
  lateral pterygoid plate, 1150–1160
  posterior clinoid process, 1112
  superior orbital fissure, 1112
Sphenoid sinus, 1107–1122, 3130–3142, 5298–5308
Spinal cord, 1178–1610, 3190–3265, 5288–5298
  anterior horn, 1197
  conus medullaris, 1616–1629
  dorsal horn, 1243
  ventral horn, 1243
Spinal lemniscus, 1107–1150
Spinalis muscle, 3237–3311, 5308–5328
Spinalis thoracis muscle, 1521–1555, 3299–3305
Spleen, 1490–1595, 3190–3265, 5338–5428
Splenic
  artery, 1555
  vein, 1539–1583, 3142, 5288–5378
  vessels, 5388
Splenium, 3190–3196
Splenius capitis muscle, 1150–1285, 3208–3242, 5238–5348
Splenius cervicis muscle, 1307–1352
Sternoclavicular disc, 1340
Sternocleidomastoid muscle, 1160–1326, 3130–3214, 5228–5358
Sternohyoid muscle, 1243–1362, 3124–3148, 5278–5288
Sternothyroid muscle, 1260–1362, 5318
Sternum, 1381–1521, 3076–3124
  manubrium, 1340–1362, 3118–3130
  suprasternal notch, 3124
Stomach, 1521–1539, 3076–3237, 5318–5388
  body, 1510
  cardiac orifice, 1504–1510
  cardiac portion, 1504
  fundus, 1490
  longitudinal ridge, 1510–1521
  pyloric antrum, 1521
  pyloric sphincter, 1555–1570
Straight sinus, 1122, 3237–3242
Stria medullaris, 1082–1095
Stria terminalis, 1082
Styloglossus muscle, 1178–1197, 5258
Stylohyoid muscle, 1160–1218, 5318
Stylopharyngeus muscle, 1178–1188
Subarachnoid space, 1028–1326, 3214–3242, 5308
Subclavian
  artery, 1307–1352, 5218–5228
  vein, 1307–1317, 3166, 5218–5358
  muscle, 1307–1326, 5178–5408
Subcutaneous fascia, 2781–2793
Sublingual duct, 1197
Sublingual gland, 1197–1212
Submandibular
  duct, 1218
  gland, 1197–1235, 3136, 5338
  lymph node, 1218
Submental vein, 1218–1243
Subpatellar adipose tissue, 3178
Subsartorial fascia, 2106–2178
Subscapularis muscle, 1293–1420, 3190–3293, 5158–5198, 5368–5448
  tendon, 5458
Substantia nigra, 1107–1122
Subtalar joint, 2767
Subtendinous space, 5438
Superficial
  abdominal vein, 1490
  cerebral vessel, 1060
  cervical vein, 1235
  epigastric vein, 1829–1861
  epigastric vessel, 1850–1897
  facial vessel, 5368
  fascia, 1420–2867, 3046–3299, 5158–5498
    feet, 5418
    thigh, 3082–3094
  gluteal vessel, 1992
  temporal vein, 1134, 5218–5238
  temporal vessel, 1060–1134, 5368
  thoracic vein, 1461
  transverse perineal muscle, 1940
Superior
  cerebellar peduncle, 1122–1134
  cerebral vessel, 1048
  colliculus, brachium, 1107
  colliculus, commissure, 1107
  frontal gyrus, 1028–1072
  frontal sulcus, 1028–1060
  gemellus muscle, 1873–1886, 5208, 5368–5408
  gluteal artery, 1812
  gluteal vein, 1829
  gluteal vessel, 1835, 2274
  mesenteric artery, 5308
  mesenteric vein, 1583–1740, 3136, 5298–5308
  oblique muscle, 1095, 3118
  peroneal retinaculum, 2757
  pulmonary vein, 1381–1388
  rectal vessel, 1829–1835
  rectus muscle, 1095

Superior (*continued*)
  sagittal sinus, 1028–1122, 3100–3248, 5298
  temporal gyrus, 1082–1112
  vena cava, 1362–1405, 3160–3178, 5268–5278
Supinator muscle, 1595–1657, 5538
Supracallosal gyrus, 1082
Supraorbital vessel, 1028–1082
Suprapatellar bursa, 2250–2274, 3154–3166, 5168, 5418–5448
Suprascapular vessel, 5178
Supraspinatus muscle, 1272–1307, 3196–3265, 5158–5198, 5368–5468
Supraspinous ligament, 1692–1750, 3282–3299
Sural vessels, 2341
Suspensory ligament, 1908–1922, 3106–3112, 5308
Sustentaculum tali, 2757

T1
  body, 1285–1307
  costal facet, 1285
  lamina, 1285
  pedicle, 1285
  spinous process, 1285–1293
  transverse process, 1285, 3214
T1–2, intervertebral disc, 1307
T2
  body, 1307–1326
  lamina, 1307
  spinous process, 1307–1317
  transverse process, 1293
T2–3, intervertebral disc, 1326
T3
  body, 1326–1340
  lamina, 1326
  spinous process, 1326–1340, 3265
  transverse process, 1326
T4
  body, 1352–1362
  lamina, 1352
  spinous process, 1352–1362
  superior articular process, 1340
  transverse process, 1340–1352
T5
  body, 1381–1388
  lamina, 1381
  spinous process, 1381–1388
  superior articular process, 1362
  transverse process, 1362
T5–6, intervertebral disc, 1388
T6
  body, 1405
  spinous process, 1405–1420
  superior articular process, 1381–1388
  transverse process, 1388–1420
T6–7, intervertebral disc, 3248
T7
  body, 1420–1436
  lamina, 1428–1436
  spinous process, 1428–1436
  vertebra, 3248
T8
  body, 1452–1461
  spinous process, 1452–1490, 3276
  transverse process, 1452
T9
  body, 1473–1490
  spinous process, 1504–1510
T9–10, intervertebral disc, 1490
T10
  body, 1504–1510
  transverse process, 1504
T11
  body, 1521–1539
  spinous process, 1549–1555
  transverse process, 1549
T12
  body, 1549–1583
  spinous process, 1583
  vertebra, 3248
  vertebral body, 3214
T12–L1, intervertebral disc, 1549–1610, 3231
Talocalcaneal joint, 2745–2757
Talofibular joint, 2740
Talonavicular joint, 2767
Talotibial joint, 2725
Talus bone, 2740–2767, 3190–3242, 5448–5478
  lateral malleolar process, 2740
  trochlea, 2725
Tapetum, 1107
Tarsal plate, 3082
Tarsometatarsal joint, 2793
Tarsometatarsal ligament, 2805
Tectum, 1107–1112
Teeth
  canine, lower, 1197
  canine, upper, 1178
  central incisor, root, 1218
  incisor, lower, 1197, 3082
  molar, lower, 3076
  premolar, lower, 1188–1197
  premolar, upper, 1178
Temporal bone, 1060–1150, 3214–3220
  articular tubercle, 3148
  auditory tube, 1134–1150
  petrous part, 1122
  styloid process, 1150–1178
  tubercle, 1134
Temporal gyrus, middle, 1107–1112
Temporal lobe, 1107–1122, 3130–3208, 5238–5278, 5358
Temporalis muscle, 1048–1160, 3106–3196, 5228–5248, 5338–5368
  tendon, 1112–1122
Temporomandibular joint, 1134
Tendon(s), see full name of tendon
Tensor fascia lata muscle, 1835–1964, 3124–3184, 5168, 5448–5478
Tensor veli palatini muscle, 3160
Tentorium, 1107–1122, 3196–3237, 5258–5338
Teres major muscle, 1352–1436, 3202–3311, 5158–5168, 5408–5478
Teres minor muscle, 1307–1362, 3214–3299, 5178, 5448–5498
Testicle, 1980–1992, 2010, 3118–3136, 5298–5328
Testicular vessel, 1657–1861, 3046–3088, 5208–5238, 5398
Thenar muscles, 3046–3088, 5208–5398
Thoracic duct, 1307–1555
Thoracoepigastric vein, 1504
Thoracoepigastric vessel, 1428–1490
Thoracolumbar fascia, 1420–1504, 3299–3311
Thoracolumbar vein, 5448
Thumb, 3046
Thymus, atrophic, 1362–1388, 3112–3136
Thyrohyoid muscle, 1235–1243, 3136, 5318
Thyroid cartilage, 1243–1260, 3136, 5308
  inferior horn, 1272
  superior horn, 1243
Thyroid gland, 1272–1307, 3148–3166, 5268–5278
  isthmus, 5288
  lateral lobe, 1285, 5318
Tibia, 2341–2725, 3166–3242, 5178, 5208, 5458–5468
  anterior border, 2394–2704
  articular cartilage, 2341–2725
  condyles, 3190
  intercondylar eminence, 2330
  lateral condyle, 2341–2353, 3196–3237, 5418–5458
  lateral surface, 2368–2716
  medial border, 2382–2668
  medial condyle, 2341–2353, 3196–3214, 5198, 5388–5428
  medial malleolus, 2725, 3196–3248, 5438–5448
  medial margin, 2680–2704
  medial surface, 2368–2716
  periosteum, 5188, 5398, 5448
  posterior surface, 2382–2704
  shaft, 3184–3225, 5178, 5408–5458
Tibial
  collateral ligament, 2324–2353, 3190, 5378
  nerve, 2178–2757
  tuberosity, 2368–2382, 3172–3184, 5438
Tibialis
  anterior muscle, 2368–2680, 3184–3214, 5458–5478
    tendon, 2668–2781, 3190–3208, 5468–5478

posterior muscle, 2368, 2477–2680, 3214–3259, 5168, 5418–5448
tendon, 2668–2767, 5438
Tibiofibular joint, 2353, 2716
Tibiofibular syndesmosis, 2692–2704
Toe
fifth, 2867–2877, 3112
first, 2853–2867, 3058–3076
fourth, 2867–2877, 3094–3112
second, 3070–3082
third, 2867, 3082–3100
Tongue, 1178–1212, 3088–3142, 5278–5288, 5308–5318
apex, 1188
intrinsic muscles, 1212–1218
root, 1212–1218, 3148
Trabeculae carnae, 1436–1490
Trachea, 1285–1362, 3154–3184, 5298
carina, 1381, 3196
Tracheal cartilage, 1317–1340
Tracheal lymph node, 1317–1381
Transverse arytenoid muscle, 1260
Transverse
cervical artery, 5218
pontine fibers, 1122–1134
sinus, 1134, 3196–3237, 5248–5278
Transversus abdominis muscle, 1570–1850, 3082–3208, 5178–5228, 5388–5428
Transversus thoracis muscle, 1510–1539
Trapezium bone, 1778, 5198
Trapezius muscle, 1150, 1307–1510, 3196–3317, 5158, 5308–5458
Trapezoid bone, 1778–1789
Triceps brachii muscle, 1388–1595, 3190–3288
lateral head, 1352–1510, 5498–5538
long head, 1340–1510, 3254–3265, 5478–5538
medial head, 1362–1510, 5488
tendon, 1504–1570, 3259–3276
Tricuspid valve, 1436–1452, 3130–3136
leaf, 1461–1473
Trigeminal nerve, 1122
nucleus of mesencephalic tract, 1122
spinal nucleus, 1150–1160
Triquetral bone, 1778–1800
Tuber cinereum, 1107
Tympanic cavity, 1134, 3166

Ulna, 1640–1789, 3076–3271, 5478–5538
coronoid process, 1595, 3237
interosseous crest, 1640–1740
medial epicondyle, 1595
radial notch, 1610
shaft, 1657–1740, 5538
styloid process, 1789
supinator crest, 1616–1629
trochlear notch, 1583–1595
Ulnar
collateral vessel, 1583
nerve, 5488
tuberosity, 1616–1629
Umbilicus, 1750, 3046–3064, 5278–5298
Uncus, 1107
Urachus, 1873
Ureter, 1640–1861
Urethra, 1908–1964, 3118, 5298
membranous, 1922
penal, 1980–2010
sphincter, 1922
Urogenital diaphragm, 1933, 3196, 3231, 5288–5298
Uvula, 1178–1188
Uvula muscle, 1160

Vagus nerve, 1212–1285
Vallate papilla, 1178–1188
Vastus intermedius muscle, 1922–2274, 3130–3214, 5168–5198, 5418–5468
aponeurosis, 2166–2178
Vastus lateralis muscle, 1908–2250, 3106–3254, 5158, 5428–5498
Vastus medialis muscle, 1951–2286, 3106–3196, 5178–5418
Vein(s), see specific vein or vessel by name
Ventral sacroiliac ligament, 1778–1812
Ventricle
fourth, 1122–1150, 3190
choroid plexus, 1150
left, 1420–1473, 3106–3184, 5318–5378
right, 1420–1490, 3100–3190, 5298–5338
third, 1095–1107, 1381–1504, 3208–3225, 5308
Vertebral artery, 1178–1293, 3178
Vertebral canal, 3231–3271
Vertebral foramen, 1778
Vertebral vein, 1260–1293, 1671, 1750
Vessel(s), see specific name
Vestibule, 1178–1188
Vitreous chamber, 1112
Vocal folds, 3148
Vocalis muscle, 1260
Vomer bone, 1150

White matter, 1028–1072

Xiphoid process, 3070–3076
cartilage, 1539–1555

Zygomatic arch, 3100–3136
Zygomatic bone, 1095–1122, 3112, 5348–5358
Zygomaticus major muscle, 1134–1178, 3094